Dynamic aspects of plant ultrastructure

Consulting Editor

Professor M. B. Wilkins
University of Glasgow

A. W. Robards: Editor

Senior Lecturer in Biology, University of York

Dynamic aspects of plant ultrastructure

London · New York · St Louis · San Francisco · Düsseldorf
Johannesburg · Kuala Lumpur · Mexico · Montreal
New Delhi · Panama · São Paulo · Singapore
Sydney · Toronto

Published by

McGraw-Hill Book Company (UK) Limited
MAIDENHEAD · BERKSHIRE · ENGLAND

07   084017   2

Text set in 10/12pt IBM Press Roman, Printed by photolithography, and bound in Great
Britain at The Pitman Press, Bath.

Contents

v

Contributors

Professor A. M. Catesson, Laboratoire de Botanique, E.N.S. Paris.

Professor E. C. Cocking, Department of Botany, University of Nottingham.

Professor J. Cronshaw, Department of Biological Sciences, University of California, Santa Barbara.

Dr B. E. S. Gunning,* Department of Botany, The Queen's University of Belfast.

Professor W. A. Jensen, Department of Botany, University of California, Berkeley.

Dr J. G. Lafontaine, Départment de Biologie, Université Laval, Faculté des Sciences, Cité Universitaire, Québec.

Professor Ph. Matile, Eidgenössische Technische Hochschule Zürich, Institut für Allgemeine Botanik.

Dr H. H. Mollenhauer, Texas A. and M. University, College Station, Texas.

Professor D. J. Morré, Department of Botany and Plant Pathology and Department of Biological Sciences, Purdue University, Lafayette, Indiana.

Dr T. P. O'Brien, Department of Botany, Monash University, Clayton, Victoria, Australia.

Dr H. Öpik, Department of Botany and Microbiology, University College of Swansea.

Professor D. J. Paolillo, Section of Genetics, Development and Physiology, Division of Biological Sciences, Cornell University, Ithaca, New York.

Professor J. Pate, Botany Department, University of Western Australia, Nedlands, Australia.

Professor J. D. Pickett-Heaps, Department of Molecular, Cellular and Developmental Biology, University of Colorado, College of Arts and Sciences, Boulder, Colorado.

Professor R. D. Preston, Astbury Department of Biophysics, The University of Leeds.

Professor E. Schnepf, Lehrstuhl für Zellenlehre, Der Universität Heidelberg.

Professor W. W. Thomson, Department of Biology, University of California, Riverside.

---

* Now Professor of Developmental Biology, Research School of Biological Sciences, Australian National University, Canberra.

Preface

The ultrastructure of plant cells is a rapidly advancing field of study. As we see with increasing clarity the finest details of cell structure, so we are made aware of the necessity for considering the dynamic and functional roles that these structures play in the living cell; hence the title of this book. It is assumed that the reader of this volume will have a basic knowledge of plant ultrastructure to the approximate level covered by many books, including my own earlier text *Electron Microscopy and Plant Ultrastructure* (McGraw-Hill, 1970). This textbook is, therefore, aimed at advanced-course students and research workers in Botany. Much that is contained here will be relevant to final-year undergraduates as well as to post-graduate students specializing in ultrastructural and molecular studies. It has been compiled to complement and extend the information available in introductory volumes, while easing the task of discerning overall trends in a specific area by searching the current literature.

In planning a volume such as this, there are many considerations which make heavy demands upon the subjective assessments of the Editor. However, my task has been made easier by the help of Professor M. B. Wilkins (Regius Professor of Botany, University of Glasgow; and Consulting Editor of this Series) whose own edited book *Physiology of Plant Growth and Development* (McGraw-Hill, 1969) has set such a high standard in this form of publication.

The most important decisions to be made in producing a multi-author book are (a) which subjects should be covered and (b) who should deal with them. In the first case, the Editor's choice has been made on the basis of rapidly advancing areas of interest, topics which have been unwarrantedly neglected, and those which, because of some special significance, merit inclusion. The unequal state of development of knowledge in each of these areas has meant that the depth of treatment has varied from chapter to chapter, but I regard this as an inevitable, and necessary, result of choosing the topics in the manner described. My choice of subjects was also influenced by the need to maintain continuity throughout the book. Fortunately, this was not too difficult as, using the criteria mentioned, many of the chapters (and contributors) chose themselves. However, I accept that other botanists would have made different selections; it is unavoidable that some topics have been omitted which another editor would have regarded as indispensable. Having decided upon the subject matter of the book, I approached the scientists who I felt would together produce the most stimulating result; I had the immense good fortune of immediate co-operation from each one of my fellow botanists, and all contributions are by the person originally contacted.

With subjects and contributors chosen, it remained to compile the book to the best possible effect. This involved asking the contributors to conform to precise details of manuscript preparation, and also to write within the general subject limits for their chapter; I am most grateful to them all for their co-operation in this respect. The main editorial duties were seen as to guide and then collate the information presented by the contributors. Once the manuscript was received from a contributor, I regarded it as important to allow freedom of style and approach; not least because it would have been impossible to produce uniformity from such diverse subjects at different levels of understanding. The chapters are not reviews, nor are they exclusively research papers: they represent something between the two. I think it important that, in many cases, the authors have not only given the answers to questions but have also raised new problems which remain to be solved. Further, electron microscopy has the most serious disadvantage that structures look static: this colours our thinking and concepts of ultrastructural activity. The title of the book, and the approach of the contributors, are meant to reinforce the need to consider ultrastructure in dynamic and functional terms, so that we may understand how cells work and not merely how they are put together. There are, of course, disadvantages to multi-author volumes, not least in variation of style, depth of coverage, and similar problems. In my opinion these are more than outweighed by the immense value of having contributions written by experts in their own fields. Another disadvantage is that the distribution of fifteen different authors across the globe inevitably leads to delays not associated with single-author volumes; this has become acutely clear during the course of production. However, authors have had the opportunity to up-date references at the proof-reading stage. The book is therefore up to the minute, although some changes of detail will doubtless have taken place here and there; that is one of the unavoidable penalties of attempting to provide a contemporary commentary close to the frontiers of a particular field of scientific study.

Finally, I should like to thank my friends and fellow scientists for their excellent contributions. Necessarily, such people always have other heavy commitments and I am grateful for their hard work in adding yet another burden to their onerous schedules. My thanks are also due to staff in my own laboratory who have helped so much in reading and correcting manuscripts and all the other jobs involved in preparing a book: Miss R. M. Dean, Miss G. S. Dolman, Mrs S. M. Jackson, and Mrs M. E. Robb. As ever, I am indebted to my wife for her help in so many respects while editing this book.

Anthony W. Robards

1

The nucleus
J. G. Lafontaine

Introduction

Important advances have recently been made, at the biochemical level, towards a better understanding of the function of both the chromosomes and the nucleolus. To a very large extent, such progress has been realized with animal cells. In many instances, the use of synchronized cell populations has helped in obtaining solutions to otherwise most difficult problems. Particularly exciting examples of recent major contributions to the study of the nucleus are the discovery of satellite DNA (reviewed in Walker *et al.*, 1969; Bostock, 1971; Flamm, 1972; Walker, 1972) and its localization in the centromeric portion of chromosomes (Jones, 1970; Pardue and Gall, 1970); the demonstration of highly redundant DNA as part of the genome of a wide variety of organisms (reviewed in Britten and Kohne, 1969); the discovery of the genetic amplification phenomenon (reviewed in Brown and Dawid, 1968; Lima-de-Faria, 1969) and the characterization of the polycistronic organizer DNA (reviewed in Birnstiel *et al.*, 1971).

At the electron microscopic level, in spite of many recent interesting findings (reviewed in Lafontaine and Lord, 1969; DuPraw, 1970; Ris and Kubai, 1970; Comings, 1972), progress in elucidating the macromolecular organization of nuclear structures has been much more modest.

The present review will, as much as possible, emphasize biochemical and ultrastructural aspects of nuclear structures in plant cells. However, in view of the fact that most current investigations of the structure and biochemical function of chromosomes and of the nucleolus are being carried out with animal cells, this chapter will forcibly draw rather heavily from such studies. In keepir₃ with the purpose and general orientation of the present volume, this chapter will include a certain amount of background information which should prove useful to advanced undergraduate or graduate students in understanding some of the current work on the organization and biochemical activities of nuclear structures.

The Interphase Nucleus

Fixation

As chromatin within the living interphase nucleus is only slightly more refringent than the ground substance or nucleoplasm, its organization *in vivo* has been very difficult to study until recent years. With the development of phase contrast and interference optics, new information has become available on the state of chromatin in living cells and, in this respect, especially interesting progress has been achieved in the case of plant cells (Bajer and Molè-Bajer, 1963). Nevertheless, much of our current understanding of the organization of nuclear structures still stems from examination of fixed material or from biochemical characterization of isolated nuclei.

Early work revealed a wide variety of images of interphase nuclei following preservation with acid or alcohol-containing fixatives, and controversies therefore arose as to the reality of the nuclear structures thus observed. For some workers, fixatives induced a sol—gel transformation giving rise to a nuclear reticulum. Others were inclined to visualize nuclear organization as colloidal in nature.

A number of classical cytologists turned to plant material for studying nuclear structures in living cells and investigating their reaction to certain experimental conditions. Staminate hair and secondary roots, among other materials, were found most appropriate for such work (Bělǎr, 1929; Dangeard, 1941). As a result of these *in vivo* studies, which generally involved painstaking efforts to minimize artefacts, it became evident that factors such as the degree of hydration, exposure to salt solutions or to dilute acetic acid would induce reversible modifications in the appearance of chromatin.

Following their successful isolation of chromosomes from resting nuclei of mammalian tissues, Ris and Mirsky (1949) carried out a more detailed analysis of some of the physico-chemical parameters involved in inducing a reversible transformation of chromatin in different cell types from the dispersed, life-like, state to the condensed state. Among other interesting findings these authors conclusively demonstrated that, in freshly prepared *Allium cepa* epidermis, the chromatin absorbs very diffusely in the u.v. range whereas the nucleolus is, on the contrary, quite distinctly delineated. Addition to dilute acetic acid was found to induce the appearance of coiled nuclear structures which evolved into thick and highly refractile chromosomes upon increasing the concentration to 45%. Examination, under u.v. microscopy, of the effect of acetic acid on interphase and prophase nuclei led to the conclusion that chromatin is in an extended state in living cells and that it becomes condensed upon fixation.

Since the early days of electron microscopy, osmium tetroxide has gained wide acceptance as, perhaps, the most satisfactory overall fixative for both histological and cytological applications. The virtues of this fixative for preserving the life-like organization of cellular components were, at that time, already well known from classical observations on cultured cells (Strangeways and Canti, 1927). As electron microscope techniques improved, striking progress was rapidly achieved in unravelling the ultrastructure of various cytoplasmic components; it soon became evident, however, that analysis of chromosome organization at the macromolecular level was beset with unforseen difficulties. Little of the regular structural pattern of chromosomes with hierarchies of coils, so familiar from light optical studies, could be recognized in ultrathin sections of mitotic chromosomes. Efforts were therefore made in various laboratories to improve the fixation of chromosomes. Undoubtedly guided by the well-established action of cations on the conformation of nucleoproteins in solutions, various authors have investigated the possible advantages of adding $CaCl_2$ to standard fixatives, especially osmium tetroxide. According to certain workers (De Robertis, 1956), the microfibrillar component of chromosomes is better preserved in the presence of calcium ions. Other studies (Robbins, 1961; Davies and Spencer, 1962; Barnicot and Huxley, 1965) drew attention to the fact that chromosomes in cultured cells change in appearance, to the point of becoming invisible, during fixation in buffered solutions of osmium tetroxide not containing added calcium. More recent investigations have revealed (Ris, 1969; Ris and Kubai, 1970) that metal-binding buffers affect the native macromolecular organization of chromosomes during fixation.

The effect of bivalent cations on the appearance of nuclear structures, following fixation, may also be studied to advantage with plant cells since, as is well known, certain species exhibit a regular and conspicuous chromatin reticulum during part of interphase. A quite diffuse Feulgen staining of the chromatin of *Vicia faba* interphase nuclei is observed following fixation in 1% osmium tetroxide dissolved in demineralized distilled water (Lafontaine, 1968). Little evidence of a chromatin reticulum may be recognized at interphase in such preparations, even when examined by means of phase contrast optics. At the ultrastructural level, interphase nuclei show, besides the nucleolus, a rather uniform distribution of fine fibrillar material indicating that the chromatin reticulum has completely merged with the surrounding nucleoplasm. As the calcium concentration is increased from 10^{-4} M to 10^{-3} M in the fixing osmium tetroxide solution, the chromatin strands begin to appear in Feulgen-stained preparations. Concomitantly, a reticulum-like structure becomes faintly recognizable within the interphase nucleus studied by electron microscopy. At salt concentrations of from approximately 5×10^{-3} M to 10^{-2} M, the degree of condensation of this interphase chromatin reticulum, as well as its macromolecular texture, correspond rather closely to that characterizing nuclei in material fixed with standard 1% osmium tetroxide solutions adjusted to pH's of 6.5–7.2 with a phosphate buffer. However, when the concentration of $CaCl_2$ reaches 0.1 M, the interphase chromatin exhibits a coarsely fibrillar texture indicative of a certain degree of clumping.

From these various observations it thus clearly appears that, in the absence of metal-binding buffers, the concentration of calcium in the fixative has a marked influence on the degree of condensation of chromatin. Even though, according to various criteria, current fixation techniques do furnish satisfactory preservation of many cell components it is, nevertheless, quite obvious that they do not necessarily give rise to images of the interphase nucleus which closely reflect the

3

organization of chromatin in the living state. One is led to assume that, in living plant cells, the interphase chromatin reticulum exists in a much more extended state than is revealed by fixation methods developed so far. Future efforts to improve this state of affairs will undoubtedly involve selecting cultured cells which, in the living state, exhibit sufficient visible nuclear organization under phase contrast microscopy so as to permit close monitoring of the effect of different fixation conditions on this organization. Certain plant cells would appear particularly well suited for such studies (Bajer and Molè-Bajer, 1963).

Types of interphase nuclei in plant cells

The wide spectrum of organization which plant interphase nuclei exhibit following fixation is undoubtedly most striking. This great diversity in appearance is known, from early observations on living plant cells, to reflect a corresponding variety in the structure of plant nuclei (discussed in Dangeard, 1947). For purpose of discussion, two extreme situations will be considered here: the reticulate and the chromocentric interphase nuclei. It is important to note, however, that there exist a whole series of intermediates between these two extreme types of interphase nuclei: some nuclei show a rather conspicuous chromatin reticulum with relatively small chromocentres, in others this reticulum is much less regular and the heterochromatic masses quite large.

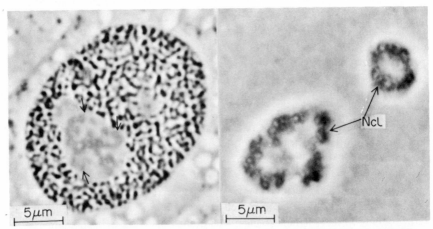

Fig. 1.1   *Triticum vulgare* interphase nucleus. The convoluted chromatin strands show a rather constant diameter and at many places touch one another, thus giving rise to an apparent reticulum. Note that many segments of these strands attach to the nuclear envelope. The nucleolus consists of a dense irregular central portion or nucleolonema embedded in a more translucent substance. At several points (arrows). the nucleolonema reaches the nucleolar surface and seems to be continuous with neighbouring chromatin strands. Phase contrast micrograph of a 0.5 μm section stained by means of the Feulgen procedure. (From Lord and Lafontaine, 1969.)

Fig. 1.2.   Phase contrast micrograph of nucleoli from isolated *Allium porrum* nucleus treated with 'Tween 80' for a few minutes. A coarse filamentous component, the nucleolonema (Ncl), is now most conspicuous and exhibits an organization strikingly similar to that seen under electron microscopy within the central, fibrillar portion of the nucleolus. (From Lord and Lafontaine, 1969.)

4

Reticulate nuclei. The plant species that most attracted the attention of classical authors for studies of the mitotic cycle showed, in most cases, interphase nuclei with a conspicuous chromatin reticulum and were characterized by relatively large chromosomes during the mitotic stages. Well-known examples of such species are *Allium cepa, Tradescantia virginica,* and *Vicia faba.*

The examination of relatively thin (0·3–0·7 μm) Feulgen-stained preparations reveals that the gross morphological organization of the interphase chromatin reticulum varies noticeably from one nucleus to another. Certain of these nuclei exhibit rather uniform chromatin strands approximately 0.3 μm in diameter which seem to form a reticulum (Fig. 1.1). As a result of their complex arrangement within the nuclear cavity and of their highly unravelled condition at interphase, these chromatin strands indeed give the impression that they anastomose or branch at many places. That such an apparent network results from fortuitous contacts between neighbouring segments of the closely packed chromatin strands within the living interphase nucleus is evident from elegant time-lapse studies of the evolution of interphase chromosomes in the endosperm of *Haemanthus* (Bajer and Molè-Bajer, 1963).

By electron microscopy, a quite similar apparent reticulum is observed with the difference that, due to the limited section thickness, the chromatin strands are much less regular in outline than in 0.5–1 μm preparations (Figs. 1.3 and 1.4). These chromatin strands, it is important to note, consist of fine fibrillar material the texture and density of which are indistinguishable from that of the heterochromatin masses (Lafontaine 1968; Lafontaine and Lord, 1969). As discussed in a later section of this chapter, recent autoradiographic observations have shown that much of the complex ultrastructural appearance of the interphase reticulum stems from the fact that segments of the chromatin strands have unravelled to various degrees and have sometimes even transformed into diffuse chromatin. Depending on the relative importance of this relaxation phenomenon, the chromatin reticulum within different interphase nuclei thus appears more or less regularly organized. The relation between these modifications of the macromolecular conformation of interphase chromatin and certain of its key biochemical activities will be discussed later on.

A futher interesting aspect of the organization of reticulate interphase nuclei concerns the mode of attachment of chromatin to the nuclear envelope. Under light microscopy, one cannot help noticing that various segments of the chromatin strands become perpendicularly oriented to the nuclear envelope as they attach to it. At the ultrastructural level, it is easily verified that a large proportion of the chromatin strands running close to the nuclear envelope are associated with it through several finger-like projections or stalks (Fig. 1.3). Numerous seemingly isolated chromatin masses, some of them well below the resolution level of the light microscope, are also found attached to the nuclear envelope. Serial sectioning would be required to verify whether these tiny chromatin masses might not correspond, as seems most likely, to the extremities of the fine stalks just referred to. Determination of the number of these sites of attachment of the interphase chromatin to the nuclear envelope would also permit verification of the recent

5

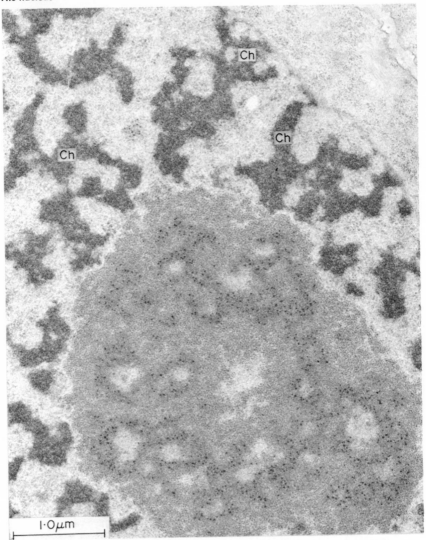

Fig. 1.3 Portion of interphase nucleus (*Allium porrum*). A number of chromatin strands (Ch) are seen to attach to the nuclear envelope. Other strands project towards the nucleolar surface and are closely associated with this organelle. In between these dense strands, the nucleoplasm consists of diffuse fibrillar and granular material. The peripheral portion of the nucleolus is composed predominantly of ribosome-like particles. More internally, the nucleolus contains a nucleolonemal reticulum. This convoluted filament is characterized by the presence of very dark particles somewhat larger than the nucleolar RNP granules. The light meshes within the reticulum show diffuse fibrillar material which is assumed to be chromatin. (Courtesy Mr P. Nadeau.)

hypothesis that they correspond to points of initiation of DNA synthesis (Comings, 1968; Comings and Kakefuda, 1968; Mizuno *et al.*, 1971).

Chromocentric nuclei. In other plants, an interphase chromatin reticulum either cannot be recognized in preparations stained by means of the Feulgen procedure or it appears as a faint meshwork almost merging with the surrounding nucleoplasm. The most striking feature of such interphase nuclei is the presence of dense chromatin masses which vary greatly in size, number, and distribution within the nuclear cavity depending on the species examined. Throughout the years, these dense chromatin structures have been referred to by different authors as prochromosomes, chromocentres and heterochromatic masses (discussed in Lafontaine, 1968). An abundant literature has dealt with the classification of such non-reticulate plant nuclei, certain authors distinguishing between prochromosomal and chromocentric nuclei and even arguing that the basic organization of the so-called prochromosomes was quite different from that of the chromatin strands characterizing reticulate plant nuclei. Critical analysis of this problem (Ris, 1945; Delay, 1948) revealed that these different chromosome forms could possibly be accounted for through differential coiling of basic strands, or chromonemata. Current biochemical and physico-chemical investigations of chromatin suggest that its specific degree of compaction in different nuclei is also most likely determined by a number of additional factors the nature of which is only beginning to be investigated (reviewed in Hearst and Botchan, 1970; Brasch, 1971).

The examination in this laboratory of a variety of plants with large hetero-chromatin masses at interphase (*Raphanus sativus, Tropaeolum majus*) has shown that these lumps consist of densely packed material, the macromolecular texture of which seems identical to that of the chromatin strands found in reticulate nuclei (*Allium porrum, Vicia faba*). The majority of these heterochromatic masses are found closely associated but not actually touching the nuclear envelope, except through short stalks similar to those observed in reticulate nuclei. The remaining non-nucleolar portion of the nuclear cavity consists of loosely dispersed fine fibrillar material. That much of this material corresponds to diffuse chromatin is clearly demonstrated by radioautography. Following tritiated thymidine incorporation, certain interphase nuclei show heavy labelling which is restricted to the large peripheral heterochromatic masses. In other interphase nuclei, these chromocentres remain unlabelled, the silver grains being instead scattered over the nucleoplasm. It would thus appear that, in such nuclei, an important portion of the chromatin is in a dispersed state and is responsible for part of the diffuse fibrillar substance observed within the nuclear cavity. It is also evident from these observations that, as has already been demonstrated (reviewed in Lima-de-Faria, 1969; Comings, 1972), heterochromatin replicates its DNA asynchronously with respect to diffuse or euchromatin.

Possible factors involved in nuclear structure diversity

Conformation of deoxyribonucleohistones. Early fractionation studies of isolated chromosomes revealed that chromatin consists predominantly of deoxyribonucleohistones (Mirsky and Ris, 1947). Following extraction of the histones, there remains a residual thread-like structure containing DNA, RNA and non-histone proteins. Further studies using chromatography and gel electrophoresis showed the existence of several histone fractions (reviewed in Murray, 1964). Certain data even led to the belief that the nature of these histone fractions varied significantly from species to species or in the different cell types of a given organism and nourished the hypothesis that cell differentiation might result from the specific interaction of a wide spectrum of histone types with different chromosome loci. Pitfalls in such an assumption soon became evident (Bloch, 1963). More refined biochemical analysis revealed the existence of a few major histone fractions only. Striking similarities were also observed between histones prepared from different organisms (reviewed in Fambrough, 1969; Smith et al., 1970).

Other biochemical studies, besides suggesting that histones play a regulatory action in gene function, have also underlined their possible influence in modulating the macromolecular conformation of chromatin. As a result, there has developed a renewed interest in the physico-chemical characterization of deoxyribonucleohistones. Such work has involved the study of native nucleohistones isolated from chromatin as well as the examination of reassociated complexes of DNA and histones. Hydrodynamic data (Gionnani and Peacocke, 1963) suggest that nucleohistones consist of single DNA molecules which, according to flow birefringence and circular dichroism experiments (Fasman et al., 1970; Simpson and Sober, 1970), are highly coiled or folded within this complex. X-ray diffraction studies of these macromolecules have likewise led to the conclusion that DNH fibres assume a supercoil configuration when maintained at high relative humidity (Richards and Pardon, 1970). Using calf thymus DNH, these authors showed that fibres that had been stretched and then allowed to form a stiff gel in water would give rise to much sharper reflexions than when the same preparations where in a relaxed state. According to the model proposed, the attachment of histones to DNA generates a supercoil the pitch and diameter of which are approximately 12 nm. In the relaxed state, DNA is assumed to be inclined to the axis of this supercoil but becomes re-aligned upon stretching.

Much interesting evidence in support of a complex conformation of deoxyribonucleohistones has also come from the examination of these macromolecules under electron microscopy. According to Bram and Ris (1971), isolated nucleohistone fibres are about 10 nm in diameter and exhibit knobby protuberances some 20–30 nm in total thickness. This irregular diameter of the DNH fibres is assumed to result from localized supercoiling or folding of the DNA double helix. Other workers are of the opinion that the nucleohistone complex, as present within chromatin, is characterized by first- and second-order coils which give rise to fibrils of correspondingly larger diameters. From densitometric measurements of whole-mount preparations of chromosomes, it has been inferred that one micron of the 23 nm wide chromatin microfibrils must contain some 56 μm of DNA (DuPraw and Bahr, 1969). The proposal was therefore made that the DNA double helix first forms a supercoil to give rise to 6–10 nm wide fibrils, the A fibrils (Fig. 1.4). Further coiling of the latter units forms the B fibrils which are some 20 nm in diameter (DuPraw, 1970).

This model of chromatin fibril structure has recently been elaborated further by Brasch (1971) as a result of this biochemical and ultrastructural study of chromatin in avian erythrocyte and liver nuclei. In this work, extraction of histone fraction I (lysine-rich) was observed to bring about a reduction in the diameter of the chromatin fibrils from 20 nm to approximately 10 nm. Sequential extraction of the other more tightly bound histone fractions gave fibrils of variable diameter down to 2.0 nm. On the basis of these observations, it has been hypothesized that these latter histones form a supercoil with DNA and thus give rise to the 6–10 nm fibrils (A fibrils) reported by many authors. This type of supercoil is also assumed to correspond to the DNH complex studied by Richards and Pardon (1970). According to Brasch, the specific role of the loosely bound,

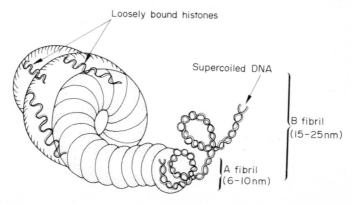

Fig. 1.4   Proposed model of chromatin fibril structure. In this model the double helix DNA molecule first assumes a supercoil configuration as a result of complexing with tightly bound histones. This nucleohistone fibre, from 6 to 10 nm in diameter, in turn forms a secondary supercoil thus giving rise to the B fibril. (Modified from Dupraw (1970) by Brash (1971).)

lysine-rich, histones would be to stabilize the A fibril into a secondary coil through interaction with several adjacent loops. As illustrated in Fig. 1.4, such secondary coiling of the nucleohistone complex accounts for the observed diameter (15–25 nm) of the B fibrils. The additional finding (Brasch, 1971) that the B fibrils transform reversibly to the A conformation when isolated nuclei are exposed to water and then brought back to saline seems to furnish strong support for the above model.

Secondary structure of chromatin. Granting, as now seems well established, that chromatin consists of basic fibrillar units some 15–25 nm in diameter, the question remains largely unanswered as to the various factors which determine its degree of compaction in different segments of the interphase chromosomes. Many observations point to the fact that condensation of chromatin reflects a specific packing of its constituent fibrillar elements and is not, therefore, the result of fortuitous aggregations of different portions of the chromosomal strands. Phase contrast investigations on isolated chicken erythrocyte and liver nuclei have shown, for instance, that chromatin first dispersed in water will recondense in the presence of saline and give rise to the original nuclear organization (Brasch, 1971; Brasch et al., 1971). Several hypotheses have been advanced to account for the particular distribution of condensed and dispersed chromatin observed in interphase nuclei of different species or in different tissues of the same organism. Since histones represent one of the major constituents of chromosomes and are known to interact readily with DNA through electrostatic bonds, it was generally believed, a few years back, that these proteins specifically controlled the degree of compaction of chromatin. The later finding that a very restricted number of histone fractions exist, and that these do not differ markedly in character amongst various animal and plant species (Bonner et al., 1968; Fambrough, 1969; Smith et al., 1970) indicates that this view is probably somewhat too simple.

The possibility remains that certain histones are more prominent in dense as compared to diffuse chromatin. A particularly favourable material for verifying this possibility is provided by the mealy bug. In the male, one haploid set of chromosomes is present as dense heterochromatic masses, in 47 per cent of the cells, while the whole chromosome complement is in a diffuse state in the female. Gel electrophoresis analysis of the histones prepared from the male and female revealed no significant difference in the nature of quantitative importance of the various fractions of these proteins (Pallotta et al., 1970). A comparative study of the histones isolated from HeLa cells interphase and metaphase chromosomes has shown that condensation of the chromosomes is not accompanied, either, by any important change in the content or nature of their different histones (Sadgopal and Bonner, 1970). However, it was noted that the cysteine residues which characterize histone III are oxidized in the condensed metaphase chromosomes, where they also mostly occur in a multimere form as a result of intermolecular disulphide bridges. It is thus conceivable, on the basis of these observations, that selective compaction of different proportions of the chromatin of interphase nuclei is controlled by similar changes in the conformation of certain protein fractions of chromosomes.

Another factor which may be involved in determining the structural configuration of certain chromosome regions is their content in highly repetitive DNA sequences. This DNA sometimes forms a satellite band in a caesium chloride density gradient and has so far been localized in the heterochromatic centromeric region of chromosomes in a few organisms (Pardue and Gall, 1970; Eckhardt and Gall, 1971) as well as in the heterochromatin of the sex chromosomes in the vole (Arrighi et al., 1970). Extensive DNA–DNA interactions between the repetitive sequences have been proposed to account for the compaction of centromeric chromatin (Botchan et al., 1971).

A further question which arises in the case of plant cells is why certain species (*Allium cepa, Vicia faba*) show a dense and highly organized chromatin reticulum in Feulgen preparations, whereas only the chromocentres are visible in the case of other species (*Phaseolus vulgaris; Raphanus sativus*). No definitive explanation exists of this most striking phenomenon. Nevertheless, it is known from early observations (reviewed in Delay, 1948) that plant species with conspicuous inter-phase chromatin reticulum are also characterized by long chromosomes. Con-vincing evidence has been furnished, in fact, that the chromosomes are 3—4 μm or more in length in these species, whereas interphase nuclei of the prochromosomal or chromocentric type possess shorter chromosomes. Certain observations show that the organization of the interphase nucleus remains of the non-reticulate type in related plant species with short chromosomes, even though the number of chromosomes may increase markedly as a result of polyploidy. Conversely, a reticulate organization is observed in species (e.g. *Crepis capillaris*) with few chromosomes, provided that the latter are sufficiently long. A second well-established characteristic of reticulate plant species is that they contain relatively large amounts of DNA per nucleus as compared to other plants, as well as to the large majority of animal species (reviewed in Lafontaine, in press). It finally emerges from recent biochemical studies that such plants with high DNA con-tents also have surprisingly large percentages of repetitive DNA (Britten and Kohne, 1969; Bendich and McCarthy, 1970; Chooi, 1971). As expected, plants of the chromocentric type exhibit a much lower percentage of redundant sequences (Ranjekar and Lafontaine, unpublished). Additional work will be required to verify whether this large proportion of repetitive DNA affects the compacity of the interphase chromatin in reticulate nuclei.

Clearly, then, no simple explanation exists of the diversified and specific organization of the interphase chromatin in different cell types. Besides the few factors just discussed, a number of others may also be involved in controlling the chromatin condensation pattern at interphase (reviewed in Hearst and Botchan, 1970; Brasch, 1971). More precise information on this problem should undoubtedly also throw some light on the well-known sequence of transforma-tion which chromosomes undergo prior to mitosis.

Replication of DNA in reticulate interphase nuclei

Modern cytological and biochemical studies have revealed that the interphase nucleus, far from being in a resting or quiescent state, as long imagined, is the most metabolically active stage of the mitotic cycle. Certain major activities of the interphase nucleus such as the synthesis of chromosomal deoxyribonucleo-histones and the transcription of RNA on chromosomes have attracted much attention during the last two decades or so. The vast body of information that has accumulated so far on various aspects of these most important subjects was reviewed by several authors in recent years (Bonner *et al.*, 1968; Darnell, 1968; Frenster, 1969; Prescott, 1970). In order to avoid undue repetition, the present section will mostly focus on the ultrastructure of plant interphase chromosomes in relation to their DNA synthetic activity.

Classical studies of dividing cells have revealed that the rod-like and well-defined mitotic chromosomes gradually transform throughout telophase. In certain plants, the chromosomes largely disappear from view except for scattered heterochromatic lumps of various sizes; in other species they give rise to a more or less conspicuous chromatin reticulum. By using the radioautographic technique in conjunction with electron microscopy, it has been possible to follow the sequence of transformations which the chromosomes of the latter species undergo throughout interphase and, most particularly, to identify nuclei of the early and late S periods (Nagl, 1968, 1970a, 1970b; Kuroiwa and Tanaka, 1971a, 1971b; Lafontaine and Lord, 1973; Lafontaine, in the press).

In the different reticulate plant species (*Allium porrum*, *Triticum vulgare*, *Vicia faba*) studied in this laboratory, the relatively small G_1 nuclei have rather irregular contours and show partly uncoiled chromosomes which are generally still aligned along the pole to pole direction. In appropriate preparations, the emerging small nucleolar bodies are seen in close association with certain of the unravelling chromosomes. By the S period, the nuclei have increased in size and become more roundish in outline. The chromosomes are now completely uncoiled and form a coarse chromatin reticulum the strands of which are approximately 0.3 μm in diameter. Following incorporation of tritiated thymidine for a 10—20 min period, two types of S nuclei become labelled. A certain proportion of these nuclei are heavily labelled and show a number of dense chromatin masses or chromocentres which, in Feulgen stained preparations, appear to be linked by a network of fine strands. At the ultrastructural level, it is observed that this partial disorganization of the reticulum results from a narrowing of the strands which now show most irregular contours. The further observation that a large proportion of the radioautographic grains are located over the nucleoplasm suggests that portions of the chromatin strands have transformed into diffuse material, presumably for DNA duplication purposes.

The second population of labelled interphase nuclei shows much more regular chromatin strands which have been transformed again into a rather conspicuous reticulum. Some of these nuclei are rather densely labelled while others exhibit only a few clusters of radioautographic grains, the majority of which are restricted to particular regions of the chromatin reticulum. In *Allium porrum*, certain of these regions are located at the surface of the nucleolus and consist of narrow twisted filaments forming more or less spherical structures. Taking into account that, in this species as well as in many other plants, heterochromatin masses are often closely associated to the nucleolar surface, it appears reasonable to infer that the regions under consideration are partially unravelled chromocentres. Since it is now also well established that heterochromatic chromosome segments are generally late replicating in a variety of organisms (reviewed in Lima-de-Faria, 1969; Prescott, 1970; Comings, 1972), one is led to the conclusion that the nuclei under consideration correspond to the end portion of the S period.

Considered together, these various observations show that much of the complex appearance of reticulate interphase nuclei, as recorded by either light or electron microscopy, stems from the fact that the chromosomes undergo important conformational changes from the G_1 to the late S periods. The observations carried out on *Allium porrum* (Lafontaine and Lord, 1973) suggest that the dense regular strands characterizing the G_1 period first transform to a diffuse state as different chromosome segments undergo replication. Judging from the ultrastructural organization of such early S nuclei, one is led to assume, in accord with different other observations (Plaut, Nash, and Fanning, 1966; Painter, Jermany and Rasmussen, 1966; Huberman and Riggs, 1968; Okada, 1968; Callan, 1972), that numerous chromosome segments unravel simultaneously for duplication purposes, thus accounting for the large proportion of the chromatin which is in a dispersed state during the first portion of the S period. The fact that a second population of the labelled nuclei exhibit a dense reticulum is best interpreted by assuming that, subsequent to replication, chromatin returns to a compact organization, a conclusion that has also been recently reached in the case of *Crepis capillaris* (Kuroiwa and Tanaka, 1971a, 1971b). These different studies with reticulate plant nuclei add further support, therefore, to the view (Hay and Revel, 1963; Tokuyasu *et al.*, 1968; Milner, 1969) that DNA duplication does not take place within dense chromatin, but is rather restricted to the immediately adjacent diffuse chromatin areas.

Organization and possible role of puff-like chromosome segments

Interphase plant nuclei frequently exhibit spherical structures possessing most interesting cytochemical and ultrastructural characteristics (Lafontaine, 1965). In *Allium cepa, Allium porrum* (Fig. 1.5) and a few other species with reticulated interphase nuclei, these structures occur at points of convergence of two chromatin strands and consist of a distinct meshwork of convoluted microfibrils embedded in an all-pervading amorphous material slightly denser than the surrounding nucleoplasm. Following double staining with uranyl acetate and lead hydroxide, these microfibrils match the chromatin in density and, in stereoscopic electron micrographs, they are actually seen to be continuous with the immediately adjacent chromosome segments. The latter observation, taken in conjunction with the loose ultrastructural organization of the spherical bodies in question strongly suggests, therefore, that they correspond to differentiated segments of the interphase chromosomes.

Further insight into the possible significance of these structures is provided by a study of their cytochemical characteristics. Although these bodies are difficult to analyse under the light microscope, their metachromatic staining with azure B indicates the presence of RNA. This conclusion is supported by electron microscopic observations of preparations either digested with ribonuclease or stained with uranyl acetate according to Bernhard's (1969) regressive technique (unpublished observations). Using this latter procedure the chromosomes remain unstained and show a totally amorphous texture whereas, on the contrary, the spherical structures under discussion maintain their finely fibrillar organization and stain as densely as the nucleolus and the perichromatin granules. That these spherules also contain proteins is revealed by their noticeably reduced electron-opacity following digestion with either pepsin or trypsin. The presence of DNA within these structures may finally be confirmed by using desoxyribonuclease. Considering the mode of action of the foregoing enzymes,

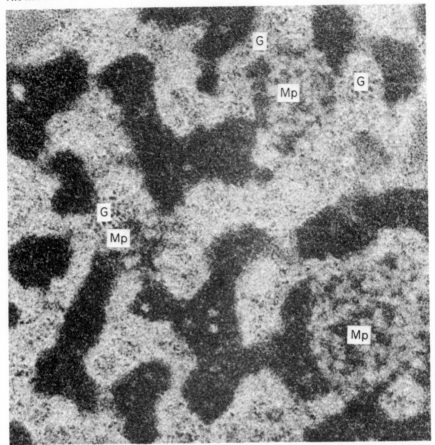

Fig. 1.5 Electron micrograph depicting a portion of a reticulate nucleus (*Allium porrum*).
It is noted that the margins of the dark chromatin segments are intimately associated with
much more diffuse fine fibrillar material which pervades the nucleoplasm. At places, portions
of these chromatin strands have unravelled into light spherical structures consisting of loose
convoluted microfibrils. Material slightly darker than the nucleoplasm permeates these
presumed micropuffs (Mp). The nucleoplasm is richly provided with diffuse fibrillar material
and also exhibits a variety of granules (G) of various sizes, some of which form clusters. The
background texture of the nucleoplasm is finely punctate and difficult to characterize.

it would appear that the latter bodies consist essentially of a meshwork of DNA
microfibrils the interstices of which are pervaded by an amorphous material containing both RNA and proteins.

   Owing to their ultrastructural organization and their apparent continuity with
more compact portions of chromatin strands, one is thus inclined to view these
spherical bodies as highly unravelled chromosome segments (Fig. 1.5). If the
above hypothesis is correct, it also follows that these chromosome segments
correspond to derepressed loci and, consequently, that they may be visualized
as micropuffs (Lafontaine, 1968; Lafontaine and Lord, 1969) actively involved

in RNA synthesis. Although direct radioautographic confirmation of such localized chromosomal RNA has not yet been obtained, the following comparison of these spherical structures with the loops of lampbrush chromosomes and puffs of salivary gland chromosomes furnishes some support for this view. First, as these latter well-known structures, the spherical bodies under discussion consist of highly distended microfibrillar elements which are continuous with the immediately adjacent, more condensed chromatin. Moreover, as the loops and puffs of giant chromosomes, these bodies contain RNA and proteins. Finally, these spherules are observed from mid-telophase to mid-prophase and thus appear at stages of the mitotic cycle during which chromosomes are generally believed to support RNA synthesis. Their number at interphase has not yet been established but, since ultrathin sections often exhibit five or six such spherules, one may roughly estimate that certain nuclei may contain up to two hundred.

Since, as argued in the preceding section, the bulk of non-nucleolar RNA synthesis in reticulate plant interphase nuclei must be taking place on portions of the chromosome strands which have dispersed throughout the nucleoplasm, the problem naturally arises concerning the significance of the micropuffs. Are they chromosome loci as actively engaged in RNA synthesis as diffuse chromatin? Could they, instead, possibly represent sites of RNA transcription which, contrary to diffuse chromatin, remain active for longer periods of interphase and thus manufacture types of RNA's which are essential throughout that stage? These questions must unfortunately remain unanswered at this time. Hopefully, the recently developed *in situ* molecular hybridization technique utilized in conjunction with radioautography under electron microscopy (Jacob *et al.*, 1971) could be of value to characterize these micropuffs further.

Ultrastructure of the nucleoplasm

From classical *in vivo* studies or examination of fixed preparations, the interphase nucleus has long been considered to contain an all-pervading amorphous substance: the nucleoplasm or karyolymph. Even though chromatin emerges as rather distinct masses, or in the form of a more or less conspicuous reticulum, in preparations stained according to the Feulgen procedure or photographed at 260 nm, the distinction between chromatin and surrounding nucleoplasm is usually much more difficult to establish in the electron microscope (Hay and Revel, 1963; Moses, 1964; Lafontaine, 1968). In spite of these difficulties, many interesting details have been resolved within the nucleoplasm, during the last decade or so. The existence of a whole spectrum of particles closely associated with or between the interphase chromatin regions has, in particular, been demonstrated in animal cells. This most interesting progress in the study of the nucleoplasm results largely from the development of new staining procedures with heavy metals (Watson, 1962; Bernhard, 1969; Monneron and Bernhard, 1969) as well as from the adaptation of well-known enzymatic digestion methods to electron microscopy (Swift, 1963; Bernhard and Granboulan, 1968).

Most investigations of the structure of plant cell nuclei have, so far, placed little emphasis on the nucleoplasm proper. One such report contained a description of its fine texture at interphase, prophase and telophase in *Vicia faba* meristematic cells (Lafontaine and Chouinard, 1963). This portion of the interphase and prophase nucleus was found essentially to consist of loosely and uniformly distributed convoluted fibrils similar to those observed within the chromosomes. A variety of dense granules, ranging from 15 to 30 nm in diameter and sometimes grouped into clusters, was also noted.

As a result of improvement in fixation and also of the utilization of Bernhard's (1969) regressive staining technique, the nucleoplasm of reticulate plant nuclei now appears much more structured than previously observed. First, as noted in a previous section of this chapter, it is presently evident that segments of the dense interphase chromatin strands loosen up at many places to the extent that their constituent fine fibrillar material merges imperceptibly with the immediately adjacent nucleoplasm. At sufficiently high magnification, local variations in the degree of diffuseness of this all-pervading fibrillar material can be noted (Fig. 1.5). Although part of this background fibrillar meshwork is presumably chromosomal in nature, differential staining by means of Bernhard's (1969) procedure reveals that it also consists of RNA and, presumably also, of ribonucleoproteins. This latter fibrillar material either appears as irregular compact masses closely appressed to the surface of the dense chromatin strands or takes the form of a loose network in between these strands. Numerous spherical structures, the presumed micropuffs described earlier, also become most conspicuous with this staining technique.

Apart from this more or less homogeneous distribution of fine fibrillar material, recent observations have also uncovered a whole spectrum of granules within the nucleoplasm of reticulate plant interphase nuclei. Certain of these are approximately 50 nm in diameter and usually form clusters close to the surface of the dense chromatin strands; others are found in the neighbourhood of micropuffs or appear to be immersed within zones of diffuse chromatin (Figs. 1.3 and 1.6A). These granules most likely correspond to the perichromatin granules of other authors (Swift, 1963; Bernhard and Granboulan, 1968; Monneron and Bernhard, 1969). Another reasonably well-defined category of particles is also observed in plant interphase nuclei. These are from 20 to 25 nm in diameter and usually escape attention in standard electron microscopic preparations unless, as is sometimes the case, they are grouped into clusters. When the density of chromatin is reduced by means of Bernhard's regressive staining technique, however, such granules are brought into sharper focus and their quite widespread distribution throughout the nucleoplasm is then much more easily established. Close examination of high magnification electron micrographs reveals a more finely punctuate texture of the nucleoplasm than that due to the presence of the 25 nm granules. This background particulate texture may well result from sharp twists and kinks in the fine underlying fibrillar meshwork.

There seems little doubt from these recent studies that the nucleoplasm is far from the structureless ground substance in which chromatin, the nucleolus and occasional inclusions were once assumed to be suspended. In functional

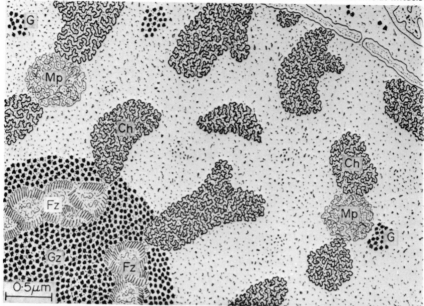

Fig. 1.6A   Diagram depicting the various morphological components of an interphase nucleus. The chromatin strands (Ch) are of rather uniform diameter except at places where they unravel to form micropuffs (Mp). Many of these strands are intimately associated with the nuclear envelope. The nucleoplasm, besides exhibiting fine fibrillar material, also contains granules (G) of various sizes.

The nucleolus, apart from vacuoles, essentially consists of two types of zones. The granular zones (Gz) extend throughout the peripheral portion of the nucleolus but are also found more centrally in the form of irregular patches. Other regions, the fibrillar zones (Fz), are characterized by a quite complex organizational pattern and, at places, are seen to be continuous with segments of chromatin projecting within the nucleolar mass.

terms, it now seems meaningless to attempt to dissociate the chromatin, especially that portion which is in a diffuse state, from the nucleoplasm since these two nuclear components are physically integrated and are presumably in constant intimate interaction.

Structure of the Chromosomes at Mitosis

General consideration

For many decades following the discovery of mitosis and meiosis, chromosomes were under intensive investigations at the light microscope level. From these studies, some concensus was reached concerning their gross morphology and certain finer details of their structural organization as determined by the presence of centromeres, secondary constrictions and nucleolar organizer regions. Other morphological characteristics were, on the contrary, the objects of much debate over the years. Foremost amongst these were the problems of the nature of the chromomeres, and that concerning the degree of strandedness of chromosomes during various stages to the cell cycle (reviewed in Kaufmann, 1948; Ris, 1957; Moses, 1964).

With the advent of electron microscopy in the early fifties, students of chromosome organization held great hopes of rapidly settling these controversial issues. It was likewise assumed that the much improved resolving power thus provided would permit rapid progress in correlating the well-documented variations in gross morphology of chromosomes during the mitotic cycle with recognizable changes in their macromolecular organization. As has already been pointed out by a number of authors, definite answers to these basic aspects of chromosome structure have proved unexpectedly difficult to obtain. It has also been clearly realized, during the last two decades, that the structural organization of chromosomes at the chromatid or subchromatid level is, unfortunately, rather poorly reflected in electron micrographs recorded from conventional ultrathin tissue sections. For these reasons, much effort has recently been devoted to devise preparation procedures more appropriate to unravelling the organizational pattern of chromosomes (reviewed in DuPraw, 1970; Ris and Kubai, 1970).

Evolution of the gross morphology of chromosomes from early prophase to telophase

Preprophase and prophase. The transition of a reticulate nucleus from interphase to early prophase involves a sequence of changes in the overall distribution of the chromatin strands, in their density and diameter which have so far not yet been fully analysed by either light or electron microscopy. The most revealing information on the almost imperceptible changes which interphase strands undergo as they gradually give rise to easily recognizable prophase chromosomes was provided by phase contrast cinematographic studies of cultured *Haemanthus katherinae* endosperm cells (Bajer and Molè-Bajer, 1963; Bajer, 1965). In this material, the first noticeable change in interphase nuclei preparing to enter mitosis is a slight increase in the density and coarseness of the slender chromatin strands. There is then observed a slight redistribution of these strands within the nuclear cavity following which the chromosomes are first seen to emerge. This subtle transformation cannot be more aptly described than by quoting these authors: 'The whole nucleus differentiates into numerous but indistinctly separated territories or areas, each of them representing a future chromosome.' During a brief period, minute spirals are formed but these soon disappear and are therefore seldom observed in fixed material. It would thus appear that the fine strands previously meandering throughout the nuclear cavity first undergo a certain degree of contraction, become slightly coiled and, following conformational changes still poorly understood, eventually give rise to the early prophase chromosomes. The later structures, in the species, are sometimes distinctly double but may also only become so slightly later. This most illustrative sequence of events has unfortunately not yet been recorded under electron microscopy.

Although not providing the unique advantages of endosperm cells, meristematic cells (*Allium porrum, Triticum vulgare*) have also been found useful for following the evolution of chromosomes during the mitotic cycle (Lafontaine,

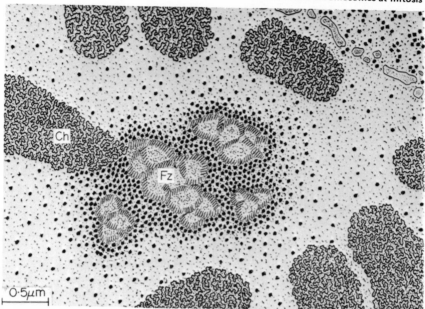

Fig. 1.6B   This diagram illustrates the organization of the nucleolus, at late prophase, when this organelle has reached an advanced stage of disorganization. At that time, it is noted that the nucleolar RNP particles have dispersed throughout the nucleoplasm but that the fibrillar zones (Fz) still show a coarse reticulum consisting of the convoluted nucleolonemal filament. The meshes or lacunae of this reticulum contain loose fibrils and a few particles.

1968; Lafontaine and Lord, 1969). Judicious use of consecutive thick (0.5–1 μm) and ultrathin sections for light and electron microscopy, respectively, greatly facilitates the task of identifying the different stages of the cell cycle and of correlating new ultrastructural details with available cytological information. In fixed material, the interphase chromatin reticulum seems more extensive than indicated by the above observations on living endosperm cells. The first sign of the evolution of an interphase nucleus, as noted in 0.5 μm preparations, consists in a slight reorganization of the coarse reticulum characterizing the late S period and a disappearance, here and there, of some of the fortuitous contacts between neighbouring segments of the chromatin strands. As in the case of *Haemanthus* late interphase nuclei, this gradual disorganization of the reticulum is accompanied by a corresponding redistribution of the chromatin strands. Once these convoluted chromatin filaments have segregated to distinct portions of the nucleus, the loosely coiled early prophase chromosomes become recognizable as such. In the course of this emergence of the early prophase chromosomes, but slightly before these become clearly discernable, twin parallel strands may be noted within the peripheral portion of the nucleus or in close association with the nuclear envelope. Such paired chromatin elements undoubtedly result from duplication of the deoxyribonucleohistones during the preceeding S period. However, it is far from clear how these twin structural units become separated from the mother interphase strands. The observation that many of the paired elements appear in intimate contact with the nuclear envelope may be most significant in this respect and may imply that this structure is involved in the process of getting daughter strands organized into the emerging prophase chromosomes.

As a result of continued spiralization of the slender chromatin strands during early prophase, the chromosomes eventually take on the form of quite regularly coiled structures. Viewed in cross section, in 0.5 μm preparations or under electron microscopy, these condensing chromosomes appear as ring-like structures with a central light core consisting of fine fibrillar material and occasional granules of the type found within the nucleoplasm. By mid-prophase, the daughter chromatids are loosely coiled around each other and, as a result, the bipartite organization of the chromosomes has thus become much more conspicuous. At that stage, the nucleolus generally exhibits quite irregular contours; segments of the dense chromosomes are then found running into these large nucleolar indentations and establishing contacts with the surface of this organelle through numerous finger-like chromatin projections. At the periphery of the nucleus, portions of chromosomes remain attached to the nuclear envelope by means of fine stalks similar to those observed at earlier stages (Sparvoli et al., 1965; Brinkley and Stubblefield, 1970; Lafontaine, unpublished).

As condensation of the mid-prophase chromosomes progresses, the light zones in between the gyres of the coiling chromonemata gradually decrease in size and, by late prophase, they have usually disappeared altogether. At that stage, the chromosomes are largely located within the peripheral portion of the nucleus which also exhibits large amounts of RNP particles originating from the dispersing nucleolus.

Other important changes undoubtedly take place during condensation of the chromosomes throughout prophase. In *Leucojum aestivium* endosperm cells, for instance, much nuclear material was found to migrate progressively to the cytoplasm during that stage. The total amount of material thus lost, according to interferometric measurements, corresponds to approximately 75 per cent of the dry mass of the preprophase nucleus (Richards and Bajer, 1961). It is not clear whether this material originates from the chromosomes or from the nucleoplasm proper. Likewise, the possible significance of this migration phenomenon is therefore still uncertain.

Metaphase and anaphase. Following dispersion of the nucleolus, the nuclear envelope takes on undulating contours and usually exhibits a number of small breaks, here and there. As the latter structure becomes more disorganized, fragments break off and begin to migrate to the periphery of the spindle area. Other portions of the envelope remain attached to the chromosomes and, according to some observations, even maintain this association throughout metaphase and anaphase (Comings and Okada, 1970). Once the chromosomes have reached the cell equator, their two constituent chromatids may easily be recognized in either longitudinal or cross-sections. They appear as dense structures consisting primarily of closely packed convoluted microfibrils and of scattered granules identical to

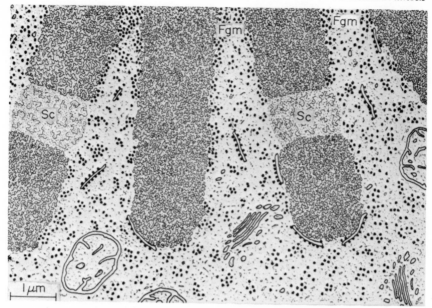

Fig. 1.6C   In this diagram, the early telophase chromosomes are still pointing to the equator of the cell and have not yet begun to unravel. Segments of membranous vesicles are moving to the surface of the chromosomes where they will attach and coalesce to initiate formation of a new nuclear envelope. A fibrillogranular material (Fgm) is observed at that stage in between the chromosomes or intimately associated with their surface in the form of a thin irregular coating. At that stage, the nucleolar secondary constrictions (Sc) consist of fine fibrillar material which typically stains much less intensely with heavy metals than the adjacent chromosome segments.

those found within chromosomes at earlier stages. A quite similar ultrastructural organization characterizes the anaphase chromosomes. Grazing sections of either metaphase or anaphase chromosomes, in 0.5 μm preparations, strongly support the classical view that these structures consist of helically arranged elements (reviewed in Kaufmann et al., 1960). However, except for their serrated contours little, if any, evidence of such an organizational pattern can be recognized within mitotic chromosomes under electron microscopy (Lafontaine and Chouinard, 1963: Barnicot and Huxley, 1965: Bajer and Molè-Bajer, 1969). Whether this masking of the internal organization of metaphase and anaphase chromosomes results from the presence of a matrix substance in between the coiled chromonemata or is simply due to the tight spiralization of these units remains to be determined. That this latter factor is certainly partly responsible for the compact organization of mitotic chromosomes is evident from measurements carried out in *Haemanthus* living endosperm cells (Bajer, 1959). In this material, a period of rapid shortening is observed a few minutes prior to disolution of the nuclear envelope. This process continues at different rates during subsequent periods of mitosis and, by telophase, the chromosomes may be as much as 75 per cent shorter than at early prophase.

Telophase. Telophase nuclei vary greatly in appearance depending on the material examined. In plants with chromocentric interphase nuclei (Lafontaine and Lord, 1969) as well as in many animal cells (Robbins and Gonatas, 1964; Stevens, 1965; Noel *et al.*, 1971) the late anaphase chromosomes become closely appressed and form a large irregular chromatin mass at each pole. Elongate membranous cisternae and numerous roundish vesicles are recognized in the immediate vicinity of this seemingly fused chromatin mass or already associated with parts of its surface. A progressive relaxation of chromatin takes place throughout telophase and, by the end of that stage, only a few chromocentres or heterochromatic masses remain visible.

In nuclei of the reticulate type, on the contrary, the chromosomes remain well aligned at the cell poles during telophase, as they uncoil, and their evolution is, therefore, much more easily analysed. Early telophase chromosomes in these species (*Allium porrum, Vicia faba*) are usually still quite compactly organized but may nevertheless show evidence of a certain degree of uncoiling (Lafontaine and Chouinard, 1963; Sparvoli *et al.*, 1965; Lafontaine and Lord, 1969). Portions of the nuclear envelope are then seen closely following the outer irregular contours of the chromosome set and, at places, invaginating in between the chromosome arms still pointing toward the cell equator. As unravelling of the chromosomes progresses futher, their complex spiral internal structure becomes apparent. In 0.5 μm-thick preparations, the impression is gained that each chromosome consists of two distinct subunits or chromonemata loosely coiled around each other. Corresponding electron micrographs show much more elaborate chromosome profiles which are difficult to interpret in terms of only two constituent subunits (Sparvoli *et al.*, 1965; Lafontaine and Lord, 1969). By early interphase, individual chromosomes are no longer recognizable as such and have transformed into complex reticulate structures. These structures eventually come to occupy all of the non-nucleolar portion of the nuclear cavity and, as a result of their proximity, touch one another at various points thus giving rise to the apparent chromatin reticulum characteristic of interphase.

Centromere and secondary constriction. Most of our present knowledge of the ultrastructure of the kinetochore or centromere comes from studies of animal cells, especially under synchronized conditions (reviewed in Brinkley and Stubblefield, 1970). According to these authors, the early prophase centromere is first recognized as a denser filamentous zone located in a small indentation on each side of the chromosome. This structure subsequently elongates during pro-phase and gives rise to a band which may be up to 0.5 μm long and is intimately associated with the surface of both sides of the prophase chromosome. At sufficient magnification, the centromere is then seen to consist of a denser axial portion 20–30 nm in diameter and, on each side, of a more translucent zone from 20–60 nm thick. By late prophase and prometaphase, the nuclear envelope has generally broken down to a noticeable extent and microtubules become attached to the centromere. The fine structural organization of the centromere is studied to better advantage at metaphase and anaphase. At these stages, they are observed to consist, in fact, of two adjacent filaments, each of which is in turn composed of two structural components: an axial denser zone and several fibrils or loops projecting more or less perpendicularly to this central element. Other somewhat different views of the organization of the centromere in animal (Jokelainen, 1967) and plant cells (Bajer, 1968; Bajer and Molè-Bajer, 1969; Bajer and Molè-Bajer, 1972) have also been presented.

Apart from the above progress in resolving the macromolecular organization of the centromere, certain information has also recently been obtained concerning some of its biophysical characteristics. These new developments concern the discovery that, in various species where this chromosomal region is heterochromatic, it contains highly repetitive DNA (Jones, 1970; Pardue and Gall, 1970; Botchan et al., 1971; Eckhardt and Gall, 1971; Bostock et al., 1972). It is not clear, as yet, whether the presence of redundant DNA represents a general characteristic of the centromere and, if so, how this property is related to its unique and most specialized functions during mitosis.

Besides the centromere and other heterochromatic segments, chromosomes also exhibit other differentiated regions known as secondary or nucleolar constrictions. Studies of these Feulgen-negative chromosome segments have developed into a most active chapter of cell biology during the last decade and have led to a whole new body of information regarding their key role in nucleologenesis. This subject will be covered in detail in a later section of this chapter.

Strandedness of the chromosomes

Much of the difficulty in interpreting available ultrastructural data on chromosomes stems not only from the fact that they are evidently most intricate structures but also from persisting uncertainties as to the number of basic strands they contain. The notion that mitotic chromosomes are multistranded was largely documented by classical authors and became widely accepted in the thirties (reviewed in Kaufman, 1948). Certain recent studies in a number of laboratories have led to a similar view (reviewed in Swift, 1965; Wolff, 1969; Prescott, 1970; Ris and Kubai, 1970). The weight of this conclusion rests on various lines of evidence. The most convincing demonstration, perhaps, of chromosome multistrandedness comes from the very elegant phase contrast observations on living endosperm (Bajer, 1965). In these cells, chromatid subunits first become apparent at late prometaphase or metaphase and are most distinct during middle anaphase. During telophase, the chromosomes loosen up and each appears to consist of at least two if not four structural units (see Fig. 1.6D). Corresponding subunits are also clearly recognized in squash preparations following unravelling of mitotic chromosomes by means of various treatments. Trosko and Wolff (1965), for instance, obtained particularly well resolved subchromatid units in *Vicia faba* metaphase chromosomes fixed in neutral formalin and subsequently uncoiled by trypsinization. The presence of a complex array of ultrastructural fibrillar elements in mitotic chromosomes which sometimes appear to be organized into larger subunits or chromonemata has also frequently been invoked in favour of the multi-strandedness thesis (Kaufmann et al., 1960; Sparvoli et al., 1965).

Labelling experiments of both plant (Taylor et al., 1957) and animal cells

23

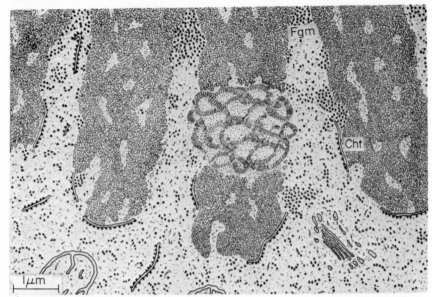

Fig. 1.6D   Diagram of part of a mid-telophase nucleus. The chromosomes have unravelled to some extent and are seen to consist of loosely coiled subunits or chromonemata (Cht). The process of nuclear envelope formation is now well under way. Of especial interest, at this stage, is the appearance within the nucleolar secondary constriction of a glomerulus-like structure, strikingly reminiscent of the convoluted nucleolonemal filament characteriz-ing the mature nucleolus. As a result, a small spherical structure, the emerging nucleolus, may now be observed. At this point in its evolution, the nucleolar body thus essentially consists of the tightly organized nucleolonema and of pervading fine fibrillar material. Note that patches of fibrillo-granular material are still evident in between the chromosomes.

(Prescott and Bender, 1963) have played a central role, during the last decade or so, in studies of the strandedness of chromosomes. When cells labelled for short periods are allowed to proceed to mitosis (M_1) and, then, to go through a second replicating period (S_2) in the absence of label, it is noted that the chromosomes reaching the second metaphase (M_2) show a semiconservative distribution of radioautographic grains (Taylor et al., 1957). Such observations constitute for certain authors a clear demonstration of the unistrandedness of chromosomes (discussed in Prescott, 1970). Several researchers, however, have reported varia-tions to this labelling pattern which lead to both chromatids being labelled at M_2 (LaCour and Pelc, 1958; Peacock, 1963). This isolabelling of metaphase chromo-somes, as it is referred to, has generally been interpreted in terms of multistranded models (Wolff, 1969).

Even though the above observations would appear to provide compelling evidence in favour of a polynemic organization of chromosomes, debate on this issue has been continuing with sustained vigour, during recent years, in view of the series of equally convincing arguments presented by proponents of the uninemic thesis. One such key argument is that, in spite of the much improved and even striking three-dimensional overall view provided by whole-mount prepa-rations, no suggestion of longitudinal split is observed under electron microscopy, along chromatids of metaphase chromosomes. According to models which have been proposed, each chromatid would consist of a single fundamental deoxyribonucleoprotein fibre (Fig. 1.4) some 25 nm in diameter (DuPraw, 1970). The folding pattern of this fibre along metaphase chromosomes is assumed to account for the apparent paired subunits observed within the chromatids following certain preparation procedures. Further support in favour of the uninemic hypothesis has come from studies utilizing X-rays to induce chromosome breaks. For that purpose, cells in known phases of the mitotic cycle are irradiated and the chromosomal aberrations thus produced are examined at the following metaphase. In experiments carried out with *Vicia faba* root tip cells, whole chromosome breaks were produced from cells irradiated during the G_1 period; on the contrary, chromatid breaks were induced if the time of irradiation coincided with the S or G_2 periods (Wolff and Luippold, 1964). Such results have generally been taken to indicate that, since the two units of the G_1 chromosomes seem to be sufficiently close to break together, they most probably correspond to the two strands of a DNA molecule. This view is also consistent with digestion experiments of lampbrush chromosomes with deoxyribonuclease (Gall, 1963) as well as with recent electron microscope observations of these structures (Miller and Beatty, 1969). Apart from the above most suggestive findings, the original demonstration (Taylor *et al.*, 1957) of the semiconservative distribution of labelled DNA among chromosomes also provides strong support to the uninemic concept. Other equally important arguments for or against this theory are often also cited (Swift, 1965; Wolff, 1969; Ris and Kubai, 1970; Thomas, 1971).

A new and most promising approach to the problem of chromosome strandedness is based on measurements of renaturation kinetics of denatured eukaryote DNA. This technique has, so far, largely been utilized to detect the presence of repetitive sequences within the genome of various organisms (reviewed in Britten and Kohme, 1969). By appropriate use of internal DNA standards during measurements of renaturation kinetics, the sensitivity of the method can apparently be improved to the point of revealing whether the various stretches of DNA within an haploid chromosome set are present in more than one copy (Laird, 1971). According to this author, most of the non-repeated DNA sequences in the mouse sperm are unique and each chromatid must therefore be considered as unineme, at least at the informational level. Further studies will be required to verify whether, in this species, the chromatids really consist of a single DNA molecule or are instead formed of two or possibly more such molecules each characterized by different sequences along the non-repeated stretches. Extension of these studies to other material should provide more decisive data, in relation to the problem of chromosome strandedness, than available so far.

Organization of the interphase nucleolus

Ultrastructure

The nucleolus has been recognized from early cytological observations to represent an essential component of eukaryote cells. Due to its variation in size and shape in different cell types and, presumably, also to its lack of continuity throughout the mitotic cycle, the function of this organelle in the economy of the cell has long remained obscure. A surprisingly wide spectrum of hypotheses were therefore advanced, at various periods, to account for its possible role within the nucleus. One view was that this organelle acts as a sort of reservoir which collects material for formation of the prophase chromosomes. This assumption was undoubtedly inspired by the observation that the nucleolus, a quite conspicuous structure at interphase, disappears at late prophase at a time when the condensing chromosomes become much more heavily stained. Another popular notion, at one time, was that the nucleolar substance contributes to formation of the spindle. Several additional assumptions are discussed in earlier reviews on the subject (Gates, 1942; Vincent, 1955).

In plant cells, the nucleolus is generally a quite conspicuous structure and, in certain species, may occupy a surprisingly large proportion of the nuclear cavity. Such large nucleoli usually result from the fusion of two or more nucleolar bodies formed at telophase: this accounts for the dumb-bell appearance of certain nucleoli, at later stages, as well as for the presence of several organizers within their mass.

The fact that different types of structure were mistakenly taken for nucleoli may account for the great variability in organization which are to be found in early descriptions of this organelle. While there is at present no doubt that plant nucleoli are characterized by a rather heterogeneous internal structure, their organizational pattern does not seem to vary appreciably with species (Lafontaine, 1968) contrary to the situation observed in animal cells (reviewed in Bernhard and Granboulan, 1968; Hay, 1968; Busch and Smetana, 1970). Following staining of $0.5-1$ μm sections with methylene blue, plant nucleoli exhibit at least two components, each segregated into zones of different densities (Fig. 1.1). Under electron microscopy, the lighter nucleolar regions are found to consist predominantly of rather loosely packed particles the density and size of which closely resemble the cytoplasmic ribosomes (Fig. 1.3). This granular material quite generally occupies the peripheral portion of the nucleolus and forms irregular patches within its mass (Lafontaine and Chouinard, 1963; Hyde, 1967; Lord and Lafontaine, 1969; Jordan and Godward, 1969). Similar material normally also surrounds the large and medium sized nucleolar vacuoles and is loosely dispersed within their cavities together with fine fibrils, from 6 to 10 nm in diameter.

The remaining portions of the plant interphase nucleolus appear somewhat more opaque in preparations stained with methylene blue, and confer a rather complex organization on this organelle. These latter zones invariably occupy more central regions of the nucleolus (Fig. 1.1). Further information on the internal structure of these more opaque nucleolar zones is obtained by exposing cells to a 0.1% aqueous solution of 'Tween 80' for brief periods so as to disperse the nucleolus. During such treatment, a complex skein-like structure, or nucleolonema (Fig. 1.2), is gradually seen to emerge from the denser nucleolar portions (LaCour, 1966; Lord and Lafontaine, 1969; Lafontaine and Lord, 1969). According to observations by LaCour (1966), one such coarse filamentous structure is present in the interphase nucleoli which have remained separate. Following fusion, however, the resulting single interphase nucleolus exhibits as many nucleolonemata as the species contains nucleolar chromosomes. In the triploid endosperm nuclei of *Scilla sibirica*, for instance, three such structures are observed whereas only two appear in nucleoli from diploid roots. In the various plant species studied by this author, the nucleolonemata are surprisingly long (35–55 μm) and approximately 1.4 μm in width. The evidence in support of this author's hypothesis, that the structures in question correspond to loops originating from specific chromosomes, will be discussed in a later section of this chapter.

At the ultrastructural level, the denser nucleolar portions under discussion consist of closely packed fibrillar material and exhibit a quite complex organization due to the presence of a convoluted nucleolonemal structure quite similar in appearance to the skein revealed under phase contrast microscopy in Tween-treated nucleoli. One striking feature of these fibrillar zones is the presence of a large number of lighter regions which appear to be closely associated with the nucleolonema. This relationship of the lighter nucleolar regions to the nucleolonema is seen to much better advantage in nucleoli, such as the one illustrated in Fig. 1.3, which exhibit particularly opaque particles within their fibrillar zones. The increased conspicuousness of the nucleolar skein, resulting from the presence of such opaque granules, brings into sharper focus the fact that the numerous light nucleolar zones in question correspond to the meshes of a coarse reticulum. A closer examination shows that their ultrastructural texture is quite similar to that of chromatin and, in favourable preparations, it is apparent that these lacunae are somehow intimately associated with chromosome segments which project within the nucleolar mass. In view of the most important role that intranucleolar DNA is presently known to play as the site of transcription of preribosomal RNA, the chromatin-containing skein is undoubtedly a key component of nucleolar architecture in plant cells. Certain cytochemical characteristics of this structure will therefore be presented further on, together with a description of its evolution during the mitotic cycle.

Biochemical composition and cytochemistry of the nucleolus

Although early attempts were made at characterizing the nucleolus by means of various staining procedures as well as by digestion with certain enzymes, the demonstration that this organelle contains RNA (Brachet, 1940; Caspersson and Schultz, 1940) furnished the first really reliable information concerning its composition. Using Unna's methyl green-pyronin staining procedure in conjunction with ribonuclease digestion of histological preparations, Brachet succeeded in showing the presence of RNA within the nucleolus. Cytospectrophotometric measurements in Caspersson's laboratory soon confirmed this important finding and also established that a large proportion of cellular RNA was actually located within this organelle. Further insight into the biochemical composition of the nucleolus awaited the development of cell fractionation techniques which eventually led to the preparation of isolated nucleoli in sufficient quantities for biochemical analysis (reviewed in Vincent, 1955; Busch and Smetana, 1970). Unfortunately, a very restricted number only of these various biochemical studies concern purified plant nucleoli, presumably due to the greater difficulty in handling plant tissues (reviewed in Birnstiel, 1967).

RNA. Biochemical studies of nucleoli isolated from various sources have shown these organelles to contain from 5 to 20 per cent RNA. The development of refined separation techniques, such as sucrose gradient sedimentation, column chromatography or polyacrylamide gel electrophoresis have made possible significant progress in the characterization of nucleolar RNA's (Birnstiel, 1967; Loening, 1968; Busch and Smetana, 1970). According to available evidence, these various classes of nucleolar RNA are typical of this organelle and are related to the different types of ribosomal RNA. This most interesting relationship between nucleolar and ribosomal RNA has been the object of intensive investigation in recent years and will be considered in the next section of this chapter.

Cytochemical studies have revealed that RNA is distributed throughout the nucleolus but that its concentration varies within its mass (Swift, 1962; Hay and Gurdon, 1967; Lord and Lafontaine, 1969). In plant cells, for instance, stainability of the nucleolus with azure B is completely removed following digestion with ribonuclease. By utilizing consecutive ultrathin and 0.5 μm preparations, it can be shown that those nucleolar areas which stain more intensely with azure B are granular in nature, while the lighter ones consist predominantly of densely packed fibrillar material interspersed with a few particles. In view of the specificity of this stain, under appropriate conditions, it is thus apparent that the former nucleolar zones contain higher concentrations of RNA. This conclusion is supported by biochemical data showing that isolated nucleoli have a much lower RNA : protein ratio (1:7) after extraction of their RNP particles (Hyde *et al.*, 1965). This ratio, as is well known, is roughly 1:2 for the nucleolar particles themselves. At the ultrastructural level, ribonuclease digestion completely disorganizes the nucleolar particles which are then no longer recognizable as such. Material also seems to be extracted from the fibrillar zones, but the nucleolonema present within these nucleolar areas remains unchanged in opacity and is therefor much more conspicuous than in untreated preparations (Lord and Lafontaine, 1969).

Proteins. It is now generally accepted that, besides DNA and RNA, nucleoli of both plant and animal cells consist mostly of proteins. According to current estimates, the protein content of nucleoli ranges from 70 to 85 per cent. The latter substances are apparently quite varied and comprise different enzymes such as RNA polymerase, nucleolar methylases, NAD pyrophosphorylase and ATPases (Busch and Smetana, 1970). Analysis of nucleolar proteins have not reached, by far, the degree of sophistication of current studies on ribosomal proteins. So far, nucleolar proteins have been fractioned into three or four broad classes according to their solubility in dilute saline and in HCL (Birnstiel *et al.*, 1964; Grogan and Busch, 1967). When nucleolar proteins are separated by polyacrylamide gel electrophoresis, several bands are obtained thus indicating their quite diversified nature.

Information on the distribution of proteins within the nucleolar mass has been sought by various means. Radioautographic studies have, for instance, furnished ample evidence that proteins are present within both the fibrillar and granular nucleolar zones. However, the results thus obtained do not indicate whether certain types of protein are located preferentially, or in larger amounts, in one or the other of these nucleolar regions (Mattingly, 1963; Chouinard and Leblond, 1967; unpublished observations).

Potentially, cytochemical techniques should be more useful for localizing certain broad classes of proteins within the nucleolar mass. Since basic proteins and, according to some authors, histones are present in non-negligible quantities within the nucleolus, attempts have been made to reveal these proteins by means of the alkaline fast green technique. Several authors have reported negative results with this method but others, using frozen-dried plant material, have, on the contrary, observed a positive reaction over nucleoli and have, therefore, taken this as evidence for the presence of histones. Selective staining of nucleolar proteins with amino black 10 B has also recently been reported but the types of proteins thus revealed remain unknown (Mundkur and Brauer, 1966).

Use of enzymatic digestion in conjunction with electron microscopy has, finally, been exploited in efforts to pinpoint the distribution of the various nucleolar proteins. Pronase, a proleolytic enzyme with a particularly wide spectrum of action, is found to extract much of the nucleolar mass, leaving behind a coarse reticulum of fibrillar material interspersed with small granules (Monneron, 1966). Much more restrictive effects are observed when more specific enzymes such as pepsin or trypsin are utilized. Following digestion of animal or plant cells with pepsin, nucleoli appear somewhat more translucent as a whole and the RNA particles characterizing the granular zones become particularly more conspicuous (Marinozzi, 1964: Lord and Lafontaine, 1969). These results are best explained by assuming that pepsin extracts an amorphous substance or matrix which pervades the nucleolar mass. Although the mode of action of pepsin is pretty well understood, as far as *in vitro* systems are concerned, it is difficult at this time to infer what types of proteins are actually extracted by this enzyme from fixed preparations.

Information on the effect of trypsin on nucleolar ultrastructure is presently quite meagre. Most authors agree that the extent of its action on cell structures varies considerably with fixation conditions. There is, nevertheless, some evidence that it disorganizes the RNP granules found in the nucleolus and, eventually, will also extract other material throughout its mass. Part of this extraction could possibly be attributed to presence of traces of ribonuclease which have been reported in commercial preparations of trypsin. The spectrum of action of trypsin being quite narrow, one is tempted to assume that this enzyme specifically hydrolyses certain classes of nucleolar proteins, especially the more basic ones. However, it is not clear that cleavage by trypsin of peptide linkages involving either arginine or lysine would extract basic proteins more selectively since, according to available biochemical data (Busch and Smetana, 1970), the percen-

29

tage of these two basic amino acids differs only slightly within the different major classes of nucleolar proteins.

The variety and quantitative importance of nucleolar proteins suggests that they fulfil a number of key functions within the nucleolus. It was already noted that certain of these proteins are enzymes, some of which are concerned with the metabolism of nucleolar RNA. Evidence has also been accumulating to the effect that many of the nucleolar proteins are complexed with RNA. It appears, for instance, that certain nucleolar proteins, presumably those which are more basic in character, interact with newly synthesized RNA and thus stabilize this RNA until it is processed further within the nucleolar mass (Vincent et al., 1966; Perry, 1967; Mundell, 1968). Since 45 S RNA has been isolated as a stable 60 to 110 S ribonucleoprotein complex, it most likely exists in this form within the fibrillar zone of the nucleolus where it is transcribed. Various preribosomal RNA intermediates also exist as ribonucleoprotein complexes of characteristic sizes and sedimentation constants which can be isolated from nucleoli of both animal (Izawa and Kawashima, 1969; Liau and Perry, 1969; Warner and Soeiro, 1967) and plant (Birnstiel et al., 1963) cells. Judging from their overall morphological characteristics, certain classes of these complexes undoubtedly originate from the granular portion of the nucleolus (Shankar Narayan and Birnstiel, 1969). The problem of the site of synthesis of nucleolar proteins will be discussed further on.

Intranucleolar DNA. It has been recognized, from early studies, that the nucleolus reacts positively to basophilic stains, but this property long remained unexplained. The observation that this organelle does not stain with the Feulgen procedure suggested, however, that its basophilia could not be due to the presence of DNA. Nevertheless, several reports appeared, over the years, showing that nucleoli contain Feulgen-positive inclusions but the significance of this particular type of intranucleolar chromatin is still not understood.

Since nucleoli form at telophase on specific chromosome segments, there were good reasons, on theoretical grounds, to assume that chromatin should be part of the nucleolar structure. Biochemical analysis of nucleoli isolated from a wide spectrum of animal (reviewed in Busch and Smetana, 1970) and plant (Stern et al., 1959; Birnstiel and Hyde, 1963; McLeish, 1968) cells have furnished ample evidence in support of such a view. Present estimates of the concentration of intranucleolar DNA range from 2 to 10 per cent depending on the material analysed. In view of its relatively low concentration and of indications that DNA is in a diffuse state within the nucleolar mass, a number of new experimental approaches were exploited, in recent years, to tackle this problem. Incorporation of labelled thymidine within the nucleolus has, for instance, been demonstrated by radioautography in both animal and plant tissues (Hay and Revel, 1963; Granboulan and Granboulan, 1964; McLeish, 1968; Lafontaine and Lord, 1973). Intranucleolar DNA has also been elegantly localized by formation of specific complexes with labelled actinomycin D (Camargo and Plaut, 1967) or, more recently, by means of molecular hybridization *in situ* with labelled ribosomal RNA (Pardue et al., 1970; Jacob et al., 1971). The regressive staining technique developed by

30

Bernhard for ultrastructural studies has, moreover, led to most suggestive results as far as the importance and distribution of chromatin within the nucleolus of animal (Bernhard, 1969; Monneron and Bernhard, 1969) and plant cells (unpublished results). Actual visualization in the electron microscope of nucleolar DNA has, finally, recently been achieved by dispersing nucleoli in detergents and staining the isolated DNA with heavy metals (Miller and Beatty, 1969).

Transcription of preribosomal RNA on organizer DNA

An impressive body of new information has accumulated, of late, concerning the role of the organizer within the nucleolar body (reviewed in Birnstiel *et al.*, 1971). This breakthrough in our understanding of the biochemical and genetic function of the nucleolar organizer can be traced to a decisive and most important discovery, that of the anucleolate mutants of the toad *Xenopus laevis* (Elsdale *et al.*, 1958). Cytological observations reveal that the absence of nucleoli in the tadpole of such mutants reflects the deletion of specific chromosome segments, the nucleolar organizers. Heterozygous mutants exhibit a single nucleolus, whereas the homozygous individuals lack nucleoli completely (Fischberg and Wallace, 1960; Kahn, 1962). Biochemical studies have established a direct relationship between presence of a nucleolus and ribosome production within the cell by demonstrating that anucleolate *Xenopus* embryos do not synthesize 28 S and 18 S ribosomal RNA in any detectable amounts, although synthesis of transfer RNA remains unaffected (Brown and Gurdon, 1964). Concurrent studies with HeLa cells showed that these ribosomal RNA's originate from larger precursor molecules (45 S RNA) which are unstable (Scherrer *et al.*, 1963) and, most revealing, which are synthesized within the nucleolar mass (Perry, 1964).

The demonstration that 45 S RNA is synthesized within the nucleolus led to the view that organizer DNA acts as template for transcription of this particular RNA. A variety of biochemical and physico-chemical techniques has been exploited to characterize this intranucleolar DNA and establish its anticipated intimate relationship to ribosomal RNA. Utilizing mutants of *Drosophila melanogaster*, Ritossa and Spiegelman (1965) were able to clearly demonstrate, by means of the DNA–RNA hybridization technique, that DNA complementary to ribosomal RNA is indeed localized in the nucleolar organizer region. It was also found that, in the wild type, 0.27 per cent of total DNA is complementary to ribosomal RNA and, moreover, that these genes are represented in equal amounts in the X and Y chromosomes. Appropriate calculations show that each of these chromosomes should therefore carry approximately one hundred stretches for each of the two ribosomal RNA's

Other studies were undertaken with mutants of *Xenopus laevis* to verify that deletion of the secondary constriction corresponds to the absence of the nucleolar organizer proper (Birnstiel, 1967). One of the most suggestive steps in this direction was the demonstration that roughly twice as much labelled ribosomal RNA would hybridize to total DNA of wild type organisms as compared to that annealing with DNA isolated from mutants showing one instead of two secondary constrictions. Significantly enough, ribosomal RNA fails to hybridize specifically with DNA from anucleolate mutants (Wallace and Birnstiel, 1966). It is also observed that 0.07 and 0.04 per cent of total DNA of wild type organisms is complementary to 28 S and 18 S RNA respectively. The observation that the RNA–DNA hybrids increase in density as annealing approaches saturation strongly suggests, as in the case of *Drosophila,* that certain segments in the DNA preparation contain several stretches capable of forming specific hybrids with both 28 S and 18 S RNA. It is, indeed, now believed that the chromosome complement, in a diploid cell of *Xenopus,* carries some 1600 alternating 28 S and 18 S RNA cistrons represented in equimolar amounts and apparently massed in the form of long polycistronic blocks (Birnstiel *et al.,* 1968).

The fact that deletion of the secondary constrictions, as observed in cytological preparations (Kahn, 1962), corresponds to removal of DNA which is homologous to ribosomal RNA can be verified by taking advantage of the observation that, in Xenopus, some 0.15–0.2 per cent of total DNA forms a satellite band in a caesium chloride density gradient (Wallace and Birnstiel, 1966). When the sedimentation patterns of DNA isolated from different individuals are compared, it is found that a satellite band appears at a density of 1.723 g/cm³ in the case of both wild types and heterozygous mutants but that it is absent in the case of homozygous toads (Birnstiel *et al,* 1966). As anticipated, this satellite DNA anneals specifically with both 28 S and 18 S ribosomal RNA. Measurements of the kinetics of association of denatured organizer DNA with ribosomal RNA indicate, moreover, that this DNA contains highly repetitive base sequences (Birnstiel *et al,* 1968). This important finding, together with the elegant electron microscopic observations of Miller and Beatty (1969) on isolated nucleolar DNA, clearly shows that the nucleolar organizer, in *Xenopus,* is a polycistronic gene consisting of several hundred tandemly arranged repeating DNA stretches separated by linkers the function of which is presently unknown.

The nucleolar cycle

Fate of the nucleolus during prophase

The ultrastructural and cytochemical organization of the interphase plant nucleolus having already been described, the evolution of this organelle will now be followed throughout other stages of the mitotic cycle (refer to Figs. 1.6 A–D) and special emphasis placed on the process of nucleolar dissolution at late prophase and that of nucleologenesis at telophase.

At early prophase, the nucleolus of plant cells shows a tendency to become slightly more angular in outline and, as during interphase, may contain one large central vacuole together with numerous small peripheral lighter zones. These latter are seen, under electron microscopy, to consist of diffuse material the texture of which is indistinguishable from that of chromatin. As prophase progresses, the general organizational pattern of the nucleolus gradually changes in several respects. A large central vacuole becomes much less frequently observed and the nucleolar surface invariably takes on extremely irregular contours as a result of the formation of pointed and angular projections which extend in between the condensing chromosomes. In appropriate preparations, it is possible to detect one, two or more segments of chromosomes which project more or less deeply within the nucleolar mass and, as in interphase, seem to be in continuity with the nucleolonema and the adjacent light chromatin zones.

By late prophase the densely stained and contracted chromosomes have become marginated but, in appropriate 0.5–1 μm preparations, one, two or more chromosome segments are still seen projecting towards the dispersing nucleolus. In stained preparations, especially under phase contrast microscopy, the nucleolar remnant consists of a dense irregular and heterogeneous core surrounded by a gradient of lighter material. At the ultrastructural level, the peripheral portion of the partly dissolved nucleolus mostly contain loose granules interspersed with fine fibrillar material. The dense central core still exhibits a nucleolonemal network but the latter is now much less clearly delineated than at early and mid-prophase (Fig. 1.6B). One also gains the impression that the light zones of chromatin, once part of this nucleolar region, are either no longer associated with the transforming nucleolonemal reticulum or have condensed and thus become difficult to recognize as such.

By the time the nuclear envelope breaks down, the nucleolonema has generally disappeared from view and is therefore assumed to have condensed back to the chromosomal axis.

Evolution of the nucleolar constriction from metaphase to early telophase

As just noted above, dispersion of the nucleolus at late prophase is accompanied by a gradual transformation of the nucleolonemal network with the result that, eventually, this structure is most difficult to identify as such on account of its greatly reduced opacity. For this reason, as well as due to the difficulty of obtaining appropriate planes of sectioning in random thin sections prepared for electron microscopy, there is at present little ultrastructural information on the eventual fate of the nucleolonema following breakdown of the nuclear envelope at late prophase. As will be argued later on, it is reasonable to assume that the nucleolonema, or what is left of it, collapses back to the main chromosomal axis and gives rise to a particular chromosome segment, the secondary constriction. It is well known from Heitz's classical studies that certain of these constrictions are intimately involved in formation of the telophase nucleolus. Although they appear Feulgen-negative in most preparations, such constrictions can be shown to contain a fine chromatin strand which is continuous with the adjacent chromosomes segments (Ohnuki, 1968).

In all cases so far observed in either animal (Hsu et al., 1967) or plant cells (Lafontaine and Chouinard, 1963; Lafontaine, 1968), secondary constrictions of metaphase and anaphase chromosomes show a homogeneous fibrillar texture. In plant cells, for instance, they are usually characterized by a density which contrasts sharply with that of the chromosomes but which closely matches that of the surrounding spindle material; as a result, such constrictions easily escape attention. By late anaphase, as the chromosome sets reach the cell poles, the density and ultrastructural texture of the secondary constrictions have not changed to any noticeable degree. Slightly later on, however, the nucleolus begins to form and important modifications of the internal organization of these chromosome segments may then be observed (Lafontaine and Lord, 1969).

Formation of the nucleolus

The emerging nucleolus, as it is first identified under electron microscopy, consists almost exclusively of a glomerulus-like structure (Fig. 1.6D) embedded in a fibrillar matrix resembling the material filling the secondary constriction at earlier stages. This tightly convoluted structure, to all appearances, corresponds to the filamentous component of the nucleolonemal reticulum characterizing the mature interphase nucleolus. In view of the attachment of the young nucleolar body to a particular chromosome segment, it is tempting to assume that the coarse intranucleolar filament in question is of chromosomal origin and must, therefore, be closely related to the nucleolar organizer. This glomerular organization of the emergent nucleolus is far from evident, however, in many plant species Small nucleolar bodies are, for instance, very difficult to recognize in plants such as *Vicia faba* on account of the presence of large amounts of a diffuse material which permeates the interchromosomal space at telophase and thus masks the contours of the young nucleolus (Lafontaine and Chouinard, 1963). The earliest stages of formation of the nucleolus and its mode of association with specific chromosome segments are, likewise, most difficult to analyse, in our experience, in plants with chromocentric interphase nuclei (*Raphanus sativus*). Not only do such species exhibit an all-pervading so-called prenucleolar substance throughout telophase, but the chromosomes themselves, except for a few chromocentres, have already completely disappeared from view at the time a nucleolar body may first be identified.

During early interphase the nucleolus gradually increases in size, and a number of important modifications of its ultrastructural organization may be observed. One eventually notes that the growing nucleolar mass has acquired a thin, but nevertheless distinctive, peripheral layer of fibrillogranular material consisting of RNP particles interspersed with fine convoluted fibrils. This material accumulates throughout the G_1 period. A second most striking aspect of the morphological evolution of the forming nucleolus is the progressive expansion of the central light nucleolonemal network into a much looser structure with correspondingly larger meshes. Although the successive stages of this loosing up process are generally difficult to detect in most species, they can be followed in some details in certain plants. In such cases, careful analysis of nucleolar growth reveals that, as the nucleolonemal thread relaxes and takes the appearance of a complex three-dimensional network, the meshes thus formed exhibit a fine fibrillar material which presents all the characteristics of diffuse chromatin.

By the late G_1 period, nucleoli have reached a certain size and, in many plant species, have already started fusing into a large and quite irregular body. The resulting single nucleolus then contains two or more distinct nucleolonemal networks separated by irregular zones of granular material.

Origin of the nucleolar substance

Role of the prenucleolar material in nucleologenesis. The problem of the formation of the nucleolus has been closely linked, over the years, to that of the nature and origin of the so-called prenucleolar substance observed in between the unravelling telophase chromosomes (refer to Moses, 1964; Birnstiel, 1967; Lafontaine, 1968, for recent reviews). This interchromosomal material was thought by many earlier cytologists to originate from a chromosomal matrix which in turn was formed from the dispersing late prophase nucleolus. Other authors assumed that, as the nuclear envelope formed anew at early telophase, interchromosomal spindle material was incorporated into the nuclear cavity and soon became involved in nucleologenesis.

In its original form, the first hypothesis implied that part of the nucleolar material is transported to the cell poles by the anaphase chromosomes and is thus continuous from one division cycle to the next. Cytochemical data to the effect that nucleolar RNA does indeed accumulate onto late prophase chromosomes have not been substantiated by later radioautographic studies and no significant amounts of labelled RNA have, so far, been observed on mitotic chromosomes. Electron microscopic examination of dispersing nucleoli has generally furnished little evidence for the accumulation of nucleolar material onto the late prophase chromosomes. In *Vicia faba,* for instance, the nucleolar RNP granules were found to diffuse throughout the nucleoplasm but few of these appeared to become associated with the chromosomes (Lafontaine and Chouinard, 1963). Unfortunately, on account of its macromolecular texture, the fibrillar component of the nucleolus is much more difficult to follow, once this organelle has dispersed, and it is still not known whether part of this material condenses onto the chromosomes. Recent biochemical studies to be discussed in more details later on, now seem to indicate that certain molecules formed in the nucleolus, the 32 and 45 S RNA's do in fact become associated with the mitotic chromosomes and most probably play an important role, during the following telophase, in formation of the nucleolus (Fan and Penman, 1971). A number of reports have also appeared to the effect that isolated anaphase chromosomes contain significant amounts of ribosomal RNA (discussed in Hearst and Botchan, 1970). No data exist, as yet, regarding the fate of this RNA during the following telophase.

The second hypothesis, that the interchromosomal substance seen at telophase corresponds to spindle material which is entrapped into the forming nucleus, has received little support from recent electron microscopic studies. In both animal and plant cells the reconstituting nuclear envelope appears closely appressed to the outer surface of the telophase chromosomes grouped at the cell poles with the result that restricted amounts of spindle material appear to be actually enclosed within the nuclear boundaries (Lafontaine and Chouinard, 1963; Stevens, 1965; Noel *et al.,* 1971). Nevertheless, as will be discussed shortly, compelling evidence now exists that nucleolar material, specially proteins synthesized in the cytoplasm, moves to the nucleus during interphase (Craig and Perry, 1971). Whether such migration of proteins occurs from the onset of formation of the nucleolus is still uncertain.

A variety of observations has been presented, throughout the years, in support of the concept that the so-called prenucleolar substance observed between the telophase chromosomes is utilized, at least in part, during formation of the nucleolus. All of these different lines of evidence are unfortunately circumstantial in nature and generally rest on certain cytochemical and ultrastructural characteristics which this material shares with the nucleolus. This substance, whether diffuse or in the form of small irregular bodies, stains with azure B and is partly digested with ribonuclease; its strong u.v. absorption at 280 nm following such extraction indicates that, as the nucleolus, it consists mostly of proteins. This interchromosomal material and the nucleolus both contain, moreover, a silver-reducing component presumably proteinaceous in nature (Tandler, 1966). At the ultrastructural level, the texture of this diffuse substance or that of the numerous small bodies scattered between the early telophase chromosomes is best described as fibrillogranular in nature and is, therefore, very similar to material found within the nucleolus. In *Allium cepa* and *Allium porrum*, these small bodies have often been observed in this laboratory to contain very opaque 25 nm particles which, in several plant species, are typically located over the skein portion of both the growing and mature nucleoli (Fig. 1.3). Using the regressive staining technique (Bernhard, 1969), it can be demonstrated, moreover, that the interchromosomal material under discussion reacts quite identically to the nucleolus and, thus, contains RNA (unpublished observations). Finally, it has generally been noted that this interchromosomal substance gradually disappears as the forming nucleolus increases in size (Lafontaine and Chouinard, 1963; Stevens, 1965; Tandler, 1966).

In spite of these highly suggestive characteristics of the interchromosomal substance, its truly prenucleolar nature remains to be demonstrated. It must be stated, however, that the various arguments offered, so far, by tenants of the view according to which the so-called prenucleolar substance is in no way related to nucleologenesis, cannot be considered as being conclusive either. The main argument put forth by these workers rests on the demonstration that the nucleolar organizer is essential for transcription of ribosomal RNA and for formation of a normal nucleolus capable of elaborating RNP particles which act as precursors to the mature cytoplasmic ribosomes. This is a most important finding, indeed. It should not detract, however, from the equally important fact that the nucleolar mass consists predominantly of proteins the synthesis of which has not yet been shown to be controlled by the organizer proper. On the contrary, relevant observations would tend to indicate that certain nucleolar components can be formed in the absence of the organizer. One should recall, at this point, that animal and plant mutants lacking nucleolar organizers are nevertheless capable of forming globular bodies which are predominantly fibrillar in texture and seem to contain less RNA than normal nucleoli (Swift and Stevens, 1966; Hay and Gurdon, 1967). Likewise, micronuclei containing no organizer chromosomes still form nucleolar-like bodies which, according to ultrastructural studies, exhibit most of the characteristics of normal organelles (Phillips and Phillips, 1969). Equally revealing is the recent demonstration that nucleolus-like bodies may form in amoebae in the absence of post-mitotic RNA synthesis (Stevens and Prescott, 1971). Such pseudonucleoli contain residual amounts of RNA and appear to consist predominantly of proteins.

Whether the nucleolar components which are formed in the absence of the organizer are related to the substance observed between the telophase chromosomes is unfortunately not known. Until relevant information becomes available, it is not too farfetched, I think, to envisage the possibility that this substance consists mostly of proteins which normally associate with RNA transcribed on the organizer during growth of the nucleolus. In the absence of such RNA synthesis, either due to deletion of the organizer or to inhibition with actinomycin D, these proteins would form, according to this view, the nucleolar-like bodies of the type just referred to.

Site of synthesis of nucleolar RNA and proteins

It is now generally accepted that a large proportion of nucleolar RNA originates from precursor 45 S molecules transcribed on a polycistronic gene, the organizer, contained within the mass of this organelle. Much information has accumulated from studies of *Xenopus laevis* mutants showing that deletion of the organizer DNA results in the absence of 45 S RNA synthesis and, consequently, of ribosome formation. Of special significance for the purpose of the present discussion is the observation that normal nucleoli are not formed following deletion of the organizer regions (reviewed in Hay, 1968). Instead, there are observed spherical bodies, the organization of which differs markedly from that of true nucleoli. Since these structures are naturally not attached to specific chromosome segments, they contain no diffuse chromatin, as is the case of nucleoli, and do not synthesize detectable amounts of preribosomal RNA. Such pseudonucleoli may consist exclusively of fine fibrillar material (Jones, 1965) but a restricted particulate component segregated in the form of a small cap may sometimes also be observed (Hay and Gurdon, 1967). The presence of both the RNP granules and of RNA in the fibrillar portion of these bodies remains unexplained.

Ultrastructural and cytochemical studies of microspores of maize stocks which also entirely lack nucleolar organizers have furnished results essentially similar to those obtained with *Xenopus*. These plant cells exhibit a number of basophilic spherical bodies consisting of fine fibrillar material and devoid of the usual peripheral RNP particles (Swift and Stevens, 1966). To explain the formation of RNA-containing spheres in the absence of nucleolar organizers, these authors come to the tentative conclusion that a close interrelationship exists between the organizer itself and other loci of chromosomal RNA synthesis.

These observations on *Xenopus laevis* and *Zea mays* mutants show that, in the absence of ribosomal RNA synthesis, the RNP nucleolar particles are not formed. This, however, does not necessarily imply that ribosomal proteins and, consequently, nucleolar proteins are not synthesized. The fact, for instance, that the pseudo-

37

nucleoli appear to consist predominantly of proteins would seem to suggest that the organizer itself is not primarily involved in the synthesis of nucleolar proteins, but that these are instead coded by RNA transcribed on other chromosomal loci. This problem of the synthesis of nucleolar proteins has only recently begun to be analysed. Since preribosomal particles are formed within the nucleolar mass, some information on the regulatory processes involved in the synthesis of certain nucleolar proteins can be gained by investigating the mechanisms controlling the formation of ribosomal proteins. From studies with *Xenopus laevis*, it appears that synthesis of ribosomal proteins does not take place at stages of embryonic development when ribosomal RNA synthesis is absent. In anucleolate mutants, no ribosomal proteins are formed, either, at stages when normal embryos are actively engaged in ribosomal protein synthesis. It is concluded, from such results, that synthesis of ribosomal proteins is coordinated with that of ribosomal RNA (Hallberg and Brown, 1969). Other lines of evidence also suggest a close relationship between the synthesis of these two ribosomal constituents. It has, for instance, been shown that cycloheximide, an inhibitor of protein synthesis, drastically interferes with transcription of 45 S RNA in cultured HeLa cells (Willems *et al.*, 1969) as well as in rat liver cells (Muramatsu *et al.*, 1970). It is not clear whether, under such experimental conditions, RNA polymerase is affected or whether normal nucleolar function is regulated in part by the availability of certain protein components which may or may not be ribosomal in nature. Experiments with L cells in culture treated with low doses of actinomycin D so as to stop selectively synthesis of ribosomal RNA have, likewise, shown that the proteins made during that period are not capable of being assembled into ribosomes when inhibition of RNA synthesis was removed (Craig, 1971).

The foregoing observations point to a close coordination between synthesis of ribosomal RNA and ribosomal proteins, but several aspects of this important problem still remain unsettled. Of special interest, now that the origin of ribosomal RNA has been partly clarified, is the question of the possible role of the nucleolus in the synthesis of both nucleolar and ribosomal proteins. A few reports have appeared indicating that the nucleolus is the site of synthesis of specific proteins (Hnilica *et al.*, 1966; Zimmerman *et al.*, 1969). According to more recent investigations, however, ribosomal proteins are not formed within this organelle as some authors had believed. The first important step in elucidating this problem was provided by the demonstration that the synthesis of nascent ribosomal proteins could be completed *in vitro* using a HeLa cell microsomal preparation (Heady and McConkey, 1970). Similar, but much more comprehensive data were subsequently obtained from studies carried out with a cell-free system obtained from L cells in culture (Craig and Perry, 1971). In this work, elongation of nascent proteins was essentially achieved by adding labelled amino acids and supernatant factors necessary for protein synthesis to a preparation of polysomes free of nuclear contaminants. Comparison of the electrophoretic profile of the proteins thus synthesized *in vitro* revealed sets of bands similar to those furnished by proteins isolated from mature L cell ribosomes. It is evident, therefore, that most, if not all, of the ribosomal proteins are synthesized on cytoplasmic polysomes. Corresponding *in vitro* studies with polysomes isolated from actinomycin D treated cells showed that ribosomal protein synthesis was not inhibited in the absence of ribosomal RNA synthesis, as certain earlier observations had suggested, but that these proteins were simply not assembled into ribosomal particles.

On the strength of available evidence one may conclude that most of nucleolar RNA is synthesized within this organelle but that, on the contrary, many of its constituent proteins originate from the cytoplasm. A problem which remains largely unsettled, however, is when exactly during the nucleolar cycle these two types of macromolecules begin to be synthesized. It may be shown by radio-autography that, no uridine incorporation takes place over the early and mid-telophase interchromosomal substance in plant cells (unpublished observations). Significant incorporation is detected only later on when the growing nucleolar bodies have reached a diameter of three microns or more. This observation is consistent with the recent biochemical studies of Fan and Penman (1971) indicating the persistence throughout the mitotic stages of important amounts of ribosomal RNA precursors (45 and 32 S RNA). These RNA species are present in the form of stable ribonucleoprotein particles tightly bound to the metaphase chromosomes. The synthesis and processing of new 45 S RNA is resumed once the nucleolus has reformed. A second key information furnished by these studies is that synthesis and processing of 45 S RNA may take place within the reformed nucleolus in the absence of post-mitotic protein synthesis. According to these authors, such renewed activity of the nucleolus indicates that the proteins necessary for nucleolar function are present prior to reformation of this organelle.

The above autoradiographic and biochemical observations, as well as those of Gaffney and Nardone (1968), clearly suggest that the first stages of nucleolar formation are not necessarily dependent on concurrent RNA and protein synthesis. Future studies will be required to verify, among other possibilities, whether part of the presynthesized nucleolar RNA and proteins might not be localized within the so-called prenucleolar material before finding its way to the mass of the forming nucleolus.

The nucleolar loops: facts and hypothesis

Since the nucleolus is the site of transcription of preribosomal RNA and since, moreover, ribosomal RNA is complementary to part of organizer DNA, it has generally been assumed that the organizer is located within the nucleolar body and must therefore represent a key structural component of this organelle. Although, as summarized earlier in this chapter, the presence of DNA has conclusively been revealed within the nucleolus of different types of cells it still is not clear whether all of this DNA represents the nucleolar organizer locus. Relevant electron microscopic observations have not yet converged, either, to a concensus as to which ultrastructural component of the nucleolus corresponds to the organizer proper.

Amongst the various morphological nucleolar components recorded in most cell types studies so far, one, the nucleolonema, has particularly attracted attention and has occasionally been assumed to represent the organizer. Although such a filamentous nucleolar structure had been observed by a number of classical cytologists, it gained status as an essential nucleolar component mostly due to later observations by Estable and Sotelo (1955). According to these two authors, the nucleolonema is a convoluted filament assuming the shape of a glomerulus immersed in an amorphous matrix. Much electron microscopic evidence has accumulated during the last two decades supporting the notion that a similar thread-like structure is part of nucleolar bodies in cells from both lower and higher organism (reviewed in Bernhard and Granboulan, 1968; Hay, 1968). Views as to the nature, origin and role of this coarse nucleolar filament are still quite divergent, however. Certain authors (Estable, 1966) envisage the nucleolonema as a self-replicating structure which fragments and attaches to the chromosomes at prophase. According to this view, one of these fragments persists throughout the mitotic stages and acts as a nucleolar organizer at telophase. It is, unfortunately, not clear from such a concept what is the actual morphological relation between the nucleolonema and the nucleolar chromosomes. Other searchers, on the basis of cytochemical and radioautographic studies, are of the opinion that the nucleolonema is a complex structure of chromosomal origin and consists of a DNA core (Lettré et al., 1966; Ghosh et al., 1969). The view has finally been put forth that, in animal cells, the nucleolonema is made of RNA and proteins and that chromatin is instead present at the periphery as well as within the mesh of the nucleolonemal network (Bernhard and Granboulan, 1968; Hay, 1968). The relationship between the nucleolonema and the nucleolar organizer region of chromosomes is seen to better advantage, perhaps, in the case of the polytene chromosomes in salivary gland cells of insects. In this material, coiled chromosomal threads approximately 0.1 μm in diameter are seen meandering throughout the inner fibrillar region of the nucleolus (Kalnins et al., 1964). As expected, incorporation of labelled uridine is first observed to take place over the nucleolar organizer region, the radioautographic grains appearing a few minutes later over the inner homogeneous nucleolar zone. Eventually, the whole nucleolar body shows presence of labelled RNA (Gaudecker, 1967).

Radioautographic studies with mammalian cells have, likewise, led to the conclusion that organizer DNA was located within the fibrillar zones of the nucleolus. Following exposure to radioactive uridine for short periods, incorporation first appears over these latter zones and is not detected over the granular nucleolar portions till some 10 minutes later. When incorporation of uridine is extended to a 30-minute period, the whole nucleolus becomes labelled (Granboulan and Granboulan, 1965). That rapidly labelled nucleolar RNA actually migrates from the fibrillar zones to the adjacent granular areas was shown by exposing mammalian cells in culture to labelled uridine during 5 minutes and, for chasing purposes, then transferring these cells to a medium containing a much higher concentration of cold uridine. During the chasing period, the radioautographic grains are first localized over the fibrillar zones of the nucleolus and appear in increasing number over the particulate portions as the former zones become progressively depleted of such grains (Geuskens and Bernhard, 1966). As a result of these most revealing observations, the hypothesis was advanced that 45 S RNA is actually transcribed within the fibrillar zones of the nucleolus and moves to the granular portions where it is involved in formation of the nucleolar RNP particles.

The obvious implication of the various studies just referred to is that derepressed organizer chromatin is present within the fibrillar portion of animal cell nucleoli. Since the nucleolonemal network seems to be a constant feature of these fibrillar zones in both animal (Bernhard and Granboulan, 1968; Hay, 1968) and plant (Lord and Lafontaine, 1969; Lafontaine and Lord, 1969) cells, the question arises as to whether this convoluted filament, the immediately adjacent fribrillar material, or both, correspond to the nucleolar organizer proper. Any tentative answer to this question must, at present, be considered purely hypothetical for lack of sufficient information. Nevertheless, a number of observations carried out on plant cells appear particularly relevant and will now be discussed in some detail. The interest of these observations stems from the important finding (LaCour, 1966) that the number of nucleolonemata per interphase nucleus is related to the ploidy of the species and, moreover, that these convoluted nucleolar filaments seem to correspond to loops attached to certain chromosomes. Each loop was therefore envisaged to form through unfolding of the nucleolar organizer segment (LaCour and Wells, 1967). As noted previously, strong support for this hypothesis is provided by studies of the nucleolar cycle in plant cells and, most particularly, of the evolution of the structure of secondary constrictions at the onset of formation of the nucleolus. The fact that the earliest observable nucleolar body consists essentially of a nucleolonema-like coarse filament seems to be best interpreted by assuming that this latter structure is in the process of looping out of the secondary constriction onto which the nucleolus is forming (Lafontaine and Lord, 1969).

Even though the nucleolonema is a constant morphological component of the nucleolus from early telophase to late prophase, certain data suggest that this filamentous structure does not correspond to the nucleolar organizer proper. Indeed, according to current thinking, the organizer should consist of diffuse chromatin, at least at those stages of the cell cycle during which it plays an active role in RNA transcription. In plants, on the contrary, the nucleolonema appears as a rather dense filament (Lord and Lafontaine, 1969) and, for this reason, it seems more logical to assume that the meshes of the nucleolonemal network represent the active component of the organizer (Figs. 1.3 and 1.6D). The corollary of this hypothesis, namely, that the nucleolonemal meshes under consideration consist predominantly of diffuse or derepressed chromatin rests on the following data:

(a) First, extraction of 0.5 μm sections with electrophoretically purified desoxyribonuclease significantly reduces the azure B staining intensity of those regions of plant-cell nucleoli which are known from electron microscopy to exhibit a fibrillar texture. The extent of this effect is such that it cannot be attributed to extraction of DNA from the nucleolonemal thread only (Lord and Lafontaine, 1969).

(b) Use of appropriate staining procedures for electron microscopy reveals that the numerous meshes of the nucleolonemal network of both animal (Bernhard, 1969; Monneron and Bernhard, 1969) and plant (unpublished observations) cells reacts identically to extranucleolar chromatin.

(c) Examination, under electron microscopy, of preparations treated with hot TCA, a procedure known to extract DNA (Douglas, 1970), also shows that the meshes of the nucleolonemal network have become as translucent as the chromosomes themselves (unpublished observations).

(d) High resolution autoradiography of plant cells (*Allium porrum*) labelled with tritiated thymidine reveals a striking localization of the silver grains over the nucleonemal skein, and most particularly over its lacunar components (Lafontaine and Lord, 1973).

(e) Pulse labelling with tritiated uridine indicates that little, if any, RNA synthesis can be detected by radioautography over the growing plant nucleolus until this organelle acquires its chromatin-like zones. By that time a layer of RNP granules is also noted at the nucleolar periphery.

(f) In the case of the mature interphase animal nucleolus, corresponding labelling experiments have clearly shown that radioautographic grains first occur over its fibrillar regions (Granboulan and Granboulan, 1965; Geuskens and Bernhard, 1966; LaCour and Crawley 1965).

Taken together, these observations are extremely suggestive and, I think, may serve as the basis for some speculation concerning nucleologenesis as well as concerning the respective role of the major morphological components of the nucleolus. First, since the earliest nucleolar body which can be identified in favourable preparations consists almost exclusively of a nucleolonema embedded in fine fibrillar material (Fig. 1.6D), it is reasonable to assume that this latter structure plays a key role in nucleologenesis. It would appear that the onset of nucleolar formation coincides with a loosening up or looping of basic elements of the secondary constriction together with the concurrent addition of appropriate material so as to give rise to the coarse glomerular filament observed at early interphase. The fine fibrillar substance found in between this nucleolar skein presumably corresponds to the low contrast material characterizing secondary constrictions at earlier stages and possibly, also, to newly added material which may play an essential role at this particular stage of nucleolar formation. Since, at this early period, no detectable incorporation of labelled uridine can be detected by radioautography over plant nucleoli (unpublished observations), such new material is either not RNA or, if so, has been synthesized at earlier stages. As mentioned earlier, there exists biochemical evidence for the availability of presynthesized RNA and proteins prior to formation of the telophase nucleolus (Gaffney and Nardone, 1968; Fan and Penman, 1971). This absence of RNA synthesis during the earliest stages of nucleologenesis indicates that the nucleolar organizer is still in a repressed state. It is conceivable, therefore, that the subsequent appearance of fine fibrillar material within the enlarging meshes of the nucleolonemal network is related to the renewed nucleolar RNA synthesis activity which is observed by early interphase. If our assumption that this fibrillar

---

material corresponds to the active portion of the nucleolar organizer is correct, one important question will then remain to be clarified; this concerns the role of the coarse nucleolonemal filament itself. One possibility which comes to mind is that this structure acts as a backbone from which DNA loops out during derepression of the organizer.

Acknowledgements

Preparation of this chapter was aided by research grants from the Ministry of Education of Quebec and the National Research Council of Canada. The Author also gratefully acknowledges the assistance of Drs A. Lord and D. Pallotta during redaction of the manuscript, and of Mr S. Gugg and Miss D. Leclerc for their expert technical help.

Bibliography

Further reading

Bahr, G. F. (1970) 'Human chromosome fibers. Consideration of DNA-protein packing and of looping patterns', *Expl Cell Res.*, **62**, 39–49.
Busch, H. and Smetana, K. (1970) *The nucleolus*. Academic Press Inc., New York.
Cantor, K. P. and Hearst, J. E. (1970) 'The structure of metaphase chromosomes. I. Electrometric titration, magnesium ion binding and circular dichroism', *J. molec. Biol.*, **49**, 213–229.
Du Praw, E. J. (1970) *DNA and chromosomes*, Holt, Rinehart and Winston, Inc., New York.
Lima-de-Faria, A. (ed.), (1969) *Handbook of molecular cytology*, North-Holland Publishing Co., Amsterdam.
Maio, J. J. and Schildkraut, C. L. (1967) 'Isolated mammalian metaphase chromosomes. I. General characteristics of nucleic acids and proteins', *J. molec. Biol.*, **24**, 29–39.
Mirsky, A. E., Burdick, C. J., Davidson, E. H., and Littau, V. C. (1968) 'The role of lysine-rich histone in the maintenance of chromatin structure in metaphase chromosomes'. *Proc. natn. Acad. Sci. U.S.A.*, **61**, 592–597.
Rogers, M. E., Loening, U. E., and Fraser, R. S. S. (1970) 'Ribosomal RNA precursors in plants', *J. molec. Biol.*, **49**, 681–692.
Sivolap, Y. M. and Bonner, J. (1971) 'Association of chromosomal RNA with repetitive DNA', *Proc. natn. Acad. Sci. U.S.A.*, **68**, 387–389.
Sponar, J., Boublík, M., Frič, I., and Šormová, Z. (1970) 'Study of DNA conformation in native, partial and reconstituted nucleohistones, using hydrodynamic methods and optical rotatory dispersion', *Biochim. biophys. Acta*, **209**, 532–540.
Stubblefield, E. and Wray, W. (1971) 'Architecture of the chinese hamster metaphase chromosome', *Chromosomoa*, **32**, 262–294.
Tanifuji, S., Higo, M., Shimada, T., and Higo, S. (1970) 'High molecular weight RNA synthesized in nucleoli of high plants', *Biochim. biophys. Acta*, **217**, 418–425.
Yasmineh, W. G. and Yunis, J. J. (1970) 'Localization of mouse satellite DNA in constitutive heterochromatin', *Expl Cell Res.*, **59**, 69–75.

References

Arrighi, F. E., Hsu, T. C., Saunders, P., and Saunders, G. F. (1970) 'Localization of repetitive DNA in the chromosomes of *Microtus agrestis* by means of *in situ* hybridization', *Chromosoma*, **32**, 224–236.
Bajer, A. (1959) 'Change of length and volume of mitotic chromosomes in living cells', *Hereditas*, **45**, 579–596.

The nucleus

Bajer, A. (1965) 'Subchromatid structure of chromosomes in the living state', *Chromosoma*, 17, 291—302.

Bajer, A. (1968)'Behavior and fine structure of spindle fibers during mitosis in endosperm', *Chromosoma*, 25, 249—281.

Bajer, A. and Molè-Bajer, J. (1963) 'Cine analysis of some aspects of mitosis in endosperm', In: *Cinemicrography in cell biology* (ed. G. G. Rose), pp. 357—409, Academic Press Inc., New York.

Bajer, A. and Molè-Bajer, J. (1969) 'Formation of spindle fibers, kinetochore orientation and behaviour of the nuclear envelope during mitosis in endosperm. Fine structural and *in vitro* studies', *Chromosoma*, 27, 448—484.

Bajer, A. S. and Molè-Bajer, J. (1972) 'Spindle dynamics and chromosome movements', *Int. Rev. Cytol.* (eds. G. H. Bourne and J. F. Danielli), Vol. 34, Suppl. 3, Academic Press, New York.

Barnicot, N. A. and Huxley, H. E. (1965) 'Electron microscope observations on mitotic chromosomes'. *Q. Jl Microsc. Sci.*, 106, 197—214.

Bĕlář, K. (1929) 'Untersuchungen an den Staubfadenhaaren von *Tradescantia virginica*', *Z. Zellforsch. Mikrosk. Anat.*, 10, 73—134.

Bendich, A. J. and McCarthy, B. J. (1970) 'DNA comparisons among barley, oats, rye and wheat', *Genetics*, 65, 545—565.

Bernhard, W. (1969) 'A new staining procedure for electron microscopical cytology', *J. Ultrastruct, Res.*, 27, 250—265.

Bernard, W. and Granboulan, N. (1968) 'Electron microscopy of the nucleolus in vertebrate cells', In: *Ultrastructure in biological systems: The nucleus* (eds. A. J. Dalton and F. Haguenau), vol. 3, pp. 81—149, Academic Press Inc., New York.

Birnstiel, M. L. (1967) 'The nucleolus in cell metabolism', *A. Rev. Pl. Physiol.*, 18, 25—58.

Birnstiel, M. L., Chipchase, M. I. H., and Hyde, B. B. (1963) 'The nucleolus, a source of ribosomes', *Biochim. biophys. Acta*, 76, 454—462.

Birnstiel, M. L. and Hyde, B. B. (1963) 'Protein synthesis by isolated nucleoli', *J. Cell Biol.*, 18, 41—50.

Birnstiel, M. L., Chipchase, M. I. H., and Flamm, W. G. (1964) 'On the chemistry and organization of nucleolar proteins', *Biochim. biophys. Acta*, 87, 111—122.

Birnstiel, M. L., Wallace, H., Sirlin, J. L., and Fischberg, M. (1966) 'Localization of the ribosomal DNA complements in the nucleolar organizer region of *Xenopus laevis*', *Natl. Cancer Inst. Monograph*, 23, 431—447.

Birnstiel, M., Speirs, J., Purdom, I., Jones, K., and Loening, U. E. (1968) 'Properties and composition of the isolated ribosomal DNA satellite of *Xenopus laevis*', *Nature, Lond.*, 219, 454—463.

Birnstiel, M. L., Chipchase, M., and Speirs, J. (1971) 'The ribosomal RNA cistrons', In: *Progress in nucleic acid research and molecular biology* (eds. J. N. Davidson and W. E. Cohn), vol. 11, pp. 351—389, Academic Press Inc., New York.

Bloch, D. P. (1963) 'The histones: Syntheses, transitions, and functions', In: *The cell in mitosis* (ed. L. Levine) pp. 205—221. Academic Press Inc., New York.

Blondel, B. (1968) 'Relation between nuclear fine structure and ^{3}H-thymidine incorporation in a synchronous cell culture', *Expl Cell Res.*, 53, 348—356.

Bonner, J., Dahmus, M. E., Fambrough, D., Huang, R. C., Marushige, K., and Tuan, D. Y. H. (1968) 'The biology of isolated chromatin', *Science, N.Y.*, 159, 47—56.

Bostock, B. (1971) 'Repetitious DNA', In: *Advances in cell biology* (eds. D. M. Prescott, L. Goldstein, and E. McConkey), vol. 2, pp. 153—223, Appleton-Century-Crofts, New York.

Bostock, C. J., Prescott, D. M., and Hatch, F. T. (1972) 'Timing of replication of the satellite and main band DNAs in cells of the kangaroo rat', *Expl. Cell. Res.*, 74, 487—495.

Botchan, M., Kram, R., Schmid, C. W., and Hearst, J. E. (1971) 'Isolation and chromosomal localization of highly repeated DNA sequences in *Drosophila melanogaster*', *Proc. natn. Acad, Sci. U.S.A.*, 68, 1125—1129.

Brachet, J. (1940) 'La détection histochimique des acides pentosenucléiques', *C.r. Séance. Soc. Biol.*, 133, 88—90.

Bram, S. and Ris, H. (1971) 'On the structure of nucleohistone', *J. molec. Biol.*, 55, 325—336.

Brasch, K. R. (1971) *The role of histone proteins and ionic environment in the structure and organization of avian liver and erythrocyte nuclei.* Ph.D. Thesis, Carleton University, Ottawa.

Bibliography

Brasch, K., Seligy, V. L., and Setterfield, G. (1971) 'Effects of low salt concentration on structural organization and template activity of chromatin in chicken erythrocyte nuclei', *Expl Cell Res.,* **65,** 61–72.

Brinkley, B. R. and Stubblefield, E. (1970) 'Ultrastructure and interaction of the kinetochore and centriole in mitosis and meiosis', In: *Advances in cell biology* (eds. D. M. Prescott; L. Goldstein and E. McConkey), vol. 1, pp. 119–185, Appleton-Century-Crofts, New York.

Britten, R. J. and Kohne, D. E. (1969) 'Repetition of nucleotide sequences in chromosomal DNA', In: *Handbook of molecular cytology* (ed. A. Lima-de-Faria), pp. 21–36, North-Holland Publishing Co., Amsterdam.

Brown, D. D. and Dawid, I. B. (1968) 'Specific gene amplification in oocytes', *Science, N.Y.,* **160,** 272–280.

Brown, D. D. and Gurdon, J. B. (1964) 'Absence of ribosomal RNA synthesis in the anucleolate mutant of *Xenopus laevis'*, *Proc. natn. Acad, Sci. U.S.A.,* **51,** 139–146.

Busch, H. and Smetana, K. (1970) *The nucleolus,* Academic Press Inc., New York.

Callan, H. G. (1972) 'Replication of DNA in the chromosomes of eukaryotes', *Proc. R. Soc. Lond. B.,* **181,** 19–41.

Camargo, E. P. and Plaut, W. (1967) 'The radioautographic detection of DNA with tritiated Actinomycin D', *J. Cell Biol,* **35,** 713–716.

Caspersson, T. O. and Schultz, J. (1940) 'Ribonucleic acids in both nucleus and cytoplasm and the function of the nucleolus', *Proc, natn. Acad. Sci. U.S.A.,* **26,** 507–515.

Chooi, W. Y. (1971) 'Comparison of the DNA of six *Vicia* species by the method of DNA-DNA hybridization', *Genetics,* **68,** 213–230.

Chouinard, L. A. and Leblond, C. P. (1967) 'Sites of protein synthesis in nucleoli of root meristematic cells of *Allium cepa* as shown by radioautography with (^{3}H) arginine', *J. Cell Sci.,* **2,** 473–480.

Comings, D. E. (1968) 'The rationale for an ordered arrangement of chromatin in the interphase nucleus', *Am. J. hum. Genet,* **20,** 440–460.

Comings, D. E. and Kakefuda, T. (1968) 'Initiation of deoxyribonucleic acid replication at the nuclear membrane in human cells', *J. molec. Biol.,* **33,** 225–229.

Comings, D. E. and Okada, T. A. (1970) 'Association of nuclear membrane fragments with metaphase and anaphase chromosomes as observed by whole mount electron microscopy', *Expl Cell. Res.,* **63,** 62–68.

Comings, D. E. (1972) 'The structure and function of chromatin', In: *Advances in human genetics* (eds. H. Harris and K. Hirshhorn), vol. 3, pp. 237–431, Plenum Press, New York.

Craig, N. C. (1971) 'On the regulation of the synthesis of ribosomal proteins in L-Cells', *J. molec. Biol.,* **55,** 129–134.

Craig, N. and Perry, R. P. (1971) 'Persistent cytoplasmic synthesis of ribosomal proteins during the selective inhibition of ribosomal RNA synthesis', *Nature, Lond.,* **229,** 75–80.

Dangeard, P. (1941) 'Recherches sur la structure des noyaux et sur l'action des fixateurs particulièrement de l'acétocarmin', *Botaniste,* **31,** 113–187.

Dangeard, P. (1947) *Cytologie végétale et cytologie générale,* Lechavallier, Paris.

Darnell, J. E. (1968) 'Ribonucleic acids from animal cells', *Bact. Rev.,* **32,** 262–290.

Davies, H. G. and Spencer, M. (1962) 'The variation in the structure of erythrocyte nuclei with fixation', *J. Cell Biol.* **14,** 445–458.

Delay, C. (1948) 'Recherches sur la structure des noyaux quiescents chez les phanérogames', *Rev. Cytol. Cytophysiol. Vég.,* **10,** 103–228.

De Robertis, E. (1956) 'Electron microscopic observation on the submicroscopic morphology of the meiotic nucleus and chromosomes', *J. Biophys. biochem. Cytol.* **2,** 785–795.

Douglas, W. H. J. (1970) 'Perchloric acid extraction of deoxyribonucleic acid from thin sections of epon-araldite-embedded material', *J. Histochem. Cytochem.,* **18,** 510–514.

DuPraw, E. J. (1970) *DNA and Chromosomes,* Holt, Rinehart and Winston, Inc., New York.

DuPraw, E. J. and Bahr, G. F. (1969) 'The arrangement of DNA in human chromosomes, as investigated by quantitative electron microscopy', *Acta Cytol.,* **13,** 188–205.

Eckhardt, R. A. and Gall, J. G. (1971) 'Satellite DNA associated with heterochromatin in *Rhynchosciara'*, *Chromosoma,* **32,** 407–427.

Elsdale, T. R., Fischberg, M., and Smith, S. (1958) 'A mutation that reduces nucleolar number in *Xenopus laevis'*, *Expl Cell Res.,* **14,** 642–643.

The nucleus

Estable, C. (1966) 'Morphology, structure and Dynamics of the nucleolonema', *Nat. Cancer Inst. Monogr*, **23**, 91−105.

Estable, C. and Sotelo, J. R. (1955) 'The behaviour of the nucleolonema during mitosis', In: *Symposium on fine structure of cells*, pp. 170−90, Noordhoff, Groningen, The Netherlands.

Fambrough, D. M. (1969) 'Nuclear protein fractions', In: *Handbook of molecular cytology* (ed. A. Lima-de-Faria), pp. 437−471, North-Holland Publishing Co., Amsterdam.

Fan, H. and Penman, S. (1971) 'Regulation of synthesis and processing of nucleolar components in metaphase-arrested cells', *J. molec. Biol.*, **59**, 27−42.

Fasman, G. D., Schaffhausen, B., Goldsmith, L., and Adler, A. (1970) 'Conformational changes associated with f-1 histone-deoxyribonucleic acid complexes. Circular dichroism studies', *Biochemistry, N.Y.*, **9**, 2814−2822.

Fischberg, M. and Wallace, H. (1960) 'A mutation which reduces nucleolar number in *Xenopus laevis*', In: *The cell nucleus* (ed. J. S. Mitchell). pp. 30−33, Butterworth, London.

Flamm, W. G. (1972) 'Highly repetitive sequences of DNA in chromosomes', *Int. Rev. Cytol.* (eds. G. H. Bourne and J. F. Danielli), vol. 32, pp. 1−51.

Frenster, J. H. (1969) 'Biochemistry and molecular biophysics of heterochromatin and euchromatin', In: *Handbook of molecular cytology* (ed. A. Lima-de-Faria), pp. 251−276, North-Holland Publishing Co., Amsterdam.

Gaffney, E. V. and Nardone, R. M. (1968) 'Nucleolar RNA synthesis in synchronous cultures of strain L-929', *Expl Cell Res.*, **53**, 410−416.

Gall, J. G. and Callan, H. G. (1962) 'H³-uridine incorporation in lampbrush chromosomes', *Proc. natn. Acad. Sci. U.S.A.*, **48**, 562−570.

Gall, J. G. (1963) 'Kinetics of deoxyribonuclease action on chromosomes', *Nature* (London), **198**, 36−38.

Gates, R. R. (1942) 'Nucleoli and related nuclear structures', *Bot. Rev.*, **8**, 337−409.

Gaudecker, B. von. (1967) 'RNA synthesis in the nucleolus of *Chironomus thummi* as studied by high resolution autoradiography', *Z. Zellforsch. mikrosk. Anat.*, **82**, 536−557.

Geuskens, M. and Bernhard, W. (1966) 'Cytochimie ultrastructurale du nucléole. III. Action de l'actinomycin D sur le métabolisme du RNA nucléolaire', *Expl Cell Res.*, **44**, 579−598.

Ghosh, S., Lettré, R., and Ghosh, I. (1969) 'On the composition of the nucleolus with special reference to its filamentous structure', *Z. Zellforsch. mikrosk. Anat*, **101**, 254−265.

Giannoni, G. and Peacocke, A. R. (1963) 'Thymus deoxyribonucleoprotein. III. Sedimentation behaviour', *Biochim. biophys. Acta*, **68**, 157−166.

Granboulan, N. and Granboulan, P. (1964) 'Cytochimie ultrastructurale de nucléole. I. Mise en évidence de chromatine à l'intérieure du nucléole', *Expl Cell Res.*, **34**, 71−87.

Granboulan, N. and Granboulan, P. (1965) 'Cytochimie ultrastructurale du nucléole. II. Etude des sites de synthèse du RNA dans le nucléole et le noyau', *Expl Cell Res.*, **38**, 604−619.

Grogan, D. E. and Busch, H. (1967) 'Studies on fractionation of saline-soluble nucleolar proteins on diethylaminoethyl-cellulose', *Biochemistry, N.Y.*, **6**, 573−578.

Hallberg, R. L. and Brown, D. D. (1969) 'Co-ordinated synthesis of some ribosomal proteins and ribosomal RNA in embryos of *Xenopus laevis*', *J. molec. Biol.*, **46**, 393−411.

Hay, E. D. (1968) 'Structure and function of the nucleolus in developing cells', In: *Ultrastructure in biological systems. The nucleus* (eds. A. J. Dalton and F. Hagueneau) vol. 3, pp. 1−79.

Hay, E. D. and Revel, J. P. (1963) 'The fine structure of the DNP component of the nucleus', *J. Cell Biol.*, **16**, 29−51.

Hay, E. D. and Gurdon, J. B. (1967) 'Fine structure of the nucleolus in normal and mutants Xenopus embryos', *J. Cell Sci.*, **2**, 151−162.

Heady, J. E. and McConkey, E. H. (1970) 'Completion of nascent HeLa ribosomal proteins in a cell-free system', *Biochem. biophys. Res. Commun.*, **40**, 30−36.

Hearst, J. E. and Botchan, M. (1970) 'The eukaryotic chromosome', *A. Rev. Biochem.*, **39**, 151−182.

Hnilica, L. S., Liau, M. C., and Hurlbert, R. B. (1966) 'Biosynthesis and composition of histones in Novikoff hepatoma nuclei and nucleoli', *Science*, N.Y., **152**, 521−523.

Hsu, T. C., Brinkley, B. R., and Arrighi, F. E. (1967) 'The structure and behaviour of the nucleolus organizers in mammalian cells', *Chromosoma*, **23**, 137−153.

Huberman, J. A. and Riggs, A. D. (1968) 'On the mechanism of DNA replication in mammalian chromosomes', *J. molec. Biol.*, **32**, 327–341.

Hyde, B. B. (1967) 'Changes in nucleolar ultrastructure associated with differentiation in the root tip', *J. Ultrastruct. Res.*, **18**, 25–54.

Hyde, B. B., Sankaranarayanan, K., and Birnstiel, M. L. (1965) 'Observations on fine structure in pea nucleoli in situ and isolated', *J. Ultrastruct. Res.*, **12**, 652–667.

Izawa, M. and Kawashima, K. (1969) 'Some properties of ribonucleoprotein particles in the isolated nucleoli of mouse ascites tumor cells', *Biochim. biophys. Acta*, **174**, 124–136.

Jacob, J., Todd, K., Birnstiel, M. L., and Bird, A. (1971) 'Molecular hybridization of ^{3}H-labelled ribosomal RNA with DNA in ultra-thin sections prepared for electron microscopy', *Biochim. biophys. Acta*, **228**, 761–766.

Jokelainen, P. T. (1967) 'The ultrastructure and spatial organization of the metaphase kinetochore in mitotic rate cells', *J. Ultrastruct. Res.*, **19**, 19–44.

Jones, K. W. (1965) 'The role of the nucleolus in the formation of ribosomes', *J. Ultrastruct. Res.*, **13**, 257–262.

Jones, K. W. (1970) 'Chromosomal and nuclear location of mouse satellite DNA in individual cells', *Nature, Lond.*, **225**, 912–915.

Jordan, E. G. and Godward, M. B. E. (1969) 'Some observations on the nucleolus in *Spirogyra*', *J. Cell Sci.*, **4**, 3–15.

Kahn, J. (1962) 'The nucleolar organizer in the mitotic chromosome complement of *Xenopus laevis*', *Q. Jl Microsc. Sci.*, **103**, 407–409.

Kalnins, V. I., Stich, H. F., and Bencosme, S. A. (1964) 'Fine structure of the nucleolar organizer of salivary gland chromosome of chironomids', *J. Ultrastruct. Res.*, **11**, 282–291.

Kaufmann, B. P. (1948) 'Chromosome structure in relation to the chromosome cycle. II', *Bot. Rev.*, **14**, 57–126.

Kaufmann, B. P., Gay, H., and McDonald, M. R. (1960) 'Organizational patterns within chromosomes', *Int. Rev. Cytol.*, **9**, 77–127.

Kemp, C. L. (1966) 'Electron microscope autoradiographic studies of RNA metabolism in *Trillium erectum* microspores', *Chromosoma*, **19**, 137–148.

Kuroiwa, T. and Tanaka, N. (1971a) 'Fine structure of interphase nuclei I. The morphological classification of nucleus in interphase of *Crepis capillaris*', *Cytologia*, **36**, 143–160.

Kuroiwa, T. and Tanaka, N. (1971b) 'Fine structure of interphase nuclei IV. The behavior of late replicating chromatin during a late portion of the S period as revealed by electron microscope radioautography', *J. Cell Biol.*, **49**, 939–942.

LaCour, L. F. (1966) 'The internal structure of nucleoli', In: *Chromosomes today* (eds. C. D. Darlington and K. R. Lewis), vol. I, pp. 150–160, Oliver and Boyd, Edinburgh and London.

LaCour, L. F. and Pelc, S. R. (1958) 'The effect of colchicine on the utilization of labelled thymidine during chromosomal reproduction', *Nature, Lond.*, **182**, 506–508.

LaCour, L. F. and Crawley, J. W. G. (1965) 'The site of rapidly labelled ribonucleic acid in nucleoli', *Chromosoma*, **16**, 124–132.

LaCour, L. F. and Wells, B. (1969) 'The origin of fibrillar particles sometimes seen in plant nucleoli', *Z. Zellforsch. mikrosk. Anat*, **97**, 358–368.

Lafontaine, J. G. (1965) 'A light and electron microscope study of small spherical nuclear bodies is meristematic cells of *Allium cepa, Vicia faba*, and *Raphanus sativus*', *J. Cell Biol.*, **26**, 1–17.

Lafontaine, J. G. (1968) 'Structural components of the nucleus in mitotic plant cells', In: *Ultrastructure in biological systems: The Nucleus* (eds. A. J. Dalton and F. Haguenau), vol. 3, pp. 151–196, Academic Press Inc., New York.

Lafontaine, J. G. and Chouinard, L. A. (1963) 'A correlated light and electron microscope study of the nucleolar material during mitosis in *Vicia faba*', *J. Cell Biol.*, **17**, 167–201.

Lafontaine, J. G. and Lord, A. (1969) 'Organization of nuclear structures in mitotic cells', In: *Handbook of molecular cytology* (ed. A. Lima-de-Faria), pp. 381–411, North-Holland Publishing Co., Amsterdam.

Lafontaine, J. G. 'Ultrastructural organization of plant cell nuclei', In: *The cell nucleus* (ed. H. Busch), Academic Press, New York (in the press).

Lafontaine, J. G. and Lord, A. (1973) 'An ultrastructural and radioautographic investigation of the nucleolonemal component of plant cell (*Allium porrum*) nucleoli', *J. Cell Sci.*, **12**, 369–383.

Laird, C. D. (1971) 'Chromatid structure: Relationship between DNA content and nucleotide sequence diversity', *Chromosoma*, **32**, 378–406.

Lettré, R., Siebs, N., and Paweletz, N. (1966) 'Morphological observations on the nucleolus of cells in tissue culture with special regard to its composition', *Natl. Cancer Inst. Monograph.*, **23**, 107–123.

Liau, M. C. and Perry, R. P. (1969) 'Ribosome precursor particles in nucleoli', *J. Cell Biol.*, **42**, 272–283.

Lima-de-Faria, A. (1969) 'DNA replication and gene amplification in heterochromatin', In: *Handbook of molecular cytology* (ed. A. Lima-de-Faria), pp. 277–325, North-Holland Publishing Co., Amsterdam.

Loening, U. E. (1968) 'RNA structure and metabolism', *A. Rev. Pl. Physiol.*, **19**, 37–70.

Lord, A. and Lafontaine, J. G. (1969) 'The organization of the nucleolus in meristematic plant cells. A cytochemical study', *J. Cell Biol.*, **40**, 633–647.

Marinozzi, V. (1964) 'Cytochimie ultrastructurale du nucléole. RNA et protéines intranucleolaires', *J. Ultrastruct. Res.*, **10**, 433–456.

Mattingly, A. (1963) 'Nuclear protein synthesis in *Vicia faba*', *Expl Cell Res.*, **29**, 314–326.

McLeish, J. (1968) 'Chemical and autoradiographic studies of intranucleolar DNA in *Vicia faba*', *Expl Cell Res.*, **51**, 157–166.

McQuade, H. A. and Atchison, A. A. (1969) 'Localization of RNA synthesis in interphase nuclei of *Allium cepa*', *Caryologia*, **22**, 7–24.

Miller, O. L. and Beatty, B. R. (1969) 'Extrachromosomal nucleolar genes in amphibian oocytes', *Genetics Suppl.*, **61**, 133–143.

Milner, G. R. (1969) 'Nuclear morphology and the ultrastructural localization of DNA synthesis during interphase', *J. Cell Sci.*, **4**, 569–582.

Mirsky, A. E. and Ris, H. (1947) 'The chemical composition of isolated chromosomes', *J. gen. Physiol.*, **31**, 7–18.

Mizuno, N. S., Stoops, C. E., and Pfeiffer, Jr, R. L. (1971) 'Nature of the DNA associated with the nuclear envelope of regenerating liver', *J. Molec. Biol.*, **59**, 517–525.

Monesi, V. (1969) 'DNA, RNA and protein synthesis during the mitotic cycle', In: *Handbook of molecular cytology* (ed. A. Lima-de-Faria), pp. 472–499, North-Holland Publishing Co., Amsterdam.

Monneron, A. (1966) 'Utilisation de la pronase en cytochimie ultrastructure', *J. Microscopie.*, **5**, 583–596.

Monneron, A. and Bernhard, W. (1969) 'Fine structural organization of the interphase nucleus in some mammalian cells', *J. Ultrastruct. Res.*, **27**, 266–288.

Moses, M. J. (1964) 'The nucleus and chromosomes: A cytological perspective', In: *Cytology and cell physiology* (ed. G. H. Bourne), 3rd ed., pp. 423–558, Academic Press Inc., New York.

Mundell, R. D. (1968) 'Studies on nucleolar and ribosomal basic proteins and their relationship to nucleolar function', *Expl Cell Res.*, **53**, 395–400.

Mundkur, B. and Brauer, B. (1966) 'Selective localization of nucleolar protein with amido black 10B', *J. Histochem. Cytochem.*, **14**, 94–103.

Muramatsu, M., Shimada, N., and Higashinakagawa, T. (1970) 'Effect of cycloheximide on the nucleolar RNA synthesis in rat liver', *J. molec. Biol.* **53**, 91–106.

Murray, K. (1964) 'The heterogeneity of histones' In: *The Nucleo-histones* (eds. J. Bonner and P. Ts'o), pp. 21–35. Holden-Day, Inc., San Francisco.

Nagl, W. (1968) 'Der mitotische und endomitotische kernzyklus bei *Allium carinatum*. I. Struktur, volumen und DNS-gehalt der kerne', *Österr. Bot. Z.*, **115**, 322–353.

Nagl, W. (1970a) 'Correlation of chromatin structure and interphase stage in nuclei of *Allium flavum*', *Cytobiologie*, **4**, 395–398.

Nagl, W. (1970b) 'The mitotic and endomitotic nuclear cycle in *Allium carinatum*. II. Relations between DNA replication and chromatin structure', *Caryologia*, **23**, 71–78.

Noel, J. S., Dewey, W. C., Abel, J. H. Jr, and Thompson, R. P. (1971) 'Ultrastructure of the nucleolus during the chinese hamster cell cycle', *J. Cell Biol.*, **49**, 830–847.

Ohnuki, Y. (1968) 'Structure of chromosomes. I. Morphological studies of the spiral structure of human somatic chromosomes', *Chromosoma*, **25**, 402–428.

Okada, S. (1968) 'Replicating units (replicons) of DNA in cultured mammalian cells', *Biophys. J.*, **8**, 650–664.

Painter, R. B., Jermany, D. A., and Rasmussen, R. E. (1966) 'A method to determine the number of DNA replicating units in cultured mammalian cells', *J. molec. Biol.*, 17, 47–56.

Pallotta, D., Berlowitz, L., and Rodriguez, L. (1970) 'Histones of genetically active and inactive chromatin in mealy bugs', *Expl Cell Res.*, 60, 474–477.

Pardue, M. L. and Gall, J. G. (1970) 'Chromosomal localization of mouse satellite DNA., *Science, N.Y.*, 168, 1356–1358.

Pardue, M. L., Gerbi, S. A., Eckhardt, R. A., and Gall, J. G. (1970) 'Cytological localization of DNA complementary to ribosomal RNA in polytene chromosomes of Diptera', *Chromosoma*, 29, 268–290.

Peacock, W. J. (1963) 'Chromosome duplication and structure as determined by autoradiography', *Proc. natn. Acad. Sci. U.S.A.*, 49, 56–113.

Pelling, C. (1959) 'Chromosomal synthesis of ribonucleic acid as shown by incorporation of uridine labelled with tritium', *Nature, Lond.*, 184, 655–656.

Perry, R. P. (1964) 'Role of the nucleolus in ribonucleic acid metabolism and other cellular processes', *Natl. Cancer Inst. Monograph.*, 14, 73–90.

Perry, R. P. (1967) 'The nucleolus and the synthesis of ribosomes', In: *Progress in nucleic acid research and molecular biology* (eds. J. N. Davidson and W. E. Cohn) vol. 6, pp. 219–257, Academic Press, Inc., New York.

Phillips, S. G. and Phillips, D. M. (1969) 'Sites of nucleolus production in cultured chinese hamster cells', *J. Cell Biol.*, 40, 248–268.

Plaut, W., Nash, D., and Fanning, T. (1966) 'Ordered replication of DNA in polytene chromosomes of *Drosophila melanogaster*', *J. molec. Biol.*, 16, 85–93.

Prescott, D. M. (1970) 'The structure and replication of eukaryotic chromosomes', In: *Advances in cell biology* (eds. D. M. Prescott, L. Goldstein, E. McConkey), vol. 1, pp. 57–117, Appleton-Century-Crofts, New York.

Prescott, D. M. and Bender, M. A. (1962) 'Synthesis of RNA and protein during mitosis in mammalian tissue culture cells', *Expl Cell Res.*, 26, 260–268.

Prescott, D. M. and Bender, M. V. (1963) 'Autoradiographic study of chromatid distribution of labelled DNA in two types of mammalian cells *in vitro*', *Expl. Cell Res.*, 29, 430–442.

Richards, B. M. and Bajer, A. (1961) 'Mitosis in endosperm. Changes in nuclear and chromosome mass during mitosis', *Expl Cell Res.*, 22, 503–508.

Richards, B. M and Pardon, J. F. (1970) 'The molecular structure of nucleohistone (DNH)', *Expl Cell Res.*, 62, 184–196.

Ris, H. (1945) 'The structure of meiotic chromosomes in the grasshopper and its bearing on the nature of "chromomeres" and "lampbrush chromosomes" ', *Biol. Bull.*, 89, 242–257.

Ris, H. (1957) 'Chromosome structure', In: *The chemical basis of heredity* (eds. W. McElroy and B. Glass) pp. 23–62, Johns Hopkins Press, Baltimore, Maryland.

Ris, H. (1969) 'The molecular organization of chromosomes', In: *Handbook of molecular cytology* (ed. A. Lima-de-Faria) pp. 221–250. North-Holland Publishing Co., Amsterdam.

Ris, H. and Mirsky, A. E. (1949) 'The state of the chromosomes in the interphase nucleus', *J. Gen. Physiol.* 32, 489–502.

Ris, H. and Kubai, D. F. (1970) 'Chromosome structure', *A. Rev. Genet.*, 4, 263–294.

Ritossa, F. M. and Spiegelman, S. (1965) 'Localization of DNA complementary to ribosomal RNA in the nucleolus organizer region of *Drosophila melanogaster*', *Proc. natn. Acad. Sci. U.S.A.*, 53, 737–745.

Robbins, E. (1961) 'Some theoretical aspects of osmium tetroxide fixation with special reference to the metaphase chromosomes of cell cultures', *J. Biophys, Biochem. Cytol.*, 11, 449–455.

Robbins, E. and Gonatas, N. K. (1964) 'The ultrastructure of a mammalian cell during the mitotic cycle', *J. Cell Biol.*, 21, 429–463.

Sadgopal, A. and Bonner, J. (1970) 'Proteins of interphase and metaphase chromosomes compared', *Biochim. biophys. Acta*, 207, 227–239.

Scherrer, K., Latham, H., and Darnell, J. E. (1963) 'Demonstration of an unstable RNA and of a precursor to ribosomal RNA in HeLa cells', *Proc. natn. Acad. Sci. U.S.A.*, 49, 240–248.

Shankar Narayan, K. and Birnstiel, M. L. (1969) 'Biochemical and ultrastructural characteristics of ribonucleoprotein particles isolated from rat-liver cell nucleoli', *Biochim. biophys. Acta*, 190, 470–485.

Simpson, R. T. and Sober, H. A. (1970) 'Circular dichroism of calf liver nucleohistone', *Biochemistry, N.Y.*, 9, 3103–3109.

The nucleus

Smith, E. L., DeLange, R. J., and Bonner, J. (1970) 'Chemistry and biology of the histones', *Physiol. Rev.*, **50**, 159–170.

Sparvoli, E., Gay, H., and Kaufmann, B. P. (1965) 'Number and pattern of association of chromonemata in the chromosomes of *Tradescantia*', *Chromosoma*, **16**, 415–435.

Stern, H., Johnston, F. B., and Setterfield, G. (1959) 'Some chemical properties of isolated pea nucleoli', *J. biophys. biochem. Cytol.*, **6**, 57–60.

Stevens, B. J. (1965) 'The fine structure of the nucleolus during mitosis in the grasshopper neuroblast cell', *J. Cell Biol.*, **24**, 349–368.

Stevens, A. R. and Prescott, D. M. (1971) 'Reformation of nucleolus-like bodies in the absence of postmitotic RNA synthesis', *J. Cell Biol.*, **48**, 443–454.

Strangeways, T. S. P. and Canti, R. G. (1927) 'This living cell *in vitro* as shown by darkground illumination and the changes induced in such cells by fixing reagents', *Q. Jl Microsc. Sci.*, **71**, 1–17.

Swift, H. (1962) 'Nucleic acids and cell morphology in dipteran salivary glands', In: *The molecular control of cellular activity* (ed. J. M. Allen), pp. 73–125, McGraw-Hill Book Company, Inc., New York.

Swift, H. (1963) 'Cytochemical studies on nuclear fine structure', *Expl Cell Res. Suppl.*, **9**, 54–67.

Swift, H. (1965) 'Molecular morphology of the chromosome', In: *The chromosome: Structural and functional aspects* (ed. C. J. Dawe), vol. I, pp. 26–49, Williams and Wilkins Inc., Baltimore.

Swift, H. and Stevens (1966) 'Nucleolar-chromosomal interaction in microspores of maize', *Natl. Cancer Inst. Monograph*, **23**, 145–166.

Tandler, C. J. (1966) 'Detection and origin of nucleolar components: A model for nucleolar RNA function', *Natl. Cancer Inst. Monograph*, **23**, 181–190.

Taylor, J. H., Woods, P. S., and Hughes, W. L. (1957) 'The organization and duplication of chromosomes as revealed by autoradiographic studies using tritium-labeled thymidine', *Proc. natn. Acad. Sci. U.S.A.*, **43**, 122–128.

Taylor, J. H. (1960) 'Nucleic acid synthesis in relation to the cell division cycle', *Ann. N.Y. Acad. Sci.*, **90**, 409–421.

Thomas, C. A. (1971) 'The genetic organization of chromosomes', *A. Rev. Genet.*, **5**, 237–256.

Tokuyasu, K., Madden, S. C., and Zeldis, L. J. (1968) 'Fine structural alterations of interphase nuclei of lymphocytes stimulated to growth activity *in vitro*', *J. Cell Biol.*, **39**, 630–660.

Trosko, J. E. and Wolff, S. (1965) 'Strandedness of *Vicia faba* chromosomes as revealed by enzyme digestion studies', *J. Cell Biol.*, **26**, 125–135.

Vincent, W. S., Baltus, E., Lovlie, A., and Mundell, R. E. (1966) 'Proteins and nucleic acids of starfish oocyte nucleoli and ribosomes', *Natl. Cancer Inst. Monograph*, **23**, 235–253.

Vincent, W. S. (1955) 'Structure and chemistry of nucleoli', *Int. Rev. Cytol.*, **4**, 269–298.

Walker, P. M. B., Flamm, W. G., and McLaren, A. (1969) 'Highly repetitive DNA in rodents', In: *Handbook of molecular cytology* (ed. A. Lima-de-Faria), pp. 52–66, North-Holland Publishing Co., Amsterdam.

Walker, P. M. B. (1972) 'Repetitive DNA in higher organisms' In: *Progress in biophysics and molecular biology* (eds. J. A. V. Butler and D. Noble), vol. 23, pp. 145–190. Pergamon, Oxford.

Wallace, H. and Birnstiel, M. L. (1966) 'Ribosomal cistrons and the nucleolar organizer', *Biochim. biophys. Acta*, **114**, 296–310.

Warner, J. R. and Soeiro, R. (1967) 'Nascent ribosomes from HeLa cells', *Proc. natn. Acad. Sci. U.S.A.*, **58**, 1984–1990.

Watson, M. L. (1962) 'Observations on a granule associated with chromatin in the nuclei of cells of rat and mouse', *J. Cell Biol.*, **13**, 162–167.

Willems, M., Penman, M. and Penman, S. (1969) 'The regulation of RNA synthesis and processing in the nucleolus during inhibition of protein synthesis', *J. Cell Biol.*, **41**, 177–187.

Wolff, S. (1969) 'Strandedness of chromosomes', In: *Int. Rev. Cytol.* (eds. G. H. Bourne, J. F. Danielli), vol. 25, pp. 279–296, Academic Press, New York.

Wolff, S. and Luippold, H. E. (1964) 'Chromosome splitting as revealed by combined X-ray and labeling experiments', *Expl Cell Res.*, **34**, 548–556.

Woodard, J., Rasch, E., and Swift, H. (1961) 'Nucleic acid and protein metabolism during the mitotic cycle in *Vicia faba*', *J. Biophys. biochem. Cytol.*, 9, 445–462.
Zimmerman, E. F., Hackney, J., Nelson, P. and Arias, I. M. (1969) 'Protein synthesis in isolated nuclei and nucleoli of HeLa cells', *Biochemistry, N.Y.*, 8, 2636–2644.

2

Mitochondria
Helgi Öpik

Introduction

The first observation of mitochondria is credited to Altmann who, in 1890, described the organelles in fixed animal cells using the name 'bioblasts', the term 'mitochondrion' being coined by Benda nine years later. Following Meves' initial report of mitochondria in plant cells in 1904, mitochondria were described in many plant tissues during the first half of this century. The advent of electron microscopy took the study of cell structure to a new level and, as early as 1947, Buchholz published some electron micrographs of mitochondria teased out of *Tsuga* eggs and maize (*Zea mays*) pollen mother cells, and dried down on films, but little internal detail was visible in these crude preparations. The internal structure of mitochondria was described in 1953 independently by Palade and Sjöstrand from ultra-thin sections of animal cells, and the same basic organization has subsequently been shown to hold for all types of mitochondria. The study of mitochondrial metabolism can be said to have originated with the elucidation of the Krebs cycle in the 1930s, although then the location of this reaction series was not yet known. Intensive work on the activities of isolated mitochondria began around 1940, and within a decade animal mitochondria had been proved to be important centres of cellular energy metabolism, carrying the complete Krebs cycle, the cytochrome electron transport chain, and the system for oxidative phosphorylation. Progress with plant mitochondria was slower due to isolation difficulties; nevertheless, by 1953 the same basic respiratory functions had also been identified in mitochondria from plant sources.

The initial examinations of mitochondrial fine structure and the pioneer work on mitochondrial activity were carried out quite separately, but studies on structure and function soon became combined. A consideration of structure is essential for an understanding of the functioning of the mitochondrion as a whole, and for an understanding of the reaction mechanisms. At the levels of structure visible by conventional electron microscopy, basic structure has been elucidated and attention is now focusing on comparative studies of mitochondria at different developmental stages, under different environmental conditions, and in different cell types. One of the most active areas of current mitochondrial research, however, is the study of mitochondrial structure at the molecular and multimolecular complex levels, almost beyond the reach of even electron microscopy, and the correlation of structure at these levels with the energy-linked reactions. Electron transport and oxidative phosphorylation are absolutely dependent on structural organization of the relevant macromolecules, and proceed accompanied by structural changes in the mitochondrial membranes. The location of many of the mitochondrial enzymes even at cruder levels, is also still under active study. On these lines, work with plant mitochondria is, unhappily, still lagging behind, and data on molecular structure and conformational changes in plant mitochondria are extremely scanty. The advances being made in these fields are, however, of such interest and importance, that rather than omit consideration of the subject, the present article will include findings from animal mitochondria where corresponding data for plant mitochondria are not available.

Another topic which has recently risen to prominence is that of genetic control of mitochondrial development and of synthesis of mitochondrial macromolecules; a whole discipline of mitochondrial genetics has sprung up. This subject will be considered in the final section of this chapter, although only a summary outline of the current picture is feasible within the present scope.

Mitochondrial structure

Since mitochondrial diameter rarely exceeds 1 μm, even when the length extends to several micrometres, light microscopy can reveal no more than the overall shape of the organelles, and for all knowledge of their internal structure electron microscopy must be relied on. The structure of a typical plant mitochondrion, based on pictures obtained with chemical fixation and thin sectioning techniques, is illustrated in Fig. 2.1A, B. An *outer membrane* of smooth outline encloses a highly convoluted *inner membrane*, with the infoldings or *cristae* of variable form and number, projecting into the central *matrix*. Except where it folds inwards to form the cristae, the inner membrane follows the contours of the outer, leaving an *intermembrane space* of fairly constant width (but see below), continuous with the *intracristal space* or *lumen*.

This general model of the mitochondrion has become so familiar that one tends to forget that, notwithstanding all the micrographs which suggest this interpretation, there is uncertainty about the true width and indeed the very existence of the intermembrane space and the intracristal lumen, in mitochondria in the living cell. Both of the common chemical fixatives, osmium tetroxide and potassium permanganate, usually produce a narrow but distinct intermembrane space. The intracristal space is usually narrow with permanganate fixation but may be much more dilated when the same tissue is osmium-fixed. Freeze-etching, assumed to stabilize organelle morphology unchanged, has shown the two envelope membranes of yeast mitochondria to be closely adpressed with no intermembrane space, while the intracristal spaces are narrow, approximately 5–10 nm across (Moor and Mühlethaler, 1963). The intracristal spaces seem to be very narrow

Mitochondria

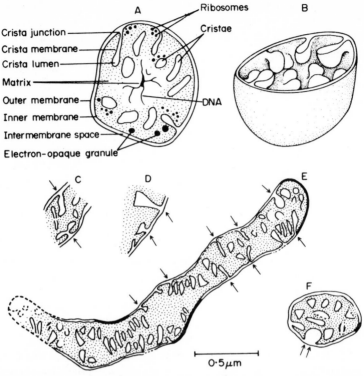

Fig. 2.1 The morphology of plant mitochondria, based on chemical fixation and thin sectioning techniques; the intermembrane space and the crista lumina may be exaggerated.
(A) Diagrammatic cross-section.
(B) Diagrammatic three-dimensional reconstruction of the membrane systems.
(C to E) Actual tracings of electron micrographs of thin sections of glutaraldehyde/osmium tetroxide fixed material, to illustrate the form of the junctions of cristae and inner membrane. C, E and F: coleoptile of rice (*Oryza sativa*); D: etiolated leaf of mung bean (*Phaseolus aureus*). The matrix is stippled; indistinct (obliquely sectioned) membranes are dotted. C, D and E show the common situation where the crista junctions (arrowed) appear narrow, even when the cristae form fairly wide folds or sacs. A wide junction as depicted in E (double arrows) is rarely found; if plate-like cristae had openings extending the whole length of the fold, this kind of picture should be seen much more frequently.

also in freeze-etched onion (*Allium cepa*) root tip (Branton and Moor, 1964). For animal tissues, too, quick-freezing techniques have yielded pictures showing no gap between the outer and inner membranes, nor between the membranes of the cristae (Malhotra, 1966; Steinert, 1969). Thus there may be much less space between the membranes *in vivo* than conventional electron microscopy indicates.

The membrane systems

It is agreed that the outer and inner membranes are of very different nature. Both are lipoprotein structures and can be resolved to display the triple-layered 'unit membrane' profile, 5–7 nm thick (see, e.g., Baker, Elfvin, Biale, and Honda, 1968, for a fine illustration of this in a sweet potato (*Ipomoea batatas*) mitochondrion).

But differences in staining density are frequently apparent, and the outer membrane is more susceptible to fixation damage, sometimes disintegrating when the inner membrane still looks intact. In mitochondrial fragments obtained by hypotonic treatment and negatively stained, or in freeze-etched *in situ* preparations, the outer membrane looks smooth, except that pits, 2.5–3 nm in diameter, were seen in the outer membranes of negatively stained mitochondria from six species of higher plants (Parsons, Bonner, and Verboon, 1965), and slits, approximately 10 x 100 nm, in yeast mitochondria after freeze-etching (Moor and Mühlethaler, 1963; Plattner and Schatz, 1969). Animal mitochondrial outer membranes have not revealed any perforations and it is possible that these are indicative of damage. The cristal membranes, on the other hand, are characterized by the appearance in negatively stained preparations, of stalked particles on the side facing the matrix; examples of plant tissues where these have been seen are *Ricinus* endosperm (Nadakavukara, 1964); squash (*Cucurbita pepo*), mung bean (*Phaseolus aureus*), beetroot (*Beta vulgaris*), potato (*Solanum tuberosum*), wheat (*Triticum* sp.), and cauliflower (*Brassica oleracea*), (Parsons *et al.*, 1965); yeast (Shinagawa, Inouye, Ohnishi, and Hagihara, 1966) and *Neurospora* (Stoeckenius, 1965). In all cases the particles have a roughly spherical 'head', with reported diameters varying from 7.5 to 11 nm, subtended by a stalk 4–6 nm long. Freeze-etching shows particles about 10 nm in diameter on the inner membranes of onion root cell mitochondria (Branton and Moor, 1964); it is not clear whether these are stalked. Similar particles coat the cristae of freeze-etched animal mitochondria (Moor, 1964).

The stalked particles have incited much controversy, some authors having considered them artifacts induced by hypotonic treatment, comparable to myelin figures. Sjöstrand, Anderson-Cedergren, and Karlsson (1964) comment that the particles seen by negative staining only appear on ageing, and are more difficult to demonstrate in the very stable heart muscle mitochondria than in the more labile mitochondria of kidney tissue. The negative staining data are to some extent reinforced by the revelation of globular subunits also by freeze-etching and by thin sectioning (Ashurst, 1965, moth nerve tissue), where hypotonic pretreatment used with negative staining is not involved. One cannot be certain, however, that what is seen after different techniques, is the same entity. Nevertheless, the regular array of particles seen in negatively stained preparations must at least represent some regular repeating pattern in the crista membranes, even if the appearance of the subunits is modified by the treatment applied.

Another difference between the outer and inner membranes lies in their osmotic behaviour. The inner membrane is an osmotic barrier to substances such as sucrose, and expands and contracts reversibly in response to changes in sucrose concentration. The outer membrane is sucrose-permeable. In a medium of high osmolarity, the inner membrane shrinks and the outer remains unchanged; in hypotonic solutions, the outer membrane swells, but irreversibly. Consideration of the enzymic content of the outer and inner membranes and of membrane structure at the molecular level will be deferred to the next section.

Are the crista membranes identical in composition with that part of the inner membrane which parallels the contours of the outer membrane? The physical continuity of the two parts of the membrane speaks in favour of identity; so does the fact that in extreme swelling the crista folds disappear, leaving finally a single smooth inner membrane sac, at least in animal mitochondria (Green and Baum, 1970). But Kellerman, Biggs, and Linnane (1969), noting that in anaerobically grown yeast, the cristae were disorganized but the inner membrane appeared normal, suggested that the two may be separate entities, and that a mitochondrion should therefore be considered to be a three-membrane system rather than a two-membrane one.

The three-dimensional shape of the cristae

A cursory examination of a few mitochondrial profiles in a thin section will give some impression of the abundance and arrangement of cristae in a cell type but the precise three-dimensional shape can be deduced only from a careful study of many profiles, and may require serial sectioning. A broad morphological division of cristae has been made into tubular invaginations ('microvilli') and plate-like folds ('true cristae'). Plant mitochondria are often stated to possess predominantly the microvillous type of cristae, in contrast to animal mito- chondria, where the cristae are usually plate-like, but this statement is too sweeping. Tubular cristae do occur in a fair number of algae and fungi, but by no means in all. As for higher plants, the present writer's impression is that really unequivocally tubular cristae are in the minority; some good examples are often found in companion cells, e.g., *Pisum sativum* (Wark, 1965), and *Phaseolus vulgaris* (Öpik, unpublished), and occasionally in other instances: e.g., aerial roots of *Chlorophytum* (Mota, 1963). It is true, however, that the cristae of plant mitochondria are much less regular than the neatly stacked folds characterizing numerous animal mitochondria. Very regular parallel cristae occur in some pollen mitochondria (Larson, 1965), in *Neurospora* (Beck, Decker, and Greenawalt, 1970), and sometimes in yeast, but irregularly oriented folds and sacs are much more common. What these should be called depends on definition; one can find in plant mitochondria the whole range: regular plates, irregular folds, swollen sacs, narrower vesicles, fingerlike tubules. The morphology is also fixation- dependent: what appear as wide folds or vesicles with osmium fixation can contract to narrow plates and tubules with permanganate. The general term 'crista' will therefore be used throughout this chapter for infoldings of the inner membrane irrespective of morphology. The shapes can be very complex: cylindrical, branched, goblet-like (Diers and Schötz, 1965); no significance can at present be assigned to the cristal shape variations. Cristae of a peculiar form have been reported as occurring in a small number of mitochondria in bean (*Phaseolus vulgaris*) roots (Newcomb, Steer, Hepler, and Wergin, 1968); plate- like, extending almost the whole length of the organelle, and with the inner crista surfaces tightly apposed, leaving no intracristal lumen. The result is a five- layered, rigid-looking structure, with a surface pattern of oblique electron opaque lines, and of unknown function.

The opening from the intracristal space into the intermembrane space (or the point of junction of the membranes, if there is no intermembrane space) must be narrow even in plate-like cristae or it would be encountered much more commonly in electron micrographs than is actually the case. A count of 50 mitochondrial profiles, from a variety of plant cells, yielded 1029 cristae, of which only 9 per cent showed the junction (Öpik, unpublished), and where the junction was visible the cristal lumen often bulged out from a finer neck (Fig. 2.1 C to E). The plate-like cristae of animal mitochondria are also said to have very small points of junction (Green and Baum, 1970).

The matrix

Earlier studies on mitochondrial structure tended to concentrate on the discrete membrane systems and to dismiss the matrix as the 'soluble phase'. When permanganate is used as the fixative, matrix structure is generally lost, but osmium fixation, particularly when preceded by glutaraldehyde, reveals structure also in the matrix. Its staining density can be higher or lower than that of the cytoplasmic background; in isolated mitochondria, matrix density can vary according to incubation conditions. Areas of higher and lower electron-opacity often occur in the same organelle. The solid material of the matrix is mainly protein, in concentrations which make a gel state probable; e.g., for liver mito-chondria, the concentration of matrix protein is calculated to be 56 per cent (Hackenbrock, 1968a). In 'texture' the matrix may look finely granular, or may give the impression of a meshwork of filaments. Some authors have viewed the electron-opaque material as the structural framework; e.g., Baker et al. (1968) think that the matrix contains 6 nm diameter electron opaque granules; others have interpreted the structure as electron-translucent granules in an electron-opaque amorphous ground mass (Weintraub, Ragetli, and John, 1966).

In 1963, Nass and Nass reported a detailed study on the presence of DNA filaments in the mitochondrial matrix of several animal tissues; since then, similar filaments have been found in many plant mitochondria (see e.g., Mikulska, Odintsova, and Turischeva, 1970). The DNA is frequently located in a very electron-translucent area of the matrix and the filaments can be fine and diffuse or coagulated into an opaque mass, from which finer strands may radiate. The filaments may be connected to the cristae or inner membrane (Fig. 2.2A). Most often one DNA-containing region is observed per mitochondrial profile; some-times several are visible, especially in elongate mitochondria but, without serial sectioning, one cannot rule out possible interconnections. Mitochondria lacking distinct DNA filaments may show electron-opaque regions of a more diffuse outline, generally one per mitochondrion, and more or less median (Fig. 2.2D), which could represent the nucleic acid in a different form. The presence of mitochondrial DNA has been amply confirmed by analytical methods; its role in mitochondrial development will be discussed in the last part of this chapter.

Ribosomes ('mitoribosomes') can be seen in the mitochondrial matrix, looking slightly smaller than the cytoplasmic ribosomes ('cytoribosomes') which is borne out by ultracentrifugation studies. The arrangement may appear random, or they can form small groups, possibly polysomes: from Neurospora, mitochondrial

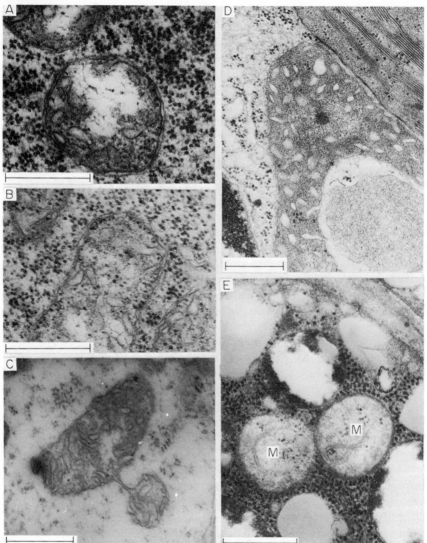

Fig. 2.2   Electron micrographs of plant mitochondria fixed in glutaraldehyde/osmium tetroxide, post-stained with uranyl acetate and lead citrate. The scale line in all cases = 0.5 μm.

(A) Mitochondrion from root tip meristem of tomato (*Lycopersicon esculentum*); the large electron-translucent region contains DNA filaments, some of which extend to the membrane and cristae; the mitochondrion contains also ribosome-like particles.

(B) Another mitochondrion from the same tissue, with ribosome-like particles just to the inside of the inner membrane.

(C) Two mitochondrial profiles joined by a narrow 'neck'; coleoptile of wheat (*Triticum sativum*).

(D) Mitochondrion with darkly staining region (see text, page 57); young leaf of mung bean (*Phaseolus aureus*).

(E) Two mitochondria (M) in coleoptile of ungerminated rice grain, each with one crista profile; ribosome-like particles are present. The 'fuzzyness' of the cristae is not due to poor focusing, since the plasmalemma (top right) is very clear.

polysomes have been isolated (Küntzell and Noll, 1967). Sometimes the mito-ribosomes appear associated with cristae or the inner membrane (Fig. 2.2B). It may be difficult to decide from mere visual observation whether some small particles in a mitochondrial section are ribosomes or not.

Occasional matrix components are electron-opaque granules about 20—40 nm in diameter, probably representing insoluble phosphate deposits. Protein crystals have been seen in lentil (*Lens culinaris*) leaf epidermal cell mitochondria (Lance-Nougarède, 1966), and some kind of crystalline structures occasionally lie in mitochondria of radicle cells of ungerminated bean (*Phaseolus vulgaris*) embryo (Öpik, unpublished).

The matrix can accordingly be seen to contain several important structures (as well as enzymes: see next section). Pihl and Bahr (1970) even suggest that the matrix rather than the membrane systems should be considered the true mito-chondrial skeleton, since in whole mounts and sections of critical-point dried, negatively stained, mitochondria (rat liver), they believe that one can see a 'matrix cord', 30—90 nm wide, twisted round and round to outline the mitochondrial shape.

The intermembrane and intracristal spaces have so far revealed no visible structure and appear electron-translucent; they are generally assumed to contain an aqueous fluid but, as discussed, their dimensions may be greatly exaggerated by fixation.

Mitochondrial activity: energy transduction

The reaction systems; activity of isolated mitochondria

Mitochondria can carry out numerous metabolic activities and must contain well over fifty enzymes. The basic metabolic function of mitochondria is, however, considered to be energy transduction, the conversion of the potential energy of oxidizable substrates to a different form, most commonly to ATP; and the enzyme systems more or less directly connected with energy transduction are the ones studied most thoroughly, and these will be considered here.

The central process in energy metabolism is electron transport. Electrons of high negative potential pass along a series of carriers, the electron transport chain, to molecular oxygen. The electron flow is dependent on a supply of electrons of negative potential from the mitochondrial substrate oxidizing systems and can be coupled to a synthesis of ATP (oxidative phosphorylation) at the expense of the fall in the negative potential of the electrons. Alternatively, the energy made available during electron transport can be used directly for ion movements through the mitochondrial membranes or for reductive reactions. The final stage of substrate oxidation is the Krebs cycle, into which the products of carbohydrate breakdown enter as pyruvate; carbon skeletons of amino acids can join at several points, and two-carbon units produced in the β-oxidation of fatty acids come in as acetyl coenzyme A. The β-oxidation system of animal cells is itself contained in mitochondria; for plants, its location is less certain. As the substrates are oxidized in the Krebs cycle, pairs of hydrogen atoms are accepted by NAD, and the resulting NADH feeds the electrons into the electron transport chain. (Only in the succinate oxidation step the coenzyme does not mediate.)

A simplified outline of the electron transport chain, where the carriers are alternatively oxidized and reduced by loss and acceptance of electrons as these pass along, is presented in Fig. 2.3. In this general form, the scheme is applicable

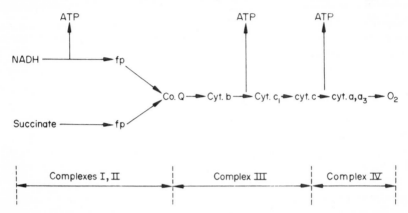

Fig. 2.3 The electron transport chain. The horizontal arrows denote the normal direction of electron flow; the vertical arrows, sites of ATP formation. Co.Q = coenzyme Q; cyt. = cytochrome, fp = flavoprotein. The cytochromes a + a_3 constitute cytochrome oxidase. The non-haeme iron and copper atoms associated with various parts of the chain are not indicated. The Complexes I to IV refer to the subdivision of the chain by Green and coworkers (see page 64).

to mitochondria from different sources, but details vary. The mammalian electron transport chain contains only one b cytochrome, but in higher plant mitochondria there are several, and these have been thought to mediate the cyanide-resistant respiration which is very appreciable in some plant tissues and their isolated mitochondria, bypassing the highly cyanide-sensitive cytochrome oxidase. Some recent work, however, indicates that the cyanide-resistant pathway branches off from the cytochrome chain at the flavoprotein level. The discovery that hydroxamic acids are very specific inhibitors of the cyanide-insensitive oxidase (Schonbaum, Bonner, Storey, and Bahr, 1971), may help in its identification. (For detailed analyses of the plant electron transport chain, see: Lance and Bonner, 1968; Erecinska and Storey, 1970; Storey, 1970a, b; Wilson and Bonner, 1970a, b, c.)

The sites of ATP synthesis can be deduced from a knowledge of the differences in redox potential between successive members of the chain. There must be a minimum difference of 0.20 volts if the transfer of a pair of electrons is to result in the formation of a high energy phosphate bond (Lehninger, 1965). The sites have also been identified experimentally by observation of changes in the degree of oxidation of the various carriers at the onset of oxidative phosphorylation, and by confining electron transport to limited sections of the chain with appropriate combinations of substrates and inhibitors.

The expected values of the ADP/O ratio (numerically equivalent to the P/O or

P/2 electron ratios), i.e., moles high energy phosphate formed per atom of oxygen utilized, are 3.0 for NADH oxidation and 2.0 for succinate (see Fig. 2.3). Together with a substrate level phosphorylation in the Krebs cycle, this gives a total of 15 molecules of ATP per 1 molecule of pyruvate oxidized, equivalent to a conservation of approximately 50 per cent of the available energy. About the same level of energy conservation is calculated for fatty acid oxidation.

With animal mitochondrial preparations, the theoretical ADP/O ratios were obtained already before 1950, but it took considerably longer to attain comparable values with plant mitochondria. Undoubtedly this was due to damage during isolation: the phosphorylation system is much more labile than the substrate oxidation systems and the electron transport chain. The possession of a high ADP/O ratio is one of the criteria for a 'good' mitochondrial preparation, and with newer techniques, plant mitochondria have given ADP/O values of the theoretical level. A ratio of 6.0 for α-ketoglutarate oxidation has been claimed in one instance (Sarkissian and Srivastava, 1969). The *uncoupling* of electron transport from phosphorylation can occur spontaneously as a result of damage, or be induced by chemicals; the classical uncoupling agent is DNP (2, 4-dinitrophenol) but many other and more potent uncouplers have been discovered.

A second criterion for evaluating the intactness of mitochondria is the RC (respiratory control) ratio. In 1955, Chance and Williams defined five states of mitochondrial activity according to what factor (substrate, ADP or oxygen) was limiting oxidation. In 'state 4' when activity was limited by lack of the phosphate acceptor ADP, though ample substrate and oxygen were present, the oxygen uptake was found to be very slow. Addition of ADP resulted in rapid oxidation, 'state 3', limited now only by the capacity of the electron transport chain. Then:

$$\text{RC ratio} = \frac{\text{rate of oxygen uptake in state 3}}{\text{rate of oxygen uptake in state 4}}$$

The higher this ratio is, the more tightly is the electron transport coupled to phosphorylation. Mammalian mitochondria may exhibit RC ratios of over 40. Early plant mitochondrial preparations manifested little or no respiratory control, but here again refinements in technique have produced improvements. Parsons *et al.* (1965) obtained mitochondrial RC ratios of 2.9–4.2 with six higher plant tissues; Ikuma and Bonner (1967) reported 5.6 for mung bean hypocotyl mitochondria; Sarkissian and Srivastava (1968) measured RC ratios of up to 10 in wheat seedling mitochondria; and Ku, Pratt, Spurr, and Harris (1968) working with mitochondria from tomato (*Lycopersicon esculentum*) fruit found that with successive cycles of ADP addition, the RC for succinate oxidation gradually reached infinity, i.e., there was finally no measurable state 4 oxidation at all. The RC mechanism is believed to regulate mitochondrial activity *in vivo*. During periods of rapid energy expenditure, rapid cellular utilization of ATP would keep up a constant supply of ADP, whereas when the rate of ATP utilization is low, mitochondrial activity is kept low by the limited production of ADP. At the moment there is still a tendency to regard plant mitochondria as more loosely coupled than animal mitochondria; future studies must show whether this loose coupling is inherent, or an isolation artifact. (Note: the '*loose* coupling' with respect to respiratory control is not to be confused with *un*coupling. Loosely coupled mitochondria have low RC ratios, i.e., they will carry out electron transport in the absence of ADP, without oxidative phosphorylation; if ADP is supplied, however, they can phosphorylate with normal ADP/O ratios. Uncoupled mitochondria are incapable of phosphorylation; even in the presence of ADP, their ADP/O = 0.)

The $Q[O_2(N)]$ values of carefully prepared plant mitochondria compare very favourably with those of animal mitochondria and may indeed exceed the latter (Ikuma and Bonner, 1967). The energy transducing system is clearly fundamentally the same in plant and animal mitochondria.

The intramitochondrial location and structural organization of the enzyme systems involved in energy transduction

Already since the 1940s, it has been recognized that there is compartmentation of activities within mitochondria, and that for reaction series as complex as electron transport, the enzymes exist not singly but as multienzyme systems, supramolecular complexes of specific structure. But even now, the locations are known with certainty for only a limited number of the mitochondrial enzymes and the precise structure of any complex at the molecular level is unknown. Both topics are currently under very active study, but the ideas to be presented in this section stem almost exclusively from work with animal (mainly mammalian) organelles. In view of their similarity in metabolism and overall structure, it is scarcely plausible that plant and animal mitochondria should show drastic differences in enzyme location and organization, but the possibility of minor variations should be kept in mind.

Some mitochondrial enzymes are very firmly membrane-bound; this includes the cytochromes (except cytochrome c) and succinic dehydrogenase. It is almost impossible to get cytochrome oxidase and succinic dehydrogenase into true solution. When solubilized, they show properties quite different from those of the membrane-bound form; soluble succinic dehydrogenase is very unstable, whereas submitochondrial membrane fragments possess reasonable stability and may still be capable of electron transport, even of oxidative phosphorylation. Other enzymes, including the remaining Krebs cycle dehydrogenases, and the enzymes of fatty acid oxidation, are easily brought into solution from both plant and animal mitochondria although completely intact isolated mitochondria retain the enzymes.

It is agreed that the electron transport chain, including succinic dehydrogenase, and the phosphorylating system, are located in the inner membrane only, being still present in preparations which have been treated to remove the more labile outer membrane. Succinoxidase and NADH oxidase have been located histochemically at the electron microscope level in the inner membranes and cristae by deposition of enzymatically produced formazan crystals (Ogawa and Barrnett, 1964). The animal mitochondrial electron transport chain is inaccessible to externally added NADH, though this is oxidized by isolated plant mitochondria, which may possess a NADH dehydrogenase on the outer membrane (Douce, Mannella and Bonner, 1973).

The siting of the remaining Krebs cycle dehydrogenases and other substrate-oxidizing enzymes is, however, still in dispute. The classical view was to regard these as soluble in the matrix. Location in the outer membrane was discounted because of the phenomenon of latency: intact animal mitochondria are unable to oxidize externally supplied substrate acids, except pyruvate: apparently the enzymes are inside the membranes which pyruvate alone can permeate; only when the structure is mildly damaged are other added acids oxidized. Again, plant mitochondria are able to oxidize externally added acids even when their RC and ADP/O ratios indicate reasonable intactness, but this could be explained by a difference in permeability. Perhaps the perforations seen in the outer membranes of plant mitochondria (pp. 55) provide channels of entry. The ready release of the dehydrogenases into solution on sonication suggested that they were in free solution and the matrix offered itself as the obvious location.

More recently, Green and his co-workers have proposed that the dehydrogenases are associated in labile binding with the outer membrane and positioned in the intermembrane space. A full discussion and references to original papers can be found in Green and Baum (1970); in brief, the claim rests on subfractionation of beef heart mitochondria, using agents such as sonication, acetone-drying, phospholipase digestion, cholate or digitonin, and separating the resulting particles by centrifugation. The authors have obtained particles which they believe to be derived from the outer membrane and which contain the Krebs cycle dehydrogenases, fatty acid oxidases and β-hydroxybutyrate dehydrogenase.

This interpretation is rejected by other authors who have used methods of subfractionation which they consider to be more gentle. The mitochondria are allowed to swell in a hypotonic medium; on replacement in a sucrose solution, the inner membrane and matrix contract, but the irreversibly expanded outer membrane remains separate, and after gentle disruption, pieces float off and can be centrifuged out (Ernster and Kuylenstierna, 1969). Outer membrane fractions so prepared lack any Krebs cycle activity. Heidrich, Stahn, and Hannig (1970) used a combination of electrophoresis and density gradient centrifugation after hypotonic treatment to separate the outer membrane fraction from the inner membrane plus matrix; both preparations appeared pure in electron micrographs. No Krebs cycle enzyme activity was detected in the outer membrane fraction, but the inner-membrane/matrix contained malic and glutamic dehydrogenases, releasable by detergents and sonic action. If this is correct, Green's supposed outer membrane fraction must have included vesicles containing matrix material. It may also be commented that the intermembrane space, inferred to be the location of these enzymes by Green, is under suspicion of not existing as such *in vivo*.

Structural organization of the inner membrane-bound enzymes.

The major constituents of mitochondrial membranes are lipid and protein. The lipid, mainly phospholipid, amounts to approximately 30 per cent by dry weight, a lower percentage than found in other cellular membranes. About

40 per cent of the membrane protein is thought to be 'structural protein' (Criddle, 1969), a somewhat ill-defined entity. The name was originally applied to a protein fraction devoid of enzyme activity and highly insoluble, but more recently it has been claimed to dissolve even in distilled water after suitable extraction methods (Branton, 1969). Many 'structural protein' preparations have probably been highly heterogeneous mixtures. Nevertheless, evidence has accumulated for a class of non-enzymatic proteins not only in mitochondria but in other cellular membranes, with molecular weights of 20 000–30 000, capable of binding to enzymes stoichiometrically and modifying enzyme properties; *Neurospora* mutants are known where differences in mitochondrial malic dehydrogenase activity can be traced to modifications in structural protein (Woodward, 1968). (This could, incidentally, be taken as evidence for a membrane-bound location of malic dehydrogenase.) The cytochromes and other insoluble mitochondrial enzymes must, however, also be regarded as integral structural components of the mitochondrial membranes.

For biological membranes, current structural models are of two basic types: the bimolecular lipid leaflet type, and the subunit (particulate) type. The first visualizes a continuous phase of lipid covered on its outer surfaces by protein; the second, discrete subunits of protein, lipid, or lipoprotein, without a continuous lipid phase. In the case of mitochondria, the evidence points strongly to a particulate substructure. According to some authors, the quantity and molecular shape of mitochondrial lipids does not allow their packing into a condensed lipid bilayer (O'Brien, 1967). Freeze-etching shows a particulate structure in mito-chondrial membranes. Branton (1969) claims that the fracture plane revealing the particles runs inside the membranes (between the two darkly staining lines of the 'unit membrane' of sectioned material). He interprets this as indication of a composite membrane of particles embedded in a lipid bilayer, for a fracture would be expected to cleave a lipid bilayer, the hydrophobic bonding of the lipid molecules being weakened at the low temperature. The mitochondrial membranes have about 2700 particles per square micrometre. Particles are visible on many other freeze-etched membranes, but Staehelin (1968) notes that the particles can be removed from other cellular membranes but not from those of mitochondria (nor from chloroplast membranes, which functionally are most closely comparable to the mitochondrial). The observation of stalked subunits in negatively stained preparations can again be quoted in support of a subunit structure.

Green's group has circulated a very detailed model for structure of the inner/ cristal membrane, based on subfractionation, and electron microscopy. The membrane has been fractionated into four functional complexes, each with a total molecular weight of approximately 5×10^5 and containing enzymes, structural protein, phospholipid, and insoluble cofactors, as follows:

Complex I: NADH–Co . Q reductase, catalysing electron transport from NADH to Co . Q.
Complex II: Succinate–Co . Q reductase, catalysing electron transport from succinate to Co . Q.

Complex III: Co . Q–cytochrome c reductase, catalysing electron transport from Co . Q to cytochrome c.

Complex IV: cytochrome c oxidase.

Complexes I, III and IV or II, III and IV in sequence constitute the electron transport chain (Fig. 2.3), Co . Q and cytochrome c being considered as mobile joining links, and the functional chain has been reconstituted *in vitro* from the complexes. The building block of the membrane is taken to be a *tripartite unit,* of which any of the above complexes is a roughly cubical basepiece, embedded in the membrane, whence project the *stalk* and the headpiece displayed in negatively stained material:

Tripartite unit = basepiece + stalk + headpiece

The basepieces are considered integral parts of the membrane, joined laterally through their phospholipid moieties, and not detachable from the membrane without destroying it, while the stalk and headpiece can be detached, leaving the membrane still intact. It is recognized that basepieces other than the four complexes listed may exist, but the detachable portion is believed to be identical for all. The only enzyme activity detected in the headpieces is ATPase, which is probably active in ATP synthesis under normal conditions. The stalk, lacking enzyme activity, is thought to act as a link for coupling electron transport in the basepieces to ATP synthesis in the headpieces. No detailed model has been proposed for the outer membrane; the Krebs cycle enzymes are considered by Green and co-workers to be in a much looser association with the membrane, not forming integral parts of it.

This four-complex model is, however, only one view. The fractionating procedures of Green's school have been criticized; isolation of particles with certain enzyme activity does not prove that they existed as such in the membrane, and the tripartite form of the subunits may also be an artifact. An entirely different model has been postulated by Sjöstrand and Barajas (1968; 1970), depending on electron microscopy of mitochondria in rat kidney cells treated in ways which, the authors claim, avoid the protein denaturation to be expected from standard dehydration procedures utilizing alcohol or acetone. In their studies, the tissues were fixed in 1% glutaraldehyde for 0.5–60 minutes, or left unfixed, and dehydrated in ethylene glycol, sometimes air-dried. Very thin 'dark grey' sections were examined, and the membranes displayed globular subunits, the average diameter of the subunits increasing with time of fixation from 6.4 nm (0.5 minutes) to 10.9 nm (60 minutes). Taking a particle diameter of 7 nm obtained with 3 minutes' fixation, and considering the known molecular weights of the mitochondrial membrane-bound enzymes, the authors have constructed three-dimensional models of subunits, which they also designate complexes I, II and III, rather confusingly, because these are quite different from the complexes of Green's school; here complex I, e.g., contains *all* the electron transport chain plus some dehydrogenases, with a total mass of 5×10^6 daltons. These complexes, with gaps between them filled by regions of lipid bilayer, are postulated to make up the membranes. Though this model is interesting, it is rather speculative, none of the complexes having been isolated and the change in apparent particle dimensions with variation in fixation time makes one wonder which, if any, is the 'correct' size. Clearly at this vital level of structure, our knowledge is still extemely incomplete.

The energized state and conformational changes

Of all the mitochondrial activities connected with energy conversion, the mechanism of oxidative phosphorylation is least understood. The earliest hypotheses were in terms of formation of high energy intermediates in sequence:

$$\text{electron transport} \rightarrow X \sim I \rightarrow X \sim P \rightarrow ADP \sim P$$

where P stands for a phosphate group and X, I, for unknown compounds. The action of various uncouplers can be interpreted as premature cleavage of one or other of the intermediates. Three protein coupling factors have been isolated, apparently responsible for the coupling of electron transport to phosphorylation at each of the three phosphorylation sites (Fig. 2.3), but the high energy intermediate compounds have defied identification; and some authors have sought alternative types of mechanism.

The chemiosmotic hypothesis of Mitchell (1961, 1969) supposes that the H^+ and OH^- ions generated during electron transport are discharged asymmetrically due to enzyme positioning, the H^+ to the outside and the OH^- to the inside of the mitochondrial membrane, which is regarded highly impermeable to both ions. The resultant proton gradient or 'proton motive force' then drives ATP synthesis by reversal of an ATPase suitably oriented in the membrane. Uncouplers are surmised to destroy the differential permeability of the membrane. Mitochondria certainly possess ATPase activity, and electric potential gradients can arise during electron transport (Lieberman and Skulachev, 1970), but the hypothesis faces objections. The direction of proton movements in fresh, undamaged mitochondria has been claimed to be the opposite from that required for Mitchell's hypothesis, an H^+ ion output only appearing after ageing (Packer and Utsumi, 1969). Projection of the ATPase bearing inner membrane particles into the matrix would place the ATPase rather far from any membrane surface. Chance, Azzi, Lee, Lee, and Mela (1969) noted electron transport and ion movements associated with mitochondrial fragments during intense sonication, which would have broken up any vesicles, and they concluded that ion movements need not represent discharge to two sides of an impermeable membrane.

Since around 1966, it has been noticed that changes in the activity of isolated (and occasionally *in situ*) mitochondria are accompanied by changes in structure which can be detected by electron microscopy with rapid fixation, and traced spectrophotometrically as changes in light scattering by the isolated active particles in suspension. A very thorough study of the reactions of mouse and rat liver mitochondria was made by Hackenbrock. These mitochondria when fixed *in situ* generally display an 'orthodox' configuration with narrow crista lumina and a matrix of only moderate electron-opacity. When fixed directly after isolation in 0.25 M sucrose, they possessed a strikingly different form termed by Hackenbrock 'condensed': the crista lumina were highly dilated, and the matrix very electron-opaque. Incubation in state 4 (see page 61) induced a gradual change to the orthodox form, taking 25 minutes to reach completion at 30 °C; the subsequent addition of ADP, giving state 3 conditions, resulted in a return to the condensed in 35 seconds (Hackenbrock, 1966). The conversion from condensed to orthodox was prevented by inhibitors of electron transport, and partial inhibition gave a proportionally slower configurational change, which thus seemed to depend directly on electron transport; uncoupling with DNP had no direct effect (Hackenbrock, 1968b). It was suggested that the ultrastructural changes reflect differences in the energy state of the mitochondrial enzymes, the orthodox configuration representing a high energy state, built up in the absence of ADP, and the condensed form, a low energy state, produced when ADP addition permits energy utilization for phosphorylation (Hackenbrock, 1968a, 1972).

Green and his colleagues have put forward a scheme for oxidative phosphory-lation as dependent on conformational changes in the tripartite membrane subunits; the visible changes are assumed to be 'gross expressions of the conformations of the repeating units' (Green, Asai, Harris, and Penniston, 1968). They, however, consider the *condensed* form to represent the energized state. It was found that the mitochondria were condensed in rat heart muscle fragments incubated in the presence of succinate, rotenone and rutamycin, which was interpreted to show that the energized state was generated by succinate oxidation, and its utilization was prevented by the rotenone (which blocks any reverse electron trans-port to NAD) and the rutamycin (inhibiting ATP synthesis); DNP caused the mitochondria to assume the orthodox form (Harris, Williams, Caldwell, and Green, 1969). Normally the energized state would be discharged when its energy is used for ATP synthesis, for ion movements, or for driving reversed electron transport, while uncouplers would discharge it with a dissipation of energy. The generation of the energized state is said to be inducible with ATP, which is hydrolysed in the process, i.e., the reaction between ATP and the energized state is reversible.

Unfortunately, a review of data from a number of sources shows the reactions of mitochondria to varied conditions to be inconsistent. Buffa, Guarriera-Bobyleva, Muscatello, and Pasquali-Ronchetti (1970) found that both isolated rat liver mitochondria, and mitochondria of chick embryo myoblasts treated and fixed *in situ*, became condensed in the presence of DNP, in direct contradiction to the above-quoted results of Harris *et al.* (1969). In insect flight muscle, too, DNP has been found to produce highly condensed mitochondria (Smith, Smith, and Yunis, 1970). In the one report on plant mitochondria that has come to the writer's attention, no differences in appearance were found in sweet potato mitochondria as isolated in 0.4 M sucrose, or incubated in various respiratory states: in all cases, the condensed configuration prevailed (Baker *et al.,* 1968). In 0.25 M sucrose, the mitochondria took on an appearance intermediate between the condensed and orthodox forms, apparently reacting merely to the osmolarity change.

Changes in visible mitochondrial configuration with change in activity are an unquestionable fact, even if not of universal occurrence. The idea of shape changes in the tripartite unit is more hypothetical; isolated particles, supposed to correspond to the tripartite units, contract in the presence of ATP (Green and Hechter, 1965). The question of whether the conformational changes are *the* primary energy conversion as supposed by Green, must be left open. Theoretically simultaneous generation of both high energy intermediates (or proton gradients) and conformational changes during electron transport would be possible. It has been argued that unless conformational changes are the primary energy conversion process, they represent an excessive dissipation of energy; but even with maximal ADP/O ratios, the energy conversion efficiency is only 50 per cent leaving a possible margin for conformational changes (Lehninger, 1965).

The contradictory results obtained by different workers are confusing. One source of this confusion may be failure to discriminate between two distinct phenomena: the conformational changes at the molecular level, and the visible changes in membranes and matrix. A particular visible change may be the consequence of a certain change in molecular conformation, but it need not be the inevitable consequence under all circumstances. There is evidence from fluorescence changes of a dye bound to mitochondrial membranes that changes in protein conformation do occur concurrently with cytochrome oxidation within a few milliseconds, whereas the gross structural changes may take many minutes for completion, and may represent 'relaxation to minimum entropy' (Chance *et al.*, 1969). This equilibrium state could be influenced by factors other than macromolecule conformation, e.g., by the medium composition.

Whether the secret of energy conservation is ultimately proved to reside in the conformational changes in mitochondrial membranes or not, it is clear that such changes are intimately connected with electron transport. The functioning mitochondrion is a highly dynamic entity, its submicroscopic structural components oscillating ceaselessly between alternate configurations, and all these molecular changes will have to be studied and understood as parts of the electron transport and energy conversion process. It is to be hoped that some relevant work will be conducted also with plant mitochondria.

Mitochondrial development

Mitochondrial changes during cellular differentiation

As cells differentiate, the electron microscopic images of the mitochondria undergo changes which, at least in some cases or in some respects, can be interpreted in terms of functional changes, and changes in mitochondrial activity have also been observed independently of structural investigations. In the survey below, morphological descriptions refer to osmium-fixed material unless stated otherwise.

In mitochondria of meristematic cells of higher plants the cristae are not particularly well developed. Meristematic tissues tend to have a partly fermentative respiration, with respiratory quotients above unity. The mitochondrial cristae are irregular, sac-like and electron-translucent regions with conspicuous DNA filaments occupy a large proportion of the matrix; ribosomes or ribosome-like particles can also be seen frequently (Fig. 2.1A, B; Öpik, 1968b). Differentiation often involves an increase in the amount of cristae per mitochondrion; the cristal density becomes particularly high in cells of high metabolic activity such as companion cells and secretory cells but remains lower in ground parenchyma. Mitochondrial size may increase; in leaf bundle sheath cells of *Amaranthus edulis* and *Atriplex lentiformis,* the mitochondria attain twice the diameter of those in adjacent palisade cells (Laetsch, 1968). The bundle sheath cells are believed to be active in transport of compounds into the vascular tissue. In the green flagellate *Prymnesium parvum* a mitochondrion lies at the base of the mechanically active flagellum and the cristae are densest at the end of the mitochondrion next to the flagellar base (Manton, 1966).

Changes in matrix opacity and the extent of dilation of the cristae commonly occur during differentiation; it is difficult to pick out any universal trends, except that senescent tissues often contain mitochondria with extremely dilated cristae and darkly staining matrix — the 'condensed configuration' (e.g., Öpik, 1965). In the light of the discussion in the preceding section an appraisal of the functional significance of such apparent changes is very difficult. The DNA filaments and their surrounding electron-translucent areas become less conspicuous as cells age and may become completely obliterated (Street, Öpik, and James, 1968; Öpik, 1965). This does not necessarily mean a disappearance of the mitochondrial DNA which can still be isolated from mature tissues; far more probably it results from a change in state influencing its reaction with fixatives and stains.

The number of mitochondria per cell increases as cells grow; accurate quantitative estimations, however, are not easy and few have been made. In maize root cap, the number of mitochondria per cell increases from 50 in the cap initials to approximately 175 in the mature cells (Clowes and Juniper, 1964).

According to Lance and Bonner (1968) differences in the respiratory oxygen uptake rates of various plant tissues are determined by the amount of mitochondria per unit mass of tissue, rather than by differences in the specific oxidative activity of the mitochondria, since they found very little variation in the $Q[O_2(N)]$ of mitochondria isolated from five tissues covering a wide range of respiration rates. Potato tuber with a tissue $Q(O_2)$ (per fresh weight) of 40, and skunk cabbage (*Symplocarpus foetidus*) spadix with a $Q(O_2)$ of 4000, had mitochondrial $Q[O_2(N)]$ values of 1180 and 1030 respectively, but the skunk cabbage tissue yielded about 40 times more mitochondria per unit weight. Increases in respiration rate which occur in seed storage tissues during germination, however, have been reported to be accompanied by increases in the mitochondrial $Q[O_2(N)]$ as well as by increases in mitochondrial material. (E.g. *Ricinus* endosperm, Akazawa and Beevers, 1957; *Arachis* cotyledons, Cherry, 1963; maize scutellum, Hanson, Vatter, Fisher, and Bils, 1959.) Where examined, the mitochondrial cristal density has simultaneously been found to increase (Cherry, 1963). Increases in mitochondrial $Q[O_2(N)]$ concomitant with increased cristal density have also been noted in differentiating maize root (Lund, Vatter, and Hanson, 1958) and developing *Arum maculatum* spadix (Simon and Chapman,

69

1961). The logical interpretation of these results is, that respiration rate rises at least partly because of increased proportions of oxidative enzymes in mitochondria, reflected as increases in cristal density and the $Q[O_2(N)]$. The discrepancy between these findings and those of Lance and Bonner (1968) may be due to the latter authors having used more or less mature tissues where the average crista density may have varied little, or some of the apparent changes of $Q[O_2(N)]$ of earlier reports could conceivably be results of the less sophisticated isolation methods employed and reflect changes in the amount of contaminating non-mitochondrial material at different developmental stages.

Mitochondria do persist in dry, dormant seed tissues of higher plants and in spores; reports to the contrary can be attributed to the difficulty of fixing and sectioning such material. In dormant spores of *Allomyces*, no mitochondria could be seen with permanganate fixation, but the organelles were demonstrable by freeze-etching (Skukas, 1968). The protoplasmic changes that occur during dehydration greatly lower membrane contrast. Additionally, the number of cristae per mitochondrion may decrease as seeds dry out; this occurs, e.g., in pea (*Pisum sativum*) cotyledons (Bain and Mercer, 1966) and radicle (Perner, 1966). The cristae which remain are characteristically narrow and the matrix is fairly light, sometimes with a halo of darker matrix material around the cristae (Fig. 2.2E). The DNA filaments may still be discernible and also granular material that could represent the mitoribosomes. (*Phaseolus vulgaris* cotyledons and radicle – Öpik, unpublished; rice (*Oryza sativa*) embryo – Öpik, 1972.) Thus all normal structural components of mitochondria are retained in dormant tissues.

Environmental control of mitochondrial development

The number and appearance of mitochondria in plant cells can be influenced by various factors. Incubation of young etiolated bean (*Phaseolus vulgaris*) leaves on 0.2 M sodium chloride solution has been found to induce a marked increase in mitochondria per cell as compared with water-treated controls (Siew and Klein, 1968), suggesting a stimulation of mitochondrial development by a demand for energy for osmoregulation. During a period of digestion in Venus' flytrap (*Dionaea muscipula*) glands, the mitochondria change configuration from orthodox to condensed, and revert to the orthodox when the process is completed (Schwab, Simmons, and Scala, 1969). When mung bean seedlings are grown for 7 days in the dark, the leaf cell mitochondria have matrices of low electron-opacity; in light-grown seedlings, the matrices are electron-opaque and the cristae are more regularly arrayed (Öpik, unpublished); the significance of these changes is yet to be evaluated.

Systematic studies on the effect of oxygen on mitochondrial development have been conducted on yeasts which are capable of growth under aerobic or anaerobic conditions. It was first reported that anaerobically grown yeast (*Torula utilis* and *Saccharomyces cerevisiae*) entirely lacked mitochondria and cytochromes, which appeared rapidly on aeration, but continued research has revealed a more complex situation. Yeast is unable to synthesize unsaturated fatty acids and ergosterol under anaerobic conditions, and in anaerobiosis, the

70

unsaturated fatty acid content of the mitochondrial lipids falls to one-sixteenth of the normal value (Watson, Haslam, and Linnane, 1970), and this apparently makes the membranes very difficult to fix, and to isolate intact. When the medium is supplemented with lipids, mitochondria can be seen with permanganate fixation (Wallace, Huang, and Linnane, 1968), and even in the lipid-depleted cells, mitochondria are revealed by carefully controlled glutaraldehyde fixation (Damsky, Nelson, and Claude, 1969). With appropriate techniques, mitochondria can also be isolated from the anaerobically grown cells (Watson *et al.*, 1970).

In all the studies quoted above, the mitochondria formed under anaerobic conditions were found to be developed very poorly with scarcely any cristae, lipid addition increasing the frequency of cristae, though not to the aerobic level, and the number of mitochondria per cell was low. Criddle and Schatz (1969), on the other hand, claim a yield of mitochondrial protein from anaerobic lipid-supplemented yeast equal to the yield from aerobic cells; and Plattner and Schatz (1969) have found distinctly cristate mitochondria in freeze-etched preparations of anaerobic lipid-depleted yeast. All authors agree, however, that the anaerobically cultured cells lack the normal cytochrome complement and are incapable of normal oxidative metabolism. The organelles produced in anaerobiosis, sometimes termed promitochondria, are converted to normal mito-chondria during oxygen adaptation (Schatz and Criddle, 1969).

The interpretation of the data is complicated by the phenomenon of glucose repression. Anaerobic growth requires a fermentable substrate and most of the results just discussed have been obtained with *Saccharomyces* cells grown on glucose. But in *Saccharomyces* glucose strongly represses aerobic metabolism and mitochondrial development even in oxygen (Yotsuyanagi, 1962). When *Saccharomyces* is cultured anaerobically on galactose, the mitochondria are morphologically better developed, though still deficient in aerobic cytochromes. In the obligately aerobic yeast *Candida parapsilosis,* which is not subject to glucose repression, lowering of aeration leads to poor development of cristae and to a depression of cytochrome synthesis which must be attributed directly to the effect of oxygen shortage (Kellerman, Biggs, and Linnane, 1969). Thus, in yeast, mitochondrial development is controlled by the extent of aerobic metabolism of the cells and suppression of this, whether by oxygen lack or presence of glucose, also suppresses mitochondrial development, especially the synthesis of cytochromes, the formation of cristae, and in some cases at least, also the number of mitochondria per cell.

Higher plants are obligate aerobes, yet some species are adapted for germination under practically anaerobic conditions, and the coleoptile of rice (*Oryza sativa*) will grow to a height of several centimetres under complete anaerobiosis, thus affording an opportunity for studying the effect of oxygen withdrawal on mitochondrial development in a higher plant. In anaerobically grown coleoptiles the capacity for aerobic respiration and cytochrome oxidase activity show a strong decrease (Table 2.1; Öpik, 1973), which appears most dramatic when expressed per organ, being to some extent due to a lowered cell number and weight. But even per cell, the activity depression is still very appreciable. Studies on fine

Mitochondria

structure show no significant differences in mitochondrial structure between the aerobically and anaerobically grown seedlings. In young coleoptiles, cristal densities are identical with the two treatments (Table 2.1), nor is there any significant difference in the appearance of the mitochondria in other respects. In older coleoptiles anaerobic cristal density per unit mitochrondrial cross-sectional area is slightly lowered (Öpik, 1973).

TABLE 2.1

The Effect of Anaerobiosis on some Oxidative Activities and Mitochondrial Development in Rice Coleoptiles (Öpik, 1973).

Growth conditions	Aerobic	Anaerobic
Days to reach 14 mm	4	5
Cells per coleoptile	13.1×10^4	6.37×10^4
Oxygen uptake (μl/hr/coleoptile)	approx. 2.7	approx 0.51
Cytochrome oxidase:		
relative OD units* per coleoptile	14.5	2.08
per cell	13.0	3.19
Average mitochondrial area	1.7	1.9
Average number of cristae per unit of		
mitochondrial area	5.3	5.5

* The per coleoptile and per cell units are not directly comparable.

Note: Anaerobic conditions were achieved under hydrogen in microbial anaerobiosis jars; growth was at $27°$ in the dark. Data are for 14 ± 2 mm-long coleoptiles except for crista density and mitochondrial area, which were measured in small, just-emerging coleoptiles. For cytochrome oxidase activity, frozen-thawed coleoptiles were incubated in a p-aminodiphenylamine reagent, the blue colour extracted into ethanol and assayed spectrophotometrically at 560 nm. Estimates of crista density are based on counts of approx. 275 mitochondria for each treatment. Mitochondrial cross-sectional area in arbitrary units.

This situation is quite different from that in yeast. Admittedly microorganisms have a great tendency for adaptive formation of enzymes which in higher organisms are formed constitutively; and lipid limitation is not so likely to affect the rice which has a lipid store in the grain. But since factors such as presence of sodium chloride can influence mitochondrial structure in a higher plant, it is surprising the lack of oxygen makes no difference. There can be no question of the rice mitochondria all being pre-formed in the ungerminated embryo, for in the dormant and newly imbibed grain, the coleoptile mitochondria have very few cristae and gradually increase in complexity both in presence and absence of oxygen. Cell number also increases under anaerobiosis without a diminishing of the number of mitochondria per cell. It may be that the rice coleoptile is a special case. It would be of interest to investigate the effect of varied oxygen tension on mitochondrial development in other higher plant organs not specially adapted for anaerobic growth. The case of the rice does demonstrate, however, that mitochondria may look perfectly normal while deficient in cytochrome oxidase, i.e., superficially normal membranes can form even during shortage of some membrane components.

Genetic control

The science of mitochondrial genetics has made great advances in the last decade; for comprehensive reviews and references, the reader is referred to Borst and Kroon (1969), Nass (1969), and 'The Control of Organelle Development', vol. 24 (1970) of the Symposia of the Society for Experimental Biology.

Mitochondrial origin. This long-disputed problem is even now not yet solved completely; much evidence, however, points to mitochondrial formation by division of pre-existing mitochondria. Divisions (and fusions) of mitochondria have been observed in living cells during the so-called pleomorphic shape changes, and in some unicellular algae, the cell possesses a single mitochondrion, which divides at each cell division (Manton, 1961). Many electron micrographs contain constricted mitochondrial profiles, suggestive of fission, or of budding of small mitochondria from larger ones. These static pictures are, however, only indirect evidence, especially when three-dimensional reconstructions from serial sections are not available, for shapes seen in single sections can be misleading. Dynamic evidence comes from radioactive tracer experiments. Luck (1963) labelled the mitochondria of a choline-requiring *Neurospora* mutant with radioactive choline and during subsequent growth in cold choline found the label distributed over all the mitochondria, with continuous decrease of label per mitochondrion. Similarly, when the mitochondria of the protozoan *Tetrahymena pyriformis* were labelled with tritiated thymidine, the average label per mitochondrion was halved at each cell division, but the total mitochondrial label in the culture was conserved over at least four generations (Parsons and Rustad, 1968). These results are compatible only with formation of mitochondria by division from the pre-existing ones. If other modes of mitochondrial reproduction exist, these must be additional to formation by division. A claim for the evagination of mitochondria from the nuclear membrane in developing egg cells of the fern *Pteridium aquilinum* has been made (Bell and Mühlethaler, 1964), on electron microscopic evidence, but has been firmly denied by other authors (Diers, 1964; Tourte, 1968), and recent work from Bell's own laboratory has failed to produce supporting evidence from radioactive labelling experiments (Sigee and Bell, 1971).

Very little information is available on the timing of mitochondrial division during the cell cycle. In the shoot meristem of *Epilobium hirsutum* mitochondrial multiplication occurs during the interphase (Anton-Lamprecht, 1967), and in the sea urchin egg, too, the maximum number per cell is observed in the interphase (Agrell, 1955).

Mitochondrial inheritance. By now nobody can deny that mitochondria have some genetic properties. Certain 'petite' mutants of yeast, and 'poky' mutants of *Neurospora,* which all contain abnormal mitochondria and are unable to respire aerobically, transmit the character cytoplasmically. Mitochondrial DNA differs from the nuclear DNA of the same species with respect to molecular weight, shape and buoyant density (the latter reflecting the base composition); the amount varies from 0.2 per cent (mouse L cells) to 50 per cent (giant amphibian egg) of the total cellular DNA; in diploid *Saccharomyces* it is approximately 15 per cent (Borst and Kroon, 1969). In some of the petite yeasts the mito-chondrial DNA has a grossly abnormal buoyant density and molecular weight (Williamson, 1970); in some cases it is said to be lost altogether (Borst and Kroon, 1969). Thus the petite condition though commonly called a mutation is very different from a chromosomal point mutation. Yeast mitochondrial mutants differing in drug resistance are known, and recombinations between these can take place. Mitochondria contain DNA polymerase, and have been shown to incorporate tritiated thymidine, e.g., in *Tetrahymena* (Parsons and Rustad, 1968), and rat liver (Parsons and Simpson, 1969). Isolated rat liver mitochondria can incorporate 5-bromodeoxyuridine into DNA in amounts indicating synthesis of long strands rather than a mere repair process (Karol and Simpson, 1968). Forked molecules, presumed to be stages in replication, have been seen in electron micrographs of DNA molecules isolated from rat liver mitochondria (Kirchner, Wolstenholme, and Gross, 1968). Such observations establish that mitochondria contain a DNA-based genetic system, but are far from proving that mitochondria are genetically autonomous. Indeed yeast, which has provided so much evidence for mitochondrial autonomy, also provides very clear evidence against absolute autonomy. In addition to the mitochondrial petites, yeast forms petites indistinguishable from those in morphology and metabolism, but with the condition determined by a nuclear gene mutation. Normal mitochondrial development in yeast thus requires a favourable genome both in the nucleus and in the mitochondria. The question is: how great is the degree of mitochondrial autonomy and precisely what is controlled by mitochondrial DNA?

One approach to this problem has been to calculate the coding capacity of the mitochondrial DNA. All higher animal mitochondrial DNA is of remarkable uniformity in size, with circular molecules about 5 μm long and a molecular weight of 10^7. Yeast mitochondrial DNA is also circular, but the molecular weight is higher, approximately 5×10^7 (Hollenberg, Borst, Thuring, and van Bruggen, 1969). For higher plants, data are fewer, but all reports to date claim the molecules to be linear, and the molecular weights often to be still higher— over 10^8 for lettuce (*Lactuca sativa*) (Wells and Birnstiel, 1969) and 1.6×10^8 for bean (*Phaseolus vulgaris*) (Borst and Kroon, 1969), though somewhat less, not over 5.8×10^7, for pea (Mikulska, Odintsova, and Turischeva, 1970).

The coding capacity of mitochondrial DNA can be derived from its molecular weight, assuming that all the mitochondrial DNA molecules of a species are identical. There is no proof of this, but in the absence of contrary evidence, it is the simplest hypothesis. There may be several molecules of DNA per mito-chondrion, but as long as they are identical, the argument is not affected. The calculated coding capacity can then be compared with estimates of the actual number of mitochondrial macromolecules to be coded.

Mitochondria are capable of protein synthesis (see below). A protein synthesizing system consists of ribosomes (constituted of two species of RNA and some fifty proteins); about twenty species each of transfer RNA and of amino acid activating enzymes, and a few more enzymes concerned with polypeptide chain initiation, growth and release. The mitochondrial ribosomes are different from the cytoplasmic, being smaller; in yeast, *Neurospora* and *Aspergillus,* the sedimentation constant is about the same for bacterial ribosomes, 70 S; in higher animals, 55–60 S (Attardi and Ojala, 1971; Noll, 1970). At least some of the transfer RNA's and activating enzymes are specific to the mitochondria. Then the mitochondria contain some 40–50 species of enzymes for their metabolic functions, and structural protein(s). It has been calculated that to code for the protein synthesizing system alone, approximately 6×10^7 daltons of DNA are required (Noll, 1970). A molecule of animal mitochondrial DNA, mass 10^7 daltons, will not suffice even for that. Yeast and pea mitochondrial DNA ($5-5.8 \times 10^7$ daltons) might just about stretch to it–but with nothing left for coding any of the proteins to be synthesized! If the larger values of the molecular weight of higher plant mitochondrial DNA are correct, its coding capacity might theoretically be enough for a complete coding of mitochondrial components. However, it cannot be emphasized too strongly that for these considerations, it is assumed firstly, that all mitochondrial DNA molecules in a cell are identical (existence of several different ones would increase the coding capacity); and secondly, that all the DNA is functional without duplication of information (redundant pieces, or duplication, would lower the coding capacity). Neither premise is proved and experimental studies of what the mitochondrial DNA actually does code, are indispensable.

Much effort has been expanded on elucidating which of the mitochondrial proteins are synthesized within the mitochondria. Isolated mitochondria incorporate radioactive amino acids mostly into an insoluble protein fraction, not clearly identified, possibly structural protein(s). Application *in vivo* of antibiotics which selectively inhibit protein synthesis by bacterial (70 S) ribosomes, but not by cytoribosomes has inhibited the synthesis of cytochromes a, a_3, b and c_1 in liver and yeast mitochondria (Firkin and Linnane, 1969; Clark-Walker and Linnane, 1967), with some inhibition of succinic dehydrogenase synthesis in yeast, i.e., inhibition was of insoluble, membrane-bound, enzymes. On the other hand, the soluble cytochrome c is definitely known to be synthesized outside mitochondria. It is therefore suggested that mitochondria synthesize (some of) their insoluble proteins, while (some of) their soluble proteins are made on the cytoplasmic ribosomes. Much more work is needed here.

Identification of the site of protein synthesis by no means yet identifies the site of the code, nor of the code for the protein synthesizing machinery itself. Nuclear messenger RNA could enter mitochondria and code intramitochondrial protein synthesis, and the coding of protein synthesis by mitochondrial messenger on cytoplasmic ribosomes is also possible. From nucleic acid hybridization experiments, it appears that the mitochondrial ribosomal and soluble RNA are in fact analogous to mitochondrial DNA, and mitochondrial yeast petites lack the normal mitochondrial RNA (Wintersberger and Viehhauser, 1968); this indicates that mitochondrial DNA codes for mitochondrial RNA. Petite mutants have also been found to lack the insoluble cytochromes a, a_3, b and c_1 (Borst and Kroon, 1969), and in a *Neurospora* mitochondrial mutant the structural protein is abnormal (Woodward, 1968). Cytochrome c is, however, coded by nuclear genes, and so are Krebs cycle dehydrogenases in *Neurospora* (Woodward, Edwards, and Flavell, 1970) and maize malic dehydrogenase (Longo and Scandalios, 1969). The emerging picture is one of dual control of mitochondrial synthesis, with the exact roles of the nuclear and mitochondrial DNA not yet disentangled, but some evidence for mitochondrial autonomy in the synthesis of mitochondrial RNA and insoluble proteins. It would be most interesting to have comparative data for the coding activities of animal, yeast and higher plant mitochondrial DNA's, with their greatly different molecular weights: would the higher plant mitochondria, with more DNA, be found to have greater autonomy? There is also the possibility that too much emphasis has been put on mitochondrial DNA as coding macromolecules, against an alternative role of regulating the rate of mitochondrial replication (Williamson, 1970).

The physical process of mitochondrial growth, involving the synthesis and orderly structural integration of so many different macromolecules, under double control from two genomes, is a fascinating problem. What can ensure that components coded by the two genomes are synthesized in correct proportions, for instance? Repressors coded by mitochondrial DNA might control the activity of nuclear genes coding mitochondrial components (Barath and Künzel, 1972). The actual structure has been considered too complex for assembly of the subunits without a structural template; alternatively or additionally, the 'homing instinct' of molecules has been invoked; functional mitochondrial subunits will automatically reconstruct themselves from their components *in vitro*, showing that the molecular properties lead to their interaction. But how do the cytochrome c molecules, synthesized extramitochondrially, 'sense' their site on the inner membrane through the intervening outer envelope? If Krebs cycle dehydrogenases are synthesized on the cytoplasmic ribosomes how'and why do they enter the mitochondria? Again, we are ignorant of whether the membranes are built of preassembled subunits or whether a skeleton membrane is first synthesized, enzymic components being added afterwards. The latter possibility is suggested by the presence of cristae in yeast and rice mitochondria deficient in cytochromes, and by the observation of sequential, rather than simultaneous, synthesis of various components of sweet potato mitochondria (Sakane and Asahi, 1971). Also, in cytochrome-deficient yeast mutants, freeze-etching has revealed no abnormalities in particle distribution on the fractured mitochondrial membranes (Packer, Williams, and Criddle, 1973). How does recombination of mitochondrial characters occur? Electron micrographs occasionally show adjacent mitochondria joined by constricted necks, which might represent conjugation canals (Noll,

1970); one such is illustrated in Fig. 2.2C. Fission and fusion during pleomorphic changes might provide an opportunity for material exchange. In discussions of the biochemical aspects of mitochondrial development, the structural aspects must not be forgotten.

Possible heterogeneity of mitochondria

There are some scattered references to heterogeneity among the mitochondrial population of a cell or tissue. Avers (1961) noted that in root epidermal cells of two grasses the average number of particles per cell which stained with the vital mitochondrial stain Janus' green was the same as staining for cytochrome oxidase, but only approximately half that number stained with tetrazolium dyes for succinic, pyruvic, citric, or isocitric dehydrogenases. In yeast also, a larger number stained for cytochrome oxidase than for succinic dehydrogenase (Avers, Pfeffer, and Rancourt, 1965) and a similar result was obtained by Ogawa and Barrnett (1964) for rat heart muscle mitochondria at the electron microscope level. These results suggest variability if not in enzyme content, then in permeability to substrates; the mitochondria of different reactivity in the above instances could represent organelles at different developmental stages.

Heterogeneity of a different kind has been reported for mitochondria from hybrid maize scutella (Sarkissian and McDaniel, 1967). On density gradient centrifugation, mitochondria from the parent strains separated out into one or two bands; a preparation from the hybrid showed all parent bands and a third additional one. These mitochondria showed 'hybrid vigour', with the activity of mitochondria from the hybrid exceeding that of either parent. (Synergism for oxidation rate was also observed when isolated mitochondria from the parent strains were mixed *in vitro*.) In the case of mitochondria isolated from hybrid barley, Grimwood and McDaniel (1970) have been able to separate out on sucrose density gradients fractions differing in the proportions of three mitochondrial malic dehydrogenase isozymes. Such results are suggestive of genetic heterogeneity. Federman and Avers (1967) introduced genetically different mitochondria into the same cell by crossing haploid strains of normal and petite yeasts, and found that the mixed population persisted for some time, with gradual elimination of the petite type of organelle. Smaller genetic differences between mitochondria in a cell may well exist and go undetected.

The ultimate origin

Advancement of our understanding of mitochondrial structure and genetics has revealed striking similarities with bacteria (Nass, 1969). The size of a mitochondrion is of the same order of magnitude as that of a bacterial cell; bacteria bear oxidative enzymes in the cell membrane. The mitochondrial protein synthesizing system is of the bacterial type and the mitochondrial DNA-containing regions resemble bacterial nucleoplasm, while the circular DNA molecules of yeast and animal mitochondria can be compared with the circular molecules of bacterial DNA—although the amount of DNA even in a higher plant mitochondrion is considerably less than in a bacterial cell. These similarities make the idea of ultimate origin of mitochondria from symbiotic prokaryotic organisms very attractive. One could perhaps postulate that in higher plant mitochondria, specialization to ATP-synthesizing machines has not proceeded as far as in the mitochondria of higher animals, the plant mitochondria retaining more DNA, more matrix, and more ancillary (i.e., non-oxidative) functions. The presence of different isozymes of the same enzyme in mitochondria and in the cytoplasmic ground substance becomes understandable on this hypothesis: the cytoplasmic enzyme would be the original 'host' enzyme. The outer membrane of mitochondria, so different from the inner, but with definite resemblances to the endoplasmic reticulum, might be derived from a host cell membrane, which generally tends to surround invading micro-organisms.

When Altmann in 1890 described his 'bioblasts' he considered these to be 'Elementarorganismen' evolved from and related to free-living forms such as bacteria. After 80 years of research, we are led towards the same conclusion.

Bibliography

Additional reading

Borst, P. and Kroon, A. M. (1969) 'Mitochondrial DNA: physicochemical properties, replication and genetic function', *Int. Rev. Cytol.*, **26**, 107–190.
'Control of organelle development', *Symp. Soc. exp. Biol.*, **24**, (1970).
Green, D. E. and Baum, H. (1970) *Energy and the mitochondrion*, Academic Press, New York and London.
Hanson, J. B. and Hodges, T. K. (1967) 'Energy-linked reactions of plant mitochondria', In: *Current topics in bioenergetics*, (ed. D. R. Sanadi), vol. 2, pp. 65–98. Academic Press, New York and London.
Lehninger, A. L. (1965) *The mitochondrion*, W. A. Benjamin Inc., New York and Amsterdam.
Nass, S. (1969) 'The significance of the structural and functional similarities of bacteria and mitochondria', *Int. Rev. Cytol.*, **25**, 55–129.
Öpik, H. (1968a) 'Structure, function and developmental changes in mitochondria of higher plant cells', In: *Plant cell organelles* (ed. J. B. Pridham), pp. 47–88. Academic Press, London and New York.
Roodyn, D. B. and Wilkie, D. (1968) *The biogenesis of mitochondria*, Methuen, London.

References

Agrell, I. (1955) 'A mitotic rhythm in the appearance of mitochondria during the early cleavages of the sea urchin egg', *Expl Cell Res.*, **8**, 232–234.
Akazawa, T. and Beevers, H. (1957) 'Mitochondria in the endosperm of the germinating castor bean: a developmental study', *Biochem. J.*, **67**, 115–118.
Altmann, R. (1890) *Die Elementarorganismen und ihre Beziehungen zu den Zellen*, Veit and Co., Leipzig.
Anton-Lamprecht, I, (1967) 'Anzahl und Vermehrung der Zellorganen im Scheitelmeristem von *Epilobium*', *Ber. dt. bot. Ges.*, **80**, 747–754.
Ashhurst, D. E. (1965) 'Mitochondrial particles seen in sections', *J. Cell Biol.*, **24**, 497–499.
Attardi, G. and Ojala, D. (1971) 'Mitochondrial ribosomes in Hela cells', *Nature New Biology, Lond.*, **229**, 133–136.
Avers, C. J. (1961) 'Histochemical localisation of enzyme activities in root meristem cells', *Am. J. Bot.*, **48**, 137–143.
Avers, C. J., Pfeffer, C. R., and Rancourt, M. W. (1965) 'Acriflavine induction of different kinds of "petite" mitochondrial populations in *Saccharomyces cerevisiae*', *J. Bact.*, **90**, 481–494.

Bibliography

Bain, J. M. and Mercer, F. V. (1966) 'Subcellular organization of the developing cotyledons of *Pisum sativum* L.', *Aust. J. biol. Sci.*, **19**, 49–67.

Baker, J. E., Elfvin, L. G., Biale, J. B., and Honda, S. I. (1968) 'Studies on ultrastructure and purification of isolated plant mitochondria', *Pl. Physiol., Lancaster*, **43**, 2001–2022.

Barath, Z. and Künzel, H. (1972) 'Induction of mitochondrial RNA polymerase in *Neurospora crassa*', *Nature New Biol.*, **240**, 195–197.

Beck, D. P., Decker, G. L., and Greenawalt, T. W. (1970) 'Ultrastructure of striated inclusion in *Neurospora*', *J. Ultrastruct. Res.*, **33**, 245–251.

Bell, P. R. and Mühlethaler, K. (1964) 'The degeneration and reappearance of mitochondria in the egg cells of a plant', *J. Cell Biol.*, **20**, 235–248.

Branton, D. (1969) 'Membrane structure', *A. Rev. Pl. Physiol.*, **20**, 209–238.

Branton, D. and Moor, H. (1964) 'Fine structure in freeze-etched *Allium cepa* L. root tips', *J. Ultrastruct. Res.*, **11**, 401–411.

Buchholz, J. T. (1947) 'Methods in the preparation of chromosomes and other parts of cells for examination with an electron microscope', *Am. J. Bot.*, **34**, 445–454.

Buffa, P., Guarriera-Bobyleva, V., Muscatello, U., and Pasquali-Ronchetti, I. (1970) 'Conformational changes of mitochondria associated with uncoupling of oxidative phosphorylation *in vitro* and *in vivo*', *Nature, Lond.*, **226**, 272–274.

Chance, B., Azzi, A., Lee, I. Y., Lee, C. P., and Mela, L. (1969) 'The nature of the respiratory chain: location of energy conservation sites, the high energy store, electron transfer-linked conformation changes, and the 'closedness' of submitochondrial vesicles', In: *Mitochondria, structure and function* (eds. L. Ernster and Z. Drahota), pp. 233–273, V FEBS meeting, Prague 1968, Symposium vol. 17.

Chance, B. and Williams, G. R. (1955) 'Respiratory enzymes and oxidative phosphorylation, I to V', *J. biol. Chem.*, **217**, 383–393, 395–407, 409–427, 429–438, 439–451.

Cherry, J. H. (1963) 'Nucleic acid, mitochondria and enzyme changes in cotyledons of peanut seeds during germination', *Pl. Physiol., Lancaster*, **38**, 440–446.

Clark-Walker, G. D. and Linnane, A. W. (1967) 'The biogenesis of mitochondria in *Saccharomyces cerevisiae*', *J. Cell Biol.*, **34**, 1–14.

Clowes, F. A. L. and Juniper, B. E. (1964) 'The fine structure of the quiescent centre and neighbouring tissues in root meristems', *J. exp. Bot.*, **15**, 622–630.

Criddle, R. S. (1969) 'Structural proteins of chloroplasts and mitochondria', *A. Rev. Pl. Physiol.*, **20**, 239–252.

Criddle, R. S. and Schatz, G. (1969) 'Promitochondria of anaerobically grown yeast. I. Isolation and biochemical properties', *Biochemistry, N.Y.*, **8**, 322–334.

Damsky, C. H., Nelson, W. M., and Claude, A. (1969) 'Mitochondria in anaerobically-grown, lipid-limited, yeast', *J. Cell Biol.*, **43**, 174–179.

Diers, L. (1964) 'Bilden sich während der Oogenese bei Moosen und Farnen die Mitochondrien und Plastiden aus dem Kern?' *Ber. dt. bot. Ges.*, **77**, 369–371.

Diers, L. and Schötz, F. (1965) 'Über den Feinbau pflanzlicher Mitochondrien', *Z. Pfl. Physiol.*, **53**, 334–343.

Douce, R., Mannella, C. A., and Bonner, W. D. Jr. (1973) 'The external NADH dehydrogenases of intact plant mitochondria', *Biochim. biophys. Acta*, **292**, 105–116.

Erecinska, M. and Storey, B. T. (1970) 'The respiratory chain of plant mitochondria. VII. Kinetics of flavoprotein oxidation in skunk cabbage mitochondria', *Pl. Physiol., Lancaster*, **46**, 618–624.

Ernster, L. and Kuylenstierna, B. (1969) 'Structure, composition and function of mitochondrial membranes', In: *Mitochondria, structure and function* (ed. L. Ernster and Z. Drahota), pp. 5–31, V FEBS meeting, Prague 1968, Symposium vol. 17.

Federman, M. and Avers, C. J. (1967) 'Fine-structure analysis of intercellular and intracellular mitochondrial diversity in *Saccharomyces cerevisiae*', *J. Bact.*, **94**, 1236–1243.

Firkin, F. C. and Linnane, A. W. (1969) 'Biogenesis of mitochondria 8. The effect of chloramphenicol on regenerating rat liver', *Expl Cell Res.*, **55**, 68–76.

Green, D. E., Asai, J., Harris, R. A., and Penniston, J. T. (1968) 'Conformational basis of energy transformations in membrane systems III. Configurational changes in the mitochondrial inner membrane induced by changes in functional states', *Archs. Biochem. Biophys.*, **125**, 684–705.

Green, D. E. and Hechter, O. (1965) 'Assembly of membrane subunits', *Proc. natn. Acad. Sci. U.S.A.*, **53**, 318–325.

Mitochondria

Grimwood, B. G. and McDaniel, R. G. (1970) 'Variant malate dehydrogenase isoenzymes in mitochondrial preparations', *Biochim. biophys. Acta*, **220**, 410–415.

Hackenbrock, C. R. (1966) 'Ultrastructural bases for metabolically linked mechanical activity in mitochondria. I. Reversible ultrastructural changes with change in metabolic steady state in isolated liver mitochondria', *J. Cell. Biol.*, **30**, 269–297.

Hackenbrock, C. R. (1968a). 'Chemical and physical fixation of isolated mitochondria in low-energy and high-energy states', *Proc. natn. Acad. Sci. U.S.A.*, **61**, 598–605.

Hackenbrock, C. R. (1968b) 'Ultrastructural bases for metabolically linked mechanical activity in mitochondria', *J. Cell Biol.*, **37**, 345–369.

Hackenbrock, C. R. (1972) 'Energy-linked ultrastructural transformations in isolated liver mitochondria and mitoplasts', *J. Cell Biol.*, **53**, 450–465.

Hanson, J. B., Vatter, A. E., Fisher, M. E., and Bils, R. F. (1959) 'The development of mitochondria in the scutellum of germinating corn', *Agron. J.*, **51**, 295–301.

Harris, R. A., Williams, C. H., Caldwell, M., and Green, D. E. (1969) 'Energized configurations of heart mitochondria *in situ*', *Science, N.Y.*, **165**, 700–702.

Heidrich, H. G., Stahn, R., and Hannig, D. (1970) 'The surface charge of rat liver mitochondria and their membranes. Clarification of come controversies concerning mitochondrial structure', *J. Cell Biol.*, **46**, 137–150.

Hollenberg, C. P., Borst, P., Thuring, R. W. J., and van Bruggen, E. F. J. (1969) 'Size, structure and genetic complexity of yeast mitochondrial DNA', *Biochim. biophys. Acta*, **186**, 417–419.

Ikuma, H. and Bonner, W. D. Jr. (1967) 'Properties of higher plant mitochondria. I. Isolation and some characteristics of tightly coupled mitochondria from dark-grown mung bean hypocotyl', *Pl. Physiol., Lancaster*, **42**, 67–75.

Karol, M. H. and Simpson, M. V. (1968) 'DNA biosynthesis by isolated mitochondria: a replicative rather than a repair process', *Science, N.Y.*, **162**, 470–473.

Kellerman, G. M., Biggs, D. R., and Linnane, A. W. (1969) 'Biogenesis of mitochondria. XI. A comparison of the effects of growth-limiting oxygen tension, intercalating agents, and antibiotics on the obligate aerobe *Candida parapsilosis*', *J. Cell. Biol.*, **42**, 378–391.

Kirchner, R. H., Wolstenholme, D. R., and Gross, N. J. (1968) 'Replicating molecules of circular mitochondrial DNA', *Proc. natn. Acad. Sci. U.S.A.*, **60**, 1466–1471.

Ku, H. S., Pratt, H. K., Spurr, A. R., and Harris, W. M. (1968) 'Isolation of active mitochondria from tomato fruit', *Pl. Physiol., Lancaster*, **43**, 883–887.

Küntzel, H. and Noll, H. (1967) 'Mitochondrial and cytoplasmic polysomes from *Neurospora crassa*', *Nature, Lond.*, **215**, 1340–1345.

Laetsch, W. M. (1968) 'Chloroplast specialization in dicotyledons possessing the C_4-dicarboxylic acid pathway for photosynthetic CO_2 fixation', *Am. J. Bot.*, **55**, 875–883.

Lance, C. and Bonner, W. D., Jr. (1968) 'The respiratory chain components of higher plant mitochondria', *Pl. Physiol., Lancaster*, **43**, 756–766.

Lance-Nougarède, A. (1966) 'Présence de structures protéiques à arrangement périodique et d'aspect cristallin dans les mitochondries de l'épiderme des jeunes feuilles de lentille (*Lens culinaris* L.)', *C.r. hébd. Séanc. Acad. Sci. Paris.*, D **263**, 246–247.

Larson, D. A. (1965) 'Fine structural changes in the cytoplasm of germinating pollen', *Am. J. Bot.*, **52**, 139–154.

Lieberman, E. A. and Skulachev, V. P. (1970) 'Conversion of biomembrane-produced energy into electric form. IV. General discussion', *Biochim. biophys. Acta*, **216**, 30–42.

Longo, G. P. and Scandalios, J. G. (1969) 'Nuclear gene control of mitochondrial malic dehydrogenase in maize', *Proc. natn. Acad. Sci. U.S.A.*, **62**, 104–111.

Luck, D. J. L. (1963) 'Genesis of mitochondria in *Neurospora crassa*', *Proc. natn. Acad. Sci. U.S.A.*, **49**, 233–240.

Lund, H. A., Vatter, A. E., and Hanson, J. B. (1958) 'Biochemical and cytological changes accompanying growth and differentiation in the roots of *Zea mays*', *J. biophys. biochem. Cytol.*, **4**, 87–98.

Malhotra, S. K. (1966) 'A study of the structure of the mitochondrial membrane system', *J. Ultrastruct. Res.*, **15**, 14–37.

Manton, I. (1961) 'Some problems of mitochondrial growth', *J. exp. Bot.*, **12**, 421–429.

Manton, I. (1966) 'Observations on scale production in *Prymnesium parvum*', *J. Cell Sci.*, **1**, 375–380.

Bibliography

Mikulska, E., Odintsova, M. S., and Turischeva, M. S. (1970) 'Electron microscopy of DNA in mitochondria of pea seedlings', *J. Ultrastruct. Res.,* **32,** 258–267.

Mitchell, P. (1961) 'Coupling of phosphorylation to electron and hydrogen transfer by a chemiosmotic type of mechanism', *Nature, Lond.,* **191,** 144–148.

Mitchell, P. (1969) 'The chemical and electrical components of the electrochemical potential of H^+ ions across the mitochondrial cristae membrane', In: *Mitochondria, structure and function* (ed. L. Ernster and Z. Drahota), pp. 219–232. V FEBS meeting, Prague 1968, Symposium vol. 17.

Moor, H. (1964) 'Die Gefrier-fixation lebender Zellen und ihre Anwendung in der Elektronenmikroskopie', *Z. Zellforsch. mikrosk. Anat.,* **62,** 546–580.

Moor, H. and Mühlethaler, K. (1963) 'Fine structure of frozen-etched yeast cells', *J. Cell Biol.,* **17,** 609–628.

Mota, M. (1963) 'Electron microscope study of the relationship between the nucleus and mitochondria in *Chlorophytum capense* (L.) Kuntze', *Cytologia,* **28,** 409–416.

Nadakavukara, M. J. (1964) 'Fine structure of negatively stained plant mitochondria', *J. Cell Biol.,* **23,** 193–195.

Nass, M. M. K. and Nass, S. (1963) 'Intramitochondrial fibres with DNA characteristics. I. Fixation and electron staining reactions. II. Enzymatic and other hydrolytic treatments', *J. Cell Biol.,* **19,** 593–611, 613–629.

Newcomb, E. H., Steer, M. W., Hepler, P. K., and Wergin, W. P. (1968) 'An atypical crista resembling a "tight junction" in bean root mitochondria', *J. Cell Biol.,* **39,** 35–42.

Noll, H. (1970) 'Organelle integration and the evolution of ribosome structure and function', In: *Control of organelle development; Symp. Soc. exp. Biol.,* **24,** 419–448.

O'Brien, J. S. (1967) 'Cell membranes–composition; structure; function', *J. theor. Biol.,* **15,** 307–324.

Ogawa, K. and Barrnett, R. J. (1964) 'Electron histochemical examination of oxidative enzymes and mitochondria', *Nature, Lond.,* **203,** 724–726.

Öpik, H. (1965) 'Respiration rate, mitochondrial activity and mitochondrial structure in the cotyledons of *Phaseolus vulgaris* L. during germination', *J. exp. Bot.,* **16,** 667–682.

Öpik, H. (1968b) 'Development of cotyledon cell structure in ripening *Phaseolus vulgaris* seeds', *J. exp. Bot.,* **19,** 64–76.

Öpik, H. (1972) 'Some observations on coleoptile cell ultrastructure in ungerminated grains of rice (*Oryza sativa,* L.)' *Planta,* **102,** 61–71.

Öpik, H. (1973) 'Effect of anaerobiosis on respiratory rate, cytochrome oxidase activity and mitochondrial structure in coleoptiles of rice (*Oryza sativa,* L.)', *J. Cell Science,* **12** (In press).

Packer, L. and Utsumi, K. (1969) 'The relation of respiration–dependent proton transfer to mitochondrial structure', *Archs Biochem. Biophys.,* **131,** 386–403.

Packer, L., Williams, M. A., and Criddle, R. S. (1973) 'Freeze-fracture studies on mitochrondria from wild-type and respiratory-deficient yeasts', *Biochim. biophys. Acta,* **292,** 92–104.

Palade, G. E. (1953) 'An electron microscope study of the mitochondrial structure', *J. Histochem. Cytochem.,* **1,** 188–211.

Parsons, D. F., Bonner, W. D., and Verboon, J. G. (1965) 'Electron microscopy of isolated plant mitochondria and plastids using both the thin-section and negative-staining techniques', *Can. J. Bot.,* **43,** 647–655.

Parsons, J. A. and Rustad, R. C. (1968) 'The distribution of DNA among dividing mitochondria of *Tetrahymena pyriformis*', *J. Cell Biol.,* **37,** 683–693.

Parsons, P. and Simpson, M. V. (1967) 'Biosynthesis of DNA by isolated mitochondria: incorporation of thymidine triphosphate-2-C^{14}', *Science, N.Y.,* **155,** 91–93.

Perner, E. (1966) 'Das endoplasmatische Reticulum in der Radicula von *Pisum sativum* während der Keimung', *Z. Pfl. Physiol.,* **55,** 198–215.

Pihl, E. and Bahr, G. F. (1970) 'Matrix structure of critical-point dried mitochondria', *Expl Cell Res.,* **63,** 391–403.

Plattner, H. and Schatz, G. (1969) 'Promitochondria of anaerobically grown yeast. III. Morphology', *Biochemistry, N.Y.,* **8,** 339–343.

Sakano, K., and Asahi, T. (1971) 'Biochemical studies on biogenesis of mitochondria in wounded sweet potato root tissue. I. Time course analysis of increase in mitochondrial enzymes. II. Active synthesis of membrane-bound protein of mitochondria', *Plant and Cell Physiol.,* **12,** 417–426, 427–436.

Mitochondria

Sarkissian, I. V. and McDaniel, R. G. (1967) 'Mitochondrial polymorphism in maize. I. Putative evidence for de novo origin of hybrid-specific mitochondria', *Proc. natn. Acad. Sci. U.S.A.*, **57**, 1262–1266.

Sarkissian, I. V. and Srivastava, H. K. (1968) 'On methods of isolation of active, tightly coupled mitochondria of wheat seedlings', *Pl. Physiol., Lancaster*, **43**, 1400–1410.

Sarkissian, I. V. and Srivastava, H. K. (1969) 'High efficiency, heterosis and homeostasis in mitochondria of wheat', *Proc. natn. Acad. Sci. U.S.A.*, **63**, 302–309.

Schatz, G. and Criddle, R. S. (1969) 'The biosynthesis of mitochondrial energy transfer components in baker's yeast', In: *Mitochondria, structure and function* (ed. L. Ernster and Z. Drahota), pp. 189–198. V FEBS meeting, Prague 1968, Symposium vol. 17.

Schonbaum, G. R., Bonner, W. D. Jr., Storey, B. T., and Bahr, J. T. (1971) 'Specific inhibition of the cyanide-insensitive respiratory pathway in plant mitochondria by hydroxamic acid', *Pl. Physiol., Lancaster*, **47**, 124–128.

Schwab, D. W., Simmons, E., and Scala, J. (1969) 'Fine structure changes during function of the digestive gland of Venus' flytrap', *Am. J. Bot.*, **56**, 88–100.

Shinagawa, Y., Inouye, A., Ohnishi, T., and Hagihara, B. (1966) 'Electronmicroscopic studies of isolated yeast mitochondria with negative staining and thin sectioning methods', *Expl Cell Res.*, **43**, 301–310.

Siew, D. and Klein, S. (1968) 'The effect of sodium chloride on some metabolic and fine structure changes during the greening of etiolated leaves', *J. Cell Biol.*, **37**, 590–596.

Sigee, D. C. and Bell, P. R. (1971) 'The cytoplasmic incorporation of tritiated thymidine during oogenesis in *Pteridium aquilinum*', *J. Cell Sci.*, **8**, 467–487.

Simon, E. W. and Chapman, J. A. (1961) 'The development of mitochondria in *Arum* spadix', *J. exp. Bot.*, **12**, 409–420.

Sjöstrand, F. S. (1953) 'Electron microscopy of mitochondria and cytoplasmic double membranes', *Nature, Lond.*, **171**, 30–32.

Sjöstrand, F. S., Andersson-Cedergren, E., and Karlsson, U. (1964) 'Myelin-like figures from mitochondrial material', *Nature, Lond.*, **202**, 1075–1078.

Sjöstrand, F. S. and Barajas, L. (1968) 'Effect of modifications in conformation of protein molecules on structure of mitochondrial membranes', *J. Ultrastruct. Res.*, **25**, 121–155.

Sjöstrand, F. S. and Barajas, L. (1970) 'A new model for mitochondrial membranes based on structural and on biochemical information', *J. Ultrastruct. Res.*, **32**, 293–306.

Skukas, G. P. (1968) 'Changes in wall and internal structure in *Allomyces* resistant sporangia during germination', *Am. J. Bot.*, **55**, 291–295.

Smith, U., Smith, D. S., and Yunis, A. (1970) 'Chloramphenicol-related changes in mitochondrial ultrastructure', *J. Cell Sci.*, **7**, 501–521.

Staehelin, L. A. (1968) 'The interpretation of freeze-etched artificial and biological membranes', *J. Ultrastruct. Res.*, **22**, 326–347.

Steinert, M. (1969) 'The ultrastructure of mitochondria', *Proc. R. Soc. B.*, **173**, 63–70.

Stoeckenius, W. (1965) 'Some observations on negatively stained mitochondria', *J. Cell Biol.*, **17**, 443–454.

Storey, B. T. (1970a and b) 'The respiratory chain of plant mitochondria VI. Flavoprotein components of the respiratory chain of mung bean mitochondria; and VIII. Reduction kinetics of the respiratory chain carriers of mung bean mitochondria with reduced nicotinamide adenine dinucleotide', *Pl. Physiol., Lancaster*, **46**, 13–20, 625–630.

Street, H. E., Öpik, H., and James, F. E. L. (1968) 'Fine structure of the main axis meristems of cultured tomato roots', *Phytomorphology*, **17**, 391–401.

Tourte, Y. (1968) 'Observations sur le comportement du noyau, des plastes et les mito-chondries au cours de la maturation de l'oosphère du *Pteridium aquilinum* L', *C.r. hebd. Séanc. Acad. Sci., Paris*, **266**, 2324–2326.

Wallace, P. G., Huang, M., and Linnane, A. W. (1968) 'The biogenesis of mitochondria. II. The influence of medium composition on the cytology of anaerobically grown *Saccharomyces cerevisiae*', *J. Cell Biol.*, **37**, 207–220.

Wark, M. C. (1965) 'Fine structure of the phloem of *Pisum sativum* II. The companion cell and phloem parenchyma', *Aust. J. Bot.*, **13**, 185–193.

Watson, K., Haslam, J. M., and Linnane, A. W. (1970) 'Biogenesis of mitochondria. XIII. The isolation of mitochondrial structures from anaerobically grown *Saccharomyces cerevisiae*', *J. Cell Biol.*, **46**, 88–96.

Weintraub, M., Ragetli, H. W. J., and John, V. T. (1966) 'Fine structural changes in isolated mitochondria of healthy and virus-infected *Vicia faba* L.', *Can. J. Bot.*, **44**, 1017–1024.
Wells, R. and Birnstiel, M. (1969) 'Kinetic complexity of chloroplastal deoxyribonucleic acid and mitochondrial deoxyribonucleic acid from higher plants', *Biochem. J.*, **112**, 777–786.
Williamson, D. H. (1970) 'The effect of environmental and genetic factors on the replication of mitochondrial DNA in yeast', In: Control of organelle development; *Symp. Soc. exp. Biol.*, **24**, 247–276.
Wilson, S. B. and Bonner, W. D., Jr. (1970a) 'Effects of guanidine inhibitors on mung bean mitochondria', *Pl. Physiol., Lancaster*, **46**, 21–24.
Wilson, S. B. and Bonner, W. D., Jr. (1970b) 'Preparation and some properties of sub-mitochondrial particles from tightly coupled mung bean mitochondria', *Pl. Physiol., Lancaster*, 25–30.
Wilson, S. B. and Bonner, W. D., Jr. (1970c) 'Energy-linked functions of submitochondrial particles prepared from mung bean mitochondria', *Pl. Physiol., Lancaster*, 31–35.
Wintersberger, E. and Viehhauser, G. (1968) 'Function of mitochondrial DNA in yeast', *Nature, Lond.*, **220**, 699–702.
Woodward, D. O. (1968) 'Functional and organizational properties of *Neurospora* mitochondrial structural protein', *Fedn. Proc. Fedn. Am. Socs exp. Biol.*, **27**, 1167–1173.
Woodward, D. O., Edwards, D. L., and Flavell, R. B. (1970) 'Nucleocytoplasmic interactions in the control of mitochondrial structure and function in *Neurospora*', In: Control of organelle development; *Symp. Soc. exp. Biol.*, **24**, 55–70.
Yotsuyanagi, Y. (1962) 'Etudes sur le chondriome de la levure. 1. Variation de l'ultrastructure du chondriome au course du cycle de la croissance aérobie', *J. Ultrastruct. Res.*, **7**, 121–140.

3

The endomembrane concept: a functional integration of endoplasmic reticulum and Golgi apparatus

D. James Morré and H. H. Mollenhauer

Introduction

The endomembrane system (Morré, Mollenhauer, and Bracker, 1971; Figs. 3.1 to 3.4A) is proposed as a developmental continuum of membranous cell components that functions along two main pathways of membrane differentiation, one to the cell surface and another to the vacuolar apparatus of the cell's interior, to account for the biogenesis and multiplication of the cytoplasmic membranes of eukaryotic cells. Included within the endomembrane system are endoplasmic reticulum, the outer membrane of the nuclear envelope and various transitional membrane elements like those of the Golgi apparatus as well as structures derived from these major components. Continuities (Bracker et al., 1971; Morré, Mollenhauer, and Bracker, 1971; Figs. 3.3 to 3.15) among the separate endomembrane components demonstrate that these cytoplasmic membranes occur as localized specializations of an interassociated system. Endomembrane components are distinguished bio-chemically from the semiautonomous organelles (with definite inner and outer membrane systems: so-called double membrane structure), such as chloroplasts and mitochondria, by the absence of DNA and cytochrome oxidase, the inability to generate ATP through respiratory- or photosynthetic-chain-linked phosphorylation, and a high degree of functional interdependence.

In this chapter, we introduce the endomembrane concept to explain specific characteristics of endomembrane structure and function. Each of the components of the endomembrane system is described and then discussed in terms of its potential role in the dynamic and integrated activities of the system as a whole.

Concepts of *membrane flow* and of *membrane differentiation* are combined to
provide a mechanism whereby membranes are transferred and transformed within
the endomembrane system along a chain of cell components in a subcellular
developmental pathway. While the *endomembrane system* is a well-established
part of the cell (although rarely discussed as a functional continuum), the
endomembrane concept is an hypothesis for which supporting information is still
being sought.

The chapter is not intended as a comprehensive review of the literature. Rather,
we stress concepts along with facts and reasoning leading to the concepts. To the
extent possible, we present original supporting data. Summary articles listed on
page 131 give the historical development of the concepts together with more complete reviews of pertinent literature.

We emphasize the role of the endomembrane system in its own biogenesis and
multiplication. Its participation in the synthesis, modification, transport and
storage of products destined for export out of the cell is a parallel and essential
endomembrane function that has been treated extensively in previous reviews
(Beams and Kessel, 1968; Mollenhauer and Morré, 1966a; Morré, Mollenhauer,
and Bracker, 1971; Schnepf, 1968, 1969a; Schramm, 1967).

Much of the structural information upon which the endomembrane concept
is based comes from plants (e.g., this chapter) and fungi (Grove, Bracker, and
Morré, 1968, 1970; C. E. Bracker and colleagues, in preparation). Yet, the concept is equally applicable to animals (Keenan, Morré, and Huang, in press; Morré,
Keenan, and Mollenhauer, 1971; Morré, Mollenhauer, and Bracker, 1971) and
most of the biochemical information upon which the concept is based derives
from animal sources. As a result, we summarize findings from rat liver and
mammary gland along with those from plants. The concepts are meant to apply
to eukaryotes generally, be they plant or animal.

To illustrate the significance of endomembrane function, we remind the reader
that major food and fibre commodities of agriculture and commerce (except for
starch) are synthesized, stored, and/or released from cells through the cooperative
action of endomembrane components. Included are fat and protein components
of milk (Keenan, Morré, and Huang, in press); cell wall materials (the celluloses,
pectins, hemicelluloses, and perhaps even lignins of wood, fruits, vegetables, and
forages) (Mollenhauer and Morré, 1966a; Northcote, 1969), latex (the source of
natural rubber), natural oils (Schnepf, this volume), hormones (Nakane and
Kawari, 1971), vitamins (Nyquist, Crane, and Morré, 1971; Nyquist, Matschiner,
and Morré, 1971), collagen and mucopolysaccharides of connective tissues in
meat (Revel and Hay, 1963; Sheldon and Kimball, 1962), egg shells (Breen and
de Bruyn, 1969), egg yolks (Bellairs, 1964) and storage lipids and proteins of
seeds (Mollenhauer and Totten, 1971).

To the extent that the endomembrane system is involved in the origin of the
plasma membrane (Morré and VanDerWoude, in press; Morré *et al.*, 1971), those
surface characteristics which determine self and non-self (especially as related to
morphogenesis), immunological specificity, oncogenic (tumour causing) transformations, etc. (Curtis, 1967; Ginsburg and Neufield, 1969; Roseman, 1970) are

also determined by activities of the endomembrane system. Additionally, dis-
functions of endomembrane components have been implicated in disease
(Berhsohm and Grossman, 1971; Bracker and Littlefield, 1973; Dales, 1971;
Dickens et al., 1968). Nearly all of the structural and reserve components of
eukaryotic cells depend on endomembrane activities for synthesis or secretion.
Yet the most important roles of the endomembrane system may be those involving synthesis, turnover and differentiation of membranes as these functions
relate to normal and abnormal processes of cell division, growth and differentiation.

The endomembrane concept

The *endomembrane system* (Figs. 3.1 to 3.4) is presented as a functional and
developmental continuum of the internal cytoplasmic membranes that characterize eukaryotic cells. The inner membrane systems of the semi-autonomous
organelles — chloroplasts, mitochondria, and possibly the nucleus as well, are
excluded from the endomembrane system, e.g., Fig. 3.5; a discussion of membrane
continuity is provided in the Appendix (p. 126). The outer membrane of the
nuclear envelope, endoplasmic reticulum (ER), Golgi apparatus and other transitional membrane elements are the major components of the system, but membrane structures derived from these major components (secretory vesicles,
microbodies (?)[1], and lysosomes) are included along with the major end
products (vacuole membrane or tonoplast, plasma membrane and the outer
membrane envelopes of both mitochondria and plastids[2]). Structural and functional features common to all components of the system are more readily
visualized by considering all the internal membranes as an interassociated unit
(Fig. 3.1; Fig. 3.2), as are the dynamics, pleomorphic aspects and three-dimensional complexity of the system. Most important, functional activities (e.g.,
Fig. 3.3) as well as the very existence of the system may depend on the integrated
activities of its major parts.

   The major parts of the system have or are expected to have many properties
in common but, more important, there are also many properties which are unique
to each part (Figs. 3.1 to 3.18; Tables 3.1 to 3.5, 3.7). The endomembrane concept differs from other cytomembrane concepts such as the unit membrane theory

---

   [1] The question of whether or not to include the microbody as an organelle or as an
endomembrane component is difficult to decide on the basis of present information. Although
the membranes of microbodies are closely associated with (perhaps even derived from) ER
(Figs. 3.5D, H), the microbody appears to function as a type of primitive respiratory organelle
although no ATP is generated as a result of microbody respiration (deDuve, 1969a).

   [2] The organelles, particularly chloroplasts and mitochondria, are thought to have arisen
through endosymbiotic invasion from prokaryotic ancestors (Margulis, 1970). According
to this notion, the outer membrane of these organelles corresponds to the membrane of an
endocytotic vacuole (originating from the host plasma membrane) while the inner membranes correspond to the plasma membrane of the endosymbiont. Carrying this argument
further, it might be expected that the outer organellar membranes would originate via a
mechanism similar to that for the origin of host plasma membrane. Connections between
ER and outer organellar membranes have been reported (Bracker and Grove, 1971; Morré,
Merritt, and Lembi, 1971; Franke and Kartenbeck, 1971; Figs. 3.5F, G).

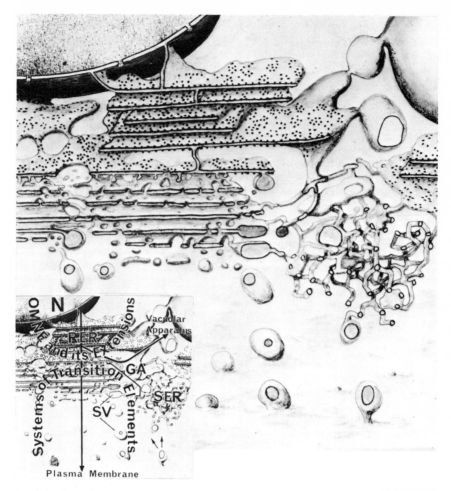

Fig. 3.1   Generalized interpretation of a portion of the endomembrane system characteristic
of eukaryotic cells. The diagram combines information from animal (rat liver), fungal (vegetative hyphae of *Pythium* spp.) and plant (onion stem). Features from each cell type are
emphasized. Some aspects (rough and smooth endoplasmic reticulum) are more typical of
the animal cell while other aspects (the Golgi and vacuolar apparatus) are more typical of the
plant and fungus. A key to labelling is provided in the inset.

   The part of the endomembrane system labelled OMNE (outer membrane of the nuclear
envelope) *and its extensions* includes rough endoplasmic reticulum (RER) as well as the
outer membrane of the nuclear envelope. These structures have ribosomes, pore apparatus,
or both associated with their membranes and function in nucleo-cytoplasmic exchanges
and/or protein biosynthesis.

   The part of the endomembrane system labelled *Systems of Transition Elements* includes
smooth endoplasmic reticulum (SER), Golgi apparatus (GA) and secretory vesicles (SV).
According to the endomembrane concept, these systems of transition elements function
along two main pathways of endomembrane differentiation (arrows): one is to the plasma
membrane and the cell surface and another is to lysosomes and the vacuolar apparatus of
the cell's interior.

(Robertson, 1964), principally in that specialization is considered to arise through progressive differentiation of the components of the system according to a defined developmental sequence. It combines the concepts of *membrane flow* (Bennett, 1956; Franke *et al.*, 1971; Morré *et al.*, 1971) and of *membrane differentiation* (Grové, Bracker, and Morré, 1968; Schnepf, 1969b) to provide a mechanism whereby membranes are transferred and transformed along a chain of cell components in a subcellular developmental pathway (Morré *et al.*, 1971).

TABLE 3.1

Endomembrane Differentiation based on Membrane Dimensions in Stem Cells of Onion (*Allium cepa*). Glutaraldehyde-osmium Tetroxide Fixation. (Measurements are approximate but taken from the same electron micrographs)*

Endomembrane component	Average membrane thickness (nm)
Outer membrane of nuclear envelope or rough endoplasmic reticulum	4–5
Outer mitochondrial and plastid membrane (microbody membrane)	4–5.5
Golgi apparatus cisternae (central plate or saccule)	5–7
'Free' secretory vesicles	8–9.5
Plasma membrane	9.5

* The thickness of the tonoplast could not be determined accurately because of dense deposits on one or both membrane surfaces but was estimated to be in the range of 6.0–7.5 nm.

Parts of the endomembrane system

The endomembrane system consists of two major parts : (a) the outer membrane of the nuclear envelope and its extensions including the portion of ER with ribosomes on its outer surface (rough ER) which functions in protein synthesis and/or nucleocytoplasmic exchanges; and (b) systems of transition elements incapable of protein synthesis but which function in one or both of two major pathways of membrane differentiation (Figs. 3.1 and 3.2). One pathway is to the cell surface as the membranes become progressively more like plasma membrane (Figs. 3.1, 3.2; 3.8, 3.9 and 3.13). The other pathway is to the cell's interior as ER, ER-like membranes or other prevacuolar structures give rise to vacuole membranes (Buvat, 1971; Fig. 3.10) and perhaps contribute constituents to the outer membranes of plastids and mitochondria (Bracker and Grove, 1971; Franke and Kartenbeck, 1971; Morré, Merritt, and Lembi, 1971; Figs. 3.5F, G) as well as to lysosomal membranes (Brandes, 1965; Essner and Novikoff, 1962; Novikoff and Shin, 1964; Fig. 3.2). The term *transition* element denotes an endomembrane component with properties intermediate between those of the outer membrane of the nuclear envelope or rough ER at one extreme and plasma membrane or tonoplast at the other extreme (see footnote 6 of Morré, Mollenhauer, and Bracker, 1971; Figs. 3.1 to 3.3, 3.6 to 3.16; Tables 3.1, 3.2, 3.4, 3.5, and 3.7). Included in this category are forms which lack ribosomes such as smooth

TABLE 3.2

Phospholipid Composition and Sterol Content of Components of the Endomembrane System of Rat Liver

Phospholipid phosphorous (% of total recovered)

Phospholipid	Microsome fractions*			Membrane fractions †		
	Rough	Smooth I	Smooth II	Rough endoplasmic reticulum	Golgi apparatus	Plasma membrane
Phosphatidyl choline	53	52	50	61	45	40
Phosphatidyl ethanolamine	22	20	20	19	17	18
Phosphatidyl serine	8	9	8	3	4	4
Phosphatidyl inositol	11	12	13	9	9	7
Sphingomyelin	6	7	9	4	12	19
Lysophosphatidyl choline	—	—	—	5	6	7
Lysophosphatidyl ethanolamine	—	—	—	—	6	6
Cholesterol/phospholipid	0.07	0.11	0.17	0.04	0.14	0.21

* Glauman and Dallner (1968).   † Keenan and Morré (1970).

The endomembrane concept

ER (Figs. 3.3 and 3.7A), the Golgi apparatus (Figs. 3.8 to 3.13), specialized membranes which function as Golgi apparatus equivalents (Figs. 3.3B and 3.14) and various types of vesicles (e.g., Figs. 3.3, 3.8, 3.9, 3.11, 3.13 and 3.14). A three-dimensional representation of endomembrane organization and the pathways of endomembrane differentiation are provided in Figs. 3.1 and 3.2.

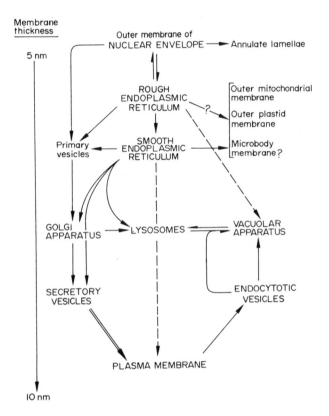

Fig. 3.2 Pathways of endomembrane flow and differentiation based on relative membrane thickness, functional associations and biochemical characteristics. The dashed lines indicate flow of constituents from ER to the plasma membrane and vacuole membranes via pathways not involving the Golgi apparatus. As a first approximation, any two membranous cell components within the endomembrane system can be related by knowing the thickness of their membranes relative to ER and plasma membrane.

Emphasis is placed on membrane differentiation in the 'outward' direction from nuclear envelope or ER to the plasma membrane or 'inward' to the vacuolar apparatus. Reversals of membrane differentiation are not excluded but the pathways may be different in the two directions. deDuve (1969b) has emphasized that materials that enter the vacuolar apparatus by endocytosis can invade a large portion of the endomembrane system but not all of it. They never seem to reach the innermost parts of the system, i.e., there is no evidence that materials (or membranes) entering a cell by endocytosis are transferred to the lumen (or membranes) of the ER directly. This inner space, corresponding approximately to the space bounded by the ER, and called endoplasmic space by deDuve, corresponds to the nuclear envelope and its extensions in our scheme.

The outer membrane of the nuclear envelope and its extensions

The outer membrane of the nuclear envelope and its extensions includes rough ER (Figs. 3.3 to 3.8) and annulate lamellae (not illustrated, see Kessel, 1968). They are structurally continuous (Fig. 3.4C) and share many properties in common (see below). Both the outer membrane of the nuclear envelope and rough ER have ribosomes on their outer or cytoplasmic surfaces (compare Figs. 3.4B and 3.6A) and, rough ER, at least, is a major site of biosynthesis of proteins and phospholipids within the endomembrane system (Table 3.6 and ref. cit.). Rough ER is continuous with the rest of the endomembrane system through regions of smooth ER (ER lacking ribosomes) (Figs. 3.3, 3.4, 3.7, 3.8, and 3.15).

Outer membrane of the nuclear envelope. The outer membrane of the nuclear envelope is the most widespread representative of the endomembrane system (specific references are provided in footnote 2 of Morré, Mollenhauer, and Bracker, 1971). The envelope consists of two adjacent membranes separated by a perinuclear space (Figs. 3.4, 3.5A and 3.8A, B). At irregular intervals, the inner and outer membranes of the envelope are joined at sites where circular or octagonal pores connect cytoplasm and nucleoplasm (Franke, 1970a; Figs. 3.4B, C, 3.5A and 3.8A) and provide for exchange of both large and small molecules (Feldherr, in press). The pores are not simple holes in the envelope but contain a complex pore apparatus (Feldherr, in press; Franke, 1970a) which modifies the properties of the membranes at the pore junction (Franke and Scheer, 1970).

In many respects, the two membranes which form the nuclear envelope are separate and distinct entities (e.g., Fig. 3.5A). The inner membrane of the nuclear envelope is closely associated with the peripheral chromatin (DNA + histone) in the interphase nucleus (Figs. 3.4C and 3.5A; Feldherr, in press). It appears to function in the control of transcription (Comings and Kakefuda, 1968; reviewed by Feldherr, in press), contains cytochrome oxidase (E. D. Jarasch, unpublished) and is in many ways analagous to the cell membranes of prokaryotes and to the inner membrane systems of eukaryotic organelles (e.g., Fig. 3.5). The outer membrane is frequently studded with ribosomes (Figs. 3.4A, B, and 3.8C), is continuous with rough ER (Fig. 3.4C) and is very much a part of the endomembrane system (Figs. 3.4 and 3.5A). The anticlinal membrane cylinder of the pore complex is a structural component of the pore complex and a remarkably stable part of the nuclear envelope (Franke and Scheer, 1970). The pore complex may restrict free exchange of constituents between the inner and outer nuclear membranes even though the two membranes are structurally continuous in the pore regions (Fig. 3.5A; Franke and Scheer, 1970). Detailed descriptions of the nuclear envelope and its functions are given by Lafontaine (chapter 1).

Some investigators consider that all endomembranes are somehow related to, originate from, or are derived from the outer membrane of the nuclear envelope (See Franke, Eckert, and Krien, 1971 for discussion). In keeping with this concept are studies which indicate that turnover of membrane proteins in the outer membrane of the nuclear envelope may be greater than those of rough ER (Franke,

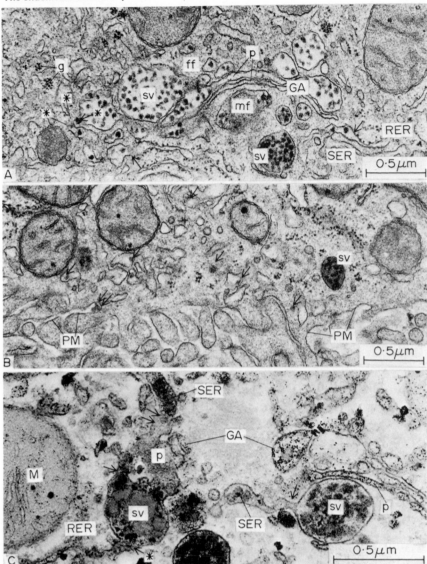

Fig. 3.3   Portions of the endomembrane system of rat liver illustrating structural and functional continuities among endomembrane components involved in the biosynthesis and export of lipoprotein particles. Scale mark = 0.5 μm.

(A) A portion of the Golgi apparatus. Lipoprotein particles are recognized as 30 to 90 nm diameter, darkly stained, spherical profiles in the secretory vesicles (sv) and peripheral elements of the smooth endoplasmic reticulum (SER). The apoproteins of the lipoprotein particles are presumed to be synthesized on polyribosomes and associated messenger RNA of the rough endoplasmic reticulum (RER) but, following periods of synthesis, lipoprotein particles first appear within the SER at or near SER—RER junctions. Regions of RER—SER continuity are shown at the single arrows. At least some of the particles appear to migrate from the SER to secretory vesicles (sv) of the Golgi apparatus (GA). The lipoprotein particles are absent from the plate-like central regions of the cisternae (p) and are thought to

Eckert, and Krien, 1971; see, however, Franke *et al.*, 1971). Although direct transfer of membrane from nuclear envelope to ER remains to be demonstrated, membranes of the Golgi apparatus, for example, appear to originate from vesicles or short tubules which are derived from specific ribosome-free sites along the outer membrane of the nuclear envelope and which subsequently fuse to form flattened cisternae (Fig 3.8A, B; Bracker *et al.*, 1971; reviewed by Beams and Kessel, 1968; Franke, Eckert, and Krien, 1971; Grove, Bracker, and Morré, 1970; Morré, Mollenhauer, and Bracker, 1971). This idea is introduced here to indicate a possible role of the outer membrane of the nuclear envelope in membrane flow and biogenesis. We shall return to the importance of this structural pattern in relation to formation of Golgi apparatus later in the chapter.

Rough ER and outer membrane of the nuclear envelope sometimes appear interconvertible, perhaps even interchangeable (Figs. 3.4, 3.5A and 3.8). For example, ER cisternae provide an alternative source of Golgi apparatus membrane by a mechanism entirely analogous to that afforded by the outer membrane of the nuclear envelope (Fig 3.8C, D; literature reviewed by Morré, Mollenhauer, and Bracker, 1971). To what extent ER and nuclear envelope, or inner and outer

---

pass directly from SER to the secretory vesicles of the forming face (ff) of the apparatus via direct tubular connections. The lipoprotein particle at the arrow marked by asterisks appears to represent a grazing section through one of these tubular connections. A clear illustration of continuities between lipoprotein particle-containing SER and secretory vesicles is provided in Fig. 3.3 C. Note that the lipoprotein particles contained in the secretory vesicles of the Golgi apparatus, especially those at the maturing face (mf), are of smaller diameter than those in SER. Note also that secretory vesicles at the mature face (mf) of the Golgi apparatus, as well as migrating secretory vesicles free in the cytoplasm (Fig. 3.3B), are characterized by a darkly staining matrix while those of the forming face have an electron transparent matrix. g = glycogen. Osmium tetroxide fixation.

(B) Portion of the cytoplasm bordering the capillary space (Space of Disse). Illustrated is a secretory vesicle (sv) free in the cytoplasm and the system of smooth endoplasmic reticulum characteristic of the cytoplasm near the cell surface. Lipoprotein particles of slightly greater diameter (50 to 100 nm) than those of the secretory vesicles occur singly or in small clusters within the cavities of the SER (arrows). These particles lack the electron dense matrix characteristic of mature secretory vesicles (see Fig. 3.3A). These differences (occurrence in small clusters, larger diameter, absence of electron dense matrix) serve to distinguish SER lipoprotein particles from those of secretory vesicles. Single particles of the SER type are found near the plasma membrane (PM) and appear to be secreted directly into the extracellular space (double arrows). These latter images have been interpreted by Morré and VanDerWoude (in press) to result from fusion of SER membranes or membrane of vesicles derived from SER directly with the plasma membrane. Adapted from Morré and VanDerWoude (in press). Osmium tetroxide fixation.

(C) Transport of lipoprotein particles from smooth ER (SER) to secretory vesicles (sv) of the Golgi apparatus involves direct tubular connections between SER and the vesicles. These connections are difficult to visualize in ordinary thin sections since the connections are smaller in diameter than the thickness of the average thin section (see Fig. 3.3A). To increase tubule visibility, tubule interiors were stained by the procedure of Goldfischer (1965) for cytochemical localization of arylsulphatases. Arylsulphatases are distributed throughout the endomembrane system at low levels in rat liver (Wilkinson, Nyquist, Merritt, and Morré (1972), and this technique revealed numerous tubular connections between lipoprotein-containing SER and secretory vesicles of the Golgi apparatus. Connections are indicated (arrows) between lipoprotein-containing SER tubules and the large secretory vesicles of the Golgi apparatus (single arrows), between these tubules and a portion of a Golgi apparatus plate (p) seen in face view (double arrows) and between a secretory vesicle and rough ER (RER: arrow with asterisk). M = Mitochondrion. (From a study by F. E. Wilkinson and D. J. Morré.)

membrane of the nuclear envelope, are interconvertible is not known. When division occurs in many types of cells, the nuclear envelope breaks down during prophase and re-forms during anaphase and early teleophase by 'precipitation' of membrane fragments on the surface of the condensed chromosomes. During

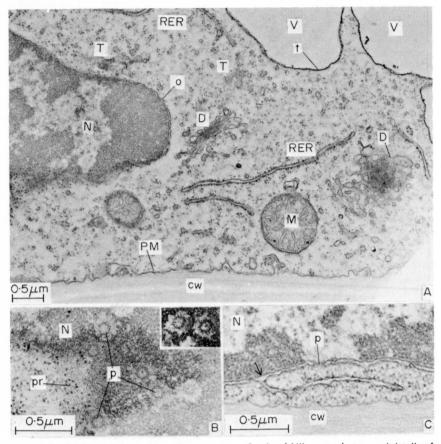

Fig. 3.4   Portions of the endomembrane system of onion (*Allium cepa*) stem and details of the outer membrane of the nuclear envelope and its extensions. Glutaraldehyde-osmium tetroxide fixation. Scale marks = 0.5 μm.

(A) Stem cell of onion showing the nucleus (N), outer (o) and inner membranes of the nuclear envelope, rough endoplasmic reticulum (RER), vacuoles (V), vacuole membrane or tonoplast (t), mitochondria (M), dictyosomes (D) of the Golgi apparatus (a stack in cross section, left, and a single cisterna from the mid-region of the stack in face view, right), plasma membrane (PM), cell wall (cw) and tubular transition elements (T). The latter may represent functional equivalents to smooth endoplasmic reticulum in this tissue.

(B) Outer membrane of the nuclear envelope in grazing section at the edge of a nucleus (N). Pores which connect inner and outer nuclear membranes and provide for nucleo-cyto-plasmic exchanges are seen in face view (p and inset). Polyribosomes (pr) are associated with the outer membrane of the nuclear envelope although perhaps less frequently than with rough endoplasmic reticulum membranes (Compare with Fig. 3.6A).

(C) An extension of the outer membrane of the nuclear envelope (arrow) morphologically indistinguishable from rough ER. p = nuclear pore in cross section, N = nucleus, cw = cell wall.

simultaneous re-formation and multiplication of the organized nuclear envelope, some of the membrane fragments appear to be derived from ER or pieces of old nuclear envelope (Barer, Joseph, and Meek, 1959; Merriam, 1961). Does ER and 'old' outer membrane of the nuclear envelope give rise only to 'new' outer membrane of the nuclear envelope or are outer and inner nuclear membranes interconvertible (and/or derived from ER) at this stage? Perhaps the differential staining reaction shown in Fig. 3.5A. applied to dividing cells, might help resolve this question. In any event, it seems doubtful that the unique features of the outer and inner membranes of the nuclear envelope (e.g., Figs. 3.4 and 3.5A) would be conserved during the process if they were completely interconvertible. Pores are apparent even at initial stages of nuclear envelope reconstitution (Merriam, 1961) but details of the process of pore formation are unknown (Feldherr, in press).

In many fungi and euglenoid flagellates, the nuclear envelope does not break down during nuclear division (Leedale, 1969). Yet, a complete nuclear envelope surrounds each of the daughter nuclei. These observations, and observations on the formation of annulate lamellae,[1] more clearly implicate the nuclear envelope in at least some aspects of its own formation.

Rough-surfaced endoplasmic reticulum (rough ER). Rough, rough-surfaced or granular ER (rough ER) in its familiar form consists of a system of plate-like or sheet-like, flattened cisternae with ribosomes attached on the cytoplasmic surface, the surface facing the cytoplasm (Figs. 3.3 to 3.9). Since not all ER membranes have associated ribosomes, those parts which do are distinguished as rough ER by this association (Porter, 1961). Rough ER is a major site of biosynthesis within the endomembrane system (Siekevitz *et al.*, 1967; Tables 3.3, 3.6 and ref. cit.) and provides structural and functional continuity between the outer membrane of the nuclear envelope and the rest of the endomembrane system (Figs. 3.3 to 3.18; Franke, Eckert, and Krien, 1971; for an explanation and examples of functional continuity, see Fig. 3.3 and the Appendix).

Specific characteristics of rough ER vary with the organism, developmental stage, history and metabolic state of the cell (Porter, 1961) and the possibility has been considered that it may differ with respect to regions within a given cell (See Leskes, Siekevitz, and Palade, 1971a, b for experimental approaches to this problem). Rough ER is frequently most extensively developed in cells engaged in synthesis of proteins for export, such as enzyme-producing cells of digestive glands (e.g., Amsterdam, Ohad, and Schramm, 1969; Jamieson and Palade, 1967; Palade, 1959; Porter, 1961; Schnepf, 1968, 1969a). In these and other cells which produce proteins for export (Fig. 3.3), rough ER lamellae, arranged in parallel to

[1] Annulate lamellae are interpreted as pore-containing extensions of the nuclear envelope (Feldherr, in press; Kessel, 1968; Scheer and Franke, 1969). Annulate lamellae appear to form from small primary vesicles derived from the nuclear envelope (Kessel, 1963). The vesicles are morphologically similar to those involved in the formation of Golgi apparatus (Morré, Mollenhauer, and Bracker, 1971). Annulate lamellae provide an example of an endomembrane component thought to be derived exclusively from the outer membrane of the nuclear envelope. Of limited occurrence, annulate lamellae are most prevalent in oocytes and are absent from most somatic cells (Kessel, 1968). Their function is unknown.

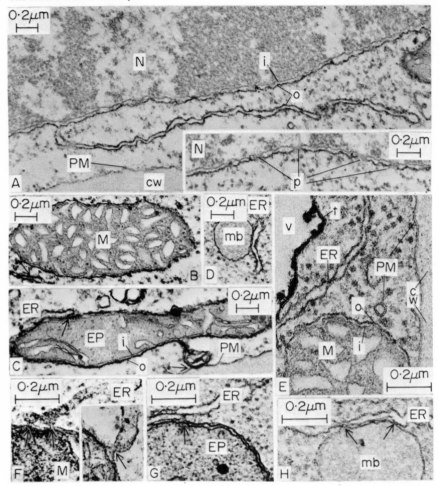

Fig. 3.5 Endomembrane differentiation in stem cells of onion. Glutaraldehyde-osmium tetroxide fixation. Sections were post-stained with alkaline lead citrate so that endoplasmic reticulum (ER) and ER-like membranes were darkly stained and plasma membrane (PM) and PM-like membranes were lightly stained. The differential staining was not due to stain penetration since all membranes in the section had equal access to the stain. Control sections subjected to conventional lead staining procedures did not show such marked differential staining of endomembranes. Scale marks = 0.2 μm.

(A) Nuclear envelope and its extensions. The outer membrane of the nuclear envelope (o) is darkly stained, resembling ER, while the inner membrane of the nuclear envelope (i) is lightly stained. This electron micrograph demonstrates that the inner and outer membranes of the nuclear envelope differ in their staining properties. The outer membrane resembles ER while the inner membrane resembles the inner membrane systems of the organelles (Fig. 3.5B,C and E). Inset shows the separation of inner and outer nuclear membrane afforded by nuclear pores (p). N = nucleus, cw = cell wall.

(B) Mitochondrion (M). The outer mitochondrial membrane stains intensely resembling ER membranes (arrow, lower left) while the inner membrane system of the mitochondrion is lightly stained.

each other in stacked arrays, are a distinguishing cytoplasmic feature. Stacking is not an essential feature of the structure of rough ER and rough ER cisternae may exist in the cell without any obvious orientation to other rough ER cisternae (Figs. 3.3 to 3.9). Rough ER may also appear tubular or fenestrated (Fig. 3.6A).

The rough ER, along with the outer membrane of the nuclear envelope, is the portion of the endomembrane system specialized for protein synthesis (Siekevitz et al., 1967; Table 3.6 and ref. cit.). This synthetic capacity resides in the ribosomes attached to the outer surface of the ER membranes (cf. Sabatini et al., 1971; Figs. 3.6A and 3.7). Since many ribosomes may participate simultaneously in the transcription of a single messenger RNA (Attardi, 1967), ribosomes are most frequently encountered on rough ER or nuclear envelope in the form of polyribosomal spirals or *polysomes* (Figs. 3.4B, 3.6A and 3.7B).

Polysomes occur free in the cytoplasm (Figs. 3.3 and 3.4) and the association with membranes does not appear to be an essential part of their function in protein synthesis (Hamada et al., 1968). Yet, the association of polysomes with membranes provides a means of compartmentalizing protein synthesis (Sabatini et al., 1971).

(C) Etioplast. Outer plastid membranes (o) stain darkly and resemble ER while the inner plastid membranes (i) and plasma membrane (PM) are only lightly stained. Approximately eight plastid profiles out of ten examined showed a fragment of a membrane element resembling ER in close association with the outer plastidal membrane. These associations have not been studied in sufficient detail to determine if the ER lumen is continuous with the space between inner and outer plastid membranes. The dark outline of the ER membrane and outer plastid membrane can be followed without interruption (arrow upper left) in contrast to a point of contact with the plasma membrane (arrow, lower right) where the membranes appear to contact but are still resolved as separate membranes.

(D) Microbody. Microbody (mb) membranes are characterized by a close physical association with a smooth (lacking ribosomes) segment of rough ER. The microbody membrane stains darkly, like ER.

(E) Endomembranes at higher magnification illustrating that even though the ER membranes and outer mitochondrial membranes (o) are more darkly stained than plasma membrane (PM) and inner mitochondrial membrane (i), the relative membrane thicknesses illustrated in Table 3.1 are still maintained (most evident in the relative width of the light space of the dark—light—dark pattern). The prominent dark—light—dark pattern of PM-like membranes is especially evident in the vesicular profile (centre right). The tonoplast (t) or vacuole (v) membrane is partly obscured in this tissue by darkly staining deposits especially prevalent on the inner surface facing the vacuole contents. cw = cell wall.

(F) Presumed connections (arrow) between ER and the outer membrane of mitochondria (M). The presumed connections appear as short, narrow (15—30 nm) tubules which extend perpendicular from the outer mitochondrial membrane to join with the ER. The tubules connect with regions of the ER lacking ribosomes but ribosomes are attached to other parts of the same cisterna to permit positive identification of the membranes as ER (Morré, Merritt, and Lembi, 1971). (From a study by D. J. Morré and C. A. Lembi.)

(G) A tubule similar to that of Fig. 3.5F between ER and the outer membrane of an etioplast (EP). As with the ER-mitochondrion connections, the tubules are narrower than the thickness of a single section so that the lumen (if one exists) is most generally obscured. Similar images are formed (with indistinct lumina) by the 20 to 60 nm diameter tubules which interconnect secretory vesicles of the Golgi apparatus and cisternae of smooth ER (Figs. 3.3A and C). (From a study by D. J. Morré and C. A. Lembi.)

(H) Region of microbody (mb) ER association showing what appears to be membrane continuity (arrow, left) and a tubular profile (arrow, right) between ER and the microbody membrane. In the absence of correlative information, luminal continuity can be neither decided nor excluded on the basis of this type of thin section analysis since the points of association are resolved only as short, narrow tubules.

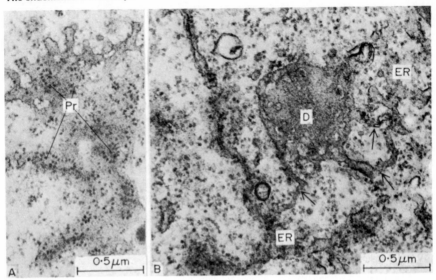

Fig. 3.6   Endomembranes of onion stem. Flattened cisternae and tubules of rough ER and dictyosomes (D) of the Golgi apparatus. Glutaraldehyde-osmium tetroxide fixation. Scale marks = 0.5 μm.

(A) Rough endoplasmic reticulum in face view shows that ribosomes of the membrane surface are arranged in polyribosomal configurations (Pr). Rough ER membranes in this tissue may also appear tubular or fenestrated (with small openings or perforations).

(B) A thick thin section shows a dictyosome (D) in face view (See also Figs. 3.4A and 3.11 inset) with a central plate-like portion and peripheral tubules of one cisternae plus some profiles of tubules from under- and overlying cisternae. The tubules marked by arrows extend from the central plate to ER. The tubules extend for several micrometres from the centre of the dictyosome cisternae but branch and change directions frequently so that they are difficult to trace for long distances within the plane of a single thin section. Because of overlap, continuity of the Golgi apparatus tubules with ER is difficult to determine from thick sections. Continuity between Golgi apparatus and ER most frequently seems to be of the intermittent tubular or vesicular bridge type (See Appendix) and is illustrated in Figs. 3.8 and 3.11.

Fig. 3.7   Endomembranes of rat liver. Both rough ER and smooth ER are prominently developed in this tissue. Osmium tetroxide fixation. Scale marks = 0.5 μm.

(A) Rough ER in cross-section (RER) continuous with smooth ER (SER) at the arrows. Double arrows mark lipoprotein particles which help to identify SER in rat liver (Compare with Fig. 3.3). Regions of continuity between RER and SER occur at junctional regions (JR). Similar appearing junctional regions are encountered in plant cells which lack a prominent system of SER (Fig. 3.4A and Fig. 3.11). The tendency for rough ER cisternae of protein-secreting cells to occur in stacked arrays is illustrated at the left of the micrograph. Glycogen is marked by an asterisk.

(B) Rough ER in face view showing numerous polyribosomal profiles of the membrane surface. A region of RER—SER continuity is shown at the arrow. The lipoprotein particle of the ER lumen (double arrow) identifies the smooth-surfaced portion (lacking ribosomes) as SER.

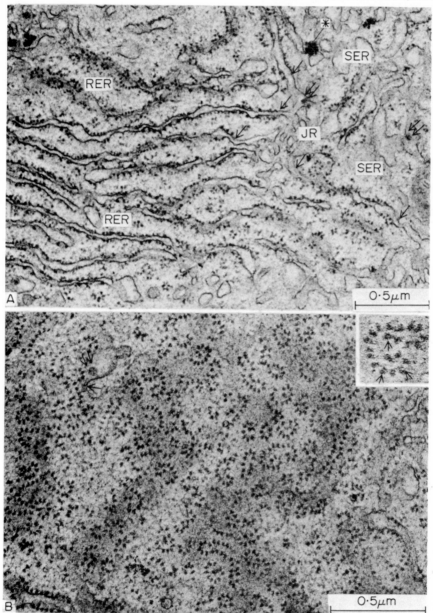

Fig. 3.7 (continued)

A polyribosomal configuration consisting of 21 ribosomes is illustrated in the inset (2 x photographic enlargement). Membrane-bound polysomes composed of up to 25 ribosomes are common in rat liver. The ribosomes marked by arrows show a lightly-staining central core which may correspond to the channels through the ribosome and membrane which permit entry of secretory proteins from their site of synthesis within the interior of the ribosome to their site of accumulation within the ER lumen (Sabatini *et al.*, 1971).

The endomembrane concept

Even though free and membrane-bound ribosomes are identical in most respects (even interchangeable, Tanaka, Takagi, and Ogata, 1970), they may engage in very different types of protein synthesis in the living cell (Bulova and Burka, 1970; Galling 1970; Ganoza and Williams, 1969; Ragnotti, Lawford, and Campbell, 1969; Redman, 1968, 1969; Sabatini *et al.*, 1971 and ref. cit., Takagi, Tanaka, and Ogata, 1970, as examples). While cytoplasmic ribosomes engage in the synthesis of proteins to be used by the cell in carrying out its metabolic functions, ribosomes attached to rough ER are thought to be engaged primarily in the synthesis of secretory proteins (Sabatini *et al.*, 1971).

Yet, a possibility often overlooked is that rough ER might synthesize other types of proteins in addition to secretory proteins. Where do membrane proteins

TABLE 3.3

Summary of Enzymes, Multienzyme Systems and Metabolic Sequences of Endoplasmic Reticulum based on studies with Isolated Microsomal Fractions from Mammalian Cells

Steroid biosynthesis
    Biosynthesis of cholesterol from squaline
        Squaline cyclohydrase
    Production of bile acids
Phospholipid biosynthesis
    Lecithin biosynthesis
        CDP-choline pathway
        Stepwise methylation of phosphatidyl ethanolamine
    Cephalin biosynthesis
        Biosynthesis of phosphatidyl ethanolamine
        Biosynthesis of phosphatidyl serine
    Phosphatidyl inositol biosynthesis
    Sphingolipid biosynthesis (?)
Mucopolysaccharide biosynthesis
    Hexosamine transferase
Detoxification of xenobiotics; steroid hydroxylation
    NADPH-O_2-linked mixed function oxidases
    NADPH-linked acyl- and alkoxylargyl hydroxylases
Other oxidases
    Cytochrome b_5-linked cytochrome c reductase
        NADH $\xrightarrow{\text{cyt } b_5}$ O_2
        NADPH $\xrightarrow{\text{cyt } b_5}$ O_2
    NADPH-linked lipid peroxidase
    Dicoumarol-sensitive nonspecific NADH/NADPH dye reductase (DT diaphorase)
Conjugating enzymes
    Glucuronide biosynthesis
Ascorbic acid synthesis
Glycogen metabolism
    Glucose-6-phosphatase
    UDP-glucose-glycogen transferase 'conversion factor'
Triglyceride (neutral fat) biosynthesis
Terpene biosynthesis
Sulphur metabolism
Transport
    Calcium-stimulated ATPase

come from, for example? The hypothesis to be developed later in the chapter is that ER participates in the formation of Golgi apparatus and other transition elements as well as plasma membrane, vacuole membranes and the outer membranes of the organelles. Rough ER may also give rise to rough ER, i.e., be responsible for its own formation (Kuriyama *et al.*, 1969; Franke *et al.*, 1971; Omura, Siekevitz and Palade, 1967; Siekevitz *et al.*, 1967). A simple but untested hypothesis which derives logically from biochemical and morphological evidence (see Fig. 3.2) is that the rough ER is the site of synthesis of its own membrane proteins as well as those for much of the rest of the endomembrane system and an important site for the synthesis of membrane lipids. We do not exclude the possibility that, among endomembranes, synthesis of proteins and certain phosphoglycerides such as lecithin are restricted to rough ER (See Bracker and Grove, 1971 for a brief review concerning outer mitochondrial membranes).

A puzzling question relevant to membrane assembly in general, is how are newly synthesized membrane constituents, especially the membrane proteins, inserted into the membranes? Presumably, ribosomes and messenger RNA templates are involved, but there is no proof of this. An interesting possibility is raised by reports of membrane-associated RNA (Chauveau *et al.*, 1962; Moulé, Rouiller, and Chauveau, 1960; see Moulé, 1968 for a review). Although most of the RNA of rough microsomes (isolated fragments of rough ER) is recovered with the

TABLE 3.4

Summary of Endoplasmic Reticulum Activities of Golgi Apparatus Fractions from Rat Liver (Morré *et al.*, 1971)

Constituent	Specific activity or amount per milligramme of protein	Ratio: Golgi apparatus to ER
Glucose-6-phosphatase	1.3 μmole/hr	0.13
NADH-cyt. c reductase	6.3 μmole/hr	0.14
NADPH-cyt. c reductase	0.6 μmole/hr	0.12
Cytochrome b_5	0.18 μmole/hr	0.30
$P_{450} + P_{420}$	0.21 mμmole	0.33
Total iron	36.4 mμatom	1.03
UDP-glucuronyl transferase (bilirubin as acceptor)	1.0 mμmole/hr	0.16
Arylsulphatase C	18.0 mμmole/hr	0.23
L-gulonolactone oxidase	87.0 mμmole/hr	0.22

ribosomes, a portion (about 5%) remains attached to the membrane and is found in both rough and smooth microsomes. The base composition of this RNA differs from that of ribosomal RNA or soluble RNA of the cytoplasm. It is heterogeneous and shows a high degree of labelling that suggests rapid turnover (Moulé, 1968). Among the functional possibilities for this RNA are: as a template or messenger RNA for the synthesis of membrane proteins, or a role in ribosome attachment. The next few years should bring some clarification of the function of this interesting RNA fraction of the ER membrane.

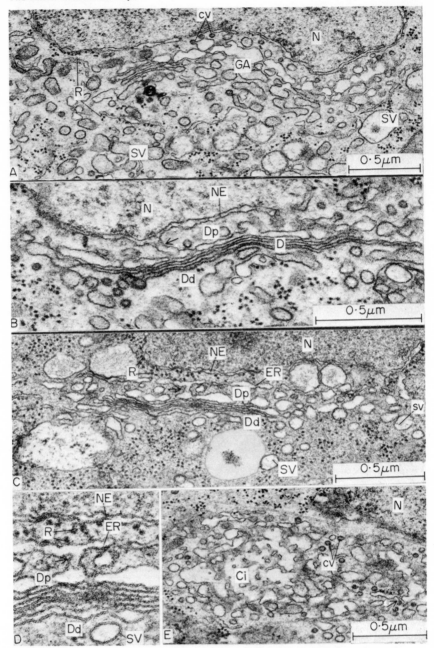

Fig. 3.8 Continuity among endomembrane components in fungal hyphae. Glutaraldehyde-osmium tetroxide fixation. Scale marks = 0.5 μm. From Bracker *et al.* (1971).

(A) Portion of a Golgi apparatus (GA) adjacent to a nucleus (N) of *Pythium aphanidermatum*. Numerous primary vesicles, some of which have alveolate 'coats' (cv), are found in the space between the nuclear envelope and the paranuclear Golgi apparatus cisternae.

Transition elements

Transfer of membranes and secretory products from rough ER (or outer membrane of the nuclear envelope) to the cell surface, to the vacuolar apparatus or to any other part of the endomembrane system is accomplished, to the extent that it occurs, within a system of membranous elements which lack ribosomes. The elements have membranes with characteristics intermediate between those of the nuclear envelope and its extensions at one extreme and those of the plasma membrane or tonoplast (vacuole membrane) at the other extreme (Tables 3.1, 3.2, 3.4, 3.5 and 3.7; Figs. 3.8, 3.9, 3.13, 3.15 and 3.17; Morré, Mollenhauer, and Bracker, 1971; Morré and VanDerWoude, in press). Included in this category are smooth ER, Golgi apparatus and various other types of tubules, vesicles and smooth cisternae which are neither rough ER nor plasma membrane or vacuole but somewhere in between. Because these elements are transitional — they function in the passage of membrane constituents or secretory products from one cell component to another (see also Fig. 3.3) — we refer to them all as transition elements.

The intermediate nature of transition elements is most readily determined from the relative thickness and staining characteristics of their membranes under favourable conditions of tissue preparation for electron microscopy (Table 3.1,

---

These vesicles appear to bleb off the nuclear envelope and are thought to give rise, through fusion, to new Golgi apparatus cisternae. R = ribosome, SV = secretory vesicle

(B) Paranuclear dictyosome (D) of the Golgi apparatus of *Pythium aphanidermatum*. The outer membrane of the nuclear envelope (NE) is continuous with the distended first cisterna of the dictyosome through a tubular connection (arrow) formed perhaps through the fusion of one or more primary vesicles. An intermittent bridge is the same phenomenon as the blebbing configuration and may represent a transient condition in which a bleb is extended in time and space so that it coalesces with the dictyosome cisterna before it is detached from the nuclear envelope. This type of connection illustrates how intermittent tubular or vesicular bridges insure transfer of materials in the absence of permanent structural continuity. N = nucleus, Dp = cisterna at the proximal pole of the dictyosome (the pole adjacent to nuclear envelope or endoplasmic reticulum). Dd = cisterna at the distal pole of the dictyosome.

(C) Transverse section of a dictyosome in a hypha of *Pythium middletonii*. The cisterna at the proximal pole (Dp) is adjacent to endoplasmic reticulum (ER). The arrow shows a tubular connection between the ER cisterna and the first dictyosome cisterna similar to that shown in B for the nuclear envelope–Golgi apparatus association. Ribosomes (R) are attached to the surface of the nuclear envelope (NE) and ER as well as found in the space between these membranes. Ribosomes are found in the cytoplasm near the distal pole (Dd) of the dictyosome, but the dictyosome is surrounded by a cytoplasmic 'zone of exclusion' in which ribosomes are scarce or absent, a characteristic of Golgi apparatus (Fig. 3.9D, E; Morré, Mollenhauer, and Bracker, 1971). N = nucleus, SV = secretory vesicle.

(D) Part of C enlarged approximately 2x to show the intercisternal connection (arrow) between ER and the first dictyosome cisterna (Dp). Note also the progressive increase in membrane thickness (shown by a gradual widening of the light space of the dark–light– dark pattern) from ER across the stacked cisternae to the distal pole (Dp) of the dictyosome and the membrane of the secretory vesicle (SV), NE = nuclear envelope, R = ribosome).

(E) Dictyosome cisterna (Ci) of *Pythium aphanidermatum* in face view. Presumably, the typical central, plate-like region of the cisterna is formed through progressive addition and coalescence of the nuclear envelope- or ER-derived primary vesicles. The arrow shows a place of apposition between ER and the dictyosome cisterna at the periphery of the dictyosome. N = nucleus, cv = coated vesicle.

Figs. 3.8, 3.9, 3.13 and 3.15 and summarized in Fig. 3.2). (Not all fixations for electron microscopy show the differences as clearly as others (Compare Figs. 3.4 and 3.5; Figs. 3.11 and 3.13).) Functional considerations (e.g., secretory proteins move from rough ER to smooth ER to Golgi apparatus to secretory vesicles to the cell surface, Fig. 3.3) help verify that interactions take place. Yet, proof of membrane differentiation and of the intermediate nature of transition elements must come from detailed biochemical and correlative studies such as those being carried out with Golgi apparatus from rat liver (p. 113ff). As discussed in the sections which follow, transition elements are thought to provide structural and functional continuity between cell components having dissimilar membranes but at the same time carry out specific functions frequently related to membrane differentiation or to the transformation of secretory products. Transition elements need not have permanent structural continuity between conjoining cell components to function in these capacities (see Appendix). Intermittent tubular or vesicular bridges will result in transfer of materials in the absence of permanent structural continuity. Specific examples of transition elements include the following:

Smooth-surfaced endoplasmic reticulum (smooth ER). Smooth, smooth-surfaced or agranular ER (ER lacking ribosomes or smooth ER) is a category of transition elements in which no sharp morphological discontinuity exists between it and the rough ER (Figs. 3.3, 3.4 and 3.7). Smooth ER provides functional and structural continuity between the rough ER and other components of the endomembrane system, e.g., Golgi apparatus (Fig. 3.3; Morré, 1971).

Smooth ER owes its identity to structural continuity with rough ER and the absence of ribosomes (Fig. 3.7A). Yet, smooth ER appears to be more than just rough ER that has lost its ribosomes (Fig. 3.3; Table 3.2). The two forms of ER (rough and smooth) are often similar (Tables 3.2 and 3.3) but may also differ in important respects (Table 3.2; Fig. 3.3; Jones and Fawcett, 1966).

The basic structural element of smooth ER, at least in animal cells, is frequently a 50 to 100 nm diameter tubule (Fig. 3.7A; Jones and Fawcett, 1966; Porter, 1961). Cisternal regions are found dispersed among and continuous with the tubules (the structural pattern is readily visualized from electron micrographs of smooth ER fractions negatively stained with phosphotungstic acid, e.g., Fig. 1 of Morré, Merritt, and Lembi, 1971) and the system is often represented by a three-dimensional lattice of ribosome-free tubules and small cisternae (Fig. 3.1). Yet, there is considerable variation in smooth ER morphology among species and cell types and within a species (Porter, 1961).

In plants and fungi, lamellar regions of ER-lacking ribosomes are frequently interspersed with lamellar regions having ribosomes (Figs. 3.4A, 3.5, 3.6, 3.8 and 3.15; Bracker and Grove, 1971). Regions of rough ER lamellae immediately adjacent to Golgi apparatus, plasma membrane, microbodies, mitochondria, chloroplasts and other cell components characteristically lack ribosomes (e.g., Figs. 3.5 and 3.8). Such regions might be aptly described as ER membranes lacking ribosomes to differentiate them from the tubular type of 'smooth ER' found in abundance in rat liver (Fig. 3.7A) and certain plant glands (Schnepf, this volume) as examples.

Although there are few definitive studies pointing to more than one functional class of smooth ER, it seems inadvisable on the basis of structural differences to regard all types of ER-lacking ribosomes as equivalents (Porter, 1961). The transitional role for smooth regions of ER lamellae adjacent to or connecting with other cell components may differ somewhat for each type of association.

In rat liver, the lipid and protein compositions of rough and smooth ER are similar, but small differences in lipid composition are found (Table 3.2). Increased amounts of sterols and sphingomyelin are encountered in smooth ER fractions relative to fractions derived from rough ER with a corresponding decline in phosphoglycerides such as lecithin. Similar changes are encountered in comparing rough ER and Golgi apparatus (or Golgi apparatus and plasma membrane) from rat liver (Table 3.2) or mammary gland (Keenan, Morré, and Huang, in press). These findings from Golgi apparatus have been interpreted as evidence for the differentiation of membranes from ER-like to plasma membrane-like during membrane flow (Keenan and Morré, 1970; Morré, Mollenhauer, and Bracker, 1971; Morré et al., 1971). A lipid composition of smooth ER fractions intermediate between that of rough ER and that of Golgi apparatus is to be expected in rat liver, since smooth ER serves as one of the transition elements between rough ER and Golgi apparatus in this tissue (Claude, 1970; Morré, Mahley, Bennett, and LeQuire, 1971; Fig. 3.3). Most of the enzymatic activities of liver microsomes (Table 3.3) are exhibited by both rough and smooth ER fragments. Specific activities may vary between the two types of ER in liver, but no enzyme activity has been found which characterizes one type of ER to the exclusion of the other.

The biochemical similarities of rough and smooth ER are consistent with current views that smooth ER is synthesized first in the form of rough ER or at least by the rough ER (e.g., Higgins and Barnett, 1971). During periods of rapid ER development in embryonic rat livers, the in vivo incorporation of leucine-C^{14} into proteins and lipids is higher in the rough ER than in the smooth ER for the first hour after injection of isotope (Dallner, Siekevitz, and Palade, 1966). Later (at about 10 hours), the situation is reversed. Considerable evidence for derivation of smooth ER from rough ER has come from experiments with drug-treated animals. This literature has been reviewed extensively (e.g., Ernster and Orrenius, 1965; Gillete et al., 1969; Jones and Fawcett, 1966) and will not be repeated here.

The fate of ribosomes after detachment from rough ER is unknown, and it is uncertain that ribosomes are actually detached. One possibility is that they do detach and either enter the cytoplasmic pool of ribosomes or re-attach to newly formed membranes. Alternatively, smooth ER may arise from the outgrowth of ribosome-free regions continuous with rough ER (junctional regions; Fig. 3.7A) so that ribosomes are never attached to smooth ER membranes.

Smooth ER shares many functional properties with rough ER including general functions in transport, storage and secretion (Porter, 1961). Yet smooth ER lacks the ability to synthesize proteins and is specialized for other activities including biogenesis of isoprenoid compounds (sterols and terpenes), ion transport and secretion and subcellular detoxification (Porter, 1961; Schnepf, this volume; Siekevitz, 1965; Wollenweber, Egger, and Schnepf, 1971).

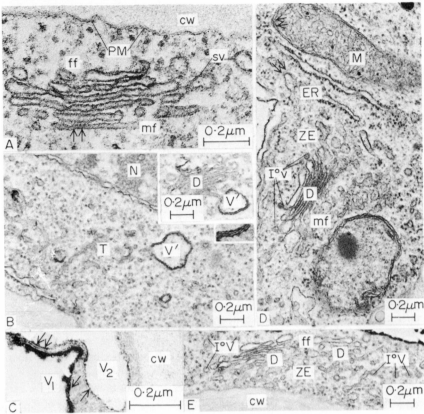

Fig. 3.9   Endomembrane differentiation within the Golgi apparatus of onion stem.
Glutaraldehyde-osmium tetroxide fixation. Scale marks = 0.2 μm.

(A) A dictyosome from a thin section post-stained as described for Fig. 3.5 to show the
change in staining from dark and ER-like at the forming face (ff) to progressively light and
more like plasma membrane (PM) toward the maturing face (mf) and secretory vesicles (sv).
Membranes at the forming face are 4 to 5 nm thick while those of secretory vesicles are
8 to 9.5 nm thick and approach the dimensions of plasma membrane. In contrast, the central
plate-like portions of cisternae have membranes which seldom exceed 7 nm in thickness.
These findings, summarized in Table 3.1 and those of Fig. 3.13 for maize roots, show that
in these dictyosomes, which may be typical of those of higher plants, membrane differentiation from an ER-like morphology to a morphology indistinguishable from plasma membrane
is most evident in (but not necessarily restricted to) membranes of the secretory vesicles
(the structures that ultimately fuse with and contribute new membrane to the plasma
membrane). These findings contrast in detail only, those from the fungus *Pythium ultimum*
(Grove, Bracker, and Morré, 1968) where both the cisternal membranes and membranes of
the secretory vesicles differentiate from ER-like to plasma membrane-like and those from
the alga studied by Brown (1969) where the entire cisterna may give rise to one large
secretory vesicle. The results with onion and maize root (Fig. 3.13) confirm a role of the
Golgi apparatus in membrane differentiation and as a system of plasma membrane-
generating transition elements. Yet, at the same time, the results raise the interesting
possibility that Golgi apparatus may participate simultaneously in more than one pathway
of membrane differentiation. The membranes of central plate-like portions of cisternae may
become closely appressed in these cells and, in the appressed regions, attract heavy and
irregular deposits of stain (double arrows). Vesicular profiles with similar appearing membranes (V', Fig. 3.9B) are encountered in the vicinity of other dictyosomes in these same

Vacuole-forming transition elements. The bounding membranes of vacuoles in both plant and animal cells seem to come from parts of pre-existing membrane systems of the cytoplasm (for a discussion of some of the possibilities, see the article by Buvat, 1971). With few exceptions, these pre-existing membranes appear to be derived ultimately from rough ER.

Direct continuity of plant vacuole membranes with rough ER as illustrated in Fig. 3.10A and 3.10B are exceptional. More usual is the intermediate involvement of some other (ER-derived) structure, an example of which is the prevacuolar body of the maize root tip (Whaley, Kephart, and Mollenhauer, 1964; Fig. 3.10C).

Components of the vacuolar apparatus of animal cells are the lysosomes (deDuve and Wattiaux, 1965; Fig. 3.3). They contain enzymes for intracellular digestion and are formed as part of a system of transition elements including smooth ER, Golgi apparatus, or both (Brandes, 1965; Essner and Novikoff, 1962; Moe, Rostgaard, and Behnke, 1965; Novikoff and Shin, 1964; Novikoff *et al.* 1971). For a discussion of the possible functional equivalence of animal lysosomes and plant vacuoles see the chapter by Matile in this volume (chapter 5).

Vesicle-forming transition elements. Vesicle-forming transition elements include Golgi apparatus cisternae (Figs. 3.3A,C, 3.8, 3.9A, 3.11, 3.12, and 3.13; Morré, Mollenhauer, and Bracker, 1971); smooth ER (Fig. 3.3B; Morré, Mahley, Bennett, and LeQuire, 1971; Morré and VanDerWoude, in press) and Golgi apparatus equivalents of fungi (Bracker, 1968; Girbardt, 1969; Grove and Bracker, 1970; Morré, Mollenhauer, and Bracker, 1971; Fig. 3.14) and other cells (Franke, Eckert, and Krien, 1971). Secretory vesicles are formed directly from the cisternae (Figs. 3.3, 3.11 inset, Fig. 3.14B) which, in effect, serve as transition elements. The transformations required to produce a secretory vesicle, although perhaps initiated within the continuous span of the vesicle-forming cisterna or even within the tubule to which they are attached, appear to be completed within the vesicle

---

cells and a role in the formation or differentiation of complex membranes is indicated for the plant Golgi apparatus. cw = cell wall.

(B) Complex membranes (V') associated with dictyosomes (D) and dictyosome tubules (T). The complex membranes show a five-layered (dark—light—dark—light—dark) pattern. The insets show one of these regions enlarged 3x. N = nucleus.

(C) A five-layered membrane (double arrows) at the region of apposition of two vacuoles (V₁ and V₂) which apparently resulted from the lateral association or fusion of the two three-layered (dark—light—dark) membranes (single arrows) of the individual vacuoles. This figure together with Fig. 3.9A (double arrows) suggests that the complex membranes of Fig. 3.9B (V') might result from lateral fusion of 'cisternal membranes'. How laterally fused cisternal membranes are formed into vesicular profiles is unresolved and the problem of differentiation of complex membranes in plant Golgi apparatus is a subject of continuing investigation. cw = cell wall.

(D) A dictyosome within a zone of exclusion (ZE). Primary vesicles (I⁰V) and coated vesicles of the Golgi apparatus zone are restricted to zones of exclusion. Alveolate or spinous coats resembling those of coated vesicles are associated, on occasion, with regions of the outer mitochondrial (M) membrane and ER (double arrows). mf = mature face of the dictyosome.

(E) Dictyosomes (D) within a zone of exclusion (ZE) showing membrane differentiation across the stacked cisternae. Primary vesicles (I⁰V) are prevalent at the forming face (ff) of onion stem dictyosomes (see also Fig. 3.9A and Fig. 3.9D) but are encountered throughout the zone of exclusion at the dictyosome periphery (see also Fig. 3.9D). cw = cell wall.

itself and sometimes not until after the vesicle has left its site of origin and has migrated to near the plasma membrane (Wood, 1967). Vesicle-forming transition elements frequently occur as part of a system of transition elements (i.e., where structural continuity with rough ER may be direct or indirect via vesicular bridges) and is discussed in more detail in relation to the Golgi apparatus in the sections which follow.

   Plasma membrane-forming transition elements. Proteins of the plasma membrane arise by synthesis somewhere within the cell's interior and subsequent

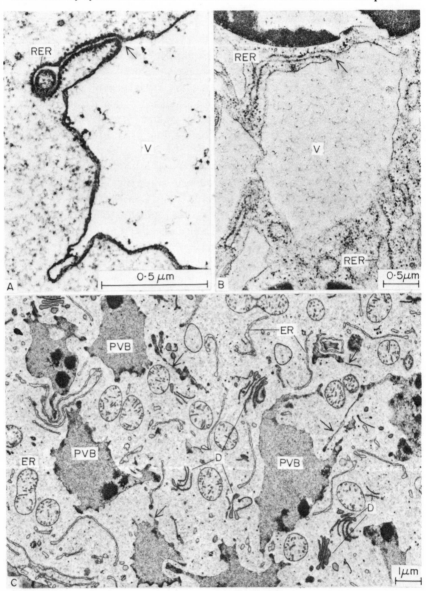

transfer to the plasma membrane (Amsterdam *et al.*, 1971; Baranick and Lieberman, 1971; Franke *et al.*, 1971; Morré and VanDerWoude, in press; Morré *et al.*, 1971; Ray, Lieberman, and Lansing, 1968). Kinetic evidence suggests that at least two pathways are operative (Franke *et al.*, 1971; Morré and VanDerWoude, in press); (a) Secretory vesicles derived from the Golgi apparatus provide a structural basis for one of the pathways (Fig. 3.3A). (b) Vesicles (Fig. 3.3B) or direct transfer (Fig. 3.15) from ER provide a second pathway (Morré and VanDerWoude, in press). Although membrane structures, conforming to ER structurally continuous with plasma membrane, have been described from cells of imaginal discs in insects (Agrell, 1966) and from mammary gland of the rat and onion stem (Morré and VanDerWoude, in press), their functional significance in terms of material transport (e.g., Fig. 3.15) cannot be ascertained without additional correlative information. In spite of the uncertainty surrounding direct ER-plasma membrane connections, ER-derived vesicles do appear to provide an alternative to Golgi apparatus-derived vesicles as vehicles for the transport of membrane constituents and secretory products to the cell surface in rat liver (Fig. 3.3; Morré and VanDerWoude, in press).

Vesicles. Vesicles are a major vehicle for providing continuity among endomembrane components where permanent structural continuity is not maintained but where compartmental integrity is required. Many types of vesicles characterize the cytoplasm of eukaryotic cells (e.g., Figs. 3.3, 3.8, 3.9 and 3.11) but only two types will be discussed — primary vesicles and secretory vesicles.

(a) *Primary vesicles.* The 30 to 60 nm diameter primary vesicles which bleb off the nuclear envelope or rough ER (Fig. 3.8A–D) and coalesce, for example to form new Golgi apparatus cisternae, are an important component of systems of transition elements (see Morré, Mollenhauer, and Bracker, 1971). These vesicles are frequently, but not necessarily, coated (distinguishable by the presence of a highly organized layer of material on the cytoplasmic surface in the form of a nap-like or hairy coating or alveolate reticulum; Figs. 3.8A and 3.9D), a property correlated with the participation of such vesicles in functional specialization, transport, and fusion (Bracker *et al.*, 1971). Primary vesicles are not restricted to fusion with forming Golgi apparatus cisternae. Associations with primary vesicles

---

Fig. 3.10   The vacuolar apparatus of plant cells.
(A) and (B) Rough endoplasmic reticulum (RER) continuous (arrows) with the membranes (tonoplast) which surround plant vacuoles (V). Glutaraldehyde-osmium tetroxide fixation. Scale marks = 0.5 μm.
(A) Immature stem cell of onion from a study by D. J. Morré and C. A. Lembi.
(B) Maturing pea cotyledon from a study by H. H. Mollenhauer and C. Totten.
(C) Portion of a maize root tip cell showing prevacuolar bodies (PVB). In early developmental stages, prevacuolar bodies are characterized by electron-dense contents. As the cells mature, the electron-dense contents are lost so that the prevacuolar bodies appear as small vacuoles. The large central vacuole formed as the cells reach maturity seems to be derived from the fusion of many smaller vacuoles as happens in a variety of plant cells (Buvat, 1971). At the stage of prevacuolar body development shown here, many of the extensions of the prevacuolar body membranes (arrows) are difficult to distinguish from endoplasmic reticulum (ER) or dictyosome (D) cisternae. Potassium permanganate fixation. (From a study by D. D. Jones and D. J. Morré.) Scale mark = 1 μm.

109

occur at the peripheries of formed Golgi apparatus cisternae (Figs. 3.8E and 3.9E) and provide one basis for suggesting that membrane components may be added not only to immature dictyosome cisternae but also to successive cisternae across the stack (Morré, Mollenhauer, and Bracker, 1971). Primary vesicles may also participate in the transfer of materials from nuclear envelope or rough ER to other endomembrane components (e.g., annulate lamellae; Kessel, 1963) or from smooth ER to other endomembrane components (e.g., plasma membrane; Morré and VanDerWoude, in press) although this latter involvement has not been studied extensively.

(b) *Secretory vesicles.* Secretory vesicles are formed at the ends of tubules which in turn are most often attached to flattened cisternae (Figs. 3.11 and 3.14b). Secretory vesicles associated with Golgi apparatus cisternae are a familiar example (see Fig. 4 of Morré, Mollenhauer, and Bracker, 1971; Mollenhauer and Morré 1966b; Morré, Keenan, and Mollenhauer, 1971; Ovtracht, Morré, and Merlin, 1969; Figs. 3.9A, 3.11 and 3.13). Once formed, the secretory vesicles are detached from the tubules and migrate, usually to the plasma membrane (but sometimes perhaps to the vacuoles, or other structures) or are retained for a time in the cytoplasm. The membranes of mature secretory vesicles, those which fuse with the plasma membrane, are morphologically similar to the plasma membrane (Table 1; Figs. 3.9A and 3.13; Morré, Mollenhauer, and Bracker, 1971 and ref. cit.). As the membranes of the vesicles fuse with the plasma membrane, the contents of the vesicles are discharged into the extracellular space, and the vesicle membranes are incorporated into the plasma membrane (Fig. 3.11). In this manner, secretory

---

Fig. 3.11 Electron micrograph of a representative outer cap cell of the maize root tip showing dictyosomes with attached secretion vesicles (D), free secretory vesicles (sv), endoplasmic reticulum (ER) and other cytoplasmic components. Dictyosomes appear to be organized into small closely associated groups, the groups separated and interconnected by ER cisternae. That dictyosomes must be functionally interconnected is deduced from the observation that all dictyosomes of a cell are at the same developmental or functional stage. Specialized junctions between ER cisternae are shown at the arrows and between ER and dictyosomes at the double arrows. Fusion of secretory vesicles with the plasma membrane (PM) has resulted in accumulations of the secretory product (Pr) outside the plasma membrane adjacent to an exterior cell wall (CW). M = mitochondrion, mb = microbody, EP = etioplast. Inset shows a dictyosome in face view (left) and a dictyosome in cross section (right) for comparison. Shown are the central plate-like region (P) of the cisternae, the system of peripheral tubules (T) and secretory vesicles (sv). The secretory vesicles are attached to the plate-like region of the cisterna via the system of peripheral tubules (arrow; see also Fig. 3.3). Root tip fixed in potassium permanganate. Scale mark = 1 μm.

ER-Golgi apparatus associations in the maize root tip are manifest in the form of distinctive junctional regions at the Golgi apparatus periphery. Similar junctions occur between ER and plasmodesmata and between sheets of ER (Mollenhauer and Morré, in preparation). At the junctions, the ER appears to branch and rejoin forming a transitional region between the conjoining cell components. The most conspicuous feature of the junctional regions in glutaraldehyde-osmium tetroxide fixed cells (see Fig. 3.4A) is the characteristic anastomotic transition from a sheet form of ER with ribosomes to a smooth tubular form of transition element lacking ribosomes. This organization of ER in maize roots is similar to that in liver (Fig. 3.7A) which, like pancreas (Jamieson and Palade, 1967), is characterized by 'transition elements' consisting of part rough and part smooth ER. In both plant and animal cells, smooth tubules seem to be associated with the periphery of the Golgi apparatus cisternae.

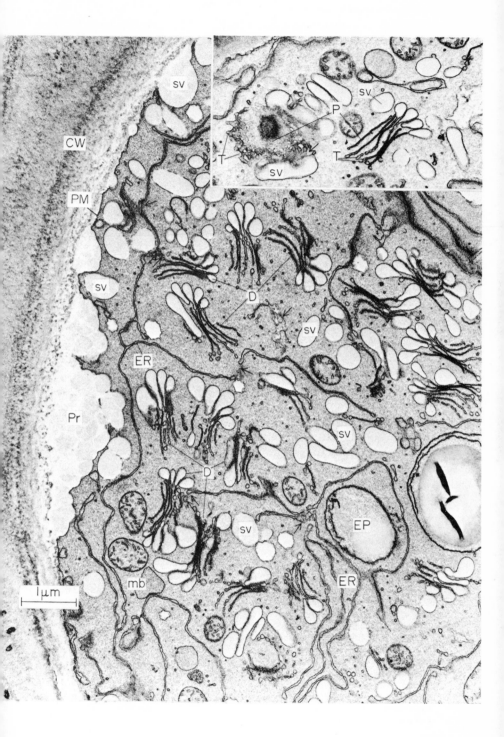

vesicles provide functional continuity between the plasma membrane and the rest of the endomembrane system.

Golgi apparatus: a system of transition elements. The basic functional unit of the Golgi apparatus is the cisterna (Morré, Mollenhauer, and Bracker, 1971). In most Golgi apparatus, the cisternae are organized into stacks called dictyosomes (Figs. 3.3A, 3.4A, 3.6B, 3.8, 3.9 and 3.11 to 3.13; Mollenhauer, Morré, and Bergman, 1967). In the Golgi apparatus of plants, invertebrates and some animal germ cells, the dictyosomes are concentrated in certain regions of the cytoplasm (Fig. 3.11) but appear as discrete cell components. In most somatic cells of higher animals, the dictyosomes are arranged end to end (as distinguished from face to face, back to face or back to back), close together, in a localized, often curved, array (Flickinger, 1969). Details of Golgi apparatus organization are given by Morré, Mollenhauer, and Bracker (1971).

Golgi apparatus cisternae, an example of a vesicle-forming transition element, lack ribosomes, are usually flattened and, most often, are continuous with a peripheral system of tubules (e.g., Figs. 3.4A, 3.6B and 3.11). In some cisternae, the plate-like regions predominate (Figs. 3.4A and 3.6B). In others, tubular elements predominate (Figs. 3.8E and 3.11 inset; see also 3.14B). Both types of cisternae may exist within a single stack or dictyosome (Cunningham, Mcrré, and Mollenhauer, 1966; Flickinger, 1969; Mollenhauer and Morré, 1966b; Mollenhauer, Morré, and Bergman, 1967). Golgi apparatus cisternae appear to arise by coalesence of ER- or nuclear envelope-derived vesicles.[1] They function in the formation of secretory vesicles, lysosomes and other types of specialized vesicles and cisternae some or all of which contribute constituents to the plasma membrane (by fusing with the plasma membrane, cf. Brown, 1969; Schnepf and Koch, 1966a, b; additional examples are provided by Beams and Kessel, 1968; Morré, Mollenhauer, and Bracker, 1971 and Schnepf, 1969a; see also Schnepf, this volume).

As a result of their unique structural and functional organization, dictyosomes frequently appear polarized with a forming face where new cisternae arise by fusion of primary vesicles and an opposite, maturing or secreting, face where cisternae are lost through production of secretory vesicles (Fig. 3.8; the situation for rat

---

[1] The evidence for this suggestion is primarily from morphological studies which show numerous ER- or nuclear envelope-derived primary vesicles aligned in the space between ER and dictyosomes where new Golgi apparatus cisternae are presumed to be formed (Fig. 3.8; see pp. 93 to 99 of Morré, Mollenhauer, and Bracker, 1971 for a general discussion and specific references). During assumed dictyosome multiplication in the fungus *Pythium aphanidermatum,* dictyosome doubling is accompanied by an extension of the forming face region where the dictyosomes interface with a region of ER-lacking ribosomes. Cisternae having twice normal diameter are observed in stages corresponding to their derivation from the coalesence of small evaginations or blebs derived from the cisternal walls of the 'extended' forming face regions. Complete dictyosomes with all cisternae of twice-normal diameter are found as are 'hybrid' dictyosomes having two stacks of cisternae with normal dimensions located side by side directly underneath forming face regions of normal dimensions and on top of a single stack of cisternae with twice-normal dimensions. The 'hybrid' structures are assumed to derive from 'division' of an 'extended' forming face region with the result that two dictyosomes are formed where only one existed previously. These and other observations have led to a proposed mechanism of cisternal formation and dictyosome multiplication based on the coalesence of ER- or nuclear envelope-derived vesicles (Footnote 7 and Fig. 14 of Morré, Mollenhauer, and Bracker, 1971).

liver (Fig. 3.3), maize roots (Figs. 3.11 and 3.13), and onion stem (Fig. 3.9) may be variations of this functional pattern in which potential regions of ER-Golgi apparatus interactions are most evident at the cisternal peripheries). Membranes of the forming face resemble those of ER on the basis of thickness, staining characteristics, and cytochemical properties (Grove, Bracker, and Morré, 1968; Figs. 3.9 and 3.13). The membranes of the 'mature' Golgi apparatus cisternae and especially those of the 'mature' secretory vesicles attached to these cisternae are plasma membrane-like (Grove, Bracker, and Morré, 1968; Figs. 3.9 and 3.13). These characteristics are essential to the functional role of secretory vesicles as vehicles for the transport of membrane and other materials destined for export to the cell surface. Thus, membranes of secretory vesicles, derived from ER or nuclear membrane, appear to be transformed during passage through the Golgi apparatus (Fig. 3.13) and destined to become plasma membrane. These dynamic concepts of endomembrane functioning involving the Golgi apparatus are diagrammed in Fig. 3.18.

The endoplasmic reticulum—Golgi apparatus—secretory vesicle—plasma membrane export route: a model for endomembrane differentiation

Biochemical studies comparing ER, Golgi apparatus and plasma membrane from rat liver provide one of the important sources of information upon which the endomembrane concept is based (Tables 3.2, 3.4 and 3.5). These studies suggest that the intermediate character of transition elements will be exemplified by the enzymatic activities, lipid composition, and progressive modification of the proteins and lipids (e.g., glycosylation) of the membranes. Additionally, some enzyme activities will be lost while others will be gained either by addition or deletion of enzyme molecules or by activation or inhibition of enzyme molecules already associated with the membranes.

Not only do the biochemical studies reflect the transitional nature of Golgi apparatus shown to us first from morphological studies (Grove, Bracker, and Morré, 1968) but, equally important, they indicate the nature of some of the biochemical changes required to effect the transformation of ER membrane into plasma membrane. We visualize the process as a gradual one with some of the changes occurring between the ER and Golgi apparatus, some occurring at the Golgi apparatus, and others occurring between the Golgi apparatus and plasma membrane, or at the plasma membrane itself. Because of the importance of membrane differentiation to the endomembrane concept, these findings from rat liver will be discussed briefly in the context of a biochemical model for endomembrane differentiation.

If ER is the source of membrane that will ultimately become plasma membrane through the production of Golgi apparatus vesicles, a conversion of membranes from ER-like to plasma membrane-like must occur at the Golgi apparatus. From morphological studies, we expect that cisternae of the forming face are more like ER and contain less of the constituents that characterize the plasma membrane

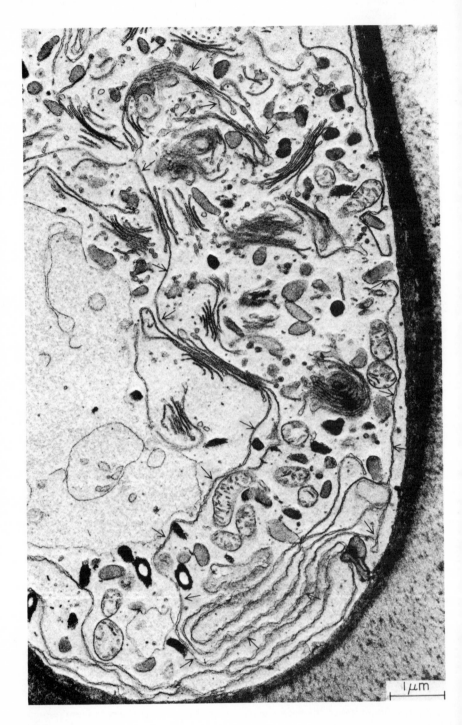

1μm

whereas cisternae at the opposite face are more like plasma membrane (specific morphological evidence for this transformation in Golgi apparatus from rat liver is given by Morré, Keenan, and Mollenhauer, 1971). This means that the composition of the membranes should reflect the gradient in morphology from ER-like to plasma membrane-like across the stacked cisternae. To test this idea fully, it would be helpful to be able to fractionate Golgi apparatus into the different classes of cisternae so that cisternae from the maturing and forming faces could be compared directly. But since the analyses summarized in Tables 3.2, 3.4, 3.5 and 3.7 are based on intact Golgi apparatus, the results are an average of all cisternae in the stack.

TABLE 3.5

Summary of Some Plasma Membrane Activities Comparing Endoplasmic Reticulum (ER)- Golgi Apparatus (GA)- and Plasma Membrane (PM)-Rich Cell Fractions from Rat Liver. From Keenan, Morré, and Huang (in press); Morré, Keenan, and Mollenhauer (1971); Morré *et al.,* (1971) and ref. cit.

Constituent	Specific activity or amount per milligramme of protein			
	ER	GA	PM	units
5'-nucleotidase (AMP)	1.9	3.2	54	μmole/hr
UDP-galactose hydrolase	7	12	47	mμmole/hr
Mg^+ +-ATPase	2	2.5	45	μmole/hr
Mg^+ +-UTPase	0.6	2.2	4.6	μmole/hr
Nucleoside diphosphatase				
TDP	0.5	1.0	2.1	μmole/hr
GDP	4.0	2.2	1.5	μmole/hr
Sialic acid	2.5	18	32	mμmole

The comparisons of intact Golgi apparatus with fractions of ER and plasma membranes from rat liver reveal several significant membrane changes which must be accomplished to achieve membrane differentiation (Tables 3.2, 3.4 and 3.5). These include a decrease in the amount of lecithin and perhaps other phosphoglycerides (Keenan, Morré, and Huang, in press) with a compensating increase in sphingomyelin. Sterols, sialic acid and specific sugars or sugar derivatives of the membranes (including glycolipids and glycoproteins) also increase (Keenan, Morré, and Huang, in press; results unpublished). Enzymatic activities characteristic of plasma membrane, i.e., plasma membrane marker enzymes, must be acquired while enzyme activities characteristic of ER membranes must be lost. Activities characteristic

Fig. 3.12   Portion of an outer cap cell of a maize root grown for 10 hr at 9° C. An extensive region of ER—Golgi apparatus continuity is outlined by the arrows. The cold treatment resulted in the conversion of the normal tubular-vesicular ER—Golgi apparatus junctions (See Fig. 3.11) into a continuous sheet-like structure in which continuity was more easily followed in a single electron microscope section. From Mollenhauer, VanDerWoude, and Morré (in preparation). Root tip fixed in potassium permanganate. Scale mark = 1 μm.

of the Golgi apparatus membranes (thiamine pyrophosphatase and specific glycosyl transferases activities) must also be acquired and lost. These interpretations are based on experimental findings (Tables 3.2, 3.4 and 3.5 and ref. cit.) and are independent of the mechanisms by which the changes take place and of whether membrane flow (physical transfer of membrane from one cell component to another discussed in the next section) is part of the mechanism.

The data comparing ER, Golgi apparatus and plasma membranes show a progressive increase in sphingomyelin with a corresponding decrease in phosphoglycerides, particularly lecithin (Table 3.2; Keenan, Morré, and Huang, in press). Other cell fractions from rat liver do not contain sphingomyelin in concentrations approaching that of either the Golgi apparatus or the plasma membrane (Table 3.2 and ref. cit.). These findings confirm the uniqueness of endomembrane components but still show the ordered progression of change indicated from Fig. 3.2. Plant membranes are not known to contain sphingomyelin and the nature of the lipid changes accompanying membrane differentiation will probably vary from organism to organism. Similarly, the cholesterol and sialic acid contents of the Golgi apparatus fraction from rat liver, although less than that found in the plasma membrane, are still much greater than those found in the ER or other cell components (Tables 3.2 and 3.5 and ref. cit.). A similar situation is true for gangliosides and total glycoprotein (Keenan, Morré, and Huang, in press; results unpublished). Again, sialic acid has not been found as a normal constituent of plant membranes and the biochemical details of membrane differentiation are expected to vary among organisms, among cells within a given organism, and among pathways within a given cell.

Several enzymatic activities characteristic of the plasma membrane of rat liver (Benedetti and Emmelot, 1968; Table 3.5 and ref. cit.) may be present in all three fractions but are concentrated in plasma membrane (ER $<$ GA $\lll$ PM). Of these, 5'-nucleotidase, Mg^{++}-ATPase and a UDP-galactose hydrolase have been determined for ER and Golgi apparatus (Table 3.5). In other examples (Mg^{++}-UTPase and nucleoside diphosphatases), the relative specific activity of each of the enzymes in the Golgi apparatus fractions is at a level intermediate between that of the two reference fractions (ER $<$ GA $<$ PM).

Endoplasmic reticulum markers show the reverse pattern (ER $>$ GA $\ggg$ PM or ER $>$ GA $>$ PM). They are present in the Golgi apparatus at levels 15 to 30 per cent of those found in ER but are lower or absent from plasma membrane (Table 3.5) or at a level intermediate between that of the two reference fractions, i.e. nucleoside diphosphatase (GDP) of Table 3.4. At least with glucose-6-phosphatase, there is evidence that the activity of the Golgi apparatus fraction is endogenous and not the result of contamination by ER (Morré *et al.*, 1971). Progressive loss of enzyme activities and constituents of the electron transport system of ER (first

---

Fig. 3.13 Membrane differentiation in dictyosomes (D) from the Golgi apparatus of outer cap cells of the maize root tip. Glutaraldehyde-osmium tetroxide fixation. Scale marks = 1μm.

(A) Membranes are stained so that plasma membrane (PM) is dark and endoplasmic reticulum (ER) is light. Membranes of secretory vesicles are lightly stained at the forming

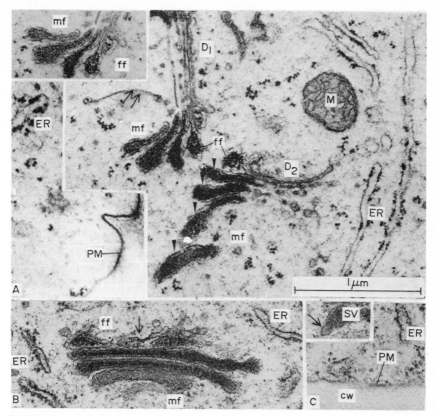

Fig. 3.13 (continued)

face (ff) and the staining intensity increases progressively across the stack (arrows) to the maturing face (mf). Marked changes in staining characteristics were not observed with membranes of the central plate-like portions of the cisterna (compare with Fig. 3.9A). Inset (top left) shows the secretory vesicles from D_1 reprinted at lower contrast to emphasize the differential staining reaction of the contents of the secretory vesicles from the forming face to the maturing face. The staining of the vesicle contents varies inversely to membrane staining in this preparation. Contents of vesicles at the forming face are most intensely stained while the membranes of these vesicles are least stained. A mature dictyosome cisterna which has separated from the stack is indicated by the double arrows.

Based on ultrastructural evidence, Mollenhauer (1971) suggested that release of secretory vesicles in the maize root cap is accompanied by the loss of the entire distal cisterna (see also, Morré, Mollenhauer, and Bracker, 1971). The amount of membrane lost to the cytoplasm in this manner is estimated to be approximately equal to that discharged to the plasma membrane in the form of secretory vesicles (Mollenhauer, 1971). M = mitochondrion.

(B) and (C) Membranes are stained as illustrated for onion stem in Figs. 3.5 and 3.9. Plasma membrane (PM) is light and endoplasmic reticulum (ER) is dark (arrow). Inset shows a secretory vesicle (sv) from the mature face (mf) of a dictyosome. All were taken from the same print. The staining pattern of the secretory product is the same as in (A) which shows that product and membrane stain independently. These figures also show that membrane differentiation from ER-like to plasma membrane-like can be expressed (within the same cell type) either as light to dark or dark to light depending upon the conditions of fixation and the procedure for post staining the sections. Irrespective of the pattern, dictyosome membranes of the forming face stain like ER while those of the maturing face stain like plasma membrane. Membrane staining seems to reflect intrinsic properties of the membrane and appears to provide a reliable criterion of membrane differentiation.

117

the enzyme activities and then the prosthetic groups (the cytochromes)) within the Golgi apparatus has been suggested (Morré *et al.,* 1971) on the basis of the Golgi apparatus/ER ratios of these constituents (Table 3.4) plus the observation that the total iron contents (haem + non-haem iron) of ER and Golgi apparatus are similar (Morré *et al.,* 1971; Table 3.4).

Only thiamine pyrophosphatase (Cheetam *et al.,* 1971) and certain glycosyl transferases (Morré, Merlin, and Keenan 1969; Schachter *et al.,* 1970) are concentrated in the Golgi apparatus. The other enzymes examined have specific activities in Golgi apparatus fractions intermediate between ER and plasma membrane (Tables 3.4 and 3.5 and ref. cit.). This contrasts with marker enzymes of the respiratory chain of mitochondria (cytochrome oxidase, succinic dehydrogenase), for example, which are low or absent from all three endomembrane components (Cheetham *et al.,* 1970).

Since Golgi apparatus contain both glycosyl transferases (Morré, Merlin, and Keenan, 1969; Schachter *et al.,* 1970) and enzymes of phospholipid biosynthesis (Morré, Keenan, and Mollenhauer, 1971; Morré, Nyquist, and Rivera, 1970), phospholipid changes and glycosylations of proteins and other molecules could take place within the endomembrane system as part of the process of endomembrane differentiation. These changes are easily visualized as localized biosynthetic

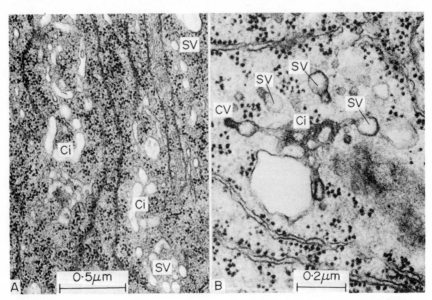

Fig. 3.14   Vesicle-producing cisternae (Ci) from subapical regions of young hyphae of fungi. Electron micrographs courtesy of S. N. Grove, Purdue University. (Adapted from Morré, Mollenhauer, and Bracker (1971).)

(A) *Aspergillus niger.* Formaldehyde-glutaraldehyde-osmium tetroxide fixation. Scale mark = 0.5 μm.

(B) *Armillaria mellea.* Glutaraldehyde-osmium tetroxide fixation. Scale mark = 0.2 μm.

These structures exist as single cisternae (Ci) and show no tendency to form stacks. Yet they appear to carry out functions usually associated with more complex systems of transition elements such as Golgi apparatus. SV = secretory vesicle, CV = coated vesicle.

events (either concomitant with or independent of membrane flow) and certain of the required enzyme systems have already been demonstrated to be present in Golgi apparatus of rat liver (Morré, Keenan, and Mollenhauer, 1971; Morré, Merlin, and Keenan, 1969; Morré, Nyquist, and Rivera, 1970; Schachte• et al., 1970).

The situation with regard to enzyme changes is less readily visualized. Since Golgi apparatus is presumably incapable of enzyme synthesis (Morré, Mollenhauer, and Bracker, 1971), the process of membrane transfer and differentiation must involve a selective transfer (with membrane flow) or alternatively, some enzymes derived from ER are progressively activated or inhibited as changes in the lipid and carbohydrate composition of the membrane occur. This is because some enzyme activities characteristic of ER are lost (but not necessarily the actual enzyme molecules) and some enzyme activities characteristic of the plasma membrane are gained (either by addition of the proteins or by activation of enzymes already associated with the membrane). In the absence of a flow mechanism, all enzymes would be transferred from sites of synthesis to sites of utilization in membrane formation via the soluble cytoplasm or via the cisternal lumina. Even with a flow mechanism, addition (from any source) or deletion of enzymes could be part of the mechanism of membrane differentiation.

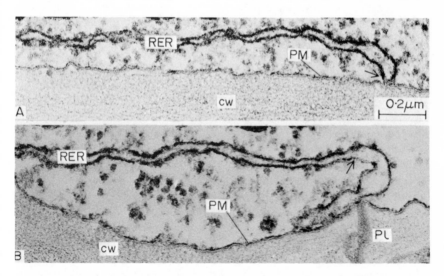

Fig. 3.15   Regions of association between rough endoplasmic reticulum (RER) and plasma membrane (PM) of onion stem. Glutaraldehyde-osmium tetroxide fixation. Scale mark = 0.2 μm.

(A) A region of close association (arrow) between dark-staining RER membrane and light-staining plasma membrane. Luminal continuity is not shown but membrane contact or continuity remains as a possibility. The region of association contains a mixture of dark- and light-staining membranes which may or may not be continuous.

(B) A structure similar to that in A near an intercellular connection or plasmodesma (pl) which tranverses the cell wall (cw). In the region of ER-plasma membrane association, the staining properties of the membrane change from thin and dark-staining (ER-like) to thicker and light-staining (PM-like). This is especially evident in the portion of the membrane from the arrow to near the plasmodesma.

The various possibilities are summarized in Fig. 3.19. They range from 'pre-programmed' membrane (originating at the ER with all enzymes already present but some inactive) to a 'shuttle' mechanism (in which new enzymes are added to the membrane as other enzymes are removed). At present there is little or no evidence to distinguish among the possibilities.

Metabolic studies

That Golgi apparatus and endoplasmic reticulum are dynamic and pleomorphic structures has been shown by phase contrast light microscopy (Dougherty and Lee, 1967; Rose and Pomerat, 1960). They change shape and position rapidly. This property of membranes known as *dynamic pleomorphism* (Novikoff *et al.*, 1962), has led to the view that cytoplasmic membranes may be subject to rapid turnover and quantitative changes as well as the high degree of interaction with other cell components as predicted by the endomembrane concept. Yet, quantitative studies of the dynamics of change and interaction are still in their infancy.

Membrane turnover

Metabolic studies of membrane turnover in plants offer little additional information concerning endomembrane interactions than has already been deduced from morphological studies. If all ER membranes are more or less in equilibrium and Golgi apparatus membranes are derived from ER, then Golgi apparatus turnover will be determined principally by the rate of ER turnover. Data with acetate-C^{14} supplied to stem explants of onion show turnover of microsomes and dictyosomes to be approximately equal, i.e., a half-life time of 1.2 (0.5 hours of label prior to addition of excess unlabelled acetate) to 5 (2 hours of label prior to addition of excess unlabelled acetate) hours (Morré, 1970). The longer (2 hr) period of labelling would be expected to permit spreading of label into constituents such as proteins that are more stable than the fatty acid residues of membrane lipids preferentially labelled by the shorter (0.5 hr) labelling time. Since different parts of the membrane, or even different parts of a phospholipid molecule are replaced at different rates (Hokin, 1968; Siekevitz *et al.*, 1967; Table 3.6), these differences in turnover with length of incubation time are not difficult to reconcile. With C^{14}-choline, labelling was rapid, but turnover was even slower than with acetate (about 43 hr half-life), due in part to ineffective chase conditions (choline uptake was found to be not proportional to external choline concentration, Morré, 1970; see also Chlapowsky and Band, 1971a). Yet, the half-life times for membrane constituents determined in this study were of the same order or shorter than those determined for membrane constituents of rat liver microsomes (Table 3.6) and for Golgi apparatus of rat liver (Franke *et al.*, 1971).

From the membrane flow hypothesis, we predict that disappearance of label from dictyosomes will be influenced in large measure by the rate of disappearance of label from ER. If ER membranes give rise to Golgi apparatus membranes and ER membranes are assembled by insertion of molecules followed by two-dimensional diffusion in the plane of the membrane (see Leskes, Siekevitz, and Palade,

TABLE 3.6

Relative Rates of Turnover of Different Constituents of Endoplasmic Reticulum Membranes Based on Studies with Isolated Microsome Fractions

Constituent	Label	Half-life time	References
Protein:	C^{14}-L-guanido-arginine*		
Total			Arias and DeLeon (1967)
Trypsin-soluble			Arias, Doyle, and Schimke (1969)
Trypsin-resistant			Argyris and Magnus (1968)
NADH-cytochrome c reductase		2–2.5 days	Ernster and Orrenius (1965)
Aminopyrine demethylase			Franke et al., (1971)
Barbiturate side chain oxidation activity			Kuriyama et al., (1969)
Cytochrome P_{450}			Omura, Siekevitz, and Palade (1967)
			Shuster and Jick (1966)
Cytochrome b_5		~ 4 days	Arias, Doyle, and Schimke (1969)
			Kuriyama et al., (1969)
Lipid:			Omura, Siekevitz, and Palade (1967)
			Omura, Siekevitz, and Palade (1967)
Total	Acetate-2-C^{14}	2.5–3.5 days	
	Glycerol-1, 3-C^{14}	~ 2 days	
Glycerol	Acetate-2-C^{14}	~ 1.7 days	
backbone	Glycerol-1, 3-C^{14}	~ 1.2 days	

* Half-life times are normalized to or derived from studies in which the label was supplied as C^{14}-L-guanido-arginine. Liver contains large amounts of an arginase that quickly converts the label-containing guanido group to $C^{14}O_2$ and water after the protein is broken down so that re-utilization of radioactivity once incorporated into protein is minimized. Half-life times based upon C^{14}-L-arginine or C^{14}-L-leucine are approximately twice as long (e.g., 5 days for total protein) since the labelled carbon atoms of the amino acids are re-utilized significantly after the proteins themselves are broken down, i.e., the 'apparent' life times are lengthened by an amount proportional to the degree to which the carbon is re-utilized.

1971b for a discussion of the latter possibility), ER and Golgi apparatus will appear to turn over as a unit in short-term labelling studies involving radioactive precursors. This is because the specific radioactivity of the Golgi apparatus will be high as long as the specific radioactivity of the ER is high. This will be true even if individual molecules are introduced and turn over in a manner independent of the assembly and turnover of the membrane as a whole. ER turnover would be the rate limiting step. This condition is met by the available plant (Morré, 1970) and animal (see Franke *et al.*, 1971) data. Thus the turnover data, although consistent with membrane flow, may say nothing about rate of dictyosome turnover *per se*. The best estimates of dictyosome turnover for plants come from the algae studied by Schnepf and Koch (1966a, b) and Brown (1969) where the process is visualized in the electron microscope and monitored in living cells by light microscopy. Here, as proximal cisternae are formed, large vesicles are discharged at the distal pole, for incorporation into the plasma membrane, with the loss of the entire distal cisterna. From these studies, dictyosomes are estimated to turn over within 20 to 40 minutes and cisternae are released at the rate of one every 1 to 4 minutes (see also Neutra and LeBlond, 1966, for similar estimates from animal cells).

Fig. 3.16 Diagram summarizing the functional similarities and differences among Golgi apparatus (top), vesicle-producing cisternae (centre) and direct ER-plasma membrane association (bottom) as systems of plasma membrane-generating transition elements. (From Morré and VanDerWoude, in press.)

Membrane flow

The membrane flow hypothesis as stated by Franke *et al.* (1971) predicts that 'the biogenesis of certain membranes is accomplished by the physical transfer of membrane material from one cell component to another in the course of their formation or normal functioning'. The concept of membrane flow was originally applied to endocytosis (Bennett, 1956) but membrane flow provides an equally attractive mechanism to account for exocytosis as well as other types of transfer associated with membrane biogenesis and differentiation. In spite of the attractiveness of the hypothesis, membrane flow remains unproven. Whether or not membrane flow occurs is critical to understanding the mechanisms of membrane differentiation, and proof of membrane flow would greatly strengthen the endomembrane concept.

Fig. 3.17 Functional polarity of maize root cap dictyosomes illustrated by prolonged exposure to osmium tetroxide (42 hrs at 40°C) according to the procedure of Friend and Murray (1965). Under these conditions, electron-opaque deposits of reduced osmium are restricted to endoplasmic reticulum (ER) cisternae at the immature or forming face of the dictyosomes (double arrows) and some of the tubules at the peripheries of intercalary cisternae (inset). In the inset, some of the osmium reactive material is seen to course through the central plate-like region of the cisterna (P) from points of connection with the peripheral tubules (T) giving the impression that some structural element of the tubules extends well

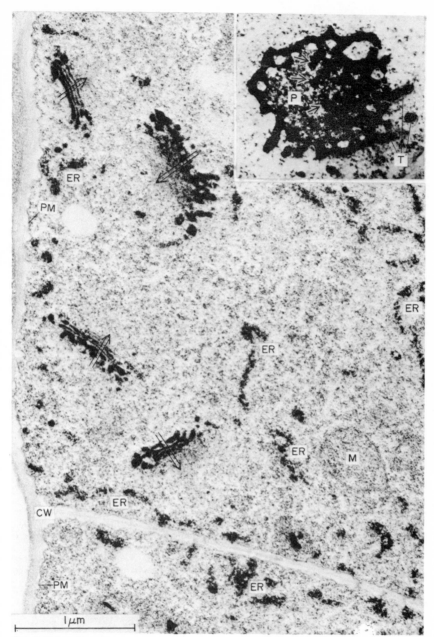

Fig. 3.17 (continued)

into the interior of the central plate (arrows). Prolonged exposure to osmium tetroxide is just one of many techniques used to demonstrate that Golgi apparatus cisternae on or near the forming face of the dictyosomes exhibit a character that is ER-like. In epidermal cells of the maize root, the differences illustrated in this figure are even more striking. In epidermal cells, the dense deposits of reduced osmium are restricted to ER, the forming face cisterna and a few tubules at the dictyosome periphery. PM = plasma membrane, M= mitochondrion. CW = cell wall.

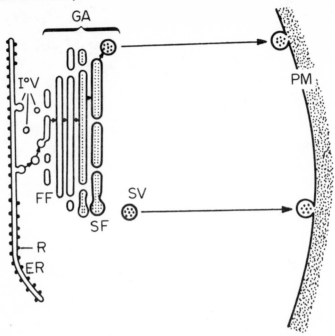

Fig. 3.18 Diagrammatic representation of endomembrane functioning in membrane flow and differentiation. The diagrammed concept involves the biosynthesis of new membranes at the rough endoplasmic reticulum (ER). R = ribosome. Small vesicles, the primary vesicles (I° V), arise by blebbing of the smooth surfaces of the rought ER adjacent to the dictyosomes and function in the transfer of materials, including membranes, from ER to the Golgi apparatus (GA). The input of new membranes occurs at a forming face (FF) of the Golgi apparatus. The membrane is progressively transformed from ER-like to plasma membrane-like as the cisternae mature. The membranes at the maturing cisternae are then utilized in the production of secretory vesicles (SV) which are released at the secreting face (SF) of the apparatus. The vesicles have membranes resembling the plasma membrane (PM); migrate to the plasma membrane and fuse with the plasma membrane to effect the discharge of secretory products to the cell's exterior and to contribute new membrane to the cell surface. Arrows denote directions of membrane flow and vesicle migration. (From Morré et al., 1971.)

In pollen tubes, fungal hyphae, rhizoids and trichomes (including root hairs), where elongation is primarily by means of tip growth, secretory vesicles appear to be a major or sole contributor to the new cell surface (Grove, Bracker, and Morré, 1970; VanDerWoude, Morré, and Bracker, 1971; Morré and VanDerWoude, in press and ref. cit.). In pollen tubes, approximately 1000 vesicles per cell per minute must be produced by dictyosomes to maintain a steady elongation rate of 6 μm per minute (Morré and VanDerWoude, in press).

One critical test of membrane flow is afforded by labelling studies with radioactive amino acids. When rats, for example, are pulse labelled with (C^{14}-guanido)-L-arginine, the order of labelling of membrane proteins of liver fractions is: (a) ER (nuclear envelope), (b) Golgi apparatus and (c) plasma membrane (Table 3.7) corresponding to the order established for migration of secretory proteins in liver (for literature, see Franke et al., 1971). In these studies, rapid turnover components of ER and Golgi apparatus appear sufficient to account for labelling of plasma membranes via a flow mechanism. However, the stepwise physical transfer of specific membrane proteins from one cell component to another via a flow mechanism remains to be demonstrated directly.[1]

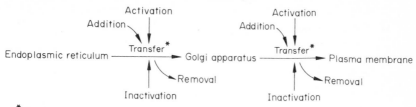

*May be either selective or nonselective

Fig. 3.19 Summary of some of the types of transformations which might account for changes in enzyme activities and protein composition of endomembrane components during membrane flow and differentiation.

TABLE 3.7

Time of Half Maximal Labelling of Membrane Proteins Following Administration of Radioactive Amino Acid $(C^{14}$-guanido-L-arginine) to Adult Male Rats. From Franke et al. (1971)

Cell fraction	Time of half maximal labelling (minutes)
Nuclear envelope	5
Endoplasmic reticulum	5—7
Golgi apparatus	10
Plasma membrane	15

Yet, as anyone who has observed living cells can testify, membranes flow. They are dynamic and pleomorphic. They move, extend and change shape with such rapidity that a major argument against membrane flow might be that membranes change too rapidly for this to be the only source of new membrane. Additionally, observations with heterokaryons suggest that surface antigens (presumably glycoproteins) spread rapidly over the cell membranes of fused cells (Frye and Edidin, 1970). Membrane flow does occur. It is only an understanding of the extent to which membrane flow will or will not explain the origin and continuity of the endomembrane system that must await further experimental findings.

[1] In preliminary studies with onion stem (Morré and VanDerWoude, in press), leucine incorporation into dictyosomes lagged behind that of incorporation into microsomes whereas incorporation into the plasma membrane fraction showed two phases not unlike the situation described for rat liver (Franke et al., 1971). The second phase of leucine incorporation into the plasma membrane fraction lagged approximately 30 min behind incorporation into dictyosomes while the initial phase more nearly paralleled incorporation into dictyosome A lag of 30 min between labelling of dictyosomes and labelling of plasma membrane is sufficient to allow for a new cisterna to be added to each stack every 5 to 6 min (4 to 5 cisternae/stack = mean life time of the dictyosome of 20 to 30 min, i.e., the time required for membranes added at the forming face to be displaced across the stack and released at the maturing face). In plant stems, the secretory product is primarily polysaccharide (Mollenhauer and Morré, 1966a), and specific labelling of membrane proteins by radioactive amino acids is assumed.

Summary

The endomembrane system denotes the functional continuum of membranous cell components consisting of the outer membrane of the nuclear envelope, endoplasmic reticulum, and Golgi apparatus, as well as vesicles and other structures derived from the major components which characterize the cytoplasm of eukaryotic cells. Documented continuity between each of the separate components leads to the view that they are all localized specializations of a single interassociated membrane system.

The endomembrane concept differs from other cytomembrane concepts in that specialization is considered to arise through progressive differentiation of the components of the system according to defined developmental sequences. It combines the concepts of membrane flow and of membrane differentiation to provide mechanisms whereby membranes are transferred and transformed along a chain of cell components in a subcellular developmental pathway.

APPENDIX
Structural, functional, and ontogenetic continuity of endomembrane components

The problem of membrane continuity is one of the most difficult structural concepts of cytology. To say that one type of membrane structure exhibits continuity with another implies physical continuity where one structure grades imperceptibly into the other without interruption of the membrane profiles. Yet one membrane may be derived from another in a manner where direct clear-line membrane continuity is not maintained. Alternatively, one membrane structure may derive materials from another without a permanent fusion of their structures. Membranes are not to be viewed as functionally independent units. They interact with each other and with non-membranous components of the cytoplasm. In one way or another, all the components of a cell interact. To aid in understanding these complex relationships among the internal membranes of eukaryotic cells, types of membrane continuity will be defined and discussed. These definitions are not meant to be mutually exclusive, and membranous cell components may exhibit any or all types of continuity with other membranous cell components at some time during development.

Structural continuity

When applied to membranes, structural continuity usually is interpreted to mean clear-line membrane continuity accompanied by luminal continuity so that adjacent compartments as well as the membranes forming the compartments are continuous in every respect. Yet, continuity among adjacent membrane components need not involve luminal continuity. A further possibility raised is that sites of contact or apposition may have functional significance in terms of the transfer of materials among endomembrane components. In addition to the types of structural continuity where both membranes and lumina are joined, other types of continuity among adjacent membrane components will be described where luminal continuity is not involved.

Luminal continuity. Luminal continuity is most readily resolved in the electron microscope when the image is formed from a cross-section through a pair of membranes lying parallel to each other but perpendicular to the plane of section, i.e., between rough ER and smooth ER (Fig. 3.3A). Under these conditions, the electron-opaque lines delimiting the membranes appear uninterrupted and the electron-translucent lumina are continuous. This type of connection is encountered between the outer membrane of the nuclear envelope and the ER (Fig. 3.4C), between different kinds of ER (Fig. 3.3A, 3.7A), between ER and vacuoles (Fig. 3.10A, B) and occasionally between ER and Golgi apparatus (Fig. 3.3C).

Another type of structural continuity in which adjacent compartments are continuous in every respect is that which results from narrow tubules 20 to 50 nm in diameter (Fig. 3.3C). Analyses from isolated cell fractions in negative stain show this type of connection to occur more frequently than can be determined solely on the basis of analysis of thin sections (Morré, Merritt, and Lembi, 1971). Because these tubular connections are often narrower than the thickness of a single thin section, the lumen is partly obscured with bridges at the sites of contact with conjoining membrane elements (Fig. 3.8B, D), and use of serial sections provides little additional information concerning their reality than is derived from single sections. Yet, in terms of transfer of membrane constituents or materials within the lumen from one compartment to another, this type of structural continuity is of considerable importance. Tubular continuity has been demonstrated between Golgi apparatus and associated secretory vesicles (Mollenhauer and Morré, 1966b; Morré, Merritt, and Lembi, 1971; Fig. 3.3C; Fig. 3.11) or Golgi apparatus equivalents (Fig. 3.14) and between the ER or nuclear envelope and Golgi apparatus (Bracker, Heintz, and Grove, 1970; Bracker *et al.,* 1971; Morré, Mahley, Bennett, and LeQuire, 1971; Figs. 3.3C and 3.8). This type of continuity is also indicated between ER and the outer membranes of mitochondria (Bracker and Grove, 1971; Franke and Kartenbeck, 1971; Morré, Merritt, and Lembi, 1971; Fig. 3.5F).

Membrane continuity without luminal continuity. This type of continuity is most frequently demonstrated in the electron microscope by showing continuity between one or both of the dark lines of the dark—light—dark pattern of an individual membrane after osmium fixation. Membrane continuity in the absence of luminal continuity has been described between certain types of specialized thickened membranes and the plasma membrane (Gunning, 1970) and between such membranes and the ER (Bulger, Griffiths, and Trump, 1966). This type of continuity has so far been encountered only in situations where one might logically expect transfer of membrane constituents under conditions where luminal continuity would prove disastrous, i.e., connections at the cell surface not involved with exo- or endocytosis (see also Morré and VanDerWoude, in press).

Contact or apposition. Sites of apposition which show no resolvable space between the appressed membranes are relatively common (Figs. 3.5C, 3.5H, 3.9B, 3.9C, 3.15; Bracker and Grove, 1971; Gunning, 1970; Porter, 1969). Observations of living cells with the light microscope show frequent collisions between membrane structures, some of which result in stable associations while most appear as collisions and no more than that. In the electron microscope, images of membrane apposition might result from fixation of contacting membranes representative of the living cell, from fixation of membranes which were brought into contact during fixation or both. In any critical evaluation of membrane contact, it is necessary to distinguish between true contact and what appears as contact through superposition of images, i.e., 'overlap'. Two adjacent membranes, close together but not contacting, may appear in contact or even to be fused because the image is formed not from the surface of the section but from its entire thickness. The depth of field of an electron micrograph is very large relative to the thickness of a single section. Any folds in the membrane or tilting of the membranes will contribute to the image. This problem can frequently be resolved by mechanically tilting the specimen so that a near perpendicular view is obtained (Willis, 1970). In routine electron microscopy, one can be immediately suspicious of regions of membrane contact where the membranes appear 'fuzzy' or 'smudged', a sure indication that the membranes are tilted relative to the plane of sectioning (cf. Fig. 3.15B).

Thread connections are the most tenuous of all reported membrane continuities. Images of thread-like continuities have been reported between ER and mitochondria where ER is in a parallel association with the mitochondrion or in an end-on relationship (Bracker and Grove, 1971). Both membranes (mitochondrial and ER) frequently project toward each other at the sites of these thread connections suggesting that the bridge is a genuine intermembrane connection and not the result of superposition of spurious electron-opaque grains in the section. Thread connections have been reported between adjacent strands of ER between ER and Golgi apparatus, between ER and the plasma membrane and between these structures and microtubules and secretory vesicles (Franke, 1970b; Franke *et al.*, 1971). Their significance has not been resolved.

Functional continuity (functional membrane interactions)

Functional continuity is most frequently established by biochemical or cytochemical studies that show vectorial transfer of materials or cell constituents from one compartment to another (Amsterdam, Ohad, and Schramm, 1969; Amsterdam *et al.*, 1971; Bracker, Heintz, and Grove, 1970; Brandes, 1965; Brown, 1969; Buvat, 1971; Claude, 1970; Essner and Novikoff, 1962; Franke *et al.*, 1971; Glauman and Dallner, 1968; Higgins and Barnett, 1971; Jamieson and Palade, 1967, 1968; Jones and Fawcett, 1966; Morré *et al.*, 1971; Morré, Jones, and Mollenhauer, 1967; Novikoff and Shin, 1964; Redman, 1968; Redman, Siekevitz, and Palade, 1966; Revel and Hay, 1963; Sabatini *et al.*, 1971; Schnepf, 1968, 1969a, b; Schnepf and Koch, 1966a, b; Sheldon and Kimball, 1962; Siekevitz *et al.*, 1967; VanDerWoude, Morré, and Bracker, 1971 as examples selected from the bibliography). Many additional examples are given by Morré Mollenhauer, and Bracker, 1971, Beams and Kessel, 1968; and other chapters in this volume. Structural continuity need not occur but is frequently involved. The variety of techniques that have been used to demonstrate functional continuity among cell components include use of isotopes in conjunction with cell fractionation (Chlapowski and Band, 1971a, b; Franke et al., 1971; Jamieson and Palade, 1967, 1968; Northcote, 1969) or autoradiography (LeBlond and Warren, 1965; Neutra and LeBlond, 1966; Pickett-Heaps, this volume; Revel and Hay, 1963), immunochemical techniques (Frye and Edidin, 1970; Luridkvist and Perlmann, 1967), cytochemistry (Novikoff et al., 1962), morphological markers in conjunction with light and electron microscopy (Bracker, Heintz, and Grove, 1970; Brown, 1969; Claude, 1970; Kessel, 1963; Morré, Jones, and Mollenhauer, 1967), etc. Any technique that permits determination of the movement or transfer of materials from one compartment to another or demonstrates that such movement or transfer has taken place is a potentially useful tool in studying functional continuity.

There are varying degrees of functional continuity involving both structured and nonstructured communication pathways. These can be further grouped into two broad categories.

Compartmentalized functional continuity. This type of continuity is usually structured and normally does not involve passage of material from one compartment to another via the cytoplasmic matrix (Jamieson, 1971). It occurs when adjacent cell components have continuous lumina (i.e., within the ER or through tubules connecting ER of Golgi apparatus to secretory vesicles, Fig. 3.3) but adjacent components need not have permanent structural continuity to have functional continuity. Intermittent tubular or vesicular bridges would ensure transfer of materials (ontogenetic functional continuity) (Morré, Mollenhauer, and Bracker, 1971). Transfer is most often vectorial (from component A to component B but not in the reverse direction) (see Franke et al. (1971) for specific references to vectorial movements of secretory proteins in mammalian cells).

Functional continuity between ER and Golgi apparatus has been established by a variety of techniques (see Beams and Kessel, 1968; Morré, Mollenhauer, and Bracker, 1971). Here, vectorial transfer of secretory proteins from ER to Golgi apparatus to secretory vesicles is accomplished via luminal continuity between conjoining cell components (Jamieson and Palade, 1967, 1968; Redman, Siekevitz, and Palade, 1966). Small vesicles which bleb from ER cisternae have been implicated as an important mechanism for providing the required luminal continuity in the absence of permanent structural continuity (Ziegel and Dalton, 1962; Jamieson and Palade, 1967). Regardless of the mechanism, at no time do the secretory proteins appear to be transported in soluble form through the cell sap.

Close physical association may be sufficient to ensure vectorial transfer without contact or continuity between the adjacent membranes and in the absence of luminal continuity. Apposition between elements of the sarcoplasmic reticulum and specialized extensions of the plasma membrane (transverse or T-tubules of the sarcolemma) in skeletal muscle are thought to be sufficient to permit vectorial conduction of excitation from the cell surface to the myofibril located within the cell's interior (Smith, 1965). The membranes do not appear to contact but remain separated by some constant minimal spacing (Smith, 1965).

Non-compartmentalized functional continuity. This type of continuity is usually non-structured and normally involves entry into the cell sap. Examples include the passage of ATP from a chloroplast or mitochondrion into the cell sap and its movement by diffusion or cytoplasmic turbulence to another cell component such as ER. This type of continuity apparently is not limited to small and intermediate sized molecules but to large molecules as well. The extent to which large molecules are transferred from one compartment to another via the cell sap or from sites of synthesis in the cell sap (e.g., free polysomes) to membranous compartments remains to be investigated.

Ontogenetic continuity

The term ontogenetic continuity is proposed to denote structural continuity in time and space during either cellular or subcellular ontogeny. If one cell component derives from or gives rise to another, parent and progeny are ontogenetically continuous. For example, the Golgi apparatus produces secretory vesicles which fuse with and become part of the plasma membrane. As a result, the plasma membrane and the Golgi apparatus are ontogenetically continuous via a common pathway of subcellular ontogeny even though they are not structurally continuous.

At the cellular level, another type of ontogenetic continuity is encountered. Membrane structures are dynamic and in a constant state of flux or movement so that connections are continuously being formed and broken. Two-cell components that are continuous at one time during cellular development may become separated at a later time. For example, connections between rough and smooth ER are observed most frequently during periods of rapid proliferation of smooth ER (Jones and Fawcett, 1966). Nuclear envelope and rough ER may be continuous during early developmental stages but these connections may be lost during later stages. Thus, cell components may exhibit ontogenetic continuity even though they are never structurally continuous or if structural continuity is lost during development or normal cell functioning. Ontogenetic continuity is a form of structural continuity where transfer of both membranes and luminal contents is afforded by transient connections or by intervening tubular or vesicular bridges between conjoining cell components.

Acknowledgements

Portions of the experimental findings reported were from studies sponsored by NSF GB 1084, 6751, 03044, 7078, 23183 and 25110. Purdue University Agricultural Experiment Station Journal Paper No. 4651. Charles F. Kettering Research Laboratory Contribution No. C-455. The authors are indebted to Professors C. E. Bracker, T. W. Keenan, and W. W. Franke for their many contributions to the endomembrane concept. Although they did not participate directly in the preparation of this manuscript, credit for those features of the endomembrane concept which prove to be correct must be shared with them while those features which prove to be incorrect must be regarded as misjudgements of the authors alone.

Bibliography

General references

Beams, H. W. and Kessel, R. G. (1968) 'The Golgi apparatus: structure and function', *Int. Rev. Cytol.*, **23**, 209–276.

Clementi, F. and Ceccarelli, B., eds. (1971) 'Advances in cytopharmacology' (*Proceedings First International Symposium on Cell Biology and Cytopharmacology, Venice, Italy, July, 1969*), Raven Press, New York.

Grove, S. N., Bracker, C. E., and Morré, D. J. (1968) 'Cytomembrane differentiation in the endoplasmic reticulum-Golgi apparatus-vesicle complex', *Science, N.Y.*, **161**, 171–173.

Grove, S. N., Bracker, C. E., and Morré, D. J. (1970) 'An ultrastructural basis for hyphal tip growth in *Pythium ultimum*', *Am. J. Bot.*, **57**, 245–266.

Jamieson, J. D. and Palade, G. E. (1967) 'Intracellular transport of secretory proteins in the pancreatic exocrine cell. I. Role of the peripheral elements of the Golgi complex', *J. Cell Biol.*, **34**, 577–596.

Jones, A. L. and Fawcett, D. W. (1966) 'Hypertrophy of the agranular endoplasmic reticulum in hamster liver induced by phenobarbital (with a review of the functions of this organelle in liver)', *J. Histochem. Cytochem.*, **14**, 215–232.

Mollenhauer, H. H. and Morré, D. J. (1966) 'Golgi apparatus and plant secretion', *A. Rev. Pl. Physiol*, **17**, 27–46.

Morré, D. J., Franke, W. W., Deumling, B., Nyquist, S. E., and Ovtracht, L. (1971) 'Golgi apparatus function in membrane flow and differentiation: origin of plasma membrane from endoplasmic reticulum', *Biomembranes* (*Proceedings Symposium on Membranes and Coordination of Cellular Activities*, Gatlinburg, Tennessee, April 5–8, 1971, ed. L. A. Mason), vol. 2, pp. 95–104, Plenum Press, New York.

Morré, D. J., Mollenhauer, H. H., and Bracker, C. E. (1971) 'The origin and continuity of Golgi apparatus', In: *Results and problems in cell differentiation. II. Origin and continuity of cell organelles* (eds. T. Reinert and H. Ursprung), pp. 82–126, Springer-Verlag, Berlin.

Northcote, D. H. (1969) 'Fine structure of cytoplasm in relation to synthesis and secretion in plant cells', *Proc. R. Soc. B*, **177**, 21–30.

Novikoff, A. B., Essner, E., Goldfischer, S., and Heus, M. (1962) 'Nucleosidediphosphatase activities of cytomembranes', In: *The interpretation of ultrastructure* (ed. R. J. C. Harris), pp. 149–192, Academic Press Inc., New York.

Schnepf, E. (1969) 'Sekretion und exkretion bei pflanzen', *Protoplasmatologia, Handbuch der Protoplasmaforschung*, **8**, 1–181.

Siekevitz, P., Palade, G. E., Dallner, G., Ohad, I., and Omura, T. (1967) 'The biosynthesis of intracellular membranes', In: *Organizational biosynthesis* (ed. H. J. Vogel, J. O. Lampen and V. Bryson), pp. 331–361, Acadmic Press Inc., New York.

Whaley, W. G., Kephart, J. E., and Mollenhauer, H. H. (1964) 'The dynamics of cytoplasmic membranes during development', In: *Cellular membranes in development* (ed. M. Locke), pp. 135–173, Academic Press Inc., New York.

References

Agrell, I. P. S. (1966) 'Continuity of membrane systems in the cells of imaginal discs', *Z. Zellforsch. mikrosk. Anat.*, **72**, 22–29.

Amsterdam, A., Ohad, I., and Schramm, M. (1969) 'Dynamic changes in the ultrastructure of the acinar cell of rat parotid gland during the secretory cycle', *J. Cell Biol.*, **41**, 753–773.

Amsterdam. A., Schramm, M., Ohad, I., Salomon, Y., and Selinger, Z. (1971) 'Concomitant synthesis of membrane protein and exportable protein of the secretory granule in rat parotid gland', *J. Cell Biol.*, **50**, 187–200.

The endomembrane concept

Argyris, T. S. and Magnus, D. R. (1968) 'The stimulation of liver growth and demethylase activity following phenobarbital treatment', *Devl Biol.*, **17**, 187–201.

Arias, I. M. and DeLeon, A. (1967) 'Estimation of the turnover rate of barbiturate side-chain oxidation enzyme in rat liver', *Mol. Pharmacol.*, **3**, 216–218.

Arias, I. M., Doyle, D., and Schimke, R. T. (1969) 'Studies on the synthesis and degradation of proteins of the endoplasmic reticulum of rat liver', *J. biol. Chem.*, **244**, 3303–3315.

Attardi, G. (1967) 'The mechanism of protein synthesis', *A. Rev. Microbiol.*, **21**, 383–416.

Barancik, L. C. and Lieberman, I. (1971) 'The kinetics of incorporation of protein into the liver plasma membrane', *Biochem, biophys. Res. Comm.*, **44**, 1084–1088.

Barer, R., Joseph, S. and Meek, G. A. (1959) 'The origin of the nuclear membrane', *Expl Cell Res.*, **18**, 179–182.

Beams, H. W. and Kessel, R. G. (1968) 'The Golgi apparatus: structure and function', *Int. Rev. Cytol.*, **23**, 209–276.

Bellairs, R. (1964) 'Biological aspects of the yolk of the hen's egg', *Adv. Morphogenesis,* **4**, 217–272.

Benedetti, E. L. and Emmelot, P. (1968) 'Structure and function of plasma membranes isolated from liver', In: *Ultrastructure in biological systems. The membranes* (eds. A. Dalton and F. Haguenau), pp. 33–120, Academic Press Inc., New York.

Bennett, H. S. (1956) 'The concepts of membrane flow and membrane vesiculation as mechanisms for active transport and ion pumping', *J. biophys. biochem. Cytol.*, **2**(Suppl.), 99–103.

Berhsohm, J. and Grossman, H. J. (1971) *Lipid storage diseases: enzymatic defects and clinical implications,* Academic Press Inc., New York.

Bracker, C. E. (1968) 'The ultrastructure and development of sporangia in *Gilbertella persicaria',* *Mycologia,* **5**, 1016–1067.

Bracker, C. E. and Grove, S. N. (1971) 'Continuity between cytoplasmic endomembranes and outer mitochondrial membranes in fungi', *Protoplasma,* **73**, 15–34.

Bracker, C. E., Grove, S. N., Heintz, C. E., and Morré, D. J. (1971) 'Continuity between endomembrane components in hyphae of *Pythium* spp.' *Cytobiol.,* **4**, 1–8.

Bracker, C. E., Heintz, C. E., and Grove, S. N. (1970) 'Structural and functional continuity among endomembrane organelles in fungi', In: *Microscopie Electronique, 1970, Proc. Septième Congrés Intern. de Microscopie Electronique,* Grenoble, France, vol. III, pp. 103–104, Soc. Française Microscopie Electronique, Paris.

Bracker, C. E. and Littlefield, L. (In preparation) 'Concept of the host-parasite interface', In: *Fungal pathogenicity. The host's response* (Proceedings 3rd Long Ashton Symposium, Bristol, England, ed. C. V. Cutting), Academic Press Inc., New York.

Brandes, D. (1965) 'Observations on the apparent mode of formation of "pure" lysosomes', *J. Ultrastruct. Res.,* **12**, 63–80.

Breen, P. C. and de Bruyn, P. P. H. (1969) 'The fine structure of the secretory cells of the uterus (shell gland) of the chicken', *J. Morph.,* **128**, 35–66.

Brown, R. M. (1969) 'Observations on the relationship of the Golgi apparatus to wall formation in the marine Chrysophycean alga, *Pleurochrysis scherffelii* Pringshein', *J. Cell Biol.,* **41**, 109–123.

Bulger, R. E., Griffiths, L. D., and Trump, B. F. (1966) 'Endoplasmic reticulum in renal interstitial cells: molecular rearrangement after water deprivation', *Science, N. Y.,* **151**, 83–86.

Bulova, S. I. and Burka, E. R. (1970) 'Biosynthesis of nonglobin protein by membrane-bound ribosomes in reticulocytes', *J. biol. Chem.,* **245**, 4907–4912.

Buvat, R. (1971) 'Origin and continuity of cell vacuoles', In: *Results and problems in cell differentiation. II. Origin and continuity of cell organelles* (eds. J. Reinert and H. Ursprung), pp. 127–157, Springer-Verlag, Berlin.

Chauveau, J., Moulé, Y., Rouiller, C., and Schneebeli, J. (1962) 'Isolation of smooth vesicles and free ribosomes from rat liver microsomes', *J. Cell Biol.,* **12**, 17–29.

Cheetham, R. D., Morré, D. J., Pannek, C., and Friend, D. S. (1971) 'Isolation of a Golgi apparatus-rich fraction from rat liver. IV. Thiamine pyrophosphatase', *J. Cell Biol.,* **49**, 899–905.

Cheetham, R. D., Morré, D. J., and Yunghans, W. N. (1970) 'Isolation of a Golgi apparatus-rich fraction from rat liver. II. Enzymatic characterization and comparison with other cell fractions', *J. Cell Biol.,* **44**, 492–500.

Chlapowski, F. J. and Band, R. N. (1971a) 'Assembly of lipids into membranes in

Acanthamoeba palestinensis. I. Observations on the specificity and stability of choline-^{14}C and glycerol-^{3}H as labels for membrane phospholipids', *J. Cell Biol.,* **50,** 625–633.

Chlapowski, F. J. and Band, R. N. (1971b) 'Assembly of lipids into membranes in *Acanthamoeba palestinensis.* II. The origin and fate of glycerol-^{3}H-labeled phospholipids of cellular membranes', *J. Cell Biol.,* **50,** 634–651.

Claude, A. (1970) 'Growth and differentiation of cytoplasmic membranes in the course of lipoprotein granule synthesis in the hepatic cell. I. Elaboration of elements of the Golgi complex', *J. Cell Biol.,* **47,** 745–766.

Comings, D. E. and Kakefuda, T. (1968) 'Initiation of deoxyribonucleic acid replication at the nuclear membrane in human cells', *J. molec. Biol.,* **33,** 225–229.

Cunningham, W. P., Morré, D. J., and Mollenhauer, H. H. (1966) 'Structure of isolated plant Golgi apparatus revealed by negative staining', *J. Cell Biol.,* **28,** 169–179.

Curtis, A. S. G. (1967) *The cell surface: its molecular role in morphogenesis,* Logos Press, London.

Dales, S. (1971) 'Involvement of membranes in the infectious cycle of vaccinia', In: *Cell membranes. Biological and pathological aspects* (eds. G. W. Richter and D. G. Scarpelli), pp. 136–144, Williams and Wilkins, Baltimore.

Dallner, G., Siekevitz, P., and Palade, G. E. (1966) 'Biogenesis of endoplasmic reticulum membranes. I. Structural and chemical differentiation in developing rat hepatocyte', *J. Cell Biol.,* **30,** 73–96.

De Duve, C. (1969a) 'Evolution of the peroxisome', *Ann. N. Y. Acad. Sci.,* **168,** 369–381.

De Duve, C. (1969b) 'The lysosome in retrospect', In: *Lysosomes in biology and pathology.* (eds. J. T. Dingle and H. B. Fell), vol. I, pp. 3–40, American Elsevier, New York.

De Duve, C. and Wattiaux, R. (1966) 'Functions of lysosomes', *A. Rev. Physiol.,* **28,** 435–492.

Dickens, F., Randle, P. J., and Whelan, W. J., eds. (1968) *Carbohydrate metabolism and its disorders,* Academic Press, New York.

Dougherty, W. J. and Lee, M. McN. (1967) 'Light and electron microscope studies of smooth endoplasmic reticulum in dividing rat hepatic cells', *J. Ultrastruct. Res.,* **19,** 200–220.

Ernster, L., Siekevitz, P., and Palade, G. E. (1962) 'Enzyme-structure relationships in the endoplasmic reticulum of rat liver. A morphological and biochemical study', *J. Cell Biol.,* **15,** 541–562.

Ernster, L. and Orrenius, S. (1965) 'Substrate-induced synthesis of the hydroxylating enzyme system of liver microsomes', *Fed. Proc.* **24,** 1190–1199.

Essner, E. and Novikoff, A. B. (1962) 'Cytological studies on two functional hepatomas. Interrelations of endoplasmic reticulum, Golgi apparatus and lysosomes', *J. Cell Biol.,* **15,** 289–312.

Feldherr, C. M. (In press) *Structure and function of the nuclear envelope.*

Flickinger, C. J. (1969) 'Fenestrated cisternae in the Golgi apparatus of the epididymus', *Anat. Rec.,* **163,** 39–54.

Franke, W. W. (1970a) 'On the universality of nuclear pore complex structure', *Z. Zellforsch. mikrosk. Anat.,* **105,** 405–429.

Franke, W. W. (1970b) 'Cytoplasmic microtubules linked to endoplasmic reticulum with cross bridges', *Expl Cell Res.,* **66,** 486–489.

Franke, W. W., Eckert, W. A., and Krien, S. (1971) 'Cytomembrane differentiation in a ciliate, *Tetrahymena pyriformis.* I. Endoplasmic reticulum and dictyosomal equivalents', *Z. Zellforsch. mikrosk. Anat,* **119,** 577–604.

Franke, W. W. and Kartenbeck, J. (1971) 'Outer mitochondrial membrane continuities with endoplasmic reticulum', *Protoplasma,* **73,** 35–41.

Franke, W. W., Kartenbeck, J., Zentgraf, H.-W., Scheer, U., and Falk, H. (1971) 'Membrane-to-membrane cross bridges. A means to orientation and interaction of membrane faces', *J. Cell Biol.,* **51,** 881–888.

Franke, W. W., Morré, D. J., Deumling, B., Cheetham, R. D., Kartenbeck, J., Jarasch, E.-D., and Zentgraf, H.-W. (1971) 'Synthesis and turnover of membrane proteins in rat liver: an examination of the membrane flow hypothesis', *Z. Naturf.* 26b, 1031–1039.

Franke, W. W. and Scheer, U. (1970) 'The ultrastructure of the nuclear envelope of amphibian oocytes: a reinvestigation. I. The mature oocyte' *J. Ultrastruct. Res.,* **30,** 288–316.

Friend, D. S. and Murray, M. J. (1965) 'Osmium impregnation of the Golgi apparatus', *Am. J. Anat.,* **117,** 135–150.

The endomembrane concept

Frye, L. D. and Edidin, M. (1970) 'The rapid intermixing of cell surface antigens after formation of mouse-human heterokaryons', *J. Cell Sci.*, **7**, 319–335.

Galling, G. (1970) 'Ribosomes in unicellular green algae. II. Dynamic alterations in the polysome patterns of *Chlorella*', *Cytobiol.*, **2**, 359–375.

Ganoza, M. C. and Williams, C. A. (1969) '*In vitro* synthesis of different categories of specific protein by membrane-bound and free ribosomes', *Proc. natn. Acad. Sci. U.S.A.*, **63**, 1370–1376.

Gillete, J., Conney, A., Cosmides, G., Estabrook, R., Fouts, J., and Mannering, G. (1969) *Microsomes and drug oxidations*, Academic Press Inc., New York.

Ginsburg, V. and Nuefeld, E. F. (1969) 'Complex heterosaccharides of animals', *Ann. Rev. Biochem.*, **38**, 371–388.

Girbardt, M. (1969) 'Die Ultrastruktur der Apikalregion von Pilzhyphen', *Protoplasma*, **67**, 413–441.

Glaumann, H. and Dallner, G. (1968) 'Lipid composition and turnover of rough and smooth microsomal membranes in rat liver', *J. Lipid Res.*, **9**, 720–729.

Glaumann, H. and Dallner, G. (1970) 'Subfractionation of smooth microsomes from rat liver', *J. Cell Biol.*, **47**, 34–48.

Golfischer, S. (1965) 'The cytochemical demonstration of lysosomal arylsulfatase activity by light and electron microscopy', *J. Histochem. Cytochem.*, **12**, 659–669.

Grove, S. N. and Bracker, C. E. (1970) 'Protoplasmic organization of hyphal tips among fungi: vesicles and Spitzenkörper', *J. Bact.*, **104**, 989–1009.

Grove, S. N., Bracker, C. E., and Morré, D. J. (1968) 'Cytomembrane differentiation in the endoplasmic reticulum-Golgi apparatus-vesicle complex', *Science, N.Y.*, **161**, 171–173.

Grove, S. N., Bracker, C. E., and Morré, D. J. (1970) 'An ultrastructural basis for hyphal tip growth in *Pythium ultimum*', *Am. J. Bot.*, **57**, 245–266.

Gunning, B. E. S. (1970) 'Lateral fusion of membranes in bacteroid-containing cells of leguminous root nodules', *J. Cell Sci.*, **7**, 307–317.

Hamada, K., Yang, P., Heintz, R., and Schweet, R. (1968) 'Some properties of reticulocyte ribosomal subunits', *Archs. Biochem. Biophys.*, **125**, 598–603.

Higgins, J. A. and Barnett, R. J. (1971) 'The biogenesis of smooth endoplasmic reticulum membranes in hepatocytes of phenobarbital treated rats', *Abstracts 11th Meeting, American Society for Cell Biology*, New Orleans, Louisiana, 17–20 November 1971, p. 125.

Hokin, L. E. (1968) 'Dynamic aspects of phospholipids during protein secretion', *Int. Rev. Cytol.*, **23**, 187–208.

Jamieson, J. D. (1971) 'Role of the Golgi complex in the intracellular transport of secretory proteins', In: *Advances in cytopharmacology (Proceedings 1st Intern. Symp. Cell Biol. and Cytopharmacology*, Venice, Italy, 7–11 July 1969, eds. F. Clementi and B. Ceccarelli), vol. 1, pp. 183–190, Raven Press, New York.

Jamieson, J. D. and Palade, G. E. (1967) 'Intracellular transport of secretory proteins in the pancreatic exocrine cell. I. Role of the peripheral elements of the Golgi complex', *J. Cell Biol.*, **34**, 577–596.

Jamieson, J. D. and Palade, G. E. (1968) 'Intracellular transport of secretory proteins in the pancreatic exocrine cell. III. Dissociation of intracellular transport from protein synthesis', *J. Cell Biol.*, **39**, 580–588.

Jones, A. L. and Fawcett, D. W. (1966) 'Hypertrophy of the agranular endoplasmic reticulum in hamster liver induced by phenobarbital (with a review on the functions of this organelle in liver)', *J. Histochem. Cytochem.*, **14**, 215–232.

Keenan, T. W. and Morré, D. J. (1970) 'Phospholipid class and fatty acid composition of Golgi apparatus isolated from rat liver and comparison with other cell fractions', *Biochemistry, N.Y.*, **9**, 19–25.

Keenan, T. W., Morré, D. J., and Huang, C.-M. (In press) 'Membranes of the mammary gland', In: *Lactation: a comprehensive treatise* (eds. B. L. Larson and V. R. Smith), Academic Press Inc., New York.

Kessel, R. G. (1963) 'Electron microscope studies on the origin of annulate lamellae in oocytes of *Necturus*', *J. Cell Biol.*, **19**, 391–414.

Kessel, R. G. (1968) 'Annulate lamellae', *J. Ultrastruct Res.*, **24**, (Suppl. 10), 1–82.

Kuriyama, Y., Omura, T., Siekevitz, P., and Palade, G. E. (1969) 'Effects of phenobarbital on the synthesis and degradation of the protein components of rat liver microsomal membranes', *J. biol. Chem.*, **244**, 2017–2026.

LeBlond, C. O. and Warren, K. B., eds. (1965) *The use of radioautography in investigating protein synthesis,* Academic Press Inc., New York.

Leedale, G. F. (1969) 'Observations on endonuclear bacteria in euglenoid flagellates', *Öst. bot. Z.,* **116,** 279–294.

Leskes, A., Siekevitz, P. and Palade, G. E. (1971a) 'Differentiation of endoplasmic reticulum in hepatocytes. I. Glucose-6-phosphatase distribution *in situ*', *J. Cell Biol.,* **49,** 264–287.

Leskes, A., Siekevitz, P., and Palade, G. E. (1971b) 'Differentiation of endoplasmic reticulum in hepatocytes. II. Glucose-6-phosphatase in rough microsomes', *J. Cell Biol.,* **49,** 288–302.

Lundkvist, U. and Perlmann, P. (1967) 'Immunochemical characterization of submicrosomal rat liver membranes', *Immunology,* **13,** 179–191.

Margulis, L. (1970) *Origin of eukaryotic cells,* Yale University Press, New Haven.

Merriam, R. W. (1961) 'Nuclear envelope structure during cell division in *Chaetopterus* eggs', *Expl Cell Res.,* **22,** 93–107.

Moe, H., Rostgaard, J., and Behnke, O. (1965) 'On the morphology and origin of virgin lysosomes in the intestinal epithelium of the rat', *J. Ultrastruct. Res.,* **12,** 396–403.

Mollenhauer, H. H. (1971) 'Fragmentation of mature dictyosome cisternae', *J. Cell Biol.,* **49,** 212–214.

Mollenhauer, H. H and Morré, D. J. (1966a) 'Golgi apparatus and plant secretion', *A. Rev. Pl. Physiol.,* **17,** 27–46.

Mollenhauer, H. H and Morré, D. J. (1966b) 'Tubular connections between dictyosomes and forming secretory vesicles in plant Golgi apparatus', *J. Cell Biol.,* **29,** 373–376.

Mollenhauer, H. H., Morré, D. J., and Bergman, L. (1967) 'Homology of form in plant and animal Golgi apparatus', *Anat. Rec.,* **158,** 313–318.

Mollenhauer, H. H. and Totten, C. (1971) 'Studies on seeds. II. Origin and degeneration of lipid vesicles in pea and bean cotyledons', *J. Cell Biol.,* **48,** 394–404.

Mollenhauer, H. H. and Whaley, W. G. (1963) 'An observation on the functioning of the Golgi apparatus', *J. Cell Biol.,* **17,** 222–225.

Morré, D. J. (1970) *In vivo* incorporation of radioactive metabolites by Golgi apparatus and other cell fractions of onion stem', *Pl. Physiol., Lancaster,* **45,** 791–799.

Morré, D. J., Franke, W. W., Deumling, B., Nyquist, S. E., and Ovtracht, L., (1971) 'Golgi apparatus function in membrane flow and differentiation: origin of plasma membrane from endoplasmic reticulum', *Biomembranes (Proceedings Symp. Membranes and Coordination of Cellular Activities,* Gatlinburg, Tennessee, 5–8 April 1971, ed. L. A. Manson) vol. 2, pp. 95–104, Plenum Press, New York.

Morré, D. J., Jones, D. D., and Mollenhauer, H. H. (1967) 'Golgi apparatus mediated polysaccharide secretion by outer root cap cells of *Zea mays.* I. Kinetics and secretory pathway', *Planta,* **74,** 286–301.

Morré, D. J., Keenan, T. W., and Mollenhauer, H. H. (1971) 'Golgi apparatus function in membrane transformations and product compartmentalization: studies with cell fractions from rat liver', In: *Advances in cytopharmacology (Proceedings 1st Intern. Symp. Cell Biol. and Cytopharmacology,* Venice, Italy, 7–11 July 1969, eds. F. Clementi and B. Ceccarelli) vol. 1, pp. 159–182, Raven Press, New York.

Morré, D. J., Mahley, R. S., Bennett, B. D., and LeQuire, V. S. (1971) 'Continuities between endoplasmic reticulum, secretory vesicles and Golgi apparatus in rat liver and intestine', *Abstracts 11th Annual Meeting, American Society for Cell Biology,* New Orleans, Louisiana, 17–20 November 1971, p. 199.

Morré, D. J., Merlin, L. M., and Keenan, T. W. (1969) 'Localization of glycosyl transferase activities in a Golgi apparatus-rich fraction isolated from rat liver', *Biochem. biophys. Res. Commun.,* **37,** 813–819.

Morré, D. J., Merritt, W. D., and Lembi, C. A. (1971) 'Connections between mitochondria and endoplasmic reticulum in rat liver and onion stem', *Protoplasma,* **73,** 43–49.

Morré, D. J., Mollenhauer, H. H., and Bracker, C. E. (1971) 'Origin and continuity of Golgi apparatus', In: *Results and problems in cell differentiation. II. Origin and continuity of cell organelles* (eds. J. Reinert and H. Ursprung), pp. 82–126, Springer-Verlag, Berlin.

Morré, D. J., Nyquist, S., and Rivera, E. (1970) 'Lecithin biosynthetic enzymes of onion stem and the distribution of phosphoryl choline cytidyl transferase among cell fractions', *Pl. Physiol., Lancaster,* **45,** 800–804.

Morré, D. J. and VanDerWoude, W. J. (In press) 'Origin and growth of cell surface components', *Developmental Biology (Proceedings 30th Symposium of the Society for Developmental Biology,* Seattle, Washington, 17–19 June 1971) Supplement 5.

The endomembrane concept

Moulé, Y. (1968) 'Biochemical characterization of the components of the endoplasmic reticulum in rat liver cell', In: *Structure and function of the endoplasmic reticulum in animal cells (Proceedings Fed, European Biochem. Soc.,* 4th., ed. F. C. Gran), pp. 1–12, Academic Press, New York.

Moulé, Y., Rouiller, C., and Chauveau, J. (1960) 'A biochemical and morphological study of rat liver microsomes', *J. biophys. biochem. Cytol.,* 7, 547–558.

Nakane, P. K. and Kawari, Y. (1971) 'Localization of growth hormone (GH) and prolactin (MtH) on ultrathin sections of epon embedded pituitary gland with peroxidase labeled antibody', *Abstracts 11th Annual Meeting, American Society for Cell Biology,* New Orleans, Louisiana, 17,–20 November 1971, p. 203.

Neutra, M. and LeBlond, C. P. (1966) 'Synthesis of the carbohydrate of mucus in the Golgi complex as shown by electron microscope radioautography of goblet cells from rats injected with glucose-H^3', *J. Cell Biol.,* 30, 119–136.

Northcote, D. H. (1969) 'Fine structure of cytoplasm in relation to synthesis and secretion in plant cells', *Proc. R. Soc. B,* 177, 21–30.

Novikoff, A. B., Essner, E., Goldfischer, S., and Heus, M. (1962) 'Nucleosidediphosphatase activities of cytomembranes', In: *The interpretation of ultrastructure* (ed. R. J. C. Harris), pp. 149–192, Academic Press Inc., New York.

Novikoff, A. B. and Shin, W.-Y. (1964) 'The endoplasmic reticulum in the Golgi zone and its relations to microbodies, Golgi apparatus and autophagic vacuoles in rat liver cells', *J. Microscopie,* 3, 187–206.

Novikoff, P. M., Novikoff, A. B., Quintana, N., and Hauw, J.-J. (1971) 'Golgi apparatus, GERL, and lysosomes of neurons in rat dorsal root ganglia, studied by thick section and thin section cytochemistry', *J. Cell Biol.,* 50, 859–886.

Nyquist, S. E., Crane, F. L., and Morré, D. J. (1971) 'Vitamin A: concentration in the rat liver Golgi apparatus', *Science, N.Y.,* 173, 939–941.

Nyquist, S. E., Matschiner, J. T., and Morré, D. J. (1971) 'Distribution of vitamin K among rat liver cell fractions', *Biochem. biophys. Acta,* 244, 645–649.

Omura, T., Siekevitz, P., and Palade, G. E. (1967) 'Turnover of constituents of the endo-plasmic reticulum membrane of rat hepatocytes', *J. Biol. Chem.,* 242, 2389–2396.

Ovtracht, L., Morré, D. J., and Merlin, L. M. (1969) 'Isolement de l'appareil de Golgi d'une glande sécrétrice de mucopolysaccharides de l'escargot (*Helix pomatia*)', *J. Microscopie,* 8, 989–1002.

Palade, G. (1959) 'Functional changes in the structure of cell components', In: *Subcellular Particles* (ed. T. Hayashi). pp. 64–83, Ronald Press, New York.

Porter, D. (1969) 'Ultrastructure of *Labyrinthula*', *Protoplasma,* 67, 1–19.

Porter, K. R. (1961) 'The ground substance in the cell', In: *The cell* (eds. J. Brachet and A. E. Mirsky), pp. 621–675, Academic Press Inc., New York.

Ragnotti, G., Lawford, G. R., and Campbell, P. N. (1969) 'Biosynthesis of microsomal nicotinamide-adenine dinucleotide phosphate-cytochrome c reductase by membrane-bound and free polysomes from rat liver', *Biochem. J.,* 112, 139–147.

Ray, T. K., Lieberman, I., and Lansing, A. I. (1968) 'Synthesis of the plasma membrane of the liver cell', *Biochem. biophys. Res. Commun.,* 31, 54–58.

Redman, C. M. (1968) 'The synthesis of serum proteins on attached rather than free ribosomes of rat liver', *Biochem. biophys. Res. Commun.,* 31, 845–850.

Redman, C. M. (1969) 'Biosynthesis of serum proteins and ferritin by free and attached ribosomes of rat liver', *J. biol. Chem.,* 244, 4308–4315.

Redman, C. M., Siekevitz, P., and Palade, G. E. (1966) 'Synthesis and transfer of amylase in pigeon pancreatic microsomes', *J. biol. Chem.,* 241, 1150–1158.

Revel, J. R. and Hay, E. D (1963) 'An autoradiographic and electron microscopic study of collagen synthesis in differentiating cartilage', *Z. Zellforsch. mikrosk. Anat.,* 61, 110–144.

Robertson, J. D. (1964) 'Unit membranes; a review with recent new studies of experimental alterations and a new subunit structure in synaptic membranes', In: *Cellular membranes in development* (ed. M. Locke), pp. 1–81, Academic Press Inc., New York.

Rose, G. G. and Pomerat, C. M. (1960) 'Phase contrast observations on the endoplasmic reticulum in living tissue cultures', *J. biophys. biochem. Cytol.,* 8, 423–430.

Roseman, S. (1970) 'The synthesis of complex carbohydrates by multiglycosyltransferase systems and their potential function in intracellular adhesion', *Chem. Phys. Lipids,* 5, 207–297.

Bibliography

Sabatini, D. D., Blobel, G., Nomura, Y., and Adelman, M. R. (1971) 'Ribosome-membrane interaction: structural aspects and functional implications', In: *Advances in cytopharmacology (Proceedings 1st Intern. Symp. Cell Biol. and Cytopharmacology,* Venice, Italy, 7–11 July 1969, eds. F. Clementi and B. Ceccarelli) vol. 1, pp. 119–129, Raven Press, New York.

Schachter, H., Jabbal, I., Hudgin, R. L., Pinteric, L., McGurie, E. J., and Roseman, S. (1970) 'Intracellular localization of liver sugar nucleotide glycoprotein glycosyl transferase in a Golgi-rich fraction', *J. biol. Chem.,* **245,** 1090–1100.

Scheer, U. and Franke, W. W. (1969) 'Negative staining and adenosine triphosphatase activity of annulate lamellae of newt oocytes', *J. Cell Biol.,* **42,** 519–533.

Schnepf, E. (1968) 'Transport by compartments', In: *Transport and distribution of matter in cells of higher plants* (eds. K. Mothes, E. Muller, A. Nelles and D. Neumann), pp. 39–49, Akademie-Verlag, Berlin.

Schnepf, E. (1969a) 'Sekretion und Exkretion bei Pflanzen', *Protoplasmatologia, Handbuch der Protoplasmaforschung,* **8,** 1–181.

Schnepf, E. (1969b) 'Membranfluss und membrantransformation', *Ber. Dtsch. Bot. Ges.,* **82,** 407–413.

Schnepf, E., and Koch, W. (1966a) 'Golgi-Apparat und Wasserausscheidung bei *Glaucocystis',* *Z. PflPhysiol.,* **55,** 97–109.

Schnepf, E. and Koch, W. (1966b) 'Uber die Entstehung der puslierenden Vacuolen von *Vacuolaria virescens* (Chloromonadophyceae) aus dem Golgi-Apparat', *Arch. Mikrobiol.,* **54,** 229–236.

Schramm, M. (1967) 'Secretion of enzymes and other macromolecules', *A. Rev. Biochem.,* **36,** 307–320.

Sheldon, H. and Kimball, F. B. (1962) 'Studies on cartilage.' III. The occurrence of collagen within vacuoles of the Golgi apparatus', *J. Cell Biol.,* **12,** 599–613.

Shuster, L. and Jick, H. (1966) 'The turnover of microsomal protein in the livers of phenobarbital-treated mice', *J. biol. Chem.,* **241,** 5361–5365.

Siekevitz, P. (1965) 'Origin and functional nature of microsomes', *Fed. Proc.,* **24,** 1153–1155.

Siekevitz, P., Palade, G. E., Dallner, G., Ohad, I., and Omura, T. (1967) 'The biosynthesis of intracellular membranes', In: *Organizational biosynthesis* (eds. H. J. Vogel, J. O. Lampen and V. Bryson), pp. 331–361, Academic Press Inc., New York.

Smith, D. S. (1965) 'The organization of flight muscle in an aphid, *Megoura viciae* (Homoptera), with a discussion on the structure of synchronous and asynchronous striated muscle fibers', *J. Cell Biol.,* **27,** 379–393.

Takagi, M., Tanaka, T., and Ogata, K. (1970) 'Functional differences in protein synthesis between free and bound polysomes of rat liver', *Biochim. biophys. Acta,* **217,** 148–158.

Tanaka, T., Takagi, M., and Ogata, K. (1970) 'Studies on the metabolism of RNA of free and membrane-bound polysomes from rat liver', *Biochim. biophys. Acta,* **224,** 507–517.

VanDerWoude, W. J., Morré, D. J., and Bracker, C. E. (1971) 'Isolation and characterization of secretory vesicles in germinated pollen of *Lilium longiflorum',* *J. Cell Sci.,* **8,** 331–351.

Whaley, W. G., Kephart, J. E., and Mollenhauer, H. H. (1964) 'The dynamics of cytoplasmic membranes during development', In: *Cellular membranes in development* (ed. M. Locke), pp. 135–173, Academic Press, Inc., New York.

Willis, R. A. (1970) 'Tilting and stereoscopy as aids to interpretation in biological electron microscopy', *Philips Bulletin,* March 1970, pp. 1–5. Analytical Equipment Dept., N. V. Philips Gloeilampenfabriken, Eindhoven, Netherlands.

Wilkinson, F. E., Nyquist, S. E., Merritt, W. D., and Morré, D. J. (1972) 'Aryl sulphatases: properties and subcellular distribution in rat liver', *Proc. Indiana Acad. Sci. for 1971,* **81,** 121–132.

Wollenweber, Von E., Egger, K., and Schnepf, E. (1971) 'Flavonoid-Aglykone in *Alnus-* Knopsen und die Feinstruktur der Drüsenzellen', *Biochem. Physiol. Pflanzen,* **162,** 193–201.

Wood, J. G. (1967) 'The relationship of nucleotidase activity to catecholamine storage sites in adrenomedullary tissue', *Am. J. Anat.,* **121,** 671–704.

Zeigel, R. F. and Dalton, A. J. (1962) 'Speculations based on the morphology of the Golgi system in several types of protein secreting cells', *J. Cell Biol.,* **15,** 45–54.

4

Ultrastructure of mature chloroplasts
William W. Thomson

Introduction

This chapter is concerned with the ultrastructure of mature chloroplasts in higher plants. Because of space limitations, no attempt has been made to review other interesting aspects of chloroplast structure such as development, senescence, algal chloroplasts, and genetic influence. The reader is referred to recent reviews where different aspects of chloroplast ultrastructure are discussed (e.g., Bogorad, 1967; Kirk and Tilney-Bassett, 1967; Walles, 1967; von Wettstein, 1967; Lefort-Tran, 1969; Kreutz, 1970; Butler and Simon, 1971; Park and Sane, 1971).

The mature chloroplast can be divided into three phases: (a) the chloroplast envelope; (b) the stroma, comprising the internal matrix material; (c) the grana-fretwork system comprising the internal membrane system. To a certain extent, this division is rather artificial as the phases are undoubtedly interrelated functionally developmentally and even probably structurally. However, each phase has structural and functional properties not found in the others and each will be discussed separately although the relationships between them have been continuously borne in mind.

The internal membrane system is complex both in function and structure, and a plethora of terms has been applied in describing its organization. In this review the terminology developed by Weier will be used (Weier, 1961; Weier and Thomson, 1962a). To avoid confusion a résumé of these terms will be presented here. Briefly, the internal membrane system consists of compartmented structures, *grana* (Fig. 4.1, G), which are interconnected to other grana by membrane-bound channels, the *frets* (Fig. 4.1, f). The grana are bounded by marginal and end membranes which isolate the internal regions of the grana from the stroma (Fig. 4.1, arrows) (Weier, 1961; Weier and Thomson, 1962a; Thomson, 1965). The *marginal* and *end membranes* are continuous with the fret membranes. The grana are subdivided into compartments by membranous complexes, the *partitions* (Fig. 4.1, p). Often these partitions can be resolved into two electron opaque layers separated by a region or A space (Fig. 4.1, p) (Heslop-Harrison, 1963) which is usually electron translucent. The electron opaque layers of the partition are continuous with the marginal membranes of the grana. The partitions compartmentalize the grana and separate electron translucent regions, the *loculi* (Fig. 4.1, 1), from each other. The loculi are confluent with the electron translucent region enclosed by the fret membrane, the fret channels, and thus also with the loculi of other grana. A granum may consist of many compartments but the simplest granum is defined as consisting of two compartments and recognized by the presence of a single partition separating two loculi. The entire complex of compartmented grana and interconnecting frets is termed the *grana-fretwork system.*

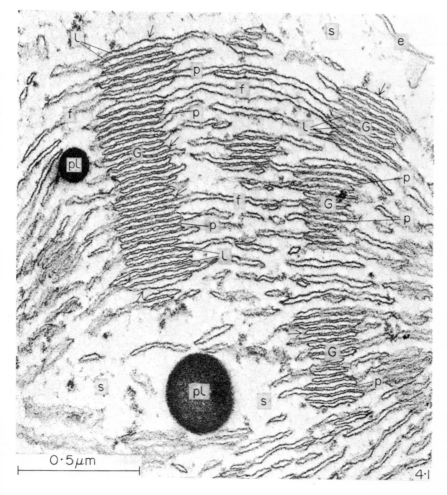

Fig. 4.1   A portion of a chloroplast from a green tomato fruit. The various components and structures of the chloroplasts are identified as: grana (G), frets (f), partitions (p), loculi (l), stroma (s), plastoglobuli (pl), envelope (e). The arrows indicate the end and marginal membranes of the grana. Fixation, osmium tetroxide, x87 000.

Size, volume, and shape

Chloroplasts in higher plants are generally about 5 μm in length and 2–3 μm in width (Granick, 1955) and are usually described as lens-shaped and, in profile views, appear bi-convex, plano-convex or concavo-convex in form. This is particularly true when chloroplasts are seen closely adjacent to the anticlinal walls of palisade parenchyma cells – a position which is thought to be the most favourable in reference to the intercellular air spaces and the absorption of CO_2 as well as in relation to light (Haberlandt, 1914). Chloroplasts in different regions in the same cell may vary considerably from the typical lenticular shape.

Transformation in the shape of the chloroplasts, including changes in the peripheral layer of stroma, were described by Senn in 1908. Renewed interest in these changes has been accelerated by the elegant phase microscopy and cinematographic studies of Wildman and his associates (Wildman, Hongladarom, and Honda, 1962; Spencer and Wildman, 1962; Wildman et al., 1964; Wildman, 1967). Of particular interest is the formation of finger-like protuberances which extend from the chloroplast (Heitz, 1936; Spencer and Wildman, 1962; Wildman, 1967). Protuberances have been observed also in electron microscopic studies and have been described for etioplasts (Laetsch and Price, 1969; Laetsch, 1969b), mature chloroplasts (Fig. 4.3, po; Laetsch, 1969a, b), chloroplasts undergoing the transitions to chromoplasts (Thomson, 1969; Nichols, Stubbs, and James, 1967), chloroplasts in tissues infected with a virus (Shalla, 1964), in manganese-deficient plants (Possingham, Vesk, and Mercer, 1964) and in chloroplasts in starvation-stressed, degenerating cells (Ragetli, Weintraub, and Lo, 1970). At the ultrastructural level there appear to be two types of protuberance. Some appear as long, thin extensions which are bounded by the chloroplast envelope and mainly contain stroma material (Weier and Thomson, 1962b; Shalla, 1964; Nichols et al., 1967). This 'type' of extension is frequently curled and, in some instances, the tip appears to reunite with the chloroplast forming small enclaves (Shalla, 1964; Nichols et al., 1967) in which other cellular constituents such as ribosomes and mitochondria are often seen (Vesk, Mercer, and Possingham, 1965). The second 'type' is generally much wider and may occur as bulges or protuberances (Figs. 4.2, 3, po). These contain stroma material but are characterized by the occurrence of many invaginations of the inner membrane of the chloroplast envelope. Chloroplasts of plants possessing high photosynthetic capacities (i.e., C_4 plants) frequently have bulges and long protuberances in which the invaginations of the inner membrane of the chloroplast envelope are extensively developed into an anastomosing network of tubules (Fig. 4.2, 3 po; and Laetsch, 1969a).

The significance of the protuberances is not clear. Their occurrence in etioplasts (Laetsch, 1969b), and the fact that proplastids are often amoeboid (Esau, 1965; Kirk and Tilney-Bassett, 1967) seems to rule out an absolute relationship to light. However, this does not eliminate the possibility that in mature chloroplasts their formation may be influenced by light. Also, their occurrence in mature chloroplasts apparently mitigates their being necessarily related to degenerative or abnormal conditions as suggested by Clowes and Juniper (1968).

Wildman and his associates (Wildman *et al.*, 1962; Spencer and Wildman, 1962; Wildman *et al.*, 1964) have reported from their light microscopic and cinephoto-micrographic studies that mitochondrial-like bodies develop from the protuberances. However, Vesk *et al.* (1965) considered the light and electron microscopic evidence for the possible interconversion of mitochondria and concluded that a definitive answer cannot be provided.'. . . the critical events lie beyond the resolution of the light microscope and the electron microscope gives only a static picture'. However, as emphasized by Wildman (1967), these changes in shape indicate that the surface of the chloroplast and the underlying stroma is dynamic and mobile.

Prior to the advent of electron microscopy, light microscopists were aware that *in vivo* chloroplasts swell or contract if the tissues are placed in hypotonic or hypertonic medium. The osmotic response of the chloroplast was probably, at that time, the strongest evidence that the chloroplast was surrounded by a semi-permeable membrane (see Weier, 1938; Weier and Stocking, 1952; Packer and Siegenthaler, 1966 for reviews). Changes in isolated chloroplasts in response to changes in the osmotic concentration of the suspending medium have been studied by a number of investigators. These studies have shown that, not only do intact chloroplasts swell and contract osmotically, but granal compartments respond in a similar manner (see Packer and Siegenthaler, 1966; Packer, Mehard, and Murakami, 1970 for reviews). It is therefore important to emphasize that, since chloroplasts and granal compartments respond osmotically, in experiments concerned with these phenomena it should be clearly distinguished whether the studies involve the whole chloroplast, the granal compartments or both.

The integrity of the chloroplasts is an important factor in *in vitro* studies (Walker, 1965, 1967; Leech, 1967) and two general categories of isolated chloroplasts are recognized: intact chloroplasts (Class I) with stroma and grana-fretwork system within an envelope, and broken chloroplasts (Class II) which lack envelopes and consist almost entirely of the grana-fretwork system (Spencer and Wildman, 1962; Spencer and Unt, 1965). Isolated intact chloroplasts appear as bright, shiny, opaque objects under phase contrast while broken chloroplasts have a dark, granular appearance (Kahn and von Wettstein, 1961; Spencer and Wildman, 1962; Leech, 1964; Spencer and Unt, 1965).

Recent studies by Harnischfeger (1970) and Harnischfeger and Gaffron (1969, 1970) on swelling and resulting chloroplast disintegration, indicate that changes in the shape and integrity of chloroplasts and the grana-fretwork system have a significant effect on the photosynthetic capacity of chloroplasts. They reported that both the whole chloroplast as well as the granal compartments within the chloroplasts are capable of swelling at the same time, and that the transition point between Class I chloroplasts and Class II chloroplasts occurs when the grana-fretwork system swells and becomes closely associated with the chloroplast envelope. At the transition point they observed a drop in the efficiency of the Hill reaction in blue light which, under certain conditions, could be reversed or delayed by increasing the osmotic concentration of the suspending medium. Since the effects on the Hill reaction were correlated with the swelling of the chloroplast and grana-fretwork system and the transition from intact to broken chloroplasts, they suggested that a physical disruption of the grana-fretwork system when swollen osmotically causes a separation of the pigment complexes within the membranes and a concomitant decrease in the efficiency of light utilization and energy transfer between complexes.

The evidence is quite convincing that chloroplasts, both *in vivo* and when isolated, undergo light-induced volume changes. The evidence has been provided through light and electron microscopic studies and with the use of Coulter counter and packed volume techniques on isolated chloroplasts (Kushida *et al.,* 1964; Hilgenberger and Menke, 1965; Packer, Barnard, and Deamer, 1967; Murakami and Packer, 1970; Nobel, 1967a, b, 1968a, b, c, 1969; Nobel *et al.,* 1969). These studies have shown that chloroplasts flatten and decrease significantly in volume in the light, and that these changes are reversed in the dark. Several studies have been directed at the mechanism of shrinkage as an osmotic consequence of ion transport (see Packer, Mehard, and Murakami, 1970 for review). However, Nobel (1968a) observed that isolated chloroplasts from illuminated leaves had twice the rate of photophosphorylation of those isolated from plants in the dark, and suggested that the light-induced changes in shape and volume may influence the photosynthetic efficiency of the chloroplasts. This is an interesting suggestion since it implies that changes in the form of the chloroplast may, to some extent, influence and regulate photosynthetic activity. Exactly how the changes in shape may influence the function of chloroplasts would be speculative at this point. However, it is probably not too extreme to point out that, along with a possible change in membranes, the increased surface-to-volume ratio in the light-induced, flattened chloroplast may be important.

The loculi of the granal compartments are also known to contract and diminish in width in the light and expand in the dark in both *in vivo* and isolated chloroplasts (Itoh, Izawa, and Shibata, 1963; Izawa and Good, 1966; Dilley, Park, and Branton, 1967; Murakami and Packer, 1970; Sundguist and Burris, 1970). A similar flattening of the compartments occurs when the pH of the suspending medium is lowered by addition of acid which is reversible by the addition of alkali (Murakami and Packer, 1970). Packer, Mehard, and Murakami (1970) have reviewed and summarized the experimental studies on this phenomenon and directly imply that the flattening of the compartments is an osmotic response due to a light-dependent efflux of osmotically active species (ion translocation) from the grana which is then reversed in the dark.

The chloroplast envelope

Before the advent of electron microscopy and, particularly, the technique of thin-sectioning, the presence or absence of a membrane around the chloroplast was seriously argued (Weier, 1938; Weier and Stocking, 1952). However, it is now well established that the chloroplast envelope consists of two membranes, each about 8–10 nm in thickness, separated by a region about 10–20 nm in width which appears electron translucent in electron micgrographs (Fig. 4.1, e). The chloroplast envelope must be considered one of the most important interfaces in the plant. It is across this envelope that photosynthetic metabolities and products enter and leave the chloroplast. Further developmental studies have shown that invaginations and vesicles formed from the inner membrane of the envelope are apparently involved in the development of the grana-fretwork system (see Kirk and Tilney-Bassett, 1967 for review).

Considering the importance of the envelope, it is surprising that more attention has not been directed at elucidating its structure, function and composition. However, with the development of techniques for isolating intact chloroplasts (Leech, 1964; Walker, 1965, 1967; Nobel, 1967b), studies have come forth particularly with regard to the transport of substrates, products of photosynthesis, and the movement of ions (Heber, 1969; Stocking and Larson, 1969; Nobel et $al.$, 1969). Also, preliminary studies on the isolation and characterization of the chloroplast envelope have been reported recently (Mackender and Leech, 1970).

Electron microscope studies on the ultrastructure of the chloroplast envelope are not abundant. However, the studies to date indicate that there are numerous variations in its general organization. Invaginations of the inner membrane of the envelope are observed in most mature chloroplasts. Apparently the most common type of invaginations are small vesicular elements, although flatter, finger-like invaginations are not unusual. In some cases, much longer invaginations, which generally lie in the peripheral region of the stroma, occur (Fig. 4.6, i). One or several of these lamellar extensions may occur and they generally run parallel to the envelope. Often the invaginations may be confluent with the inner envelope membrane in two or more places. Schotz and Diers (1967), from serial sections, have reported that the invaginations are plate-like structures which enclose small enclaves and pockets of stroma.

Weier and Thomson (1962b) earlier described small single membrane bodies which were apparently isolated between the inner and outer envelope membranes and the studies of Schotz and Diers (1965, 1967) support this finding. These bodies contain a matrix which, after $KMnO_4$ fixation, appears similar to the stroma and Schotz and Diers (1967) have suggested that they may develop from the small enclaves formed by invaginations which connect in two or more places with the inner envelope membrane. Both Weier and Thomson (1962b) and Schotz and Diers (1965, 1967) have observed bodies in the cytoplasm similar to those between the two membranes of the envelope. These investigators have advanced the suggestion that these cytoplasmic particles are stroma particles, formed and elaborated by pocketing of the chloroplast envelope and which are emitted to the cytoplasm.

One of the most fascinating variations in the ultrastructure of the chloroplast envelope has been described in chloroplasts of Zea $mays$ and chloroplasts of some other plants (Fig. 4.3, 4, pr) which have a high photosynthetic capacity and a low CO_2 compensation value, i.e., C_4 plants (Shumway and Weier, 1967; Rosado-Alberio, Weier, and Stocking, 1968; Laetsch, 1968, 1969a, b). In these chloroplasts,

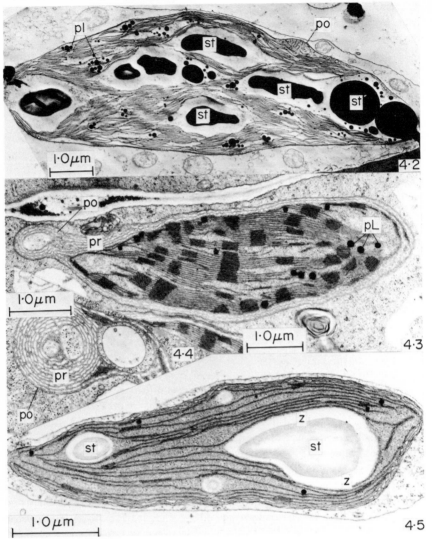

Fig. 4.2   A chloroplast from the cactus, *Echinocactus* with round to oblong starch grains (st), many plastoglobuli (pl), and a surface protuberance (po). Fixation: osmium tetroxide. x15 600.

Fig. 4.3   Mesophyll chloroplast from Bermuda grass (*Cynodon dactylon* L.)—a plant with a high photosynthetic capacity. Note the extensive peripheral reticulum (pr) between the chloroplast envelope and the grana-fretwork system. The peripheral reticulum also occurs in the protuberance (po). Numerous plastoglobuli (pl) are also present. Fixation: glutaraldehyde and osmium tetroxide. x20 000.

Fig. 4.4   A large protuberance (po) from the surface of a chloroplasts of Bermuda grass (*Cynodon dactylon* L.). Note the extensively developed peripheral reticulum (pr) within the protuberance. Fixation: glutaraldehyde and osmium tetroxide. x22 250.

invaginations of the inner envelope membrane in the form of an elaborate anastomosing system of tubules or peripheral reticulum have been described. The tubules apparently connect with the grana-fretwork system (Rosado-Alberio et al., 1968). The function of the peripheral reticulum is not known although Rosado-Alberio et al. (1968) and Laetsch (1969a, b) have suggested that it may be involved in the transport of materials to and from the grana. However, it is somewhat doubtful that the peripheral reticulum is a general feature of chloroplasts of plants with high photosynthetic capacities since Downton, Bisalputra, and Tregunna (1969) did not observe it in *Atriplex rosea,* a plant with a low CO_2 compensation value. More comparative studies on other genera and on plants in different physiological states are needed before a clearer understanding of the significance and function of the peripheral reticulum is obtained.

Detailed studies on the substructure of the envelope membranes have not been abundant and it is probably fair to say that this aspect of chloroplast ultra-structure has been relatively ignored. In most early electron microscopic studies the membranes appeared as opaque lines each about 10 nm in width (Mercer, 1960). However, Weier and Thomson (1962b) reported that, after $KMnO_4$ fixation, the membranes often appeared unequal in thickness. Heslop-Harrison (1963), also using $KMnO_4$ fixed material, described each membrane of the chloroplast envelope as having a unit membrane (Robertson, 1959, 1964) pattern of electron opacities — that is, each membrane was composed of two electron opaque layers, about 2 nm in thickness, separated by an electron translucent region approximately 1 nm in width. In contrast, Weier and his associates (Weier et al., 1965a) briefly noted that the envelope membranes were composed of spherical subunits, although a micrograph clearly demonstrating this was not provided.

Very little is known about the interspace between the two membranes of the chloroplast envelope. Recently, Sabnis, Gordon, and Galston (1970) using histochemical techniques for the fine structural localization of a nucleoside phosphatase have reported the occurrence of a Mg^{++}-dependent ATPase between the two membranes of the envelope. The enzyme appeared to be light activated and these investigators have suggested that it may be involved in the light dependent contraction of chloroplasts. Clarification as to the nature and function of this and other enzymes that may reside in the envelope depends on studies with intact, isolated chloroplasts and on purified preparations of the envelope.

Stroma

The stroma (Fig. 4.1, s) has long been identified as proteinaceous (Granick, 1961) and it is apparent from a number of studies that the major constituent is Fraction I protein (Moyse, 1967) consisting mostly of the enzyme, ribulose diphosphate

---

Fig. 4.5 A chloroplast of a bundle sheath cell of *Zea mays* L. Note the reduced number and small size of the grana. Starch grains (st) are evident with a clear zone (z) between the grains and the stroma. Fixation: glutaraldehyde and osmium tetroxide. x31 000.

carboxylase. (However, see Kawashima and Wildman's (1970) excellent review on Fraction I protein for a discussion of this point.) In electron micrographs of thin sections of chloroplasts the main bulk of the stroma, presumably Fraction I protein, has a granular appearance. The size and the distribution of the granules vary according to the fixative used in preparing the material. After $KMnO_4$ fixation the granules are usually rather uniform in size and distribution and measure about 4.5—5.5 nm in diameter. When osmium tetroxide is used as the primary fixative, the size of the granules varies considerably, and frequently the distribution is irregular with clusters of granules often observed. After fixation with glutaraldehyde followed by osmium tetroxide, the bulk of the stroma material appears granular and rather evenly distributed.

Gunning and his associates (Gunning, 1965a; Gunning, Steer, and Cochrane, 1968; Steer et al., 1968) have described a fibrillar spherulite 8—9 nm in width and up to 200 nm in length in the stroma of Avena chloroplasts after aldehyde-osmium tetroxide fixation. Gunning et al. (1968) have suggested that these spherulites or 'stromacentres' are possibly fibrillar aggregates of Fraction I protein. It is particularly interesting that although 'stromacentres' have not been described in normal chloroplasts of most other plants, similar, semi-crystalline arrays of granules and fibrils have been induced in chloroplasts of beans by treatment of the leaves with the oxidant air pollutants, ozone (Fig. 4.7, arrows) and peroxyacetyl nitrate (Thomson, Dugger, and Palmer, 1965, 1966). Crystalline structures also were observed in the stroma of isolated chloroplasts by Perner (1962), and similar structures have been shown to occur in the stroma of chloroplasts in leaves that had been plasmolysed or in chloroplasts subjected to centrifugation (Wrischer, 1967; Shumway, Weier, and Stocking, 1967; Gunning et al., 1968; Bradbeer et al., 1970). More recently Ragetli, Weintraub, and Lo (1970) have described exceedingly large, pseudocrystalline bodies in the stroma of chloroplasts of starvation stressed cells.

Both Shumway et al. (1967) and Gunning et al. (1968) have pointed out that the formation of these semi-crystalline arrays was probably related to the increased dehydration of the stroma due to the removal of water. This explanation could also explain the formation of the semi-crystalline arrays in oxidant-treated plants (Thomson et al., 1965, 1966), since one of the first symptoms of this type of injury is a 'water logging' of the intracellular spaces due to the loss of water from the cells (see Middleton, 1961 for review). As Gunning et al. (1968) have pointed out, this explanation does not answer the question as to the presence of the 'stromacentres' in Avena in which there is no reason to suggest that the hydration state of the chloroplast is significantly altered or different from most other plants.

As discussed by Gunning et al. (1968) although the semi-crystalline arrays and 'stromacentres' have many similarities there are some important differences. The 'stromacentres' observed in Avena (Gunning 1965a, b; Gunning et al., 1968; O'Brien and Thimann, 1967) and the semi-crystalline arrays of fibrils observed in oxidant treated, Phaseolus chloroplasts are generally arranged in clusters of hexagonal and curvilinear arrays. However, the semi-crystalline arrays induced by water stress in *Phaseolus* (Gunning *et al.*, 1968) and the crystalline arrays observed by Perner (1962) and Shumway *et al.* (1967) in isolated, intact chloroplasts, are arranged in linear and angular arrays. The striking similarity between the semi-crystalline arrays and the 'stromacentres' in the width of the fibrils, longitudinal periodicity, their location in the stroma, and the overall organization, suggest they are probably composed of much the same material, with Fraction I protein being a likely candidate (Gunning *et al.*, 1968). The variation in the organization of the semi-crystalline arrays and the occurrence of the 'stromacentre' in normal *Avena* chloroplasts may be due to differences in the major component of these aggregates in different species (Gunning *et al.*, 1968).

If the major component or sole constituent of these arrays is Fraction I protein, then the differences must be subtle, since, as Kawashima and Wildman (1970) have pointed out, there is a great deal of biochemical, biophysical, and immunological similarity between Fraction I protein isolated from different plants. Although attractive, the argument that these arrays are mainly if not entirely composed of Fraction I protein needs further support.

Starch

Starch grains are one of the commonly observed bodies in the stroma of chloroplasts in leaves exposed to the light. The grains appear roughly oval in shape in most electron micrographs although they may vary from oblong to round (Figs. 4.2, 5, 6, st). Starch grains are easily stained for light microscope studies but a reliable positive stain has not been developed for material prepared for electron microscopy. Recently, Palevitz and Newcomb, (1970) have used enzymic treatments of thin sections to study the structure and composition of starch in sieve-tube plastids. The application of this approach to starch grains in other plastids including chloroplasts would be valuable in identifying the bodies and elucidating aspects of their structure and development.

Although starch grains are not positively identified in thin sections, what are assumed to be starch grains show wide variations in staining characteristics, in particular to osmium and potassium permanganate. The grains usually stain rather lightly or appear electron translucent in micrographs (Fig. 4.5, st, Buttrose, 1960). Frequently, however, they appear uniformly electron opaque (Fig. 4.2, st, Frey-Wyssling and Mühlethaler, 1965). This may happen even though in both cases the material was prepared by identical methods. This difference may be a result of different opacities and hydration states of starch grains which determine the ability of osmium and permanganate to penetrate into the grain. The question of whether the electron opaque material within the grains represents a reaction product or simply an accumulation of heavy metal within the grain has not been answered, although the latter seems the more probable. Probably the most common image of starch grains, particularly after the use of osmium tetroxide, is a gradation of opacity with the internal region being more electron translucent and towards the edge of the grain there is a gradual although not intense increase in opacity (Fig. 4.6, st). After osmium fixation, granules often occur in association with the surface of the starch grains with the result that the edge of the grain has a granulated, opaque boundary (Fig. 4.6, arrows). Whether the granules are involved in the development of degradation of the grains is not known.

Starch grains are closely bounded by stroma in many electron micrographs, although frequently a light, electron translucent zone exists between the stroma and the starch grain (Fig. 4.5, z). This may reflect differential shrinkage of the stroma and starch grains during fixation and dehydration. The grains occur in the stroma of chloroplasts and do not appear to be associated closely with the grana-fretwork system. There seems to be little evidence that starch grains in chloroplasts are associated with membranes as has been described in amyloplasts (Buttrose, 1960; Badenhuizen, 1962, 1959) and for plastids of the meristematic cells of potato tuber buds (Marinos, 1967). Neither is there any compelling evidence that the starch grains in chloroplasts form within vacuoles (Caporali, 1959; Buvat, 1963).

Starch grains may increase in number and size to such an extent that the chloroplasts become extended and often very irregular in outline. There is some indication that the grana-fretwork system may become reduced in extent, making it difficult to differentiate these plastids from amyloplasts. Recently, Grob and Rufener (1969) reported that when the autotrophic water plant *Spirodela* was grown under continuous illumination in a medium containing glucose, the grana-fretwork system was seriously reduced and that there was a correlated increase in starch. Returning the plants to a glucose-free medium resulted in a regeneration of the grana-fretwork system and a decline in starch.

The possibility that starch accumulation coupled with a concomitant decrease in the grana-fretwork system has a regulatory influence on the photosynthetic capacity of chloroplasts is interesting and should be investigated in greater detail (see Wildman, 1967 for a pertinent comment on this point). However, in some cases other factors may be involved. Ballantine and Forde (1970) observed considerable variation in the development of grana and starch in the chloroplasts of soybean plants grown under different light and temperature conditions. For example, they reported that the chloroplasts in the palisade parenchyma cells of plants grown under high light/low temperature conditions had a reduced grana-fretwork system and lacked starch. However, under high light/high temperature conditions the chloroplasts of the palisade parenchyma cells had small grana and starch grains. In contrast, under low light/high temperature conditions, the chloroplasts in the spongy mesophyll had a well-developed grana-fretwork system and large starch grains. It is interesting that when plants were transferred from the low light/high temperature condition to one of high light/low temperature, there was an increase in starch and a disorganization of the grana-fretwork system in the spongy mesophyll chloroplasts.

The accumulation of starch may vary considerably between chloroplasts in different tissue in the same leaf (Ballantine and Forde, 1970). Starch accumulates readily in the bundle sheath chloroplasts of many plants with high photosynthetic capacities and less so in mesophyll chloroplasts (Downton and Tregunna, 1968; Laetsch, 1969a, b; Crookston and Moss, 1970). However, the suggestion that the presence of large amounts of starch in the bundle sheath chloroplasts can be used to characterize plants with high photosynthetic capacities has been criticized by Black and Mollenhauer (1971) who observed starch grains in both bundle sheath and mesophyll chloroplasts of several plants with high photo-synthetic capacities. Laetsch (1968, 1969a, b) has also noted that mesophyll chloroplasts in high photosynthetic capacity plants may accumulate starch. He has suggested that the large amount of starch in the bundle sheath chloroplasts is related to their position as a sink in a gradient of photosynthetically fixed carbon from the mesophyll cells. Laetsch (1969a) has also pointed out that the bundle sheath chloroplasts are much larger than the mesophyll chloroplasts in plants with high photosynthetic capacities and thereby have greater potential for starch synthesizing and storage. From a similar point of view he has noted that the bundle sheath chloroplast in *Zea mays* (Fig. 4.5) and sugar cane have a high ratio of stroma to internal membranes since these chloroplasts lack grana or only possess small grana and are characterized by a series of parallel membranes ('stroma lamellae') (Hodge, McLean, and Mercer, 1955; Laetsch, Stetler, and Vlitos, 1966; Weier, Stocking, and Shumway, 1966). It should be pointed out that reasonably quantitative and comparative determinations of the ratio of stroma to internal membranes have not been reported for any chloroplasts except possibly for the obvious difference between the bundle sheath and mesophyll chloroplasts in such plants as corn. Further, Black and Mollenhauer (1971) have reported that the mesophyll chloroplasts in high photosynthetic capacity plants may be equal or larger in size than chloroplasts in the bundle sheath. Also, the observations of Ballantine and Forde (1970) that the chloro-plasts may have large starch accumulation and a well-developed grana-fretwork system seems to argue against the ratio of stroma to internal membrane being of primary importance in the ability of chloroplasts to accumulate starch. Detailed studies correlating starch accumulation with changes in the extent and organization of the grana-fretwork system would be helpful in clarifying the possible interrelationships that may exist.

Plastoglobuli

A characteristic feature of chloroplasts fixed with osmium tetroxide is the occurrence of round, electron opaque bodies in the stroma (Figs. 4.1, 2, 3, 6, pl). These plastoglobuli are not present after $KMnO_4$ fixation, although irregular, electron opaque outlines, 'star-bodies', have been observed (Weier, 1961; Weier and Thomson, 1962a; Heslop-Harrison, 1963). These are probably the result of the reaction of $KMnO_4$ only at the surface of the plastoglobuli, and during dehydration in organic solvents the internal, unreacted material is extracted with the consequence that the surface deposits collapse, forming the irregular 'star-bodies'.

The plastoglobuli appear free in the stroma and may occur singly or in clusters. Some affinity with the grana-fretwork membranes may exist since in isolated, washed chloroplasts the plastoglobuli occur associated with the grana-fretwork

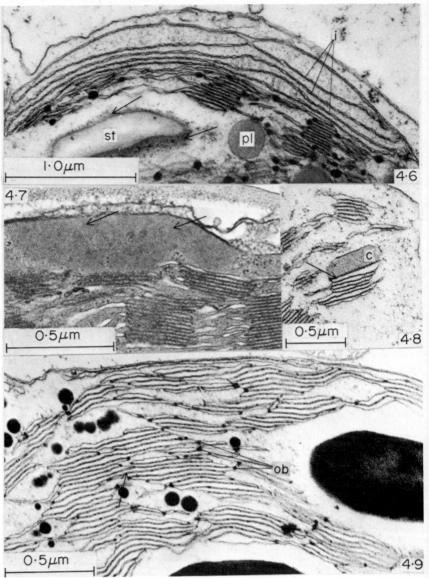

Fig. 4.6   A portion of a chloroplast from a green Valencia orange (*Citrus sinensis* L.). Long invagination of the chloroplast envelope (i) occur frequently in these chloroplasts. Note the presences of dense granules (arrows) near the surface of the starch grain (st) and the rather large plastoglobuli (pl). Fixation: osmium tetroxide. x45 500.

Fig. 4.7   A portion of a chloroplast from the leaf of cotton plant (*Gossipium hirsutum* L.) which had been treated with the oxidant air pollutant, ozone. Note the crystalline structure in the peripheral stroma (arrows). Fixation: glutaraldehyde and osmium tetroxide. x74 000.

Fig. 4.8   A crystalline body (c) in close association with a granum (arrow) in a spinach chloroplast. Fixation: osmium tetroxide. x42 500.

systems under conditions in which they might be expected to have been washed free.

The size of the plastoglobuli in normal photosynthesizing chloroplasts generally ranges from 10 to 500 nm in diameter (Leech, 1968) although it is not unusual to find plants in which the plastoglobuli are quite large (Falk, 1960). The size and number tend to increase with aging (Lichtenthaler, 1966, 1968; Dodge, 1970), and in some chloroplast to chromoplast transitions the plasto-globuli increase in number and size and become the dominant feature of the mature chromoplasts (Thomson, 1966; Kirk and Tilney-Bassett, 1967; Butler and Simon, 1971).

Evidence from several types of study indicates that the plastoglobuli are not artifacts. Greenwood, Leech, and Williams (1963) have isolated plastoglobuli and found that they have the same electron microscopic characteristic as those of *in vivo* chloroplasts. Further, globules of the same size and distribution can be observed in freeze-etched preparations (Mühlethaler, Moor, and Szarkowski, 1965), and in chloroplasts in tissue that was fixed with osmium tetroxide but dehydrated by air drying rather than with solvents (Fig. 4.10, pl). It is interesting that the globules appear to be less opaque in the air-dried material. This suggests that the organic solvents usually used in dehydration have some role in the deposition of osmium within the globules.

It is generally suggested that plastoglobuli represent a reservoir of excess lipid (Greenwood *et al.*, 1963; Bailey and Whyborn, 1963). However, little or no chlorophyll or carotenoid pigments have been found in globules from photo-synthesizing green leaves (Greenwood *et al.*, 1963). With aging and during the transition of chloroplasts to chromoplasts, the composition of the globules changes, and carotenoid pigments accumulate in them (Lichenthaler, 1969a, b).

A particularly interesting observation has been made by Adams and Strain (1969) in a study of the drought-deciduous desert plant *Cercidium.* They found that chloroplasts in the rather ephemeral leaves which appear after a rainfall contained starch and are ultrastructurally similar to those of other higher plants. The chloroplasts in the green bark tissue, which evidently provides a major source of photosynthetically fixed carbon to the plant, resemble those in the ephemeral leaves and other plants, except they lacked starch and have numerous large plastoglobuli. Adams and Strain have suggested that the plastoglobuli may represent a form in which photosynthetically fixed carbon is stored in these chloroplasts subsequent to further utilization in the carbon economy of the plant. Thus, these plastoglobuli may have the same general role as starch in most other plants. This is an attractive hypothesis, particularly in regard to plants where limited water may be available for metabolic processes and from the point of view of efficient energy conservation and utilization.

Fig. 4.9   A portion of a chloroplast from a cactus (*Echinocactus*) showing elaborately branched grana with compartments extending between grana (arrow). Note the presence of small osmiophilic bodies (ob) along the margins of the grana. Fixation: osmium tetroxide. x62 700.

Phytoferritin

Hyde *et al.* (1963) reported the occurrence and isolation of an iron-protein complex from pea embryos and termed it phytoferritin because of the similarity in
structure and composition to ferritin present in animal tissues. The phytoferritin
particles appear electron opaque even in material not fixed in osmium or
poststained with heavy metals (Hyde *et al.,* 1963). This intrinsic electron
opacity is apparently due to the high concentration of iron in the particles.
Phytoferritin may exist in either crystalline arrays consisting of particles aligned
in parallel, curved, or straight rows, as a more irregular paracrystalline
aggregation, or as clusters of particles (Robards and Humpherson, 1967).
Hyde *et al.* (1963) reported from negative stained images of isolated phytoferritin
that the particles measured about 10 nm in diameter. In thin sections of material
they reported electron opaque particles about 5.5 nm in diameter which, when
arranged in a crystalline fashion, have a centre-to-centre spacing of about 10
to 10.5 nm (Hyde *et al.,* 1963; Robards and Humpherson, 1967). The electron
opaque particle apparently represents the iron-containing core and the electron
translucent region around the particles the proteinaceous shell (Robards and
Robinson, 1968).

Phytoferritin has been found most commonly in proplastids and differentiating
plastids or in senescing chloroplasts and developing chromoplasts (Hyde *et al.,*
1963; Robards and Humpherson, 1967; Robards and Robinson, 1968; Marinos,
1967; Newcomb, 1967; Catesson, 1966, 1970; Barton, 1970). Similar particles
have been observed in the stroma of chloroplasts in tissue infected by a virus
(Cronshaw, Hoefert, and Esau, 1966). Since phytoferritin occurred in proplastids
and differentiating plastids, and because in mature chloroplasts, iron-containing
compounds such as ferredoxin are essential for photosynthesis, Hyde *et al.* (1963)
suggested that phytoferritin represented a reservoir of stored iron in the form of
a non-toxic, iron-containing protein. Support for this view has come from
experiments in which chloroplasts of iron-deficient plants which contained no
phytoferritin on treatment with excess iron, accumulated heavy deposits of
phytoferritin (Seckback, 1968). Phytoferritin has been reported in mature
chloroplasts of *Tradescantia* (Sprey, 1965). Even in mature chloroplasts it may
also represent a storage product if sufficient iron is available.

Crystalline bodies

Schimper (1885) reported the occurrence of what he interpreted as protein
crystals in chloroplasts in several plants. At the electron microscope level,
crystalline bodies have been reported in chloroplasts of coconut palm leaves (Price,
Martinez, and Warmke, 1966), *Macadamia* leaves (Price and Thomson, 1967), and
the chloroplasts of leaves of *Phaseolus* (Wrischer, 1967). Crystalline bodies are
also observed in spinach chloroplasts (Fig. 4.8, c). These crystalline bodies are
closely associated with the grana (Fig. 4.8, arrow) and are composed apparently
of granules aligned in linear rows (Fig. 4.8, c). Whether the crystalline bodies

are related to those described by Schimper is not known. They do not seem to be related to viruses, since symptoms of a viral infection were not observed in the plants investigated. However, viral related inclusions have been reported in plants showing no visual symptoms of the infection (McWhorter, 1965).

The crystalline bodies in the chloroplasts of the coconut palm are quite large, measuring about 1 μm in width and appear to be enclosed within a membrane (Price *et al.,* 1966). The bodies are composed of hexagonal particles about 12 nm in diameter. Price and Thomson (1967) observed that the crystalline bodies in *Macadamia* chloroplasts were variable in size and closely associated with the granal and fret membranes. They found two types of line patterns, with the more common type being a series of parallel electron dense lines alternating with electron translucent regions. The electron opaque lines had a centre-to-centre spacing of about 16 nm. Small electron opaque strands extending across the translucent region connecting the electron opaque lines suggested that there was a subunit structure to the crystalline body. The other image consisted of a hexagonal, honeycombed pattern and the centre-to-centre spacing between each honeycomb unit was about 16 nm. Price and Thomson (1967) have suggested that the parallel line pattern is observed when the section runs parallel to the columns and the honeycomb pattern is observed when the section is across the columns.

The significance of the crystalline bodies is not known. Price and Thomson (1967) suggested that the bodies may be developmentally or functionally related to the granal and fret membranes or that they were possibly a storage form of excess protein or lipoprotein. Wrischer (1967) found that the crystalline bodies were removed when the thin sections were treated with pronase, which is an indication of their proteinaceous nature. It is interesting to note that amorphous and crystalline bodies have also been described in proplastids (Marinos, 1967; Newcomb, 1967; Srivastava and O'Brien, 1966). The studies by Marinos on the proplastids of potato tuber buds and Newcomb (1967) on proplastids of root also provided histochemical and enzymatic removal evidence that these are proteinaceous bodies.

Ribosomes

A characteristic feature of chloroplasts is the presence of ribosomes in the stroma (see Swift and Wolstenholme, 1969; Lefort-Trans, 1969). These are particularly evident after staining with uranyl acetate, but they are not observed after permanganate fixation. In thin sections the ribosomes measure about 17 nm in diameter and are thus smaller than ribosomes in the hyaloplasm which measure about 25 nm in diameter (see also mitochondrial ribosomes, p. 57). The ribosomes frequently appear as individual particles distributed throughout the stroma. However, Brown and Gunning (1966) have described clusters of ribosomes in developing chloroplasts and Bartels and Weier (1967) have described a helically arranged polysome in the proplastids of *Tritium* which lies free in the stroma. More recently, Falk (1969) has provided rather convincing electron microscopic evidence for the occurrence of excentric spiral and circular whorls of polysomes in chloroplasts of *Phaseolus*. The polysomes were attached to the surface of the fret and the end membranes of the grana. High magnification micrographs revealed a thin, 1–1.5 nm, strand extending between some of the ribosomes. Falk's observations are supported by recent combined electron microscopic and experimental studies on isolated granal membranes by Philippovich *et al.* (1970) and by other biochemical studies which indicate that some of the chloroplast RNA is associated with these membranes (Dyer and Leech, 1968).

DNA

As with ribosomes, the evidence from cytological, electron microscopic, autoradiographical, and biochemical studies has clearly established the presence of DNA in chloroplasts. At the ultrastructural level the DNA appears as an irregular mesh of 2.5 nm fibrils located in clear zones in the stroma (Ris and Plaut, 1962; Kislev, Swift, and Bogorad, 1965). Swift and his collaborators (see Kislev, Swift, and Bogorad, 1966; Swift and Wolstenholme, 1969 for reviews) have found that these fibrils bind uranyl ions and that the fibrils were removed with DNAase treatment of the tissue. While these studies were mainly on proplastids or developing chloroplasts, however, clear zones containing 2.5 nm fibrils have been observed in mature chloroplasts of several different plants (e.g., Gunning, 1965b; Yokomura, 1967) and based on the ultrastructural similarity to the DNA sites observed by Swift and his co-workers, these regions are usually interpreted as sites of DNA. Herrmann and Kowallik (1970b) have recently shown that the stroma may also mask the presence of DNA fibrils and that they can be visualized after removal of the stroma by enzymic extraction. Several DNA sites may be observed in the same chloroplast and Herrmann (1969, 1970) and Herrmann and Kowallik (1970a) have reported that the number of sites is correlated with the size of the chloroplast. They have suggested that the increased number of DNA sites relative to the size of the chloroplasts may indicate a polyploidy or polyvalency of the chloroplast DNA.

Chloroplast DNA spread on surface films from ruptured spinach chloroplasts consisted of linear and looped filaments. The filaments measured up to 150 μm in length, and some seemed to connect with plastid membranes (Woodcock and Fernandez-Moran, 1968). The filaments do not appear circular (Woodcock and Fernandez-Moran, 1968; Swift and Wolstenholme, 1969). However, Kroon (1969) has pointed out that the evidence is not conclusive on this point. The observation of Woodcock and Fernandez-Moran that the filaments may emanate from plastid membranes is interesting since, in thin sections, the DNA sites appear to be free in the stroma.

Grana-fretwork system

The structure and organization of the internal membrane system has proved to be a particularly fascinating subject to cytologists. Meyer (1883) and Schimper (1885) recognized that the internal region of chloroplasts was divided into two phases: the stroma; and numerous greenish granules, the grana. The presence of grana was a serious point of controversy until the mid-1930s when Doutreligne (1935),

Weier (1936), and Heitz (1936) established that they were not artifacts. Also
during the 'thirties, polarizing light microscope studies indicated that chloroplasts
(presumably the grana) had a lamellar substructure (see Thomas, 1960; Frey-
Wyssling and Mühlethaler, 1965 for discussion).

Confirmation of the presence of grana and their lamellar nature was provided in
the early published electron micrographs of chloroplasts (Granick and Porter,
1947; Steinmann, 1952a, b; Steinmann and Sjöstrand, 1955; Cohen and Bowler,
1953). It was also apparent in these early studies that the grana were connected
to each other by a series of membranes. Serial sections, 'face view' sections, and
studies on isolated chloroplasts have supplied substantial evidence that the granal
and intergranal membranes form an integral membrane system (Weier and Thomson,
1962a; Weier et al., 1963; 1965b; Heslop-Harrison, 1963, 1966; Paollilo and Falk,
1966; Paollilo and Reighard, 1967; Wehrmeyer, 1963a, b, 1964a, b). Thus, the
grana-fretwork system can be viewed as a complete membrane system with two
primary regions of differentiation and possibly function, the grana and the
interconnecting fret membranes.

General organization of the grana-fretwork system

Weier et al. (1966) has defined the grana by the presence of partitions. By this
criterion a granum may consist of only two compartments separated by a partition,
or may be a multicompartmented structure with many partitions and loculi. If
the grana are composed of several compartments, the common impression from
many published micrographs and two- and three-dimensional representations is that
the grana are precisely arranged, compartmented cylinders. However, grana are
frequently observed which approximate different geometrical forms (Weier, 1961;
Weier and Thomson, 1962a). The length of the compartments may vary in an
individual granum, and often one or more compartments and partitions may
extend from one granum to another (Fig. 4.9, arrow; Weier, 1961; Weier and
Thomson, 1962a; Thomson, 1965). Frequently, where compartments extend from
granum to granum the entire complex may appear branched (Fig. 4.9; Weier and
Thomson, 1962a). Serial sections often reveal that what appears to be separate
grana in one section are actually components of the same granum (e.g., Weier and
Thomson, 1962a; Thomson, 1965), and such studies indicate that the grana
are often quite large structures extending throughout large regions of the
chloroplast and may have branched, bowed, curves and possibly spiralled
configurations.

Another general impression, derived from many published electron micrographs
of chloroplasts, is that the grana are aligned mainly with the partitions running
approximately parallel to the long axis of the chloroplasts. These views are usually
obtained by photographing chloroplasts that lie adjacent to a long, flat cell wall
such as along the anticlinal walls of palisade parenchyma cells. However, in
chloroplasts in other positions in the cell, as well as in chloroplasts that do not
conform to the more lenticular shape, the grana are often aligned in other planes
(Weier, 1961; Weier and Thomson, 1962a; Thomson, 1965). Often 'face view',
slanting images as well as apparently median longitudinal sections of grana are

found in the same section of a chloroplast, and a chloroplast may also have some grana with partitions aligned parallel to the long axis of the chloroplast and others with partitions aligned perpendicular to this axis (Weier, 1961; Thomson, 1965). The variation in the alignment of the grana suggests that the grana are capable of movement within the chloroplast (Weier and Stocking, 1962; Thomson, 1965). It is well known that chloroplasts move within the cell in relation to light (Haupt, 1966). However, whether changes in alignment of the grana within the chloroplasts also occur in relation to light or other factors is not known. It may well be that the evolutionary differentiation of the grana-fretwork system as opposed to the more lamellar system in many algal cells (Kirk and Tilney-Bassett, 1967) is related to the increased flexibility and selective advantage of the grana-fretwork type of organization in an environment where physical and climatic factors have a greater range and flux.

The nature and organization of the intergranal membranes have been a focal point of much discussion. In many of the early studies on the ultrastructure of chloroplasts these structures were interpreted as large flattened discs which extended throughout large regions of the stroma (Eriksson et al., 1961; Menke, 1962). However, evidence accumulated from studies of serial sections, 'face views' of grana, and of isolated and shadowed or negative stained preparations of the grana-fretwork system (Weier and Thomson, 1962a; Weier et al., 1963; Heslop-Harrison, 1963, 1966; Wehrmeyer, 1964a; Paollilo, 1970), indicates that the intergranal membranes, the frets, were more comparable to flexible channels. The frets are often branched and extend to, and connect with, two or more grana, with the result that the fretwork appears as highly fenestrated sheets. Close examinations of the connection of the frets with the grana revealed that they were often connected with two or more compartments of the same granum (Weier and Thomson, 1962a; Heslop-Harrison, 1963). The number of frets connecting to an individual compartment varies (Thomson, 1965; Paolillo and Falk, 1966; Paolillo and Reighard, 1967; Paolillo, Mackay, and Graffius, 1969; Paolillo, 1970), and tangential and serial sections indicate that frets are arranged in a spiral fashion around the grana (Weier and Thomson, 1962a; Heslop-Harrison, 1963; Paolillo and associates – see above). Paolillo and his associates, from a series of detailed studies of serial sections of grana in chloroplasts of different plants, have concluded that the spiral follows a right-to-left helix around the grana (Paolillo, 1970) and that several helices may occur on the same granum. The functional significance of the fretwork is still not understood although Weier et al. (1966) have suggested it may be related to the exchange of materials between the grana and stroma.

As previously pointed out, the grana are bounded on all sides that are in contact with the stroma by a membrane which, depending on the fixative, is approximately 8.5 nm in thickness (Weier, 1961; Weier and Thomson, 1962a; Weier et al., 1965a). The fret membranes also measure about 8.5 nm in thickness and are continuous with the marginal and end membranes where they connect with the grana (Weier and Thomson, 1962a; Weier et al., 1965b; Heslop-Harrison, 1963, 1966; Paolillo, 1970). Since the fret membranes are continuous with the end and marginal membranes of the grana, and because they are identical in thickness, it would appear, on morphological grounds, that these membranes are structurally and functionally identical (see section on substructure for further discussion).

The fret, marginal, and end membranes are also continuous with the partitions. However, the partitions are apparently a complex of two membranes which in many instances are separated by a region, the A-space, which usually measures about 2.4 nm in thickness. The partitions usually appear to extend across the granum from one margin to the other. However, in many instances a small gap between the end of the partition and the marginal membrane of the granum occurs (Weier and Thomson, 1962a; Diers and Schotz, 1966). In these regions the loculi of adjacent compartments are confluent. The two membranes of the partition are continuous with, and may be invaginations or overlaps of, the marginal membrane (Menke, 1962; Wehrmeyer, 1964b; Wehrmeyer and Robbelen, 1965; Diers and Schotz, 1966; Heslop-Harrison, 1966). However, the width of the partitions including the A-space is less than the sum of the thicknesses of two marginal membranes (Weier et al., 1965a). Assuming that the partitions form by the apposition of vesicles (von Wettstein, 1958, 1959; Mühlethaler and Frey-Wyssling, 1959), or by infoldings of the marginal membrane, the diminution in the total width of the partition during their formation would suggest that the membranes become differentiated during the formation of the partition. This differentiation may be due to a rearrangement of the structural components of the original membrane, to the subtraction or addition of material during partition formation or a combination of these possibilities.

One danger in interpreting the partition membranes as highly differentiated from the marginal and fret membranes is indicated by the studies of Izawa and Good (1966). They found that the grana-fretwork organization in isolated chloroplasts was transformed into lamellar sheets on the removal of salts from the isolating medium. On the addition of salts, grana-like structures were reformed from the dispersed lamellae. If the grana reformed randomly from the lamellar membrane, it would provide evidence that there was little difference between granal and fret membranes. However, it is possible that grana are reformed from specialized and differentiated portions of the lamellar sheets which were originally granal and partition membranes. Whatever the answer, these observations by Izawa and Good provide evidence that the grana-fretwork system is an integral membrane system with the grana representing morphological and thus probably functional regions of specialization.

The probability that the partition membranes are differentiated structurally and compositionally raises the question as to whether the partitions are specialized in function. In this regard the A-space has received considerable attention. The A-space may appear electron translucent (Fig. 4.1, p) or electron opaque depending on the fixation, dehydration and staining procedures used (Heslop-Harrison, 1963, 1966; Schidlovsky, 1965; von Wettstein, 1967; Heslop-Harrison, 1963; Weier et al., 1963, 1965a). In many instances when the region between the two partition membranes appears electron opaque (Fig. 4.11, arrows) it is difficult to determine whether this actually represents a discrete and distinct region or a zone of close appression between the two membranes. Thus, whether an A-space occurs as a universal feature of the partition complex is still unresolved. However, it has been suggested several times that the A-space may possibly represent a cementing layer between the two membranes of the partition (Kahn and von Wettstein, 1961; Thomson, 1965; Heslop-Harrison, 1966). In those instances where only a dark band is observed, it may also be equivalent to the A-space and represent a zone of interdigitation and union between the two membranes. Further, since the partition complex is not disrupted by osmotic treatments, Weier and Benson (1966, 1967) have suggested that the partition complex, and particularly the A-space, is a hydrophobic region, where chlorophyll molecules may be preferentially localized.

The question of the localization of chlorophyll in the grana-fretwork is a primary question when considering the structural and functional relationships in chloroplasts. As Weier and Benson (1967) pointed out, the presence of chlorophyll in the grana is fully supported by their green colouration in the light microscope, as described by Meyer in 1883, and the red fluorescence of the grana (Wildman, Hongladarom, and Honda, 1964). Using a tetrazolium salt as a Hill reagent, Weier *et al.* (1966) have reported the localization of the reduced formazan reaction product in the partition and particularly along the A-space. These observations support the suggestion by Weier and Benson that chlorophyll is present in the partitions. Other indirect observations support this suggestion. Ongun, Thomson, and Mudd (1968) observed that after osmium fixation, the A-space was seldom present in *in vivo* and isolated spinach chloroplasts and that 69 per cent of the chlorophyll was extracted from the isolated chloroplasts during the dehydration in preparation for electron microscopy. However, in $KMnO_4$-fixed material the A-space was easily seen, and they found that only 8 per cent of the chlorophyll was extracted during dehydration. Weier, Shumway, and Stocking (1968) have reported that when care is taken to prevent the removal of chlorophyll and hydrophobic lipids either by low temperature dehydration or air drying the A-space is maintained in the partition (Figs. 4.10, 12, 13, arrows; Schidlovsky, 1965; Heslop-Harrison, 1966; Thomson, unpublished observations). These observations in aggregate, i.e., when chlorophyll is retained the A-space is present and when it is removed the A-space is diminished or lost, are indicative that chlorophyll is a major component of the A-space in the partition.

Several lines of evidence indicate that chlorophyll is also present in the fret membranes. Lintilhac and Park (1966) reported that isolated chloroplast membranes which they considered to be fret (stromal lamellae) membranes had a red fluorescence characteristic of chlorophyll. Howell and Moudrianakas (1967a) using a tetrazolium salt as a Hill reagent, have reported that the formazan reaction product occurs along the entire membrane system. More recently Nir and Seligman (1970) in experiments with diaminobenzidine as an electron donor in photosynthesis have shown that the osmiophilic reaction product occurs along the membranes of both the grana and frets. These experiments would indicate that chlorophyll is present in both grana and fret membranes. The results of Howell and Moudrianakas and Nir and Seligman also suggest that the chlorophyll is active in photosynthesis. However, generalizations based on localization of a reaction product must be made cautiously, since the degree of movement before deposition may be quite large (Shumway and Park, 1969).

Recently, Murakami and Packer (1970) have reported that not only do the loculi contract and flatten in the light, but that the intergranal membranes, presumably the partitions, also decreased in width by 13–23 per cent. The addition of H^+ in the dark produced a similar decrease in this width. These investigators (see Packer, Mehard, and Murakami, 1970) have interpreted the decrease in membrane thickness as a reflection of a conformational change in the membrane due to ion binding (protonation). Interesting as these results are, the A-space is not observable in Murakami and Packer's (1970) published micrographs. Determining whether the decrease in partition thickness is relative to a change in the A-space or the partition membranes is an important consideration. It would also be interesting to ascertain whether the end, marginal and fret membranes change in thickness in the dark and the light.

The loculi and fret channels are known to swell when isolated grana-fretwork systems (Class II chloroplasts) are placed in hypotonic medium (Jacobi and Perner, 1961; Kahn and von Wettstein, 1961; Weier *et al.*, 1963; Weier *et al.*, 1965; von Wettstein, 1967). Not only do these types of experiments indicate that the loculus and fret channels are hydrophilic and possibly similar in nature, but that end, marginal, and fret membranes are semipermeable and, if not identical, probably closely related. The nature of the loculi is poorly understood. Although they appear electron translucent in most electron micrographs, they appear quite electron opaque in other situations (see for example, Hodge, McLean, and Mercer, 1955). Von Wettstein (1967) has shown the loculus (i.e., the intradisc space) to be electron opaque after 'heavy' staining with lead and uranyl ions. Differences in the electron opacity of the loculus are also related to fixation and dehydration procedures. For example, Heslop-Harrison (1966) and Israel and Steward (1967) have shown that after aldehyde fixation, alone and with subsequent poststaining of the thin sections, the loculus is electron opaque while the membranes of the grana have an electron translucent appearance. Similarly, this pattern of contrast between the loculus and the granal membranes is apparent in chloroplasts fixed with osmium tetroxide, but dehydrated in air instead of organic solvents (Figs. 4.10, 12, 13, l; Schidlovsky, 1965). Whether the differences in the electron opacity of the loculus reflect differences in technique or some underlying constitutive differences is not yet known.

The evidence that the loculi swell or contract osmotically is indicative of the presence of hydrophilic components within the loculi (see Weier and Benson, 1966, 1967 for references and discussion). Whatever these and the possible other constituents of the loculi are, is unknown. Heslop-Harrison (1966) has emphasized, however, that the materials within the loculi are important in the organization of the grana and that they probably have considerable functional significance.

In many electron micrographs osmiophilic bodies are often observed along the margins of the grana at the junction of the partition with the marginal membranes (Figs. 4.6, 9, 11, ob). These do not appear to be plastoglobuli, and it has been suggested that these 'bodies' result from the movement and accumulation of

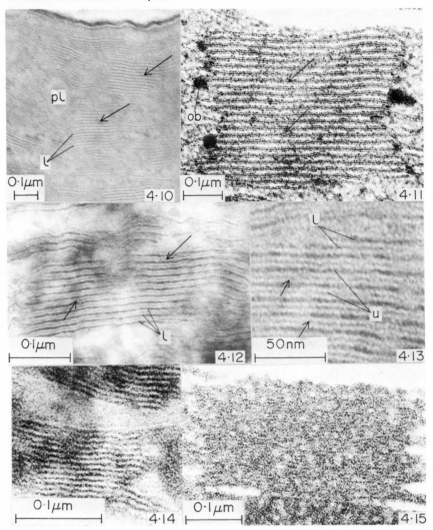

Fig. 4.10 A portion of a chloroplast of a bean leaf (*Phaseolus vulgaris* L.) which was fixed in osmium tetroxide but air-dried rather than dehydrated with organic solvents. Note the reversal in contrast as compared to the granum in Fig. 4.11 which was dehydrated with organic solvents. The loculi, Fig. 4.10 l, and the A-space appear electron-opaque. Plastoglobuli (pl) are easily identified in the air-dried material. x87 000.

Fig. 4.11 A granum in a chloroplast of tobacco. Note the presence of an electron-opaque band (arrows) in the centre of the partitions and the occurrence of osmiophilic bodies (ob) along the margins of the grana. Fixation: glutaraldehyde and osmium tetroxide. x140 000.

Fig. 4.12 A granum in a chloroplast from a section of a bean leaf (*Phaseolus vulgaris* L.) which was fixed with osmium vapours. Note the negative contrast of the membranes as compared to the electron-opaque loculi (l) and A-spaces (arrows). x220 000.

more loosely associated membrane lipids at these points (Frey-Wyssling and Kreutzer, 1958a, b). Some support for this conclusion comes from studies on chloroplast-to-chromoplast transition in which these osmiophilic bodies are particularly evident along the margins of the grana (Thomson, 1966). Since the grana-fretwork system is undergoing degradation during the transition these bodies may represent accumulation of membrane lipids released from the granal membranes which become loosely associated with the grana during membrane disintegration.

These osmiophilic bodies are absent from osmium-fixed, air-dried material (Figs, 4.10, 12) and from material dehydrated at low temperatures (Weier et al., 1968). Since extraction and movement of lipids does not occur to any large extent during these preparative procedures, the absence of osmiophilic bodies would tend to support the suggestion that they are artifacts in more conventionally prepared material. Care, however, must be taken before this suggestion is completely accepted, particularly since Sabnis, Gordon, and Galston (1969) have presented evidence and arguments to the effect that these bodies may not be lipid or enzymic sites but represent localized sites having different affinities for metal ions.

The substructure of the granal and fret membranes

Probably no other aspect of the fine structure of chloroplasts has drawn as much attention and debate as the ultrastructure of the granal and fret membranes. One obvious reason for this is centred on the recognition that this membrane is the site of the primary photochemical processes in photosynthesis (Trebst, Tsujimoto, and Arnon, 1958; Park and Pon, 1961, 1963; Park, 1966) and the possible correlation between structure and function at the membrane level. However, a general agreement as to the molecular organization of these membranes is lacking.

This lack of agreement is directly related to the absence of consistent common denominators in observations obtained using a variety of different techniques and approaches (Staehelin and Probine, 1970). Secondly, and probably more importantly, all models and interpretations, no matter what techniques are used, are only tentative until more theoretical and factual information is available concerning the physical and chemical parameters underlying the different techniques and resulting observations upon which these interpretations are formed.

Fig. 4.13   A high magnification view of a granum in a chloroplast of a bean leaf (*Phaseolus vulgaris* L.). The tissue was fixed with osmium tetroxide and dehydrated by air-drying. Note the negative contrast of the membranes and the apparent globular subunits in the partition membranes (u). The loculi (l) and A-space (arrows) appear electron-opaque when the tissue is prepared in this manner. x500 000.

Fig. 4.14   A granum from a chloroplast of a bean leaf (*Phaseolus vulgaris* L.). The tissue was fixed in osmium tetroxide and partially dehydrated in acetone before dehydration was completed by air-drying. Note the intermediate pattern of densities in the granum between that which was dehydrated in organic solvents (Figs. 4.1, 4.9, 11) as contrasted with grana in chloroplasts dehydrated by air drying (Figs. 4.10, 12, 13). x310 000.

Fig. 4.15   A granum from a chloroplast of tobacco. The tissue was fixed in glutaraldehyde and then extracted with acetone prior to post fixation with osmium. Note the absence of definable membrane structure. x202 000.

Several investigators have published micrographs which apparently indicate that the granal—fret membranes are tripartite structures composed of two electron opaque layers separated by a more electron translucent region (Mühlethaler 1960; Sitte, 1962; Heslop-Harrison, 1966; von Wettstein, 1967). From these types of images it has been suggested (Mühlethaler, 1960; Mühlethaler, Moor, and Szarkowski, 1965) that the structure of the granal—fret membranes is analogous to the unit membrane concept proposed by Robertson (1959, 1964). That is, the central electron translucent layer represents a bimolecular lipid leaflet, and the opaque layers on both sides of the electron translucent layer represent the accumulation of an electron opaque 'stain' along the polar ends of the lipids and in association with protein molecules which 'coat' the membrane on both sides.

Hohl and Hepton (1965) and particularly Weier and his associates in an extensive series of studies (Weier et al., 1965, 1966; Weier and Benson, 1966, 1967) of chemically fixed chloroplasts have presented evidence that the granal—fret membranes are composed of globular subunits. The studies of Weier and his colleagues were mainly done on aldyhyde-fixed material, post-fixed with osmium, $KMnO_4$, or Dalton's chrom-osmium solution. They reported that the globules are composed of a 3.7 nm electron translucent core surrounded by an electron opaque rim of approximately 2.8 nm in width. The fret, marginal, and end membranes are composed of a two-dimensional series of these globular subunits. Each membrane of the partition complex is also composed of a two-dimensional series of subunits. Weier and Benson (1967a, b), to explain the diminution in the width of the partition membranes, have suggested that the apposition of the subunits along the A-space induces a reorientation of the molecular components of the subunits, which results in a decrease in total membrane width.

Benson (1964, 1966), and Weier and Benson (1966, 1967) have developed the concept, based on the above observations, as well as chemical, physical and compositional considerations, that the globular subunits represent lipoprotein complexes in which the hydrophobic, hydrocarbon chains of the polar lipids are tightly associated with or inserted into hydrophobic intersticies in the protein. Based on this concept the electron opaque rims would represent the accumulation of the reaction products of osmium and $KMnO_4$ along the more hydrophilic surface of the proteins in association with the hydrophilic polar groups of the lipid

It is interesting, however, that a 'negative' image of chloroplast membranes is obtained after fixation with osmium, or glutaraldehyde followed by osmium, when the material is dehydrated in air rather than with organic solvents (Figs. 4.10, 11, 12; Schidlovsky, 1965; Heslop-Harrison, 1966). The granal and fret membranes appear electron translucent while the stroma, loculus and A-space are electron opaque. At high magnification (Fig. 4.13, u) the membranes appear to be composed of globular subunits of the same approximate size and periodicity as the subunits observed by Weier et al. (1965, 1966), after chemical fixation and organic solvent dehydration.

An evaluation of the meaning of opacity patterns of membranes in chemically fixed, dehydrated, embedded, thin sectioned, and stained material in terms of possible molecular organization is dependent on the understanding of the effects of these chemicals and procedures on the organization and components of the membranes (lipids, protein, water, etc.); the sites of action of the fixative (osmium, $KMnO_4$); how and where the heavy metals accumulate; and to what degree the heavy metals and the components of the membrane contribute to the opacity patterns observed. These problems have remained as constant factors limiting the acceptance of any one interpretation of membrane structure based on electron microscopic observations. During the past few years numerous reviews and articles have appeared which are mainly concerned with this aspect (e.g., Frey-Wyssling and Mühlethaler, 1965; Stoeckenius and Engelman, 1969; Korn, 1966; Branton, 1969; Staehelin and Probine, 1970; Hendler, 1971). Therefore, only a limited discussion will be developed here.

Of primary concern has been the question of whether the fixatives, osmium and $KMnO_4$, react with the polar lipids in membranes (see Stoeckenius and Engelman, 1969; and Staehelin and Probine, 1970; Hendler, 1971 for discussion). Fleischer, Fleischer, and Stoeckenius (1967) found that mitochondrial membranes extracted of 95 per cent of their lipids still had a tripartite appearance in electron micrographs, after fixation with osmium. These observations would tend to indicate that the reaction of osmium with lipids contributes little to electron opacity patterns of membranes. However, chloroplast membranes in leaf tissue fixed in glutaraldehyde and extracted in acetone and subsequently fixed with osmium have a uniform grey appearance in electron micrographs (Fig. 4.15; Swanson and Thomson, unpublished observations). No evidence of opacity patterns suggestive of a tripartite or subunit structure was observed. In contrast to the studies of Fleischer *et al.* on mitochondria, these observations suggest that the presence of lipids and the probable reaction of osmium with them are requisites for the opacity patterns in more typical electron micrographs of chloroplast membranes.

Korn and Weisman (1966) have shown after fixation with osmium the extraction of lipids occurred during the dehydration and embedding of amoeba. From these results Korn (1966) has pointed out that little certainty can be given to the original localization and position of lipids in membranes from opacity patterns in micrographs. However, with isolated chloroplasts, Ongun *et al.* (1968) found that retention of lipids in the membranes was necessary to preserve the membrane ultrastructure. They further reported that if the fixation time was sufficiently long complete fixation of the phospholipids and glycolipids could be achieved. Nevertheless, these results do not guarantee that the density distributions observed after the use of osmium or $KMnO_4$ reflect the original position of lipids in the membrane. In this respect the 'negative' images produced after osmium fixation and air drying are particularly interesting. Similar 'negative' images or 'reverse' contrast patterns of membranes are observed in freeze-dried and freeze-substituted tissues (Sjöstrand and Baker, 1958; Grunbaum and Welling, 1960; Pease, 1966), and in tissues dehydrated by 'inert' dehydration using ethylene glycol (Pease, 1966; Sjöstrand and Barajas, 1968; Sjöstrand, 1969). It thus appears that the more 'typical' electron opacity patterns are, to some degree, a function of the non-polar solvents such as ethanol and acetone used in the dehydration of the tissue.

These dehydrating agents are good solvents for membrane lipids and they also are denaturing agents of proteins. An explanation of the opacity patterns observed in chloroplast membranes after their use may be related to these properties. If, as pointed out above, the membrane lipids are essential factors underlying the more 'typical' patterns of electron opacity of granal membranes and if the hydrophobic, hydrocarbon chains which contain the olefinic groups – the most likely sites of reaction of osmium and $KMnO_4$ with membrane lipids (Staehelin and Probine, 1970; Korn, 1966) – are localized in hydrophobic regions such as in hydrophobic interstices of protein, these reactive sites would normally be unavailable for reaction since these regions, being hydrophobic, would tend to exclude water and with it the aqueous solutions of osmium and $KMnO_4$. If these sites are not readily available, the amount of heavy metals deposited in the membranes would be seriously reduced. Since contrast in the electron microscope is a function of the differential between scattering of electrons in one part as opposed to another, a reduction in the heavy metal concentration in the membrane with the consequential decrease in electron scattering could explain the 'negative' image in the air-dried material. Thus the more typical patterns of electron opacity observed after organic solvent dehydration may well be due to the dislocation of the reactive sites (double bonds) from the hydrophobic regions through solubilization of the lipids in the organic solvents and/or the disruption of the hydrophobic regions through denaturation of the proteins. On dislocation, the reactive sites become available, and if osmium or $KMnO_4$ is present in the tissue, as it most probably is since little care is taken to remove them in most instances, then accumulations of heavy metals and increased electron opacity of the membranes would be expected.

It is interesting from this point of view that when tissues are fixed in osmium and dehydrated in a graded series of acetones and after each step some of the tissue is air-dried, that a transition from a 'negative' to the 'positive' image of the chloroplasts membranes occurs after treatment with 50 and 70% acetone (Fig. 4.14). Further, Swanson and Thomson (unpublished results) have found that extraction and thus movement of the polar membrane lipids occurs at these concentrations from untreated and glutaraldehyde-fixed material. In summary, the evidence would indicate that contrast patterns observed in 'typical' or 'positive' electron micrographs of chloroplast membranes reflect the location of lipids at the time of their reaction with osmium, and not necessarily their original position or orientation within the membrane.

The different contrast patterns, i.e., tripartite (Heslop-Harrison, 1966), globular (Weier *et al.*, 1965) or single lines (von Wettstein, 1967) may well result from the degree of movement of the lipids before reacting with the heavy metal fixatives and the completeness of the reaction. For example, according to the model of Weier and Benson (1966, 1967) the chloroplast membranes are constructed of a polymeric, two-dimensional array of lipoprotein subunits in which the hydrophobic, hydrocarbon chains of the lipids are buried in hydrophobic internal regions of globular proteins. If this model is correct, then the olefinic groups of the unsaturated fatty acids would also be located in relatively hydrophobic regions in the proteins. As pointed out previously, because of the hydrophobic nature of these regions, aqueous solutions of heavy metal fixatives would not have ready access to the reactive sites. However, Weier and his associates (Weier *et al.*, 1965, 1966; Weier and Benson, 1966, 1967) have described globular subunits consisting of an electron translucent core surrounded by electron opaque rims. Weier and Benson (1966, 1967) have not provided a satisfactory explanation for this pattern of opacity since, according to their model and its theoretical basis, the most likely sites for the reaction of the heavy metal fixatives would either be unavailable or if the reaction and accumulation of heavy metals did occur with the olefinic groups the core of the subunit should appear electron opaque. Two explanations can be given: either the heavy metal fixatives react with other reactive sites, resulting in an accumulation of heavy metal at the surface of the lipoprotein subunits, or the hydrocarbon chains of the lipids are dislocated from the internal hydrophobic regions in the protein to the surface of the subunit where the heavy metal fixatives react with the olefinic groups.

Some reaction and accumulation of osmium and $KMnO_4$ probably does occur at the surface of the subunits, which probably explains the globular structures observed in high magnification views in the 'negative' images of air-dried material. However, the reduced contrast in the air dried material would indicate that it is not extensive. The more probable situation is that organic solvents used in dehydration bring about a disruption of the hydrophobic lipoprotein inter-action and a dislocation of olefinic reactive sites with the result that osmium and $KMnO_4$ react with the lipid moieties and accumulate at the surface of the lipoprotein subunit.

Weakening of the hydrophobic interaction due to denaturation of the protein by the organic solvents, or because of the shift from an aqueous to a non-aqueous situation, and the solubilization of the lipids by the solvents would be some of the factors responsible for the shift in position of the lipids. Extensive deformation and reorientation of the lipids and protein by the solvents and the fixatives might well lead to other density patterns (Lenard and Singer, 1966).

Whether the chloroplast membranes are composed of particles has been a focal point of much discussion. Frey-Wyssling and Steinmann (1953) observed that the isolated, shadowed granal membranes had a granular appearance in electron micrographs. These observations led them to suggest a particulate nature for the granal membranes. Re-emphasis in this possibility came with the observations of Park and Pon (1961) that shadowed preparations of isolated granal membranes dried on grids had a particulate structure. These particles measured about 10 nm in depth and 15.5 by 18.5 nm along the sides, and each particle was composed apparently of four 6.5 nm subunits (Park and Biggins, 1964; Park, 1965). These particles have been interpreted by Park and his associates as being buried in the granal membranes. The particles were frequently arranged in a paracrystalline fashion, and Park and Drury (1967) noted that the paracrystalline pattern is more evident in the winter and under long night conditions than in the summer under short nights. The term 'quantasomes' was applied to these particles (Park and Biggins, 1964) with the suggestion that they were possibly morphological equivalents of physiological units.

The question of whether these particles are on or in the membranes has been raised several times. Studies on negative stained preparations of granal membranes have revealed the presence of particles on the surface of the membranes (Murakami, 1964, 1968; Heslop-Harrison, 1966; von Wettstein, 1967; Howell and Moudrianakis, 1967a, b). These particles have the same approximate dimensions and substructure of the quantasomes, and Heslop-Harrison (1966) noted that the particles were released from the membranes by washing and that they were also similar in appearance to Fraction I protein. Howell and Moudrianakis (1967a, b) reported that some of the particles could be freed from the membrane by washing in buffer and that these particles appeared to be similar to Fraction I protein. They also found that the remaining particles could be released from the membrane by washing with EDTA and that these particles were a Ca^{++} dependent ATPase and were probably a coupling factor protein. Paracrystalline arrays of the particles were observed, and Howell and Moudrianakis (1967a, b) have suggested that these particles (i.e., Fraction I and the ATPase) were actually the particles viewed by Park and his associates in shadowed preparation of isolated grana, and they seriously questioned the quantasome concept and the interpretation that they were localized within the membrane. More recently Arntzen, Dilley, and Crane (1969) and Park and Pfeifhofer (1968) have reported that negatively stained particles occurred on the surface of granal membranes and that these could be removed by washing and treatment with EDTA. The evidence thus appears quite conclusive that particles occur on the granal membranes and that these consist of Fraction I protein and a Ca^{++} dependent coupling factor.

However, studies of granal membranes using freeze-etching techniques indicate that the membranes may indeed be composed of particulates. Mühlethaler (1966, 1967) and his associates (1965) described particles associated with the granal membranes in frozen-etched preparations of spinach chloroplasts. These workers originally interpreted the particles as being attached or partially embedded in the membranes. On the outside, i.e., the stroma side, the particles existed in groups of four approximately 12 nm in diameter and 6 nm in width, while on the inside the particles were scattered, and measured about 6 nm in diameter. Serious criticisms of the interpretation that the particles were on the outside of membrane have been developed by Branton and Park and their associates (Branton and Park, 1967; Branton, 1966, 1967, 1969; Park and Branton, 1967; Park and Pfeifhofer, 1969a, b). These investigators have reported that freeze-etched preparations of chloroplast membranes reveal two classes of particles on the fracture faces of the membranes. The large particles measure about 17.5 nm in diameter and the smaller particles 11 to 12 nm in diameter. These investigators have presented experimental and theoretical evidences that the fracture plane occurs along the central, hydrophobic region of the membrane (see Branton, 1969; and also Staehelin and Probine, 1970, for discussion) and concluded that the particles occur within the granal membranes. Branton and Park (1967) have proposed that since the two types of particles occur on different fracture faces that the particles are asymmetrically placed in the membrane. In their model they indicate that the small particles are distributed in that part of the membrane adjacent to the stroma and supposedly the A-space in the partition and the large particles in that part of the membrane adjacent to the loculus. Recently, Arntzen et al. (1969) and Park and Pfeifhofer (1969a, b) have shown that these two types of particles occur in granal membranes from which the negative stained particles, i.e., Fraction I protein and the Ca^{++} dependent coupling factor, have been removed by washing and EDTA treatment, which provides strong evidence that 17.5 and 11 nm particles revealed by freeze-etching are within the membrane and also not related to the particles which are observed on the membranes by negative staining.

The studies by Arntzen et al. (1969) are particularly interesting, since they found that, after digitonin treatment and differential centrifugation of chloroplast membranes, they could separate a fraction at 10 000 g which was enriched in photosystem II activity and another fraction at 144 000 g which had only photosystem I activity. Freeze-etching studies of the 10 000 g, photosystem II enriched fraction revealed large 17.5 nm particles on most of the fracture faces while the 144 000 g, photosystem I fraction had only small 11 nm particles. The particles in the two fractions corresponded in size to those observed in intact chloroplast membranes. On the basis of these studies they suggested that the membrane is binary in that it is composed of two structurally and functionally different components.

By deep etching, Park and Shumway (1968) have observed paracrystalline arrays of particles consisting of from two to four subunits which they interpret as representing the quantasome observed in air-dried shadowed material by Park and Biggins (1964). Recently, Park and Pfeifhofer (1969a, b) have termed the 17.5 nm particles quantasome cores, suggesting they represent a portion of a quantasome. They claim these paracrystalline arrays occur on the locular surface of the granal membranes. They have noted also that the particles occurring on the stromal surface, within the membrane, and on the locular surface of the granal membranes may be intimately related.

In the discussion of the existence of particles or subunits revealed either by chemical fixation or freeze-etching techniques, it is still not certain whether these are discrete structural units or represent localized periodicity of membrane components within the continuum of the membrane. Although the studies by Park and his associates (Park and Biggins, 1964) and Arntzen et al. (1969) are highly suggestive, Arntzen et al. have pointed out that it is still not possible 'to assign a specific functional role to the particles per se, . . .'. In consideration of the freeze-etching observations, the occurrence of two and possibly three types of 'particles' (i.e., the 11 nm, 17.5 nm, and the paracrystalline arrays described by Park and Shumway, 1968) on and in the granal membranes are probably, in essence, not artifacts.

Recent studies using fractionation techniques and freeze-etching also have provided new clues as to the nature and function of the end, marginal, and fret membranes (see Park and Sane, 1971 for review). Remy (1969) published micrographs of freeze-etched chloroplasts which indicate that only the 11 nm particles occur in these membranes while both the 17.5 and 11 nm particles occur in the partition membranes. Similar observations have been made by Sane, Goodchild, and Park (1970). The latter investigators, using mechanical means of fractionation, also isolated two membrane fractions from the chloroplasts which differed in photochemical activity. One fraction was active only in those reactions related to photosystem I while the other fraction showed activity for both photosystem I and II. Thin sections and freeze-etching studies revealed that the fraction with only photosystem I activity consisted of vesicles with 11 nm particles on the fracture planes, while the fraction with photosystem I and II consisted of grana with both the 11 and 17.5 nm particles occurring, but on different fracture faces. These results are quite similar to those obtained by Arntzen *et al.* (1969) using digitonin treatment; however, Sane *et al.* have concluded that the fraction containing only photosystem I activity and 11 nm particles is derived from the end and fret membranes. These results and conclusions would indicate, as discussed previously, that although chlorophyll occurs throughout the grana-fretwork system the end, marginal, and fret membranes are differentiated structurally and functionally from the partition membranes in the grana.

Summary

This chapter has been concerned almost entirely with the ultrastructure of 'normal', mature chloroplasts in higher plants. Even within the content of the article many papers and important contributions were undoubtedly omitted. Many of these were omitted because of the lack of space but others were undoubtedly overlooked inadvertently. Nevertheless, an attempt has been made to indicate the variability and complexity of chloroplast ultrastructure, the conflicts and limitations in interpretation, and the many areas where basic information is lacking.

Acknowledgements

The author must acknowledge the financial support of a grant from the National Science Foundation (GB 8199) during the preparation of this review. I also would like to thank the Department of Biology, University of York, England, for the hospitality and stimulating environment provided during the sabbatical year in which this article was completed. In this regard particular thanks must be given to Dr Rachel Leech with whom I had many stimulating discussions about chloroplasts.

Bibliography

Adams, M. S. and Strain, B. S. (1969) 'Seasonal photosynthetic rates in stems of *Cercidum floridum* Benth', *Photosynthetica*, 3, 55–62.

Arntzen, C. J., Dilley, R. A., and Crane, F. L. (1969) 'A comparison of chloroplast membrane surfaces visualized by freeze-etch and negative staining techniques; and ultrastructural characterization of membrane fractions obtained from digitonin-treated spinach chloroplasts', *J. Cell Biol.*, 43, 16–31.

Badenhuizen, N. P. (1959) 'Chemistry and biology of the starch granule', *Protoplasmalogia*, 2, 1–74.

Badenhuizen, N. P. (1962) 'The development of the starch granule in relation to the structure of the plastid', *Proc. K. ned Akad. Wet.*, 65, 123–135.

Bailey, J. L. and Whyborn (1963) 'The osmiophilic globules of chloroplasts: II Globules of spinach beet chloroplasts', *Biochim. biophys. Acta*, 78, 163–174.

Ballantine, J. E. and Forde, B. J. (1970) 'The effects of light intensity and temperature on plant growth and chloroplast ultrastructure in soybean', *Am. J. Bot*, 57, 1150–1159.

Bartels, P. G. and Weier, T. E. (1967) 'Particle arrangements in proplastids of *Triticum vulgare* L. seedlings', *J. Cell Biol.*, 33, 243–253.

Barton, R. (1970) 'The production and behaviour of phytoferritin particles during senescence of *Phaseolus* leaves', *Planta*, 94, 73–77.

Benson, A. A. (1964) 'Plant membrane lipids', *A. Rev. Pl. Physiol.*, 15, 1–16.

Benson, A. A. (1966) 'On the orientation of lipids in chloroplasts and cell membranes', *J. Am. Oil Chem. Soc.*, 48, 265–270.

Black, C. C., Jr and Mollenhauer, H. H. (1971) 'Structure and distribution of chloroplasts and other organelles in leaves with various rates of photosynthesis', *Pl. Physiol., Lancaster*, 47, 15–23.

Bogorad, L. (1967) 'Chloroplast structure and development', In: *Harvesting the sun* (eds. San A. Pietro, F. A. Greer, and T. J. Army), pp. 191–210, Academic Press, New York and London.

Bradbeer, J. W., Clijsters, H., Gyldenholm, A. O., and Edge, H. J. W. (1970) 'Plastid development in primary leaves of *Phaseolus vulgaris*', *J. exp. Bot.*, 21, 525–533.

Branton, D. (1966) 'Fracture faces of frozen membranes', *Proc. natn. Acad. Sci. U.S.A.*, 55, 1048–1056.

Branton, D. (1967) 'Structural units of chloroplast membranes', In: *Le chloroplaste* (ed. C. Sironval), pp. 48–54, Masson et Cie, Paris.

Branton, D. (1969) 'Membrane structure', *A. Rev. Pl. Physiol.*, 20, 209–238.

Branton, D. and Park, R. B. (1967) 'Subunits in chloroplast lamellae', *J. Ultrastruct. Res.*, 19, 283–303.

Brown, F. A. M. and Gunning, B. E. S. (1966) 'Distribution of ribosome-like particles in *Avena* plastids', In: *Biochemistry of chloroplasts* (ed. T. W. Goodwin), pp. 365–373, Academic Press, New York and London.

Butler, R. D. and Simon, E. W. (1971) 'Ultrastructural aspects of plant senescence', In: *Advances in gerontological research 111* (ed. B. Strehler), Academic Press, New York and London.

Buttrose, M. S. (1960) 'Submicroscopic development and structure of starch granules in cereal endosperms', *J. Ultrastruct. Res.*, 4, 231–257.

Buvat, R. (1963) 'Electron microscopy of plant protoplasm', *Int. Rev. Cytol.*, 14, 41–155.

Caporali, L. (1959) 'Recherches sur les infrastructures des cellules radicularies de *Len culinaris* et particulierement sur l'evolutions des leucoplaste', *Annls. Sci. nat. (Bot.) (11) Biol. Vegetale.*, 20, 215–247.

Catesson, A-M. (1966) 'Presence de phytoferritin dans le cambium et les tissus conducteurs de la tige de sycomore *Acer pseudoplantanus*', *C.r. hebd. Séanc. Acad. Sci., Paris*, 262, 1070–1073.

Catesson, A-M. (1970) 'Evolution des plastis de pomme au cours de la maturation au fruit. Modifications ultrastructurales et accumulation de ferritine', *J. Microscopie*, 9, 949–974.

Clowes, F. A. L. and Juniper, B. E. (1968) *Plant cells*, Blackwell Scientific Publications, Oxford and Edinburgh.

Cohen, M. and Bowler, E. (1953) 'Lamellar structure of the tobacco chloroplast', *Protoplasma*, 42, 414–416.

Ultrastructure of mature chloroplasts

Cronshaw, J., Hoefert, L., and Esau, K. (1966) 'Ultrastructural features of *Beta* leaves infected with beet yellows virus', *J. Cell Biol.*, 31, 429–434.

Crookston, R. K. and Moss, D. N. (1970) 'The relation of carbon dioxide compensation and chlorenchymatous vascular bundle sheaths in leaves of dicots', *Pl. Physiol., Lancaster,* 46, 564–567.

Diers, L. and Schotz, F. (1966) 'Uber die dreidimensionale gestaltung des Thylakoidsystem in den Chloroplasten', *Planta,* 70, 322–366.

Dilley, R. A., Park, R. B., and Branton, D. (1967) 'Ultrastructural studies of light-induced chloroplast shrinkage', *Photochem. Photobiol.,* 6, 407–412.

Dodge, J. D. (1970) 'Changes in chloroplast fine structure during the autumnal senescence of *Betula* Leaves', *Ann. Bot.,* 34, 817–824.

Doutreligne, J. (1935) 'Note sur la structure des chloroplasts', *Proc. K. ned. Acad. Wet.,* 38, 886–896.

Downton, W. J. S. and Tregunna, E. B. (1968) 'Carbon dioxide compensation – its relation to photosynthetic carboxylation reactions, systematics, of the Gramineae, and leaf anatomy', *Can. J. Bot.,* 46, 207–215.

Downton, W. J. S., Bisalputra, T., and Tregunna, E. B. (1969) 'The distribution and ultra-structure of chloroplasts in leaves differing in photosynthetic carbon metabolism. II. *Atriplex rosea* and *Atriplex hastata* (Chenopodiaceae)', *Can. J. Bot.,* 47, 915–919.

Dyer, T. A. and Leech, R. M. (1968) 'Chloroplast and cytoplasmic low-molecular-weight ribonucleic acid components of the leaf of *Vicia faba* L.', *Biochem. J.,* 106, 689–698.

Eriksson, G., Kahn, A., Walles, B., and von Wettstein, D. (1961) 'Zur makromolecularn Physiologie der Chloroplasten', *Ber. dt. bot. Ges.,* 74, 221–232.

Esau, K. (1965) *Plant anatomy* (2nd ed.), Wiley and Sons, New York, London, and Sidney.

Falk, H. (1960) 'Magnoglobuli in Chloroplasten von *Ficus elastica* Roxb.', *Planta,* 55, 525–532.

Falk, H. (1969) 'Rough thylakoids: polysomes attached to chloroplast membranes', *J. Cell Biol.,* 42, 582–587.

Fleischer, S., Fleischer, B., and Stoeckenius, W. (1967) 'Fine structure of lipid-depleted mitochondria', *J. Cell Biol.,* 32, 193–208.

Frey-Wyssling, A. and Kreutzer, E. (1958a) 'The submicroscopic development of chromo-plasts in the fruit of *Capsicum annum* L.', *J. Ultrastruct. Res.,* 1, 397–411.

Frey-Wyssling, A. and Kreutzer, E. (1958b) 'Die submikroskopische Entwicklung der Chromoplasten in der Bluten von *Ranunculus repens* L.', *Planta,* 51, 104–114.

Frey-Wyssling, A. and Mühlethaler, K. (1965) *Ultrastructural plant cytology,* Elsevier, New York.

Frey-Wyssling, A. and Steinmann, E. (1953) 'Ergebnisse der Feinbau-analyse der Chloro-plasten', *Naturforsch. Ges. Zurich. Vierteljahressche.,* 98, 20–29.

Granick, S. (1955) 'Plastid structure, development and inheritance', In: *Encyclopedia of plant physiology* (ed. W. Ruhland), vol. 1, pp. 507–564, Springer, Berlin.

Granick, S. (1961) 'The chloroplasts: Inheritance structure and function', In: *The cell* (eds. J. Brachet and A. E. Mirsky), vol. 2, pp. 489–602, Academic Press, New York and London.

Granick, S. and Porter, K. R. (1947) 'The structure of the spinach chloroplast as interpreted with the electron microscope', *Am. J. Bot.,* 34, 545–550.

Greenwood, A. D., Leech, R. M., and Williams, J. P. (1963) 'The osmiophilic globules of chloroplasts. I. Osmiophilic globules as a normal component of chloroplasts and their isolation and composition in *Vicia faba'*, *Biochim. biophys. Acta.,* 78, 148–162.

Grob, E. C. and Rufener, J. (1969) 'Influence of sugar containing nutrients on ultrastructure and photosynthesis activity of *Spirodela oligorrhiza* chloroplasts', In: *Progress in photosynthesis.* (ed, H. Metzner), vol. 1, pp. 55–62.

Grunhaum, B. W. and Wellings (1960) 'Electron microscopy of cytoplasmic structures in frozen-dried mouse pancrease', *J. Ultrastruct. Res.,* 4, 73–80.

Gunning, B. E. S. (1965a) 'The fine structure of chloroplast stroma following aldehyde osmium-tetroxide fixation', *J. Cell Biol.,* 24, 79–93.

Gunning, B. E. S. (1965b) 'The greening process in plastids. I. The structure of the prolamellar body', *Protoplasma,* 60, 111–130.

Gunning, B. E. S., Steer, M. W., and Cochrane, M. P. (1968) 'Occurrence, molecular structure, and induced formation of the "stromacentre" in plastids', *J. Cell Sci.,* 3, 445–456.

Haberlandt, G. (1914) Drummond, M. (Trans), *Physiological plant anatomy*, Macmillan, London.

Harnischfeger, G. (1970) 'Changes in appearance, volume and activity during the early stages of disintegration in isolated chloroplasts', *Planta*, 92, 164–177.

Harnischfeger, G. and Gaffron, H. (1969) 'Transient color sensitivity of the Hill reaction during the disintegration of chloroplasts', *Planta*, 89, 385–388.

Harnischfeger, G. and Gaffron, H. (1970) 'Transient light effects in the Hill reaction of disintegrating chloroplasts *in vitro*', *Planta*, 93, 89–105.

Haupt, W. (1966) 'Phototaxis in plants', *Int. Rev. Cytol.*, 19, 267–299.

Heber, U. (1969) 'Conformational changes of chloroplasts induced by illumination of leaves *in vivo*', *Biochim. biophys. Acta.*, 180, 302–319.

Heitz, E. (1936) 'Untersuchungen uber den Bau der Plastiden. I. Die gerichteten Chlorophyllscheiben der Chloroplasten', *Planta*, 26, 134–163.

Hendler, R. W. (1971) 'Biological membrane ultrastructure', *Physiol. Rev.*, 51, 66–97.

Herrmann, R. G. (1969) 'Are chloroplasts polyploid?', *Expl Cell Res.*, 55, 414–416.

Herrmann, R. G. (1970) 'Multiple amounts of DNA related to the size of the chloroplasts. I. An autoradiographic study', *Planta*, 90, 80–96.

Herrmann, R. G. and Kowallik, K. V. (1970a) 'Multiple amounts of DNA related to the size of chloroplasts. II. Comparison of electron microscopic and autoradiographic data', *Protoplasma*, 69, 365–372.

Herrmann, R. G. and Kowallik, K. V. (1970b) 'Selective presentation of DNA-regions and membranes in chloroplasts and mitochondria', *J. Cell Biol.*, 45, 198–202.

Heslop-Harrison, J. (1963) 'Structure and morphogenesis of lamellar systems in grana containing chloroplasts. I. Membrane structure and lamellar architecture', *Planta*, 60, 243–260.

Heslop-Harrison, J. (1966) 'Structural features of the chloroplast', *Sci. Prog.*, 54, 519–541.

Hilgenherger, H. and Menke, W. (1965) 'Lichtabhangige, anisotrope Veranderungen des chloroplasten Volumens in lebenden Zellen', *Z. Naturf.*, 20b, 699–701.

Hodge, A. J., McLean, J. D., and Mercer, F. V. (1955) 'Ultrastructure of the lamellae and grana in the chloroplasts of *Zea Mays* L.', *J. biophys. biochem. Cytol.*, 1, 605–614.

Hohl, A. R. and Hepton, A. (1965) 'A globular subunit pattern in plastid membranes', *J. Ultrastruct. Res.*, 12, 542–546.

Howell, S. H. and Moudrianakis, E. N. (1967a) 'Hill reaction site in chloroplast membranes: Non-participation of the quantasome particle in photoreduction', *J. Molec. Biol.*, 27, 323–333.

Howell, S. H. and Moudrianakis, E. N. (1967b) 'Function of the "quantasome" in photosynthesis, structure and properties of membrane bound particles active in the dark reactions of photophosphorylation', *Proc. natn. Acad. Sci. U.S.A.*, 58, 1261–1268.

Hyde, B. B., Hodge, A. J., Kahn, A., and Birnstiel, M. L. (1963) 'Studies on phytoferritin 1. Identification and localisation', *J. Ultrastruct. Res.*, 9, 248–258.

Israel, H. W. and Steward, F. C. (1967) 'The fine structure and development of plastids in cultured cells of *Daucus carota*', *Ann. Bot.*, 31, 1–18.

Itoh, M., Izawa, S., and Shibata, K. (1963) 'Shrinkage of whole chloroplasts upon illumination', *Biochim. biophys. Acta*, 66, 319–327.

Izawa, S. and Good, N. E. (1966) 'Effect of salts and electron transport on the conformation of isolated chloroplasts. II. Electron microscopy', *Pl. Physiol., Lancaster*, 41, 544–552.

Jacobi, G. and Perner, E. (1961) 'Strukturelle und biochemische Probleme der Chloroplastenisolierung', *Flora, Jena*, 150, 209–226.

Kahn, A. and Wettstein, D. von (1961) 'Macromolecular physiology of plastids. II. Structure of isolated spinach chloroplasts', *J. Ultrastruct. Res.*, 5, 557–574.

Kawashima, N. and Wildman, S. G. (1970) 'Fraction I protein', *A. Rev. Pl. Physiol.*, 21, 325–358.

Kirk, J. T. O. and Tilney-Bassett, R. A. E. (1967) *The plastids; their chemistry, structure, growth and inheritance*, Freeman, London and San Francisco.

Kislev, N., Swift, H., and Bogorad, L. (1965) 'Nucleic acid of chloroplasts and mitochondria of Swiss chard', *J. Cell Biol.*, 25, 327–344.

Kislev, N., Swift, H., and Bogorad, L. (1966) 'Studies of nucleic acids in chloroplasts and mitochondria in Swiss chard', In: *Biochemistry of chloroplasts* (ed. T. W. Goodwin), vol. 1, pp. 355–363. Academic Press, New York and London.

Korn, E. D. (1966) 'Structure of biological membranes', *Science, N.Y.,* **153,** 1491–1498.

Korn, E. D. and Weisman, R. A. (1966) 'Loss of lipids during preparation of amoebae for electron microscopy', *Biochim. biophys. Acta.,* **116,** 309–316.

Kreutz, W. (1970) 'X-ray structure research on the photosynthetic membrane', In: *Advances in botanical research* (ed. R. D. Preston), vol. 3, pp. 53–169.

Kroon, A. M. (1969) 'DNA and RNA from mitochondria and chloroplasts (biochemistry)', In: *Handbook of molecular cytology* (ed. Lima de Faria), pp. 943–971, North Holland Publ. Co., Amsterdam and London.

Kushida, H., Itoh, M., Izawa, S., amd Shibata, K. (1964) 'Deformation of chloroplasts on illumination in intact spinach leaves', *Biochim. biophys. Acta.,* **79,** 201–203.

Laetsch, W. M. (1968) 'Chloroplast specialization in dicotyledons possessing the C_4-dicarboxylic acid pathway of photosynthetic CO_2 fixation', *Am. J. Bot.,* **55,** 875–883.

Laetsch, W. M. (1969a) 'Specialized chloroplast structure of plants exhibiting the dicarboxylic acid pathway of photosynthetic CO_2 fixation', In: *Progress in photosynthetic research* (ed. H. Metzner), vol. 1, pp. 36–46.

Laetsch, W. M. (1969b) 'Relationship between chloroplast structure and photosynthetic carbon fixation pathways', *Sci. Prog. Oxf.,* **57,** 323–351.

Laetsch, W. M. and Price, I. (1969) 'Development of the dimorphic chloroplasts of sugar cane', *Am. J. Bot.,* **56,** 77–78.

Laetsch, W. M., Stetler, D. A., and Vlitos, A. J. (1966) 'The ultrastructure of sugar cane chloroplasts', *Z. PflPhysiol.,* **54,** 472–474. .

Leech, R. M. (1964) 'The isolation of structurally intact chloroplasts', *Biochim. biophys. Acta.,* **79,** 637–639.

Leech, R. M. (1967) 'Comparative biochemistry and comparative morphology of chloroplasts isolated by different methods', In: *The biochemistry of chloroplasts* (ed. T. W. Goodwin), vol. 1, pp. 65–74, Academic Press, New York.

Leech, R. M. (1968) 'The chloroplast inside and outside the cell', In: *Plant cell organelles* (ed. J. B. Pridham), pp. 137–162, Academic Press, New York and London.

Lefort-Tran, M. (1969) 'Cytochemistry and ultrastructure of chloroplasts', In: *Handbook of molecular cytology* (ed. A. Lima-De-Faria), pp. 1047–1076, North Holland Publ. Co., Amsterdam and Holland.

Lenard, J. and Singer, S. J. (1966) 'Protein conformation in cell membrane preparations as studied by optical rotatory dispersion and circular dichroism', *Proc. natn. Acad. Sci. U.S.A.,* **56,** 1828–1835.

Lichtenthaler, H. K. (1966) 'Plastoglobuli und Plastidenstruktur', *Ber. dt. bot. Ges.,* **79,** 82–88.

Lichtenthaler, H. K. (1968) 'Plastoglobuli and the fine structure of plastids', *Endeavour,* **27,** 144–149.

Lichtenthaler, H. K. (1969a) 'Die Plastoglobuli von Spinat, ihre Grosse, und Zusammensetzung wahrend der Chloroplastendegeneration', *Protoplasma,* **68,** 315–326.

Lichtenthaler, H. K. (1969b) 'Die Plastoglobuli von Spinat, ihre Grosse, Isolierung und Lipochinonzusammensetzung', *Protoplasma,* **68,** 65–77.

Lintilhac, P. M. and Park, R. B. (1966) 'Localization of chlorophyll in spinach chloroplast lamellae by fluorescence microscopy', *J. Cell Biol.,* **28,** 582–585.

Mackender, R. O. and Leech, R. M. (1970) 'Isolation of chloroplast envelope membranes', *Nature, Lond.,* **228,** 1347–1349.

Marinos, N. G. (1967) 'Multifunctional plastids in the meristematic region of potato tuber buds', *J. Ultrastruct. Res.,* **17,** 91–113.

McWhorter, F. P. (1965) 'Plant virus inclusions', *A. Rev. Phytopath.,* **3,** 287–312.

Menke, W. (1962) 'Structure and chemistry of plastids', *A. Rev. Pl. Physiol.,* **13,** 27–44.

Mercer, F. (1960) 'The submicroscopic structure of the cell', *A. Rev. Pl. Physiol.,* **11,** 1–24.

Meyer, A. (1883) *Das Chlorophyllkorn in chemischer, morphologischer und biologischer Beziehung,* Leipzig, A. Felix.

Middleton, J. T. (1961) 'Photochemical air pollution damage to plants', *A. Rev. Pl. Physiol.,* **12,** 431–448.

Moyse, A. (1967) 'Les chloroplastes: activite photosynthetique et metabolisme des glucides', In: *Biochemistry of chloroplasts* (ed. T. W. Goodwin), vol. 2, pp. 91–129. Academic Press, New York and London.

Mühlethaler, K. (1960) 'Die Strukter der Grana- und Stromalamellen in Chloroplasten', *Z. wiss. Mikrosk.*, **64**, 444–452.

Mühlethaler, K. (1966) 'The ultrastructure of the plastid lamellae', In: *Biochemistry of chloroplasts* (ed. T. W. Goodwin), vol. 1, pp. 49–64, Academic Press, New York and London.

Mühlethaler, K. (1967) 'L'ultrastructure de la lamellae des chloroplaste', In: *Le chloroplaste* (ed. C. Sironval), pp. 42–47, Masson et Cie, Paris.

Mühlethaler, K. and Frey-Wyssling, A. (1959) 'Entwicklung und Struktur der Proplastiden', *J. biophys. biochem. Cytol.*, **6**, 507–512.

Mühlethaler, K., Moor, H., and Szarkowski, J. W. (1965) 'The ultrastructure of chloroplast lamellae', *Planta*, **67**, 305–323.

Murakami, S. (1964) 'Microstructure of chloroplast lamellae', *J. Electron Micro.*, **13**, 234–236.

Murakami, S. (1968) 'On the nature of the particles attached to the thylakoid membrane of spinach chloroplasts', In: *Comparative biochemistry and biophysics of photosynthesis* (eds. K. Shibata, A. Takamiya, A. T. Jagendorf, and R. C. Fuller), pp. 82–83, University of Tokyo Press, Tokyo.

Murakami, S. and Packer, L. (1970) 'Protonation and chloroplast membrane structure', *J. Cell Biol.*, **47**, 332–351.

Newcomb, E. H. (1967) 'Fine structure of protein-storing plastids in bean root tips', *J. Cell Biol.*, **33**, 143–163.

Nichols, B. W., Stubbs, J. M., and James, A. T. (1967) 'The lipid composition and ultrastructure of normal developing and degenerating chloroplasts', In: *Biochemistry of chloroplasts* (ed. T. W. Goodwin), vol. 2, pp. 677–690. Academic Press, New York and London.

Nir, I. and Seligman, A. M. (1970) 'Photoxidation of diaminobenzidine (DAB) by chloroplast lamellae', *J. Cell Biol.*, **46**, 617–620.

Nobel, P. S. (1967a) 'Relation of swelling and photophosphorylation to light-induced ion uptake by chloroplasts *in vitro*', *Biochim. biophys. Acta*, **131**, 127–140.

Nobel, P. S. (1967b) 'A rapid technique for isolating chloroplasts with high rates of endogenous photophosphorylation', *Pl. Physiol., Lancaster*, **42**, 1389–1394.

Nobel, P. S. (1968a) 'Chloroplasts shrinkage and increased photophosphorylation *in vitro* upon illuminating intact plants of *Pisum sativum*', *Biochim. biophys. Acta*, **153**, 170–182.

Nobel, P. S. (1968b) 'Energetic basis of light-induced chloroplast shrinkage *in vivo*', *Plant and Cell Physiol.*, **9**, 499–509.

Nobel, P. S. (1968c) 'Light-induced chloroplast shrinkage *in vivo* detectable after rapid isolation of chloroplasts from *Pisum sativum*', *Pl. Physiol., Lancaster*, **43**, 781–787.

Nobel, P. S. (1969) 'Light-induced changes in ionic content of chloroplasts in *Pisum sativum*', *Biochim. biophys. Acta*, **172**, 134–143.

Nobel, P. S., Chang, P. T., Wang, C., Smith, S. S., and Barcus, D. E. (1969) 'Initial ATP formation, NADP reduction, CO_2 fixation and chloroplast flattening upon illuminating pea leaves', *Pl. Physiol., Lancaster*, **44**, 655–661.

O'Brien, Y. P. and Thimann, K. V. (1967) 'Observations on the fine structure of the oat coleoptile. II. The parenchyma cells of the apex', *Protoplasma*, **63**, 417–422.

Ongun, A., Thomson, W. W., and Mudd, J. B. (1968) 'Lipid fixation during preparation of chloroplasts for electron microscopy', *J. Lipid Res.*, **9**, 416–424.

Packer, L., Barnard, A. C., and Deamer, D. W. (1967) 'Ultrastructural and photometric evidence for light-induced changes in chloroplast structure *in vivo*', *Pl. Physiol., Lancaster*, **42**, 283–293.

Packer, L., Mehard, C. W., and Murakami, S. (1970) 'Ion transport in chloroplasts and plant mitochondria', *A. Rev. Pl. Physiol.*, **21**, 271–304.

Packer, L. and Siegenthaler, P-A. (1966) 'Control of chloroplast structure by light', *Int. Rev. Cytol.*, **20**, 97–124.

Palevitz, B. A. and Newcomb, E. H. (1970) 'A study of sieve element starch using sequential enzymatic digestion and electron microscopy', *J. Cell Biol.*, **45**, 383–398.

Paolillo, D. J. Jr (1970) 'The three-dimensional arrangement of intergranal lamellae in chloroplasts', *J. Cell Sci.*, **6**, 243–255.

Paolillo, D. J. Jr and Falk, R. H. (1966) 'The ultrastructure of grana in mesophyll plastids of *Zea mays*', *Am. J. Bot.*, **53**, 173–180.

Paolillo, D. J. Jr, MacKay, N. C., and Graffius, J. R. (1969) 'The structure of grana in flowering plants', *Am. J. Bot.*, **56**, 344–347.

173

Ultrastructure of mature chloroplasts

Paolillo, D. J. Jr and Reighard, J. A. (1967) 'On the relationship between mature structure and ontogeny in the grana of chloroplasts', *Can. J. Bot.*, **45**, 773–782.

Park, R. B. (1965) 'Substructure of chloroplast lamellae', *J. Cell Biol.*, **27**, 151–161.

Park, R. B. (1966) 'Subunits of chloroplast structure and quantum conversion in photosynthesis', *Int. Rev. Cytol.*, **20**, 67–95.

Park, R. B. and Biggins, J. (1964) 'Quantasome: size and composition', *Science, N.Y.*, **144**, 1009–1011.

Park, R. B. and Branton, D. (1967) 'Freeze-etching of chloroplasts from glutaraldehyde fixed leaves', *Brookhaven Symp. Biol.*, **19**, 341–351.

Park, R. B. and Drury, S. (1967) 'The effects of daylength on quantasome structure and chloroplast phytochemistry in *Spinacea oleracea* L. var. early hybrid 7', In: *Le chloroplaste* (ed. C. Sironval), pp. 328–334, Masson et Cie, Paris.

Park, R. B. and Pfeifhofer, A. O. (1968) 'The continued presence of quantasomes in ethylenediaminetetraacetate washed chloroplast lamellae', *Proc. natn. Acad. Sci., U.S.A.*, **60**, 337–343.

Park, R. B. and Pfeifhofer, A. O. (1969a) 'Ultrastructural observations on deep-etched thylakoids', *J. Cell Sci.*, **5**, 299–311.

Park, R. B. and Pfeifhofer, A. O. (1969b) 'The effect of ethylenediaminetetracetate washing on the structure of spinach thylakoids', *J. Cell Sci.*, **5**, 513–519.

Park, R. B. and Pon, M. G. (1961) 'Correlation of structure with function in *Spinacea oleracea* chloroplasts', *J. molec. Biol.*, **3**, 1–10.

Park, R. B. and Pon, M. G. (1963) 'Chemical composition and substructure of lamellae isolated from *Spinacea oleracea* chloroplasts', *J. molec. Biol.*, **6**, 105–114.

Park, R. B. and Sane, P. V. (1971) 'Distribution of function and structure in chloroplast lamellae', *A. Rev. Pl. Physiol.*, **22**, 395–430.

Park, R. B. and Shumway, L. K. (1968) 'The ultrastructure of fracture and deep etch faces of spinach thylakoids', In: *Comparative biochemistry and biophysics of photosynthesis* (eds. K. Shibata, A. Takamiya, A. T. Jagendorf, and R. C. Fuller), pp. 57–66, University of Tokyo Press, Tokyo.

Pease, D. C. (1966) 'The preservation of unfixed cytological details by dehydration with "inert" agents', *J. Ultrastruct. Res.*, **14**, 356–378.

Perner, E. (1962) 'Elekronenmikronskopische befunde uber kristallgitter strukturen in Stroma isolieter Spinatchloroplasten', *Port. Acta. biol.*, **A6**, 359–372.

Philippovich, I. I., Tongur, A. M., Alina, B. A., and Oparin, A. I. (1970) 'Localization and conformation of polyribosomes bound to chloroplast lamellae', *Expl Cell Res.*, **62**, 399–406.

Possingham, J. V., Vesk, M., and Mercer, F. V. (1964) 'The fine structure of leaf cells of manganese-deficient spinach', *J. Ultrastruct. Res.*, **11**, 68–83.

Price, J. L. and Thomson, W. W. (1967) 'Occurrence of a crystalline inclusion in the chloroplasts of *Macademia* leaves', *Nature, Lond.*, **214**, 1148–1149.

Price, W. C., Martinez, A. P., and Warmke, H. E. (1966) 'Crystalline inclusions in chloroplasts of the coconut palm', *J. Ultrastruct. Res.*, **14**, 618–621.

Ragetli, H. W. J., Weintraub, M., and Lo, E. (1970) 'Degeneration of leaf cells resulting from starvation and excision. I. Electron microscopic observations', *Can. J. Bot.*, **48**, 1913–1922.

Remy, R. (1969) 'Etude de la structure des lamelles chloroplastiques de *Vicia faba* L. par la technique du cryodecapaqe', *C.r. hebd. Séanc Acad. Sci., Paris* **268**, 3057–3060.

Ris, H. (1961) 'Ultrastructure and molecular organization of genetic systems', *Can. J. Genet. Cytol.*, **3**, 95–120.

Ris, H. and Plaut, W. (1962) 'Ultrastructure of DNA-containing areas in the chloroplast of *Chlamydomonas*', *J. Cell Biol.*, **13**, 383–391.

Robards, A. W. and Humpherson, P. G. (1967) 'Phytoferritin in plastids of the cambial zone of willow', *Planta*, **76**, 169–178.

Robards, A. W. and Robinson, C. L. (1968) 'Further studies on phytoferritin', *Planta*, **82**, 179–188.

Robertson, J. D. (1959) 'The ultrastructure of cell membranes and their derivatives', *Biochem. Soc. Symp.*, **16**, 3–43.

Robertson, J. D. (1964) 'Unit membranes. A review with recent studies of experimental alterations and a new subunit structure in synaptic membranes', pp. 1–82; In: Soc. Study Develop. Growth Symp. 22, *Cellular membranes in development*, Academic Press, New York.

Rosadac-Alberio, J., Weier, T. E., and Stocking, C. R. (1968) 'Continuity of the chloroplast membrane system in *Zea mays* L.', *Pl. Physiol., Lancaster,* **43**, 1325–1331.

Sabnis, D. D., Gordon, M., and Galston, A. W. (1969) 'A site with an affinity for heavy metals on the thylakoid membranes of chloroplasts', *Pl. Physiol., Lancaster,* **44**, 1355–1363.

Sabnis, D. D., Gordon, M., and Galston, A. W. (1970) 'Localization of adenosine triphosphatase activity on the chloroplast envelope in tendrils of *Pisum sativum*', *Pl. Physiol., Lancaster,* **45**, 25–32.

Sane, P. V., Goodchild, D. J., and Park, R. B. (1970) 'Characterization of chloroplast photosystem 1 and 2 separated by a non-detergent method',*Biochim. biophys. Acta,* **216**, 162–178.

Schidlovsky, G. (1965) 'Contrast in multilayer systems after various fixations', *Lab. Invest.,* **14**, 1213–1233.

Schimper, A. J. W. (1885) 'Untersuchungen uber die Chlorophyllkorper und die ihnen homologen Gebilde', *Jb. wiss Bot.,* **16**, 1–247.

Schotz, F. and Diers, L. (1965) 'Elektron mikroskopische untersuchungen uber die abgabe von plastidenteilen ins plasma', *Planta,* **66**, 269–292.

Schotz, F. and Diers, L. (1967) 'Differentiation processes within the limiting double membrane of chloroplasts', In: *Le chloroplaste* (ed. C. Sironval), pp. 21–29, Masson et Cie, Paris.

Seckback, J. (1968) 'Studies on the deposition of plant ferritin as influenced by iron supply to iron-deficient beans', *J. Ultrastruct. Res.,* **22**, 413–423.

Shalla, T. (1964) 'Assembly and aggregation of tobacco mosaic virus in tomato leaflets', *J. Cell Biol.,* **21**, 253.

Shumway, L. K. and Park, R. B. (1969) 'Light-dependent reduction of some tetrazolium and ditetrazolium salts by chloroplasts', *Expl Cell Res.,* **56**, 29–32.

Shumway, L. K. and Weier, T. E. (1967) 'The chloroplast structure of iojap maize', *Am. J. Bot.,* **54**, 773–780.

Shumway, L. K., Weier, T. E., and Stocking, R. C. (1967) 'Crystalline structures in *Vicia faba* chloroplasts', *Planta,* **76**, 182–189.

Sitte, P. (1962) 'Zum chloroplasten-Feinbau bei *Elodea*', *Port. Acta biol.,* **A6**, 269–278.

Sjöstrand, F. S. (1969) 'Morphological aspects of lipoprotein structure', In: *Structural and functional aspects of lipoprotein in living systems* (eds. E. Tria and A. M. Scance), pp. 73–128, Academic Press, New York and London.

Sjöstrand, F. S. and Baker, R. F. (1958) 'Fixation by freeze-drying for electron microscopy of tissue cells', *J. Ultrastruct. Res.,* **1**, 239–246.

Sjöstrand, F. S. and Barajas, L. (1968) 'Effect of modifications in conformation of protein molecules of mitochondrial membranes', *J. Ultrastruct. Res.,* **25**, 121–155.

Spencer, D. and Unt, H. (1965) 'Biochemical and structural correlations in isolated spinach chloroplasts under isotonic and hypotonic conditions', *Aust. J. biol. Sci.,* **18**, 197–210.

Spencer, D. and Wildman, S. G. (1962) 'Observations on the structure of grana-containing chloroplasts and a proposed model of chloroplast structure', *Aust. J. biol. Sci.,* **15**, 599–610.

Sprey, B. (1965) 'Beitrage zur makromolekularen Organization der Plastiden. I', *Pfl. Physiol.,* **53**, 255–261.

Srivastava, L. M. and O'Brien, T. P. (1966) 'On the ultrastructure of cambium and its vascular derivatives', *Protoplasma,* **61**, 277–293.

Staehelin, L. A. and Probine, M. G. (1970) 'Structural aspects of cell membranes', In: *Advances in botanical research* (ed. R. D. Preston), vol. 3, pp. 1–52.

Steer, M. W., Gunning, B. E. S., Graham, T. A., and Carr, D. J. (1968) 'Isolation, properties and structure of Fraction I protein from *Avena sativa* L.', *Planta,* **79**, 254–267.

Steinmann, E. (1952a) 'An electron microscope study of the lamellar structure of chloroplasts', *Expl Cell Res.,* **3**, 367–372.

Steinmann, E. (1952b) 'Contribution to the structure of chloroplasts', *Experientia,* **8**, 300–301.

Steinmann, E. and Sjöstrand, F. S. (1955) 'The fine structure of chloroplasts', *Expl Cell Res.,* **8**, 15–23.

Stocking, C. R. and Larson, L. (1969) 'A chloroplast cytoplasmic shuttle and the reduction of extraplastid NAD', *Biochim. biophys. Res. Commun.,* **37**, 278–282.

Ultrastructure of mature chloroplasts

Stoekenius, W. and Engelman, D. M. (1969) 'Current models for the structure of biological membranes', *J. Cell Biol.*, 42, 613–646.

Sundquist, J. E. and Burris, R. H. (1970) 'Light-dependent structural changes in the lamellar membranes of isolated spinach chloroplasts: measurement by electron microscopy', *Biochim. biophys. Acta.*, 223, 115–121.

Swift, H. and Wolstenholme, D. R. (1969) 'Mitochondria and chloroplasts: nucleic acids and the problem of biogenesis (genetics and biology)', In: *Handbook of molecular cytology* (ed. Lima-de-Faria), pp. 972–1046. North Holland Publ. Co., Amsterdam.

Thomas, J. B. (1960) 'Chloroplast structure', In: *Encyclopedia of plant physiology* (ed. W. Ruhland), vol. 5, pp. 511–565.

Thomson, W. W. (1965) 'The ultrastructure of *Phaseolus vulgaris* chloroplasts', *J. exp. Bot.*, 16, 169–176.

Thomson, W. W. (1966) 'Ultrastructural development of chromoplasts in Valencia oranges', *Bot. Gaz.*, 127, 133–139.

Thomson, W. W. (1969) 'Ultrastructural studies on the epicarp of ripening oranges', *Proc. First Int. Citrus Sym.,.* 3, 1163–1169.

Thomson, W. W., Dugger, W. M. Jr, and Palmer, R. L. (1965) 'Effects of peroxyacetyl nitrate on ultrastructure of chloroplasts', *Bot. Gaz.*, 126, 66–72.

Thomson, W. W., Dugger, W. M. Jr, and Palmer, R. L. (1966) 'Effects of ozone on the fine structure of the palisade parenchyma cells of bean leaves', *Can. J. Bot.*, 44, 1677–1682.

Trebst, A. V., Tsujimoto, H. Y., and Arnon, D. I. (1958) 'Separation of light and dark phases in the photosynthesis of isolated chloroplasts', *Nature, Lond.*, 187, 351–355.

Vesk, M., Mercer, F. V., and Possingham, J. V. (1965) 'Observations on the origin of chloroplasts and mitochondria in the leaf cells of higher plants', *Aust. J. Bot.*, 13, 161–169.

Walker, D. A. (1965) 'Correlation between photosynthetic activity and membrane integrity in isolated pea chloroplasts', *Pl. Physiol., Lancaster*, 40, 1157–1161.

Walker, D. A. (1967) 'Photosynthetic activity of isolated pea chloroplasts', In: *Biochemistry of chloroplasts* (ed. T. W. Goodwin), vol. 2, pp. 53–69.

Walles, B. (1967) 'Use of biochemical mutants in analyses of chloroplasts morphogenesis', In: *Biochemistry of chloroplasts* (ed. T. W. Goodwin), vol. 2, pp. 633–653, Academic Press, New York and London.

Wehrmeyer, W. (1963a) 'Qualitative and quantitative Untersuchungen uber den Bau der Stromamembranen der ausdifferenziesten Chloroplasten von *Spinacea oleracea*', *Z. Naturf.*, 18b, 60–66.

Wehrmeyer, W. (1963b) 'Uber Membranbildungsprozesse in Chloroplasten. I. Zur Morphogenese der Grana-membranen', *Planta*, 59, 280–295.

Wehrmeyer, W. (1964a) 'Uber Membranbildungsprozesse im Chloroplasten. II. Zur Entstehung der Grana durch Membranuberschiebung', *Planta*, 63, 13–30.

Wehrmeyer, W. (1964b) 'Zur klarung der strukturellen Variabilitat der Chloroplastengrana des Spinats in Profil under Aufsicht', *Planta*, 62, 272–293.

Wehrmeyer, W. and Robbelen, G. (1965) 'Raumliche Aspekte zur Membranschichtung in den Chloroplasten einer Arabidopsismutahte unter Auswertung von serienschnitten', *Planta*, 64, 312–329.

Weier, T. E. (1936) 'The structure of the non-starch containing beet chloroplast', *Am. J. Bot.*, 23, 645–652.

Weier, T. E. (1938) 'The structure of the chloroplast', *Bot. Rev.*, 4, 497–530.

Weier, T. E. (1961) 'The ultramicro structure of the starch free chloroplasts of fully expanded leaves of *Nicotiana rustica*', *Am. J. Bot.*, 48, 615–630.

Weier, T. E. and Benson, A. A. (1966) 'The molecular nature of chloroplast membranes', In: *The Biochemistry of chloroplasts* (ed. T. W. Goodwin), vol. 1, pp. 91–113, Academic Press, New York and Londor

Weier, T. E. and Benson, A. A.   7) 'The molecular organization of chloroplast membranes', *Am. J. Bot.*, 54, 389–40?

Weier, T. E., Engelbrecht, A. H. ʀ., Harrison, A., and Risley, E. B. (1965a) 'Subunits in the membranes of chloroplasts of *Phaseolus vulgaris*, *Pisum sativum* and *Aspidistra* sp.', *J. Ultrastruct. Res.*, 13, 92–111.

Weier, T. E., Shumway, L. K., and Stocking, C. R. (1968) 'The organization of chloroplast membranes of *Vicia faba* and *Zea mays* after processing to retain chlorophyll and hydrophobic lipids in the chloroplasts', *Protoplasma*, 66, 339–355.

Weier, T. E. and Stocking, C. R. (1952) 'The chloroplast structure, inheritance and enzymology', *Bot. Rev.,* **18,** 14–15.

Weier, T. E. and Stocking, C. R. (1962) 'The cup plastid of *Nicotiana rustica', Am. J. Bot.,* **49,** 24–32.

Weier, T. E., Stocking, C. R., Bracker, C. B., and Risley, E. B. (1965b) 'The structural relationships of the internal membrane systems of *in situ* and isolated chloroplasts of *Hordeum vulgare', Am. J. Bot.,* **52,** 339–352.

Weier, T. E., Stocking, C. R., and Shumway, L. K. (1966) 'The photosynthetic apparatus in chloroplasts of higher plants', *Brookhaven Symp. Biol.,* **19,** 353–374.

Weier, T. E., Stocking, C. R., Thomson, W. W., and Drever, H. (1963) 'The grana as structural units in chloroplasts of mesophyll of *Nicotiana rustica* and *Phaseolus vulgaris', J. Ultrastruct. Res.,* **8,** 122–143.

Weier, T. E. and Thomson, W. W. (1962a) 'The grana of starch-free chloroplasts of *Nicotiana rustica', J. Cell Biol.,* **13,** 89–108.

Weier, T. E. and Thomson, W. W. (1962b) 'Membranes of mesophyll cells of *Nicotiana rustica* and *Phaseolus vulgaris* with particular reference to the chloroplast', *Am. J. Bot.,* **49,** 807–820.

Wettstein, D. von (1958) 'The formation of plastid structure', *Brookhaven Symp. Biol.,* **11,** 138–159.

Wettstein, D. von (1959) 'Developmental changes in chloroplasts and their genetic control', In: *Developmental cytology* (ed. D. Rudnick), pp. 123–160, Ronald Press Co., New York.

Wettstein, D. von (1967) 'Chloroplast structure and genetics', In: *Harvesting the sun, photosynthesis in plant life* (eds. A. San Pietro, F. A. Green and T. J. Army), pp. 153–190, Academic Press, New York and London.

Wildman, S. G. (1967) 'The organization of grana-containing chloroplasts in relation to location of some enzymatic systems concerning photosynthesis, protein synthesis and ribonucleic acid synthesis', In: *Biochemistry of chloroplasts* (ed. T. W. Goodwin), vol. 2, pp. 295–319, Academic Press, New York and London.

Wildman, S. G., Hongladarom, T., and Honda, S. I. (1962) 'Chloroplasts and mitochondria in living plant cells: cinephotomicrographic studies', *Science, N. Y.,* **138,** 434–436.

Wildman, S. G., Hongladarom, T., and Honda, S. I. (1964) 'Chloroplasts and mitochondria in living plant cells: Cinephotomicrographic studies', *Science, N. Y.,* **138,** 434–436.

Woodcock, C. L. F. and Fernandez-Moran, H. (1968) 'Electron microscopy of DNA conformations in spinach chloroplasts', *J. molec. Biol.,* **31,** 627–631.

Wrischer, M. (1967) Kristalloide in Plastidenstroma. I. Electronmikroskopisch-Cytochemische Untersuchungen', *Planta,* **75,** 309–318.

Yokomura, E. (1967) 'An electron microscopic study of DNA-like fibrils in chloroplasts', *Cytologia,* **32,** 361–377.

5

Lysosomes
Ph. Matile

Introduction

The changing concept of lysosomes

The term 'lysosome' represents a combination of 'dissolution' (Gr. *lysis*) and 'body' (Gr. *soma*). It was originally introduced for designating subcellular particles which contain a set of hydrolytic enzymes potentially capable of breaking down the cellular constituents. Hence, lysosomes were conceived as safety devices that allow the storage of digestive enzymes within a living cell.

The history of the lysosome, most vividly narrated by de Duve (1969), reflects the profound changes of viewpoints in recent cell biology: after a more descriptively oriented research, 'lysosomology' is now running through a renaissance characterized by the consideration of various dynamic aspects.

Initially lysosomes were conceived as distinct organelles of rat liver homogenates characterized by the presence of several acid hydrolases in a latent form. The membrane-bounded bodies present in appropriate subcellular fractions were identified as *the* lysosomes of rat liver cells. Thus, lysosomes seemed to possess physical, biochemical and morphological properties as well defined as those which characterize mitochondria.

In the past decade this concept of lysosomes has been progressively modified. Research into the problems of ontogeny and function of lysosomes has yielded a complex terminology which reflects the complexity of structures that are involved in what may be called the lysosomal system of a cell. Within this system the classical lysosomes merely represent a transient stage of a process. This extends from the synthesis of hydrolases and primary lysosomes by endoplasmic reticulum and Golgi complexes, through the digestion of endogenous or exogenous material in secondary lysosomes, to the formation of residual bodies whose content is eventually excreted into the extracellular space. Several membrane systems participate in this process; growth, fission, coalescence and, probably, differentiation of membranes occur. Morphologically, the lysosomal system is no longer confined to a group of uniformly structured and individualized organelles. Small wonder that de Duve, who originally coined the term lysosome, now suggests the use of 'vacuome' as a more adequate term for designating the totality of membrane systems involved in the lysosomal process. The vacuome includes the endoplasmic reticulum, Golgi complexes, the whole family of lysosomes, vesicles produced by the plasmalemma through pinocytosis and phagocytosis and, lastly, the extracellular space.

Terminological considerations

De Duve's comprehensive term vacuome is familiar to the plant cytologist. It was, in fact, introduced by Dangeard in 1916 (see Dangeard, 1956) and designates the totality of plant cell compartments that have in common the ability of accumulating stains such as neutral red. It will be seen that this convergence of terminology in animal and plant cytology is meaningful because the plant vacuome *is* lysosomal in nature.

However, as the heading of this chapter indicates, the term lysosome has been introduced in plant cytology despite the fact that most, if not all, of the respective structures are identical with classical organelles termed otherwise. It would therefore be clumsy to abandon these classical terms for the benefit of a fashionable one which, as mentioned above, is already in danger of being abandoned in animal cytology. A reasonable and simple solution to the problems of terminology may be proposed as follows: the term lysosome should be used in a biochemical sense exclusively, that is, the totality of membrane-bounded structures containing lysosomal enzymes is termed the *lysosomal cell compartment;* however, the morphologist dealing with this compartment should continue to use the classical terms for designating the individual types of its constituents. It is not even feasible to use the term lysosome with a functional meaning because it will be seen that the digestive function attributed to lysosomal enzymes never represents the exclusive function of the lysosomal compartment. The proposed terminological concept will produce some minor difficulties when the dynamic aspects of the lysosomal compartment are considered: one type of organelle may differentiate into another and the transient stages will perhaps be left without appropriate terms.

The lysosomal cell compartment

Plant cells contain a variety of hydrolases capable of digesting such vital cytoplasmic constituents as protein and nucleic acids, or of hydrolysing such important metabolites as esters of phosphoric acid. It is inconceivable that these enzymes are localized in the cytoplasm in a free form. On the contrary, it must be assumed that the digestive enzymes are sequestered in specific membrane-bounded structures, that is, in a lysosomal cell compartment.

Relevance of experimental approaches

Whether cell compartments are lysosomal in nature or not can be discovered by employing enzyme cytochemistry or cell fractionation techniques. Enzyme cytochemistry has the advantage of yielding direct information about the association of reaction products of hydrolases with distinct structures. This technique possesses many inherent difficulties and, furthermore, is restricted to a few lysosomal enzyme activities, mainly acid phosphatase. An adequate biochemical characterization of lysosomes will therefore always involve the isolation of organelles. This is difficult, since the extraction of most plant cells and tissues yields almost no latent or sedimentable hydrolase activities. The example of *Acetabularia* illustrates the situation. Schweiger (1966) has demonstrated conclusively that acid ribonuclease is completely soluble in extracts obtained by centrifuging cells whose rhizoids had been cut off. Should this enzyme nevertheless be associated with a lysosomal compartment it must be hypothesized that extraction had destroyed completely the corresponding structures. In the case of *Acetabularia* cell fractionation *in vivo* has solved the problem. Winkenbach (personal communication) using stratified cells has been able to demonstrate that RNase is concentrated in the centripetal stratum occupied by the large vacuole.

The case of *Acetabularia* typifies the impossibility of isolating constituents of the plant lysosomal compartment by employing routine methods of cell extraction and factionation. Hence, the information about the lysosomal compartment has been obtained from a few organisms and tissues whose lysosomes can be extracted and isolated by specific techniques.

Types of lysosome

Vacuoles. Vacuoles isolated from a number of different organisms contain a variety of hydrolases in a concentrated form (Table 5.1). The spectrum of these enzymes ranges from peptidases and nucleases through phospho- and other esterases to various glycosidases. Most of them have an acid pH-optimum, but a few exceptions demonstrate that this is not necessarily a common property of lysosomal enzymes. Acid phosphatase, although a prominent vacuolar enzyme, may exceptionally be absent as is the case in phosphatase-repressed yeast cells (Matile and Wiemken, 1967). A few isolates have been found stable to such an extent that the latency of vacuolar enzymes could be demonstrated (Pujarniscle, 1968; Iten and Matile, 1970; Matile, 1971).

Lysosomal structures isolated from rootlets of *Cucurbita* seedlings (Coulomb, 1968), leaf tissue of the fern *Asplenium fontanum* (Coulomb, 1969) and tomatoes (Heftmann, 1971) are probably identical with small vacuoles as can be judged from the techniques employed.

It should be noted that the extraction and isolation of vacuoles always involves the rupture of a considerable proportion of these fragile organelles. Their lysosomal nature can therefore be derived from high specific hydrolase activities rather than from fractions of total activities recovered in the isolates. This explains the fundamental difficulties of evaluating the quantitative distributions of hydrolases among the structures of the lysosomal compartment.

The only important, but still missing, piece of evidence for the recognition of the lysosomal nature of plant vacuoles concerns the central vacuole of parenchymatous cells whose isolation is difficult, if possible at all. This evidence has not, as yet, been provided by the cytochemists who have also concentrated on meristems demonstrating the vacuolar localization of acid phosphatase (Fig. 5.1A) in various species of embryos and seedlings (e.g., Poux, 1963a, 1970; Berjak, 1968, 1971; Gahan and

TABLE 5.1

Vacuoles as Lysosomes: Enzyme Localization Based on Cell Fractionation Work

Object	Structures isolated	Lysosomal enzymes	References
Saccharomyces cerevisiae	Vacuoles from lysed protoplasts	Endopeptidases, RNase, acetyl esterase amino-peptidase (1 out of 4 isozymes) Invertase Exo—β1.3 glucanase	Matile and Wiemken (1967) Matile, Wiemken, and Guyer (1971) Beteta and Gascón (1971) Cortat (personal communication)
S. carlsbergensis	Vacuoles from lysed protoplasts	Acid phosphatase	Lloyd (personal communication)
Coprinus lagopus	Vacuoles from vegetative hyphae and fruiting bodies	Acid and alkaline endopeptidases, RNase, acid phosphatase, β-glucosidase, chitinase	Iten and Matile (1970)
Neurospora crassa	Vacuoles from macroconidia	Endopeptidases, RNase, acid and alkaline phosphatases, phosphodiesterase, amino-peptidase, invertase	Matile (1971)
Zea mays	Meristematic vacuoles from root tips	Endo- and exopeptidases, RNase, DNase, phosphatase, phosphodiesterase, acetyl-esterases, β-amylase, α- and β-glucosidase, β-galactosidase	Matile (1966, 1968, unpublished results)
Solanum tuberosum	Vacuoles from dark-grown shoots	Acid phosphatase, phosphodiesterase, RNase, carboxylicesterase	Pitt and Galpin (1973)
Hevea brasiliensis	Vacuoles (lutoids) from latex	Endopeptidase, RNase, phosphatase, β-glucosidase, β-galactosidase, β-N-acetylglucosaminidase	Pujarniscle (1968)
Chelidonium majus	Vacuoles from latex	Endopeptidase, RNase, phosphatase	Matile, Jans, and Rickenbacher (1970)

181

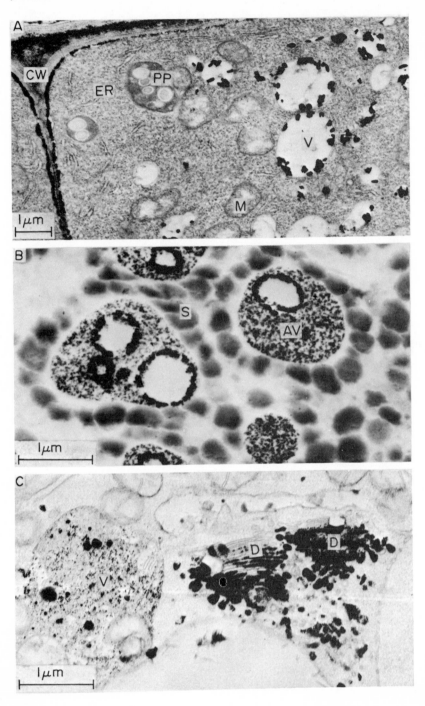

McLean, 1969; Halperin, 1969). The lysosomal nature of large vacuoles must therefore be derived indirectly from the ontogenetical relationship with meristematic vacuoles.

It is noteworthy that Dauwalder et al. (1969), having observed the vacuolar localization of β-glycerophosphatase — as well as of inosinediphosphatase — and thiamine pyrophosphatase (enzymes normally found in the Golgi) activity in radicular cells, have doubted the lysosomal nature of vacuoles because the reaction obviously was substrate unspecific. This conclusion reflects the widely spread erroneous idea that lysosomes are characterized by the presence of specific acid β-glycerophosphatase activity. It must be emphasized that plant acid phosphatases are extremely unspecific. The corresponding enzyme from tobacco leaves, purified by Shaw (1966), catalyses the hydrolysis of a wide variety of phosphate esters as well as of anhydrous bonds of phosphoric acid.

Aleurone grains. Aleurone grains represent modified vacuoles occurring in reserve tissues of seeds. They characteristically contain specific reserve proteins (protein bodies) and phytic acid. Poux (1963b, 1965) has been able to demonstrate cytochemically acid phosphatase activity in aleurone grains of wheat embryo and cotyledonary cells of *Cucumis* and *Linum* (Fig. 5.1B). Aleurone grains isolated from cotton (Yatsu and Jacks, 1968), *Vicia faba* (Morris et al., 1970), barley (Ory and Henningsen, 1969), and hempseed (St Angelo et al., 1969) contain protease; and various hydrolases including protease, RNase and acid phosphatase are associated with the corresponding organelle isolated from pea cotyledons (Matile, 1968b). A specific phytase has been localized in the aleurone grains isolated from ungerminated barley seeds (Ory and Henningsen, 1969).

Spherosomes. Evidently, the spherosomes represent a confused class of organelles, not only with regard to their belonging to the lysosomal compartment, but also from the morphological viewpoint. The only property of spherosomes that is generally accepted is their high triglyceride content. Spherosomes are therefore merely regarded as the oil storing organelle of plant cells. Indeed, Yatsu et al. (1971) have recently isolated spherosomes from oleaginous (peanut and cotton seed) and non-oily tissues (onions and cabbages) and have demonstrated the virtual absence of acid phosphatase, the most prominent of lysosomal enzymes. In contrast, at the light microscope level, acid phosphatase and several other hydrolases have been cytochemically demonstrated in spherosomes (e.g., Walek-Czernecka, 1965; Holcomb

---

Key to the labelling of electron micrographs:
V = vacuole, T = tonoplast, CS = cell sap, AV = aleurone vacuole, S = spherosome,
D = dictyosome, M = mitochondria, ER = endoplasmic reticulum, PL = plasmalemma,
PP = proplastid, CW = cell wall.

Fig. 5.1   Cytochemical localization of acid phosphatase.
(A) Gomori-reaction product in vacuoles and in cell walls of meristematic root cells of *Cucumis sativus.* (Courtesy of N. Poux.)
(B) *Triticum vulgare:* In embryo cells acid phosphatase activity is localized in the aleurone vacuoles but not in the spherosomes. Note the accumulation of reaction product around the globoids (G). (Courtesy of N. Poux.)
(C) Two dictyosomes in a cell of *Euglena gracilis* containing abundant Gomori-reaction product. (Courtesy of D. Brandes.)

et al., 1967). These contradictory findings may be due to the occurrence of distinct types of spherosomes in the same tissue as suggested by Sorokin (1967), whereby the isolation of spherosomes by flotation selects only a class of low density organelles. Nevertheless, at least one type of hydrolase seems to be associated even with spherosomes from oleaginous tissues: lipase activity has been detected in oil droplets isolated from *Ricinus* endosperm (Ory *et al.,* 1968), Douglas fir seeds (Ching, 1968), tobacco endosperm (Spichiger, 1969) and maize scutellum (Matile, unpublished). Whether lipase is a common enzyme of spherosomes – associated with the triglycerides accumulated in this organelle, as the proteases are associated with the reserve proteins in aleurone grains – has not yet been established. In any case there are considerable species specific differences in the lysosomal enzyme content of spherosomes. For example, a spectrum of hydrolases comparable with that of vacuoles (endopeptidases, phosphatase, nucleases and acetyl-esterase) has been demonstrated in spherosomes isolated from the oleaginous tobacco endosperm (Matile and Spichiger, 1968). Hence, it seems to be justifiable to classify sphero-somes as lysosomes, although in extreme cases the set of hydrolases may be reduced to a single enzyme such as lipase. There is, perhaps, a type of spherosome which contains no lysosomal enzymes at all.

Another oddity of spherosomes concerns the surrounding membrane which, in electron micrographs of glutaraldehyde-osmium fixed specimens, shows a single electron opaque layer instead of the normal triple-layered structure of vacuoles and aleurone grains (Fig. 5.5.A,B,D,E). It will be seen that this anomalous membrane structure is explained by the development of spherosomes (Schwarzenbach, 1971a) and, in turn, explains some odd features of spherosomal dynamics.

It, should be stressed that not all vacuole-like organelles belong to the lysosomal compartment. Microbodies also have a single surrounding membrane and have occasionally been confused with small vacuoles. They definitely do not represent lysosomes. Enzymes characterizing microbodies are related to peroxide metabolism (key enzyme: catalase), wherefore the term 'peroxisome' has been introduced. A specific type of peroxisome occurring in the reserve tissues of lipid storing seeds is the glyoxysome which contains enzymes of the glyoxylic acid cycle in addition to catalase and other enzymes (reviews: Beevers, 1969, de Duve, 1969).

These remarks seemed to be necessary with regard to recent findings on vacuolar localization of peroxydase activity (e.g., Poux, 1969; Czaninski and Catesson, 1969). This localization may represent an artifact due to non-enzymatic reactions of compounds present in the cell sap; in any case the formation of reaction product in vacuoles seems to be independent in pH and not affected by cyanide (Czaninski and Catesson, 1970). Cell fractionation studies, however, suggest the association of a small percentage of the total peroxidase activity with membranes, possibly tonoplasts (Parish, personal communication).

It should perhaps be noted that peroxidase is definitely absent from peroxisomes!

The extracellular space

The lysosomal nature of the space occupied by the cell wall is suggested by many cytochemical studies showing abundant Gomori-reaction product external to the plasmalemma (e.g., Figier, 1968; Zee, 1969; Halperin, 1969; Poux, 1970). Fig. 5.1A gives an example of extracellular acid phosphatase. Gahan and McLean (1969) have been able to extend this observation to acetyl-esterase.

Indeed, incubations of washed slices of various tisues of corn seedlings in the presence of suitable chromophoric substrates yield reaction products formed by hydrolases localized external to the permeability barrier. By employing detergents to destroy the semipermeability of membranes it is possible to evaluate the fractions of total hydrolase activities present in the free space: in segments of corn root tips more than half of the total activities of leucyl-aminopeptidase, phosphatase, β-glucosidase and β-galactosidase seem to have an extracellular localization (Matile, unpublished).

The phenomenon of extracellularly localized hydrolases is also well known in fungi: e.g., in yeasts acid phosphatase, invertase, glucanase and other enzymes are constituents of the cell wall (see Lampen, 1968). Certain fungi are capable of extracellularly digesting macromolecules such as proteins (Matile, 1965), DNA (Cazin et al., 1969) and RNA (Terukatsu et al., 1968) by secreting the corresponding hydrolases.

The phenomenon of hydrolase secretion has also been observed in roots (Chang and Bandurski, 1964) and in barley aleurone layers (e.g., Jacobsen and Varner, 1967; Taiz and Jones, 1970). It would be beyond the scope of this review to present all of the evidence with regard to hydrolases of the extracellular space. The few examples mentioned above suffice to consider its lysosomal nature in the purely biochemical sense of this term.

Thus, the lysosomal compartment of plant cells seems to be divided into an intracellular space (vacuoles, aleurone grains and spherosomes) and an extracellular one. In contrast to animal cells (see de Duve, 1969) the vacuome of plant cells seems to be discontinuous with the extracellular space: most probably vacuoles are neither discharged into the extracellular space nor provided with exogenous macromolecules through phagocytic activity. Therefore, it seems logical to distinguish an intracellular lysosomal compartment comprising the membrane systems of the vacuome and an exoplasmic compartment comprising the extracellular space and the structures involved in the secretion of hydrolases. While the organelles of the endoplasmic compartment represent lysosomes in a literal sense, the extracellular space may be thought of as an inside-out lysosome which surrounds the protoplast. Its limiting membrane is the plasmalemma which separates the cytoplasm from the extracellular hydrolases. Secretory vesicles containing hydrolases en route to the extracellular space (Gahan and McLean, 1969; Matile et al., 1965) may be regarded as its potential constituents.

Dynamics of the lysosomal compartment

Origin and development of vacuoles

Vacuolation. Meristematic cells of vegetative points, or of the proliferating zones of root tips, lack vacuole profiles. The enormous increase in volume upon expansion of these cells is primarily due to the formation of a large central vacuole whose development from small vacuoles has been extensively studied in radicular systems. Vital staining with neutral red shows a system of reticulated vacuoles in the youngest cells; this network is eventually transformed into numerous globular vacuoles which inflate, fuse, and finally coalesce into the single large vacuole of the expanded cells (see Buvat, 1971).

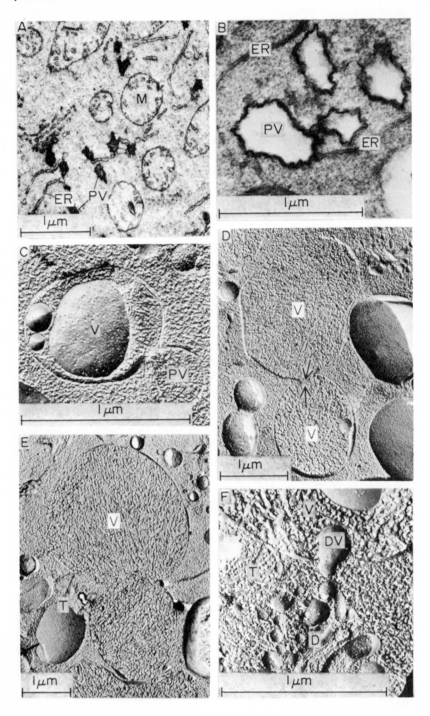

The fine structural analysis of vacuolation has yielded convincing evidence for the origin of the smallest vacuoles, subsequently termed 'provacuoles', from profiles of the endoplasmic reticulum (ER). Indeed, fractions of provacuoles isolated from maize root tips contain enzymes that are regarded as membrane constituents of the ER (Matile, 1968a). Whether provacuoles originate from local dilatations of the ER cisternae (Buvat and Mousseau, 1960; Poux, 1962; Bowes, 1965; Mesquita, 1969) or by vesiculation of the ER is not yet fully clarified. In any case, local inflations of the ER, or stellate vacuoles still being connected with the reticulum, have never been observed in frozen-etched root meristems (Matile and Moor, 1968). In addition, Fineran (1970a) has demonstrated by comparative chemical and freeze fixation that 'the irregular shapes of vacuoles in thin sections are apparently caused by shrinkage during fixation. When shrinkage is severe, portions of the tonoplast become apposed and superficially resemble profiles of the ER'. In the light of recent observations made by Berjak (1971) the vesiculation theory appears to be correct: acid phosphatase activity seems to be present in the vesicles being pinched off from the ER rather than in the cisternae of the reticulum itself (Fig. 5.2A,B). Provacuoles would thus represent a product of ER differentiation rather than a product of mere dilation, and it would not be necessary to consider the reticulum as a constituent of the lysosomal compartment.

According to this view, the differentiation of ER into provacuoles involves the synthesis of lysosomal enzymes. This process apparently continues as provacuoles develop into larger vacuoles. Indeed, several classes of isolated vacuoles, differing in size and density, are characterized by distinct proportions of hydrolases activities (Matile, 1968a). The capacity of the tonoplast to synthesize protein is also suggested by the presence in isolated vacuoles of RNA, and morphological examination (Wiemken, 1969) suggests the association of ribosomes with this membrane (Fig. 5.5C). Moreover, the membranes of root tip vacuoles assume specific properties suggesting further differentiation upon development: certain ER-specific enzymes decrease in activity, whereas transaminase activity appears in the largest vacuoles that can be isolated from root tips (Matile, 1968a).

Morphologically, the differentiation of provacuoles into vacuoles is characterized by the continuous inflation and by the extensive fusion of vacuoles (Matile and Moor, 1968); profiles of fusing vacuoles are encountered at all stages of vacuolation. Provacuoles of submicroscopic size (0.1 to 0.3 μm), as well as larger vacuoles, show the same conspicuous phenomenon of membrane coalescence initiated by a protrusion from one vacuole deforming another adjacent vacuole (Figs. 5.2,C–E; 5.3).

In contrast to the above interpretation of vacuolation, it has been proposed that provacuoles of meristems represent persisting organelles which are multiplied by

Fig. 5.2 Origin and development of vacuoles in the radicular meristem of maize.
(A,B) Acid phosphatase localized in provacuoles (PV) pinched off from the endoplasmic reticulum. (Courtesy of P. Berjak.)
(C) Fusion of a provacuole (PV) with a small vacuole.
(D, E) Fusion of larger vacuoles. Note the protrusion of one vacuole deforming the adjacent vacuole (arrows).
(F) Engulfment of a dictyosome-vesicle (DV) by the tonoplast.

division (Barton, 1965). This hypothesis is based on the observation of differentially contrasted membranes of the ER and small vacuoles (uranyl-acetate); the same observations could, however, also support the hypothesis of membrane differentiation presented above. There are, however, organisms possessing a persistent system of vacuoles. The analysis of the budding cycle of *Saccharomyces* (Wiemken *et al.*, 1970) has shown that initial budding cells are characterized by the presence of numerous small vacuoles which are distributed among mother cell and bud; as budding proceeds these vacuoles inflate and fuse to form a few large vacuoles at late budding. In turn, upon initiation of budding these large vacuoles undergo extensive shrinkage and fragmentation (Fig. 5.4). Hence, the vacuome of yeast exhibits cyclic dynamics of fusion and fission, inflation and shrinkage; in contrast to the unidirectional process of vacuolation in higher plant cells the development of yeast vacuoles represents a cyclic generation of a persistent organelle. Similar phenomena of vacuolar dynamics may also occur in certain differentiated higher plant cells. Guard cells of stomata contain numerous tiny vacuoles when the stoma are closed, but only one large vacuole when they are open (Guyot and Humbert, 1970).

Role of dictyosomes. The persistence of yeast vacuoles poses interesting problem concerning the synthesis of lysosomal enzymes that are concentrated in the cell sap (Matile and Wiemken, 1967). In growing cells the corresponding proteins are either synthesized in ribosomes located at the tonoplast, or another structure is involved in the import of enzymes synthesized elsewhere. Such a structure would represent a primary or pure lysosome: it has not so far been detected in yeast cells, but morphological and cytochemical observations in other organisms suggest that vesicles produced by dictyosomes could be involved in the intracellular translocation of hydrolases.

Indeed, the incorporation of Golgi vesicles into vacuoles (Fig. 5.2F) has been observed both in thin sections and freeze-etching preparations of root tip cells (Matile and Moor, 1968; Berjak and Villiers, 1970). Moreover, acid phosphatase activity has been localized cytochemically in dictyosome cisternae and vesicles (e.g., Brandes and Bertini, 1964; Nougarède and Pilet, 1967; Dauwalder *et al.*, 1969 Poux, 1970) in a variety of plants (Fig. 5.1C), and the same enzyme activity has in fact been found to be associated with isolated Golgi membranes (Ray *et al.*, 1969). It therefore seemed logical to conceive the Golgi vesicles as structures involved in the intracellular translocation of lysosomal enzymes. The corresponding hypothesis of lysosome formation involves the synthesis of hydrolases at ribosomes attached to the rough ER (the nascent polypeptides being released into the cisternae) and the transformation of ER elements in dictyosomes (see Morré *et al.*, 1970); at the end of this process the lysosomal enzymes are packed in dictyosome vesicles to be carried to the vacuoles (Brandes and Bertini, 1964; Coulomb *et al.*, 1972). This hypothesis closely resembles the interpretation of zymogen production and secretion in the pancrease exocrine cell (Palade *et al.*, 1962). Lysosome formation in plant cells would represent a process of internal dictyosome-mediated enzyme secretion.

A considerable weakness of the above hypothesis is its foundation on only one hydrolase: acid phosphatase, an enzyme activity which in dictyosomes could as

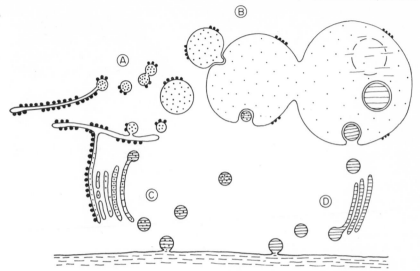

Fig. 5.3 Vacuolation in higher plant cells. Vesiculation of the ER and genesis of provacuoles (A), inflation and fusion of vacuoles (B); dictyosome mediated secretion into vacuoles and extracellular space.

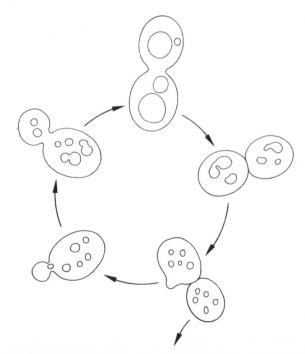

Fig. 5.4 Vacuolar dynamics in the course of the budding cycle of baker's yeast. (Adapted from Wiemken et al., 1970)

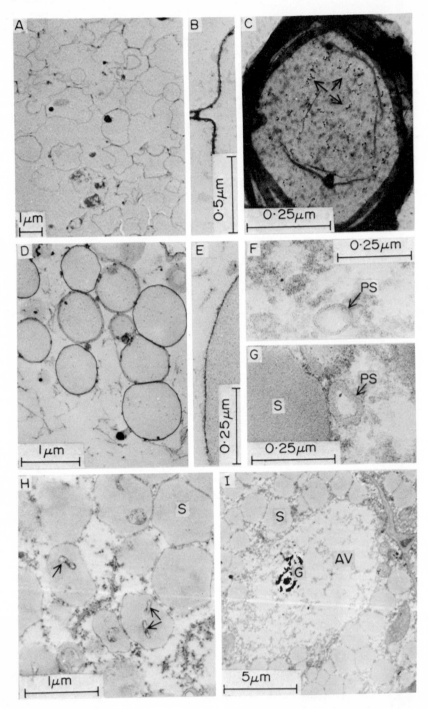

well have a specific non-lysosomal function in polysaccharide metabolism (Dauwalder *et al.,* 1969). In fact, the synthesis and secretion into the cell wall of polysaccharides represents a prominent function of the Golgi apparatus (see Mollenhauer and Morré, 1966). It could therefore be hypothesized that the incorporation of dictyosome vesicles into vacuoles represents a Golgi-mediated internal secretion of carbohydrates (Matile and Moor, 1968), and Berjak (personal communication) has in fact been able to show that monosaccharides are produced upon incubating isolated root vacuoles. It is feasible that polysaccharides imported via dictyosome vesicles into the vacuole are hydrolysed by corresponding lysosomal enzymes, whereby the monosaccharides formed would contribute to the osmotic pressure of the cell sap. As the term indicates, vacuoles represent an empty space whose spherical shape must be osmotically stabilized. Whether the necessary accumulation of micromolecules is due to intracellular digestion of osmotically inactive macromolecules or caused by the presence of tonoplast associated permeases is unknown. Yet, should it transpire that plant lysosomal enzymes are glycoproteins, as apparently are the corresponding enzymes of animal cells (Goldstone and Koenig, 1970), then the role of a Golgi-mediated step in their synthesis would be meaningful (Fig. 5.3).

Origin and development of aleurone grains and spherosomes

Aleurone grains have a single surrounding membrane (e.g., Yatsu, 1965; Jones, 1969a; Buttrose, 1971; Jacobsen *et al.*, 1972). At early stages of development they may be indistinguishable from vacuoles (e.g., Buttrose, 1963; Engleman, 1966) whose origin from the endoplasmic reticulum has already been discussed. In cotyledons of legumes the initial cell expansion is associated with the formation of large vacuoles which, upon subdivision, differentiate into protein bodies (Öpik, 1968; Briarty *et al.*, 1969). In other reserve tissues ER-derived provacuoles may immediately differentiate into aleurone vacuoles (Khoo and Wolf, 1970).

The development of aleurone grains is associated with the synthesis and accumulation of protein reserves (Bain and Mercer, 1966a). Indeed, Graham and Gunning (1970), employing specific fluorescent antibodies, have demonstrated

---

Fig. 5.5 Lysosomal membrane structure.

(A) Vacuoles isolated from yeast protoplasts showing (B) a triple layered membrane structure. (Courtesy of Wiemken and Rickenbacher.)

(C) An isolated yeast vacuole, contrasted positively with uranyl-acetate, showing membrane associated polysomes (arrows). (From Wiemken, 1969.)

(D) Spherosomes isolated from tobacco endosperm.

(E) At high resolution the spherosomal membrane exhibits a single electron-opaque contour.

(F, G) Development of spherosomes. Prospherosomes (PS) in endosperm cells of the developing *Ricinus* seed.

(H) Spherosomes of the *Ricinus* endosperm at a later stage of development. Note the cores bounded by the inner leaflet of the initial unit membrane of the prospherosome (arrows).

(I) Spherosomes and an aleurone vacuole containing a precipitating globoid (G) in an endosperm cell of *Ricinus.*

(F-I). (Courtesy of A. Schwarzenbach.)

that the reserve proteins legumin and vicilin are localized exclusively in the aleurone bodies of bean cotyledon cells. Similar results have been obtained in studies with organelles isolated from a variety of species (see Altschul *et al.,* 1966). Masses of rough endoplasmic reticulum associated with developing aleurone grains seem to be involved in the synthesis of these proteins (Öpik, 1968; Briarty *et al.,* 1969). In actual fact, autoradiographic pulse-chase experiments have demonstrated the initial association of labelled protein with ER filaments; later on the tracer moved into the protein bodies (Bailey *et al.,* 1970).

Nothing is known about the mechanism of protein transport from the ER into aleurone vacuoles. Ultrastructural data favouring a process of internal secretion via ER-derived granules containing the reserve proteins is not available. If the newly synthesized proteins are released into the cytoplasm rather than into the ER cisterna it would be necessary to postulate a specific mechanism responsible for the transport across the tonoplast resulting in the accumulation of reserve protein in the aleurone vacuole. On the other hand, electron micrographs presented by Khoo and Wolf (1970) suggest the association of polysomes with membranes of developing aleurone vacuoles. It is perhaps reasonable to assume that the lysosomal hydrolases present in mature aleurone grains (see p. 183) are synthesized in ribosomes attached to the tonoplast. In contrast to the enzymically inactive reserve proteins these enzymes would thereby not occur in the ground cytoplasm. As seed maturity is approached the gradual desiccation of the cells eventually results in the precipitation and, in the case of globulins, crystallization of reserve protein and accumulated phytic acid (Fig. 5.6).

Spherosomes originate also from the endoplasmic reticulum (Frey-Wyssling *et al.,* 1963). The prospherosomes pinched off from the ER show a distinct development. Schwarzenbach (1971a) has been able to demonstrate that the triglycerides accumulated in spherosomes are deposited in the hydrophobic central layer of the triple-layered membrane of prospherosomes (Fig. 5.5F–H; 5.7). At intermediate stages of lipid accumulation the inner electron opaque layer can still be observed in the centre of spherosomes. Hence, the development of spherosomes can be conceived as an extreme case of membrane differentiation which explains the anomalous structure of the spherosomal membrane (Fig. 5.5D,E). It is interesting to note that the enzymes involved in the synthesis of triglycerides from acetate are associated with spherosomes isolated from maize seedlings (Semadeni, 1967).

Autophagic activity of vacuoles

The autolytic breakdown of cytoplasm was the function first attributed to lysosomes; in other words, the digestive enzymes were thought to be ineffective in cell metabolism as long as they are present in a latent form, autolysis taking place only after the rupture of lysosomal membranes had occurred (see de Duve, 1969). Hence, autolysis must be related to cell death.

The phenomenon of metabolic lability of cellular constituents suggests, however, the occurrence of digestive processes (autophagy) within the living cell. Turnover of proteins, nucleic acids and other cytoplasmic components implies the simultaneous synthesis and degradation in individual cells, and the compartmentation of these opposed metabolic processes appears to be meaningful in providing the spatial limitation of cellular lytic processes. It is necessary to point out, however, that the lytic activity depends not only on the activity of digestive enzymes but also on the transport of cytoplasmic material into the lysosomal compartment.

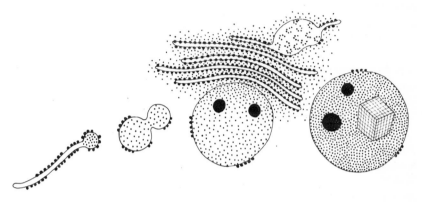

Fig. 5.6   Origin and development of aleurone grains.

The involvement of vacuoles in autophagy can be deduced from the presence of cytoplasmic material in the cell sap as shown by practically all electron micrographs of plant cells. Matile *et al.* (1965), Sievers (1966), Thornton (1968), Wardrop (1968), Zandonella (1970) and Mesquita (1972) have particularly pointed to this phenomenon, which was first interpreted by Poux (1963c) in terms of a digestive process localized in vacuoles.

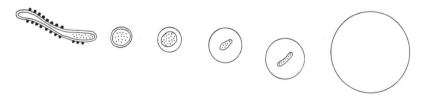

Fig. 5.7   Origin and development of spherosomes. (Adapted from Schwarzenbach, 1971a.)

The vacuolar membrane exhibits remarkable dynamic properties during the sequestration of cytoplasmic material into the cell sap. Invaginations of the tonoplast produced by local growth of the membrane result in cytoplasmic protuberances extending into the vacuolar lumen. Upon completion of these invaginations intravacuolar vesicles are pinched off from the tonoplast. Scars of both intravacuolar vesicles and tonoplast, indicate the places where membrane fission had occurred (Fig. 5.8A–E). It is assumed that the membranes surrounding the cytoplasmic exclaves are eventually ruptured (Fig. 5.11A) whereby the phagocytized material is exposed to the attack of digestive enzymes present in the cell sap (Matile and Moor, 1968). The phenomenon of invaginating tonoplasts and formation of intravacuolar vesicles has recently been observed in various species (e.g., Iten and Matile, 1970; Fineran, 1970b, 1971; Griffith, 1971).

Nothing is known about the forces that cause the engulfment of cytoplasmic material by the vacuole. Observations of Fineran (1970b) on fine structural peculiarities of the invaginating tonoplast suggest the involvement of membrane differentiation; convex fracture faces of vacuoles show a much higher density of globular particles at sites of invaginations as compared with inactive regions of the tonoplast (Fig. 5.8F). Treatments of cells inducing the autophagic activity of vacuoles, e.g., the short-term exposure of yeast cells to elevated sublethal temperatures, result in an immediate onset of tonoplast invagination and appearance of intravacuolar vesicles (Matile and Moor, unpublished). Hence the rapid growth of the tonoplast apparently does not depend on the preceding synthesis of membrane components.

Observations on autophagy in meristem cells suggest that random portions of cytoplasm, eventually including ribosomes, mitochondria and other membranous material may be sequestered into the lysosomal compartment (Fig. 5.9A). An example of this is the differentiation into sieve tubes, laticifers or tracheids (Bouck and Cronshaw, 1965; Esau, 1967; O'Brien and Thimann, 1967; Marty, 1970). In these cases the autophagic activity results in the elimination of a large proportion of cytoplasm; in tracheids it is eventually completely digested. In contrast, it has been observed in certain instances that organelles may be selectively engulfed by vacuoles. Villiers (1971a) has reported that proplastids and mitochondria, and sometimes only fractions of these organelles, are incorporated into vacuoles of growing *Fraxinus* embryos, from seeds that have been subjected to prolonged dormancy (Figs. 5.11B, 5.9B). Similar observations on a variety of organelles have been made by Fineran (1971). Proplastids present in shoot apices of *Kalanchoe* and *Bryophyllum* show peculiar inclusions arising from the accumulation of material within the cisternae of the inner membrane system; these vacuole-like inclusions are subsequently transferred to the periphery of the proplastids and selectively engulfed by the tonoplast (Gifford and Stewart, 1968). The presence in vacuoles of large membrane whirls and myelin-like bodies is possibly due to the selective sequestration of ER membranes (e.g., Thomas and Isaac, 1967; Bowes, 1969). Another example of apparent selectivity of autophagy has been detected in vacuoles of *Euphorbia characias* laticifers engulfing (indigestible!) latex globules (Marty, 1971). Hence, it seems that the autophagic activity of vacuoles represents a specifically regulated process but the factors involved in its induction and control are completely unknown.

An apparently distinct mechanism of autophagy has been reported by Villiers (1971b), Coulomb and Buvat (1968) and recently by Mesquita (1972). In dormant embryos of *Fraxinus* Villiers (1967) has observed Gomori-positive vacuoles containing cytoplasmic material. However, unlike normal vacuoles they possess a double membrane surrounding the sequestered cytoplasm. The fine structural analysis of their origin and development has shown that portions of cytoplasm may be surrounded and eventually completely enclosed by elements of the ER.

Upon dilation of the cisterna a vacuole with a single surrounding membrane and containing one or several intravacuolar vesicles is formed (Fig. 5.9C). The comparison of this mechanism with the autophagic activity of vacuoles described above reveals a rather close similarity of the two processes. It appears that autophagy and vacuolation may either occur simultaneously, or vacuolation via ER-derived pro-vacuoles may precede the autophagic activity of vacuoles. Indeed the extensive vacuolation in differentiating laticifers of *Euphorbia characias* seems to involve both types of autophagy and development of vacuoles (Marty, 1970).

It is important to realize the significance of the membrane dynamics associated with cellular lytic processes. In any case, during autophagic activity the strict compartmentation of digestive enzymes is never abolished. The portions of cytoplasm that are *en route* to the lysosomal compartment are always first wrapped and sealed into a membrane (tonoplast or endoplasmic reticulum) before they eventually become exposed to the hydrolases. This behaviour emphasizes the importance of the spatial limitation of lytic processes in the living cell and leads to the conclusion that the destruction of the lysosomal compartmentation is a feature of cell death.

Origin of the lysosomal exospace

In higher plants the formation of the extracellular space coincides with cell-plate formation after mitotic cell division. Structural analysis has pinpointed the prominent role of Golgi complexes. Numerous dictyosome vesicles that line up in the equatorial plane at late anaphase eventually coalesce to form the middle lamella of the new cell wall (Frey-Wyssling *et al.*, 1964). The newly formed plasma membranes therefore represent the product of fused Golgi membranes. The secretory activity of dictyosomes continues as the cell plate grows in thickness. This activity has generally been observed in growing cell walls and also in root caps in connection with the production and secretion of slime (see Mollenhauer and Morré, 1966). It has been established that Golgi-mediated secretion serves primarily for the deposition of polysaccharides constituting the matrix of cell walls. If, however, the hypothesis of Golgi-mediated secretion of hydrolases into the vacuome is correct (see p. 188), the secretion of dictyosome vesicles into the exospace could also account for the presence of hydrolases in the free space. Indeed, Gahan and McLean (1969) have demonstrated cytochemically acetyl-esterase activity in the Golgi complex, as well as in the cell wall; however, the reaction product was also present in ER filaments lying in close proximity to the plasmalemma. In any case, enzyme secretion by the coalescence of the homologous membranes of dictyosome vesicles with the plasmalemma would be plausible whereas ER elements containing secretory products are presumably differentiated (see Morré *et al.*, 1970) in dictyosomes before discharge into the exospace occurs. Indirect evidence favouring a correlation between vesiculation of dictyosomes and RNase secretion in barley aleurone layers treated with gibberellic acid has been reported by Jones and Price (1970).

There are, however, organisms showing conspicuous secretion without the involvement of morphologically distinguishable Golgi complexes. In baker's yeast, budding is initiated by the secretion of ER-derived vesicles that are discharged into the wall at the location of new buds (Moor, 1967). The recent isolation of these vesicles (Cortat *et al*, 1972) has yielded considerable evidence favouring the idea of secretion of both cell-wall polysaccharides (mannan) and extracellular hydrolases (glucanases). The ability of these secretory vesicles to fuse with the plasmalemma (Sentandreu and Northcote, 1969) suggests the differentiation of the initial ER membrane into a plasmalemma-homologous membrane. In *Saccharomyces cerevisiae* such a process may take place in a dictyosome-like structure which, in contrast to normal dictyosomes, is reduced to a single flattened vesicle that has occasionally been encountered in the region of ER vesiculation (Moor, personal communication).

Still another process of hydrolase secretion seems to involve ER-derived vesicles whose membranes do not coalesce with the plasmalemma. In hyphae of *Neurospora* cultured in media containing a proteinaceous source of nitrogen, secretory granules containing endopeptidases are discharged into the extracellular space by the invaginating plasmalemma which engulfs these vesicles (Matile *et al.,* 1965).

These few examples demonstrate again the strict compartmentation of lysosomal hydrolases. Secretory mechanisms are apparently designed to avoid mixing of digestive enzymes with the cytoplasm.

Functions of the lysosomal compartment

Biochemical differentiation

Turnover and autophagy. The metabolic lability of cellular constituents implies their intracellular digestion and simultaneous synthesis. It is difficult to visualize the functional significance of this turnover phenomenon under steady state conditions: e.g., in an exponentially growing population of cells. Perhaps the metabolic lability is due to the replacement of 'worn out' enzyme molecules by active ones. In any case, turnover is extremely low in exponentially growing, as compared with differentiating or resting, cells. In baker's yeast the corresponding rates of protein turnover are unmeasurably low in growing cells; conversely, about 1 per cent of total protein is turned over per hour in resting cells (see Halvorson, 1960), and rates as high as 7–15 per cent per hour have been estimated in anaerobically grown cells adapting to aerobic metabolism (Fukuhara, 1967). These findings demonstrate quite clearly the significance of cellular lytic processes in biochemical differentiation which take place if cells adapt their metabolism to changed environmental conditions.

The facultative anaerobes among yeasts are excellent organisms for the study of biochemical differentiation. In particular, *Saccharomyces* is characterized by a diauxic growth pattern in glucose media; that is to say, glucose has a repressive effect on respiration and is therefore first fermented to ethanol. Thereafter the culture goes through a stationary growth phase during which the cells differentiate to oxydative metabolism of ethanol. A few examples of enzyme activities changing dramatically during this adaptation period and in the course of other growth phases also characterized by biochemical differentiation are given in Fig. 5.10A (see Beck and von Meyenburg, 1968; Wiemken, 1969). The role of cellular lytic processes in biochemical differentiation is furthermore indicated by the ability of resting yeast cells to synthesize inducible enzymes. This is done at the expense of the amino acid pool which, in turn, is replenished from the degradation products of endogenous protein (Halvorson, 1960).

The exact establishment of the apparent correlation between biochemical differentiation and lysosomal activity is difficult, As stated above, lysosomal activity always implies both digestive enzyme and autophagic activity. In *Saccharomyces* a correlation with digestive enzyme activities but not with autophagy was established (Wiemken, 1969). In exponentially growing cells that were fully adapted to fermentative metabolism the activities of vacuolar hydrolases such as endopeptidases and RNase are comparatively low. As the culture enters the inter-exponential lag phase, during which the cells differentiate to oxydative metabolism, these activities increase dramatically and thereafter decrease as the second exponential growth phase commences (Fig. 5.10B). Attempts to correlate biochemical differentiation with the autophagic activity of vacuoles have so far failed. Autophagy undoubtedly does occur in yeast cells, but the corresponding activity of the tonoplast was conspicuous only after exposing the cells to sublethal concentrations of acridine orange, which is known to cause drastic protein degradation (see Matile, 1970).

Morphological and biochemical observations on differentiation in *Euglena* also suggest a prominent role of the lysosomal cell compartment. If these facultative photoautotrophs are kept in darkness and in media deprived of organic nutrients, autophagic metabolism is induced. Brandes *et al.* (1965) have been able to demonstrate the formation of autophagic vacuoles containing cytoplasmic organelles such as mitochondria at various stages of degradation (Fig. 5.11C). In addition, these authors have reported increased activities of hydrolases (e.g., endopeptidase) concurrent with autophagy, during carbon starvation in *Euglena gracilis* (Bertini *et al.*, 1965).

Morphological differentiation. Similar studies on biochemical differentiation in higher plants are difficult to carry out because, in contrast to populations of micro-organisms, the organs are composed of different types of cells that may behave distinctly. Nevertheless, it may be hypothesized that, among other processes, intracellular digestion plays a role in the adaptation to ever-changing environmental conditions. The most conspicuous phenomena of differentiation in higher plant cells are, however, always connected with the vectorial development of meristematic cells into the various types of specialized cells. It has already been pointed out that morphologically the significance of lytic processes is indicated by the autophagic activity of vacuoles observed in various differentiating systems (see p. 194). Observations of Fineran (1971), on the specific engulfment in root meristem cells of various structures (endoplasmic reticulum, nuclear envelope, mitochondria, amyloplasts) by the tonoplast, suggest the association of differentiation with the replacement or renewal of distinct cytoplasmic

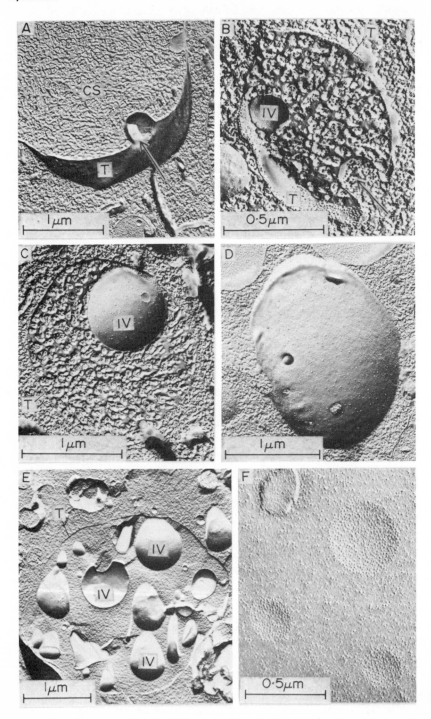

organelles. The accompanying biochemical events are, however, unknown, and corresponding experimental approaches are perhaps not feasible.

Storage and mobilization

Aleurone grains. Not only reserve proteins are accumulated in aleurone grains but also, apparently, the appropriate enzymes necessary for their mobilization. For example, edestinase, an acid endopeptidase, is localized in the aleurone grains of *Cannabis sativa* seeds together with its substrate, the reserve protein edestin (St Angelo *et al.*, 1968, 1969). This remarkable concurrence implies that the build-up of a protein deposit in the aleurone vacuoles of developing seeds is continually interfered with by the action of peptidases. Since this is probably not the case a sequential, rather than simultaneous, deposition of reserve proteins and proteases may be assumed, the enzymes being incorporated into the aleurone grains towards the end of seed maturation. Anyhow, their presence in the dormant seed allows the immediate onset of protein degradation upon imbibition.

The morphological correlate of mobilization is represented by the gradual disappearance of the protein matrix and, in certain species, crystalloids, combined with a considerable swelling of aleurone vacuoles (Nieuwdorp and Buys, 1964; Horner and Arnott, 1965; Bain and Mercer, 1966b; vaan der Eb and Nieuwdorp, 1967; Treffry *et al.*, 1967; Briarty *et al.*, 1970; Jones and Price, 1970). The tonoplast of aleurone vacuoles remains continuous throughout the period of protein mobilization. This is important with regard to the unspecificity of the proteolytic enzymes involved. The cytoplasm is unaffected and the persistence (or appearance of a new population) of ribosomes (Nougarède and Pilet, 1964) reflects the synthetic capacity of reserve cells in germinating seeds. In fact, *Vicia faba* cotyledons continue to synthesize protein whilst the content in total protein decreases sharply (Treffry *et al.*, 1967). In this and similar examples of cotyledons which eventually differentiate into green leaves it is not surprising that the mobilization of the reserve protein is connected with the partial reincorporation of amino acids into chloroplasts and other cellular constituents. However, protein synthesis is perhaps a common capacity of reserve cells, even in cases where the process of mobilization ends with cell death. In barley aleurone layers *de novo* synthesis of enzymes such as α-amylase at the expense of reserve proteins has been demonstrated (Filner and Varner, 1967). This process

---

Fig. 5.8   Autophagic activity of vacuoles.
(A–D) Formation of intravacuolar vesicles (IV) containing cytoplasmic material by the invaginating tonoplast.
(A) Beginning invagination (arrow).
(B) Cross fractured small vacuole with invaginating tonoplast (arrow).
(C,D) Scars on both intravacuolar vesicle and tonoplast indicate the sites of membrane fission.
(E) Autophagic vacuole containing numerous intravacuolar vesicles in a hymenial cell of the senescing gill of *Coprinus lagopus*. (Courtesy of W. Iten and H. Moor.)
(F) Convex fracture face of a tonoplast showing a dense population of globular particles at sites of beginning invagination. (Courtesy of B. A. Fineran.)

is induced by gibberellin-like hormones produced by the embryo. As shown by Paleg and Hyde (1964) the morphological changes accompanying the mobilization of vacuolar reserves are, in fact, drastically stimulated by the treatment of isolated barley aleurone layers with gibberellic acid. Proteolysis of stored protein, resulting in large volume increases of aleurone vacuoles and hence aleurone cells

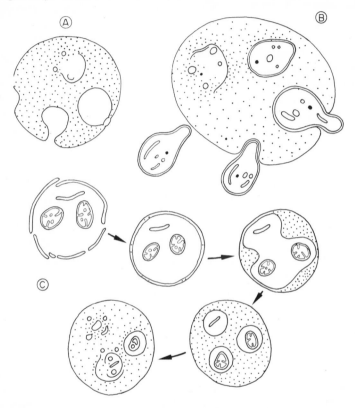

Fig. 5.9 Formation of autophagic vacuoles.
(A) Sequestration and intracellular digestion of a random portion of cytoplasm.
(B) Specific engulfment of an organelle. (See Villiers, 1971.)
(C) Endoplasmic reticulum sequestering cytoplasmic material and subsequent formation of an autophagic vacuole.

(Jones, 1969b), is therefore a prerequisite for the subsequent synthesis of hydrolases. These are eventually secreted into the dead tissue of the starchy endosperm (Varner and Chandra, 1964). The dependency of these processes on hormonal induction may imply the initial activation of existing or synthesis of new proteases. Morphological observations reported by Jones (1969b) suggest that mobilization of protein is accompanied by the production of ER-derived provacuoles which fuse with aleurone grains. These structures could be involved in the incorporation of proteolytic enzymes into the aleurone vacuoles. Fluctuations of proteolytic activities observed in germinating seeds whose aleurone grains

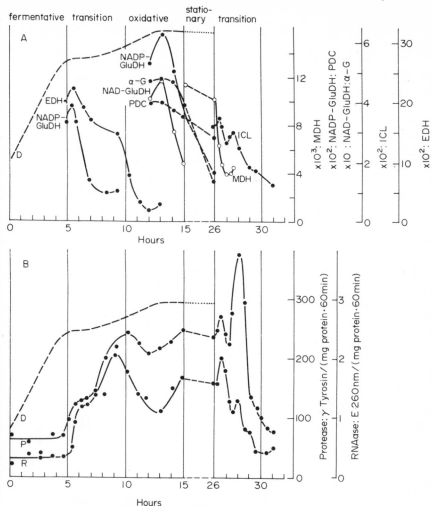

Fig. 5.10   Correlation between biochemical differentiation and lysosomal enzyme activity in the course of the diauxic growth of *Saccharomyces cerevisiae* on glucose (D: dry cell matter). After 26 hours of cultivation the stationary phase cells were exposed to fresh glucose medium. (From Wiemken, 1969.)

(A) Changes of total enzyme activities during transition phases and stationary growth phase. EDH: Ethanol dehydrogenase; NAD— and NADP—Glu DH: glutamate dehydrogenases; PDC pyruvate decarboxylase; α-G: α-glucosidase; ICL: isocitrate lyase; MDH: malate dehydrogenase.

(B) Specific activities of lysosomal enzymes. P: acid protease; R: acid ribonuclease.

are equipped with proteases upon seed development could indicate the occurrence of similar phenomena in other species (Fig. 5.12A).

Evidently, only a fraction of the amino acids produced upon proteolysis in aleurone vacuoles are reused in the storage tissue, the bulk of them being exported into the developing embryo. However, the synthetic activity counteracting

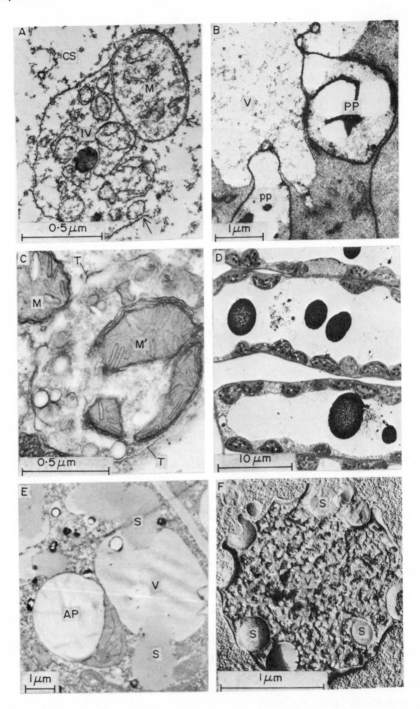

the protein mobilization as well as the transport of amino acids depends on the structural integrity of the cells and, hence, on the strict compartmentation of the lytic process. It is noteworthy that aleurone vacuoles persist after the mobilization of their own reserves. They gradually fuse and eventually coalesce into a single central vacuole. In species whose cotyledons differentiate into green leaves it represents the central vacuole of the parenchyma cells (Treffry *et al.*, 1967; Simola, 1969) whereas, in species whose reserve cells eventually die, mobilization is extended to the cytoplasm. Most probably this process is comparable with autophagy observed in senescent cells (see p. 208). At the very end of mobilization in pea cotyledons the vacuoles occupy practically the entire cell volume (Hinkelmann, 1966), and the spectrum of lysosomal hydrolases localized in aleurone vacuoles of this tissue provides a biochemical basis for the proposed process of autodigestion (Matile, 1968b).

Summarizing the story of the aleurone grain, this organelle appears as a transient differentiation product of vacuoles. It serves as a temporary storage site for reserves whose recall upon germination requires hydrolysis within a lysosomal cell compartment. Surprisingly, such differentiation of vacuoles is not limited to seeds. The synthesis of a specific type of reserve protein has been detected in various tissues of potato, tomato and other solanaceous plants. It is characterized by its powerful inhibitory effect on several proteolytic enzymes (Ryan, 1966), and is synthesized only under specific environmental or onto-genetical conditions (see Shumway *et al.*, 1970). The accumulation of this so-called chymotrypsin inhibitor I protein in cells of excised tomato leaves has been convincingly correlated with the appearance of electron opaque material (Fig. 5.11D) in the vacuoles (Shumway *et al.*, 1970). Observations of Greyson and Mitchell (1969) and of Shumway *et al.* (1972) on similar vacuolar structures associated with floral and vegetative apices of various species, suggest that the transient accumulation of protein reserves prior to organ development may be a widespread phenomenon. The specific circumstances leading to the formation and disappearance of these vacuolar inclusion bodies suggest their functional significance in the storage of nitrogen preceding the development of apical buds, fruits or tubers. The inhibitory effect of this reserve protein on plant proteases seems to be important with regard to its stability in the lysosomal compartment.

Fig. 5.11 Autophagic activity of vacuoles.
(A) Ruptured intravacuolar vesicle (IV) containing a mitochondrion and other membraneous material. (Courtesy of B. A. Fineran.)
(B) Specific engulfment of proplastids by the tonoplast in the embryo of *Fraxinus* seeds germinating after prolonged dormancy. (Courtesy of T. A. Villiers.)
(C) Upon carbon starvation autophagic vacuoles containing mitochondria and vesicular material are formed in *Euglena gracilis*. (Courtesy of D. Brandes.)
(D) Storage and mobilization: Electron-opaque globules in the vacuoles of tomato leaf palisade cells. The corresponding leaves had accumulated 565 μg of chymotrypsin inhibitor I protein per millilitre of leaf juice. (Courtesy of L. K. Shumway and C. A. Ryan.)
(E) Engulfment of spherosomes by vacuoles in a *Ricinus* embryo cell. (Courtesy of A. Schwarzenbach.)
(F) Uptake of spherosomes into the vacuole of *Saccharomyces cerevisiae*. (Courtesy of H. Moor.)

On the other hand, the degradation of this vacuolar protein seems to require a specific peptidase which is not inhibited.

Spherosomes. Spherosomes are common organelles of most plant cells, but the studies on spherosomal dynamics have been concentrated on storage tissues of seeds where they are particularly numerous. In barley aleurone layers they occupy a considerable fraction of the cell volume (Jones, 1969b), and in certain oleaginous tissues spherosomes are by far the most dominant organelle (Grieshaber, 1964).

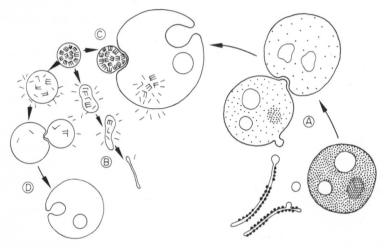

Fig. 5.12   Mobilization in aleurone grains and spherosomes. For explanation see text.

Observations on the fate of spherosomes in germinating seeds have shown their disappearance (Paleg and Hyde, 1964; Treffry *et al.*, 1967) which is apparently caused by the mobilization of triglycerides which results in a gradual decrease of the spherosomal volume (Nieuwdorp and Buys, 1964; vaan der Eb and Nieuwdorp, 1967). Recently, Mollenhauer and Totten (1971a) have reported on the transformation of spherosomes into flattened saccules occurring in cotyledons of pea and bean seeds during the first few days of germination (Fig. 5.12B). The biochemical correlate of these structural changes most probably concerns saponification of the triglycerides accumulated in spherosomes by the action of lipases and the concomitant release of fatty acids. Indeed, the flattened saccules formed upon lipid mobilization are characterized by a much higher buoyant density than the spherosomes (Mollenhauer and Totten, 1971b). It is interesting to note that the transformation of spherosomes into saccules may occur before completion of seed maturation (Mollenhauer and Totten, 1971a). This behaviour, and the presence of lipase in spherosomes of dormant seeds (see p. 184), suggests the incorporation of the mobilizing enzyme into the triglyceride storing organelle upon seed maturation.

That mobilization of triglycerides proceeds within the spherosomes is further indicated by their apparent affinity to glyoxysomes. In the endosperm of germinating castor beans spherosomes are surrounded by glyoxysomes (Vigil, 1970) which are known to be the seat of β-oxidation of fatty acids (released from spherosomes) and of the glyoxylic acid cycle involved in gluconeogenesis (see Beevers, 1969).

The transformation of spherosomes into saccules may be conceived as the reversal of spherosome development. It is, however, rather unlikely that the mobilization of triglycerides stored in the hydrophobic middle layer of the initial membrane of prospherosomes (see p. 192) results in the reconstitution of a normal triple-layered membrane, because the accumulation of lipid is associated with an enormous increase in volume and, hence, growth of the surrounding half-membrane. In any case, the membrane anomaly will not allow the fusion of spherosomes with other organelles of the lysosomal cell compartment. This is particularly interesting with regard to spherosomes lacking lipase (Jacks *et al.*, 1967). With such a modification, spherosomes should not even be transformed into saccules, and the mobilization of triglycerides might occur within a lipase-containing compartment. Indeed, morphological observations suggest that the mobilization may proceed in vacuoles, whereby the impossibility of membrane fusion is indicated in this case by the engulfment of spherosomes by the tonoplast (Fig. 5.12C). Schwarzenbach (1971b) has detected this phenomenon in the embryo of germinating *Ricinus* seeds (Fig. 5.11E). This example of *Ricinus* suggests that both mechanisms of triglyceride mobilization – lipolysis within spherosomes, and autophagic digestion within vacuoles respectively – may occur in the endosperm and embryo of the same species. Whether the spherosomes of the *Ricinus* embryo in fact lack lipase is not known, but the storage of trigly-cerides in the absence of the mobilizing enzyme could be meaningful with regard to storage of lipids that are not directly used upon germination. Spherosomes present in tissues other than reserve tissues of seeds or 'storage granules' of baker's yeast may represent such stable forms of lipid deposits. Indeed, in *Saccharomyces cerevisiae* the mobilization of spherosomes involves the engulf-ment by the tonoplast (Fig. 5.11F) and subsequent dissolution within the vacuole (see Matile *et al.*, 1969).

Still another activity of spherosomes has been observed in the endosperm of tobacco seeds. This tissue contains only the reserve oil, as the reserve protein is stored in the embryo. Hence, the spherosomes seem to represent the only organelle of the lysosomal compartment. They are characterized by the presence of a number of digestive enzymes (including lipase) otherwise encountered in vacuoles (Matile and Spichiger, 1968; Spichiger, 1969). Upon germination they behave similarly to aleurone vacuoles in showing inflation and fusion (Fig. 5.12D). Obviously the anomalous membrane of tobacco spherosomes (Fig. 5.5D,E) allows the coalescence with membranes of the same type. Even after lipolysis has been largely completed, the organelles retain their spherical shape and are involved in the reduction of cytoplasm by means of autophagic activity (Matile and Spichiger, unpublished results).

Exoplasmic space. The example of aleurone grains and spherosomes demon-strates the importance of compartmentation in the mobilization of cellular

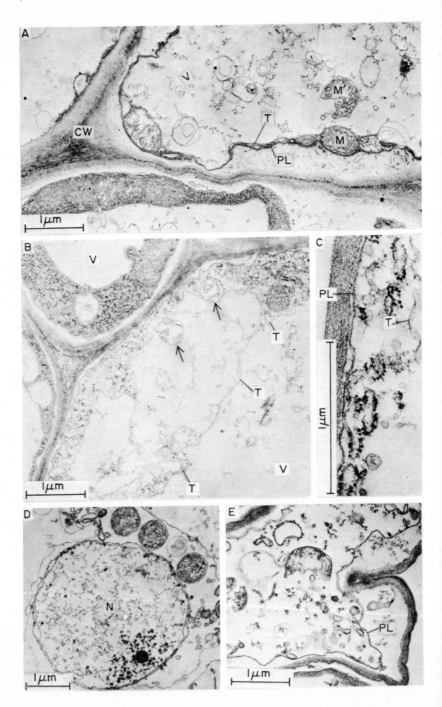

reserves. In constrast, the dead cells of the starchy endosperm of graminean caryopses provide an example of mobilization of extracellular reserves which is accomplished by hydrolases synthesized and secreted by the living cells of the aleurone layer and the scutellum. The spectrum of these extracellular lysosomal enzymes, ranging from α-amylase to RNase and protease (Chrispeels and Varner, 1967; Jacobson and Varner, 1967), reflects the nature of constituents to be dissolved in the starchy endosperm.

One enzyme among this group of secreted hydrolases, glucanase, seems to have a specific function with regard to the wall of aleurone cells. As noticed by Paleg and Hyde (1964) these very thick walls are eroded upon germination, and it is likely that this phenomenon is caused by the action of glucanase on cell wall glucans. Indeed, Jones (1971) has demonstrated the gibberellic acid induced secretion of glucanase which is correlated with extensive cell wall degradation (Taiz and Jones, 1970). Hence, the aleurone cell wall seems to function partially as a repository of carbohydrate reserves whose mobilization involves hydrolase secretion into the extracellular space. There is, in fact, a classical example illustrating this phenomenon: the cell wall mannan of the endosperm of date seeds is mobilized upon the secretion of enzymes by the haustorium of the embryo (see Keusch, 1968).

These examples may represent extreme cases of cell wall mobilization connected with germination. A number of recent observations suggest, however, that extracellular hydrolases could play a prominent role in cell wall metabolism generally. Phenomena such as plastic deformation of cell walls in extension growth and abscission have been related to secreted hydrolases. It would be beyond the scope of this chapter to present the evidence in detail. In any case, the partial *in vitro* autolysis of isolated primary cell walls (Lee *et al.*, 1967) clearly demonstrates that cell walls should not be conceived as metabolically stable products of the protoplast. It would rather seem that they are subjected to turn-over as are the cellular constituents. The membrane dynamics involved in hydrolase secretion have not been elucidated in higher plants (see p. 195), although some data are available from fungi. The secretion of β-1, 3-glucanases in budding *Saccharomyces cerevisiae* by means of ER-derived vesicles has already been mentioned (p. 196). Here the focused discharge of secretory granules into the wall results in the local intensification of glucan degradation, eventually allowing the extrusion of the bud. The occurrence of extracellular glucanases, even in stationary yeast cells, causes a continuous degradation of this cell wall polysaccharide which is probably balanced by a continuous synthetic activity

---

Fig. 5.13 Senescence and autolysis in the mesophyll of the wilting corolla of *Ipomöa purpurea.*

(A) Intensified autophagic activity. Note the presence of mitochondria (M') membranes and ribosomes in the cell sap.

(B) Shrinkage of the vacuole inflation of mitochondria (arrows) and dilution of the cytoplasm.

(C) Rupture of the tonoplast.

(D) Partially degraded nucleus after rupture of the tonoplast.

(E) Chaotic appearance of an autolysing cell.

(see Cortat *et al.*, 1972). Similar phenomena of turnover of cell wall constituents involving hydrolases of the lysosomal exospace may also occur in other fungi and higher plants.

In addition to the function in digesting certain nutrients extracellularly (see p. 185), secreted hydrolases probably play an important role in parasitic fungi. These organisms are able to dissolve the cell walls of their hosts (e.g., Calonge *et al.*, 1969). On the other hand, chitinase produced by bean leaves (Abeles *et al.*, 1970) has no substrate in the plant itself but may represent a potential extracellular antifungal agent in attacking cell walls of invading parasitic fungi. An example showing the involvement of the lysosomal cell compartment of the infected host cell has, in fact, been presented by Pitt and Coombes (1968). Finally, the possible role of hydrolase secretion in the interaction between the higher plant parasite *Cuscuta* and its host have been cytochemically demonstrated by Tripodi (1970).

Admittedly, so far as ultrastructural dynamics are concerned, only the process of secretion of hydrolase should be considered. Nevertheless, it seemed important to cover the biochemical aspect of the secreted enzymes in order to illustrate the fundamental significance of the division of the lysosomal compartment into an intracellular and an extracellular space.

Senescence and autolysis

Senescence may be conceived as unbalanced differentiation eventually resulting in cell death. The significance of lytic processes taking place in senescing tissues is reflected by the decreasing contents of protein, nucleic acids, and other cellular constituents. In turn, senescing cells continue to synthesize protein (e.g., digestive enzymes), but synthesis only partially counterbalances the degradation. Hence, the gradual decrease of cytoplasmic material is due to changes in the relative rates of synthesis and digestion, and the existence of a lysosomal cell compartment seems to represent an organizational prerequisite for the regulation of senescence (see Matile and Winkenbach, 1971).

The fine structural analysis of senescent cells has revealed the involvement of vacuoles. Bain and Mercer (1964) observed extensive vacuolation in pears during ripening, while Berjak and Villiers (1970) have been able to correlate the same phenomenon with hypersecretory activity of the dictyosomes in senescing root cap cells. The possible significance of this Golgi-mediated secretion into vacuoles has already been discussed (see p. 188–191). Several authors, concentrating on the fate of cytoplasmic organelles in senescent cells, have presented electron micrographs which show conspicuous autophagic activities of vacuoles (e.g., Bain and Mercer, 1964; Butler, 1967; Treffry *et al.*, 1967). This activity is obviously responsible for the gradual disappearance of ribosomes and membranes (Shaw and Manocha, 1965; Butler, 1967), and for related phenomena observed in senescing organs.

Apart from the increase in the vacuolar volume at the expense of the cytoplasm, the structure of organelles and the density of the cytoplasm exhibit a more or

less normal appearance during the period of intensified autophagy which characterizes senescing cells (Fig. 5.13A). At late senescence only a thin layer of cytoplasm is left between plasmalemma and tonoplast. However, when the development has reached a certain stage, the membrane systems of the ER, mitochondria
and plastids swell and the simultaneous shrinkage of the vacuole results in a
diluted appearance of the cytoplasm (Matile and Winkenbach, 1971). This
phenomenon is probably caused by permeability changes of the membranes
(Fig. 5.13B). It introduces the final stage of senescence: *autolysis.* The rupture
of the tonoplast (Treffry *et al.,* 1967; Butler, 1967; Matile and Winkenbach, 1971)
results in the mixing of lysosomal hydrolases with the cytoplasm (Berjak and
Villiers, 1970, 1972) to produce an entirely chaotic appearance of the protoplast
(Fig. 5.13C–E). The plasmalemma seems to resist temporarily the attack by
digestive enzymes, whereas mitochondria, plastids, ribosomes, and nuclei are
rapidly degraded. At the very end of autolysis only the empty cell wall is left.
A summarizing scheme of structural changes during senescence and autolysis is
given in Fig. 5.14. In illuminated flax cotyledons rupture of the tonoplast and
subsequent progressive deterioration of cell components have been observed as
early as 6 hours after treatment with the herbicide paraquat (Harris and Dodge,
1972).

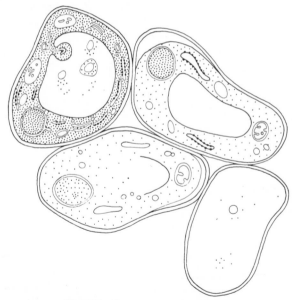

Fig. 5.14   Senescence and autolysis.

Whether the loss of compartmentation causes cell death or vice versa is difficult
to decide. As indicated by its hormonal control (and in certain cases, hormone
induced reversal), senescence seems to be a strictly programmed process.
The programme includes cell death, and only the exact definition of death would

allow an exact statement. Hence, the destruction of compartmentation may be seen as only one of a whole set of events accompanying cell death.

The corolla of the morning glory *Ipomöa* is characterized by its rapid wilting and concomitant decline in cytoplasmic constituents. Autophagic activity is, however, not strictly correlated with the macroscopic expression of wilting. The intensification of this process which characterizes senescent cells has already commenced in the mesophyll cells of the mature flower bud. In contrast, in the epidermal cells, autophagic activity is conspicuous when the wilting corolla is rolled up, whereas sieve tubes and companion cells show little change throughout the early phase of wilting. The delayed senescence in the phloem is probably meaningful with regard to the withdrawal of breakdown products from the corolla. In the majority of mesophyll cells autolysis is initiated upon the onset of wilting in the early afternoon of the flowering day. This period is also characterized by the dramatic increase of RNase and DNase activity. The corresponding proteins are presumably synthesized shortly before disorganization of the cytoplasm occurs (Winkenbach, 1970).

The fruiting bodies of the basidiomycete *Coprinus* are known for their rapid autolysis. A distinct role of the lysosomal compartment in autolysis appears from the occurrence of chitinases in vacuoles of the senescing gills. Dissolution of fruiting bodies is induced by the release of these hydrolases, probably from autolysing hyphae at the edge of gills (Iten and Matile, 1970). In *Coprinus* autolysis has a functional significance in spore release. There is no doubt, however, that in fungi lytic processes are also involved in phenomena such as sporogenesis (Chen and Miller, 1968). In asci of *Ceratocystis fimbriata* Wilson *et al.* (1970) have observed the development of a lysosomal compartment in the cytoplasm surrounding the ascospores. On the one hand the degradation of cytoplasm in vacuoles could provide the nutrients for the developing ascospores; on the other, the discharge of lysosomes into the exospace may be important in the breakdown of the wall of asci to facilitate spore release.

Concluding remarks

The discovery of a lysosomal cell compartment has provided the cytological basis for a better understanding of phenomena which had already been recognized by the classical plant physiologists. As early as 1897 Pfeffer appreciated the importance of protein degradation, which he visualized as a continuous process taking place in developing, differentiated, and senescing cells. Although Pfeffer appears as the prophet of turnover he was not in a position to explain how protein digestion can proceed within a living cell without affecting the enzymes involved in synthetic processes. We now have a preliminary answer: compartmentation of metabolism. The lysosomal compartment seems to represent the organizational prerequisite of turnover. And turnover appears, in turn, to represent a key for the understanding of the phenomena of adaptation and differentiation.

We are, however, far from really appreciating the entire structural and bio-chemical complexity, of this compartment. What we derive from static electron micrographs is certainly only an elementary idea of its dynamics. In addition, exact correlations between morphological observations and biochemical changes are difficult to make.

Despite insufficient knowledge about the nature of membranes generally, we particularly note the absence of plausible explanations for the membrane dynamics observed. The process of tonoplast invagination seems to depend, not only on the growth of the membrane in area, but also on the nature of the cytoplasmic material which is engulfed. This can be concluded from the exclusive sequestration of distinct organelles in certain instances, while biochemically it is indicated by the distinctive degradation rates of different cytoplasmic components. Whether the activity of the tonoplast is induced by the occurrence of denatured protein in the adjacent cytoplasm, or by membranes of injured or damaged organelles, the process of sequestration must be subjected to control. Even in those auto-phagic processes which accompany the differentiation into specialized cells such as sieve tube, tracheids and laticifers, the control of intracellular digestion is indicated by the maintenance of organization (order) until the developmental goal is reached. The same applies to senescence.

Another property of membranes surrounding lysosomal cell organelles is also puzzling: their stability in the presence of hydrolases. The rupture of the tonoplast in autolysing cells suggests that these membranes are in a state of dynamic stability which is metabolically maintained by the living cytoplasm.

Non-lysosomal functions of the lysosomal organelles have only been referred to briefly; e.g., in connection with the accumulation of reserves in aleurone vacuoles. It is, however, evident that phenomena such as the deposition of sugars, organic acids, amino acids, phenolic compounds and alkaloids in the cell sap depend on transport mechanisms associated with the tonoplast. In certain cases accumulation in vacuoles is reversible. Moreover, the re-use of the breakdown products depends on the transport of these micromolecules across the tonoplast into the cytoplasm. Hence, the tonoplast is apparently far from being an inert dialysis membrane which is responsible for the separation of vacuolar hydrolases from the cytoplasmic proteins. It seems, rather, that it has retained a structural complexity and functional dynamism both of which remain to be elucidated.

Acknowledgements

The author gratefully acknowledges the criticism of Dr Roger Parish and the help of Miss Ruth Rickenbacher in preparing the manuscript.

Bibliography

Introductory reading

De Duve, Ch. (1969) 'The lysosome in retrospect', In: *Lysosomes in biology and pathology* (eds. J. T. Dingle and H. B. Fell), vol. 1, pp. 3–40, North Holland Publ. Co., Amsterdam and London.

References

Abeles, F. B., Bosshart, R. P., Forrence, L. E., and Habig, W. H. (1970) 'Preparation and purification of glucanase and chitinase from bean leaves', *Plant Physiol.*, **47**, 129–134.

Altschul, A. M., Yatsu, L. Y., Ory, R. L., and Englemann, E. M. (1966) 'Seed proteins', *A. Rev. Pl. Physiol.*, **17**, 113–136.

Bailey, C. J., Cobb, A., and Boulter, A. (1970) 'A cotyledon slice system for the electron autoradiographic study of the synthesis and intracellular transport of the seed storage protein of *Vicia faba*', *Planta*, **95**, 103–118.

Bain, J. M. and Mercer, F. V. (1964) 'Organization resistance and the respiration climacteric', *Aust. J. biol. Sci.*, **17**, 78–85.

Bain, J. M. and Mercer, F. V. (1966a) 'Subcellular organization of the developing cotyledons of *Pisum sativum* L. in germinating seeds and seedlings', *Aust. J. biol. Sci.*, **19**, 49–67.

Bain, J. M. and Mercer, F. V. (1966b) 'Subcellular organization of the cotyledons in germinating seeds and seedlings of the developing cotyledon of *Pisum sativum*. L', *Aust. J. biol. Sci.*, **19**, 69–84.

Barton, R. (1965) 'Electron microscopic studies on the origin and development of the vacuoles in root tip cells of *Phaseolus*', *Cytologia*, **30**, 266–273.

Beck, Ch. and Meyenburg, H. K. von (1968) 'Enzyme pattern and aerobic growth of *Saccharomyces cerevisiae* under various degrees of glucose limitation', *J. Bact.*, **96**, 479–486.

Beevers, H. (1969) 'Glyoxysomes of castor bean endosperm and their relation to gluconeogenesis', *Ann. N.Y. Acad. Sci.*, **168**, 313–324.

Berjak, P. (1968) 'A lysosome-like organelle in the root cap of *Zea mays*', *J. Ultrastruct. Res.*, **23**, 233–242.

Berjak, P. (1972) 'Lysosomal compartmentation: Ultrastructural aspects of the origin, development and function of vacuoles in *Lepidium sativum*', *Ann. Bot.*, **36**, 73–81.

Berjak, P. and Villiers, T. A. (1970) 'Ageing in plant embryos. I. The establishment of the sequence of development and senescence in the root cap during germination', *New Phytol.*, **69**, 929–938.

Berjak, P. and Villiers, T. A. (1972) 'Ageing in plant embryos. III. Acceleration of senescence following artificial ageing treatment', *New Phytol.*, **71**, 513–518.

Bertini, F., Brandes, D., and Buetow, D. E. (1965) 'Increased acid hydrolase activity during carbon starvation in Euglena gracilis', *Biochim. biophys. Acta*, **107**, 171–173.

Beteta, P. and Gascon, S. (1971) 'Localization of invertase in yeast vacuoles', *FEBS Letters*, **13**, 297–301.

Bouck, G. B. and Cronshaw, J. (1965) 'The fine structure of differentiating sieve tube elements', *J. Cell Biol.*, **25**, 79–96.

Bowes, B. G. (1965) 'The origin and development of vacuoles in *Glechoma hederacea* L.', *La Cellule*, **65**, 359–364.

Bowes, B. G. (1969) 'Electron microscopic observations on myelin-like bodies and related membranous elements in *Glechoma hederacea* L', *Z. Pflphysiol.*, **60**, 414–417.

Brandes, D. and Bertini, F. (1964) 'Role of Golgi apparatus in the formation of cytolysomes', *Expl. Cell Res.*, **35**, 194–217.

Brandes, D., Buetow, D. E., Bertini, F., and Malkoff, D. B. (1965) 'Role of lysosomes in cellular lytic processes. I. Effect of carbon starvation in *Euglena gracilis*', *Expl molec. Pathol.*, **3**, 583–592.

Briarty, L. G., Coult, D. A., and Boulter, D. (1969) 'Protein bodies of developing seeds of *Vicia faba*', *J. exp. Bot.*, **20**, 358–372.

Briarty, L. G., Coult, D. A., and Boulter, D. (1970) 'Protein bodies of germinating seeds of *Vicia faba*. Changes in fine structure and biochemistry', *J. exp. Bot.*, **21**, 513–524.

Butler, R. D. (1967) 'The fine structure of senescing cotyledons of cucumber', *J. exp. Bot.*, **18**, 535–543.

Buttrose, M. S. (1963) 'Ultrastructure of the developing aleurone cells of wheat grain', *Aust. J. biol. Sci.*, **16**, 768–774.

Buttrose, M. S. (1971) 'Ultrastructure of barley aleurone cells as shown by freeze-etching', *Planta*, **96**, 13–26.

Buvat, R. (1971) 'Origin and continuity of cell vacuoles', In: *Origin and continuity of cell organelles* (eds. J. Reinert and H Ursprung), pp. 127–157, Springer-Verlag, Berlin, Heidelberg and New York.

Buvat, R. and Mousseau, A. (1960) 'Origine et évolution du système vacuolaire dans la racine de "*Triticum vulgare*"; relation avec l'ergastoplasme', *C.r. hebd. Séanc Acad. Sci., Paris*, 3051–3053.

Calonge, F. D., Fielding, A. H., Byrde, R. J. W., and Akinrefon, O. A. (1969) 'Changes in ultrastructure following fungal invasion and the possible relevance of extracellular enzymes', *J. exp. Bot.*, **20**, 350–357.

Cazin, J. Jr, Konzel, T. R., Lupan, D. M., and Burt, W. R. (1969) 'Extracellular deoxyribonuclease production by yeasts', *J. Bact.*, **100**, 760–762.

Chang, Ch. W. and Bandurski, R. S. (1964) 'Exocellular enzymes of corn roots', *Plant Physiol.*, **39**, 60–64.

Chen, A. Whei-Chu and Miller, J. J. (1968) 'Proteolytic activity of intact yeast cells during sporulation', *Can. J. Microbiol.*, **14**, 957–963.

Ching, Te May (1968) 'Intracellular distribution of lipolytic activity in the female gametophyte of germinating Douglas fir seeds', *Lipids*, **3**, 482–488.

Chrispeels, M. J. and Varner, J. E. (1967) 'Gibberellic acid-enhanced synthesis and release of α-amylase and ribonuclease by isolated barley aleurone layers', *Plant Physiol.*, **42**, 398–406.

Cortat, M., Matile, Ph., and Wiemken, A. (1972) 'Isolation of glucanase containing vesicle from budding yeast', *Arch. Mikrobiol.*, **82**, 189.

Coulomb, Ph. (1968) 'Etude préliminaire sur l'activité phosphatasique acid des particules analogues aux lysosomes des cellules radiculaires jeunes de la courge (*Cucurbita pepo* L. Cucurbitacée)', *C. r. hebd. Séanc Acad. Sci., Paris*, **267**, 2133–2136.

Coulomb, Ph. (1969) 'Phytolysosomes dans les frondes d'*Asplenium fontanum;*(Filicinées, Polypodiacées); isolement sur gradients et dosage de quelques enzymes', *C. r. hebd. Séanc, Acad. Sci., Paris*, **269**, 2543–2546.

Coulomb, C. and Buvat, R. (1968) 'Processus de régénérence cytoplasmique partielle dans les cellules de jeunes racines de *Cucurbita pepo*', *C. r. hebd. Séanc. Acad. Sci., Paris*, **267**, 843–844.

Coulomb, Ph., Coulomb, C., and Coulen, J. (1972) 'Origine et fonctions des phytolysosomes dans le méristème radiculaire de la courge (*Cucurbita pepo* L. Cucurbitacée)', *J. Microscopie*, **13**, 263–280.

Czaniiski, Y. and Catesson, A. M. (1969) 'Localisation ultrastructurale d'activités peroxydasiques dans les tissus conducteurs végétaux au cours du cycle annuel', *J. Microscopie*, **8**, 875–888.

Czaninski, Y. and Catesson, A. M. (1970) 'Activités peroxydasiques d'origines diverses dans les cellules d'Acer pseudoplatanus (Tissus conducteurs et cellules en culture)', *J. Microscopie*, **9**, 1089–1102.

Dangeard, P. (1956) 'Le vacuome de la cellule végétale. Morphologie', In: *Protoplasmatologia* vol. III D, p. 1.

Dauwalder, M., Whaley, W. G., and Kephart, J. E. (1969) 'Phosphatases and differentiation of the Golgi apparatus', *J. Cell Sci.*, **4**, 455–497.

De Duve, Ch. (1969) 'Evolution of the peroxisome', *Ann. N.Y. Acad. Sci.*, **168**, 369–381.

Eb vaan der, A. A. and Nieuwdorp, P. J. (1967) 'Electron microscopic structure of the aleurone cells of barley during germination', *Acta bot. neerl.*, **15**, 690–699.

Engleman, E. M. (1966) 'Ontogeny of aleurone grains in cotton embryo', *Am. J. Bot.*, **53**, 231–237.

Esau, K. (1967) 'Minor veins in beta leaves: structure related to function', *Proc. Amer. phil. Soc.*, **111**, 219–233.

Figier, J. (1968) 'Localisation infrastructurale de la phosphomonoestérase acide dans la stipule de *Vicia faba* L. au niveau du nectaire', *Planta*, **83**, 60–79.

Filner, P. and Varner, J. E. (1967) 'A simple and unequivocal test for de novo synthesis of enzymes: density labeling of barley α-amylase with $H_2 O^{18}$', *Proc. natn. Acad. Sci., U.S.A.*, **58**, 1520–1526.

Fineran, B. A. (1970a) 'An evaluation of the form of vacuoles in thin sections and freeze-etch replicas of root tips', *Protoplasma*, **70**, 457–478.

Fineran, B. A. (1970b) 'Organization of the tonoplast in frozen-etched root tips', *J. Ultrastruct. Res.*, **33**, 574–586.

Fineran, B. A. (1971) 'Ultrastructure of vacuolar inclusions in root tips', *Protoplasma*, **72**, 1–18.

Lysosomes

Frey-Wyssling, A., Grieshaber, E., and Mühlethaler, K. (1963) 'Origin of spherosomes in plant cells', *J. Ultrastruct. Res.*, 8, 506–516.

Frey-Wyssling, A., López-Sáez, J. F., and Mühlethaler, K. (1964) 'Formation and development of the cell plate', *J. Ultrastruct. Res.*, 10, 422–432.

Fukuhara, H. (1967) 'Protein synthesis in non-growing yeast. Respiratory adaption system', *Biochim. biophys. Acta*, 134, 143–164.

Gahan, P. B. and McLean, J. (1969) 'Subcellular localization and possible functions of acid β-glycerophosphatases and naphtol esterases in plant cells', *Planta*, 89, 126–135.

Gifford, E. M. and Stewart, K. D. (1968) 'Inclusion of the proplastids and vacuoles in the shoot apices of *Bryophyllum* and *Kalanchoë*', *Am. J. Bot.*, 55, 269–279.

Goldstone, A. and Koenig, H. (1970) 'Lysosomal hydrolases as glycoproteins', *Life Sci.*, 9, (II), 1341–1350.

Graham, T. A. and Gunning, B. E. S. (1970) 'The localisation of legumin and vicilin in bean cotyledon cells using fluorescent antibodies', *Nature, Lond.*, 228, 81–82.

Greyson, R. I. and Mitchell, K. R. (1969) 'Light and electron microscope observations of a vacuolar structure associated with the floral apex of *Nigella damascena*', *Can. J. Bot.*, 47, 597–601.

Grieshaber, E. (1964) 'Entwicklung und Feinbau der Sphärosomen in Pflanzenzellen', *Vierteljahrschr. Naturforsch. Ges. Zürich*, 109, 1–23.

Griffith, D. A. (1971) 'Hyphal structure in *Fusarium oxysporum* (Schlecht) revealed by freeze-etching', *Arch. Microbiol.*, 79, 93–101.

Guyot, M. and Humbert, C. (1970) 'Les modifications du vacuome des cellules stomatiques d'*Anemia rotundifolia* Schrad', *C. r. hebd. Séanc Acad. Sci., Paris*, 270, 2787–2790.

Halperin, N. (1969) 'Ultrastructural localization of acid phosphatase in cultured cells of *Daucus carota*', *Planta*, 88, 91–102.

Halvorson, H. O. (1960) 'The induced synthesis of protein', *Adv. Enzymol.*, 22, 99–156.

Harris, N. and Dodge, A. D. (1972) 'The effect of Paraquat on flax cotyledon leaves: changes in fine structure', *Planta*, 104, 201–209.

Heftmann, E. (1971) 'Lysosomes in tomatoes', *Cytobios*, 3, 129–136.

Hinkelmann, W. (1966) *Licht- und elektronenmikroskopische Untersuchungen über den Abbau der Reserveproteine in keimenden Erbsen.* Ph. thesis, Technische Hochschule Braunschweig.

Holcomb, G. E., Hildebrandt, A. C., and Evert, R. F. (1967) 'Staining and acid phosphatase reactions of spherosomes in plant tissue culture cells', *Am. J. Bot.*, 54, 1204–1209.

Horner, H. T. and Arnott, H. J. (1965) 'A histochemical and ultrastructural study of *Yucca* seed proteins', *Am. J. Bot.*, 52, 1027–1038.

Iten, W. and Matile, Ph. (1970) 'Role of chitinase and other lysosomal enzymes of *Coprinus lagopus* in the autolysis of fruiting bodies', *J. gen. Microbiol.*, 61, 301–309.

Jacks, T. J., Yatsu, L. Y., and Altschul, A. M. (1967) 'Isolation and characterization of peanut spherosomes', *Plant Physiol.*, 42, 585–597.

Jacobsen, J. V. and Varner, J. E. (1967) 'Gibberellic acid-induced synthesis of protease by isolated aleurone layers of barley', *Plant Physiol.*, 42, 1596–1600.

Jacobsen, J. V., Knox, R. B., and Pyliotis, N. A. (1972) 'The structure and composition of aleurone grains in the barley aleurone layer', *Planta*, 101, 189–209.

Jones, R. L. (1969a) 'The fine structure of barley aleurone cells', *Planta*, 85, 359–375.

Jones, R. L. (1969b) 'Gibberellic acid and the fine structure of barley aleurone cells. I. changes during the lag-phase of α-amylase synthesis', *Planta*, 87, 119–133.

Jones, R. L. (1969c) 'The effect of ultracentrifugation on fine structure and α-amylase production in barley aleurone cells', *Plant Physiol.*, 44, 1428–1438.

Jones, R. L. (1971) 'Gibberellic acid-enhanced release of β-1, 3-glucanase from barley aleurone cells', *Plant Physiol.*, 47, 412–416.

Jones, R. L. and Price, J. M. (1970) 'Gibberellic acid and the fine structure of barley aleurone cells III. Vacuolation of the aleurone cell during the phase of ribonuclease release', *Planta*, 94, 191–202.

Keusch, L. (1968) 'Die Mobilisierung des Reservemannans im keimenden Dattelsamen', *Planta*, 78, 321–350.

Khoo, U. and Wolf, M. J. (1970) 'Origin and development of protein granules in maize endosperm', *Am. J. Bot.*, 57, 1042–1050.

Lampen, J. O. (1968) 'External enzymes of yeast: their nature and formation', *Antonie van Leeuwenhoeck*, **34**, 1–18.

Lee, S., Kivilaan, A., and Bandurski, R. S. (1967) '*In vitro* autolysis of plant cell walls', *Plant Physiol.*, **42**, 968–972.

Marty, F. (1970) 'Rôle du système membranaire vacuolaire dans la différenciation des lacticifères d'*Euphorbia characias* L.', *C. r. hebd. Séanc Acad. Sci., Paris*, **271**, 2301–2304.

Marty, F. (1971) 'Vésicules autophagiques des laticifères différenciés d'*Euphorbia characias* L.', *C. r. hebd. Séanc Acad. Sci., Paris.* **272**, 399–402.

Marty, F. (1972) 'Distributions des activités phosphatasiques acides au cours du processus d'autophagie cellulaire dans les cellules du méristème radiculaire d'*Euphorbia characias* L.', *C. r. hebd. Séanc. Acad. Sci., Paris*, **274**, 206–209.

Matile, Ph. (1965) 'Intrazelluläre Lokalisation proteolytischer Enzyme von *Neurospora crassa*. I. Funktion und subzelluläre Verteilung proteolytischer Enzyme', *Z. Zellforsch. mikrosk. Anat.*, **65**, 884–896.

Matile, Ph. (1966) 'Enzyme der Vakuolen aus Wurzelzellen von Maiskeimlingen. Ein Beitrag zur funktionellen Bedeutung der Vakuole bei der intrazellulären Verdauung', *Z. Naturf.*, **21b**, 871–878.

Matile, Ph. (1968a) 'Lysosomes of root tip cells in corn seedling', *Planta*, **79**, 181–196.

Matile, Ph. (1968b) 'Aleurone vacuoles as lysosomes', *Z. PflPhysiol.*, **58**, 365–368.

Matile, Ph. (1970) 'Recent progress in the study of yeast cytology', Proc. 2nd Int. Symp. Yeast Protoplasts, Brno, *Acta Facultatis Medicae Universitatis Bruniensis*, **37**, 17–23.

Matile, Ph. (1971) 'Vacuoles, lysosomes of Neurospora', *Cytobiol*, **3**, 324–330.

Matile, Ph. and Moor, H. (1968) 'Vacuolation: Origin and development of the lysosomal apparatus in root tip cells', *Planta*, **80**, 159–175.

Matile, Ph. and Spichiger, J. (1968) 'Lysosomal enzymes in spherosomes (oil droplets) of tobacco endosperm', *Z. PflPhysiol.*, **58**, 277–280.

Matile, Ph. and Wiemken, A. (1967) 'The vacuole as the lysosome of the yeast cell', *Arch. Mikrobiol.*, **56**, 148–155.

Matile, Ph. and Winkenbach, F. (1971) 'Function of lysosomes and lysosomal enzymes in the senescing corolla of the morning glory (*Ipomoea purpurea*)', *J. exp. Bot.*, **22**, 759–771.

Matile, Ph., Jans, B., and Rickenbacher, R. (1970) 'Vacuoles of Chelidonium latex: lysosomal property and accumulation of alcaloids', *Biochem. Physiol. Pflanzen*, **161**, 447–458.

Matile, Ph., Jost, M., and Moor, H. (1965) 'Intrazelluläre Lokalisation proteolytischer Enzyme von Neurospora crassa. II. Identifikation von proteasehaltigen Zellstrukturen', *Z. Zellforsch. mikrosk. Anat.*, **68**, 205–216.

Matile, Ph.., Moor, H., and Robinow, C. F. (1969) 'Yeast cytology', In: *The yeasts* (eds. A. H. Rose and J. S. Harrison), vol. 1, pp. 219–302, Academic Press, London and New York.

Matile, Ph., Weimken, A., and Guyer, W. (1971) 'A lysosomal aminopeptidase isozyme in differentiating yeast cells and protoplasts', *Planta*, **96**, 43–53.

Matile, Ph., Balz, J. P., Semademi, E., and Jost, M. (1965) 'Isolation of spherosomes with lysosome characteristics from seedlings', *Z. Naturf.*, **20b**, 693–698.

Mesquita, J. F. (1969) 'Electron microscope study of the origin and development of the vacuoles in root-tip cells of *Lupinus albus* L.', *J. Ultrastruct. Res.*, **26**, 242–250.

Mesquita, J. F. (1972) 'Ultrastructure de formations comparable aux vacuoles autophagiques dans les cellules des racines de l'*Allium cepa* L. et du *Lupinus albus*, L.', *Cytologia*, **37**, 95–110.

Mollenhauer, H. H. and Morré, D. J. (1966) 'Golgi apparatus and plant secretion', *A. Rev. Pl. Physiol.*, **17**, 27–46.

Mollenhauer, H. H. and Totten, C. (1971a) 'Studies on seeds. II. Origin and degradation of lipid vesicles in pea and bean cotyledons', *J. Cell Biol.*, **48**, 395–405.

Mollenhauer, H. H. and Totten, C. (1971b) 'Studies on seeds. III. Isolation and structure of lipid-containing vesicles', *J. Cell Biol.*, **48**, 533–541.

Moor, H. (1967) 'Endoplasmic reticulum as the initiator of bud formation in yeast (*S. cerevisiae*)', *Arch. Mikrobiol.*, **57**, 135–146.

Morré, D. J., Mollenhauer, H. H., and Bracker, C. E. (1970) 'Origin and continuity of Golgi apparatus', In: *Origin and continuity of cell organelles* (eds. J. Reinert and H. Ursprung), pp. 82–126. Springer-Verlag, Berlin, Heidelberg, and New York.

Morris, G. F. I., Thurman, D. A., and Boulter, D. (1970) 'The extraction and chemical composition of aleurone grains (protein bodies) isolated from seeds of *Vicia faba'*, *Phytochemistry*, **9**, 1707–1714.

Nieuwdorp, P. J. and Buys, M. C. (1964) 'Electron microscopic structure of the epithelial cells of the scutellum of barley II. Cytology of the cells during germination', *Acta bot. neerl.*, **13**, 559–565.

Nougarède, A. and Pilet, P. E. (1964) 'Infrastructure des cellules du scutellum du *Triticum vulgare* Vill. au cours des premières phases de la germination', *C. r. hebd. Séanc Acad. Sci., Paris*, **258**, 2641–2644.

Nougarède, A. and Pilet, P. E. (1967) 'Activité et localisation, au niveau des infrastructures, de la phosphomonoestérase acide dans la racine et dans les jeunes feuilles du *Lens culinaris* L.', *C. r. hebd. Séanc Acad. Sci., Paris*, **265**, 663–666.

O'Brien, T. P. and Thimann, K. V. (1967) 'Observation on the fine structure of the oat coleoptile. III. Correlated light and electron microscopy of the vacuolar tissues', *Protoplasma*, **63**, 443–478.

Öpik, H. (1968) 'Development of cotyledon cell structure in ripening *Phaseolus vulgaris* seeds', *J. exp. Bot.*, **19**, 64–76.

Ory, R. L. and Henningsen, K. W. (1969) 'Enzymes associated with protein bodies isolated from ungerminated barley seeds', *Plant Physiol.*, **44**, 1488–1498.

Ory, R. L., Yatsu, L. Y., and Kircher, H. W. (1968) 'Association of lipase activity with the spherosomes of *Ricinus communis'*, *Archs. Biochem. Biophys.*, **123**, 255–264.

Palade, G. E., Siekevitz, P., and Caro, L. G. (1962) 'Structure, chemistry and function of the pancreas exocrine cell', *Ciba foundation Symp. on the exocrine pancreas 1962*, 23–49.

Paleg, L. and Hyde, B. (1964) 'Physiological effects of Gibberellic acid. VII. Electron microscopy of barley aleurone cell', *Plant Physiol.*, **39**, 673–680.

Pitt, D. and Coombes, C. (1968) 'The disruption of lysosome-like particles of *Solanum tuberosum* cells during infection by *Phytophtora erythroseptica* Pethybr', *J. gen. Microbiol.*, **53**, 197–204.

Pitt, D. and Galpin, M. (1973) 'Isolation and properties of lysosomes from dark-grown potato shoots', *Planta*, **109**, 233–258.

Poux, N. (1962) 'Nouvelles observations sur la nature et l'origine de la membrane vacuolaire des cellules végétales', *J. Microscopie*, **1**, 55–66.

Poux, N. (1963a) 'Localisation de la phosphatase acide dans les cellules méristematiques de blé *(Triticum vulgare* Vill.)', *J. Microscopie*, **2**, 485–489.

Poux, N. (1963b) 'Localisation des phosphates et de la phosphatase acide dans les cellules des embryos de blé (*Tr. vulg.* Vill.) lors de la germination', *J. Microscopie* **2**, 557–568.

Poux, N. (1963c) 'Sur la présence d'enclaves cytoplasmiques en voie de dégénerescence dans les vacuoles des cellules végétales', *C. r. hebd. Séanc Acad. Sci., Paris*, **257**, 736–738.

Poux, N. (1965) 'Loacalisation de l'activité phosphatasique acide et des phosphates dans les grains d'aleurone. I. Grains d'aleurone referment a la fois globoides et cristalloides', *J. Microscopie*, **4**, 771–782.

Poux, N. (1969) 'Localisation d'activité enzymatiques dans les cellules du méristème radiculair de *Cucumis sativus* L. II. Activité peroxidasique', *J. Microscopie*, **8**, 855–866.

Poux, N. (1970) 'Localisation d'activité enzymatiques dans les cellules du méristème radiculaire de *Cucumis sativus* L. III. Activité phosphatasique acide', *J. Microscopie*, **9**, 407–434.

Pujarniscle, A. (1968) 'Caractère lysosomale des lutoïdes du latex d'*Hevea brasiliensis* Mül. Arg', *Physiol. Vég.*, **6**, 27–46.

Ray, P. M., Shininger, T. L., and Ray, M. M. (1969) 'Isolation of β-glucan synthetase particles from plant cells and identification with Golgi membranes', *Proc. natn. Acad. Sci., U.S.A.*, **64**, 605–612.

Ryan, C. A. (1966) 'Chymotrypsin inhibitor I from potatoes: Reactivity with mammalian, plant, bacterial, and fungal proteinases', *Biochemistry, N.Y.*, **5**, 1592–1596.

Schwarzenbach, A. M. (1971a) 'Observations on spherosomal membranes', *Cytobiol.*, **4**, 145–147.

Schwarzenbach, A. M. (1971b), *Aleuronvakuolen und Sphärosomen im Endosperm von Ricinus communis während der Samenreifung und Keimung*. Ph.D. thesis No. 4645, Swiss Federal Institute of Technology, Zürich.

Schweiger, H. G. (1966) 'Ribonuclease-Altivität in *Acetabularia'*, *Planta*, **68**, 247–255.

Semadeni, E. G. (1967) 'Enzymatische Charakterisierung der Lysosomenaequivalente (Spherosomen) von Maiskeimlingen', *Planta*, **72**, 91–118.

Sentandreu, R. and Northcote, D. H. (1969) 'The formation of buds in yeast', *J. gen. Microbiol.*, **55**, 393–398.

Shaw, J. G. (1966) 'Acid phosphatase from tobacco leaves', *Archs. Biochem. Biophys.*, **117**, 1–9.

Shaw, M. and Manocha, M. S. (1965) 'Fine structure in detached senescing wheat leaves', *Can. J. Bot.*, **43**, 747–755.

Shumway, L. K., Rancour, J. M., and Ryan, C. A. (1970) 'Vacuolar protein bodies in tomato leaf cells and their relationship to storage of chymotrypsin inhibitor I protein', *Planta*, **93**, 1–14.

Shumway, L. K., Cheng, V., and Ryan, C. A. (1972) 'Vacuolar protein in apical and flower-petal cells', *Planta*, **106**, 279–290.

Sievers, A. (1966) 'Lysosomen-ähnliche Kompartimente in Pflanzenzellen', *Naturwissenschaften*, **53**, 334–335.

Simola, L. K. (1969) 'Fine structure of *Bidens radiata* cotyledons, with special reference to formation of protein bodies, spherosomes and chloroplasts', *Ann. Acad. Sci. fenn. A, IV Biologica: 156.*

Sorokin, H. P. (1967) 'The spherosomes and the reserve fat in plant cells', *Amer. J. Bot.*, **54**, 1008–1016.

Spichiger, J. U. (1969) 'Isolation und Charakterisierung von Sphärosomen und Glyoxisomen aus Tabakendosperm', *Planta*, **89**, 56–75.

St Angelo, A. J., Yatsu, L. Y., and Altschul, A. M. (1968) 'Isolation of edestin from aleurone grains of *Cannabis sativa*', *Archs. Biochem. Biophys.*, **124**, 199–205.

St Angelo, A. J., Ory, R. L., and Hansen, H. J. (1969) 'Localization of an acid proteinase in hempseed', *Phytochemistry*, **8**, 1135–1138.

Taiz, L. and Jones, R. L. (1970) 'Gibberellic acid, β-1, 3-glucanase and the cell walls of barley aleuron layers', *Planta*, **92**, 73–84.

Terukatsu, A., Uchida, T., and Egami, F. (1968) 'Studies on extracellular ribonucleases of *Ustilago sphaerogena*. Purification and properties', *Biochem. J.*, **106**, 601–607.

Thomas, P. L. and Isaac, P. K. (1967) 'An electron microscope study of intravacuolar bodies in the uredia of wheat stem rust and in hyphae of other fungi', *Can. J. Bot.*, **45**, 1473–1478.

Thornton, R. M. (1968) 'The fine structure of *Phycomyces*. I autophagic vesicles', *J. Ultrastruct. Res.*, **21**, 269–280.

Treffry, T., Klein, S., and Abrahamsen, M. (1967) 'Studies of fine structural and biochemical changes in cotyledons of germinating soy beans', *Aust. J. biol. Sci.*, **20**, 859–868.

Tripodi, G. (1970) 'Localization of tryptophan rich proteins and β-glycerophosphatase activity in *Cuscuta* haustorial cells', *Protoplasma*, **71**, 191–196.

Varner, J. E. and Chandra, G. R. (1964) 'Hormonal control of enzyme synthesis in barley endosperm', *Proc. natn. Acad. Sci. U.S.A.*, **52**, 100–106.

Vigil, E. L. (1970) 'Cytochemical and developmental changes in microbodies (glyoxysomes) and related organelles of castor bean endosperm', *J. Cell Biol.*, **46**, 435–454.

Villiers, T. A. (1967) 'Cytolysomes in long-dormant plant embryo cells', *Nature, Lond.*, **214**, 1356–1357.

Villiers, T. A. (1971a) 'Lysosomal activities of the vacuole in damaged and recovering plant cells', *Nature, New Biol.*, **233**, 57–58.

Villiers, T. A. (1971b) 'Cytology studies in dormancy. II. Pathological ageing changes during prolonged dormancy and recovery upon dormancy release', *New Phytol*, **71**, 145–152.

Walek-Czernecka, A. (1965) 'Histochemical demonstration of some hydrolytic enzymes in the spherosomes of plant cells', *Acta Soc. Botan. Polon.*, **34**, 573–588.

Wardrop, A. B. (1968) 'Occurrence of structures with lysosome-like function in plant cells', *Nature, Lond.*, **218**, 978–980.

Wiemken, A. (1969) *Eigenschaften der Hefevakuole*. Ph. D. thesis, No. 4340, Swiss Federal Institute of Technology, Zürich.

Wiemken, A., Matile, Ph., and Moor, H. (1970) 'Vacuolar Dynamics in synchronously budding yeast', *Arch. Mikrobiol.*, **70**, 89–103.

Wilson, C. L., Stiers, D. L., and Smith, G. G. (1970) 'Fungal lysosomes or spherosomes', *Phytopathology*, **60**, 216–227.

Winkenbach, F. (1970) 'Zum Stoffwechsel der aufblühenden und welkenden Korolle der Prunkwinde *Ipomoea purpurea*', *Ber. schweiz. bot. Ges.*, **80**, 374–406.

Lysosomes

Yatsu, L. Y. (1965) 'The ultrastructure of cotyledonary tissue from *Gossypium hirsutum*', *J. Cell Biol.*, **25**, 193–200.

Yatsu, L. Y. and Jacks, T. J. (1968) 'Association of lysosomal activity with aleurone grains in plant seeds', *Arch. Biochem. Biophys.*, **124**, 466–471.

Yatsu, L. Y., Jacks, T. J., and Hensarling, T. P. (1971) 'Isolation of spherosomes (oleosomes) from onion cabbage, and cottonseed tissues', *Plant Physiol.*, **48**, 675–682.

Zandonella, P. (1970) 'Infrastructure du tissu nectarigène floral de *Beta vulgaris* L.: le vacuome et la dégradation du cytoplasme dans les vacuoles', *C. r. hebd. Séanc Acad. Sci., Paris*, **271**, 70–73.

Zee, S. Y. (1969) 'The localization of acid phosphatase in the sieve elements of Pisum', *Aust. J. biol. Sci.*, **22**, 1051–1054.

6

Plant microtubules
J. D. Pickett-Heaps

Introduction

The introduction of aldehyde fixation for electron microscopy (Sabatini, Bensch, and Barrnett, 1963) has considerably extended our knowledge of cell ultrastructure, and of immediate importance to botanists was the demonstration that microtubular systems could be preserved reliably within plant cells (Ledbetter· and Porter, 1963). These organelles had been discovered earlier in plants, but were usually accompanied by some cytoplasmic disruption following osmium fixation (see p. 222). Microtubules now seem intimately involved in many interesting cellular processes, particularly mitosis, cytokinesis, and some aspects of differentiation. Much has been written on these organelles which all too often reveals the electron microscopist's urge to attribute to every structure or association of organelles he sees, various functions based on restricted observations. Space limitations have obliged me to concentrate here mainly on well-documented observations from which carefully considered conclusions may perhaps be drawn. In critically reviewing the cytology of cell wall formation, O'Brien (1972) has demonstrated how often a shaky edifice of apparent knowledge has been erected upon the uncritical interpretations of comparatively few isolated ultrastructural observations. The literature on microtubules is certainly no better, and justifies Brown's (1960) complaint that: 'at first, (light) microscopes were used with the same exuberant and indiscriminate enthusiasm with which electron microscopes are being used now, and their use led to the same accumulation of sporadic observations'.

Electron microscopy is still fundamentally limited, revealing structure only in dead, embedded cells or in specially prepared fragments of cells or their replicas, and so considerable imagination is required to visualize the behaviour of organelles *in vivo*. Sporadic observations give little real insight in this context, and many interesting experiments often lack a firm foundation for the interpretation of their results. For example, how can one really assess the effects of the application of drugs (e.g., colchicine) to dividing cells, without having a definitive account of the ultrastructure of the normal mitotic nucleus? Constructive speculation usually only follows careful, detailed studies of developmental sequences (e.g., mitosis, differentiation, etc.) in which the spatial distribution and other characteristics of organelles can be correlated with overall changes in the cell's behaviour and structure; therefore I will repeatedly emphasize such work. Most valuable is the comparison of behaviour *in vivo* (e.g., during mitosis) with concurrent ultra-structural examination of the same or similar cells. Treatment of cells with chemical or physical agents known to affect organelles may allow direct testing of hypotheses formulated from purely descriptive work. Finally, ideas may be refined further by a comparison of similar structures in different cells which probably have much in common. For example, the widely held belief that centrioles function in the assembly and/or function of the mitotic apparatus appears untenable — to thinking botanists, anyway — in view of their absence from plant spindles (see Pickett-Heaps, 1971a). The mysteries of mitosis have always stimulated the formulation of ingenious and varied hypotheses (e.g., see Mazia, 1961); on this subject, my own limited comments will concern only microtubules, drawing where possible some tentative inferences and conclusions. Throughout I shall try also to distinguish clearly between observations and subsequent interpretations to enable readers to draw their own (perhaps contrary) conclusions.

The biochemical and biophysical nature of microtubules

In understanding the behaviour and functions of microtubules, some knowledge of their chemistry is most helpful; Newcomb (1969: pp. 256–257) has reviewed this subject and I will not deal exhaustively with it. Microtubules now appear to be composed of protein subunits (Kiefer, Sakai, Solari, and Mazia, 1966) polymerized in a regular manner (see Newcomb's review, p. 255) and association/disassociation of the subunits may possibly be a quite simple process (e.g., dependent upon the presence of ions or other factors). The biophysics and biochemistry of the subunits are being intensively investigated at the moment. The general feeling (as usual!) that this protein resembles actin (Renaud, Rowe, and Gibbons, 1968; also Newcomb, 1969), is not supported by some recent biochemical results (e.g., Stephens, 1970), and since evidence to suggest that microtubules could exert force by an actin/myosin-like interaction is not extensive, this analogy may be misleading. More important is the finding that several classes of microtubular structures in the cell can be recognized by differences in their chemical properties (e.g., by their digestibility with various enzymes, their appearance after various methods of fixation, their susceptibility to drugs or lack of nutrients, etc.) — (see Behnke and Forer, 1967; Burton, 1968; Tilney and Gibbins, 1968). Behnke and Forer (1967) conclude that in cranefly spermatids four classes of tubules can be recognized: firstly, accessory tubules and central tubules of the flagellum; secondly, cytoplasmic microtubules; thirdly, and fourthly, the 'A' and 'B' tubules comprising the outer nine doublets of the flagellum. Although possibly assembled from similar subunits (Tilney and Gibbins, 1968), these microtubules are not structurally and chemically equivalent. That the nine

doublets in the cilia are easily distinguishable from other cytoplasmic microtubules is particularly significant when considering the function of the centriole
(see later). Such results are supported by biochemical data; for example,
Shelanski and Taylor (1967, 1968) have shown that of the components of
sperm flagella, only the protein of the central pair of tubules binds colchicine
(see below). However, as McIntosh (personal communication) points out, the
procedure necessary to solubilize the outer tubules would probably also destroy
colchicine-binding ability in the protein derived from the inner tubules.

As might be expected of linear polymers of protein subunits, groups of microtubules can be related to birefringence *in vivo*; this is clear in mitotic cells (Inoue
and Sato, 1967 — see Newcomb, 1969), and some other structures composed
mostly of microtubules (e.g., the axonemes of *Heliozoa*: Tilney and Porter,
1965). Other filamentous structures, however, are also birefringent (e.g., Wohlman
and Allen, 1968); nevertheless, microtubules seem responsible for most of the
birefringence of spindle fibres (Kane and Forer, 1965; Rebhun and Sander,
1967), a property most valuable in elucidating indirectly the movements of
microtubules *in vivo*. Ultra-violet light is capable of destroying such birefringence,
leading to the development of u.v.-microbeams which, when used to create areas
of reduced birefringence in the spindle *in vivo,* allow the study of some movements in the spindle fibres themselves (Inoue, 1964; Forer, 1965). This treatment
has surprisingly little effect on the normal functioning of the spindle, unless the
u.v.-microbeam is directed at regions where assembly of spindle fibres seems to
be occurring (see later). Recently, Bajer and Mole-Bajer (1970) and Bajer (1971)
report, as expected, that such areas of reduced birefringence correspond with
regions of disorganized spindle microtubules although the extent of spindle
disruption is seldom clear-cut, and is always variable, dependent upon the
experimental equipment and routine used. Finally, we need note briefly some
other reagents or techniques used to affect microtubules. Colchicine (and
many analogues) has long been known to disrupt the mitotic spindle, destroying
its birefringement fibres *in vivo* (Gaulden and Carlson, 1951; Kihlman, 1966;
Inoue and Sato, 1967; Deysson, 1968), and treated cells suffer loss of their
spindle microtubules (Inoue and Sato, 1967; Pickett-Heaps, 1967a; Brinkley,
Stubblefield and Hsu, 1967). Borisy and Taylor (1967a, b) complemented these
results by showing that a specific protein bound tritiated colchicine in various
cells known to contain large numbers of microtubules. Treatment with colchicine
and its analogues is now frequently used to study the effect on cells of loss of
microtubular systems. Treatment with low temperatures (approximately $0°$ C) or
hydrostatic pressure is also usually effective in reversibly destroying both spindle
birefringence and cytoplasmic microtubules (Inoue, 1964; Tilney, Hiramato, and
Marsland, 1966; Tilney and Porter, 1967). However, some microtubules resist
this former treatment (Behnke, 1970).

Heavy water (deuterium oxide — D_2O) has the rather remarkable effect of
increasing the birefringence and volume of the spindle *in vivo,* and also the number
of cytoplasmic microtubules (e.g., Inoue and Sato, 1967); moreover, this
treatment 'freezes' the mitotic spindle (and other microtubule-based systems;

Tilney and Gibbins, 1969) until the D_2O is washed out, whereupon mitosis resumes apparently unaffected (see also Bal and Gross, 1964; Burgess and Northcote, 1969). These results are summarized by Inoue and Sato (1967). D_2O also confers upon the spindle an increased resistance to the disruptive effects of colchicine, cold, etc. (Marsland and Zimmerman, 1965). More recently, vinblastine and other related alkaloids have been shown to disrupt the mitotic spindle (Malawista, Sato, and Bensch, 1968) and other microtubule-containing systems; meanwhile birefringement crystals (Malawista and Sato, 1969) apparently form from the sequestered microtubule subunits (Bensch and Malawista, 1969).

Some of these observations strongly suggest that microtubular proteins are conserved within the cell as pools of subunits which can associate reversibly to form the highly labile microtubules anywhere they are needed. Inoue and Sato (1967) visualize microtubules in an equilibrium with their monomeric subunits. Thus the remarkable and almost instantaneous affect of D_2O may result from a displacement of the equilibrium towards association of subunits; conversely, colchicine or cold treatment displace it the other way. This pool is formed prior to mitosis since inhibitors of protein synthesis do not affect the cell's entry into prophase, by which time the protein subunits have already been synthesized (Inoue and Sato, 1967; Robbins and Shelanski, 1969). Inoue and Sato (1967) also consider that chromosomal movement itself could be achieved by a carefully controlled dissolution of microtubules since low, marginally effective concentrations of colchicine induce a contraction in spindle fibres (Inoue, 1952).

Rosenbaum, Moulder, and Ringo (1969) have elegantly demonstrated that the microtubular systems of the flagellum are labile and also in an analogous equilibrium with a pool of cytoplasmic subunits. They found that if one flagellum is amputated from the biflagellate green alga *Chlamydomonas,* it soon regenerates whilst the remaining one shortens until both are of an intermediate length, whereupon both slowly grow back to their full length. Only this later phase of elongation was sensitive to cycloheximide, an inhibitor of protein synthesis. If both flagella are amputated in cycloheximide, both then elongate a small amount, indicating that a small pool of subunits exists even with fully extended flagella. Furthermore, colchicine treatment did not affect the shortening of the one flagellum, but it prevented the extrusion of the second. Thus, withdrawal and extrusion are separate processes, implying that assembly and disassembly of microtubular systems are also distinct.

Mitosis in higher plant cells

The spindle of higher plant cells both *in vitro* (e.g., Inoue, 1964 *et al.*; Bajer and Allen, 1966a) and fixed by crude early fixatives (e.g., Wilson, 1898; Sato, 1958, 1959), was clearly fibrillar. In particular, fibrous elements were attached to chromosomes at their leading point of movement. However, electron microscopy of the spindle initially proved frustrating since fixation with permanganate (Porter and Machado, 1960) or osmium tetroxide (e.g., Lafontaine and Chouinard, 1963) failed to reveal equivalent fibres except on rare occasions (Esau and Gill, 1965; Harris and Bajer, 1965). The advent of glutaraldehyde-based fixation, however, resolved this discrepancy between light and electron microscope images when Ledbetter and Porter (1963) preserved microtubular systems in the phragmoplast. There can now be little doubt that bundles of microtubules are equivalent to spindle fibres observed by light microscopy. Mole-Bajer and Bajer (1968) have followed individual dividing cells *in vitro* up to the moment of fixation, and then throughout fixation, dehydration and embedding; they were then able to section those areas of the spindle previously observed when live (e.g., Bajer, 1968a, b, c *et al.*). Such elegant work established firstly that the preparative methods did not degrade cytoplasmic structure (at least up to the resolution of the light microscope) except for some almost imperceptible shrinkage (Jensen and Bajer, 1969), and also that the distribution of spindle fibres coincided with microtubules. Later, Bajer and Jensen (1969) demonstrated that a striation in a section observed by Nomarski optics with the light microscope could be resolved into a group of three or four microtubules under the electron microscope. With some exceptions (e.g., Behnke and Forer, 1966) most workers now accept that spindle fibres are equivalent to microtubular systems, a conclusion supported by experimental treatments which affect both (see above). However, we must allow the possibility that the spindle fibre traction apparatus may well contain more than one component (Forer, 1966).

Since many elegant experiments, particularly those using micromanipulation (see Mazia's review, 1961; Nicklas and Staehly, 1967), suggest that spindle fibres are connected firmly to the chromosome at its leading point of movement to the pole, microtubules are surely therefore involved in the mysterious and fascinating movements of mitosis (but see Forer, 1966, for a dissenting view). Furthermore, Bajer (1968a) has followed neocentric movement (i.e., away from the kinetochore) of chromosome arms *in vivo*. Subsequent electron microscopical examination revealed bunches of microtubules impinging on the chromosomes at the region of neocentric activity.

Let us now examine each state of mitosis in higher plants, following the sequential distributions of microtubules within the cell.

Preprophase

An early naive and unsuccessful attempt to discover a centriolar equivalent in plants led me to examine 'preprophase' cells which displayed two concurrent and preliminary manifestations of premitotic nuclear activity: condensation of chromatin, and dispersal of the nucleoli, both distinct from what is seen in prophase. Rather astonishingly, many microtubules were also seen closely packed several units deep in a clearly defined band of restricted extent encircling the cell near the wall. The plane of this 'preprophase band' was usually centrally situated, and perpendicular to the long axis of various meristematic cells, thus coinciding with the anticipated plane of cell division (Pickett-Heaps and Northcote, 1966a). That this preprophase band did predict the position of the future cell plate was shown unambiguously by its location prior to those highly asymmetric cell divisions involved in the formation of stomatal initials and subsidiary cells of the stomatal complex in wheat leaf epidermis (Pickett-Heaps and Northcote, 1966b); it is emphasized that the plane of these asymmetric cell divisions is *very highly predictable* from the position of the premitotic cell in the morphogenetic pattern of the epidermis (Stebbins and Shah, 1969). These results implied that the preprophase band might be involved in establishing the plane of cell division. This most important subject has profound morphological implications and so has long fascinated botanists; therefore the phenomenon will be examined at some length here. Preprophase bands have now been reported in a variety of plants (Burgess and Northcote, 1967; Deysson and Benbadis, 1968; Esau and Gill, 1969; Evert and Deshpande, 1970; Pickett-Heaps, 1969a; Burgess, 1970c).

Some controversy has arisen over the function and significance of this preprophase band. Originally, we suggested that its constituent microtubules might be incorporated into the forming prophase spindle (Pickett-Heaps and Northcote, 1966a, p. 117) while its position appeared to be influenced by unknown factors that control the plane of cell division, particularly those perhaps emanating from the Guard Mother Cell (Pickett-Heaps and Northcote, 1966b). Later, Burgess and Northcote (1967) said that in *Phleum* root tips, the band was always symmetrically situated even though asymmetric mitoses were known to occur, but no indication was given as to how the future plane of division of their observed premitotic cells could be ascertained (as can be done with stomatal complexes). Nevertheless, the orientation of the spindle or cytokinetic apparatus could easily change during or after mitosis (e.g., during cytokinesis in elongated procambial cells which are sometimes divided longitudinally after a transverse mitosis). Burgess and Northcote (1967) conclude that these microtubules are engaged firstly in the premitotic positioning of the nucleus, and secondly in determining its orientation and hence the orientation of the spindle. The second of these possibilities is untestable at present; whether the nucleus has an axis which can be said to be oriented is problematical. The first is probably not true. Highly asymmetric and precisely predictable cell divisions occur in the green alga *Chara* in the absence of (i.e., without requiring) preprophase bands (see p. 231); furthermore, during formation of the eight-celled stomatal complex of *Commelina,* preprophase bands appear *after* the nuclei and other organelles are polarized (Pickett-Heaps, 1969a). Burgess (1970a) later confirmed this conclusion with *Dactylorchis*. Centrifugation of cells undergoing asymmetric mitoses also shows the strength of attachment of one spindle pole to the Guard Mother Cell, whether the preprophase band is present or not (Pickett-Heaps, 1969b). Observations on caffeine-treated wheat seedlings (Pickett-Heaps, 1969c) reaffirmed the persistent tendency of the preprophase band to predict the position of the cell plate. This treatment prevents cytokinesis, and when the resultant multinucleate or polyploid cells subsequently enter preprophase, the position of their preprophase band(s) was not related to the size or number of nuclei but rather to the expected plane(s) of division; when formation of the stomatal complex had been interrupted, the preprophase band predicted another attempt at asymmetric division. Similar results were obtained in untreated tissues forming abnormal complexes (Pickett-Heaps, 1969a, c).

However, an agent that predicts something is *not* necessarily involved in causing that event to happen (ask any meteorologist!), a misinterpretation of my conclusions stated by Burgess (1970a: p. 260). Neither did we ever suggest that the preprophase band has any influence on the position of the future cell plate (cf. Hepler and Newcomb, 1967: p. 512; Burgess, 1970c). What then is the significance of these microtubules? I conclude now that they represent one response of the premitotic cell to the factors inducing polarization (Pickett-Heaps, 1969a, b, c). Ledbetter and Porter (1963) noted at an early stage the disappearance of wall microtubules during mitosis. Careful scrutiny of my own sections of preprophase cells always revealed some microtubules lying between the band and the nucleus, contrary to the results of Burgess and Northcote, 1968: p. 11 (but confirmed by Burgess, 1970c). I interpret these to be moving directly into the forming spindle, probably in association with microtubule subunits (Pickett-Heaps, 1969a), this being even more clearly shown following centrifugation of stomatal complexes (Pickett-Heaps, 1969b). Burgess (1970a) questions this interpretation, and certainly the form of curved microtubules undergoing such movement would be difficult to recognize in two-dimensional sections. Nevertheless, numerous profiles of tubules are demonstrably present where described. I believe two other observations also support this idea: firstly, intact microtubules move from the wall into the spindle at preprophase in *Spirogyra* (see p. 231); secondly 'phragmosomes' occur frequently in premitotic vacuolated plant cells (Esau and Gill, 1965). These strands of cytoplasm predict the plane of division, and in *Commelina*, they contain numerous microtubules running between nucleus and preprophase band (see particularly Pickett-Heaps, 1969a; Figs. 9—9c); indeed Burgess' (1970c) own work later also showed microtubules clearly running from the band to the nuclear envelope. High-voltage electron microscopy of thick sections should resolve this issue. Thus, the preprophase band probably represents that pool (see earlier) of structural material, the wall microtubules and subunits, destined for re-utilization in the spindle. Such an interpretation has been strengthened by the treatment of preprophase cells with D_2O which increases the number of tubules in the band (Burgess and Northcote, 1969).

Burgess and Northcote (1968) have also suggested that elements of 'smooth' endoplasmic reticulum are involved in microtubular aggregation and disaggregation, in particular, at the preprophase band. In my opinion, their conclusions are not justified. Endoplasmic reticulum permeates the cell; the basis for attributing to some elements a specific interaction and function associated with the band is not made clear, and other published micrographs of these preprophase microtubules (see earlier references) do not support their claim. Furthermore, Burgess (1970a) later states that '. . . the transport of the (microtubule) subunit material via the endoplasmic reticulum seems to be the simplest explanation of all the known facts concerning microtubule redistribution in cells'. While forming microtubules are indeed often associated with membranes (e.g., at the cell plate), this statement can hardly be taken seriously. To give but one example: considerable redistribution of microtubules occurs during mitosis within the closed spindles of many fungi and algae (see pp. 233 and 239); these spindles contain no endoplasmic reticulum whatsoever.

Prophase

Bajer and Mole-Bajer (1969) provide the clearest account of the formation of prophase spindles by correlating observations *in vitro* with electron microscopy using *Haemanthus* endosperm. As prophase chromosomes condense, a 'clear' zone of growing thickness envelopes the nucleus (see particularly Bajer, 1968a: Fig. 3a) which eventually becomes enclosed in a uniform, birefringent sheath (Mazia, 1961: p. 201). Sometimes the forming spindle may be multi-polar (Wilson, 1898; Bajer and Mole-Bajer, 1969: Fig. 3), but usually it soon becomes bipolar. When examined ultrastructurally, this clear zone initially contains randomly organized microtubules which proliferate and then become more oriented along the future spindle axis. Their origin is not obvious.

Some wheat meristematic cells are similar, with extranuclear microtubules appearing close to the nucleus, and radiating from future poles (Pickett-Heaps and Northcote, 1966a). Esau and Gill's (1969) beautiful micrographs of the prophase-metaphase transition (their Figs. 7–10) show that, as the preprophase band disperses, compact spindle poles are formed containing increasing numbers of microtubules which extend tangentially to the nucleus, and these are analogous to the 'polar caps' described by light microscopists for various plant cells.

Prometaphase

This fascinating stage of mitosis is marked in higher plants by the rapid breakdown of the nuclear envelope and the first interaction of spindle fibres with chromosomes. In *Haemanthus,* the nucleus now becomes increasingly flattened across the spindle axis (Bajer and Mole-Bajer, 1969) as the nuclear envelope is subjected to intense activity near the poles, a process described as 'boiling', with this membrane being 'pushed and pulled' in adjacent areas before it breaks up. Electron microscopy shows that microtubules are probably involved in both these movements, although there is doubt as to whether they are attached directly to the nuclear envelope. However, microtubules appear to push some pieces of nuclear envelope deep into the spindle (see Metaphase below). The nucleus is clearly being invaded by microtubules which become increasingly oriented along the spindle axis (Bajer, 1968b). Meanwhile, kinetochores are forming on the chromosomes, and these begin to elaborate kinetochore fibres consisting of rapidly increasing numbers of microtubules which mingle with the continuous fibres that traverse the spindle (see also Bajer, 1968a, b). That kinetochores act as organizing or assembling centres for spindle fibres *in vitro* is indicated by their vulnerability to u.v.-microbeams which destroy both the existing kinetochore fibres and the kinetochore's capability of forming further new ones (Inoue, 1964). Microtubules appear irregularly at different kinetochores during prometaphase once the nuclear envelope is ruptured. Apparently, continuous fibres may appear to attach to kinetochores (Bajer and Mole-Bajer, 1969: p. 468), but details of this particular process are very unclear.

Metaphase

Metakinesis is the process of alignment of the kinetochores into one plane in the centre of the spindle, with sister kinetochores directed towards opposite poles. The arms of chromosomes, however, tend to align parallel to the spindle axis. Several complex phenomena are observed in endosperm. Large and small organelles, if situated near spindle fibres, are eliminated polewards at the speed of an anaphase chromosome, but slowing as they near the pole (e.g., Bajer, 1967, 1968b). The paired chromosomes may oscillate about the metaphase plate, apparently in some sort of equilibrium. The kinetochore fibres are roughly parallel, and often undulate or oscillate slightly while stretching the daughter kinetochores apart (Bajer, 1965a). Although the groups of kinetochore fibres converge, individual elements in each fibre diverge at the poles, the appearance of the spindle being modified by the degree of flattening to which it is subjected. Using live *Nephrotoma* spermatocytes, Forer (1965) has shown that areas of reduced birefringence formed by u.v.-microbeams in metaphase spindles immediately move to the pole — while the chromosomes remain stationary; if formed during anaphase, such areas move at the same speed polewards as the chromosomes. These experiments clearly indicate the dynamic state of this apparent metaphase equilibrium and suggest that either the spindle fibres themselves, or some structural organization within them, are constantly moving polewards.

An ultrastructural examination of *Haemanthus* endosperm at metaphase reveals microtubules where expected; in particular, large bundles (about 75–150: Bajer and Mole-Bajer, 1969) of kinetochore tubules run along the spindle axis, intermingling with the continuous microtubules which traverse the metaphase plate of chromosomes. In rare cases, microtubule bundles may extend from sister kinetochores to the same pole during prometaphase, an unstable situation *in vitro* which is rectified by the breakdown and reformation of one set of fibres towards the opposite pole, thus establishing the stable metaphase configuration. Elements of endoplasmic reticulum (or remnants of the nuclear envelope) also extend along the spindle axis (see below).

In wheat meristems, Pickett-Heaps and Northcote (1966a) found endoplasmic reticulum also intimately associated with some tubules during prometaphase and metaphase, and so suggested that this was involved in control and/or function of microtubules. An invasion of the spindle by endoplasmic reticulum had been reported earlier by Porter and Machado (1960), and as mentioned above, Burgess and Northcote (1968) suggest that these membranes function in microtubule aggregation/disaggregation and transport. Wilson (1970) suggests that polar aggregations of endoplasmic reticulum function in the place of centrioles (which presupposes the need for centrioles, a questionable postulate — Pickett-Heaps, 1969d). However, I believe that 'the importance of the endoplasmic reticulum in mitosis' (Burgess and Northcote, 1968) and all these suggestions are probably invalidated by the frequent occurrence of intranuclear spindles (see later) in which complex movements of spindle microtubules are accomplished in the complete absence of endoplasmic reticulum. Bajer (1968a: p. 273) agrees with this viewpoint, and the metaphase movement of endoplasmic reticulum into the spindle is probably one side-effect of the invasion of the spindle by microtubules from the poles (see, for example, Bajer and Mole-Bajer, 1969: Fig. 9). Indeed even in higher plants, metaphase spindles may be almost devoid of endoplasmic reticulum (Esau and Gill, 1969: Fig. 16).

Colchicine, as expected, prevents formation of the metaphase spindle; the paired chromosomes remained scattered within the cell.

Anaphase

Anaphase separation of chromosomes usually involves both shortening of each half spindle and elongation of the whole spindle (Mazia, 1961) and this is generally accomplished rapidly in comparison to the duration of metaphase. Observations *in vitro* indicate that the kinetochore fibres (and any closely associated granules) are transported polewards at the same speed that the kinetochore moves. Comparison of Forer's (1965, 1966) results, however, indicates the complexity of anaphase movements, and many disagree with this interpretation. Individual kinetochore fibres diverge at the poles, intermingling with other fibres (Bajer and Allen, 1966a); these remain quite birefringent as they shorten, but what happens to them at the pole is obscure. As telophase progresses, the kinetochore fibres fade and disappear while a residual birefringence of a far more diffuse interzonal fibre apparatus remains between the chromosomes.

Electron microscopy does little to dispel 'the despairing sense of mystery . . . (at) the swift precise movements of the mitotic apparatus' (Mazia, 1961); it merely confirms the presence of microtubules where anticipated. Those shortening microtubules running from the kinetochores to the pole, frustratingly never seem to end at any specific organelle or structure. Microtubules intermingle at the poles (Bajer, 1968b) but never appear to thicken during chromosomal movement (but two populations of microtubules of different sizes have been reported – Allen and Brown, 1966; Moor, 1967). The interzone initially contains few scattered microtubules, oriented along the spindle axis often in groups or sheets which soon seem to initiate phragmoplast formation (see later). These observations probably apply to other higher plant cells too (Esau and Gill, 1969; Pickett-Heaps and Northcote, 1966a).

The effect of colchicine on anaphase spindles is revealing; the two groups of chromosomes become disorganized, and dissolution of the interzonal spindle causes them to approach one another. This often results at telophase in the formation of fused hour-glass nuclei (Hindmarsh, 1953; Pickett-Heaps, 1967a).

Telophase and cell-plate formation

As chromosomes disperse inside the reforming nuclei, the birefringent interzonal fibrils running along the spindle axis proliferate markedly (Inoue, 1964), and soon a thin membranous layer, the cell plate, appears in the middle of them. The region of highly oriented cytoplasm containing these fibres, the phragmoplast, then extends laterally outwards and in it the newly forming edge of the expanding cell plate is laid down; meanwhile, the fibres adjacent to the fully formed plate slowly fade and disappear. The plate may, however, sometimes grow inwards. These events, carefully analysed *in vitro* by Bajer (1965a, b *et al.*), and Bajer and Allen (1966b) in *Haemanthus*, involve yet more complex and puzzling interzonal movements. Localized irradiation of the phragmoplast by a u.v.-microbeam (Inoue, 1964) causes localized and permanent prevention of formation of fibres, indicating that the phragmoplast contains their assembly sites. The cell plate itself forms from small elements which fuse on the interzonal fibres and then the resultant thickenings apparently slide along the fibres, lining up in the centre of the cell; sometimes, they move laterally as well, movements which Bajer (personal communication) thinks may help fuse the scattered elements of the cell plate. The thickenings also fuse to form the undulating membranous structure which soon consolidates, becoming smooth and flat and extending through the expanding system of phragmoplast fibres, which also apparently hold apart the nuclei. A little later in telophase, the spindle often shortens a little, whereupon trailing chromosome arms are deflected into the daughter nuclei which approach one another slightly; many larger granules are also eliminated away from the phragmoplast. These manifestations of activity last as long as the fibres persist. In more flattened cells (Bajer, 1968c), other subtle, transverse movements occur as proliferating fibrils, either growing from the interzone or extending from each pole during anaphase elongation, interact in the interzone before the cell plate appears. These transverse movements seem possibly to be a consequence of mechanical restraints being imposed upon spindle elongation. Regions of crossed interzonal fibrils are formed (Bajer, 1968c: Fig. 1b) and transport of particles polewards remains parallel to the adjacent fibres.

Electron microscopy of *Haemanthus* again reveals microtubules exactly where expected. The kinetochore tubules increasingly diverge at the poles, as they shorten and disappear with reformation of the nuclear envelope (Bajer, 1968a). Formation of the phragmoplast is a most mysterious process; it commences as interzonal microtubules initially clump together at regions containing amorphous dense material and later, small vesicles (Hepler and Jackson, 1968). Traces of this amorphous material can sometimes be found in early anaphase and it increases in amount and lines up across the cell, while microtubules, identical to those of the spindle, proliferate enormously. The cell plate is initiated by fusion of vesicles, forming flattened cisternae of increasing size. Some microtubules, presumably derived from the continuous interzonal system, traverse this region temporarily, and many others end in the dense amorphous material. Endoplasmic reticulum is always present, but its role is uncertain (see above); the source of cell-plate vesicles is controversial (O'Brien, 1972) and irrelevant to this discussion, although often they are derived from Golgi bodies.

Cell-plate formation in other higher plants (Allen and Bowen, 1966; Esau and Gill, 1965; Pickett-Heaps and Northcote, 1966a, b; Hepler and Newcomb, 1967;

Cronshaw and Esau, 1968; Evert and Deshpande, 1970) appears to involve the main features mentioned above. Particularly striking is the morphogenetic role of the phragmoplast in establishing curved or hemispherical cell walls (Pickett-Heaps and Northcote, 1966b; Heslop-Harrison, 1968; Pickett-Heaps, 1969a). Microtubules seem to be directing vesicles into the new wall, whose formation, however, is a very complicated and little understood process (O'Brien, 1972). Allen and Bowen (1966) suggested that the electron opaque aggregate found at the plate may be involved in microtubule assembly, a suggestion supported by several authors. Endoplasmic reticulum is also present near the cell plate (Porter and Machado, 1960 *et al.*), but for reasons given earlier, I doubt that these membranes function in microtubule assembly (cf. Burgess, 1970a).

Treatment of telophase spindles with colchicine confirms the impressions that follow from the observations reported above. Thus, disappearance of microtubules at early telophase coincides with cessation of transport within the phragmoplast, and approach (and often fusion) of daughter nuclei (Hindmarsh, 1953; Pickett-Heaps, 1967a; Bajer, 1968c). Cell-plate formation does not occur, or else it may be interrupted to leave a portion of crosswall in the cell (Pickett-Heaps, 1967a).

These results imply that the phragmoplast and cell plate are initiated and probably controlled by the interzonal fibre apparatus. This concept, however, is an oversimplification and obscures what seems to be a genuine autonomy or individuality in the behaviour of the phragmoplast. We should, for example, remember the following observations. In the multinucleate syncytium of *Haemanthus* endosperm, individual uninucleate cells are formed quite suddenly by the simultaneous formation of phragmoplasts between *all* nuclei (not just daughter nuclei); furthermore, small phragmoplasts may also appear in isolated lumps of cytoplasm (Bajer, 1968c and personal communication), and in cells treated with chloral-hydrate, they may form at prophase or metaphase (Mole-Bajer, 1965). During the formation of stomatal complexes, the phragmoplast extends far away from daughter nuclei, curving spatially to form the precise shape of cell required, often traversing a vacuole in the process (Pickett-Heaps, and Northcote, 1966b; Pickett-Heaps, 1969a). Thirdly, the phragmoplast may grow for remarkably long distances away from daughter nuclei, as for example, in the longitudinal division of cambial cells (Esau and Gill, 1965: Fig. 1).

Mitosis and cytokinesis in some algae and fungi

Cell biologists have paid remarkably little attention to the study of algae, those organisms so diverse in their structure, behaviour, evolutionary development and modes of reproduction. Indeed, most recent and otherwise excellent textbooks and reviews on 'plant' ultrastructure mention them little if at all. Yet these organisms have already provided some fascinating insights into successful variations and adaptations of cellular systems that have survived over millions of years of evolution, and I am confident that further study of them will be most profitable. Cell division is an excellent case in point. The work summarized above has given us some idea of changes in the distribution of microtubules that accompany cell division in higher plants. Some intriguing comparisons can now be made as I briefly summarize equivalent work on algae, separating mitosis and cytokinesis in each species. Nuclear division in some fungi will then be mentioned as further examples of mitotic diversity.

For reasons which should become obvious, I have divided the algae into two classes. The first includes those species in which mitosis, and sometimes cytokinesis, have obvious affinities with cell division in higher plants, and therefore probably warrant my classification as 'Advanced Mitotic Systems'. The second I shall term 'Unusual (Primitive?) Mitotic Systems'. Both classes contain regrettably few representatives, symptomatic of the lack of attention these organisms have been accorded.

The discerning reader will soon note that these cell divisions vary appreciably in detail from the account given for higher plants. Indeed, cytokinetic mechanisms differ greatly, often featuring prominent systems of (cytoskeletal?) microtubules oriented transversely across the spindle axis. These differences present an intriguing picture of structural diversity; I believe they will soon prove very important both in illustrating an ultrastructural phylogeny in organisms, and perhaps in even providing a glimpse of how various mitotic and cytokinetic systems evolved (Pickett-Heaps, 1969d).

Advanced mitotic systems in algae

Chara. Mitosis. In most respects, dividing cells of *Chara* resemble those of higher plants (Pickett-Heaps, 1967c), apart from the characteristic adherence of nucleolar material to chromosomes during division. As mentioned earlier, certain cells undergo highly asymmetric and predictable divisions (Pickett-Heaps, 1967c: Figs. 4, 6), but preprophase bands of microtubules are never formed. The nuclear envelope disperses during mitosis, and centrioles are not associated with vegetative spindles (i.e., the spindles are open and acentric, as in higher plants). On the other hand, preceding spermatogenesis, large numbers of mitoses form spermatogenous filaments, during which centrioles appear *de novo,* apparently near the nuclear envelope, whereupon these organelles then become attached to the poles. Thus the spindles become typically centric (Pickett-Heaps, 1968b: Fig. 14; Turner, 1968). This observation is not consistent with any important role of the centriole in spindle formation (Pickett-Heaps, 1969d).

Cytokinesis. Formation of the thin cell plate in *Chara,* while similar to that of higher plants, involves only few microtubules and its position is very precisely controlled (e.g., during formation of the antheridia; Pickett-Heaps, 1968a).

Spirogyra. Mitosis. Fowke and Pickett-Heaps (1969a) have described several interesting aspects of mitosis in this filamentous alga. During preprophase, the delicate cytoplasmic strands supporting the nucleus in the centre of the vacuole, thicken considerably and fill with microtubules. Meanwhile, the population of wall microtubules is being depleted, remaining thus until division is over. By prophase, these strands have thinned out and contain no microtubules; their cytoplasm has apparently been transformed into two conspicuous polar masses

231

containing thousands of longitudinally oriented microtubules, at each end of the enlarged nucleus. At prometaphase, the nuclear envelope at the poles becomes pitted and finally ruptures, allowing microtubules to invade the nucleus. Diffuse chromosomes, heavily coated with nucleolar material, form typical metaphase and anaphase figures; most continuous microtubules are grouped into a cylinder surrounding the chromosomes. The nuclear envelope, although broken up, remains around the spindle for some time.

The formation of this spindle has close parallels with that of higher plants (e.g., prometaphase rupture of the nuclear envelope); in particular, spindle microtubules appear to be derived directly from wall microtubules (cf. p. 243).

Cytokinesis. The cell membrane invaginates during anaphase at the interzone, forming a narrow annular constriction bisecting the cell. No microtubules are associated with it at this stage. However, this cleavage eventually impinges on the cylindrical array of the numerous interzonal microtubules persisting between daughter nuclei, whereupon a 'cell plate' appears at the inner edge of the furrow, containing vesicles, dense amorphous material (cf. the phragmoplast), and short microtubules. The apparent sequential utilization of these two systems for achieving cytokinesis (Fowke and Pickett-Heaps, 1969b) perhaps indicates a sequence of events leading to evolution of the phragmoplast (Pickett-Heaps, 1969d).

Closterium. Mitosis and Cytokinesis. *Closterium* and *Spirogyra* are both members of the Conjugales but their mitotic figures show minor differences (Pickett-Heaps and Fowke, 1970c). In *Closterium,* the numerous tiny chromosome do not become coated with nucleolar material. The spindle is open, and continuou microtubules are not arrayed cylindrically as in *Spirogyra.* Cytokinesis is completed by annular furrowing without any involvement of microtubules.

Post-Telophase Activity. Some extraordinary and fascinating events follow division, during which spindle components are apparently reutilized. Firstly, spindle microtubules do not disperse at telophase; instead, they congregate around the reformed daughter nuclei, and become focused upon one small polar region in which a diffuse globular structure materializes. This has been called the 'microtubule centre' (MC). It then migrates along the cell in the cytoplasm lying in a lobe of the chloroplast; large numbers of microtubules extend back from it to the nucleus from which it came. The nucleus now becomes deformed, extending a projection ensheathed in the microtubules running to the MC. Meanwhile, the MC lodges in the cleavage developing in the single chloroplast. Undergoing remarkable deformation, the whole nucleus now moves along the microtubules towards the MC. Finally, as the chloroplast is cleaved, the nucleus is inserted between the two halves.

Another system of microtubules concurrently appears near the newly formed crosswall dividing the parent cell. As this crosswall splits and starts stretching into the shape of the new semicell, these randomly organized microtubules adopt the 'hooplike' configuration typical of higher plant cells, but *only* along the new, expanding cell wall.

These complicated events re-establish the beautiful lunate symmetry of the interphase cell which is bisected at each division. I believe they allow some interesting inferences. In particular, open plant spindles may well contain diffuse, polar microtubule-organizing-centres (MTOCs) as postulated previously (Pickett-Heaps, 1969d), which in this case aggregate after mitosis to form the MC. Secondly, the microtubules must function in intracellular transport (i.e., of the nucleus). Thirdly, some role of microtubules in morphogenesis of the expanding wall seems indicated. Before they become thus involved however, these forming microtubules are oriented transversely between daughter nuclei (after cytokinesis), a configuration typical of other algae (see below) — perhaps thereby indicating how wall microtubules may have evolved from cytokinetic microtubules (Pickett-Heaps and Fowke, 1970c).

Preliminary results using colchicine treatment (Pickett-Heaps, unpublished work) confirm that microtubules are involved in nuclear movement and cell expansion; however, division of the chloroplast appears unaffected.

Chlamydomonas. Mitosis. Cell division in this well-known unicell is surprisingly complicated (Johnson and Porter, 1968). The spindle is 'closed' (i.e., the nuclear envelope does not disperse during mitosis) but has polar 'fenestrae' or openings. Connections between basal bodies and flagella are severed and the cytoplasm rotates 90 degrees to the cell axis during mitosis. A 'metaphase band' of four microtubules runs a short distance around the cell close to an invagination of the plasmalemma near the nucleus; I do not believe this has anything in common with the preprophase band of higher plants. As Johnson and Porter (1968: p. 422) point out, it persists throughout division and has a much more specific complement of tubules.

Cytokinesis. At telophase, a localized 'internuclear' system of transversely oriented, parallel microtubules appears amongst elements of endoplasmic reticulum between daughter nuclei. During mitosis, the two pairs of basal bodies remain separate from the spindle; at telophase, they become associated with a second array of 'cleavage microtubules' which grows across the cell in the same plane but roughly perpendicular to the internuclear microtubules. The cleavage furrow is initiated between the pairs of basal bodies and extends through these cleavage microtubules which radiate and then converge across the diameter of the cell. A further set of microtubules extends from the basal body region around the cell's periphery. The cleavage furrow eventually bisects the chloroplast and pyrenoid (Goodenough, 1970). If the two daughter cells divide again, the basal bodies may not replicate first, whereupon the resultant tetrad contains only one in each cell, associated as before with the cytokinetic apparatus.

The function of these microtubular arrays is not clear. Cleavage in mammalian cells does not require microtubules, but instead a ring of (contractile ?) microfilaments lines the ingrowing membrane (e.g., Selman and Perry, 1970; Schroeder, 1970 *et al.*); such microfilaments were not reported by Johnson and Porter, who suggested that perhaps the tubules facilitate growth of the furrow by altering the nature of the cytoplasm and/or directing its flow. I feel that these complex microtubular systems may also orient the cleavage precisely with respect to daughter nuclei, basal bodies and the single chloroplast, since in several algae (*Spirogyra, Bumilleria, et al.*) the cleavage furrow cuts through cytoplasm and chloroplast apparently without the need for either microtubular or micro-filamentous systems.

Kirchneriella. Mitosis. In this alga, typical of many Chlorococcales, the nucleus divides twice in succession and then cytokinesis forms four autospores which are later released from the mother cell wall.

Rudimentary (i.e., short and incompletely formed) centrioles, apparently absent from interphase cells, appear at prophase surrounded by diffuse material (Pickett-Heaps, 1970b). They move to the future poles of the spindle as the nucleus becomes ensheathed in extranuclear microtubules, which are in turn enclosed by a 'perinuclear envelope' of endoplasmic reticulum which persists during mitosis. Polar fenestrae appear in the nuclear envelope during prometaphase, through which the centriole-complexes and microtubules migrate into the nucleus. Later, after considerable elongation of the anaphase spindle, the reforming nuclear envelope is intercalated between each daughter nucleus and its attendant centriole-complex.

Cytokinesis. Incomplete cytokinesis follows the first mitosis, and during the 'Secondary Mitoses' this cleavage may be partly resorbed. Finally, the cell is partitioned into four by reconstitution of this 'Primary Septum' and the formation of two new 'Secondary Septa' (Pickett-Heaps, 1970b). The same cytokinetic system is used throughout.

After telophase, nuclei migrate to the centre of the cell and centriole-complexes move around to the inner edge of the wall. Next, two somewhat distinct arrays of transverse microtubules appear in the plane of future cleavage. One encircles the cell near the wall and the other extends randomly fan-wise through the cytoplasm. Then the septum, a cleavage membrane formed at least partly from an invagination of the plasmalemma, grows through these microtubules and so is probably spatially oriented by them. The peripheral chloroplast is often deeply indented adjacent to these microtubules before the septum impinges upon it.

This alga clearly demonstrates how spindle formation follows the ingress of organizing centres and associated microtubules into the nucleus; at telophase, both are separated from the nucleus by the contracting nuclear envelope. Furthermore, transverse arrays of microtubules feature prominently in cytokinesis. Some other members of the Chlorococcales are similar to *Kirchneriella* during division (Pickett-Heaps, unpublished data, e.g., *Scenedesmus,* which has persistent centrioles, *Ankistrodesmus* and *Tetraedron*). However, numerous mitoses and several partial cell cleavages take place in *Tetraedron* before the cell finally splits up into numerous tiny autospores, and it thereby displays an increasing tendency towards the coenocytic state (cf. p. 235). These algae use the same basic cytokinetic system as *Kirchneriella*, which may become tilted along the cell axis to cause a more longitudinal division of the cell (e.g., in *Ankistrodesmus*).

Hydrodictyon. Mitosis. The cytoplasm in the large coenocytic vegetative cells of *Hydrodictyon* can cleave up to form tiny flagellated zooids (see below) at almost any stage of growth. Consequently, centrioles are persistent, and they remain just outside the polar fenestrae (Marchant and Pickett-Heaps, 1970) during

234

mitosis. In other respects, including the acquisition of the perinuclear envelope, the mitotic nucleus resembles that in *Kirchneriella.*

Cytokinesis. A vegetative cell may grow to contain thousands of nuclei following successive mitoses. 'Cytokinesis' occurs really during formation of zooids when the cytoplasm cleaves up into tiny uninucleate fragments which become motile. This cleavage proceeds in two distinct steps (Marchant and Pickett-Heaps, 1971). First, microtubules proliferate just under the tonoplast and then a cleavage fissure grows through these tubules, cutting off the 'vacuolar envelope'; an extremely thin layer of cytoplasm, bounded by the tonoplast and one membrane of the cleavage furrow; this encloses the whole central vacuole. Next, cleavage microtubules proliferate between all the nuclei, and further membrane furrowing through these tubules cleaves the cytoplasm into hundreds or thousands of zooids. Cleavage in polyhedra (characteristic coenocytic cells that form after germination of azygotes or zygotes) is similar, except that the large mass of cytoplasm may be non-vacuolated (Marchant and Pickett-Heaps, in preparation).

The giant coenocytic cells of this alga (also a member of the Chlorococcales) thereby demonstrate beautifully how mitosis and cytokinesis have been entirely divorced from one another, whilst the cell has still retained the structural features characteristic of cell division in other members of its group (cf. *Kirchneriella*). Furthermore, the basic cytokinetic mechanism which utilizes transverse microtubules appears to have been modified during evolution so that it now forms the vacuolar envelope in addition to cleaving the cytoplasm around nuclei.

Stigeoclonium. Very brief mention of some of my unpublished observations will preface the next section on *Oedogonium.* Vegetative cells of *Stigeoclonium* sp. possess persistent centrioles and centric spindles. Cytokinesis is particularly interesting. After telophase, the spindle collapses and daughter nuclei approach one another; between them, randomly organized microtubules proliferate in a plane perpendicular to the spindle axis. Amongst these transverse microtubules, many vesicles collect which then appear to fuse, forming a crosswall growing from the centre outwards. Thus, the cell plate, apparently analogous to that of higher plants, is laid down using an entirely different cytokinetic apparatus containing transverse microtubules.

Oedogonium. The Oedogoniales have a quite extraordinary method of cell division. Before and during mitosis, a doughnut-shaped ring of soft material is laid down at one end of the cell, attached around its outer edge to the wall. After mitosis, the wall ruptures circumferentially at this ring, whose material is then quite rapidly stretched until it becomes cylindrical. Those not familiar with this extraordinary behaviour are advised to consult botanical textbooks (e.g., Fritsch, 1935: p. 298 *et seq.*), as space limitations preclude further discussion here on its subtleties.

Mitosis. During prophase-metaphase, the nuclear envelope is drawn out, particularly at the poles, into ramifying sheets or tubules of doubled membrane which are closely associated with microtubules, an unusual phenomenon which also occurs in artichoke (Yeoman, Tulett, and Bagshaw, 1970). During mitosis, the spindle is essentially closed (Pickett-Heaps and Fowke, 1969, 1970a) but proliferating microtubules appear in the nucleus by prometaphase. Prominent kinetochores, containing six to seven layers, form during prophase. These are at first randomly oriented and dispersed within nuclei but they soon line up perpendicular to the spindle axis and then move to the metaphase configuration (Pickett-Heaps and Fowke, 1970a) apparently following the elaboration of their kinetochore-microtubular systems. These elongating bundles of tubules run into (and probably form) polar evaginations of the nuclear envelope, which diminish only slowly in size during anaphase; the microtubules do not attach to any obvious structure within them. Considerable elongation of the anaphase spindle is followed by its total collapse at telophase when daughter nuclei flatten against one another (cf. *Stigeocolonium*).

Cytokinesis. Between these daughter nuclei, a planar array of transverse microtubules, interspersed with vesicles, bisects the cell (as in *Stigeocolonium*) and cuts the peripheral chloroplast. However, a crosswall is *not* formed immediately (Pickett-Heaps and Fowke, 1969, 1970b; Hill and Machlis, 1968). Instead, the cell wall now ruptures (probably due to a carefully controlled increase in turgor) and then this septum moves up the cell like a diaphragm, until it reaches the lower lip at the broken edge of the older cell wall. Only then is the crosswall finally created. This therefore represents an elegant adaptation of a cytokinetic apparatus apparently common in algae, and one vital to the success of the cell's unique form of cell expansion.

Ulva. Mitosis. In this multicellular alga, the premitotic nucleus moves to the inside wall of the thallus along with other organelles (Lovlie and Braten, 1970). The spindle is mostly closed, and some microtubules appear in the nucleus before polar fenestrae develop at metaphase; however, during anaphase the stretched nuclear envelope opens at the interzone. Extranuclear centrioles remain to one side of the spindle axis away from the fenestrae (and therefore do not constitute a focal point of the spindle microtubules). Mitosis appears normal in other respects.

Cytokinesis. The cell is partitioned by a furrow which grows across the cell from the inner wall between daughter nuclei. Microtubules are not associated with this furrow.

Prymnesium. Mitosis and cytokinesis. Manton (1964a) has described cell division in this small brown unicellular flagellate. Cytoplasmic microtubules appear in prophase often near the basal bodies which, along with other organelles, have already replicated. By metaphase, the pairs of basal bodies are widely separated, lying near but not at the pole. The spindle is open throughout and, surprisingly, Manton never observed microtubules attached to chromosomes. Anaphase elongation stretches the entire cell and after telophase the cell membrane constricts at the interzone, cleaving the elongated cell.

Unusual (primitive ?) mitotic systems in algae

Lithodesmium. Mitosis and meiosis. In four papers, Manton, Kowallik, and von Stotsch (1969a, b; 1970a, b) describe this most interesting example of cell division from a centric diatom, carefully correlating ultrastructure with behaviour *in vivo*. All interphase cells possess just outside the nuclear envelope, a characteristic, layered structure called the spindle precursor. During prophase, this precursor splits down the middle and then a system of parallel microtubules grows between the two halves. By metaphase, these halves have been separated to opposite sides of the nucleus, now devoid of its envelope, and the microtubule system sinks into the nucleus. Midway between the poles now established by the spindle precursor, the microtubules are arranged in bundles, often hexagonally close-packed. Transverse sections of the spindle allow the microtubules to be counted; the figures indicate that two separate microtubular systems (i.e., half spindles) are interdigitated for a short length in the middle of the cell. A few tubules (up to 10 per cent) may run the whole length of the spindle. Anaphase, being rapid *in vivo,* was difficult to analyse ultrastructurally, chromosomes are diffused and featureless, and their attachment to the spindle is vaguely defined. In fact, microtubules seem to touch the chromosomes laterally rather than terminally. Once the chromosomal masses are separated, considerable elongation of the spindle and cell occurs, during which the continuous microtubular system may become S-shaped.

The second meiotic division is very interesting for two reasons. Firstly, it is preceded by the formation *de novo* of centrioles from material which collects near the poles of the previous miotic division and which also contributes to the new spindle precursor. The significance of this will be mentioned later (p. 241). Secondly, this spindle is consistently smaller than those of the previous divisions, and contains a reduced number of tubules.

Cytokinesis. The cell membrane constricts rapidly after telophase and daughter cells then remain briefly attached by a cord containing microtubules (approximately equal in number to the 10 per cent presumed to run from pole to pole). Finally, the cells separate and the cord is soon resorbed.

To compare this spindle with the preceding examples is difficult and probably premature. As with *Gyrodinium* and *Euglena* (see below), we can only guess that the mechanisms and structures involved in mitosis have retained primitive features whose deeper significance will probably remain obscure until more data are available concerning a variety of such unusual cells.

Dinoflagellates. Mitosis and cytokinesis. The ultrastructure of dividing cells has been described for two species of Dinoflagellates, *Woloszynskia* (Leadbeater and Dodge, 1967) and *Gyrodinium* (Kubai and Ris, 1969). The chromosomes, permanently condensed during interphase, possess a characteristic banded structure, and their DNA is not complexed with protein (see references above). These unusual features, allied with a most atypical form of nuclear division, confirm the impression that these organisms have retained during evolution several interesting primitive characteristics.

Classical mitotic stages (prophase and telophase) are not easily recognizable. The flagellar bases duplicate before division and are never associated with the nucleus. The early division nucleus ('prophase' is probably inappropriate) develops several invaginations and channels containing cytoplasm, aligned with the 'spindle' axis; these contain numerous microtubules also oriented along the same axis. However, the nuclear envelope remains completely intact throughout division, and at no time do these microtubules penetrate into the nucleoplasm proper or make any contact with the chromosomes. Instead, the paired chromatids, now usually V-shaped, apparently contact the nuclear envelope at the apex of their V (Kubai and Ris, 1969). During 'anaphase' the nucleus elongates and then constricts into two while still interpenetrated by the channels containing microtubules which now extend between daughter nuclei. The nucleolus persists during mitosis and is partitioned equally between daughter nuclei. Little comment has been made on cytokinesis which apparently proceeds by furrowing.

There can be little similarity between the mitotic mechanisms of dinoflagellates and those described earlier for other algae and higher plants. The nuclear envelope, being attached to chromosomes, would appear to be instrumental in separating them during its own expansion, as may be the case in Procaryotes (Kubai and Ris, 1969); the microtubular systems probably confer upon the nucleus the required polarity with respect to the cleavage furrow and flagellar bases.

Euglena. Mitosis and Cytokinesis. Unfortunately the little information available concerning this organism merely sharpens our curiosity. Leedale (1968) reports that an early sign of incipient division is the replication of the pellicle, as a new ridge forms between all the older ridges. The nucleus resembles that of Dino-flagellates, containing conspicuous interphase chromosomes, but microtubules appear within the dividing nucleus whose envelope remains intact throughout division; rather astonishingly, the few microtubules present make no clear contact with chromosomes but rather become appressed to the elongating endosome (or nucleolus) which is partitioned between daughter nuclei (cf. Dino-flagellates). The means by which chromosomal separation is achieved has not been clarified. The times spent in different stages of mitosis differ radically from those of normal mitotic systems and Leedale also notes that this mitosis is insensitive to colchicine. (Whether colchicine really gets into the cell is not so clear, however. The cell furrows at cytokinesis.

Such fragmentary evidence suggests that these organisms possess a primitive spindle apparatus that will make a fascinating comparison with that of Dinoflagellates.

Nuclear division in fungi

Recently, an increasing number of interesting reports concerning mitosis and meiosis in various fungi demonstrate that these organisms also display a considerable variation in their spindle structure, all, however, utilizing micro-tubules. I will, for convenience, divide these examples into types of spindles with similar characteristics.

Open, acentric spindles. The fungus *Basidiobolus* (Tanaka, 1970) has open, acentric (i.e., perhaps 'advanced') spindles; however, this organism also possesses persistent nucleolar material (as in *Spirogyra*) which is perhaps a primitive characteristic (Pickett-Heaps, 1970c).

Closed, acentric spindles. These are reported in the plasmodia of several myxomycetes: *Clastoderma* (McManus and Roth, 1968), *Didymium* (Schuster, 1964), and *Physarum polycephalum* (mitosis and meiosis; Aldrich, 1967; Goodman and Ritter, 1969) and two other species of *Physarum* (Aldrich, 1967). In *P. flavicomum*, centrioles appear after meiosis; furthermore, whilst plasmodia contain closed, acentric spindles, the myxamoebae have open, centric spindles (Aldrich, 1969), an observation which surely emphasizes the unimportance of both centrioles and nuclear envelope in classical mitosis.

Closed, centric spindles. An interesting variation in mitotic structure is provided by the phycomycetes *Catenaria* (Ichida and Fuller, 1968), *Blastocladiella* (Lessie and Lovett, 1968), and *Saprolegnia* (Heath and Greenwood, 1970). Their spindles are entirely closed, but polar centrioles are present, situated very close to, *but outside* the nuclear envelope.

Centrosomal spindles. The spindles of several basidiomycetes are apparently organized by characteristic amorphous organelles, the centrosomes, from which spindle microtubules radiate. Girbardt (1968) clearly showed that in *Polystictus* the centrosomes (termed Globular Ends) separate at prophase to establish the microtubular systems traversing the spindle. Other tubules radiating from them into the nearby cytoplasm, are associated with oscillatory movements of the whole spindle. In *Coprinus,* Lu (1967) recorded microtubules extending between centrosomes and chromosomes (see also Lerbs and Thielke, 1969), and centrosomal spindles have also been reported for *Schizophyllum* (Raudaskoski, 1970) and *Armillaria* (Motta, 1969).

Spindles containing centrosomal plaques. Some ascomycetes contain a (so far) unique structural variant in closed spindles, a characteristic dense plaque which constitutes part of (or maybe partly separated from) the nuclear envelope, and from which microtubules extend into the nucleus and sometimes into the cytoplasm. The term 'centrosomal plaque' (Zickler, 1970) is to be preferred over 'centriolar plaque' (Robinow and Marak, 1966) in describing these structures (Pickett-Heaps, 1969d). In *Saccharomyces,* microtubules extend the length of elongating spindles between two plaques, but Robinow and Marak (1966) could not ascertain how chromosomes were attached to the spindle. Moor (1967) confirmed the existence of such plaques in freeze-etched yeast and they have now been reported in *Aspergillus* (Robinow and Caten, 1969) and *Neottiella* (Westergaard and von Wettstein, 1970). Zickler's (1970) beautiful micrographs of mitosis in several Ascomycetes illustrate their role(s) in spindle formation very clearly. The main dense component of the plaque, situated just outside the nuclear envelope, is associated with randomly organized cytoplasmic tubules. An adjacent, separate intranuclear layer marks the termination site of numerous highly oriented spindle tubules, some connected to chromosomes; in certain fungi, part of the plaque is detached from the nuclear envelope, extending into the cytoplasm. In *Ascobolus,* duplication and separation of plaques occurs during prophase, and also bundles of microtubules unmistakably connect the plaque with chromosomes (Wells, 1970).

Mitosis — summary

The ubiquity of microtubules in spindles generally confirms their vital importance in mitosis. Let us briefly summarize these results by considering two specific aspect concerning microtubules.

The importance of organizing centres and centrioles in mitosis

Centrioles have long been considered vitally involved in organizing the spindle and recently they are very often specifically implicated in the assembly of spindle microtubules. These ideas have gained wide acceptance despite the absence of centrioles in most plant cells, an unpalatable fact often attributed to the negligence or incompetence of botanical cytologists! I believe combined evidence from many sources now virtually demolishes this confusing dogma, at least till some of its protagonists convincingly explain away the evidence to the contrary.

The true role of centrioles surely is concerned with flagellum formation. The doublet flagellar tubules (different chemically to cytoplasmic microtubules) are continuous with two of the triplet tubules of the centriole and they maintain along their length the same spatial configuration as that established by the centriole. This supports the idea that the centriole acts as a specific 'seed' or determinant of the spatial organization of the flagellum. This concept is further strengthened since, invariably to my knowledge, centrioles, as basal bodies, are required for flagellum formation. Furthermore, many plant cells normally devoid of centrioles, produce them only when required during differentiation into flagellated cells (e.g., during zoosporogenesis in *Oedogonium*: Pickett-Heaps, 1971b). This point will not be further laboured (see Pickett-Heaps, 1969d, 1971a).

What of centrioles during division? Taken *in toto,* the observations described above make it quite impossible to ascribe to the centriole any vital role in spindle formation without recourse to devious and unconvincing additional hypotheses. For example, some spindles do not utilize centrioles (or basal bodies) even when they are available nearby (e.g., *Chlamydomonas,* Dinoflagellates). Other spindles acquire the centrioles only during a transition of flagellated cells (e.g., spermatogenesis in *Chara*), or else the centrioles are situated outside the intact spindle (e.g., *Catenaria* and *Blastocladiella*). Finally, of course, centrioles are simply not needed in most dividing plant cells, and anyway eucaryotic cells appear to be able to assemble and disassemble cytoplasmic microtubules without necessarily involving the centriole.

Why then should the centriole be passively attached to the spindle? I have discussed this at length elsewhere (Pickett-Heaps, 1969d), and now consider most revealing the recent observations on centrioles forming *de novo* inside centrosomes or centrosome-like structures, as in diatoms (p. 237). *Labyrinthula* (Perkins, 1970), and during spermatogenesis in primitive vascular plants (see Pickett-Heaps, 1971a). These results somewhat lift the veil of mystery surrounding genesis of centrioles, and suggest to me that the centriole is one highly structured 'Microtubule-Organizing-Centre' (MTOC) involved in forming flagellar tubules, being formed as necessary from more diffuse MTOCs elsewhere in the cell — for example, those involved in organization of the spindle (Pickett-Heaps, 1971a). So in many (e.g., mammalian) cells, the information required to form the centriole could be permanently expressed whilst the centriole itself remains dormant until flagellum formation is required. In cells such as *Kirchneriella,* it is perhaps expressed temporarily during mitosis. This explanation in simple terms absolves the centriole from any functioning in the spindle, and also renders unnecessary those attempts to attribute to mitotic structures such as centrosomes and centrosomal plaques, some structural or functional similarity to the centriole, contrary to the evidence supplied (e.g., Motta, 1967; Lerbs and Thielke, 1969; Wells, 1970). Finally, we may note that sometimes when centrioles are clearly formed *de novo* in large numbers, during zoosporogenesis in *Oedogonium* (Pickett-Heaps, 1971b), they condense from diffuse material, and, concurrently, other MTOCs also are formed, the Rootlet Templates later involved in assembly of flagellar rootlet microtubules (see p. 246).

Where then are situated the sites in the spindle that organize microtubules? The kinetochores and phragmoplast are almost certainly thus involved (see earlier) as are the centrosomes, centrosomal plaques and some other clearly defined polar structures (e.g., the spindle precursor in *Lithodesmium*), and also the amorphous material often seen around centrioles. However, open plant spindles are more difficult to explain, but perhaps their poles do contain more diffuse forms of MTOCs almost impossible to detect (Pickett-Heaps, 1969d); this has recently received some support from work on *Closterium* where a 'centrosome' or MC (see earlier) forms at the spindle pole *after* division for a purpose not directly related to mitosis (p. 232). Since microtubules normally form and disperse around the cell, similar MTOCs are probably distributed around for other purposes as well. Furthermore, Hepler and Jackson (1969) showed that the herbicide isopropyl *N*-phenyl carbamate induces the formation of multipolar spindles in dividing cells of *Haemanthus* without apparently affecting spindle microtubules; these results clearly imply that some ill-defined but nevertheless vital organizing influence or centre at the poles of the spindle had been disrupted by this treatment.

Involvement of microtubules in mitotic movements

Clearly, spindle microtubules or fibres are involved in mitotic movement, but defining their mode of action still remains the goal of many scientists. Since hypotheses solidly supported by factual observation remain scarce, my comments can be brief.

Microtubules undoubtedly possess some structural rigidity. Thus, upon elongation, bundles of them probably exert pushing forces (e.g., on active kinetochores, or granules in the spindle) which, for example, might account for metakinesis in very simple terms (Pickett-Heaps and Fowke, 1970a). This rigidity also could provide a cytoskeletal framework over which mitotic movements could take place. Furthermore, spindle elongation during anaphase could be attributed to extension of the continuous fibre systems (e.g., by active elongation of microtubules at the poles).

However, anaphase movement of chromosomes seems to require the induction of tension in kinetochore fibres. So far, the means by which this is achieved is quite obscure. Carefully controlled dissolution of microtubules at the poles could perhaps induce the required tension (Inoue and Sato, 1967), but supporting evidence remains slender and this cannot explain many of the transport properties of the spindle. Many (e.g., Bajer and his colleagues) believe that the kinetochore tubules may be transported to the poles. This suggests that perhaps shear forces are generated on these microtubules (Östergren, Mole-Bajer, and Bajer, 1960). Recently, McIntosh, Hepler, and van Wie (1969) have suggested that microtubules have polarity, and they postulate further that microtubules have sidearms which could be instrumental in generating shear force, for example between a kinetochore tubule and an adjacent 'antiparallel' continuous tubule in the spindle; this would explain also how microtubules elsewhere could be associated with movement of organelles (e.g., the postmitotic migration in *Closterium* of nuclei ensheathed in microtubules). However, problems remain even with this attractive model and I consider that the morphological evidence produced for such sidearms in spindle tubular systems (e.g., Hepler, McIntosh, and Cleland, 1970; Wilson, 1969) is not yet very convincing. Indeed, interlinking by sidearms is clearest in structures containing regular arrays of microtubules which probably move minimally if at all with respect to one another (e.g., Axonemes of *Heliozoa*: Roth, Pihlaja, and Shigenaka, 1970; Axostyles: Grimstone and Cleveland, 1965; see also Tucker, 1968). The hypothetical nature of this important topic renders it unsuitable for further discussion here (see for example, Bajer and Mole-Bajer, 1971).

Cytoplasmic microtubules

Cytoplasmic streaming

While Ledbetter and Porter (1963) tentatively suggested an involvement of microtubules in cytoplasmic streaming (a view later reiterated, usually without additional evidence, by many authors), Porter (1966) later noted a lack of experimental support for this idea which is now refuted by strong circumstantial evidence. In doses sufficient to disrupt microtubules in *Chara*, colchicine does not affect streaming. Conversely, the drug Cytochalasin B, which apparently does not affect microtubules, stops streaming in *Nitella* and many other organisms (Wessells *et al.*, 1971). Cytoplasmic streaming can now be clearly associated with microfilamentous structures (Nagai and Rebhun, 1966; O'Brien and Thimann, 1966; see Newcomb, 1969) which are affected by Cytochalasin B (Wessells *et al.*, 1971).

However, two minor qualifications need be recorded. Firstly, one particular saltatory (i.e., linear) motion of cytoplasmic particles is sensitive to colchicine (Freed and Lebowitz, 1970), and such movement exists on the spindle. Secondly, microtubular arrays often appear to channel and direct movements of cytoplasmic constituents; for example, in the movement of melanin granules in melanophores (Bickle, Tilney, and Porter, 1966); these arrays could also conceivably influence streaming too (e.g., Sabnis and Jacobs, 1967).

Relationship of microtubules to cell walls

Ledbetter and Porter (1963) first reported a close correspondence between the orientation of (newly deposited) wall microfibrils just external to the plasmalemma, and the adjacent cytoplasmic microtubules; the latter normally encircle the cell transversely to the long axis both in higher plants and many algae. This provocative observation immediately suggested to them that microtubules could be involved in the control of microfibril orientation and thus in control over cell shape (see Newcomb, 1969). Incidentally, such microfibrils do not necessarily represent the cellulose component of the wall, a popular misconception (see O'Brien, 1972). This observation has been widely confirmed, and most workers agree with Ledbetter and Porter's inference. Nevertheless, the precise role of the microtubules is far from clear. They are certainly not a prerequisite for microfibril deposition. For example, they are not present near the plasmalemma of pollen tubes (Rosen, 1968) or fungi generally (Bracker, 1967; Grove and Bracker, 1970); neither can the orientation of cellulose microfibrils in the stalk of differentiating slime molds (Hohl, Hammoto, and Hemmes, 1968) be correlated with nearby cytoplasmic microtubules. Some higher plant cells do not have microtubules near thickening walls (e.g., epidermal cells of coleoptiles) and on many occasions, microtubules do not lie parallel to microfibril orientation in primary or secondary walls (see O'Brien, 1972). In elongating root hairs, microtubules are axially oriented (Newcomb and Bonnett, 1965) but may be correlated here with similarly oriented microfibrils (Newcomb, 1969). Whilst vegetative cells of *Chara* (Pickett-Heaps, 1967b) and *Spirogyra* (Fowke and Pickett-Heaps, 1969a) possess typical transverse wall microtubules, at least two other filamentous algae with cylindrical cells have wall microtubules oriented longitudinally; one, *Oedogonium* (Pickett-Heaps and Fowke, 1969), utilizes a unique method of cell elongation after division (see earlier), but the other, *Stigeoclonium* (Pickett-Heaps, unpublished data), to the best of my knowledge, does not; detailed investigations of their wall structure might prove profitable. During the rapid expansion of semicells that follows cell division in *Closterium,* wall microfibrils appear somewhat randomly oriented in contrast to the strikingly regular array of adjacent transverse cytoplasmic microtubules (Pickett-Heaps and Fowke, 1970c). During conjugation in *Closterium,* transverse wall microtubules appear near the soft papilla wall only during early stages of expansion, disappearing soon thereafter (Pickett-Heaps and Fowke, 1971).

Treatment of growing plant tissues with colchicine has long been known to cause tumour-like swellings and rotund cell profiles (Green, 1969). This treatment also causes growing cylindrical cells of *Nitella* and *Chara* to round out, concurrent with a loss in the oriented microfibril deposition (Green, 1969); disruption of their beautifully structured reproductive organs (e.g., the spiral cells around oogonia) is also profound (Pickett-Heaps, unpublished data). Furthermore, when such treated cells were mechanically restrained from rounding up, the orientation of newly deposited microfibrils remained random; therefore the randomization was not due to the cells rounding up (Green, 1969). Removal of colchicine is soon followed by restoration of organized microfibril deposition.

Plant microtubules

Such variable and often conflicting observations make it difficult to suggest how wall microtubules might function, and why, if they seem so important, they are not universally present near fibrous walls. My feeling is that they act somehow in stabilizing the shape (and even perhaps the plasmalemma itself – O'Brien, personal communication) of large, actively growing cells, permitting the slower processes of wall deposition to proceed without disruption. Wall microtubules might also represent a comparatively recent ultrastructural development during evolution; this would explain why many (primitive) cells do not need them. Indeed *Closterium* might even demonstrate how transverse 'cytokinetic' microtubular systems could have become converted during cell expansion into typical 'wall' microtubules (Pickett-Heaps, 1969d; Pickett-Heaps and Fowke, 1970c).

We need also to recall here that, after division, cell expansion in the constricted desmids may form an exceedingly intricate and beautiful semicell whose shape is the mirror of the inherited older semicell (e.g., see Waris, 1950). The walls of such cells, often thick and ornamented, contain massive arrays of oriented microfibrils, yet neither in *Micrasterias* (Kiermayer, 1968, 1970) nor in *Cosmarium* (Pickett-Heaps, results in preparation) are microtubules associated with the expanding wall. This example should particularly temper our inclination to regard microtubules as vitally important in the morphogenesis of fibrillar cell walls. O'Brien (1972) excellently summarizes our current state of ignorance in this whole matter.

Xylem cells

The wall of xylem cells constitutes a particularly favourable object for studying wall deposition. Hepler and Newcomb (1964) first showed that microtubules were specifically situated over the plasmalemma lining growing xylem wall thickenings in redifferentiating wound vessel members of *Coleus*; as usual, these organelles also matched the orientation of wall microfibrils. Whilst Hepler and Newcomb's important results have now been widely confirmed (see Newcomb, 1969; Roberts, 1969), unfortunately, I am the only investigator so far who has considere it important to establish what happens at the *initiation* of xylem wall differentiation (Pickett-Heaps, 1966). So what follows applies to wheat coleoptiles, and remains to be confirmed elsewhere, but it establishes more forcefully the importance of microtubules in this system. Initially, the large, incipient xylem cells have transversely oriented microtubules distributed along their wall fairly evenly (i.e., as in many meristematic cells). The next stage, distinct but difficult to find, is marked by the unmistakable grouping of microtubules into regularly spaced bands along the *undifferentiated* wall. Sometimes, these bands are later seen between slight regular ridges in the wall (Pickett-Heaps and Northcote, 1966c); I now believe this is a temporary condition that confuses the issue somewhat. Next, secondary wall deposition starts and thereafter, until the cells die, microtubules persist over the thickenings (Hepler and Newcomb, 1964 *et al.*). In similar fashion, a larger grouping of microtubules lines the wall between young stomatal guard cells before and during its considerable increase in thickness (Pickett-Heaps, 1965; Kaufman, Petering, Yocum, and Baic, 1970), and they are associated with regular wall thickenings that form in shield cells of antheridia in *Chara* (Pickett-Heaps, 1968a).

The obvious implications of these observations led directly to a test of the hypothesis that bands of microtubules control the site of secondary wall deposition. Colchicine treatment of differentiating xylem cells soon showed that gross distortion and disruption of wall deposition followed disorganization of cytoplasmic microtubule bands (Pickett-Heaps, 1967a); microfibrils in the abnormal thickenings, however, were not markedly randomized during the short time of the experiment.

So, in differentiating xylem cells, wall microtubules may be functioning to a pronounced degree in a similar role to wall microtubules in expanding meristematic cells. As before, to suggest a more specific function is very difficult. Perhaps they could also specifically channel wall materials (e.g., in vesicles) into the thickening (Pickett-Heaps, 1968c; Robards, 1968). Again, O'Brien (1972) points out other difficulties in interpreting these results. Furthermore, although colchicine causes abnormal development in regenerating wound vessel members of *Coleus* too, many normal cells are formed if the tissues were treated with proline before exposure to colchicine (Roberts, 1969), which is rather difficult to explain.

Spermatogenesis

In animal cells, microtubular systems appear vital both in accomplishing the profound changes in cell morphology that characterize spermatogenesis, and also in forming a cytoskeletal framework which maintains the structural integrity of the mature sperm, organisms which are usually extremely elongated and active. The spermatozooids of *Sphagnum* (Manton, 1957), *Fucus* (Manton and Clarke, 1956), and *Pteridium* (Manton, 1959) were also shown in early work to contain a sheath of tubules, since confirmed for other plants (e.g., Diers, 1967; Norstog, 1967; Rice and Laetsch, 1967; Carothers and Kreitner, 1968; Moestrup, 1970). However, detailed observations of spermatogenesis have only been made on *Polytrichum* (Paolillo, Kreitner, and Reighard, 1968a, b), *Chara* (Pickett-Heaps, 1968b), and *Nitella* (Turner, 1968).

In *Polytrichum,* the androcyte early in differentiation forms two basal bodies (different from one another, with triplet tubules of different lengths) and a nearby 'Multi-Layered-Structure' (MLS). Two adjacent but distinct groups of microtubules, each a close-packed single layer, appear between the MLS and the two basal bodies. These microtubules elongate, and the nucleus begins to adhere to them, meanwhile condensing. Complicated movements follow in the orientation of the basal bodies and in the precise disposition of these tubules, whose number is fairly consistent in a given species. Meanwhile, the mito-chondrial mass splits in two, one half enveloping the plastid. These two halves are then rearranged on either side of the elongating nucleus. The growing bands of microtubules are obliged to circle around inside the cell, with the greatly attenuated nucleus remaining close pressed to them. The cytoplasm also condenses drastically and the mature sperm consists of an extremely long, thin spiral nucleus with a cytoplasmic appendage at one end containing the plastid and mitochondrio[]; the flagellar apparatus is inserted in the apical cytoplasm at the other end.

The accounts of spermatogenesis in the closely related species, *Chara* and *Nitella,* are essentially similar (Turner, 1968; Pickett-Heaps, 1968b). Wall microtubules, present initially in spermatogenous cells, soon disappear. Centrioles are formed *de novo* before differentiation and subsequent mitoses are then typically centric (p. 231); when cell division ceases, the centrioles move to the edge of the cell and soon extrude flagella. Near them, a flat layer of microtubules appears which eventually numbers up to twenty; these are initially very short and in *Chara*, they are inserted in a dense homogeneous structure which perhaps represents a MTOC (Pickett-Heaps, 1969d). These 'Manchette' microtubules also elongate very considerably, becoming spirally arranged in the cell as in *Polytrichum*. In *Chara* (Pickett-Heaps, 1968b), first the nucleus becomes attached to them during this elongation; then the plastids neatly line up on them next to the nucleus and, finally, the mitochondria also become linearly arranged along them, on the other side of the nucleus near the flagellar apparatus. These events are accompanied by extreme nuclear and cytoplasmic condensation. I also noted a set of microtubules possibly distinct from the Manchette tubules proper (visible also in Turner's micrographs of *Nitella*); thus these organisms may have two sets of tubules as in *Polytrichum*. Finally, a pore in the wall releases the long, thin, coiled sperm (Turner, 1968).

Spermatogenesis therefore illustrates how important microtubular systems can be in altering the shape of a cell and the disposition of its organelles. Turner's (1970) account of spermatogenesis when affected by colchicine strongly confirms this impression. For example, in treated cells the nucleus condensed as normal but never elongated or coiled. Branched flagella with abnormal complements of tubules were sometimes formed if colchicine was briefly applied early in differentia[]tion, and long-term treatment followed by recovery caused the elaboration of very atypical microtubular systems in grossly misshapen sperm cells. Thus microtubules are undoubtedly essential for this type of morphogenesis.

Rootlet and cytoskeletal microtubules

Many algal cells not enclosed by a rigid wall, possess numerous peripheral microtubules which may be cytoskeletal. Manton and her colleagues described some of these tubules as 'rootlets', being specific structural arrays associated with the flagellar apparatus (Hoffman and Manton, 1962, *et al.*; Manton, 1964b; see references given by Ringo, 1967). In *Chlamydomonas* (Ringo, 1967) four sets of three to four tubules emanate around the cell from the complex infrastructure surrounding the two basal bodies. In the multiflagellated zoospores and spermatozooids of *Oedogonium,* analogous microtubules extend from between the very numerous, circularly arranged basal bodies (Hoffman and Manton, 1962, 1963; Hoffman, 1970), and Hoffman (1970) has reaffirmed their role in transmitting the stresses induced by flagellar motion throughout the cytoplasm. During zoosporogenesis in *Oedogonium,* 'templates' for these rootlet tubules and the centrioles are both created *de novo* at the same time, and both these presumed MTOCs later concurrently extrude their respective microtubular systems (Pickett-Heaps, 1971b). Unfortunately, what happens to flagellar rootlet systems during division (e.g., in *Chlamydomonas*) has not been reported; I suspect that they might replicate along with the basal bodies. Certainly the region around basal bodies seems to constitute a complicated morphogenetic centre which produces at least one set of cleavage microtubules in *Chlamydomonas* (see Johnson and Porter, 1968).

Other less regularly arranged individual microtubules may also extend around the cell periphery near rootlet tubules (e.g., Hoffman and Manton, 1962, 1963; Hoffman, 1970; Pickett-Heaps, 1971b). When zoospores of *Oedogonium* elongate and differentiate back into germlings (i.e., when they lose their flagella), large numbers of such tubules proliferate near the wall and through the cytoplasm along the cell axis, particularly in the developing holdfast (Pickett-Heaps, 1972a, b). In *Volvox*, too, similar microtubules also proliferate during the cell elongation that accompanies both differentiation of sperm and inversion in vegetative colonies (Pickett-Heaps, 1970a). These tubules could obviously exert control over the shape of the cell (cf. Porter, 1966).

In zoospores of some *Blastocladiales*, similar microtubules have been shown to ensheath the nucleus, these emanating from an electron dense region enveloping the end of the basal body (Fuller and Calhoun, 1968; see also Fuller and Reichle, 1968). An intriguing variation of these structures has been reported in zoospores of *Phytophthora* (Reichle, 1969). The two long rootlets, planar arrays of tubules, run in opposite directions from the flagellar apparatus, underlying a groove in the cell. Perpendicular to these tubules emanate single microtubules which run rib-like around the cell's periphery.

Perhaps the only close analogy reported for these tubules in higher plant cells occurs in generative cells of germinating pollen (Burgess, 1970b). Prior to division, long bundles of tubules are dispersed around the cell periphery. As the nucleus goes into division, these diminish and, simultaneously, the cell itself becomes highly convoluted.

In *Euglena*, distinctive microtubules are also very precisely associated with ridges in the pellicle but their significance is not obvious. Elsewhere, several highly organized layers of tubules encircle the canal. These in turn are ensheathed by endoplasmic reticulum; the reservoir (into which the contractile vacuoles empty) is also enclosed by a single layer of tubules. Such microtubules could be cytoskeletal and/or associated with movement of the reservoir and canal (see review by Buetow, 1968).

A very specialized type of microtubular cytoskeleton is found in zoospores of the Hydrodictyaceae. These small uninucleate, flagellated cells are formed following cytoplasmic cleavage of a mother cell (p. 234); after a period of activity, they lose their flagella and then adhere to one another in a certain pattern to form colonies whose shape is characteristic of the species. In *Pediastrum* (Millington and Gawlik, 1970), one or two arrays of regularly spaced, parallel microtubules appear after cleavage near the plasmalemma of each zoospore, and when the cells begin to aggregate, these microtubular systems are often strikingly lined up from one cell to another, but they disappear before the colonial cells secrete fibrous walls; Millington and Gawlik briefly mention the possible involvement of these microtubules in the alignment of cells during aggregation. Analogous arrays of microtubules are also formed in zoospores of *Hydrodictyon* after cleavage (Marchant and Pickett-Heaps, 1972a), and we have good evidence from colchicine experiments that they are indeed very important in establishing the pattern of cellular adhesion and thus the pattern of vegetative nets (Marchant and Pickett-Heaps, work in progress). Moreover, these microtubular arrays are not seen in gametes of *Hydrodictyon*, which do not form colonies (Marchant after Pickett-Heaps, 1972b).

Bibliography

Selected references for further reading

For further general reading, see the references of: Porter (1966); Inoue and Sato (1967); Bajer (1968a); Bajer and Mole-Bajer (1969); Newcomb (1969); Pickett-Heaps (1969d); Bajer and Mole-Bajer (1971); Pickett-Heaps (1971a); O'Brien (1972).

References

Aldrich, H. C. (1967) 'The ultrastructure of meiosis in three species of *Physarum*', *Mycologia*, **59**, 127–147.

Aldrich, H. C. (1969) 'The ultrastructure of mitosis in myxamoebae and plasmodia of *Physarum flavicomum*', *Am. J. Bot.*, **56**, 290–299.

Allen, R. D. and Bowen, C. C. (1966) 'Fine structure of *Psilotum nudum* cells during mitosis', *Carylogia*, **19**, 299–342.

Bajer, A. (1965a) 'Behaviour of chromosomal spindle fibres in living cells', *Chromosoma*, **16**, 381–390.

Bajer, A. (1965b) 'Cine micrographic analysis of cell plate formation in endosperm', *Expl Cell Res.*, **37**, 376–398.

Bajer, A. (1967) 'Notes on ultrastructure and some properties of transport within the living mitotic spindle', *J. Cell Biol.*, **33**, 713–720.

Bajer, A. (1968a) 'Behaviour and fine structure of spindle fibres during mitosis in endosperm', *Chromosoma*, **25**, 249–281.

Bajer, A. (1968b) 'Chromosome movement and fine structure of the mitotic spindle', In: *Aspects of cell motility* (Cambridge Univ. Press) XXIInd Symp. Soc. Exp. Biol., pp. 387–310.

Bajer, A. (1968c) 'Fine structure studies on phragmoplast and cell plate formation', *Chromosoma*, **24**, 383–417.

Bajer, A. (1971) 'Influence of UV-microbeam on spindle fine structure and anaphase chromosome movements' (Oxford chromosome conference), *J. Hered. Supplement* (in press).

Bajer, A. and Allen, R. D. (1966a) 'Structure and organisation of the living mitotic spindle of *Haemanthus* endosperm', *Science, N.Y.*, **151**, 572–574.

Bajer, A. and Allen, R. D. (1966b) 'Role of phragmoplast filaments in cell-plate formation', *J. Cell Sci.*, **1**, 455–462.

Bajer, A. and Jensen, C. (1969) 'Detectability of mitotic spindle microtubules with the light and electron microscopes', *J. Microscopie*, **8**, 343–354.

Bajer, A. and Mole-Bajer, J. (1969) 'Formation of spindle fibres, kinetochore orientation, and behaviour of the nuclear envelope during mitosis in endosperm', *Chromosoma*, **27**, 448–484.

Bibliography

Bajer, A. and Mole-Bajer, J. (1970) 'Effects of UV-microbeam irradiation on fine structure of the mitotic spindle', *VIIth Int. Cong. Elect. Microscop. (Grenoble)*, pp. 267–268.

Bajer, A. and Mole-Bajer, J. (1971) 'Architecture and function of the mitotic spindle', In: *Progress in cell biology* (ed. E. Du Praw) (in press).

Bal, K. A. and Gross, P. R. (1964) 'Suppression of mitosis and macromolecular synthesis in onion roots by heavy water', *J. Cell Biol.*, **23**, 188–193.

Behnke, O. (1970) 'A comparative study of microtubules of disc-shaped blood cells', *J. Ultrastruct. Res.*, **31**, 61–75.

Behnke, O. and Forer, A. (1966) 'Some aspects of microtubules in spermatocyte meiosis in a crane fly (*Nephrotoma suturalis* Loew): intranuclear and intrachromosomal microtubules', *C.r. Trav. Lab. Carlsberg*, **35**, 437–455.

Behnke, O. and Forer, A. (1967) 'Evidence for four classes of microtubules in individual cells', *J. Cell Sci.*, **2**, 169–192.

Bikle, D., Tilney, L. G., and Porter, K. R. (1966) 'Microtubules and pigment migration in the melanophores of *Fundulus heteroclitus* L.', *Protoplasma*, **61**, 322–345.

Bensch, K. G. and Malawista, S. E. (1969) 'Microtubular crystals in mammalian cells', *J. Cell Biol.*, **40**, 95–107.

Borisy, G. G. and Taylor, E. W. (1967a) 'The mechanism of action to colchicine. Binding of Colchicine-H^3 to cellular protein', *J. Cell Biol.*, **34**, 525–533.

Borisy, G. G. and Taylor, E. W. (1967b) 'The mechanism of action of colchicine. Colchicine binding to sea urchin eggs and the mitotic apparatus', *J. Cell Biol.*, **34**, 535–548.

Bracker, C. E. (1967) 'Ultrastructure of fungi', *A. Rev. Phytopath.*, **5**, 343–374.

Brinkley, B. R., Stubblefield, E., and Hsu, T. C. (1967) 'The effects of colcemid inhibition and reversal on the fine structure of the mitotic apparatus of Chinese hamster cells *in vitro*', *J. Ultrastruct. Res.*, **19**, 1–18.

Brown, R. (1960) *Plant physiology* (ed. F. C. Steward), vol. 1A, p. 3, Academic Press Inc., New York and London.

Buetow, D. (1968) *The biology of Euglena*, vol. I; Academic Press Inc., New York and London.

Burgess, J. (1970a) 'Microtubules and cell division in the microspore of *Dactylorchis fuschii*', *Protoplasma*, **69**, 253–264.

Burgess, J. (1970b) 'Cell shape and mitotic spindle formation in the generative cell of *Endymion non-scriptus*', *Planta*, **95**, 72–85.

Burgess, J. (1970c) 'Interactions between microtubules and the nuclear envelope during mitosis in a fern', *Protoplasma*, **71**, 77–89.

Burgess, J. and Northcote, D. H. (1967) 'A function of the preprophase band of microtubules in *Phleum pratense*', *Planta*, **75**, 319–326.

Burgess, J. and Northcote, D. H. (1968) 'The relationship between the endoplasmic reticulum and microtubular aggregation and disaggregation', *Planta*, **80**, 1–14.

Burgess, J. and Northcote, D. H. (1969) 'Action of colchicine and heavy water on the polymerisation of microtubules in wheat meristem', *J. Cell Sci.*, **5**, 433–451.

Burton, P. R. (1968) 'Effects of various treatments on microtubules and axial units of lung-fluke spermatozoa', *Z. Zellforsch. mikrosk. Anat.*, **87**, 226–248.

Carothers, Z. B. and Kreitner, G. L. (1968) 'Studies of spermatogenesis in the Hepaticae', *J. Cell Biol.*, **36**, 603–616.

Cronshaw, J. and Esau, K. (1968) 'Cell division in leaves of *Nicotiana*', *Protoplasma*, **65**, 1–24.

Deysson, G. (1968) 'Antimitotic substances', *Int. Rev. Cytol.*, **24**, 94–148.

Deysson, G. and Benbadis, M. (1968) 'Etude ultrastructurale de la mise en place de l'appareil fusorial dans les cellules méristématiques des végétaux supérieurs', *C. r. Séanc. Soc. Biol.*, **162**, 601–604.

Diers, L. (1967) Der feinbau des spermatozoids von *Sphaerocarpos donnellii* Aust.', (Hepaticae), *Planta*, **72**, 119–145.

Esau, K. and Gill, R. H. (1965) 'Observations on cytokinesis', *Planta*, **67**, 168–181.

Esau, K. and Gill, R. H. (1969) 'Structural relations between nucleus and cytoplasm during mitosis in *Nicotiana tabacum*', *Can. J. Bot.*, **47**, 581–591.

Evert, R. F. and Deshpande, B. P. (1970) 'An ultrastructural study of cell division in the cambium', *Am. J. Bot.*, **57**, 942–961.

Forer, A. (1965) 'Local reduction of spindle fibre birefringence in living *Nephrotoma suturalis*

(Loew) spermatocytes induced by ultraviolet microbeam irradiation', *J. Cell Biol.*, **25**, 95–117.

Forer, A. (1966) 'Characterization of the mitotic traction system and evidence that birefringent spindle fibres neither produce not transmit force for chromosome movement', *Chromosoma*, 19, 44–98.

Fowke, L. C. and Pickett-Heaps, J. D. (1969a) 'Cell division in *Spirogyra* I. Mitosis', *J. Phycol.*, 5, 240–259.

Fowke, L. C. and Pickett-Heaps, J. D. (1969b) 'Cell division in *Spirogyra* II, Cytokinesis', *J. Phycol.*, 5, 273–281.

Freed, J. J. and Lebowitz, M. M. (1970) 'The association of a class of saltatory movements with microtubules in cultured cells', *J. Cell Biol.*, 45, 334–354.

Fritsch, F. E. (1935) *Structure and reproduction of the Algae*, vol. I, Cambridge University Press.

Fuller, M. S. and Calhoun, S. A. (1968) 'Microtubule-kinetosome relationships in the motile cells of the Blastocladiales', *Z. Zellforsch. mikrosk. Anat.*, 87, 526–533.

Fuller, M. S. and Reichle, R. E. (1968) 'The fine structure of Monoblepharella sp. zoospores', *Can. J. Bot.*, 46, 279–283.

Gaulden, E. S. and Carlson, J. G. (1951) 'Cytological effects of colchicine on the grasshopper neuroblast *in vitro* with special reference to the origin of the spindle', *Expl Cell Res.*, 2, 416–433.

Girbardt, M. (1968) 'Ultrastructure and dynamics of the moving nucleus', *Symp. Soc. exp. Biol.*, 22, 249–260.

Goodenough, U. W. (1970) 'Chloroplast division and pyrenoid formation in *Chlamydomonas reinhardi*', *J. Phycol.*, 6, 1–6.

Goodman, E. M. and Ritter, H. (1969) 'Plasmodial mitosis in *Physarum polycephalum*', *Arch. Protistenk.*, 111, 161–169.

Green, P. B. (1969) 'Cell morphogenesis', *A. Rev. Pl. Physiol.*, 20, 365–394.

Grimstone, A. V. and Cleveland, L. R. (1965) 'The fine structure and function of the contractile axostyle of certain flagellates', *J. Cell Biol.*, 24, 387–400.

Grove, S. N. and Bracker, C. E. (1970) 'Protoplasmic organization of hyphal tips among fungi: Vesicles and Spitzenkörper', *J. Bact.*, 104, 989–1009.

Harris, P. and Bajer, A. (1965) 'Fine structure studies on mitosis in endosperm metaphase of *Haemanthus katherinae* Bak', *Chromosoma*, 16, 624–636.

Heath, I. B. and Greenwood, A. D. (1970) 'Centriole replication and nuclear division in *Saprolegnia*', *J. gen. Microbiol.*, 62, 139–148.

Hepler, P. K. and Jackson, W. T. (1968) 'Microtubules and early stages of cell-plate formation in the endosperm of *Haemanthus katherinae* Baker', *J. Cell Biol.*, 38, 437–446.

Hepler, P. K. and Jackson, W. T. (1969) 'Isopropyl *N*-phenylcarbamate affects spindle microtubule orientation in dividing endosperm cells of *Haemanthus katherinae* Baker', *J. Cell Sci.*, 5, 727–743.

Hepler, P. K., McIntosh, J. R., and Cleland, S. (1970) 'Intermicrotubule bridges in mitotic spindle apparatus', *J. Cell Biol.*, 45, 438–444.

Hepler, P. K. and Newcomb, E. H. (1964) 'Microtubules and fibrils in the cytoplasm of *Coleus* cells undergoing secondary wall deposition', *J. Cell Biol.*, 20, 529–533.

Hepler, P. K. and Newcomb, E. H. (1967) 'Fine structure of cell plate formation in the apical meristem of *Phaseolus* roots', *J. Ultrastruct. Res.*, 19, 498–513.

Heslop-Harrison, J. (1968) 'Synchronous pollen mitosis and the formation of the generative cell in massulate orchids', *J. Cell Sci.*, 3, 457–466.

Hill, G. J. C. and Machlis, L. (1968) 'An ultrastructural study of vegetative cell division in *Oedogonium borisanum*', *J. Phycol.*, 4, 261–271.

Hindmarsh, M. M. (1953) 'The effect of colchicine on the spindle of root tip cells', *Proc. Linn. Soc. N.S.W.*, 77, 300–306.

Hoffman, L. R. (1970) 'Observations on the fine structure of *Oedogonium*. VI. The striated component of the compound flagellar 'roots' of *O. cardiacum*', *Can. J. Bot.*, 48, 189–196.

Hoffman, L. R. and Manton, I. (1962) 'Observations on the fine structure of the zoospore of *Oedogonium cardiacum* with special reference to the flagellar apparatus', *J. exp. Bot.*, 13, 443–449.

Hoffman, L. R. and Manton, I. (1963) 'Observations on the fine structure of *Oedogonium*. II. The spermatozoid of *O. cardiacum*', *Am. J. Bot.*, 50, 455–463.

Hohl, H. R., Hammoto, S. T., and Hemmes, D. E. (1968) 'Ultrastructural aspects of cell elongation, cellulose synthesis, and spore differentiation in *Acytostelium leptosomum*, a slime mold', *Am. J. Bot.*, **55**, 783–796.

Ichida, A. A. and Fuller, M. S. (1968) 'Ultrastructure of mitosis in the aquatic fungus *Caternaria anguillulae', Mycologia*, **60**, 141–155.

Inoue, S. (1952) 'The effect of colchicine on the microscopic and submicroscopic structure of the mitotic spindle', *Expl Cell Res. Suppl.*, **2**, 305–314.

Inoue, S. (1964) 'Organization and function of the mitotic spindle', In: *Primitive motile systems in cell biology*, pp. 549–596, Academic Press Inc., London and New York.

Inoue, S. and Sato, H. (1967) 'Cell motility by labile association of molecules', *J. gen. Physiol.*, **50**, 259–292.

Jensen, C. and Bajer, A. (1969) 'Effects of dehydration on the microtubules of the mitotic spindle', *J. Ultrastruct. Res.*, **26**, 367–386.

Johnson, U. G. and Porter, K. R. (1968) 'Fine structure of cell division in *Chlamydomonas reinhardi'*, *J. Cell Biol.*, **38**, 403–425.

Kane, R. E. and Forer, A. (1965) 'The mitotic apparatus. Structural changes after isolation', *J. Cell Biol.*, **25**, 31–39.

Kaufman, P. B., Petering, L. B., Yocum, C. S., and Baic, D. (1970) 'Ultrastructural studies on stomata development in internodes of *Avena sativa'*, *Am. J. Bot.*, **57**, 33–49.

Kiefer, B., Sakai, H., Solari, A. J., and Mazia, D. (1966) 'The molecular unit of the microtubules of the mitotic apparatus', *J. molec. Biol.*, **20**, 75–79.

Kiermayer, O. (1968) 'The distribution of microtubules in differentiating cells of *Micrasterias denticulata* Breb', *Planta*, **83**, 223–236.

Kiermayer, O. (1970) 'Elektronenmikroskopische Untersuchungen zum Problem der Cytomorphogenese von *Micrasterias denticulata* Breb', *Protoplasma*, **69**, 97–132.

Kihlman, B. A. (1966) *Actions of chemicals on dividing cells.* Prentice-Hall, Inc., Englewood Cliffs, New Jersey.

Kubai, D. F. and Ris, H. (1969) 'Division in the dinoflagellate *Gyrodinium cohnii* (Schiller)', *J. Cell Biol.*, **40**, 508–528.

Lafontaine, J. G. and Chouinard, L. A. (1963) 'A correlated light and electron microscope study of the nucleolar material during mitosis in *Vicia faba'*, *J. Cell Biol.*, **17**, 167–201.

Leadbeater, B. and Dodge, J. D. (1967) 'An electron microscope study of nuclear and cell division in a dinoflagellate', *Arch. Mikrobiol.*, **57**, 239–254.

Ledbetter, M. C. and Porter, K. R. (1963) 'A "microtubule" in plant cell fine structure', *J. Cell Biol.*, **19**, 239–250.

Leedale, G. F. (1968) 'The nucleus in *Euglena'*, In: *The biology of Euglena* (ed. by D. Buetow), vol. 1, pp. 185–242, Academic Press Inc., New York and London.

Lerbs, V. and Thielke, Ch. (1969) 'Die Enstehung der Spindel während der Meiose von *Coprinus radiatus'*, *Arch. Mikrobiol.*, **68**, 95–98.

Lessie, P. E. and Lovett, J. S. (1968) 'Ultrastructural changes during sporangium formation and zoospore differentiation in *Blastocladiella emersonii'*, *Am. J. Bot.*, **55**, 220–236.

Lovlie, A. and Braten, T. (1970) 'On mitosis in the multicellular alga *Ulva mutabilis* Foyn', *J. Cell Sci.*, **6**, 109–129.

Lu, B. C. (1967) 'Meiosis in *Coprinus lagopus*; A comparative study with light and electron microscopy', *J. Cell Sci.*, **2**, 529–536.

Malawista, S. E. and Sato, H. (1969) 'Vinblastine produces uniaxial, birefringent crystals in starfish oocytes', *J. Cell Biol.*, **42**, 596–599.

Malawista, S. E., Sato, H., and Bensch, K. G. (1968) 'Vinblastine and griseofulvin reversibly disrupt the living mitotic spindle', *Science, N.Y.*, **160**, 770–772.

Manton, I. (1957) 'Observations with the electron microscope on the cell structure of the antheridium and spermatozoid of *Sphagum'*, *J. exp. Bot.*, **8**, 382–400.

Manton, I. (1959) 'Observations on the microanatomy of the spermatozoid of the bracken fern *(Pteridium aquilinum)'*, *J. biophys. biochem. Cytol.*, **6**, 413–418.

Manton, I. (1964a) 'Observations with the electron microscope on the division cycle in the flagellate *Prymnesium parvum* Carter', *J. R. microsc. Soc.*, **83**, 317–325.

Manton, I. (1964b) 'Observations on the fine structure of the zoospore and young germling of *Stigeoclonium'*, *J. exp. Bot.*, **15**, 399–411.

Manton, I. and Clarke, B. (1956) 'Observations with the electron microscope on the internal structure of the spermatozoid of *Fucus'*, *J. exp. Bot.*, **7**, 416–432.

Plant microtubules

Manton, I., Kowallik, K., and von Stosch, H. A. (1969a) 'Observations on the fine structure of the spindle at mitosis and meiosis in a marine centric diatom (*Lithodesmium undulatum*) I. Preliminary survey of mitosis in spermatogonia', *J. Microscopie*, 89, 295–320.

Manton, I., Kowallik, K., and von Stosch, H. A. (1969b) 'Observations on the fine structure of the spindle at mitosis and meiosis in a marine centric diatom (*Lithodesmium undulatum*) II. The early mitotic stages in male gametogenesis', *J. Cell Sci.*, 5, 271–298.

Manton, I., Kowallik, K., and von Stosch, H. A. (1970a) 'Observations on the fine structure of the spindle at mitosis and meiosis in a marine centric diatom (*Lithodesmium undulatum*) III. The later stages of meiosis I in male gametogenesis', *J. Cell Sci.*, 6, 131–157.

Manton, I., Kowallik, K., and von Stosch, H. A. (1970b) 'Observations of the fine structure of the spindle at mitosis and meiosis in a marine centric diatom (*Lithodesmium undulatum*) IV. The second meiotic division and conclusion', *J. Cell Sci.*, 7, 407–443.

Marchant, H. J. and Pickett-Heaps, J. D. (1970) 'Ultrastructure and differentiation of *Hydrodictyon reticulatum*. I. Mitosis in the coenobium', *Aust. J. biol. Sci.*, 23, 1173–1186

Marchant, H. J. and Pickett-Heaps, J. D. (1971) 'Ultrastructure and differentiation of *Hydrodictyon reticulatum*. II. Formation of zooids within the coenobium', *Aust. J. biol. Sci.*, 24, 471–486.

Marchant, H. J. and Pickett-Heaps, J. D. (1972a) 'Ultrastructure and differentiation of *Hydrodictyon reticulatum*. III. Formation of daughter nets', *Aust. J. Biol. Sci.*, 25, 265–278.

Marchant, H. J. and Pickett-Heaps, J. D. (1972b) 'Ultrastructure and differentiation of *Hydrodictyon reticulatum*. IV. Conjugation of gametes and development of zygospores and azygospores', *Aust. J. Biol. Sci.*, 25, 279–291.

Marsland, D. and Zimmerman, A. M. (1965) 'Structural stabilization of the mitotic apparatus by heavy water, in cleaving eggs of *Arbacia punctulata*', *Expl Cell Res.*, 38, 306–313.

Mazia, D. (1961) 'Mitosis and the physiology of cell division', In: *The cell* (ed. J. Brachet and A. E. Mirsky), vol. 3, pp. 77–412, Academic Press Inc., New York and London.

McIntosh, J. R., Hepler, P. K., and Van Wie, D. G. (1969) 'Model for mitosis', *Nature*, 224, 659–663.

McManus, M. A. and Roth, L. E. (1968) 'Ultrastructure of the somatic nuclear division in the plasmodium of the myxomycete *Clastoderma debaryanum*', *Mycologia*, 60, 426–436.

Millington, W. F. and Gawlik, S. R. (1970) 'Ultrastructure and initiation of wall pattern in *Pediastrum boryanum*', *Am. J. Bot.*, 57, 552–561.

Moestrup, O. (1970) 'On the fine structure of the spermatozoids of *Vaucheria sesculicaria* and on the later stages of spermatogenesis', *J. mar. biol. Ass. U.K.*, 50, 513–523.

Mole-Bajer, J. (1965) 'Telophase segregation of chromosomes and amitosis', *J. Cell Biol.*, 25, 79–93.

Mole-Bajer, J. and Bajer, A. (1968) 'Studies of selected endosperm cells with the light and electron microscope. The technique', *Cellule*, 67, 257–265.

Moor, H. (1967) 'Der Feinbau der Mikrotubuli in Hefe nach Gefrierätzung', *Protoplasma*, 64, 89–103.

Motta, J. J. (1967) 'A note on the mitotic apparatus in the rhizomorph meristem of *Armillaria mellea*', *Mycologia*, 59, 370–375.

Motta, J. J. (1969) 'Somatic nuclear division in *Armillaria mellea*', *Mycologia*, 61, 873–886.

Nagai, R. and Rebhun, L. I. (1966) 'Cytoplasmic microfilaments in streaming *Nitella* cells', *J. Ultrastruct. Res.*, 14, 571–589.

Newcomb, E. H. (1969) 'Plant microtubules', *A. Rev. Pl. Physiol.*, 20, 253–288.

Newcomb, E. H. and Bonnett, H. T. (1965) 'Cytoplasmic microtubule and wall microfibril orientation in root hairs of radish', *J. Cell Biol.*, 27, 575–589.

Nicklas, R. B. and Staehly, C. A. (1967) 'Chromosome micromanipulation. I. The mechanics of chromosome attachment to the spindle', *Chromosoma*, 21, 1–16.

Norstog, K. (1967) 'Fine structure of the spermatozoid of *Zamia* with special reference to the flagellar apparatus', *Am. J. Bot.*, 54, 831–840.

O'Brien, T. P. (1972) 'The cytology of cell wall formation in some eukaryotic cells', *Bot. Rev.*, 38, 87–118.

O'Brien, T. P. and Thimann, K. V. (1966) 'Intracellular fibres in oat coleoptiles and their possible significance in cytoplasmic streaming', *Proc. natn. Acad. Sci. U.S.A.*, 56, 888–894.

Östergren, G., Mole-Bajer, J., and Bajer, A. (1960) 'An interpretation of transport phenomena at mitosis', *Ann. N. Y. Acad. Sci.*, 90, 381–408.

Paolillo, D. J., Kreitner, G. L., and Reighard, J. A. (1968a) 'Spermatogenesis in *Polytrichum juniperinum.* I. The origin of the apical body and the elongation of the nucleus', *Planta,* 78, 226–247.

Paolillo, D. J., Kreitner, G. L., and Reighard, J. A. (1968b) 'Spermatogenesis in *Polytrichum juniperinum.* II. The mature sperm', *Planta,* 78, 248–261.

Perkins, F. O. (1970) 'Formation of centriole and centriole-like structures during mitosis and meiosis in *Labyrinthula.* An electron microscope study', *J. Cell Sci.,* 6, 629–654.

Pickett-Heaps, J. D. (1965) 'The function of cell organelles in growth and differentiation in cells of the wheat seedling. Thesis, University of Cambridge, England.

Pickett-Heaps, J. D. (1966) 'Incorporation of radioactivity into wheat xylem walls', *Planta,* 71, 1–14.

Pickett-Heaps, J. D. (1967a) 'The effects of colchicine on the ultrastructure of dividing wheat cells, xylem wall differentiation and distribution of cytoplasmic microtubules', *Devl. Biol.,* 15, 206–236.

Pickett-Heaps, J. D. (1967b) 'Ultrastructure and differentiation in *Chara* sp. I. Vegetative cells', *Aust. J. biol. Sci.,* 20, 539–551.

Pickett-Heaps, J. D. (1967c) 'Ultrastructure and differentiation in *Chara* sp. II. Mitosis', *Aust. J. biol. Sci.,* 20, 883–894.

Pickett-Heaps, J. D. (1968a) 'Ultrastructure and differentiation in *Chara* sp. III. Formation of the antheridium', *Aust. J. biol. Sci.,* 21, 255–274.

Pickett-Heaps, J. D. (1968b) 'Ultrastructure and differentiation in *Chara fibrosa.* IV. Spermatogenesis', *Aust. J. biol. Sci.,* 21, 665–690.

Pickett-Heaps, J. D. (1968c) 'Xylem wall deposition; radioautographic investigations using lignin precursors', *Protoplasma,* 65, 181–205.

Pickett-Heaps, J. D. (1969a) 'Preprophase microtubules and stomatal differentiation in *Commelina cyanea',* *Aust. J. biol. Sci.,* 22, 375–391.

Pickett-Heaps, J. D. (1969b) 'Preprophase microtubules and stomatal differentiation; some effects of centrifugation on symmetrical and asymmetrical cell division', *J. Ultrastruct. Res.,* 27, 24–44.

Pickett-Heaps, J. D. (1969c) 'Preprophase microtubules bands in some abnormal mitotic cells of wheat', *J. Cell Sci.,* 4, 397–420.

Pickett-Heaps, J. D. (1969d) 'The evolution of the mitotic apparatus: an attempt at comparative ultrastructural cytology in dividing plant cells', *Cytobios,* 3, 257–280.

Pickett-Heaps, J. D. (1970a) 'Some ultrastructural features of *Volvox* with particular reference to the phenomenon of inversion', *Planta,* 90, 174–190.

Pickett-Heaps, J. D. (1970b) 'Mitosis and autospore formation in the green alga *Kirchneriella lunaris',* *Protoplasma,* 70, 325–347.

Pickett-Heaps, J. D. (1970c) 'The behavior of the nucleolus during mitosis in plants', *Cytobios,* 6, 69–78.

Pickett-Heaps, J. D. (1971a) 'The autonomy of the centriole: fact or fallacy?', *Cytobios* (in press).

Pickett-Heaps, J. D. (1971b) 'Reproduction by zoospores in *Oedogonium* I. Zoosporogenesis', *Protoplasma,* 72, 275–314.

Pickett-Heaps, J. D. (1972a) 'Reproduction by zoospores in *Oedogonium* II. Emergence of the zoospore and the motile phase', *Protoplasma,* 74, 149–167.

Pickett-Heaps, J. D. (1972b) 'Reproduction by zoospores in *Oedogonium* III. Differentiation of, and mitosis in the germling', *Protoplasma,* 74, 169–193.

Pickett-Heaps, J. D. and Fowke, L. C. (1969) 'Cell division in *Oedogonium* I.', *Aust. J. biol. Sci.,* 22, 857–894.

Pickett-Heaps, J. D. and Fowke, L. C. (1970a) 'Cell division in *Oedogonium* II. Nuclear division in *O. cardiacum',* *Aust. J. biol. Sci.,* 23, 71–92.

Pickett-Heaps, J. D. and Fowke, L. C. (1970b) 'Cell division in *Oedogonium* III. Golgi bodies, wall structure and wall formation in *O. cardiacum',* *Aust. J. biol. Sci.,* 23, 93–113.

Pickett-Heaps, J. D. and Fowke, L. C. (1970c) 'Mitosis, cytokinesis and cell elongation in the desmid *Closterium littorale',* *J. Phycol.,* 6, 189–215.

Pickett-Heaps, J. D. and Fowke, L. C. (1971) 'Conjugation in the desmid *Closterium littorale',* *J. Phycol.,* 7, 37–50.

Pickett-Heaps, J. D. and Northcote, D. H. (1966a) 'Organization of microtubules and endoplasmic reticulum during mitosis and cytokinesis in wheat meristems', *J. Cell Sci.,* 1, 109–120.

Pickett-Heaps, J. D. and Northcote, D. H. (1966b) 'Cell division in the formation of the stomatal complex of the young leaves of wheat', *J. Cell Sci.*, **1**, 121–128.

Pickett-Heaps, J. D. and Northcote, D. H. (1966c) 'Relationship of cellular organelles to the formation and development of the plant cell wall', *J. exp. Bot.*, **17**, 20–26.

Porter, K. R. (1966) 'Cytoplasmic microtubules and their functions', In: *Principles of biomolecular organization* (eds. G. E. W. Wolstenholme and M. O'Connor), pp. 308–345, Ciba Found. Symp., J. & A. Churchill, London.

Porter, K. R. and Machado, R. D. (1960) 'Studies on the endoplasmic reticulum IV. Its form and distribution during mitosis in cells of onion root tip', *J. biophys. biochem., Cytol.*, **7**, 167–180.

Raudaskoski (1970) 'Occurrence of microtubules and microfilaments, and origin of the septa in dikaryotic hyphae of *Schizophyllum commune*', *Protoplasma*, **70**, 415–422.

Rebhun, L. l. and Sander, G. (1967) 'Ultrastructure and birefringence of the isolated mitotic apparatus of marine eggs', *J. Cell Biol.*, **34**, 859–883.

Reichle, R. E. (1969) 'Fine structure of *Phytophthora parasitica* zoospores', *Mycologia*, **61**, 30–51.

Renaud, F. L., Rowe, A. J., and Gibbons, I. R. (1968) 'Some properties of the protein forming the outer fibres of cilia', *J. Cell Biol.*, **36**, 79–90.

Rice, H. V. and Laetsch, W. M. (1967) 'Observations on the morphology and physiology of *Marsilea* sperm', *Am. J. Bot.*, **54**, 856–866.

Ringo, D. L. (1967) 'Flagellar motion and fine structure of the flagellar apparatus in *Chlamydomonas*', *J. Cell Biol.*, **33**, 543–571.

Robards, A. W. (1968) 'On the ultrastructure of differentiating secondary xylem in willow', *Protoplasma*, **65**, 449–464.

Robbins, E. and Shelanski, M. (1969) 'Synthesis of a colchicine-binding protein during the HeLa cell life cycle', *J. Cell Biol.*, **43**, 371–373.

Roberts, L. W. (1969) 'The initiation of xylem differentiation', *Bot. Rev.*, **35**, 201–250.

Robinow, C. F. and Caten, C. E. (1969) 'Mitosis in *Aspergillus nidulans*', *J. Cell Sci.*, **5**, 403–431.

Robinow, C. F. and Marak, J. (1966) 'A fibre apparatus in the nucleus of the yeast cell', *J. Cell Biol.*, **29**, 129–151.

Rosenbaum, J. L., Moulder, J. E., and Ringo, D. L. (1969) 'Flagellar elongation and shortening in *Chlamydomonas*', *J. Cell Biol.*, **41**, 600–619.

Rosen, W. G. (1968) 'Ultrastructure and physiology of pollen'. *A. Rev. Pl. Physiol.*, **19** 435–462.

Roth, L. E., Pihlaja, D. J., and Shigenaka, Y. (1970) 'Microtubules in the Heliozoan axopodium' *J. Ultrastruct. Res.*, **30**, 7–37.

Sabatini, D. D., Bensch, K. G., and Barrnett, R. J. (1963) 'Cytochemistry and electron microscopy', *J. Cell Biol.*, **17**, 19–34.

Sabnis, D. D. and Jacobs, W. P. (1967) 'Cytoplasmic streaming and microtubules in the coenocytic marine alga, *Caulerpa prolifera*', *J. Cell Sci.*, **2**, 465–472.

Sato, S. (1958) 'Electron microscope studies on the mitotic figure. I. Fine structure of the metaphase spindle', *Cytologia*, **23**, 383–394.

Sato, S. (1959) 'Electron microscope studies on the mitotic figure. II. Phragmoplast and cell plate', *Cytologia*, **24**, 98–106.

Schroeder, T. E. (1970) 'The contractile ring. I. Fine structure of dividing mammalian (HeLa) cells and the effect of cytochalasin B', *Z. Zellforsch. mikrosk. Anat.*, **109**, 431–449.

Schuster, F. (1964) 'Electron microscope observations on spore formation in the true slime mold *Didymium nigripes*', *J. Protozool.*, **11**, 207–216.

Selman, G. G. and Perry, M. M. (1970) 'Ultrastructural changes in the surface layers of the newt's egg in relation to the mechanism of its cleavage', *J. Cell Sci.*, **6**, 207–227.

Shelanski, M. L. and Taylor, E. W. (1967) 'Isolation of a protein subunit from microtubules', *J. Cell Biol.*, **34**, 549–554.

Shelanski, M. L. and Taylor, E. W. (1968) 'Properties of the protein subunit of central-pair and outer-doublet of sea urchin flagella', *J. Cell Biol.*, **38**, 304–315.

Stebbins, G. L. and Shah, S. S. (1960) 'Developmental studies of cell differentiation in the epidermis of monocotyledons. II. Cytological features of stomatal development in the Gramineae', *Devl. Biol.*, **2**, 477–500.

Stephens, R. E. (1970) 'On the apparent homology of actin and tubulin', *Science, N. Y.*, **168**, 845–847.

Tanaka, K. (1970) 'Mitosis in the fungus *Basidiobolus ranarum* as revealed by electron microscopy', *Protoplasma,* **70,** 423–440.

Tilney, L. G. and Gibbins, J. R. (1968) 'Differential effects of antimitotic agents on the stability and behaviour of cytoplasmic and ciliary microtubules', *Protoplasma,* **65,** 167–179.

Tilney, L. G. and Gibbons, J. R. (1969) 'Microtubules in the formation and development of the primary mesenchyme in *Arbacia punctulata'*, *J. Cell Biol.,* **41,** 227–250.

Tilney, L. G., Hiramoto, Y., and Marsland, D. (1966) 'Studies on the microtubules in *Heliozoa.* III', *J. Cell Biol.,* **29,** 77–95.

Tilney, L. G. and Porter, K. R. (1965) 'Studies on microtubules in *Heliozoa.* I', *Protoplasma,* **60,** 317–344.

Tilney, L. G. and Porter, K. R. (1967) 'Studies on the microtubules in *Heliozoa* II', *J. Cell Biol.,* **34,** 327–343.

Tucker, J. B. (1968) 'Fine structure and function of the cytopharyngeal basket in the ciliate *Nassula'*, *J. Cell Sci.,* **3,** 493–514.

Turner, F. R. (1968) 'An ultrastructural study of plant spermatogenesis; Spermatogenesis in *Nitella'*, *J. Cell Biol.,* **37,** 370–393.

Turner, F. R. (1970) 'The effects of colchicine on spermatogenesis in *Nitella'*, *J. Cell Biol.,* **46,** 220–234.

Waris, H. (1950) 'Cytophysiological studies on *Micrasterias* I. Nuclear and cell division', *Physiologia Pl.,* **3,** 1–15.

Wells, K. (1970) 'Light and electron microscopic studies of *Ascobolus stercorarius.* I. Nuclear division in the ascus', *Mycologia,* **62,** 761–790.

Wessells, N. K., Spooner, B. S., Ash, J. F., Bradley, M. O., Luduena, M. A., Taylor, E. L., Wrenn, J. T., and Yamada, K. M. (1971) 'Microfilaments in cellular and developmental processes', *Science, N.Y.,* **171,** 135–143.

Westergaard, M. and Von Wettstein, D. (1970) 'The nucleolar cycle in an Ascomycete', *C.r. Trav. Lab. Carlsberg,* **37,** 195–237.

Wilson, E. B. (1898) *The cell in development and heredity,* Macmillan Co., New York.

Wilson, H. J. (1969) 'Arms and bridges on microtubules in the mitotic apparatus', *J. Cell Biol.,* **40,** 854–859.

Wilson, H. J. (1970) 'Endoplasmic reticulum and microtubule formation in dividing cells – a postulate', *Planta,* **94,** 184–190.

Wohlman, A. and Allen, R. D. (1968) 'Structural organization associated with pseudopod extension and contraction during cell locomotion in *Difflugia'*, *J. Cell Sci.,* **3,** 105–114.

Yeoman, M. M., Tulett, A. J., and Bagshaw, V. (1970) 'Nuclear extensions in dividing vacuolated plant cells', *Nature,* **226,** 557–558.

Zickler, D. (1970) 'Division spindles and centrosomal plaques during mitosis and meiosis in some Ascomycetes', *Chromosoma,* **30,** 287–304.

7

Plant cell walls*
R. D. Preston

Introduction

There appears to be no obvious exception to the rule that all plant cell walls contain material which is crystalline, both in the sense of showing optical anisotropy and of yielding an X-ray diagram; and that this crystalline material is embedded in substances which are not crystalline, at least in the X-ray sense. The crystallites, as they must be called since they are submicroscopic in size, are often aligned more or less parallel to each other and frequently in specific preferred directions relative to the cell axis.

The structures which have thus been developed are important for several reasons. As a solid, or semi-solid, envelope surrounding almost every plant cell, and separating it from (or joining it to) its neighbours or the environment, the cell wall forms one of the diagnostic features of plants preventing, if no other feature in the biochemistry of plants were not to do so, any rapid change in cell shape and the development of muscle. It imposes upon plants at most a slow change in cell dimensions at rates and of a kind clearly dictated in part by the mechanical properties of the wall. This is often a reciprocal relation since changes in cell dimensions inevitably lead to changes in wall structure. Whether the interaction between cell shape, rate of growth, cell wall structure and cell wall mechanics is somehow influenced by the biochemical processes inside the cell is no longer in doubt; the wall is certainly not inaccessible to the internal machinery of the cell. Yet the details of the interaction are far from being well known.

The parts of the cell wall are put together in the region of the plasmalemma. Even though the substrates and the enzymes may come from elsewhere (and even though it is barely possible that in some few species some of the assembly is done elsewhere) it is only at the plasmalemma that the finished wall comes into being. It is therefore at the plasmalemma that the instruction is given and obeyed at least for the construction of the crystallites and their orientation. It therefore always seemed likely that wall structure might be relevant for plasmalemma structure and this is now known to have been the case. Both for an

*Since the completion of this article, Professor Preston has published a comprehensive account of the subject in *Physical Biology of Plant Cell Walls*, Chapman & Hall, 1974.

understanding of cell and plant growth, and for an appreciation of plasmalemma structure and function, it is vital to know all that can be known of wall structure. It turns out, too, that wall structure has taxonomic implications.

Nowadays, of course, matters of practical industrial importance cannot be neglected either, and here the cell wall scores heavily. In the form of timber, plant cell walls have been exploited as building materials since prehistoric times, and as pulp for paper since before the invention of printing, and the exploitation is accelerating in spite of even the most modern synthetic competitors. Here, knowledge of structure has long taken the industry out of its homely craft days. As cotton hairs, as linen, ramie, hemp, jute, and sisal fibres, plant cell walls equally find a multiplicity of uses. Materials extracted from cell walls are used as pastes and glues and as food. Cell walls are dissolved and reprecipitated to form some so-called 'man-made' fibres; and the beautiful structure of wood is even ravished by tearing apart and reconstitution as chipboard and hardboard. Cell walls are clearly also a matter of concern in the ripening of fruit, in the quality of potatoes (fresh, cooked or dehydrated), in the expression of oil from oil seeds, in the invasion of plants by pathogenic organisims, and so forth. For all these purposes, complete structural determinations and full understanding of all features of wall organization are required.

In the following pages attention is confined to green plants. The general structure of cellulose – the crystallin polysaccharide of most green plants – will first be examined, and followed by a consideration of some plants in which cellulose is replaced by a different polysaccharide. Following this the differentiation between primary and secondary wall layers in higher plants (and their analogues in some lower plants) will be examined, and the question of biosynthesis will be opened. Finally, some examination will be made of the relationships between primary wall structure, cell growth, and metabolism.

The organization of cellulosic cell walls

The walls of most green plants consist of cellulose in association with other polysaccharides and polysaccharide derivatives mostly, like cellulose, 1,4-linked (Fig. 7.1), and with the constituent sugars in the β-form (the pectic compounds being an exception). Substances such as lignin which become incorporated in the wall – or are synthesized in the wall – at a later stage of development, will not be specifically considered here, but the non-cellulosic polysaccharides will be considered in the suitable place. The proportion of cellulose on a weight-to-weight basis varies widely, from 1 to 10% in the primary walls of higher plant cells and in walls of some red algae, through about 50% in the thick secondary walls of some higher plant cells, to 80% or more in some green algae. This point will be taken up later, however, in connection with chemical differentiation.

Cellulose may be defined in either of two ways, and this poses a problem which is more than a problem in semantics, since the two definitions refer to substances which are often chemically different. The cellulose of the laboratory is the residue of the wall remaining when all other soluble polysaccharides have been removed by treatments which just fail to affect the integrity of the cellulose. On hydrolysis, this residue yields sugars other than glucose to the extent of some 15% with cell walls of the xylem of some woody species, and up to 50% in some algae (Cronshaw, Myers, and Preston, 1959). The material in the wall which gives rise to the X-ray diagram is also called cellulose but this material is solely a polyglucan. With algae such as *Cladophora* and *Chaetomorpha* the two definitions coincide (Cronshaw, Myers, and Preston, 1958; Frei and Preston, 1961) since the residue after extraction yields glucose only on hydrolysis. In other cases, when reference is made to the cellulose yielding an X-ray diagram in relation to the cellulose actually in the hand, it must always be recognized that the term cellulose is being used in two senses. It was suggested some time ago that the cellulose of crystallography, consisting entirely of β-glucose residues joined by 1,4 links (Fig. 7.1) (and the celluloses extracted from plants like *Cladophora* and *Chaetomorpha*), should be called 'eucellulose' to differentiate this from the compounds extracted from most other plants' 'cellulose', which contains a family of polysaccharides (Myers and Preston, 1959). This distinction has not been widely accepted, and the general term 'cellulose' will be retained here, with closer definition as occasion demands. The molecular chains of cellulose are very long, comprising up to 15 000 glucose residues (7.7 μm).

Fig. 7.1   Cellobiose residue in a cellulose chain. ● = carbon, ○ = oxygen, positions of H shown by —

All celluloses yield X-ray and electron diffraction diagrams with arcs corresponding to almost the same spacings. This is exemplified here by an electron diffraction diagram (Fig. 7.2) of the highly oriented cellulose of a lamella of the wall of *Chaetomorpha*. From such diagrams the 'unit cell' — the unit volume containing all the features of structure — has been calculated and the sections of the appropriate chains of cellulose inserted in appropriate ways. The unit cell proposed by Meyer and Misch (1936) is represented in Fig. 7.3, corresponding to a packing together of cellulose chains in parallel array, regularly spaced at the distances given. Water cannot enter this unit cell, so that the X-ray diagram is unaffected by the humidity of the specimen.

Treatment with caustic alkali solutions induces a swelling, during which water enters and destroys the lattice. At solutions of sufficiently high concentrations (approximately 17% with higher plants) then, on washing out the alkali, the separated chains of cellulose come together in a different lattice, again monoclinic. This regenerated cellulose, also produced when cellulose is dissolved and reprecipitated, is called cellulose II to distinguish it from native cellulose, cellulose I. When cellulose II is heated at 300°C the lattice becomes orthorhombic and the cellulose is called cellulose IV. The X-ray diagram is similar to, but distinguishable from, that of cellulose I but is closely like that of polyglucuronic acid; this is relevant to an older statement that the cellulose of coltsfoot (*Tussilago farfara*) (mistakenly translated as 'hoof' cellulose in Meyer (1942)) is in the form of cellulose IV. It has long been recognized that, while the structure of cellulose I is acceptable in outline, the details are certainly wrong. In particular, the evidence that the central chain in the unit cell (and therefore half the chains) is 'upside down' is weak, and the reasons for a displacement of this chain

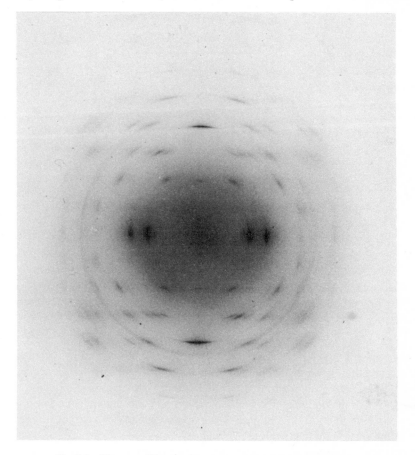

Fig. 7.2   Electron diffraction diagram of *Chaetomorpha* cellulose.

by $b/4$ relative to the corner chains are not strong either. It has been suspected that the unit cell may be four times as large as that depicted in Fig. 7.3 and this has been confirmed by Nieduszynski and Atkins (1970) who have presented eight possible eight-chain cells (Fig. 7.4). The X-ray diagram is not sufficiently rich to allow at present a final choice of one cell. The unit cell could indeed be only a 'statistical' unit cell. To some extent the details of the unit cell are, however, irrelevant; it is sufficient for present purposes that the crystallite of cellulose consists of parallel arrays of long chains of a β-1,4-linked glucan, regularly spaced apart by known distances.

X-ray and electron-diffraction diagrams have it in common that arcs lying along the equator (parallel to the shorter edge of the page in Fig. 7.2) are very much broader radially than are those lying on the meridian. This means that the distances over which the inter-chain spacing is constant are much greater parallel to

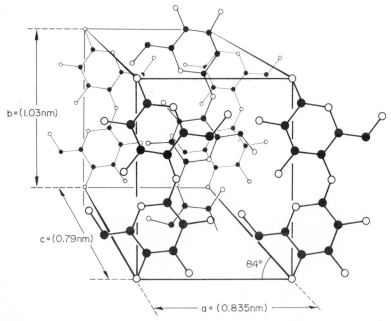

Fig. 7.3 Modified Meyer and Misch unit cell of cellulose. The modification arises from the hydrogen bonding between —OH on C4 with the ring oxygen of the next glucose residues in the chain. ● = carbon. ○ = oxygen, hydrogen omitted.

the chains than perpendicular to them. The effective crystallites are therefore long and thin, recalling the 'micelles' deduced from polarizing microscopy by von Nägeli. Hengstenberg and Mark (1928) concluded from this X-ray line broadening that in ramie fibres the crystallites are some 5.0 nm wide and at least 60 nm long, and sizes of this order in higher plant celluloses have been confirmed by low angle X-ray scattering (Heyn, 1958).

It was not altogether a surprise, therefore, that the first electron micrographs of untreated plant cell walls (Preston, Nicolai, Reed, and Millard, 1948; Frey-Wyssling, Mühlethaler, and Wyckoff, 1948) presented cellulose in the form of long thin threads immediately called microfibrils (Fig. 7.5). At that time, and this has been confirmed many times since then, the microfibrils were presented in each of two sizes. In higher plants (and now in many lower plants) the microfibrils in metal-shadowed material were about 10.0 nm wide and in some lower plants (*Valonia*, now also *Cladophora, Chaetomorpha, Glaucocystis* and others) about 20–30 nm wide, and about one-half as thick (Preston, 1951). These were overestimates on account of the added thickness of the metal deposits, but a reasonable estimate of each size might well be 8.0 nm and 18–28 nm.

The size of microfibrils is of importance for a number of reasons, particularly in questions of biosynthesis, which will be taken up later. More recent estimates, by different methods, have presented figures of a much lower order. It is vital in considering the situation thus presented to keep the two sizes of microfibrils originally proposed quite separate; confusion between the two sizes has undoubtedly led to serious error.

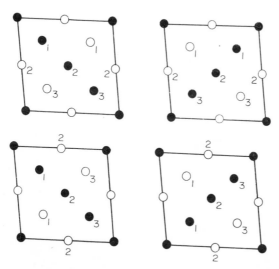

Fig. 7.4 The eight, eight-chain unit cells of cellulose, after Nieduszynski and Atkins (1970), viewed along the *b* axis. ● and ○ represent chains in antiparallel arrangement. The figures represent multiples of *b*/4 by which the chain is displaced *upwards*, along the chain direction, relative to the corner chains. There are four similar unit cells with *downward* displacements.

When cellulose microfibrils of higher plants are examined in the electron microscope after negative staining, the clear area which is unstained – representing the part into which the stain has not penetrated – is 3.5–4.0 nm wide. This has led to the idea that microfibrils consist of 'elementary' fibrils – a term originally devised by Frey-Wyssling – measuring some 3.5 nm x 3.5 nm in section. The evidence favouring this view has been persuasively summarized by Mühlethaler (1967). Deductions made from single negatively stained microfibrils have been criticized by Colvin (1963) on two grounds: firstly, that negative staining in any case is bound to give a value less than the true value unless precisely the correct amount of stain is present and, secondly, because the stain is likely to penetrate the outer paracrystalline layers of a microfibril and reveal only the inner crystalline core which is about the size required. The contention of Mühlethaler (1967) that a positive staining procedure (with chromium salts) supports the 3.5 nm concept, because the microfibril still shows the dimensions revealed by negative staining, falls short of being convincing since this staining too is likely to give a low value; for the intensity of stain falls off toward the edges of a microfibril and it becomes a matter of opinion at what point the microfibril should be taken to terminate.

In an attempt to avoid this type of criticism, measurements have been made of the centre to centre distance between the dense areas revealed when sections of walls either 'parallel' or 'perpendicular' to microfibril length are negatively stained. With higher plant celluloses this gives again the value 3.5 nm. Unfortunately, the investigators have overlooked a simple geometric feature discussed at length elsewhere (Preston, 1971). A section will be at least 50 nm thick, so that if the microfibrils are 10 nm thick there will be at least five of them in the thickness of the section. These will not be in exact register; let us assume they are hexagonally packed (Fig. 7.6A). Then in the electron microscope all five layers are focused in one plane. The distance between the dark lines in negative staining will be either 5.0 nm (Fig. 7.6A) or 3.3 nm (Fig. 7.6B) depending upon

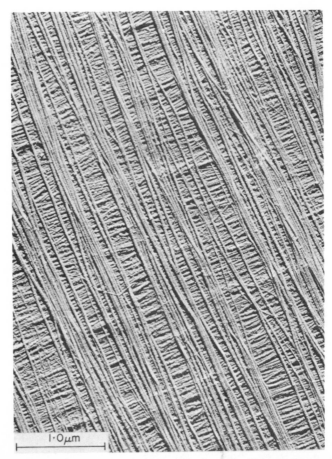

Fig. 7.5   Electron micrograph of two wall lamellae from the side wall of a cell of *Chaetomorpha melagonium* (Frei and Preston 1961). Shadowed Pd-Au; magnification ×30 000.

the accuracy with which the section is cut. Since the microfibrils are in fact probably packed more nearly at random, the minimal distance in practice will be less than this and this may well account for the reports by other workers

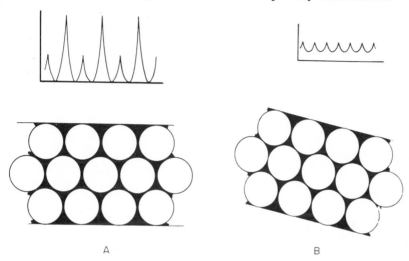

Fig. 7.6   Diagrammatic representation of the end view of microfibrils (white circles) with the interstices filled by an absorbing substance (negatively stained) and, above, the absorption distribution when viewed parallel to the long edge of the page.

(A) Plane of section normal to line of sight.

(B) Plane of section tilted 12.5 degrees from line of sight.

Note that in (A) the separation between absorption maxima is one half, and in (B) one third, of microfibril diameter.

(e.g., Kramer, 1970) of elementary fibrils even narrower than 3.5 nm. Figures 7.7A and 7.7B present an optical simulation demonstrating the reality of this effect. A similar criticism can be made of sections taken 'normal' to microfibril length (Preston, 1971). If the 10 nm measurement in higher plants (shadow-cast) is an overestimate and 8.0 nm is a better figure, then the expected values from negative staining are 4.0 nm and 2.7 nm. *The fact that under these conditions a value of 3.5 nm is observed proves that the microfibril is about 10 nm in diameter.* The elementary fibril therefore disappears and the term should be dropped. The close agreement with higher plants between the value of 3.5 nm and the crystallite diameter determined by low-angle scattering (Heyn, 1966; Stirling, 1957) means that, at best, the elementary fibril corresponds to the crystalline core. For higher plants, therefore, the model of structure presented by the present writer (Preston, 1962; Preston and Cronshaw, 1958; Dennis and Preston, 1961) (Fig. 7.8) still stands.

A natural extension of the elementary fibril hypothesis has been that the approximately 20 nm microfibrils of the algae mentioned above are fasciations of many 3.5 nm fibrils. Even superficially this concept does not make sense. As pointed out just twenty years ago (Preston, 1951), the radial breadth of the lateral arcs in the X-ray diagram of cellulose is much greater with higher plants

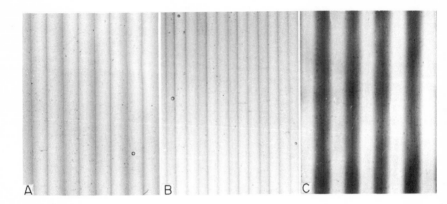

Fig. 7.7   (A) Optical simulation of Fig. 7.6(A); (B) optical simulation of Fig. 7.6(B), using perspex rods 1 cm diameter, whose full diameter is revealed when only a single layer of rods is imaged (C).

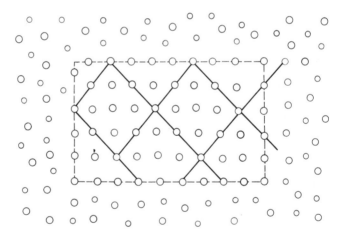

Fig. 7.8   Diagrammatic representation of a cellulose microfibril in cross-section, about 10 nm wide and 5 nm thick. Large circles represent the glucan chains, small circles polymers of other sugars. Dotted lines enclose the central crystalline core, solid lines outline the eight-chain unit cell. Note that the core accommodates only two whole unit cells.

than with the algae concerned. This means that the effective width of the crystallite must be greater with 20 nm microfibrils than with 10 nm microfibrils. A recent examination by low-angle X-ray scattering and high-angle-line broadening of both higher plant and *Chaetomorpha* celluloses has shown that the crystallite widths are about 5.0 nm and about 17 nm respectively; and these were determined under conditions guaranteed to give a minimal figure (Nieduszynski and Preston, 1970). Here again, gross underestimates are yielded by negative staining (Fig. 7.9). The conclusion is inescapable that the model presented in Fig. 7.8 stands also for these larger microfibrils.

The demonstration by Keller (1959) that in polyethylene crystals the chains are folded, and the similar finding of Bitiger and Husemann (1964) for precipitated cellulose derivatives, has led to the suggestion that, in natural cellulose microfibrils also, the chains are folded (Manley, 1964; Dolmetsch and Dolmetsch, 1962; Marx-Figini and Schultz, 1966). As will be discussed later, however, it is known that cellulose formation in plant cell walls is not comparable with precipitation; and the concept that folding would explain the regular antiparallel chains of the Meyer-Misch unit cell must probably now be abandoned

Fig. 7.9 Electron micrograph of a wall lamella of *Glaucocystis,* negatively stained. Note that the single overlying microfibril appears to be 2—3 times as wide as the underlying microfibrils which are closely packed both side by side and one over another. Magnification ×171 000.

since this unit cell is no longer acceptable (Nieduszynski and Atkins, 1970). More recently, through an ingenious series of determinations of the chain length in a series of thin wall sections, Müggli (1968), and Müggli, Elias, and Mühlethaler (1969) have finally obtained evidence which denies the folded chain concept. Nevertheless, Manley (1971) has reiterated, in a modified form, his concept that cellulose microfibrils consist of units 3.5 nm x 2.0 nm in section — which he unfortunately now calls protofibrils — within which the chains are regularly folded. His evidence for the size of the protofibrils is electron micro-scopic and is subject to the correction for overlapping of images discussed above; his evidence for folding in natural cellulose is indirect and weak.

Non-cellulosic cell walls

It was recognized a long time ago that, with some green plants, the presence of cellulose in the cell wall was questionable. The plants concerned are all algae. Oddly enough, the earliest workers considered the walls to be cellulosic (von Nägeli, 1844, *Caulerpa*; Leitgeb, 1888, *Acetabularia*; Ernst, 1903, *Codium*). Correns (1894) appears to have been the first to question the presence of cellulose in *Caulerpa,* but it was only as late as 1913 that Mirande confirmed that cellulose is absent from the walls of *Caulerpa, Acetabularia,* and *Codium.* The confusing situation which developed is exemplified by the fact that the trabeculae of *Caulerpa* are still described by von Nägeli's term '*Zellulosebalken'* in one of the more recent textbooks of botany (Strasburger, 1962). Final proof of the absence of cellulose, and identification of the skeletal polysaccharide 'replacing' cellulose, came with the work of Frei and Preston (1961, 1964, 1968).

The algae concerned almost all fall in the old order Siphonales (recently reconstituted by Christensen, 1962). They may be divided into three groups. The families Bryopsidaceae, Caulerpaceae, Udotaceae, and Dichotomosiphonaceae contain β-1,3-linked xylan (Mackie and Percival, 1959; Iriki and Miwa, 1960) as the only structural polysaccharide (Frei and Preston, 1961, 1964a). The families Codiaceae and Dasycladaceae (including the well-known *Acetabularia*), on the other hand, contain β-1,4-linked mannan (Iriki and Miwa, 1960; Miwa, Iriki, and Susuki, 1961; Love and Percival, 1964), again as the only structural polysaccharide (Frei and Preston, 1961, 1968). The third group consists of the Bangiales, an order in the red algae, in which the cell wall proper contains β-1,3-linked xylan and the cuticle β-1,4-linked mannan (Frei and Preston, 1964b). This leaves on one side the special case of *Halicystis.* This is the gametophytic phase of the sporophyte *Derbesia* but the walls of the two parts of the alterna-tion are quite different. *Halicystis* walls contain both crystalline β-1,3-linked xylan and crystalline cellulose (though probably in the form of cellulose II) while the crystalline polysaccharide of *Derbesia* is β-1,4-linked mannan.

The organization of the walls of the xylan seaweeds has been examined in detail by Frei and Preston (1964a). They are highly crystalline when wet (Fig. 7.10). When conditioned to atmospheres of R.H. below 65%, however, they yield an X-ray diagram in which the arcs are fewer and diffuse. Unlike cellulose, therefore, water must form an essential component of the crystallite, to the extent perhaps of 30% (the water content of the wall at R.H. 65%, a point at which the wall becomes fully crystalline). Boiling with hot water followed by chlorination removes from the wall polysaccharides which, on hydrolysis, yield in the main glucose with small amounts of galactose and detectable amounts of mannose, arabinose and uronic acid, to a total weight equivalent to about 40% of the initial dry weight of the wall. The residue, which may be called holoxylan by analogy with holocellulose, yields on hydrolysis xylose and glucose in the ratio about 4:1. It is completely soluble in alkali, again unlike cellulose, in concentration ranging from 10% to 18% depending upon species. The residue which remains after treatment of a wall with alkali of a strength which just fails completely to dissolve it is still microfibrillar, gives the X-ray diagram of the whole wall, and hydrolyses to give xylan with only a trace of glucose. There is therefore no doubt that the diagram of Fig. 7.10 is to be referred to a xylan, not a gluco-xylan.

Fig. 7.10   X-ray diagram of the β-1,3-linked xylan in inner wall layers of *Penicillus dumetosus,* wall untreated but conditioned to high relative humidity.

The xylan occurs as microfibrils (Fig. 7.11) within which the chains occur as triple helices (probably right-handed) (Atkins, Parker, and Preston, 1969) stabilized by a series of triads of hydrogen bonds holding the helices together (Fig. 7.12). The rather flat helical coiling of the constituent xylan chains makes the microfibrils, and the walls observed in section, negatively birefringent (i.e., with the direction of light vibration corresponding to the greater refractive index at right angles to the length) in contrast to the positive birefringence of cellulose. Among wall components of other plants, both chitin and pectin are also negatively birefringent, but with them the sign of birefringence is related to the presence of side groups on chains which lie straight and parallel.

The degree of polymerization of the xylan of *Penicillus dumetosus* has recently been examined in terms of the nitrate derivative prepared under mild conditions and at low temperatures (Mackie, 1969; Mackie and Sellen, 1971). Xylan chains are present in the wall with a degree of polymerization in excess of 10 000, many times greater than that given by earlier workers (Iriki, Susuki, Nisizawa, and Miwa, 1960).

Fig. 7.11   Electron micrograph of xylan microfibrils in the inner wall layers of *P. dumetosus;* the micrograph is in correct register with respect to Fig. 7.10. Shadowed Pd-Au; Magnification ×34 000.

Apart from these differences, the architecture of the walls of the filaments resembles that of higher plants. There is invariably an outer layer − analogous to the primary wall of higher plants − in which the microfibrils tend toward trans-verse orientation. Within this, and deposited upon it secondarily, is a layer − analogous to a secondary wall − in which the microfibrils lie almost longitudinally. The same principles of structure therefore apply here as in higher plants in spite of the different chemistry.

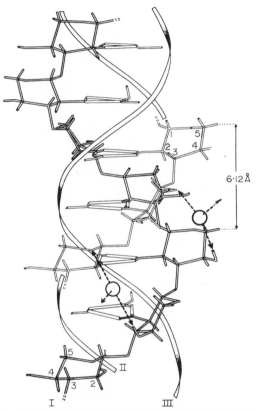

Fig. 7.12 The right-handed triple helix of β-1,3-linked xylan. Chain I fully represented, chain II partly schematic, chain III schematic. Each helix has six xylose residues per turn in a pitch of 18.36Å (1.836 nm). Stoichiometric water molecules and their hydrogen bonds are represented by circles and dotted lines. Cyclic triads of hydrogen bonds uniting I, II and III axially are shown as thicker rods.

The mannan seaweeds are even simpler in the chemical sense (Frei and Preston, 1961, 1968). Hydrolysates of whole walls contain little saccharide material other than mannose. Hot water extracts contain variable amounts of mannose together with small or very small proportions of other sugars, mainly galactose and glucose, to the total extent of 10%. Treatment with alkali achieves further physical separation into a soluble part and a residue, but no chemical separation; all fractions yield on hydrolysis mannose with a trace of glucose.

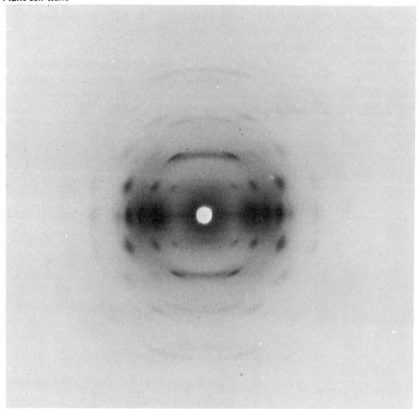

Fig. 7.13 X-ray diagram of the β-1,4-linked mannan in the (untreated) wall of *Batophora*.

Love and Percival (1964) have tentatively suggested that the glucose residues are linked to the mannan. These walls appear to be unique in the simplicity of their chemical constitution. The physical state of the mannan is, however, demonstrably not uniform even within a single plant (Frei and Preston, 1968). The residue remaining when any of these walls is treated with alkali retains intact the original membranous structure; when placed in water, the inner wall layers of some filaments (*Dasycladus, Batophora,* and *Cymopolia*) swell and dissolve while outer layers (and the whole wall in *Codium* and *Acetabularia*) remain intact. This may be associated with differences in molecular chain length. The degree of polymerization of the mannan of *Codium fragile* has been examined by Mackie and Sellen (1969) by converting under mild conditions to mannan trinitrate and using light scattering, osomometry, gel permeation chromatography, and viscosity techniques. Degrees of polymerization range from about 20 to above 10 000, 95% of the mannan lying between 100 and 2500 and with a peak at about 600. Most of this mannan has therefore degrees of polymerization much higher than that reported for higher plant mannan (Meier, 1958; Wolfrom, Laver, and Paten, 1961; Iriki and Miwa, 1960; Timell, 1957).

These mannan walls are highly crystalline, and the crystallites of inner wall layers are well oriented (Fig. 7.13). The unit cell is orthorhombic and the relative disposition of the chains within it differs somewhat from that of cellulose on account of the different hydrogen bonding (Frei and Preston, 1968). Nevertheless, electron microscopic examination of the whole wall, or of any residue remaining after the extraction processes used with cellulose walls, fails to reveal the presence of microfibrils, even though the extraction procedures do not reduce the crystallinity. Instead, the material appears granular. A milder extraction process, based on successive treatments by a detergent and urea solutions, however, gave some preparations of *Codium* and of *Acetabularia* cell walls which were undoubtedly microfibrillar (Mackie and Preston, 1968). It is not yet certain that all the crystalline oriented mannan is microfibrillar, and the current evidence is that it is not. Nevertheless, the mannan algae are in principle no exception to the rule that plant cell walls are microfibrillar in construction. The microfibrils are clearly sensitive to treatment which leaves cellulose microfibrils entirely unchanged.

Primary/secondary wall differentiation

The walls of meristematic cells of higher plants are thin and relatively free of sculpturing, and subsequent differentiation involves an increase in cell dimensions, usually anisotropic. With most cells of higher plants, and indeed of all plants higher than the Thallophyta, the wall remains thin during this increase in cell volume and remains relatively undifferentiated. Such walls are referred to as primary. With some cell types – such as collenchyma cells, and in some algae – the walls of growing cells become thick and increase in thickness during growth. These can hardly be called primary walls in the strict sense but will nevertheless be dealt with also in this section as functionally similar to the primary wall as formally defined.

At about the time when cell extension ceases, a second wall layer begins to be deposited upon the older one through wall synthesis which may continue for days before the cell becomes moribund or mature. This often involves a very considerable increase in wall thickness, in a layer which is called the 'secondary' wall. Secondary deposition does not occur invariably – the cells of some mature parenchyma are without a secondary wall, for example – and thickening of the wall is not always a sign of secondary thickening, as in some algae.

The primary walls of higher plant cells differ from the secondary walls later deposited upon them functionally, physically and chemically. These two classes of wall are therefore treated separately below, calling attention to the situation in the algae where appropriate. The structure of growing cell walls inevitably involves details relevant to the growth processes, but consideration of growth processes will be delayed to the next section. Because of the difficulties of terminology in discussing both high plants and algae, the term 'growing' and 'non-growing' will be used as more non-commital than 'primary' and 'secondary'.

Growing cell walls

Growing cell walls were classically regarded as containing a high proportion of pectic substances (Buston, 1935; Griffioen, 1938, 1939). Allsopp and Misra (1940) working on the cambia of ash, elm, and pine, found pectic compounds to be present in the walls to the extent of 21.6%, 9.91%, and 16.6% respectively on a dry weight basis. Pectin was determined by the decrease in the yield of CO_2 on extraction with 0.5% ammonium oxalate, following the method of Dickson, Otterson, and Link (1930). Allsopp and Misra also found cellulose to be present in amounts corresponding to 20.2%, 20.7% and 25.1% though the presence of up to 25% xylan in the cellulose may mean that these figures need reducing somewhat. There was at that time no satisfactory method of determining hemicellulose but it seems fairly safe to accept a figure of 21.9% mannan in conifer cambium. Similarly, in coleoptiles of *Zea mays* the pectin content of the primary walls was estimated as about 10% (Wirth, 1946; Nakamura and Hess, 1938; Thimann and Bonner, 1933) with a cellulose content of about 40% and hemicellulose 40%. The high content of pectic compounds was very satisfactory at a time when it was regarded as likely that indolyl acetic acid (IAA) exercised its influence on growth, even if indirectly, through these compounds.

More recently the high pectic content has been called into question. Bishop, Bayley, and Setterfield (1958), using refined extraction procedures, have been unable to detect more than 1% pectin in *Avena* coleoptile walls. They estimate the cellulose content as 25% and the hemicellulose content as 38%. All these figures are on a dry weight basis; on a volume basis in the fresh cell wall the proportions are likely to be about 10% cellulose and 14% hemicellulose.

The composition of growing walls of the green algae *Valonia*, *Cladophora*, and *Chaetomorpha* – which are not, strictly speaking, primary walls, since they thicken – is very different (Cronshaw, Myers, and Preston, 1958; Frei and Preston, 1961), being closer to that of the secondary walls of unlignified cells of higher plants. The major difference lies in the proportion of cellulose (40–80% by dry weight); again, however, the pectic content is low.

The physical structure of growing cell walls in elongating cells is rather uniform. The mean orientation of the crystallites, as judged either by polarization microscopy or by X-ray diffraction, is approximately transverse to cell length (Bonner, 1936 (coleoptiles); Preston and Wardrop, 1949a (conifer cambium); Green, 1958; Probine and Preston, 1961 (*Nitella* internodal cells); Frei and Preston, 1964 (some marine seaweeds); Stirling and Spit, 1957 (*Asparagus* fibres); Veen, 1970 (parenchyma cells of *Pisum* stems); Setterfield and Bayley, 1958 (coleoptiles)). In *Nitella* and conifer cambium, the crystallite orientation in fact lies in a slow helix round the cell (Preston and Wardrop, 1949a; Probine and Preston, 1961), as perhaps it does in all primary walls. This presented a puzzle which will be discussed in a later section. Polarization microscopy in particular is very sensitive to orientation, and only the low birefringence observed at that time revealed that the orientation was far from perfect.

The situation was clarified for higher plants by Roelofsen and Houwink (1951, 1953) in a series of electron microscope observations of individual lamellae in the walls of a number of growing cells. They showed that, in innermost wall lamellae neighbouring upon the plasmalemma, the cellulose

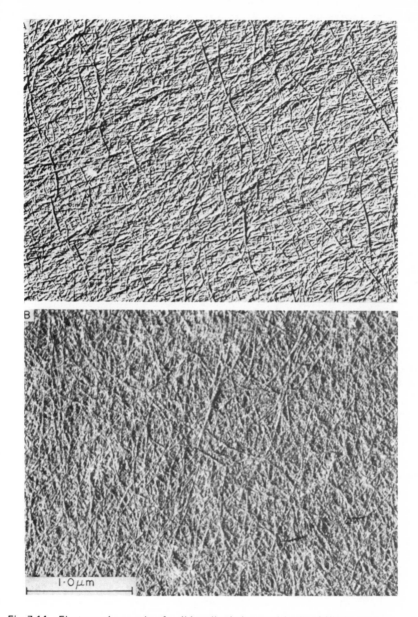

Fig. 7.14 Electron micrographs of wall lamellae in internodal cells of *Nitella opaca,*
shadowed Pd-Au, magnification x27 900. Axis of cell parallel to longer edge of page.

(A) Innermost lamella; most microfibrils lie almost transversely though many lie at other
azimuths.

(B) Outermost coherent lamella; The microfibrils show a tendency toward axial
orientation.

microfibrils, while dispersed over a wide angle, showed a nett preferred orientation lying more or less transversely to the axis of growth. Lamellae further from the plasmalemma showed either greater disorientation or a change in the nett orientation, and in outermost lamellae the microfibrils lay either at random or with a tendency to axial orientation. The observations show that wall lamellae must be passively extended during growth and led Roelofsen to propose the term 'multi-net growth'. This change in orientation through a growing wall has since then been amply confirmed both for higher plants (e.g., Stirling and Spit, 1957; Setterfield and Bayley, 1958; Veen, 1970) and lower plants. The details in lower plants are more clearly visible than they are in higher plants, since the cells used are mostly large, with thick walls and are more easily handled. Both Probine and Preston (1961), and Limaye, Roelofsen, and Spit, (1962) showed for *Nitella* that this change in orientation from inner to outer lamella is marked in elongating (or elongated) internodal cells (Fig. 7.14A,B). A most striking case occurs in *Chaetomorpha* and *Cladophora*, two marine green algae (Frei and Preston, 1961). The lateral walls of the cells of these filamentous plants are finely lamellate. The lamellae contain parallel, densely packed microfibrils, the direction of which in the main alternates through an angle somewhat less than 90 degrees from one lamella to the next (with a third lamella in a third direction present in some species) (Fig. 7.5). Of these two major directions one lies in a slow helix round the cell and the other in a steep helix. In passing from innermost to outermost lamellae the flatter helix becomes steeper, as it might be expected to do, and as expected in terms of the multi-net growth hypothesis. The fact that the steeper helix becomes flatter is a complication which will be taken up later, but is not contradictory to the hypothesis.

The primary walls of higher plant cells are not always so simply constructed as are those upon which Roelofsen based his growth hypothesis. Setterfield and Bayley (1958) have shown conclusively that with *Avena* coleoptiles, onion roots, and young petioles of celery, the walls contain, in addition to the more or less transversely oriented microfibrils, 'ribs' of longitudinally oriented microfibrils. These ribs are of two kinds and neither of them is to be confused with secondary thickenings. In one of these kinds, the rib extends throughout the whole wall (Fig. 7.15), and on the inner face of the wall passes over on the flanks of the rib into the 'transverse' set of microfibrils. These ribs are not associated with the corners at which cells meet. The other kind occurs only at the *outer* side of the wall and appears to increase in amount as the cell elongates; they are sometimes, though not always, associated with cell corners. These recall similar observations made upon epidermal and guard cell walls (Bayley, Colvin, Cooper, and Martin-Smith, 1957), and also the situation in elongating collenchyma cells in which the thickened primary walls at cell corners contain mostly microfibrils in longitudinal array (Anderson, 1927; Majumdar and Preston, 1941). The situation in collenchyma walls may well be an extreme example of a common condition in primary walls. The conclusion reached by Setterfield and Bayley that cellulose must be deposited in the wall at regions far removed from the cytoplasm seems justified. Probine (1965) has reached the same conclusion. Green (1958)

(*Nitella*) and Ray (1967) (*Avena* coleoptiles, pea stems), however, using auto-radiographic methods on cells fed with glucose-³H were unable to detect synthesis except at the wall/cytoplasm interface, though Ray (1967) provides clear evidence of the synthesis of hemicelluloses throughout the wall, enhanced by IAA. Failure to record synthesis of cellulose from glucose-³H throughout the wall cannot, however, be regarded as conclusive since glucose-³H is not a good precursor of cellulose.

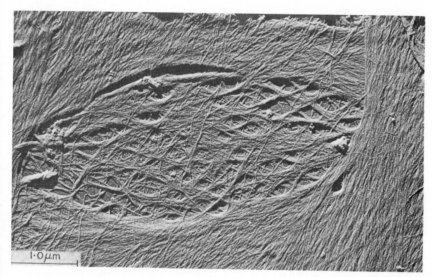

Fig. 7.15   Electron micrograph of primary pit field in the wall of a parenchyma cell of a wheat coleoptile. Shadowed Pd-Au; magnification x21 000, cell axis approximately parallel to longer edge of page. Note the rib of longitudinal microfibrils on a wall in which the bulk of them lie almost transversely. (Micrograph by Adamson and Adamson.)

Disturbances of wall structure in the primary walls of higher plants commonly occur within, and in the neighbourhood of, primary pit fields (Fig. 7.15). The microfibrils tend to run around the pit field, and within it they form a meshwork with numerous large perforations. These undoubtedly represent the locus of plasmodesmata. It is generally considered that, in life, these perforations are lined with plasmalemma confluent with the plasmalemmata of the two bordering cells (since primary pit fields correspond in position when two cells are apposed) with elements of the endoplasmic reticulum passing medianly through (e.g., López-Sáez, Giménez-Martín, and Risueño, 1966). Robards (1968, 1971) has recently given reason to believe that endoplasmic reticulum passes through the plasmodesmata in a form structurally similar to a microtubule. It was at one time believed that the presence of plasmodesmata on (double) cell walls could occur only if the wall was a division wall. It is now known that they can also form spontaneously when two cells grow into contact, as with the tyloses in angiosperm vessels.

In some algae (Probine and Preston, 1961; Frei and Preston, 1964, 1968) and in some fungi (Middlebrook and Preston, 1952), elongating hyphal walls are thin and cessation of growth is associated with a thickening wall layer, and the terms primary and secondary layer apparently apply. Here again, the orientation of the microfibrils in the primary wall (now not always cellulose) shows the normal tendency toward the transverse plane but with enormous angular dispersion, so that in some plants (e.g., *Nitella* (Probine and Preston 1961) (Fig. 7.14A), *Bryopsis* (Frei and Preston, 1964a)) longitudinal microfibrils are abundantly present.

Growing cell walls are clearly highly complex structures and it is not to be expected that the microfibril orientations observed can receive explanation through any simple mechanical device. The structures are precisely determined as part of specific cell differentiation processes involved in the control of growth. Further consideration of possible orientation mechanisms must, however, be postponed until the consequences of growth have been reviewed and the matter of wall biosynthesis comes into question.

Proteins in growing cell walls. There remains the question of the recent confirmation that structural proteins may be an integral component of wall architecture. The idea that proteins may significantly be involved in the structure of meristematic cell walls dates from observations made by Tupper-Carey and Priestly in 1924. This presented a problem which could not be resolved until recently, since no marker was available such as could distinguish between cell-wall-protein and cytoplasmic contaminants. The discovery in plant cells of a protein containing *trans*-4-L-hydroxyproline which is not involved in the normal turnover of cell proteins (Pollard and Steward, 1959), however, quickly led to developments ranging from the first statement of faith (Preston, 1961), to a suggestion that the bulk of this protein is in the wall (Dougal and Shimbayashi, 1960; Lamport and Northcote, 1960a), to the final proof (Lamport, 1965; Thompson and Preston, 1967), a proof which, in spite of disclaimers which will be mentioned later (p. 296), still stands. Perhaps the clearest evidence comes from observations made with *Chaetomorpha* filaments emptied of protoplasm by the natural process of sporulation leaving walls containing material with an amino acid composition closely similar to that reported for higher plants (Thompson and Preston, 1967) (Table 7.1). This is further supported by the observation that treatment of these walls with pronase catastrophically reduces the coherence of the walls (Thompson and Preston, 1968). The protein content varies from 1% or less, on a wall dry weight basis, for primary walls of higher plants, to as much as 6% in the growing walls of some algae, and it is always firmly bound to the wall (Lamport, 1965). In some algal walls the protein also contains hydroxylysine, another amino acid in addition to hydroxyproline also found in the animal structural protein collagen (Thompson and Preston, 1967) (Table 7.1), though with *Nitella* neither of these amino acids is present (Table 7.1).

The protein is firmly bound in the wall apparently by attachment to the polysaccharides in the wall because a number of glycopeptides containing hydroxyproline have been isolated from enzymic digests of tomato cell walls (Lamport, 1967a) in agreement with an earlier suggestion (Lamport, 1962).

276

TABLE 7.1

Amino Acid Composition of Various Cell Walls (Residues/10^5 g protein)

	Nicotiana tabacum*	Lycopersicon esculentum*	Acer pseudoplatanus*	Solanum tuberosum*	Phaseolus vulgaris*	Pisum sativum (epicotyl)	sativum (root)	Rose*	Ginkgo biloba*	Centaurea cyanus*	Nitella†	Phenolacetic acid –water extraction†	Chaetomorpha Sporulated†	Cladophora†	Codium wall†	Codium Cytoplasm
Hydroxyproline	181	142	182	85	58	64	102	52	50	43	0	37	64	46	108	26
Aspartic acid	49	50	59	61	104	71	81	70	87	69	140	109	145	130	72	107
Threonine	54	51	35	51	38	51	54	64	50	69	35	45	49	39	60	68
Serine	92	93	94	82	90	93	86	74	51	78	53	57	55	34	68	75
Glutamic acid	38	61	65	85	77	86	84	102	84	116	48	108	106	117	88	114
Proline	61	60	57	60	80	69	61	85	42	54	106	69	80	52	53	54
Glycine	48	85	51	88	66	112	95	105	146	50	123	107	117	146	87	99
Alanine	40	39	56	67	45	68	63	65	66	90	95	70	74	65	97	102
Valine	65	61	49	62	83	61	44	60	64	70	39	34	41	55	91	70
Methionine	4	4	2	15	9	trace	—	9	trace	20	trace	6	10	0	7	3
Isoleucine	20	39	26	42	31	44	45	36	39	41	30	30	22	26	31	40
Leucine	33	41	44	73	57	74	73	89	61	74	38	59	65	46	61	73
Tyrosine	29	18	15	28	24	11	15	12	23	trace	15	28	27	35	20	22
Phenylalanine	14	20	24	37	24	26	34	19	37	28	23	20	23	20	25	40
Lysine	91	81	77	66	86	66	69	89	52	72	108	75	63	71	61	52
Histidine	23	18	21	17	19	10	9	9	26	29	12	7	4	13	10	6
Arginine	21	11	18	47	25	16	11	16	14	24	36	25	25	0	20	25
Cysteic acid	nd	nd	10	—	—	—	3	—	—	—	85	65	1	64	23	22

* Lamport (1965).   † Thompson and Preston (1967).

277

More recently Lamport (1967b) has isolated Hyp-O-arabinosides from tomato cell walls. His data indicate that the link with Hyp is between its hydroxyl group and Cl of arabinose (Lamport, 1969). This clearly suggests the presence in the wall of a hydroxyproline-rich glycoprotein. Mackie (*private communication*) has also found evidence for the presence of Hyp-rich glycoproteins in the cell walls of the marine alga *Codium*. This is an organism which contains β-1,4-linked mannan in place of cellulose and Mackie has evidence for a link between the protein and this structural mannan.

Most recently Lamport *et al.* (1971) have reported the amino acid sequences of three peptides resulting from the tryptic digestions of cell walls in which the Hyp-O-arabinoside link has been cleaved by acid. The sequences are: Ser hyp hyp hyp hyp thr hyp hyp val tyr lys; ser hyp hyp hyp hyp ser hyp lys; ser hyp hyp hyp hyp lys. He has also now (*private communication*) recovered a fourth: ser hyp hyp hyp hyp val(?) lys lys. This is an exciting development which comes close to proof that the amino acids extractable from cell walls are derived from a protein and do not occur as side groups on a polysaccharide. Its significance will be discussed later (p. 297).

Non-growing cell walls

It seems almost invariable that at some point during the extension of a growing cell the wall begins to thicken fairly rapidly in a new layer called, in higher plants, a secondary layer. At about this time the cell ceases to extend; but it is not certain that cessation of growth coincides precisely with the onset of secondary layer formation (p. 300). Secondary wall layers normally differ from primary wall layers both chemically and physically.

Perhaps the most striking chemical difference lies in the much higher proportion of cellulose (or other skeletal polysaccharide in those plants which do not have cellulose), often of the order of 40%, and the greater range of hemicelluloses. Some cells, as in the xylem, also synthesize and deposit large amounts of non-polysaccharide material. It is evident that the onset of secondary layer formation heralds a switch in the biochemical processes of the cell from predominantly protein synthesis to predominantly carbohydrate synthesis. This normally leads to the death of the cell.

The hemicelluloses of secondary wall layers are built from a number of sugar residues of which the most important are: D-xylose, D-mannose, D-glucose, D-galactose, L-arabinose and, to a lesser extent, L-rhamnose, L-glucose and various O-methylated neutral sugars. Like the primary layers, the secondary layers also contain polymers of D-glucuronic acid (and/or the derivative 4-O-methyl-D-glucuronic acid) and D-galacturonic acid. Many members of the Phaeophyceae contain alginic acids (poly-L-guluronic acid, poly-D-mannuronic acid and poly-D-glucuronic acid). A typical analysis of secondary walls in xylem would be: 40–45% cellulose, 24–30% lignin, 3–4% hexuronic acid, 1% galactan, 1% araban, and with mannan and xylan 4%, 18% and 10%, 7% for angiosperms and gymnosperms respectively. Many of the hemicelluloses of land plants are essentially, like cellulose, linear polymers, though they may carry short side chains, often formed of sugars differing from those of the main chain. Reviews have recently been presented by Timell (1964, 1965) to which reference may be made for further details.

The molecular chains of xylan in wood cell walls have been shown by the use of polarized infra-red spectroscopy to lie parallel to the microfibrils (Marchessault and Liang, 1962). This is possibly true of all hemicelluloses and for the cells of other tissues — and perhaps also for primary walls where it may have special significance (p. 297). Extracted xylan (either as a true xylan or as 4-O-methyl glucurono xylan) can be crystallized and the xylan backbone then has a threefold screw axis (one xylan residue being rotated 120 degrees about the chain axis, relative to the next residue) (Marchessault and Liang, 1962), whereas cellulose has a twofold axis.

Except in vessel elements of angiosperms, the orientation of the microfibrils constituting the secondary walls differs from that in the corresponding primary walls in lying more nearly parallel to the longer axis of the cell. This is true for all elongated cells of higher plants and for all filamentous algae so far studied in which a 'secondary' wall layer is deposited upon a 'primary' layer. *Valonia, Cladophora,* and *Chaetomorpha* must be excluded here because the thickened cell walls continue to extend and the situation is not comparable. The physical properties of the secondary layer are, therefore, very different from those of the primary layer and are such as to call a halt to extension growth if this has not previously ceased through some other mechanism.

With fibres and tracheids in normal wood the secondary wall commonly consists of three superposed layers termed by I. W. Bailey (a terminology now widely accepted) the S_1, S_2, S_3 layers, S_1 being the outermost, next to the primary wall. Within these layers the cellulose microfibrils lie helically round the cell (Preston, 1934; Phillips, 1941; Preston, 1946, 1947) (Fig. 7.16). The helix is slow in S_1 and S_3 and fast in S_2. S_3 is absent in some species. Preston (1934) first demonstrated a linear relationship between cell length and the cotangent of the helical angle (the angle between the helical winding and the cell length). This was shown subsequently (Preston, 1948) to take the form

$$L = a + b \cot \theta \qquad (1)$$

where a and b are constants, an equation now known to apply to S_1, S_2 and S_3 individually (Preston and Wardrop, 1949b) and to other cell types. L and θ refer to individual cells. Since the length of tracheids ($\bar{L}$) averaged over any one annual ring of a conifer increases from inside to outside in a tree trunk, it follows that

$$\bar{L} = a + b \; \overline{\cot \theta} \qquad (2)$$

and the helices steepen across the annual rings (Preston, 1934). This means that the outer wood in a tree trunk is stronger in tension, with a smaller extension for a given load and a higher Young's Modulus, than is inner wood. Quantitative relations between Young's Modulus and microfibrillar angle has been given by Cowdrey and Preston (1966) and Cave (1969). The relevance to tree stability is obvious.

Similar relations of length to helical angle have been reported for parenchyma cells in *Avena* coleoptiles (Preston, 1938), and in a modified form for *Cladophora* (Astbury and Preston, 1940). Somewhat the same qualitative behaviour has also been reported by Veen (1970). The significance of eq. (1) is not clear except in the most general terms. *a* is formally the length of a tracheid for which $\theta = 90$ degrees. Values have been reported of $1600\,\mu m$ and $570\,\mu m$ for the S_2 layer of *Pseudotsuga douglasii* and *Pinus radiata* respectively (Preston and Wardrop, 1949b), and of $568\,\mu m$ and $500\,\mu m$ for *Picea sitchensis* and for *Abies* (Preston, 1947)

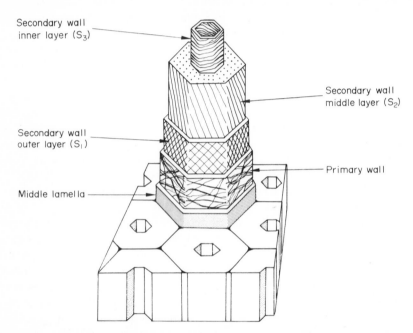

Fig. 7.16   Diagrammatic representation of wall structure in a tracheid or fibre. The oblique lines on the faces of the wall lamellae represent the run of the microfibrils.

This minimal length does not occur in any batch of material; if it corresponds to anything it represents a tracheid which would be produced by the procambium if this behaved like a cambium! It would be very satisfactory if the equation meant that the orientation is such that some feature of structure remained constant. Two obvious possibilities are that (a) the number of helical turns per cell (*n*) is constant, (b) the length of the helical winding (*l*) is constant. Neither of these holds, however; *n* constant would imply $a = 0$, and *l* constant would demand a linear relationship between *L* and $\cos \theta$ whereas the actual relationship is curvilinear. The equation does imply, however, that the inclination of the microfibrils at any part of the cell is a function of the whole length of the cell and therefore that the orienting mechanism is a function of the whole cell surface acting as a unit. The fibres of angiosperm xylem behave in a similar way. Among these we have the interesting case of apple wood in which a virus in the stock uncouples length from microfibrillar angle in the scion (Nelmes and Preston, 1968a). The normal steepening of the helix from inside to outside does not then occur and this, coupled with a disturbed lignin synthesis also induced by the virus, renders the wood rubbery.

In fibres of monocotyledons the differentiation of the wall is still more extensive. This was first demonstrated by Bailey and Kerr (1935) for *Pandanus*, and later examined in detail for a variety of bamboo fibres by Preston and Singh (1950, 1952). The secondary walls contain many layers alternating in microfibrillar angles, the helices in each kind of lamella increasing in pitch from outside to inside of the wall and again showing the relation with fibre length given by eq. (1) (Preston and Singh, 1950, 1952).

The microfibrillar angle dealt with so far was derived from polarization and X-ray studies, and therefore refers to an angle averaged over the whole layer thickness. Under the electron microscope the layers themselves appear complex. S_1 layers carry two crossed sets of microfibrils, each lying helically round the cell but in opposite senses (Wardrop, 1954; Meier, 1955; Frei, Preston, and Ripley, 1957), with some microfibrils also lying almost longitudinally (Frei *et al.*, 1957). Within the S_2 layer the helical angle also appears to vary slightly from lamella to lamella (Frei *et al.*, 1957).

The layers of the secondary wall in tracheids and fibres are differentiated chemically as well as physically. This has been demonstrated most elegantly by Meier (1961) for tracheids of spruce. The cellulose content increases from about 33% in the primary wall (including the middle lamella) to 55% in S_1 and 64% in S_2 and S_3, with a parallel decrement in the percentage of total hemicelluloses. Among the individual hemicelluloses, galactan decreases from the primary wall (plus middle lamella) (16%) to S_1 (8%) and S_2 and S_3 (0%). Araban behaves somewhat similarly, but both gluco-mannan and glucurono-arabino-xylan increase from the primary wall to S_1, and gluco-mannan further increases from S_1 to S_2 and S_3. The changing composition from outside to inside of the walls of parenchyma cells has also been examined (Nelmes and Preston, 1968b).

In most cells of higher plants the secondary layer of the wall is not uniformly present over the whole wall surface. It is absent in small localized regions, the pits, occupying the region of the primary pit fields in the primary wall upon which it is deposited. These are mostly round or elliptical in outline at the boundary between secondary and primary wall. The perforations in the secondary wall which result are straight-sided in parenchyma cells, when the pit is said to be *simple*. In the thicker cell walls of fibres, tracheids and vessels successive wall layers strongly overarch the original area and the pit is said to be *bordered*. The membrane of the pit in simple pits corresponds to the (perhaps modified) primary walls of the two contiguous cells together with the middle lamella, and it may or may not be permeable to liquids. With bordered pits there is usually a secondary differentiation into a thickened central part — the 'torus' — and a surrounding, supporting 'margo' (Fig. 7.17). In conifer tracheids, in which bordered pits occur almost exclusively on radial longitudinal walls, the margo consists of a meshwork of fibrils with interstices between them available for flow and whose dimensions have been measured both on electron micrographs and by kinetic methods (Petty and Preston, 1969). The two values tally and it is incontrovertible that the bordered pits act as the major path for flow of water from root to leaves. Under a polarizing microscope the borders of bordered pits appear as spherulites, in harmony with the impression given in the electron microscope that the cellulose microfibrils run round the pit parallel to the pit border. The mechanism whereby the border is deposited is not completely known; it is not known either why the plasmalemma at the pit fields fails to produce a secondary layer.

Once formed, the secondary wall is commonly not resorbed by the cell. This may, however, be due to the circumstance that a cell usually dies when a thick secondary layer has been completed. In the algae, where thick, highly organized walls can be found as in the Cladophoraceae, while an active protoplast is

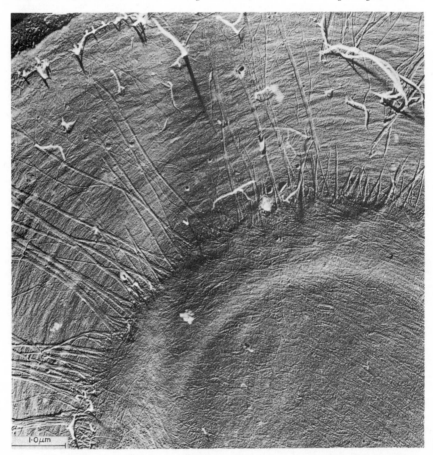

Fig. 7.17   Electron micrograph of carbon replica of part of a bordered pit membrane in *Pseudotsuga taxifolia*. Shadowed Pd-Au; magnification x22 000. Note the torus, lower right, containing a mesh of microfibrils and supported by the margo of radially arranged microfibril aggregates.

retained, the wall can in parts be resorbed. In preparation for the ejaculation of swarmers, for example, a small hole can be 'drilled' through a very compact wall (Frei and Preston, 1961). Primary walls, on the other hand, are labile (p. 300).

Biosynthesis of cell walls

The plant cell wall is thus a delicately balanced assembly of a whole family of polysaccharide and other material, and the problem of biosynthesis is consequently a highly complex matter. In examining the biochemical pathways which the various components may tread during synthesis, the overriding importance that the components must be assembled in specific spatial arrangements must be kept constantly in mind. The following discussion of biosynthesis deals only with the polysaccharide moiety of the wall and takes the biochemistry first. An attempt will then be made to express the biochemistry in terms of the cell, and to consider mechanisms whereby the necessary physical ordering may be achieved, as far as this is possible. It will become clear that neither the biochemistry nor the biophysics is as conclusive as could be wished; it is equally clear that current lines of investigation promise something spectacular over the next few years.

There seems to be general agreement over the whole range of wall polysaccharide synthesis that nucleosyl phosphate sugars act as precursors by transglycosilation, and that, in some way, a lipid is also involved. Sugar nucleotides certainly appear to be, from the thermodynamic standpoint, more promising as precursors than the other possible candidates such as sugar phosphates or oligosaccharides. The free energy of hydrolysis (ΔG°) for UDP-D-glucose is -7600 cal/mole as against a value of -4350 cal/mole for the α-1,4-link in glycogen and of -4850 cal/mole for α-D-glucose phosphate (Burton and Krebs, 1953).

The standard technique in testing the efficiency of precursors and enzymes has come to lie in the mixing of radioactive precursors with crude enzyme preparations obtained from a variety of plants by centrifuging homogenates in such a way as to obtain an active fraction. There is no guarantee that the enzyme is active in producing only one polysaccharide or, indeed, that the extract does not contain hydrolases as well as synthetases. The product is then isolated in some way (sometimes involving the addition of carrier polysaccharide of the type under investigation — which has certainly led to difficulties), identified as closely as possible, and its radioactivity measured. The enzymes are commonly found to be stabilized to some degree by dithiothreitol (which is usually therefore added), so -SH groups may be involved in the active centre. In the following, the extract used will be classed by the force of centrifugation at which the pellet was obtained, e.g., the 10 000 g fraction means the pellet obtained when an homogenate was centrifuged at 10 000 g.

Pectic compounds

An extract from mung bean (*Phaseolus aureus*) sedimenting between 500 g and 10 000 g has been shown to incorporate galacturonic acid-^{14}C from UDP-D-galactose-^{14}C (cf. Fig. 7.18A) into a polygalacturonic acid to the extent of 65% (Villemez, Lin, and Hassid, 1965) when other nucleosyl galacturonic acids are ineffective as precursors (Villemez, Swanson, and Hassid, 1966). This is one of the few investigations reported so far implicating nucleosyl sugars in the synthesis of pectin. According to Kauss and Swanson (1969), methylation of the poly-galacturonic acids is a separate and subsequent step though the enzymes involved

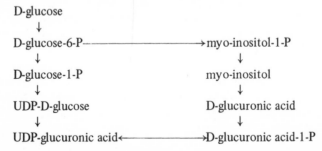

Fig. 7.18 (A) Schematic drawing of a molecule of UDPG, scale 0.1 nm = approximately 8 mm. (B) Replacement of the unit bracketed in (A) by this unit gives GDPG. (C) Myo-inositol.

in the two steps appear to be located together in a structural component of the cell. Consequently lipid membranes appear also to be involved, possibly for maintaining the structural integrity of the whole synthesizing complex (Kauss, Swanson, Arnold, and Odzuck, 1969). According to Loewus and his colleagues, myo-inositol (Fig. 7.18C) may also be involved in the synthesis of polyuronic acids, giving a pathway of synthesis outlined by them:

D-glucose
↓
D-glucose-6-P ——————————————→ myo-inositol-1-P
↓   ↓
D-glucose-1-P   myo-inositol
↓   ↓
UDP-D-glucose   D-glucuronic acid
↓   ↓
UDP-glucuronic acid ←————————————→ D-glucuronic acid-1-P

This has been extended by Roberts, Deshusses, and Loewus (1968) by showing that an extract from the root tips of *Zea mays* incorporates ^{14}C into wall polysaccharides from myo-inositol-2-[14]C. Enzyme hydrolysis of the ethanol-insoluble material then yields D-galacturonic acid, D-glucuronic acid and 4-O-methyl glucuronic acid, together with D-xylose and a trace of L-arabinose. This seems clearly to establish myo-inositol as an intermediate, though clearly it may still be one of a sequence of compounds of which UDP-glucose may be another.

Hemicelluloses

Neish and his co-workers (Altermatt and Neish, 1956; Neish, 1958) showed some time ago that D-glucose is a precursor for D-xylan by demonstrating that, when wheat plant tissues are fed with uniformly ^{14}C-labelled glucose, the xylan subsequently produced is also uniformly labelled. In a similar way, they showed that D-glucuronolactone-1-^{14}C induces the formation of a xylan with ^{14}C at Cl of the xylose residues by loss of the C6 atom (Slater and Beevers, 1958). The intervention of nucleosyl sugars has again been demonstrated, both by synthesis from UDP-glucose (Fig. 7.18A), with UDP-glucuronic acid as intermediate (Strominger and Mapson, 1957), and from UDP-xylose (Pridham and Hassid, 1966; Bailey and Hassid, 1966). The xylan is, of course, 1 → 4 linked. The material synthesized from UDP-xylose, however, gives on hydrolysis L-arabinose as well as D-xylose (1:4). It has also been claimed that, under some conditions, a mixture of polysaccharides is produced when extracts from *P. aureus* are treated with a mixture of UDP-glucose and UDP-mannose, and that one of these is a glucomannan. As we have seen above, there is also evidence that hemicellulose polysaccharides can be synthesized from myo-inositol. This has been confirmed by Roberts *et al.* (1968) who have shown that extracts from duckweed (*Lemma fibrosa*) synthesize polysaccharides which, on hydrolysis, yield labelled D-xylose and D-apiose.

Cellulose

Cellulose is more difficult to characterize than are other wall polysaccharides, since not all 1,4-linked glucans may be regarded as cellulose. It is therefore not yet certain that the synthesis of cellulose has been achieved.

Glaser (1958) was among the first to demonstrate, using *Acetabacter xylinum*, the synthesis from UDP-glucose ^{14}C of an alkali insoluble polysaccharide which he claimed to be cellulose. This synthesis was later supported by Brummond and Gibbons (1965) who claimed UDP-glucose as the most effective precursor of the various sugar derivatives they tried. Unfortunately they did not test GDP-glucose (Fig. 7.18B), which had already been claimed by Barber, Elbein, and Hassid (1964) as the only effective precursor with a 20 000 g extract from *P. aureus*. Identification of cellulose by Barber *et al.* was claimed not only on the basis of alkali insolubility but on the presence of radio-glucose, -cellobiose, -cellotriose and -cellotetraose in an hydrolysate. Feingold, Neufeld, and Hassid (1958) had earlier claimed that mung bean extracts synthesize a β-1,3-linked glucan from UDP-glucose, similar to laminarin, callose and paramylon, and at one time it began to appear that precursors had been found specific for 1,3- and 1,4-linked polysaccharides. The situation is now, however, much less clear. Ordin and Hall (1967) have shown that extracts from the oat plant (*Avena sativa*) can use either UDP-glucose or GDP-glucose in the synthesis of an alkali insoluble polysaccharide which they regard as cellulose. With UDP-glucose, treatment of the polysaccharide with *Streptomyces* cellulase yields mostly cellobiose with a small amount of mixed β-1,4-, β-1,3-linked trisaccharides. Hydrolysis of the GDP-glucose product yielded cellobiose only. They concluded (Ordin and Hall, 1968) that at least three polysaccharides are produced, one of which is cellulose and another laminarin. According to them substrate concentration has a marked effect on the end product. With concentrations of up to 5 mM UDP-glucose, the resulting alkali-insoluble polysaccharides yield on hydrolysis equal amounts of laminaribiose and cellobiose; at concentrations greater than 5 mM only traces of cellobiose appear. On addition of dithiothreitol (UDP-glucose at 0.1 mM) the percentage of cellobiose is increased so that the enzyme involved may carry an active -SH group. A re-investigation of the system by Flowers, Batra, Kemp, and Hassid (1968), using extracts from each of *P. aureus* and *Lupinus albus*, maintains, however, the situation that UDP-glucose yields only 1,3-linked polysaccharides, though with oat extracts a β-1,4-linked glucan is also produced as Ordin and Hall (1967) had claimed.

Batra and Hassid (1969) have attempted to clarify the situation by examining the reaction of synthesized polysaccharides to a highly purified exo-β(1,3)-D-glucanase extracted from a basidiomycete. The polysaccharides synthesized from UDP-glucose were found to hydrolyse to the extent of about 90% in 24 hours, whereas the GDP-glucose product remained unaffected. The UDP-glucose product is therefore in the main a 1,3-linked polysaccharide. Nevertheless, Stafford and Brummond (1970) have recovered UDP-glucose as a possible cellulose precursor by showing that the addition of glucan to the mixture (using *L. albus* extracts) can cause a tenfold increase in the presence in the product of β-1,4 links. These workers, incidentally, appear to be the only ones using acetolysis of the product and they claim to detect cellobiose octa-acetate.

Most recently, Robinson and Preston (1972) have re-examined the 2% alkali-insoluble product using a 40 000 g extract of *P. aureus* and each of UDP-glucose and GDP-glucose but without any carrier cellulose. In each case the alkali-insoluble product yields a clear X-ray diagram which is not that of cellulose I, cellulose II, or laminarin. The signs are that this product may be a mixture of the lower molecular weight oligosaccharides and that the mixture is the same whether UDP-glucose or GDP-glucose is used as precursor. The conditions are such, moreover, that little material other than this crystalline material can have been synthesized, and the earlier claims that cellulose is synthesized by this system are not substantiated.

As with other wall polysaccharides, a lipid appears to be involved in the synthesis. This was first established by Colvin (1961) for *Acetobacter xylinum*, and has since then received support from Villemez and Clark (1969) (with a 50 000 g extract of *P. aureus*) and Pinsky and Ordin (1969) (with a similar extract from *A. sativum*). The lipid appears to be tightly bound to the enzyme.

The biosynthesis of cellulose is therefore still in a confused state. Neither the demonstration of a 1,4-link, nor the presence of cellobiose in a hydrolysate, can be regarded as proof of the presence of cellulose in a polysaccharide mixture, and failure to dissolve in weak alkali solutions is no proof either. Clearly the relatively large amounts of cellulose usually added as a carrier may be confusing the issue. While it is now clear that nucleosyl sugars can act as precursors for a variety of polysaccharides, it seems certain that the *in vitro* synthesis of cellulose in the form required for cell wall incorporation has not been achieved. Current attempts to purify the putative synthetase (Lin and Hassid, 1970) may perhaps change the picture.

Localization of polysaccharide synthesis

In general, the involvement of lipids in wall polysaccharide synthesis suggests that the synthetase particles are bound to cell membranes. Mollenhauer, Whaley, and Leech (1961) were the first to claim, on the basis of electron microscopic observation of root caps, that vesicles developed from Golgi bodies are involved in wall deposition in such a way that the membrane of the vesicles contributes to the plasmalemma. This was soon supported by Drawert and Mix (1962) for *Microsterias rotata,* and by Sievers (1963) for roots of *Zea mays.* Subsequently, Northcote and Pickett-Heaps (1966) found themselves in full support by show-ing that, when wheat root tips are exposed to tritiated glucose, the label appears first in Golgi vesicles in root cap cells; a subsequent chase with cold glucose transferred the label rapidly to the wall. With cells behind the root cap, however, label appeared in the wall though with relatively no localization in Golgi bodies. There may well be, therefore, two sources of polysaccharide, only one of which is sited in Golgi vesicles. Harris and Northcote (1971), by extracting from pea roots a fraction rich in Golgi bodies, have confirmed that pectic substances and hemicelluloses are carried, and probably synthesized, in the Golgi.

Ray, Shininger, and Ray (1969) have recently introduced a much-needed gentler method of isolation of particles which may carry synthetases. Grinding and pellet formation is avoided by initial velocity centrifugation through a gradient to give an enriched zone which is then layered on a gradient for isopycnic centrifugation. They found synthetase activity to be located in one sharp peak which could utilize UDP-D-glucose, GDP-D-glucose or UDP-D galactose as donors in the ratio 5:2:1. The product from the first two give some cellobiose on hydrolysis, suggesting synthesis of a 1,4-linked glucan. Electron microscopic observation suggested that the synthetase particles were associated with Golgi vesicles. Villemez, McNab, and Albersheim (1968), also using gentler methods of extraction, had reached the conclusion that the synthetase particles were associ-ated with the plasmalemma.

From the biochemical standpoint, therefore, there is still no agreement about specific precursors for specific polysaccharides, no certainty about the poly-saccharides which are produced, and no concensus about the location of the putative synthetase particles. It is, therefore, relevant to consider the situation in terms of the nature of the wall which needs to be synthesized, and in terms of the

287

whole, undisrupted cell.

Known details of wall organization, particularly in the Cladophoraceae, drove the present writer to the unavoidable conclusion, presented in 1963 and published one year later (Preston, 1964), that cellulose synthesis could occur only by synthetase granules closely ordered on the plasmalemma (Fig. 7.19). This was

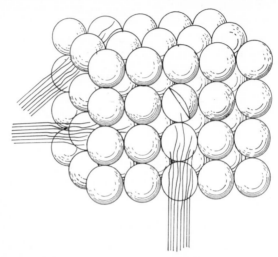

Fig. 7.19   Schematic diagram showing the relationships between three arrays of granules deduced to be associated with the plasmalemma in *Cladophora*. The granules, some 30 nm in diameter, are synthetases which are stimulated to synthesis on contact with microfibril ends (shown as aggregates of lines). The plasmalemma lies parallel to the plane of the page. Three arrays of granules are shown, to represent the geometry with three microfibril directions; three arrays are, however, not necessary.

rapidly followed by the demonstration of hexagonally close-packed particles on the plasmalemma of yeast by Moor and Mülethaler (1963), using their newly developed method of freeze-etching. These particles were associated with wall fibrils and, though these are not cellulosic, clearly pointed the way. Since that time, all plasmalemmae examined in this way have been shown to carry granules (*see* Staehelin and Probine, 1970), and Preston and Goodman (1968) have considered some of the implications of the synthetic mechanism involved. Particles on the plasmalemma of *Chaetomorpha* have now been seen (Barnett and Preston, 1970) (Fig. 7.20) precisely in the array postulated by Preston (1964), and have been shown to be associated with fibrillar bodies which can only be cellulose microfibrils (Robinson and Preston, 1971) (Fig. 7.21). What began as speculation, however informed, in 1964 has become credible in 1971; what were called 'hypothesomes' have assumed reality. Harris and Northcote (1971) have recently agreed that the synthesis of cellulose can only be understood in terms of synthetase particles on the plasmalemma.

It was already clear in 1963 (and, indeed, in principle long before that) that the plasmalemma provides the most likely locus for the cellulose synthesizing machinery. Colvin, Bayley, and Beer (1957) had shown that the microfibils of

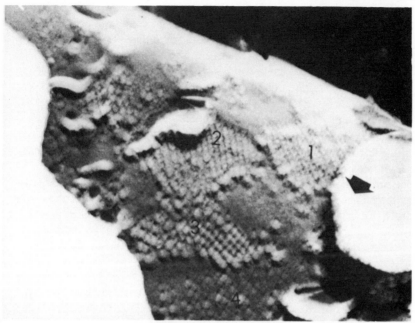

Fig. 7.20 Electron micrograph of freeze-etch replica of the surface of a swarmer of *Cladophora rupestris* immediately below the wall. Magnification x30 000. Large arrow shows direction of shadowing with Pd-Au. Areas of close-packed granules about 30 nm diameter appear in three ranks; 1, outermost, 2, median and 3, innermost. The area marked 4 represents a surface from which the granules have been stripped.

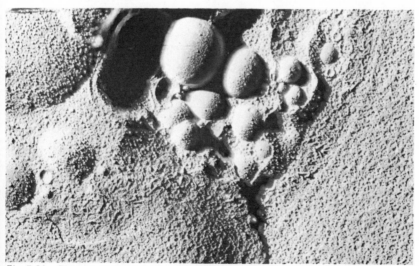

Fig. 7.21 Electron micrograph of a replica of an oblique freeze-etch of a swarmer of *Chaetomorpha melagonium*. Magnification x40 000, shadowed Pd-Au. Note, left and bottom, the region just outside the plasmalemma carrying granules with 'tails'. These may be synthetase particles attached to, or producing, cellulose microfibrils (replica by D. Robinson).

289

the (extracellular) cellulose of *Acetobacter xylinum* must be produced by end synthesis, and Preston (1958) had derived the same conditions for the cellulose of coherent cell walls and had foreshadowed the 'ordered granule' hypothesis. The necessity for end synthesis (equivalent to the necessity for a starter molecule in enzyme kinetics) seems firmly to place the synthetase granules where microfibril ends are certain to lie, namely in contact with the cell wall. It is to this locality that precursors must be channelled, whether by Golgi vesicles, by the endoplasmic reticulum or by microtubules. The interplay between these organelles may indeed be complex; both Pickett-Heaps and Northcote (1966) and Robards and Humperson (1967) have already suggested that microtubules may direct cytoplasmic vesicles to the wall. If nucleosyl sugars are donors for polysaccharide synthesis, they will then donate the sugar moiety as directed by the presence of the appropriate enzyme to the starter molecule; and it seems likely that acid polysaccharides (and perhaps hemicelluloses) will then be synthesized in Golgi vesicles. Although 'artifical' synthesis of cellulose cannot yet be claimed with certainty it seems highly likely that the location of this synthesis occurs at the plasmalemma.

It may, therefore, be of significance that a number of workers have reported the presence in cell walls of membrane systems clearly derived from the cytoplasm, whether of uncertain origin (though perhaps Golgi) (Bowes, 1969) or lomasomes (Bowes and Butcher, 1967; Marchant and Robards, 1968) or of the plasmalemma itself (Myers, Preston, and Ripley, 1956; Wardrop and Foster, 1964; Wardrop, 1965). A siting of synthetases on these membranes would harmonize with the reported synthesis of wall materials in the wall away from the plasmalemma.

Brown, Frank, Kleinig, Falk, and Sitte (1970) appear to be the only investigators to have claimed synthesis of cellulose in Golgi bodies. It could be that their organism, *Pleurochrysis scherfelii*, is exceptional; but, although they appear to have demonstrated the presence of a glucan which is possibly 1,4-linked, proof that this polysaccharide is cellulose is lacking. It seems highly unlikely that the complex wall organization observed in higher plants and in many algae could be reached through synthesis in this organelle.

Physical properties and cell extension

After a cell has left the meristem and therefore ceased to divide, it normally increases in area. Attention is confined here to this period of extension and to cells which elongate during extension. It is profitable first to examine the physical structure of the wall at the onset of growth and during growth insofar as this is known. The way is then open to enquire whether the physical properties of the wall are of consequence for growth and, if they are, how far in these terms can the effect, for instance, of growth-promoting substances be understood. Many cells have been shown to exhibit 'spiral growth' in which one end rotates with respect to the other as the cell elongates, and perhaps all cells grow in this fashion. This has been traced to the mechanical properties of the growing wall and depends upon the fact that wall structure is such that the axis of maximum stress does not coincide with the axis of maximum strain. This should be remembered, but will not be taken up here; it is responsible for the fact that the steeper helix in *Chaetomorpha* walls becomes flatter during growth (p. 274). Nor is it proposed to consider cells in which growth is localized as in, for instance, the tips of root hairs. Both these topics have been covered by Roelofsen (1959, 1965). Cells will be considered here as though they grow straight, and slight complications due to spiral growth will be ignored.

Early considerations of changes in wall structure during growth were limited by the techniques available. They were based in the main on observations under the polarizing microscope, used in a somewhat primitive way and all too often with little understanding. This technique enabled most workers to conclude that in elongating cells the aggregates of cellulose chains comprising the wall (now called microfibrils, a term we will use although out of its chronological context) lie either transversely (e.g., Bonner, 1936) or in a slow helix (Preston, 1938; Preston and Wardrop, 1949). This orientation, whether judged by polarization microscopy or X-ray diffraction used with whole tissues, remained apparently constant during growth even though mechanical stretching by only 10% caused a reorientation (Bonner, 1936). Inability to examine individual lamellae in the primary wall led to the inevitable conclusion that the wall grows and is not passively extended, even though Preston (1938) had shown by observation of single walls of parenchyma cells of oat coleoptiles, that the (helical) orientation depended upon cell length and was therefore possibly dependent upon growth, an observation recently supported by Veen (1970). It was at that time impossible to decide whether the wall is deposited by 'apposition' (layering of new material *on* the old) or by 'intussusception' (deposition of new material *in* the old). Microscopical evidence was strongly in favour of the latter, with the implication that cell growth was paralleled, in part at least, by wall growth.

The more detailed observations allowed by the electron microscope soon set the balance heavily the other way. The work of Roelofsen and Houwink (1951, 1953), already described (p. 272) showed clearly that (a) in newly deposited lamellae of the primary wall the microfibrils invariably lie almost transversely, irrespective of the elongation which the cell had undergone prior to deposition, (b) elongation after deposition causes a displacement away from the transverse, (c) as a consequence, the further a lamella is from the plasmalemma in an elongating cell the more it has been extended and the further the orientation is from the transverse. Clearly the wall is deposited by apposition and the wall extends, at least in part, passively. This concept has come to be known as the 'multinet growth hypothesis', the first word of which is perhaps unfortunate. It has been supported by many workers for many cell types (e.g., Probine and Preston (1961) and Vasumati, Roelofsen, and Spit (1962) for *Nitella*; Frei and Preston (1961) for *Chaetomorpha* – not strictly a primary wall; Veen (1970) for pea stems). Both the X-ray diagram of a cell or tissue and the image in a polarizing microscope give an orientation averaged over the whole wall and the explanation of the apparent constancy of orientation in a wall which is extended is now straightforward. The strain (relative extension) across the wall varies as shown diagrammatically in Fig. 7.22. During growth, as a new lamella is deposited on the inside, and elongation continues, each lamella moves to the right and the strain curve – and therefore the nett orientation – remains constant. As the lamellae come under greater strain they naturally become thinner and presumably in the end fracture; significantly, the outermost lamellae of the thick growing wall of *Chaetomorpha* filaments are always fractured and curled back to the nearest crosswall as collars (Frei and Preston, 1961).

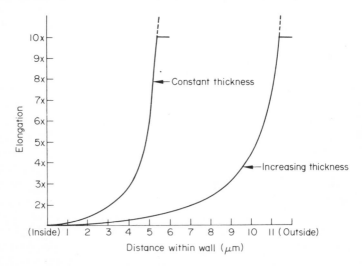

Fig. 7.22   The distribution of strain across a wall which, initially uniform, has extended in length 10x while either maintaining the wall thickness constant at 6 μm or increasing it, by apposition. By the multinet growth hypothesis the strain represents the amount of reorientation of the constituent microfibrils.

Growth in isolated cells always appears to occur most rapidly along a line of mechanical weakness in the wall, at right angles to the preferred direction of microfibril orientation unless other complications intervene. One should not be misled by the fact that cells in tissues do not behave in this way. Cambial cells, in which the inner wall microfibrils lie almost transversely (Preston and Wardrop, 1949) expand rapidly in the transverse direction to produce either tracheids in conifers or vessels in angiosperms; here the mechanical resistance to be overcome lies in the surrounding tissues and only to a minor extent in the walls of the cambium.

Changes of orientation implicit in the multinet growth hypothesis do not, however, present the whole story. In parenchyma cells of oat coleoptiles, and as already mentioned, onion roots and celery petioles, Setterfield and Bayley (1958) have demonstrated the presence of ribs of longitudinally oriented microfibrils some of which extend through the whole thickness of the primary wall (cf. Fig. 7.15) and some of which occur only on the *outside* of the cell. Similar structures have been seen in epidermal cells (Bayley, Colvin, Cooper, and Martin-Smith, 1957), in parenchyma cells of the epicotyl of pea seedlings (Probine, 1965), and in young collenchyma cells (Beer and Setterfield, 1958) though these lie in the thickened ribs in which the microfibrils already lie longitudinally (Anderson, 1927; Majumdar and Preston, 1941). Under all these circumstances the authors considered that they had evidence for cellulose biosynthesis throughout the whole wall and therefore for intussusception. This would be in harmony with the inclusion of cytoplasmic membranes in the cell wall already referred to, if the synthetase particles are membrane-bound.

While remembering that biochemical changes in the wall may well have a part to play in cell growth, it is necessary, therefore, to enquire whether mechanical properties of the wall form in any sense a growth control mechanism; and, if they do, which mechanical property is the more important and how can it be related to the various factors, including growth substances, which affect the rate of cell growth. The basic concept would be that the cell wall yields in some way to stress induced by the internal turgor of the cell. It should be remembered that in a cylindrical cell the stress is not uniform. If the internal turgor is P and the radius of the cell is r then the longitudinal stress, T_l, stretching the cell, is numerically equal to $Pr/2$ while the transverse stress, T_t, broadening the cell is twice as great, Pr. If the mechanical properties of the wall are isotropic, the cell will therefore extend more rapidly transversely than longitudinally so that a cylindrical cell would tend to become spherical. Conversely, a spherical cell, with an isotropic distribution of stress, would extend into a cylinder if the mechanical properties of the wall were anisotropic. How a cell actually reacts depends upon the anisotropy both of the stress and of the mechanical properties of the wall.

The wall can yield to stress in various ways. When a rod length l is placed under a longitudinal stress S (force per unit sectional area) and extends by a length Δl in such a way that the length l is recovered on removal of the stress, the extension is said to be 'elastic' and the behaviour can be described by a modulus of elasticity, $S/(\Delta l/l)$, which is a constant, Young's Modulus (E). If the rod is extended beyond the elastic limit, some of the extension is not recovered on removal of stress and this component of extension is called *plastic* extension. Both elastic and plastic extensions are virtually instantaneous and neither extension can be increased without increase in stress. Beyond this, many materials, including cell walls, undergo a slow, steady increase in extension under constant stress, a phenomenon called *creep*. Clearly the boundary between plastic extension and creep is indeterminate. A phenomenon related to creep may be observed if the material is instantaneously stretched and the new length is then held constant. The stress then decays in a phenomenon called 'stress relaxation'.

Though rough correlations between cell growth rates and E have been recorded (Frey-Wyssling, 1948 (roots); Probine and Preston, 1961, 1962 (internodal cells of *Nitella*)), it is clear that this cannot in any sense explain growth; for continued growth would then necessitate continued increase in stress and therefore turgor pressure and this does not occur. For this reason, and because they were working with tissues instead of isolated cells, the findings of Burström, Uhrström, and Wurschler (1967) on the relation between E and cell growth are difficult to interpret. It may, nevertheless, be significant that in, for instance, growing internodal cells of *Nitella* which elongate enormously and yet remain cylindrical, the transverse Young's Modulus E_t is 4–5 times greater transversely than longitudinally (E_l) (Probine and Preston, 1962). Remembering that the transverse stress is twice as great transversely as longitudinally, the growing wall is relatively twice as strong transversely. It is perhaps not a coincidence that at the end of growth of a *Nitella* internodal cell the ratio E_t/E_l has fallen to 2, the same ratio as the stresses. The greater transverse strength is of course due to the predominantly transverse arrangement of the microfibrils and other wall constituents.

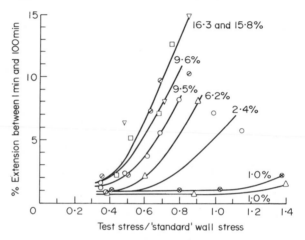

Fig. 7.23   The rate of creep of a number of longitudinal wall strips from *Nitella* internodal cells as a function of stress. Each curve corresponds to one wall strip and the figures alongside represent the percentage growth in length in the cell during the 24 hours prior to removal of the strip.

Creep is clearly a better candidate for the role mechanical properties may play in cell growth, for while the creep rate also increases with stress, it is maintained at constant stress and, moreover, changed at constant stress by changes in the internal bonding in the wall. Probine and Preston (1961, 1962) examined the creep rate of wall strips isolated from internodal cells of *Nitella opaca* whose previous growth rate was known. It had already been shown by Green (1954) that these internodal cells grow uniformly over the whole length of the cell. Probine and Preston (1962) demonstrated that wall strips stressed transversely do not creep noticeably even at stresses on the verge of breaking them, while longitudinally stressed strips creep at rates which parallel the cell growth rate (Fig. 7.23). They also showed that incorporation of Ca ions decreased both growth rate and creep, and removal of Ca ions increased both. Although creep rates and growth rates are not equal — even if allowances are made for the multiaxial application of stress in one case and uniaxial stress in the other – it is clear that something like creep may be involved in growth. The difference in the two rates is perhaps to be explained by removal of the wall, in preparing an experimental strip, from the biochemical machinery of the cell. The situation is also, of course, complicated by the fact that, since the wall extends following the miltinet growth concept (Probine and Preston, 1961), the stress distribution is not in either case uniform over the thickness of the wall. A mathematical treatment of the relationship between growth, cell wall creep, and wall structure in *Nitella* has been presented by Probine and Barber (1966).

Haughton, Sellen, and Preston (1968, 1969) and Haughton and Sellen (1969) have examined the viscoelastic properties of growing walls by stress relaxation in internodal cell walls of *Nitella* (cellulose), and in filaments of *Penicillus* (1,3-xylan, p. 267) and *Acetabularia* (1,4-mannan, p. 266). The decay of stress is found to be approximately a linear function of log (time) as found also by Wallace (1967) for *Chara*. If a reasonable mechanical model of the wall is a non-linear one (as suggested by the hysteresis in stress/strain curves) consisting of a spring and a non-Newtonian dash-pot in series, then this means that the same number of bonds break and reform in different positions in equal logarithmic time intervals. The cellulose microfibrils of cell walls have a tensile modulus of the order of 10^5 $N/mm^2/cm^2$ (Treloar, 1960; Cowdrey and Preston, 1966) at least 100 times the value observed for these algae by stress relaxation methods, so that stress relaxation must occur by slipping of microfibrils and a yielding of the matrix. Just as the stress decays linearly with log (time) so the creep rate in creep experiments decreases linearly with log (time) (Probine and Preston, 1962; Cleland, 1968), and to a very low level. During the growth of living cells, on the other hand, large deformations in the wall occur, far beyond deformations at which walls would break in tensile tests. The outcome would seem to be that, during growth, the wall deforms through the breaking and re-making of bonds within the matrix and between the matrix and the cellulose microfibrils: but that the breaking and making occurs far more extensively than it would by stress alone and that therefore, during growth, there is active biological activity within the wall itself.

Both Cleland (1971) and Cleland and Haughton (1971) have examined stress relaxation with oat coleoptiles and have found that, both for coleoptiles without pretreatment and after preconditioning by taking through a series of stress/strain cycles, stress again relaxes at a rate which is linear with log (time) so that the same considerations apply here. Pretreatment with IAA reduces the relaxation modulus in a marked way. Their data are very similar to those of Yamamoto, Shinozaki, and Masuda (1970) though their analysis is different; in neither case, however, do the data yet allow any decision on the molecular basis for stress relaxation or growth except in a most general way. Cleland and Haughton (1971) do conclude, however, that wall extension occurs by a series of extension steps, each involving viscoelastic extension, preceded or accompanied by IAA-dependent biochemical changes in the wall. These changes must involve a breaking of bonds between wall components.

The bonds involved in this making and breaking could be any or all of: hydrogen bonds; COO-bonds which might be methylated or salt-linked through Ca or Mg ions (known to depress both growth and creep rates); —S=S— and other bonds between chains of the proteins in the wall as postulated by Lamport (1965); or bonds between protein and carbohydrate. The concept that proteins are involved has been strongly reinforced by the demonstration by Thompson and Preston (1968) that in the algae *Chaetomorpha* and *Cladophora,* the tensile moduli of-the wall are catastrophically reduced when the wall is treated by either the protease pronase or by the —S=S— bond-breaking agent dithiothreitol. The work of Olsen, Bonner, and Morré (1965) showing that in oat coleoptiles removal of wall proteins had no effect on wall extensibility has been repeated by Cleland (1967) with the opposite result. Although Cleland uses a somewhat unfortunate method for separating elastic from plastic extension, and dries his isolated cell walls and stores them in ethanol before use (which must induce an irreversible change in wall bonding), his results nevertheless appear clear-cut and acceptable.

In attempting to assess the significance of a wall protein for bond making and breaking it needs to be remembered that Steward and his collaborators (e.g., Israel, Salpeter, and Steward, 1968; Steward, Mott, Israel, and Ludford, 1970) have consistently denied the existence of a Hyp-protein in the wall. Note must always be taken of the absence of hydroxyproline from wall hydrolysates of *Nitella* (Thompson and Preston, 1967) and *Valonia* (Steward *et al.,* 1970) though *Nitella* walls yield many amino acids (Table 7.1) and *Valonia* walls may contain not more than 2% protein (Steward, 1970). It could be that Hyp-proteins are important for growth in some cells but not in others; though we would then be in the situation that a factor is responsible only in some cells for a phenomenon which is common to all. In any event, this work enters a *caveat* into discussions of growth and a controversy which badly needs to be resolved. The following statements are based upon the assumptions that the many claims for the intervention of a wall protein in growth are acceptable, that wall-bound hydroxyproline is proved, if not universally, and that this amino acid is incorporated in a wall protein.

There is biochemical evidence that the hydroxyproline containing protein in the wall is directly associated with wall extension. In both carrot callus (Steward, Pollard, Patchett, and Witkop, 1958) and oat coleoptiles (Cleland, 1963a, 1967), hydroxyproline (Hyp) has been found to inhibit growth although externally fed Hyp does not enter the protein. Cleland (1967) has shown that in oat coleoptiles only the L isomer inhibits growth and then, if the treatment is preceded by an IAA treatment, only after a certain amount of IAA-induced growth has occurred; sucrose increases the inhibition. Cleland concludes that free Hyp may inhibit elongation by blocking the formation or utilization of an Hyp-protein which must be incorporated in the wall during auxin-induced elongation. On the other hand, Hyp-proteins increase markedly in the cell wall of pea epicotyl parenchyma cells during the transition from a rapidly elongating tissue to the non-elongating condition (Cleland and Karlsnes, 1967). We have therefore the odd situation that a protein apparently necessary for growth can, perhaps in excess, contribute to cessation of growth. The protein in growing walls must then be maintained in a delicate balance. There is, in any case, a lack of correlation between the rate of synthesis of Hyp-protein and the growth rate (Cleland, 1968). One part of the Hyp-protein is synthesized only when IAA and sugar are present. Sugar is known to increase cell wall synthesis, and IAA is said to increase the rate of deposition of non-cellulosic wall polysaccharides (which are intussuscepted) and to change the pattern of wall development (Ray, 1967). Indeed, 8% sucrose in the presence of auxin causes the orientation of newly deposited microfibrils to change from the normal, almost transverse direction to the longitudinal in pea epicotyl parenchyma cells, leading to a lateral expansion of the cells rather than elongation (Veen, 1970). Again, Galston, Baker, and King (1953) demonstrated that with pea epicotyls, benzimidole over a narrow range of concentration will inhibit cell elongation but encourage lateral expansion. This has been associated by Probine (1965) with the deposition of longitudinal arrays of microfibrils, this time on the outside of the wall. Cleland (1968b) concludes with some reason that the Hyp-protein might be required to ensure correct orientation of the hemicelluloses among the cellulose. In this regard, the sequencing of oligopeptides from the wall protein by Lamport (already referred to, p. 278) is a welcome and exciting development. It is not yet known whether the protein chains occur singly or in groups. In either case there will be a tendency for the sequence of 4 Hyp residues to be coiled in a helix (not the α helix) which will be achieved unless external constraints impose another conformation. In that case (Fig. 7.24) the -OH's of each Hyp will lie on the outside of the helix and might allow bonding with 4 arabinosides. This might be important for cell wall containing little protein, though in the tomato cells from which the peptides are derived a protein content of 20% (Lamport, 1969) would appear to present no problem. Notionally, the arabinosides could lie parallel to the helix (Fig. 7.24A) or perpendicular to it (Fig. 7.24B).

The way may now be open for the structural investigation necessary to determine the precise organization of the protein glycoside. If the arabinoside lies perpendicular to the helix, and if the arabinosides must lie parallel to the microfibrils as Marchessault and Liang found for xylan, then the protein must lie across the microfibrils. This would confer the possibility that one protein chain could cross-link several microfibrils.

One difficulty with the idea that a protein is involved directly in wall coherence and extension has always been that the small amount of protein assumed present in primary walls on the basis of amino acid analysis, seldom more than a few percent, has to account not only for extension but also for a number of enzymes (including a peroxidase) also bound in the wall. There would seem barely enough to go round. It may now be significant that Shannon, Kay, and Lew (1966) have reported that some horseradish peroxidases contain hydroxyproline and arabinose and galactose. There is therefore the possibility that the Hyp-protein in the wall doubles up as a peroxidase, the Hyp-galactose-arabinose component acting as an extension subunit, as mentioned by Lamport (1969) who has obtained from a peroxidase hydroxyproline-O-arabinosides.

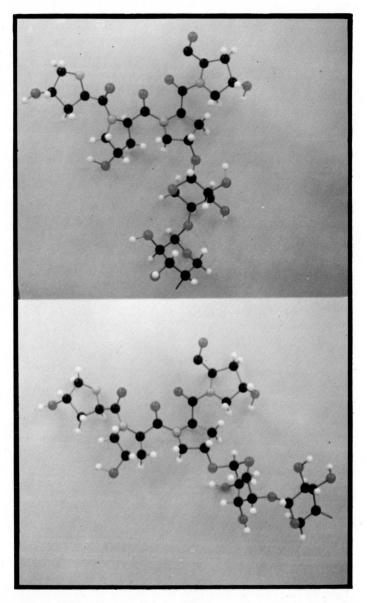

Fig. 7.24   Models of a polypeptide of four hydroxyproline residues with a section of araban attached to one —OH group.

Black, carbon; red, oxygen; blue, nitrogen; white, hydrogen

   (A) araban approximately parallel to polypeptide helix
   (B) araban approximately perpendicular to helix.

The involvement of the integrity of wall polysaccharides themselves also in growth is strongly indicated by the accumulating evidence of turnover of wall polysaccharides during growth. The more recent evidence refers, unfortunately, to pea seedlings and pea epicotyls (Matchett and Nance, 1962; Machlachlan and Young, 1962; Margerie and Lenoël, 1961), and therefore to tissues which are too complex to allow strict localization of the place of turnover. More recently, Machlachlan and Duda (1965), again with pea epicotyl, have shown that the major part of the pectic compounds are stable during growth and only the glucose- or galactose-containing hemicelluloses turn over. It is not clear that the cellulose itself ever turns over, so that the biochemical evidence, like the creep and stress relaxation evidence, points strongly to the importance of hemicelluloses in growth, as well as the wall protein.

The association of IAA-stimulated growth with wall plasticity originally advanced by Heyn (1931), and for so long ignored because others thought (erroneously) that a direct association between wall and auxin was envisaged (for which not enough auxin would be available), has since that time been supported by ample evidence (see review by Setterfield and Bayley, 1961). There can no longer be any doubt that IAA is involved in biochemical reactions leading to the formation or utilization of substances which control bond breaking and making in the cell wall, and lead to sustained creep under the influence of wall tension. It appears possible, however, that the part played by tension may be more complex than might be thought. Cleland (1968) has shown that the degree of wall loosening (i.e., the increase in 'plasticity' when the wall is mechanically stretched) varies in oat coleoptiles with turgor pressure up to at least 6 atm, but that the cell does not extend below a certain critical turgor pressure, P_c, of about 6 atm (natural turgor is greater than this). He concludes that turgor is involved in the straining or breaking of 'strong' links (presumably primary valences) which alone would not allow growth. Only when 'weak' links are also broken does wall tension induced by turgor effect cell extension. The whole matter is still more complicated by the fact that wall synthesis is also stimulated by turgor pressure, providing a kind of negative feed-back such as might be expected to be necessary in such a delicately balanced operation.

It is said that protein synthesis is needed for the induction of wall loosening (Black, Bullock, Chantler, Clarke, Hanson, and Jolley, 1967; Courtney, Moire, and Key, 1967) in harmony with the notion that enzyme activity is involved. There is, however, as yet no certainty about the nature of the enzymes. The action of IAA is reversible (Cleland, 1968) so that the enzyme cannot be cellulase (especially since cellulose is stable in the wall) even though Fan and Machlachlan (1967) have shown with decapitated oat coleoptiles that IAA induces an increase in cellulase content by a factor of 12 to 16 times. Indeed, inhibition of cellulase formation by puromycin or actinomycin D in the presence of auxin greatly enhances elongation. Fan and Machlachlan show that under their conditions increase in cellulase induces cessation of elongation and an increase of lateral expansion. An older idea, ably championed by Bennet-Clark (1956), that pectic enzymes are involved, must now be disregarded, partly because in oat coleoptiles (for which it was introduced) the amount of pectic substances is much lower than was at one time thought, and partly because there is no evidence for an effect of auxin on either pectinase or pectinmethylesterase (Jansen, Jang, Albersheim, and Bonner, 1960; Jansen, Jang, and Bonner, 1960). The only remaining competitors would appear to be hemicellulases or proteases. Only proteases are known to affect wall plasticity (Thompson and Preston, 1968).

In conclusion, it may be said that the primary wall of growing plant cells is a highly complex entity in a delicate balance which can be affected by both exogenous and endogenous factors. The wall can extend and the cell remain coherent only if (a) the making and breaking of bonds can be initiated and kept in balance, (b) turgor pressure is great enough to set wall tension above a certain critical value, and (c) the permeability of the plasmalemma will allow the necessary flux, mainly of water. A newly deposited wall layer, with pseudo-transversely oriented microfibrils, will not come under immediate stress. As it does come under stress, longitudinal strain imposed by the anisotropy of existing older layers will open out the meshwork, and perhaps at this stage hemicellulase and proteins are incorporated. This may represent the first function of turgor pressure. Extension and reorientation will at first increase resistance to further longitudinal creep, in part at least on account of microfibrillar reorientation. Beyond that, however, decreased thickness of the lamella as it 'moves' outward through the wall, coupled by failures leading to rearing of the lamella such as have been seen on the outer face of the wall in *Chaetomorpha* (Frei and Preston, 1961), will reduce the ability to withstand stress. It therefore seems likely that the load imposed on the wall by turgor must in the main be borne by the intermediate lamellae in the wall; development of outer wall ribs of longitudinal microfibrils will serve only to move the load-bearing lamellae further out in the wall. It is therefore in lamellae some distance removed from the plasmalemma that bond breakage becomes important for growth. Extension factors produced under the influence of IAA will then need to diffuse through younger lamellae to these older load-bearing lamellae. At any one instant, bond breaking will be strictly localized, and this may be why the cell extends longitudinally and the wall creeps longitudinally; since, on account of the more or less transverse orientation of the microfibrils, many more bonds would need simultaneously to be broken to allow lateral slip between microfibrils. Lowering of turgor pressure might not only reduce the incorporation of the necessary hemicellulose and proteins but might reduce the availability to these lamellae of extension factors, effects which might to some extent be offset by a reduction in wall deposition as a negative feed-back mechanism. The sum total of the growth mechanism may very tentatively be summarized as in Fig. 7.25.

Cessation of elongation would follow if any of the following factors intervene.

(a) A secondary wall is deposited with longitudinally oriented microfibrils. The effect would not be immediate since a finite thickness of secondary wall would be needed to offset the anisotropy of the primary wall. Transverse extension would be limited if *either* the secondary wall attained more than a critical thickness, *or* the chemical nature of the non-cellulosic material was such as to

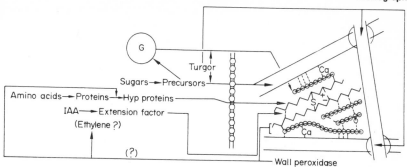

Fig. 7.25 Schematic summary of possible associations of wall extension with growth factors. ———, araban, arabanogalactan or other hemicelluloses, ⋀⋀ wall protein; Dotted lines, hydrogen bonds; S, sulphide bond.

be resistant to the extension factor. The situation examined by Veen (1970), and reported above (p. 291), may represent an intermediate situation though the explanation of his findings is not clear. His parenchyma cells before treatment with IAA and 8% sucrose were presumably elongating and changing little in transverse dimension, by the mechanism discussed above. Treatment with 8% sucrose might reduce longitudinal extension both by reducing turgor pressure and through a longitudinal strengthening of the wall through the observed deposition of longitudinally arranged microfibrils. Both of these effects would also, however, reduce the tendency to transverse expansion since the longitudinal microfibrils are deposited on a wall which was already strong enough laterally. Perhaps some mechanism is invoked which is not normally effective in growth — even a massive synthesis and utilization of cellulase such as that found by Fan and Machlachlan (1967).

(b) The bonding in the wall became so intense that localized weakening became ineffective. This might be the consequence of the increased content of Hyp protein in walls toward the end of growth (Cleland, 1967).

(c) The synthesis of the extension factor ceased, as on treatment with Hyp or with inhibitors of resporation or protein synthesis.

(d) 'Trivial' factors intervened, e.g., loss of turgor, cell senescence, etc.

Bibliography

Allsopp, A. and Misrah, P. (1940) 'The constitution of the cambium, the new wood and the mature sapwood of the common ash, the common elm and the scotch pine', *Biochem. J.*, **34**, 1078.

Altermatt, H. A. and Neish, A. C. (1956) 'The biosynthesis of cell wall carbohydrates III. Further studies on formation of cellulose and xylan from labelled monosaccharides in wheat plants', *Canad. J. Biochem. Physiol.*, **34**, 405.

Anderson, D. B. (1927) 'Uber die Struktur der Kollenchymzellwand auf Grund mikrochemischer Untersuchungen', *Sber. Akad. Wiss. Wien Math. — Naturwissensch. Cl.*, **136**, 429.

Astbury, W. T. and Preston, R. D. (1940) 'The structure of the cell wall in some species of the filamentous green alga *Cladophora*', *Proc. R. Soc.*, B **129**, 54.

Atkins, E. D. T., Parker, K. D., and Preston, R. D. (1969) 'The helical structure of the β-1,3-linked xylan in some siphoneous green algae. *Proc. R. Soc.*, B **173**, 209.

Plant cell walls

Atkins, E. D. T. and Parker, K. D. (1969) 'The helical structure of a β-D-1,3-Xylan', *J. Polyme Sci.*, C **28**, 69.

Bailey, I. W. and Kerr, T. (1935) 'The visible structure of the secondary wall and its significance in physical and chemical investigations of tracheary cells and fibers', *J. Arnold Arbor.*, **16**, 273.

Bailey, R. W. and Hassid, W. Z. (1966) 'Xylan synthesis from uridine-diphosphate-D-xylose by particulate preparations from immature corncobs', *Proc. natn. Acad. Sci. U.S.A.*, **56**, 1586.

Barber, G. A., Elbein, A. D., and Hassid, W. Z. (1964) 'The synthesis of cellulose by enzyme systems from higher plants', *J. biol. Chem.*, **239**, 2672.

Barnett, J. R. and Preston, R. D. (1970) 'Arrays of granules associated with the plasmalemma in swarmers of *Cladophora*', *Ann. Bot.* **34**, 1011.

Batra, K. K. and Hassid, W. Z. (1969) 'Determination of linkages in β-D-Glucanase from *Phaseolus aureus* by exo-β-(1→3)-D-Glucanase. *Pl. Physiol., Lancaster*, **44**, 755.

Bayley, S. T., Colvin, J. R., Cooper, F. P., and Martin-Smith, C. A. (1957) 'The structure of the primary epidermal cell wall of *Avena* coleoptiles', *J. biophys. biochem. Cytol.*, **3**, 171.

Beer, M. and Setterfield, G. (1958) 'Fine structure in thickened primary walls of collenchyma cells of celery peticles', *Am. J. Bot.*, **45**, 571.

Bennet-Clark, T. A. (1956) 'The kinetics of auxin-induced growth', In: *The chemistry and mode of action of plant growth substances* (eds. R. L. Wain and F. Wightman), Butterworth's, London.

Bishop, C. T., Bayley, S. T., and Setterfield, G. (1958) 'Chemical constitution of the primary cell walls of *Avena* coleoptiles', *Pl. Physiol., Lancaster*, **33**, 283.

Bittiger, H. and Husemann, E. (1964) 'Elektronenmikroskopische Abbildung von Einmolekülkristallen bei Cellulosetricarbanilaten', *Makromol. Chem.*, **75**, 222.

Black, M., Bullock, C., Chantler, E. C., Clarke, R. A., Hanson, A. D., and Jolley, G. M. (1967) 'Effect of inhibitors of protein synthesis on the plastic deformation and growth of plant tissues', *Nature, Lond.*, **215**, 1289.

Bonner, J. (1936) 'Zum Mechanismus der Zellstreckung auf Grund der Micellarlehre', *Jb. wiss. Bot.*, **82**, 377.

Bowes, B. G. (1969) 'The fine structure of wall modifications and associated structures in callus tissue of *Andrographis paniculate* nees', *New Phytol.* **68**, 619.

Bowes, B. G. and Butcher, R. (1967) 'Electron microscopic observations on cell wall inclusions in *Andrographis paniculata* callus', *Zeit. PflPhysiol.*, **58**, 86.

Brown, R. M., Franke, W. W., Kleinig, H., Falk, H., and Sitte, P. (1970) 'Scale formation in chrysophycean algae', *J. Cell Biol.*, **45**, 246.

Brummond, D. G. and Gibbons, A. P. (1965) 'Enzymatic cellulose synthesis from UDP-(¹⁴C)-glucose by *Lupinus albus*', *Biochem. Z.*, **342**, 308.

Burström, H. G., Uhrström, I., and Wurscher, R. (1967) 'Growth, turgor, water potential and Young's modulus in pea internodes', *Physiologia Pl.*, **20**, 213.

Burton, K. and Krebs, H. A. (1953) 'The free-energy changes associated with the individual steps of the tricarboxylic acid cycle, glycolysis and alcoholic fermentation and with the hydrolysis of the pyrophosphate groups of adenosinetriphosphate', *Biochem. J.*, **54**, 94.

Buston, H. (1935) 'Cell wall constituents of plants. XXV. Observations on the nature, distribution and development of certain cell-wall constituents of plants', *Biochem. J.*, **29**, 196.

Cave, I. D. (1969) 'The longitudinal Young's modulus of *Pinus radiata*', *Wood Sci. Technol.*, **3**, 40.

Christensen, T. (1962) 'Systematisk Botanik, 2 Alger', In: *Botanik* 2, Munksgaard, Copenhagen.

Cleland, R. (1963a) 'Hydroxyproline as an inhibitor of auxin-induced cell elongation', *Nature, Lond.*, **200**, 908.

Cleland, R. (1963b) 'Independence of effects of auxin on cell wall methylation and elongation', *Pl. Physiol. Lancaster*, **38**, 12.

Cleland, R. (1963c) 'The occurrence of auxin-induced pectin methylation in plant tissues', *Pl. Physiol., Lancaster*, **38**, 738.

Cleland, R. (1967a) 'Inhibition of cell elongation in *Avena* coleoptile by hydroxyproline', *Pl. Physiol., Lancaster*, **42**, 271.

302

Cleland, R. (1967b) 'Auxin and the mechanical properties of the cell wall', *Ann. N. Y. Acad. Sci.*, **144**, 3.

Cleland, R. and Karlsnes, Anne M. (1967) 'A possible role of hydroxyproline-containing proteins in the cessation of cell elongation', *Pl. Physiol., Lancaster*, **42**, 669.

Cleland, R. (1968a) 'Auxin and wall extensibility: reversibility of auxin-induced wall-loosening process', *Science, N. Y.*, **160**, 192.

Cleland, R. (1968b) 'Hydroxyproline formation and its relation to auxin-induced cell elongation in the *Avena* coleoptile', *Pl. Physiol. Lancaster*, **43**, 1625.

Cleland, R. (1971) 'The mechanical behaviour of isolated *Avena* coleoptile walls subjected to constant stress: properties and relation to cell elongation', *Pl. Physiol., Lancaster*, **47**, 805.

Cleland, R. and Haughton, P. M. (1971) 'The effect of auxin on stress relaxation in isolated *Avena* coleoptiles', *Pl. Physiol., Lancaster*, **47**, 812.

Colvin, J. Ross (1961) 'Synthesis of cellulose from ethanol-soluble precursors in green plants', *Can. J. Biochem. Physiol.*, **39**, 1921.

Colvin, J. Ross (1963) 'The size of the cellulose microfibril', *J. Cell Biol.*, **17**, 105.

Colvin, J. Ross, Bayley, S. T., and Beer, M. (1957) 'The growth of cellulose microfibrils from Acetobacter xylinum', *Biochim. biophys. Acta*, **23**, 652.

Correns, C. (1894) 'Ueber die Membran von *Caulerpa*', *Ber. dt. bot. Ges.*, **12**, 355.

Courtney, J. S., Morré, D. J., and Key, J. L. (1967) 'Inhibition of RNA synthesis and auxin-induced cell wall extensibility and growth by actinomycin D', *Pl. Physiol., Lancaster*, **42**, 434.

Cowdrey, D. R. and Preston, R. D. (1966) 'Elasticity and microfibrillar angle in the wood of Sitka spruce', *Proc. R. Soc.*, B **166**, 245.

Cronshaw, J., Myers, A., and Preston, R. D. (1958) 'A chemical and physical investigation of cell walls of some marine algae', *Biochim. biophys. Acta*, **27**, 89.

Dennis, D. T. and Preston, R. D. (1961) 'Constitution of cellulose microfibrils', *Nature, Lond.*, **191**, 667.

Dickson, A. D., Otterson, H., and Link, K. P. (1930) 'A method for the determination of uronic acids', *J. Am. chem. Soc.*, **52**, 775.

Dolmetsch, H. and Dolmetsch, H. (1962) 'Anzeichen für eine Kettenfaltung des Cellulosemoleküls', *Kolloid-Zeit.*, **185**, 106.

Dougal, D. K. and Shibayashi, K. (1960) 'Factors affecting growth of tobacco callus tissue and its incorporation of tyrosine', *Pl. Physiol., Lancaster*, **35**, 396.

Drawert, H. and Mix, M. (1962) 'Zur Funktion des Golgi-Apparates in det Pflanzenzelle', *Planta*, **58**, 448.

Ernst, A. (1903) '*Siphoneen*-Studien', *Beih. bot. Zbl.*, **13**, 115.

Fan, D.-F. and Machlachlan, G. A. (1967) 'Massive synthesis of ribonucleic acid and cellulase in pea epicotyl in response to indoleacetic acid, with and without concurrent cell division', *Pl. Physiol., Lancaster*, **42**, 1114.

Feingold, D. S., Neufeld, E. F., and Hassid, W. Z. (1958) 'Synthesis of a β-1,3-linked glucan by extracts of *Phaseolus aureua* seedlings', *J. biol. Chem.*, **233**, 783.

Flowers, H. M., Batra, K. K., Kemp, J., and Hassid, W. Z. (1968) 'Biosynthesis of insoluble glucans from uridine-diphosphate-D-glucose with enzyme preparations from *Phaseolus aureus* and *Lupinus albus*', *Pl. Physiol., Lancaster*, **43**, 1703.

Frei, Eva and Preston, R. D. (1961a), 'Cell wall organization and wall growth in the filamentous green algae *Cladophora* and *Chaetomorpha*. I. The basic structure and its formation', *Proc. R. Soc.*, B **154**, 70.

Frei, Eva and Preston, R. D. (1961b) 'Cell wall organization and wall growth in the filamentous green algae *Cladophora* and *Chaetomorpha*. II. Spiral structure and spiral growth', *Proc. R. Soc.*, B **155**, 55.

Frei, Eva and Preston, R. D. (1964a) 'Non-cellulosic structural polysaccharides in algal cell walls. I. Xylan in siphoneous green algae', *Proc. R. Soc.*, B **160**, 293.

Frei, Eva and Preston, R. D. (1964b) 'Non-cellulosic structural polysaccharides in algal cell walls. II. Association of xylan and mannan in *Porphyra umbilicalis*', *Proc. R. Soc.*, B **160**, 314.

Frei, Eva and Preston, R. D. (1968) 'Non-cellulosic structural polysaccharides in algal cell walls. III. Mannan in siphoneous green algae', *Proc. R. Soc.*, B **169**, 127.

Frei, Eva, Preston, R. D., and Ripley, G. W. (1957) 'The fine structure of the walls of conifer tracheids. VI. Electron microscope investigations of sections', *J. exp. Bot.*, **8**, 139.

Plant cell walls

Frey-Wyssling, A. (1948) *Growth Symposium,* **12,** 151.

Frey-Wyssling, A., Mülethaler, K., and Wyckoff, R. W. G. (1948) 'Mikrofibrillenbau der Pflanzlichen Zellwände', *Experientia,* **4,** 475.

Galston, A. W., Baker, R. S., and King, J. N. (1953) 'Benzimidazole and the geometry of cell growth', *Physiologia Pl.,* **6,** 863.

Glaser, L. (1958) 'The synthesis of cellulose in cell-free extracts of *Acetobacter xylinum'*, *J. biol. Chem.,* **232,** 627.

Green, P. B. (1954) 'The spiral growth pattern of the cell wall in *Nitella axillaris'*, *Am. J. Bot.,* **41,** 403.

Green, P. B. (1958) 'Structural characteristics of developing *Nitella* internodal cell walls', *J. biophys. biochem. Cytol.,* **4,** 505.

Griffioen, K. (1938) 'On the origin of lignin in the cell wall', *Rec. trav. bot. Néerl.,* **35,** 322.

Griffioen, K. (1939) 'Changes in the composition of the needles of *Pinus austriaca* Link during the ageing-process', *Rec. trav. bot. Néerl.,* **36,** 347.

Harris, P. J. and Northcote, D. H. (1971) 'Polysaccharide formation in plant Golgi bodies', *Biochim. biophys. Acta,* **237,** 56.

Haughton, P. M., Sellen, D. B., and Preston, R. D. (1968) 'Dynamic mechanical properties of the cell wall of *Nitella opaca'*, *J. exp. Bot.,* **19,** 1.

Haughton, P. M. and Sellen, D. B. (1969) 'Dynamic mechanical properties of the cell walls of some green algae', *J. exp. Bot.,* **20,** 516.

Haughton, P. M., Sellen, D. B., and Preston, R. D. (1969) 'The viscoelastic properties of algal cell walls', *Proc. VIth Int. Seaweed Symp.,* **6,** 453.

Hengstenberg, J. and Mark, H. (1928) 'Über Form und Grösse der Mizelle von Zellulose und Kautschuk', *Zeit. Kristallogr.,* **69,** 271.

Heyn, A. N. J. (1931) 'Der Mechanismus der Zellstreckung', *Rec. trav. bot. Néerl.,* **28,** 113.

Heyn, A. N. J. (1958) 'Small particle X-ray scattering by fibers, size and shape of micro-crystallites', *J. Appl. Phys.,* **26,** 519.

Heyn, A. N. J. (1966) 'The microcrystalline structure of cellulose in cell walls of cotton, ramie and jute fibers as revealed by negative staining of sections', *J. Cell Biol.,* **29,** 181.

Iriki, Y. and Miwa, T. (1960) 'Chemical nature of the cell wall of the green algae, *Codium, Acetabularia* and *Halicoryne'*, *Nature, Lond.,* **185,** 178.

Iriki, Y., Susuki, T., Nisizawa, K., and Miwa, T. (1960) 'Xylan of siphonaceous green algae', *Nature, Lond.,* **187,** 82.

Israel, H. W., Salpeter, M. W., and Steward, F. C. (1968) 'The incorporation of radioactive proline into cultured cells. Interpretation based on radioautography and electron microscopy', *J. Cell Biol.,* **39,** 698.

Jansen, E. F., Jang, R., Albersheim, P., and Bonner, J. (1960) 'Pectic metabolism of growing cell walls', *Pl. Physiol., Lancaster,* **35,** 87.

Jansen, E. F., Jang, R., and Bonner, J. (1960) 'Binding of enzymes to *Avena* coleoptile cell walls', *Pl. Physiol., Lancaster,* **35,** 567.

Kauss, H. and Swanson, A. L. (1969) 'Cooperation of enzymes responsible for polymerisation and methylation in pectin biosynthesis', *Z. Naturforsch.,* **24,** 28.

Kauss, H., Swanson, A. L., Arnold, R., and Odzuck, W. (1969) 'Biosynthesis of pectic substances: Localisation of enzymes and products in a lipid-membrane complex', *Biochim. biophys. Acta,* **192,** 55.

Keller, A. (1959) 'The morphology of crystalline polymers', *Makromol. Chem.,* **34,** 1.

Kramer, D. (1970) 'Fine structure of growing cellulose fibrils of *Ochromonas malhamensis* Pringsheim (syn. *Poteriochromonas stipitata* Scheffl)', *Z. Naturforsch.,* **25b,** 1017.

Lamport, D. T. A. (1962) 'Hydroxyproline of primary cell walls', *Fed. Proc.,* **21,** 398.

Lamport, D. T. A. (1965) 'The protein component of primary cell walls', In: *Advances in botanical research II* (ed. R. D. Preston), Academic Press.

Lamport, D. T. A. (1967a) 'Evidence for a hydroxyproline-O-glycosidic cross link in the plant cell-wall glycoprotein extensin', *Nature, Lond.,* **26,** 608.

Lamport, D. T. A. (1967b) 'Hydroxyproline-O-glycosidic linkage of the plant cell wall glycoprotein extensin', *Nature, Lond.,* **216,** 1322.

Lamport, D. T. A. (1969) 'The isolation and partial characterization of hydroxyproline-rich glycopeptides obtained by enzymic degradation of primary cell walls', *Biochemistry, N.Y.,* **8,** 1115.

Lamport, D. T. A., Katone, K., and Roerig, S. (1971) 'Amino acid sequence of hydroxyproline-rich tryptic peptides from acid-stripped primary cell walls', *Abstr. Amer. Soc. Biochem.*, San Francisco.

Lamport, D. T. A. and Northcote, D. H. (1960a) 'Hydroxyproline in primary cell walls of higher plants', *Nature, Lond.*, 188, 665.

Lamport, D. T. A. and Northcote, D. H. (1960b) 'The use of tissue cultures for the study of plant-cell walls', *Biochem. J.*, 76, 52P.

Leitgeb, H. (1888) 'Die Incrustation der Membran von *Acetabularia*', *Sber. Akad. Wiss. Wien. Math.-Naturwissensch.*, 96, 13.

Limaye, V. J., Roelofsen, P. A., and Spit, B. J. (1962) 'Submicroscopic structure of the cellulose in the cell walls of *Spirogyra* and *Nitella*', *Acta bot. neerl.*, 11, 225.

Lin, T.-Y. and Hassid, W. Z. (1970) 'Solubilization and partial purification of cellulose synthetase from *Phaseolus aureus*', *J. biol. Chem.*, 245, 1922.

López-Sáez, J. P., Giménez-Martín, G., and Risueño, M. C. (1966) 'Fine structure of the plasmodesm', *Protoplasma*, 61, 81.

Love, J. and Percival, E. (1964) 'The polysaccharides of the green seaweed *Codium fragile*', Part III. A β-1,4-linked mannan', *J. Chem. Soc.*, pt. 3, 3345.

Machlachlan, G. A. and Duda, C. T. (1965) 'Changes in concentration of polymeric components in excised pea-epicotyl tissue during growth', *Biochim. biophys. Acta*, 97, 288.

Machlachlan, G. A. and Young, M. (1962) 'Breakdown and synthesis of cell walls during growth', *Nature, Lond.*, 195, 1319.

Mackie, I. M. and Percival, E. (1959) 'The constitution of xylan from the green seaweed *Caulerpa filiformis*', *J. chem. Soc.*, p. 1151.

Mackie, W. (1969) 'The degree of polymerisation of xylan in the cell wall of the green seaweed *Penicillus dumetosus*', *Carbohydrate Res.*, 9, 247.

Mackie, W. and Preston, R. D. (1968) 'The occurrence of mannan microfibrils in the green algae *Codium fragile* and *Acetabularia crenulata*', *Planta*, 79, 249.

Mackie, W. and Sellen, D. B. (1969) 'The degree of polymerization and polydispersity of mannan from the cell wall of the green seaweed *Codium fragile*', *Polymer*, 10, 621.

Mackie, W. and Sellen, D. B. (1971) 'Degree of polymerization and polydispersity of xylan from the cell wall of the green seaweed *Penicillus dumetosus*', *Biopolymers*, 10, 1.

Majumdar, G. P. and Preston, R. D. (1941) 'The fine structure of the walls of collenchyma cells in *Heracleum sphondylium L.*', *Proc. R. Soc. B*, 130, 201.

Manley, R. St. J. (1964) 'Fine structure of native cellulose microfibrils', *Nature, Lond.*, 204, 1155.

Marchant, R. and Robards, A. W. (1968) 'Membrane systems associated with the plasmalemma of plant cells', *Ann. Bot.*, 32, 457.

Marchessault, R. H. and Liang, C. Y. (1962) 'The infrared spectra of crystalline polysaccharides. VIII. Xylans', *J. Polymer Sci.*, 59, 357.

Margerie, C. and Lenoël, P. D. (1961) 'Cinétique de la biosynthèse de la cellulose dans les racines de blé', *Biochim. biophys. Acta*, 47, 275.

Marx-Figini, M. and Schultz, G. V. (1966) 'Uber die Kinetik und den Mechanismus der Biosynthese der Cellulose in den höheren Pflanzen. (Nach versuchen an den Samenhaaren der Baumwolle)', *Biochim. biophys. Acta*, 112, 81.

Matchett, N. H. and Nance, J. F. (1962) 'Cell wall breakdown and growth in pea seedling stems', *Am. J. Bot.*, 49, 311.

Meier, H. (1955) 'Über den Zellwandabbau durch Holzvermorschungspilze und die submikroskopische Struktur von Fichtentracheiden und Kirkenholzfasern', *Holz Roh- u. Werkst.*, 13, 323.

Meier, H. (1958) 'On the structure of cell walls and cell wall mannans from ivory nuts and from dates', *Biochim. biophys. Acta*, 28, 229.

Meier, H. (1961) 'The distribution of polysaccharides in wood fibers', *J. Polymer Sci.*, 51, 11.

Meyer, K. H. (1942) High polymers, vol. IV. *Natural and synthetic high polymers*, Interscience, New York.

Meyer, K. H. and Misch, L. (1936) 'Positions des atomes dans le nouveau modèle spatial de la cellulose. (Sur la constitution de la partie cristallisée de la cellulose VI)', *Helv. chim. Acta*, 20, 232.

Middlebrook, Mavis J. and Preston, R. D. (1952) 'Spiral growth and spiral structure. III. Wall structure in the growth zone of *Phycomyces*', *Biochim. biophys. Acta*, 9, 32.

Miwa, T., Iriki, Y., and Susaki, T. (1961) *Colloques internationaux du centre national de la recherche scientifique No. 103. Chimie et physico-chimie des principes immédiats tirés des algues*, p. 135, Editions du Centre National de la recherche scientifique, Paris.

Mollenhauer, H. H., Whaley, W. G., and Leech, J. H. (1961) 'A function of the Golgi apparatus in outer rootcap cells', *J. Ultrastruct. Res.*, **5**, 193.

Moor, H. and Mühlethaler, K. (1963) 'Fine structure in frozen-etched yeast cells', *J. Cell Biol.*, **17**, 609.

Muggli, R. (1968) 'Fine structure of cellulose elementary fibrils', *Cellulose Chem. Technol.*, **2**, 549.

Muggli, R., Elias, H.-G., and Mühlethaler, K. (1969) 'Zum Feinbau der Elementarfibrillen der Cellulose', *Makromol. Chem.*, **121**, 290.

Mühlethaler, K. (1967) 'Ultrastructure and formation of plant cell walls', *A. Rev. Pl. Physiol.*, **18**, 1.

Myers, A. and Preston, R. D. (1959) 'Fine structure in the red algae. II. The structure of the cell wall of *Rhodymenia palmata*', *Proc. R. Soc.*, B. **150**, 447.

Myers, A., Preston, R. D., and Ripley, G. W. (1956) 'Fine structure in the red algae. I. X-ray and electron microscope investigation of *Griffithsia flosculosa*', *Proc. R. Soc.*, B **144**, 450.

Nägeli, C. von (1844) '*Caulerpa prolifera Ag*', *Z. wiss. Bot. Schleiden Nägeli*, **1**, 134.

Nakamura, Y. and Hess, K. (1938) 'Zur Kenntnis der chemischen Zusammensetzung von Mais-Koleoptilen', *Ber. dt. chem. Ges.*, **71**, 145.

Neish, A. C. (1958) 'The biosynthesis of cell wall carbohydrates. IV. Further studies on cellulose and xylan in wheat', *Can. J. Biochem. Physiol.*, **36**, 187.

Nelmes, B. J. and Preston, R. D. (1968a) 'Cellulose microfibril orientation in rubbery wood', *J. exp. Bot.*, **19**, 519.

Nelmes, B. J. and Preston, R. D. (1968b) 'Wall development in apple fruits: a study of the life history of a parenchyma cell', *J. exp. Bot.*, **19**, 496.

Nieduszynski, I. A. and Atkins, E. D. T. (1970) 'Preliminary investigation of algal cellulose', *Biochim. biophys. Acta.*, **222**, 109.

Nieduszynski, I. A. and Preston, R. D. (1970) 'Crystallite size in natural cellulose', *Nature, Lond.*, **225**, 273.

Northcote, D. H. and Pickett-Heaps, J. D. (1966) 'A function of the Golgi apparatus in polysaccharide synthesis and transport in the rootcap cells of wheat', *Biochem. J.*, **98**, 159

Olson, A. C., Bonner, J., and Morré, J. (1965) 'Force extension analysis of *Avena* coleoptile cell walls', *Planta*, **66**, 126.

Ordin, L. (1958) 'Effect of water stress on the cell-wall metabolism of plant tissue', In: *Radioisotopes in scientific research* (ed. R. C. Estermann), vol. 4, p. 553, Pergamon, Oxford.

Ordin, L. and Hall, M. A. (1967) 'Studies on cellulose cynthesis by a cell-free oat coleoptile enzyme system: inactivation by airborne oxidants', *Pl. Physiol., Lancaster*, **42**, 205.

Ordin, L. and Hall, M. A. (1968) 'Cellulose synthesis in higher plants from UDP glucose', *Pl. Physiol., Lancaster*, **43**, 473.

Petty, J. A. and Preston, R. D. (1969) 'The dimensions and number of pit membrane pores in conifer wood', *Proc. R. Soc.*, B, **172**, 137.

Phillips, E. W. J. (1941) 'The inclination of fibrils in the cell wall and its relation to the compression of timber', *Emp. For. J.*, **20**, 74.

Pickett-Heaps, J. D. and Northcote, D. H. (1966a) 'Organization of microtubules and endoplasmic reticulum during mitosis and cytokinesis in wheat meristems', *J. Cell Sci.*, **1**, 109.

Pickett-Heaps, J. D. and Northcote, D. H. (1966b) 'Cell division in the formation of the stomatal complex of the young leaves of wheat', *J. Cell Sci.*, **1**, 121.

Pinsky, A. and Ordin, L. (1969) 'Role of lipid in the cellulose synthetase enzyme system from oat seedlings', *Pl. Cell Physiol.*, **10**, 771.

Pollard, J. K. and Steward, F. C. (1959) 'The use of C^{14}-proline by growing cells: its conversion to protein and to hydroxyproline', *J. exp. Bot.*, **10.**, 17.

Preston, R. D. (1934) 'The organization of the cell wall of the conifer tracheid', *Phil. Trans. R. Soc. Ser.*, B **224**, 131.

Preston, R. D. (1938) 'The structure of the walls of parenchyma in *Avena* coleoptiles', *Proc. R. Soc.*, B **125**, 372.

Preston, R. D. (1946) 'The fine structure of the wall of the conifer tracheid. I. The X-ray diagram of conifer wood', *Proc. R. Soc.*, B **133**, 327.

Preston, R. D. (1947) 'The fine structure of the wall of the conifer tracheid. II. Optical properties of dissected cells in *Pinus insignis*', *Proc. R. Soc.*, B **134**, 202.

Preston, R. D. (1948) 'The fine structure of the wall of the conifer tracheid. III. Dimensional relationships in the central layer of the secondary wall', *Biochim. biophys. Acta,* **2,** 370.

Preston, R. D. (1951) 'Fibrillar units in the structure of native cellulose', *Faraday Soc. Disc.* No. 11, p. 165.

Preston, R. D. (1958) 'Wall organization in plant cells', *Int. Rev. Cytol.,* **8,** 33.

Preston, R. D. (1961) 'Cellulose-protein complexes in plant cell walls', In: *Macromolecular complexes* (ed. M. V. Edds), p. 229, The Ronald Press Co., N.Y.

Preston, R. D. (1962) 'The electron microscopy and electron diffraction analysis of natural cellulose', In: *The Interpretation of ultrastructure* (ed. R. J. C. Harris), *Symp. Int. Soc. Cell Biol.,* **I,** 325.

Preston, R. D. (1964) 'Structural and mechanical aspects of plant cell walls with particular reference to synthesis and growth', In: *The Formation of wood in forest trees* (ed. M. Zimmermann), p. 169, Academic Press Inc., N.Y.

Preston, R. D. (1971) 'Negative staining and cellulose microfibril size', *J. Microscopy,* **93,** 7.

Preston, R. D. and Cronshaw, J. (1958) 'Constitution of the fibrillar and non-fibrillar components of the walls of *Valonia ventricosa',* *Nature, Lond.,* **181,** 248.

Preston, R. D. and Goodman, R. N. (1968) 'Structural aspects of cellulose microfibril biosynthesis', *Jl. R. microsc. Soc.,* **88,** 513.

Preston, R. D. and Singh, K. (1950) 'The fine structure of bamboo fibres. I. Optical properties and X-ray data', *J. exp. Bot.,* **1,** 214.

Preston, R. D. and Singh, K. (1952) 'The fine structure of bamboo fibres. II. Refractive indices and wall density', *J. exp. Bot.,* **3,** 162.

reston, R. D., Nicolai, E., Reed, R., and Millard, A. (1948) 'An electron microscope study of cellulose in the wall of *Valonia ventricosa',* *Nature, Lond.,* **162,** 665.

Preston, R. D. and Wardrop, A. B. (1949a) 'The submicroscopic organisation of the walls of conifer cambium', *Biochim. biophys. Acta.,* **3,** 549.

Preston, R. D. and Wardrop, A. B. (1949b) 'The fine structure of the wall of the conifer tracheid. IV. Dimensional relationships in the outer layer of the secondary wall', *Biochim. biophys. Acta.,* **3,** 585.

Pridham, J. B. and Hassid, W. Z. (1966) 'A preliminary study on the biosynthesis of hemicelluloses', *Biochem. J.,* **100,** 21P.

Probine, M. C. (1965) 'Chemical control of plant cell wall structure and of cell shape', *Proc. R. Soc.,* B **161,** 526.

Probine, M. C. and Barber, N. P. (1966) 'The structure and plastic properties of the cell wall of *Nitella* in relation to extension growth', *Aust. J. biol. Sci.,* **19,** 439.

Probine, M. C. and Preston, R. D. (1961) 'Cell growth and the structure and mechanical properties of the wall in internodal cells of *Nitella opaca.* I. Wall structure and growth', *J. exp. Bot.,* **12,** 261.

Probine, M. C. and Preston, R. D. (1962) 'Cell growth and the structure and mechanical properties of the wall in internodal cells of *Nitella opaca.* II. Mechanical properties of the walls', *J. exp. Bot.,* **13,** 111.

Ray, P. M. (1967) 'Radioautographic study of cell wall deposition in growing plant cells', *J. Cell Biol.,* **35,** 659.

Ray, P. M., Shininger, T. L., and Ray, M. M. (1969) 'Isolation of β-glucan synthetase particles from plants cells and identification with Golgi membranes', *Proc. natn. Acad. Sci. U.S.A.,* **64,** 605.

Robards, A. W. (1968) 'A new interpretation of plasmodesmatal ultrastructure', *Planta,* **82,** 200.

Robards, A. W. (1971) 'The ultrastructure of plasmodesmata', *Protoplasma,* **72,** 315–323.

Robards, A. W. and Humpherson, P. G. (1967) 'Microtubules and angiosperm bordered pit formation', *Planta,* **77,** 233.

Roberts, R. M., Deshusses, J., and Loewus, F. (1968) 'Inositol metabolism in plants. V. conversion of myo-inositol to uronic acid and pentose units of acidic polysaccharides in root-tips of *Zea mays',* *Pl. Physiol., Lancaster,* **43,** 979.

Robinson, D. G. and Preston, R. D. (1971) 'Fine structure of swarmers of *Cladophora* and *Chaetomorpha.* I. The plasmalemma and Golgi apparatus in naked swarmers', *J. Cell Sci.,* **9,** 581–601.

Robinson, D. G. and Preston, R. D. (1972) 'Polysaccharide synthesis in mung bean roots – an X-ray investigation', *Biophys. Biochim. Acta,* **273,** 336.

Plant cell walls

Roelofsen, P. A. (1959) 'The plant cell wall', *Encyclop. Plant Anat.* vol. 3, Borntraeger, Berlin.

Roelofsen, P. A. (1965) 'Ultrastructure of the wall in growing cells and its relation to the direction of the growth', In: *Advances in botanical research* (ed. R. D. Preston), vol. II, Academic Press, London and New York.

Roelofsen, P. A. and Houwink, A. L. (1951) 'Cell wall structure of staminal hairs of *Tradescantia virginica* and its relation with growth', *Protoplasma*, **40**, 1.

Roelofsen, P. A. and Houwink, A. L. (1953) 'Architecture and growth of the primary cell wall in some plant hairs and in the *Phycomyces sporangiophore*', *Acta bot. neerl.*, **2**, 218.

Setterfield, G. and Bayley, S. T. (1958) 'Arrangement of cellulose microfibrils in walls of elongating parenchyma cells', *J. biophys. biochem. Cytol.*, **4**, 377.

Setterfield, G. and Bayley, S. T. (1961) 'Structure and physiology of cell walls', *A. Rev. Pl. Physiol.*, **12**, 35.

Shannon, L. M., Kay, E., and Lew, S. Y. (1966) 'Peroxidase isozymes from horseradish roots. I. Isolation and physical properties', *J. biol. Chem.*, **241**, 2166.

Sievers, A. (1963) 'Beteiligung des Golgi-Apparates bei der Bildung der Zellwand von Wurzelhaaren', *Protoplasma*, **56**, 188.

Slater, W. G. and Beevers, H. (1958) 'Utilization of D-glucuronate by corn coleoptiles', *Pl. Physiol., Lancaster*, **33**, 146.

Staehelin, A. S. and Probine, M. C. (1970) 'Structural aspects of cell membranes', In: *Advances in botanical research III* (ed. R. D. Preston), Academic Press, London and New York.

Stafford, L. E. and Brummond, D. O. (1970) 'β-(1→4)-D-glucan synthesis from UDP-(^{14}C)-glucose by particulate and solubilized enzyme preparations from *Lupinus albus*', *Phytochem.*, **9**, 253.

Stirling, C. (1957) 'Cellulose micelles in developing fibers of asparagus', *Acta bot. neerl.*, **6**, 458.

Stirling, C. and Spit, B. J. (1957) 'Microfibrillar arrangement in developing fibers of asparagus', *Am. J. Bot.*, **44**, 851.

Steward, F. C., Mott, R. L., Israel, H. W., and Ludford, P. M. (1970) 'Proline in the vesicles and sporelings of *Valonia ventricosa* and the concept of cell wall protein', *Nature, Lond.*, **225**, 760.

Steward, F. C., Pollard, J. K., Patchett, A. A., and Witkop, B. (1958) 'The effects of selected nitrogen compounds on the growth of plant tissue cultures', *Biochim. biophys. Acta*, **28**, 308.

Strasburger, E. (1962) *Lehrbuch det Botanik*. 28. Auflage neu bearbeitet, Gustav Fischer Verlag, Stuttgart.

Strominger, J. L. and Mapson, L. W. (1957) 'Uridine diphosphoglucose dehydrogenase of pea seedlings', *Biochem. J.*, **66**, 567.

Thimann, K. V. and Bonner, J. (1933) 'The mechanism of the action of the growth substance of plants', *Proc. R. Soc.*, B **113**, 126.

Thompson, E. W. and Preston, R. D. (1967) 'Proteins in the cell walls of some green algae', *Nature, Lond.*, **213**, 684.

Thompson, E. W. and Preston, R. D. (1968) 'Evidence for a structural role of protein in algal cell walls', *J. exp. Bot.*, **19**, 690.

Timell, T. E. (1957) 'Vegetable ivory as a source of a mannan polysaccharide', *Can. J. Chem.*, **35**, 333.

Timell, T. E. (1964) 'Wood hemicelluloses: Part I', *Adv. Carbohydrate Chem.*, **19**, 247.

Timell, T. E. (1965) 'Wood hemicelluloses: Part II', *Adv. Carbohydrate Chem.*, **20**, 409.

Treloar, L. G. R. (1960) 'Calculations of elastic moduli of polymer crystals: III. Cellulose', *Polymer*, **1**, 290.

Vasumati, L., Roelofsen, P. A., and Spit, B. J. (1962) 'Submicroscopic structure of the cellulose in the cell-walls of *Spirogyra* and *Nitella*', *Acta bot. neerl.*, **11**, 225–227.

Veen, B. W. (1970) 'Orientation of microfibrils in parenchyma cells of pea stem before and after longitudinal growth', *Proc. K. ned. Akad. Wet.*, C **73**, 113.

Villemez, C. L. and Clark, A. F. (1969) 'A particle bound intermediate in the biosynthesis of plant cell wall polysaccharides', *Biochem. biophys. Res. Commun.*, **36**, 57.

Villemez, C. L., Lin, T.-Y., and Hassid, W. Z. (1965) 'Biosynthesis of the polygalacturonic acid chain of pectin by a particulate enzyme preparation from *Phaseolus aureus* seedlings', *Proc. natn. Acad. Sci. U.S.A.*, **54**, 1626.

Villemez, C. L., McNab, J. M., and Albersheim, P. (1968) 'Formation of plant cell wall polysaccharides', *Nature, Lond.,* **218,** 878.

Villemez, C. L., Swanson, A. L., and Hassid, W. Z. (1966) 'Properties of a polygalacturonic acid — synthesizing enzyme system from *Phaseolus aureus* seedlings', *Archs. Biochem. Biophys.,* **116,** 446.

Wallace, W. J. (1967) 'Rheological behaviour of *Chara australia*', *Aust. J. biol. Sci.* **20,** 527.

Wardrop, A. B. (1954) 'Observations on crossed lamellar structures in the cell walls of higher plants', *Aust. J. Bot.,* **2,** 154.

Wardrop, A. B. (1965) 'Cellular Differentiation in Xylem', In: *Cellular ultrastructure of woody plants* (ed. W. A. Côté), p. 61, Syracuse University Press, Syracuse.

Wardrop, A. B. and Foster, R. (1964) 'A cytological study of the cat coleoptile', *Aust. J. Bot.,* **12,** 135.

Wirth, P. (1946) 'Membranwachstum während der Zellstreckung', *Ber. schweiz. bot. Ges.,* **56,** 175.

Wolfrom, M. L., Laver, M. I., and Paten, D. L. (1961) 'Carbohydrates of the coffee bean. II. Isolation and characterization of a mannan', *J. org. Chem.,* **26,** 4533.

Yamamoto, R., Shinozaki, K., and Masuda, Y. (1970) 'Stress-relaxation properties of plant cell walls with special reference to auxin action', *Pl. Cell Physiol.,* **11,** 947.

8

Ultrastructure of cultured plant cells
E. C. Cocking

Introduction

The dynamics of ultrastructure are more readily revealed in cultured cells than they are in the whole plant tissue because cells in culture can be more readily subjected to nutritional and environmental influences than individual cells of the tissues of whole plants. Included in the term cell culture are: *callus cultures* — clumps formed on static media; *suspension cultures* — single plant cell protoplasts, single cells or small cell clumps in agitated liquid medium; *plate cultures* — individually identifiable single plant cell protoplasts, or cells, or small cell clumps on agar plates and *microcultures* — one or more plant cell protoplasts, or cells, grown in chambers which facilitate microscopic observation.

Suspension cultures and callus

Suspension cultures frequently contain only a few single cells, together with a collection of cell aggregates of varying sizes and numbers of cells. As described by Roberts and Northcote (1970) in the case of sycamore, the suspension culture cells usually divide by transverse walls to form short chains of cells after a few weeks of culture. These chains then often break up to give groups of small isodiametric cells. The extent to which the behaviour of such cells — and this is, of course, reflected in their ultrastructure — is related to the fact that they have been isolated from the influence of other cells can best be assessed by first considering the sequence of events leading to their initial formation. As so aptly stated by Sinnott (1960) 'the fate of a cell is a function of its position'. Yeoman (1970) has summarized the main ways in which callus originates; suspension cultures are formed by the shaking of callus in liquid media. The initiation and development of calluses involves vigorous cell division, and this may be induced in various tissues of the plant body either by mechanical wounding, the action and reaction of microorganisms or insects, or by subjecting explants of plant tissue to media containing various stimulating chemicals. As callus growth begins there is a marked change in the metabolism of the cells of the tissue, and this frequently means that quiescent resting cells are converted into more active cells in which the metabolism is directed towards synthesis. This change is reflected in the ultrastructure of the cells. When callus formation is initiated from quiescent parenchymatous cells the storage materials in these cells such as starch, inulin, lipid, carotene, protein and phytoferritin begin to disappear. There is an increase in the amount of cytoplasm associated with this disappearance of food reserves. In the case of Jerusalem artichoke tubers, merely slicing the tissue causes the parenchyma cells to become activated, particularly if the slices are cultured in suitable media (Fowke and Setterfield, 1968). The storage parenchyma cells of the tuber are large and thin walled with the bulk of the lumen filled by a large single vacuole, as a result of which the cytoplasm is restricted to a very thin layer adpressed to the wall.

The handling of the artichoke tissue for electron microscopy was difficult, particularly in relation to satisfactory fixation, since the tonoplast and the plasmalemma of the cells were readily broken; this was the case even after initial fixation in glutaraldehyde containing the salt mixture of Zetterquist (Pease, 1964); and this highlights the need for a more systematic study of adequate fixation of highly vacuolated higher plant cells for electron microscopy. Satisfactory embedding can also be a problem since infiltration is sometimes very difficult, but the use of the low viscosity electron beam stable cross-linked styrene methacrylates has greatly helped in this connection (Mohr and Cocking, 1968), and it is clear from recent work on cultured cells that such low viscosity embedding media are finding increased usage (Davey and Street, 1971). Freeze-etching can help to solve some of these difficulties, but in this instance the added difficulty of adequate penetration of antifreeze agent into living cells can cause major problems. This difficulty can be overcome by initial fixation in aldehydes such as formaldehyde and glutaraldehyde, but deleterious effects of the fixative are often observed (Willison and Cocking, 1969).

Fowke and Setterfield (1968) noted that the artichoke tuber cells lacked much of the fine structure usually found in actively metabolizing cells. Dictyosomes were very infrequent and microtubules were only very occasionally observed in small groups lying near the wall. Only a few indistinct profiles of endoplasmic reticulum were to be seen in any cell, and ribosomes, although present, were randomly scattered and showed little association with endoplasmic reticulum membranes or with each other. The mitochondria were found in clusters, and frequently mitochondria and plastids were found concentrated in cytoplasmic thickenings around nuclei. When tuber slices were aged in water for 24 hr fine structural changes were observed which seemed to set the stage for the later active growth and expansion of the cells to form a callus under the influence of 2,4-D; this essentially centred on the amount and extent of the endoplasmic reticulum. The number of ribosomes distinctly increased, becoming mainly bound to the endoplasmic reticulum or clustered in polysomal aggregates. The cells of tissue which were expanding under the influence of 2,4-D showed, however, very significant changes. After 48 hr there was approximately a threefold increase in the thickness of the cytoplasmic layer and planimeter analysis of areas in electron micrographs proved that this increase was statistically highly significant — the cytoplasm appeared to have about the same electron opacity as the cytoplasm of the cells before the induction of callus formation — and as a result Fowke and Setterfield concluded that this increase in thickness of the cytoplasm represented a net synthesis of new cytoplasm. In these cells numerous cytoplasmic strands crossing the large central vacuole were also present whereas previously they had been conspicuously absent. The numbers of dictyosomes, microtubules, ribosomes, polysomes, mitochondria and plastids increased significantly. Overall, these results are in general agreement with those of Israel and Steward (1966) who found from studies on thin sections that the transition from quiescence to rapid division in carrot root cells is accompanied by a substantial increase in cytoplasm. These fine structural details are fully consistent with the physiological observations that protein synthesis is an essential process for continued expansion in plant cells. Active dictyosomes appear to be associated with the activity associated with cell expansion and this involves the elaboration of wall polysaccharides. As pointed out, however, by Fowke and Setterfield, the obvious structural relationships between dictyosomes, vesicles and walls, so clearly shown by O'Brien (1967) in his studies on the fine structure of cells of the oat coleoptile, were not evident. Bagshaw *et al.* (1969) and Tulett *et al.* (1969) have paid specific attention to the changes in the mitochondrial and plastid complex associated with callus growth from slices of the dormant tuber of the Jerusalem artichoke. These studies, together with those of several other workers, including Sunderland and Wells (1968) and Israel and Steward (1967), highlight the situation in higher plants: that there exist organelles which are largely autonomous, such as chloroplasts, and to a lesser extent mitochondria, and that these organelles may respond directly to whatever nutritional factors the cells are subjected, quite apart from the effects such factors have on the overall growth and development of the cell as largely controlled by the nucleus. Israel and Steward (1967) showed that the ability of carrot explants to grow under the stimulation of growth-producing systems (particularly isolated fractions of coconut milk) is frequently closely paralleled by the ability of the cultured cells to turn green in the light. As emphasized by these workers, however, the situation is different from that in the cells of whole plants in which plastid differentiation is taking place within developing foliar organs, and, moreover, in cultured cells the chloroplasts are developing in a situation in which exogenous sugar (usually sucrose) is being supplied. These workers showed that, starting from unspecialized quiescent cells of the storage root, which contain chromoplasts and plastids, the entire development of the chloroplast could be traced. They detected the pre-thylakoidal body which can be completely formed in the dark in plastids that develop under the cell division stimulus of coconut milk and that, with graded exposures to light, this body is progressively transformed into thylakoids as the dark grown plastid develops into a chloroplast. Complex interpretation resulted from their use of different fixatives and stains. The lamellae and grana of the light grown plastids became so complex and highly ordered that they appeared more complex and more highly organized than the plastids in many leaves.

From what has been said so far about the structure of cultured cells, particularly callus, it will be evident that it is still true, as pointed out by White in 1967, that the electron microscope has not been used enough in the study of plant cell cultures. Even when it has, the results are frequently very limited and quantitative analysis from them very difficult, especially when overall changes in the nature and number of organelles are being linked with physiological modification. Perhaps the future lies with the more effective use of stereological principles for morphometry in electron microscopic cytology (see Weibul, 1969). Comparisons between the cells of cultured callus and cells of the tissues of the plant body are also difficult because callus does not correspond strictly with any tissue of the plant body (Street, 1969).

A distinct advantage for studying the dynamics of the fine structure of dividing suspension culture cells is the presence of the large vacuole which, as we shall see later, makes it much easier to observe and interpret at the fine structural level (and also, incidentally, at the light microscope level) the formation of the phragmosome during division. Sinnott and Bloch (1941) clearly showed that the early stages in cell division, and the behaviour of both nucleus and cytoplasm in the period when the plane of future division is being determined, can be studied much more profitably in dividing cells which are much larger (often several hundred micrometres in diameter) and more vacuolate than in the small and only very slightly vacuolate cells in terminal meristems of root and shoot. While it is true that vacuolate dividing cells are present in shoot tips of the main groups of vascular plants, as we shall see later, the introduction of partially synchronized suspension callus cells with a reasonably high mitotic index has greatly facilitated the fine structural analysis of division in vacuolate cells (Roberts and Northcote, 1970).

Differentiation in higher organisms has been described as 'the outcome of a form of dialogue between an unchanging genome and a labile cytoplasm, the cytoplasm at any one period determining the patterns of gene expression and itself suffering consequent modification in the due course of time' (Heslop-Harrison, 1967). This dynamic developmental sequence is paralleled by a dynamic ultrastructural sequence which is well exemplified by the changes taking place in the formation of embryoids in culture, and by the growth and development of plants from these; and the growth, regeneration and division of isolated plant cell protoplasts and their aggregation (Fig. 8.1). The fate and behaviour of a cell or isolated plant cell protoplast is a function of its position. The isolation of cells results in their separation from the symplasm of the plant body; plasmodesmata are broken and may often, under plasmolysing conditions, be self-sealing; and in the case of protoplasts isolated from the symplasm of the plant there is good evidence that the plasmalemma is active at pinocytosis (Willison *et al.*, 1971), a situation, as we shall see later, which is probably not completely foreign to cells in roots (Wheeler and Hanchey, 1971).

An interesting observation of Davey and Street (1971) is that many of the outer walls of cells at the surface of cell aggregates of *Acer pseudoplatanus* cells cultured in suspension show extensive, continuous ridge-like thickenings which line the original cell wall and project into the cytoplasm. Gunning,

Pate, and Briarty (1968) had previously observed cells with similar ridge-like thickenings in the cell walls of cells in minor leaf veins, and there seems to be at least a structural parallel between these two instances, and it is interesting that Bowes (1969) has also observed similar ridge-like thickenings in cultured callus tissue of *Andrographis paniculata*. There is no physiological evidence, however, that these thickenings, either in *Acer* or in *Andrographis*, are involved in short-distance membrane-mediated transfer of solutes into the cellular aggregates from the bathing culture medium. The detailed fine structural studies of

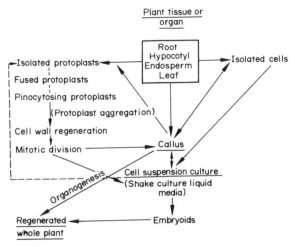

Fig. 8.1 Schematic viewpoint of the behaviour of higher plant cells in culture.

Davey and Street (1971) provided some evidence that, when the cells enter the stationary phase in batch-propagated suspension cultures, there was enhanced formation of crystal-containing microbodies. Earlier Sutton-Jones and Street (1968) had also shown that microbodies in cultured *Acer* cells became more numerous as the cells aged. These crystalloids, microbodies or peroxisomes are probably catalase deposits and, as pointed out by Davey (1970), in the leaf cells of plants in which microbodies have been identified, photosynthesis and photorespiration are intimately related. Glycolate, which is a major product of photosynthetic carbon dioxide fixation, would seem likely to be the substrate for photorespiration in such microbodies. The glycolate oxidase and a glycolate-glyoxylase system could serve to remove excess reducing power generated during photosynthesis. It should be noted that Frederick and Newcomb (1969) were able to demonstrate catalase in crystalloids present in the microbodies of leaves by using electron microscopical cytochemical methods specific for catalase. While it is easy to correlate the presence of microbodies with such metabolic interrelationships in green *Atropa* calluses, this is more difficult in *Acer* cells in culture and most suspension culture cells in which chloroplasts are either absent or not differentiated sufficiently to be active photosynthetically.

From a careful perusal of the limited amount of detailed fine structural study which has been carried out on suspension culture cells, it is clear that the actual *techniques* of electron microscopy, including fixing and embedding, do not offer insuperable difficulties in the furthering of our knowledge in this field. The main difficulty is the one of obtaining representative samples from what is usually a very heterogeneous population of cell aggregates. There are also difficulties encountered in batch propagation since medium composition varies considerably with associated effects on metabolism and ultrastructure; there are also problems associated with cell and aggregate interactions when cultures are inoculated at high cell densities. The use of media which have been 'conditioned' (Stuart and Street, 1971) allows cultures in liquid suspension to be initiated at densities as low as 2×10^3 cells/ml, and this may allow a more critical assessment of environmental factors on cell growth and, correspondingly, on changes in the ultrastructure of cells. It is, of course, now possible to stabilize media composition over prolonged periods by the use of chemostat and turbidostat systems (Wilson, King, and Street, 1971). By this means it is possible to establish steady states of growth and metabolism, thereby achieving uniform cell populations, and coupled with this, the ability to modify metabolism and differentiation patterns and to observe the related ultrastructural changes. Such methods are applicable not only to suspension culture of single cells and cell aggregates but also to the growth and development of embryoids (see Fig. 8.1). These approaches do not in themselves solve the problem of adequate sampling for electron microscopy, particularly when aggregates of a wide spectrum of different sizes are present, but the approach in which cell separating enzymes are added to batch propagated suspensions resulting in a high percentage of small cell aggregates may prove particularly useful, and may possibly be extended to continuous culture systems.

For the present, however, the most noticeable advances have come from fine structural studies (coupled where appropriate with optical studies) on suspension cultures such as those of sycamore (*Acer*), which produce moderately small aggregates in culture, and which may be fairly readily partially synchronized. As previously mentioned, a particularly useful attribute of these cultured cells is that the cells that are dividing are large and vacuolated and are more suited to the unravelling of the details of mitosis than the smaller unvacuolated (or only very slightly vacuolated) cells of meristems.

Division in suspension culture cells and naked cells

As a prelude to mitosis in sycamore suspension callus cells which have been partially synchronized to give a culture with a mitotic index of 15 per cent, Roberts and Northcote (1970) observed that the nucleus of the cell moves up against the cell wall and the numerous starch-filled plastids segregate — roughly half congregating at each end of the nucleus (Fig. 8.2). The pattern of division in these highly vacuolated cells is very similar to the pattern of division in vacuolate dividing cells in the shoot tips of the main groups of vascular plants which were extensively studied earlier by Sinnott and Bloch (1941) with the major geometrical difference that, in the cultured dividing cells, the

phragmoplast moves centrifugally outwards along the phragmosome and usually fuses first with the mother cell wall at the side of the cell where the nucleus was placed before mitosis. The cell plate then grows across the cell to fuse with the mother wall at the other side;

whereas, in the vacuolate dividing cells of shoot tips, the nucleus which has been flattened against the wall at interphase moves to the centre of the cell. This contrasts with the situation in the sycamore cells of Roberts and Northcote

where it is suspended by strands of cytoplasm and there undergoes mitosis. In early prophase these cytoplasmic strands tend to become aggregated into a more or less continuous diaphragm, the phragmosome . . . the phragmosome occupies the position where the division wall will later be laid down.

This pattern of cell division in vacuolate pith cells of meristematic shoot tips is shown semi-diagrammatically in Fig. 8.3. As we have seen, the expanding cell

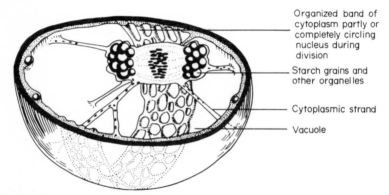

Organized band of
cytoplasm partly or
completely circling
nucleus during
division

Starch grains and
other organelles

Cytoplasmic strand

Vacuole

Fig. 8.2 Diagrammatic picture of a sycamore callus cell at metaphase. The organization of the reticulate band of cytoplasm which girdles the nucleus is depicted. (From Roberts and Northcote, 1970.)

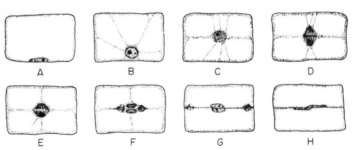

A   B   C   D

E   F   G   H

Fig. 8.3 Semi-diagrammatic drawings of cell division in vacuolate pith cells of meristematic shoot tip in longisection. Cells are dividing at right angles to shoot axis. A, interphase; B, early prophase, the nucleus becoming spherical and cytoplasmic strands penetrating the vacuole; C, prophase with phragmosome well developed; D, metaphase; E, anaphase; F and G, telophases, with kinoplasmic fibrils, and cell plate developing in the position of the phragmosome; H, two daughter cells in interphase. (From Sinnott and Bloch, 1941.)

plate follows exactly the position occupied by the phragmosome and, as stressed by Sinnott and Bloch, the 'first visible evidence of the polarity of the cell is thus provided by the cytoplasmic configuration rather than by events in the nucleus'. Since such a limited amount of detailed fine structural work and critical light microscopy (particularly the use of Nomarski differential interference optics which allows 'optical sectioning') has been carried out on cultured higher plant cells, it is difficult to assess the extent to which a centrifugal movement of the phragmoplast along the phragmosome will be found to be a general feature of cytokinesis. When isolated protoplasts from leaves, fruits or callus (see Fig. 8.1) are being cultured and are undergoing cell wall regeneration, a noticeable feature which is frequently a prelude to division, is that the nucleus, together with plastids, if present, moves to the centre of the cell, where it is suspended by strands of cytoplasm, and there undergoes mitosis — a situation for these cultured cells very similar to that found in the vacuolate pith cells of meristematic shoot tips.

It is a very unusual condition for a higher plant cell in culture to be without a cell wall, and it is to be expected that such cultured isolated protoplasts will possess certain novel features. In both the division of cultured cells and in the division of the cells of plant meristems, the cell plate which later forms into the new wall is formed by the fusion of numerous membrane-bound vesicles that have come to be aligned in the plane of the plate midway between the telophase nuclei. Such is not the case when isolated protoplasts (which possess no cell wall) are undergoing regeneration. Cell wall formation is not in this instance associated with cytokinesis. The general pattern of wall regeneration by isolated protoplasts is that the culture of protoplasts, in the simplest of balanced salt media with added plasmolyticum to ensure osmotic stability, results in activity at the plasmalemma which ultimately ensues in wall formation. So far the situation has been extensively investigated in isolated tomato fruit protoplasts (Pojnar, Willison, and Cocking, 1967) and in isolated leaf mesophyll protoplasts. The initial activity at the plasmalemma results, in the case of fruit protoplasts, in the formation of a scaly membranous envelope (Cocking, 1970); and, in the case of leaf protoplasts, in the formation of a rather electron opaque loose material which forms into an envelope. Both these appear to arise just outside the surface of the plasmalemma by a process of reverse pinocytosis involving Golgi body activity somewhat reminiscent of the deposition of pectic polysaccharide material into the growing primary wall of the root (Northcote and Pickett-Heaps, 1966). Insufficient work has as yet been done to establish at the fine structure level the detailed pathway of synthesis of this first-formed envelope material, but it is evident that the protoplast first forms this envelope at its surface as a prelude to the synthesis of cellulose fibrils which are later formed in the surface of the plasmalemma and between the plasmalemma and the envelope material so that, ultimately, a composite layered wall of mainly randomly oriented cellulose fibrils and envelope material is produced (Fig. 8.4). While it is known that the wall of a callus cell, which is of course usually non-lignified, resembles the wall of a cambial cell both in the quantities of the main poly-saccharide fractions and in the nature and amounts of the individual sugars of which they are composed (Table 8.1), it is not yet clear that a similar relationship holds for walls formed around regenerating protoplasts, and the walls of the cell types from which the protoplasts were originally isolated (see Fig. 8.1); fine structural studies need to be supplemented with direct analysis and radio-isotope incorporation type experiments. It is clear, however, that a fine structural comparison of the locule tissue of the tomato fruit reveals no structure in the

TABLE 8.1

Composition of Polysaccharide Components of Sycamore Callus and Cambial Tissues (from Northcote, 1969)

Substance	Callus, cell-wall material (%)	Cambium cell-wall material (%)
Pectin	15	15
Hemicellulose	40	45
α cellulose	30	37
	Percentage of whole dry tissue extracted with 80% ethanol	
Glucose (from α cellulose)	7.6	11.5
Galactose	6.4	8.1
Mannose	0.8	0.8
Arabinose	8.2	6.3
Xylose	4.0	2.9
Rhammose	1.0	1.1
Galacturonic acid	10.3	11.6

cell wall comparable to the 'membrane pile' envelope initially formed around regenerating protoplasts (compare Mohr and Cocking (1968) with Cocking (1970)). During the division of cells, membrane-bound particles accumulate to form the cell plate (for details of the early pioneering work in this field see Porter and Caulfield, 1958; Manton, 1961) and it would seem that the isolated protoplast recapitulates at its surface some of the phenomena associated with the continuing synthesis of the cell wall at the cell plate during the division of cultured and cambial cells. Of course, cell wall regeneration in single isolated protoplasts is unidirectional in comparison to the situation in cell wall synthesis as a result of cell division. In the case of regenerating tobacco leaf mesophyll it may be significant that when contiguous protoplasts are regenerating a new cell wall (a situation which is the closest parallel to cell wall synthesis in association with cell plate formation) a composite wall is formed (Fig. 8.5) which, in many respects, is similar to the cell wall arising from the cell plate during mitotic division (compare Fig. 8.6 and Fig. 8.5). In Fig 8.6 it is possible to trace the relationship between the middle lamella of the new cell wall, formed as a result of division in a regenerating protoplast, with the new cell wall formed *de novo* around the protoplast. Here the middle lamella is seen to link up with the dense envelope formed at the outer edge of the regenerated protoplast, and this suggests that in Fig. 8.5 the wall formed by contiguous protoplasts has a middle lamella analogous to the situation in walls of tissue cells generally. What is also evident (see Pojnar and Cocking (1968) for an earlier light microscopic study of this phenomenon) is that cell aggregates can be formed by regenerating protoplasts without cells having to divide. Cells within such aggregates readily divide, and it could well be that this phenomenon, which is often associated with the culture of protoplasts at high density, may prove of considerable importance in relation to the problem of regeneration of plants from cells in culture. This is because in what are generally described as suspension cultures of higher plant cells, which actually consist of single cells and cell clumps, studies of dynamics of growth have led to the general conclusion that mitoses are largely restricted to cells in clumps with a negligible number of cell divisions occurring in single isolated cells (see Torrey, Reinert, and Merkel, 1962; Street and Henshaw, 1963). Clearly therefore, the ability of protoplasts readily to aggregate to form clumps of cells, may greatly stimulate the division potential of cultures and also their regenerative capabilities. A schematic plan of aggregation and division in isolated tobacco leaf protoplasts is shown in Fig. 8.7.

Normally, in cultured cells, mitosis and cytokinesis appear like two parts of one process, and this is also the situation in the root meristem and other somatic tissues of higher plants. As a result, the region of the cell in which cytokinesis commences is still occupied by the remains of the mitotic spindle and, as we have already seen under these circumstances, this leads to the development of the cell plate, and the future plasmodesmata can be seen at very early stages of cell plate formation as continuous cytoplasmic strands which traverse the cell plate. As pointed out by Robards (1970), plasmodesmata are not only produced when a growing cell divides; it is clear in the case of tyloses of xylem vessel elements that plasmodesmata can also form spontaneously between cells that have grown in contact with one another. The reported presence or absence of plasmodesmata in callus tissue is somewhat variable. Butcher and Bowes (1969) in their study of the growth and fine structure of callus cultures derived from *Andrographis paniculata* were unable to identify plasmodesmata conclusively in either meristematic or vacuolated callus tissue; these authors were themselves surprised at this apparent absence of plasmodesmata since, as they pointed out, plasmodesmata are normally abundant in multicellular tissues; moreover, Halperin and Jensen (1967) were able, in a study of ultrastructural changes during the growth and embryogenesis in carrot cultures, to detect plasmodesmata in suspension culture aggregates. Both groups of workers used somewhat similar staining procedures (uranyl acetate and lead citrate) for the electron microscopic observations on thin sections of their embedded material. The failure of Butcher and Bowes to detect plasmodesmata could have been due to the very high electron opacity of the walls of their *Andrographis* cultures. This difficulty in the satisfactory detection of plasmodesmata is not restricted, however, to *Andrographis* cultures since, in earlier work, several workers were unable to detect plasmodesmata in tobacco callus; later it was reported that plasmodesmata were, in fact, present, their earlier

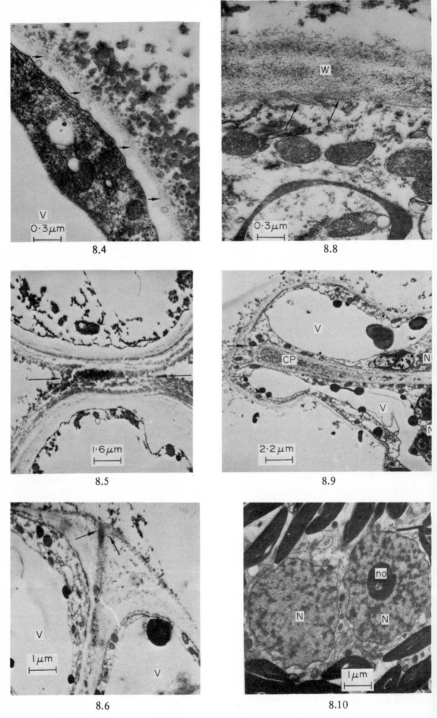

8.4

8.8

8.5

8.9

8.6

8.10

Fig. 8.4   Outer region of a regenerating tobacco leaf protoplast. The isolated leaf protoplasts had been cultured for sixteen days as described by Power *et al.* (1971). Proptoplasts
fixed in glutaraldehyde and osmium tetroxide were stained during dehydration with uranyl
acetate and embedded in a butyl-methyl methacrylate, styrene mixture. Sections were
stained with lead citrate. Rather loose electron-opaque material (probably pectin) is clearly
visible (see schematic plan, Fig. 8.7); the much less electron-opaque material of the fibrillar
cellulose of the wall (arrows) is first detectable between the plasmalemma and the electron-
opaque material. A portion of the large central vacuole (V) of the regenerating protoplast is
visible.

Fig. 8.5   Region of two contiguous protoplasts cultured as described in Fig. 8.4. A common
wall has been formed and, as a result, the electron-opaque material is acting as a sort of
cementing material between the regenerating protoplasts—this region (arrows) can possibly
be regarded as a sort of middle lamella but its origin is clearly different from the normal
middle lamella of cultured cells.

Fig. 8.6   Region of the outer region of a regenerating dividing tobacco leaf protoplast. The
protoplast was cultured as described in Fig. 8.4. The relationship between the new cell wall
which has arisen between daughter cells and the wall formed *de novo* around the regenerating mother protoplast is clearly visible. The cell plate (middle lamella) of the wall between
daughter cells has the same electron-opacity and appearance as the material formed around
the regenerating protoplast and links up with it in this region (arrows). The much less
electron-opaque fibrillar cellulose of both the wall formed *de novo* around the protoplast
and also formed as a result of division is also visible. V = vacuoles.

Fig. 8.8   Outer region of a tobacco leaf protoplast cultured as described in Fig. 8.4. Much
of the electron-opaque material has broken away but the cellulose fibrils of the newly
formed wall (W) around the protoplast are clearly visible. Microtubules are visible just
beneath the plasmalemma. Several mitochondria are also present (M).

Fig. 8.9   Low magnification view of region of a dividing regenerating protoplast. The new
cell plate (CP) has not as yet joined up with the wall formed *de novo* around the protoplast
but a channel is visible through which this will join up later (arrow). The nucleus (N) of
each daughter cell is just visible. V = vacuoles.

Fig. 8.10   Region of spontaneously fused tobacco leaf protoplast which had been cultured
for three days as described by Power *et al.* (1971). In this binucleate regenerating protoplast
systrophy has taken place and the two nuclei (N), one with a nucleolus visible (no), are
surrounded by chloroplasts. Chromatin masses are present in both nuclei and the nuclei
appear to be entering mitosis synchronously.

reported absence being due to fixation difficulties. The possibility that plasmodesmata may form spontaneously between cells that have grown in contact with
one another can now be explored experimentally using isolated protoplasts
which are regenerating a common cell wall which, as we have already seen, largely
equates with the wall developing from the cell plate which results from the
division of cells (see also Pojnar and Cocking, 1968). It is of some interest that,
in their study of the division of cultured sycamore cells, using electron microscopy
Roberts and Northcote (1970) were unable to find evidence for the presence of
a mother cell wall within which each divided protoplast forms its own new cell
wall, as was suggested by the earlier light microscopic studies of Lamport (1964).
This is relevant to the basic question of the expected nature of the wall that will
be formed between the isolated protoplasts in culture which are contiguous and
which are regenerating a wall. It is also relevant to the conditions under which
plasmodesmata will be formed between these protoplasts. It would seem likely
that regenerating dividing tobacco leaf protoplasts behave rather similarly in
culture to the dividing sycamore cells observed electron microscopically by
Roberts and Northcote.

It is, of course, possible for the division of the cytoplasmic portion of the
protoplast (cytokinesis) to be variously correlated in time with mitosis, and in
some plant cells no cytokinesis follows mitosis. This is sometimes the case in
the fertilized egg of higher plants where, even before cell division begins in the
fertilized egg, the development of the endosperm has taken place. This commences
with the division of the primary endosperm nucleus and, quite frequently, several
of these mitotic divisions can occur before cytokinesis takes place, so that a
multinucleate coenocytic state results as an intermediate stage. The removal of
cell wall from cultured cells in order to produce protoplasts may result in a

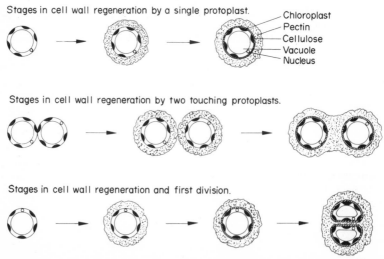

Stages in cell wall regeneration by a single protoplast.

Chloroplast
Pectin
Cellulose
Vacuole
Nucleus

Stages in cell wall regeneration by two touching protoplasts.

Stages in cell wall regeneration and first division.

Fig. 8.7 Schematic diagram of wall regeneration and division in isolated tobacco leaf
protoplasts.

similar imbalance between mitosis and cytokinesis as suggested from the studies of Eriksson and Jonasson (1969) on the division of isolated *Happlopapus* protoplasts. These workers obtained some evidence that nuclear division (mitosis) was taking place while the isolated protoplasts were remaining as protoplasts and before any cell wall regeneration resulted in their becoming cells. It would seem more likely that this nuclear division was by a simple constriction (amitosis). The possibility does exist, however, that such cultured protoplasts may undergo mitosis without cytokinesis. It would also seem likely, as has been discussed previously, that once wall regeneration has been initiated mitotic and cytokinetic temporal relationships may become more closely similar until the situation is near to that found normally in cultured cells, in root meristems, and in other somatic tissues of higher plants. Why this pattern should change is of considerable biological interest, and it is evident that structural relationships between the wall formed *de novo* around regenerating protoplasts and the new cell plate are of major importance.

During anaphase in the mitotic division of many cultured cells microtubules increase in number and are particularly noticeable at the edges of the cell plate. Microtubules are found on both sides of the young cell plate in dividing tobacco leaf cells (Cronshaw and Esau, 1968) and in dividing cambial cells of *Ulmus* and *Tilia* (Evert and Deshpande, 1970). They are also particularly numerous just beneath the plasmalemma when isolated protoplasts are regenerating a wall *de novo* at their surface (Fig. 8.8). Microtubules are also, in general, particularly evident when cells are actively synthesizing a primary cell wall, and at this stage they are arranged in a fairly random manner, under the plasmalemma. As discussed by Northcote (1969), microtubules are seen to be arranged in the cytoplasm over the sites of the secondary thickening of the cell wall. It would seem, however, that the detection of microtubules is rather variable in cultured cells. This may be related to the nature of the growth medium, and particularly to the level of auxins present in it. It is, however, impossible to generalize since, although Butcher and Bowes (1969) only rarely observed microtubules lying in the cytoplasm adjacent to the plasmalemma when their *A. paniculata* callus was grown on a basic medium containing 6.0 mg/litre of 2,4-D, Cronshaw (1967) described numerous microtubules in the callus cultures of *Nicotiana tabacum* growing in a medium containing 2.0 mg/litre of 1AA. What may, however, be the more important consideration in this aspect of the fine structure of cultured cells is the temperature of fixation of the cells with glutaraldehyde, and this is very rarely recorded in publications. At Nottingham, it has been our experience with isolated protoplasts which are regenerating cell walls that fixation, with varying concentrations of glutaraldehyde at 4°C, results in the apparent complete absence of microtubules; this puzzled us greatly during our earlier studies on cell wall regeneration in protoplasts. It was not until the introduction of fixation with glutaraldehyde at room temperature (approximately 18°C) that microtubules were readily detected lying beneath the plasmalemma in regenerating protoplasts (see Fig. 8.8). Preston and Goodman (1968) have come to the conclusion that microtubules are not directly associated with the microfibrillar synthesis that manifests itself as the primary cellulosic wall of plants, but it is still possible that microtubules are concerned (as suggested by Northcote (1969)) with the 'organized movement of the vesicles that carry material into the wall and if this is so then one of the functions of the microtubules is concerned with the deposition of the matrix material of the wall'. In this connection it would be particularly pertinent to see whether microtubules were evident when isolated protoplasts were initially forming what could be largely matrix type material at their surface (see Fig. 8.4), or whether microtubules only formed later, when microfibril synthesis had commenced.

Binucleate cells in cultured cells are sometimes observed and this could have resulted from an inhibition of cytokinesis. Experimentally binucleate, trinucleate, and tetranucleate cells can be induced in root tip cells by means of caffeine (Giménez-Martín et al., 1968), but this experimental induction of the multi-nucleate condition by this means has not as yet been extended to cultured cells. Gonzalez-Fernandez et al. (1964) have observed that in binucleate cells the nuclei, formed in a mitosis which is not followed by later cytokinesis, develop synchronously. This is also the case when nuclei with a marked difference in their chromosome complement are together in the same cytoplasm (Giménez-Martín et al., 1964).

Experimentally it is possible to follow comparable nuclear events in cultured multinucleate cells by utilizing protoplasts which have been fused together either spontaneously or by sodium salt treatment (Power, Frearson, and Cocking, 1971; Withers and Cocking, 1972). A prerequisite for spontaneous fusion is the main-tenance of contact of adjacent protoplasts via plasmodesmata during the enzyme treatment during which the cell walls of the cultured cells are being degraded by cellulase action. In such intra-species fusion bodies the number of nuclei can be from two to twenty or more, and it is observed that, after culture for a few days, there is pronounced systrophy in most of the multinucleate protoplasts. The nuclei mass together in the centre of the cell and are surrounded by the chloro-plasts, and in such protoplasts, containing two or more nuclei, some nuclear division was synchronous and concomitant with the initiation of wall regeneration (see Figs. 8.9, 8.10). A wide spectrum of different nuclear behaviour was evident in these intra-species fusion bodies during the mitotic division, with the possibility that nuclear fusion could occur during these divisions (see Fig. 8.11).

As we have already seen, callus tissue consists of rather parenchymatous-like generally undifferentiated type cells. Differentiated cells can, however, be induced in such calluses by transferring the callus to a medium usually containing less 2,4-dichlorophenoxy acetic acid, or none at all, or by varying the level of sucrose in the medium. Various other means can be used to induce differentiation, but the overall objective would seem to be to alter the ratio of the concentrations of exogenous and endogenous growth factors and nutrients in regions of the callus tissue (see Northcote, 1969, for a fuller discussion of these effects). Xylem and phloem are differentiated within the callus and form as small nodules, and in these nodules the xylem occurs near the centre and the phloem, which is situated towards the periphery, is separated from the xylem by a meristematic region.

324

With respect to subsequent organogenesis from such differentiated callus it is unfortunate, as has been emphasised by Halperin (1969), that

no adequate experimental system exists at present in which it is not only possible to control the time and place of initiation of organ primordia but also to isolate them for biochemical and ultrastructural studies during successive stages of development.

Of course, organ formation resulting from very small clumps of cells in suspension or from plate cultures may be helpful in this connection, but as yet too few

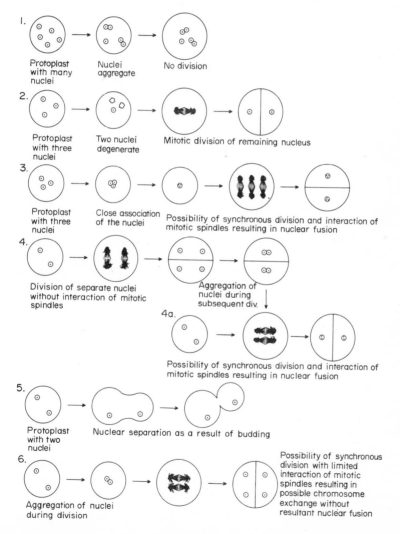

Fig. 8.11   Possible nuclear behaviour during culture of spontaneously fused protoplasts. (Reproduced with permission, J. B. Power, original.)

studies have been carried out to enable any generalization to be made. The beginnings of a comprehensive fine structural investigation of organogenesis in higher plant tissue cultures by Bowes (1970) is clearly a step in the right direction. Halperin and Jensen (1967) noted in suspension cultures of wild carrot that

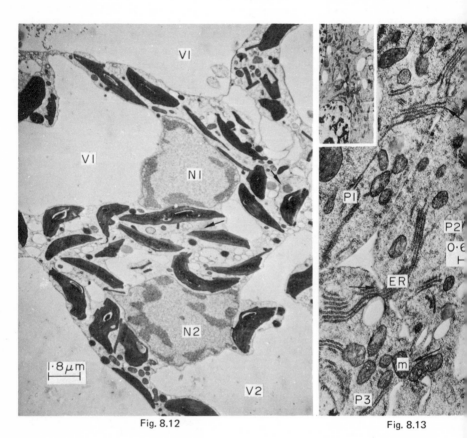

Fig. 8.12   Fig. 8.13

Fig. 8.12   Electron micrograph showing two tobacco leaf protoplasts undergoing fusion following treatment with sodium nitrate. The fusion is in an advanced state, only small remnants of the original plasmalemmae (arrows) remaining between the cytoplasms of the two protoplasts. The vacuoles V_1 and V_2 are, however, still separate. The section is of glutaraldehyde/osmium fixed material stained during dehydration with uranyl acetate embedded in methacrylate styrene and post stained with lead citrate. (N_1, N_2, nuclei of the two protoplasts.) (Reproduced with permission, L. A. Withers, original.)

Fig. 8.13   Electron micrograph of a region of spontaneous fusion in the oat root tip following enzymatic digestion of the cell walls. Broad cytoplasmic bridges (arrows) can be seen joining the three protoplasts, (P_1, P_2, P_3) elements of endoplasmic reticulum (ER) and mitochondria (M) traversing these connections. The inset, a lower magnification view of the region, shows the relationship of the three protoplasts, the nuclei of two of them being visible. The material was prepared for electron microscopy as described for Fig. 8.12. (Reproduced with permission, L. A. Withers, original.)

mitoses were localized in clumps, that few single cells divided, and that embryos which were formed were derived from such cell clumps which perpetuated themselves by growth and fragmentation. These workers observed that in smaller clumps released from parent clumps there was definite evidence at the ultrastructural level for a polarity, with each daughter clump bearing large vacuolate cells at one end and smaller meristematic cells at the other. As we have seen earlier, the ability of regenerating cultured cell protoplasts to aggregate together to form small clumps of cells may greatly help in our future understanding of the fine structural events associated with initiation of organ primordia. It may prove possible to 'assemble' aggregates from protoplasts at different stages of vacuolation and differentiation, thus imitating some of the observed early polarity of the daughter clumps of cultured wild carrot. These ultrastructural studies on cultured cells will also focus attention, not only on the activity of the plasma-lemma in cultured cells, whether it be in relation to spontaneous or sodium salt induced fusion (see Figs. 8.12, 8.13), but also on the pinocytotic activity of cultured cell protoplasts. As discussed by Clowes and Juniper (1968), Bradfute, Chapman-Andresen, and Jensen (1964) conducted an enquiry into whether pinocytosis (it is now better to use the more general term endocytosis – for a discussion of this point see Willison, Grout, and Cocking, 1971) ever takes place in higher plants, and concluded that there was no evidence for it at all. Until recently, evidence for pinocytosis in higher plants (with the exception of work on isolated protoplasts, in which it is readily possible to expose the plasmalemma to marker molecules such as viruses and ferritin (Cocking, 1970)) has been circumstantial. It is encouraging that Wheeler and Hanchey (1971) have been able to detect, in thin section, electron opaque crystals when oat seedling roots were exposed to uranyl acetate. These workers have been able to follow the pinocytotic uptake of uranyl acetate in the oat root cap with the uranyl acetate becoming sequestered into vacuoles.

From what has been written about the fine structure of cultured cells it will have become evident that the amount of detailed systematic study is very limited. Most workers, at best, have been content merely to characterize their material at the fine structural level, without systematically attempting to correlate it with physiological activity. All this is somewhat surprising, but is perhaps a reflection of the fact that many plant tissue culture workers have in the past often limited their horizons to the actual growth and behaviour of plant cells and tissues in culture. More plant physiologists and those wishing to link together biochemistry and fine structure, need to work with cultured plant cells.

Bibliography

Further reading

Northcote, D. H. (1969) 'Growth and differentiation of plant cells in culture', *Symp: Soc. gen. Microbiol.*, **19**, 333.
Pasternak, C. A. (1970) *Biochemistry of differentiation*, Wiley–Interscience, New York.

Ultrastructure of cultured plant cells

Steward, F. C., Mapes, M. O., and Ammirato, P. V. (1969) 'Growth and morphogenesis in tissue and free cell cultures', In: *Plant physiology* (ed. F. C. Steward), vol. V.B, chapter 8, Academic Press, New York and London.

Street, H. E. and Öpik, H. (1970) *The physiology of flowering plants,* Edward Arnold (Publishers) Ltd., London.

References

Bagshaw, V., Brown, R., and Yeoman, M. M. (1969) 'Arrangement and structure in dormant and cultured tissue from artichoke tubers', *Ann. Bot.,* **33,** 217–226.

Bowes, B. G. (1969) 'The fine structure of wall modifications and associated structures in callus tissue of *Andrographis paniculata* Nees', *New Phytol.,* **68,** 619–626.

Bowes, B. G. (1970) 'Preliminary observations on organogenesis in *Taraxacum officinale* tissue cultures', *Protoplasma,* **71,** 197–202.

Bradfute, O. E., Chapman-Andresen, C., and Jensen, W. A. (1964) 'Concerning morphologica evidence for pinocytosis in plants', *Expl Cell Res.,* **36,** 207–210.

Butcher, D. N. and Bowes, B. G. (1969) 'The growth and fine-structure of callus cultures derived from *Andrographis paniculata* Nees', *Z. PflPhysiol.,* **61,** 385–387.

Clowes, F. A. L. and Juniper, B. E. (1968) *Plant Cells,* Botanical monographs, vol. 8, Blackwell Scientific Publications, Oxford.

Cocking, E. C. (1970) 'Virus uptake, cell wall regeneration and virus multiplication in isolated plant protoplasts', *Int. Rev. Cytol.,* **28,** 89–124.

Cronshaw, J. (1967) 'Tracheid differentiation in tobacco pith cultures', *Planta,* **72,** 78–90.

Cronshaw, J. and Esau, K. (1968) 'Cell division in leaves of *Nicotiana'*, *Protoplasma,* **65,** 1–24.

Davey, M. R. (1970) *Growth and fine structure of cultured plant cells,* PhD Thesis, University of Leicester.

Davey, M. R. and Street, H. E. (1971) 'Studies on the growth in culture of plant cells, IX. Additional features of the fine structure of *Acer pseudoplatanus* L. cells cultured in suspension', *J. exp. Bot.,* **22,** 90–95.

Eriksson, T. and Jonasson, K. (1969) 'Nuclear division in isolated protoplasts from cells of higher plants grown *in vitro'*, *Planta,* **89,** 85–89.

Evert, R. F. and Deshpande, B. P. (1970) 'An ultrastructural study of cell division in the cambium', *Am. J. Bot.,* **57,** 942–967.

Frederick, S. E. and Newcomb, E. H. (1969) 'Cytochemical localization of catalase in leaf microbodies (peroxisomes)', *J. Cell Biol.,* **43,** 343–353.

Fowke, L. C. and Setterfield, G. (1968) 'Cytological responses in Jerusalem artichoke tuber slices during aging and subsequent auxin treatment', In: *Biochemistry and physiology of plant growth substances,* Proceedings 6th Int. Conference on Plant Growth Substances (eds. F. Wightman and G. Setterfield), p. 581. Runge Press, Ottawa.

Giménez-Martín, G., Gonzalez-Fernandez, A., and López- Sáez,J. F. (1964) 'Bimitosis', *Phyton B. Aires,* **21,** 77–84.

Giménez-Martín, G., López-Sáez, J. F., Morend, P., and Gonzalez-Fernandez, A. (1968) 'On the triggering of mitosis and the division cycle', *Chromosoma,* **25,** 282–296.

Gonzalez-Fernandez, A., López-Sáez,J. F., and Giménez-Martín, G. (1964) 'Inhibition of cytokinesis, bimitosis and polymitosis', *Phyton B. Aires,* **21,** 157–165.

Gunning, B. E. S., Pate, J. S., and Briarty, L. G. (1968) 'Specialised "transfer cells" in minor veins of leaves and their possible significance in phloem translocation', *J. Cell Biol.,* **37,** C7–12.

Halperin, W. (1969) 'Morphogenesis in cell cultures', *A. Rev. Pl. Physiol.,* **20,** 395.

Halperin, W. and Jensen, W. A. (1967) 'Ultrastructural changes during growth and embryogenesis in carrot cell cultures', *J. Ultrastruct. Re ,* **18,** 428–443.

Heslop-Harrison, J. (1967) 'Differentiation'; *A. Rev. Pl. Physiol.,* **18,** 325–348.

Israel, H. W. and Steward, F. C. (1966) 'The fine structure of quiescent and growing carrot cells: its relation to growth induction', *Ann. Bot.,* **30,** 63–79.

Israel, H. W. and Steward, F. C. (1967) 'The fine structure and development of plastids in cultured cells of *Dancus carota'*, *Ann. Bot.,* **31,** 1–18.

Lamport, D. T. A. (1964) 'Cell suspension cultures of higher plants, isolation and growth energetics', *Expl Cell Res.*, **33**, 195–206.

Manton, I. (1961) 'Observations on phragmosomes', *J. exp. Bot.*, **12**, 108–113.

Mohr, W. and Cocking, E. C. (1968) 'A method for preparing highly vacuolated senescent and damaged plant tissue for ultrastructural study', *J. Ultrastruct. Res.*, **21**, 171–181.

Northcote, D. H. (1969) 'Fine structure of cytoplasm in relation to synthesis and secretion in plant cells', *Proc. R. Soc. B.*, **173**, 21–30.

Northcote, D. H. (1969) 'Growth and differentiation of plant cells in culture', *Symp. Soc. gen. Microbiol.*, **19**, 333–349.

Northcote, D. H. and Pickett-Heaps, J. D. (1966) 'A function of the Golgi apparatus in polysaccharide synthesis and transport in the root-cap cells of wheat', *Biochem J.*, **98**, 159–167.

O'Brien, T. P. (1967) 'Observations on the fine structure of the oat coleoptile I. The epidermal cells of the extreme apex', *Protoplasma*, **63**, 385–416.

Pease, D. C. (1964) *Histological techniques for electron microscopy*, 2nd ed., Academic Press Inc., New York.

Pojnar, E., Willison, J. H. M., and Cocking, E. C. (1967) 'Cell wall regeneration by isolated tomato-fruit protoplasts', *Protoplasma*, **64**, 460–480.

Pojnar, E. and Cocking, E. C. (1968) 'Formation of cell aggregates by regenerating isolated tomato-fruit protoplasts', *Nature*, **218**, 289.

Porter, K. R. and Caulfield, J. B. (1958) 'The formation of the cell plate during cytokinesis in *Allium cepa*', 4th Int. Conf Electron Microscop, Berlin (ed. W. Bacemann), vol. 2, pp. 503–509, Springer-Verlag, Berlin.

Power, J. B., Frearson, E. M., and Cocking, E. C. (1971) 'The preparation and culture of spontaneously fused tobacco leaf spongy mesophyll protoplasts', *Biochem. J.*, **123**, 29 p.

Preston, R. D. and Goodman, R. N. (1968) 'Structural aspects of cellulose microfibric biosynthesis', *Jl. R. microsc. Soc.*, **88**, 513–527.

Robards, A. W. (1970) *Electron microscopy and plant ultrastructure*, p. 151 (European Plant Biology Series), McGraw-Hill, London.

Roberts, K. and Northcote, D. H. (1970) 'The structure of sycamore callus cells during division in a partially synchronized suspension culture', *J. Cell Sci.*, **6**, 299–321.

Sinnott, E. W. (1960) *Plant morphogenesis*, McGraw-Hill Book Co., New York.

Sinnott, E. W. and Bloch, R. (1941) 'Division in vacuolate plant cells', *Am. J. Bot.*, **28**, 225–232.

Street, H. E. and Henshaw, G. G. (1963) 'Cell division and differentiation in suspension cultures of higher plant cells', *Symp. Soc. exp. Biol.*, **17**, 234.

Street, H. E. (1969) 'Growth in organised and unorganised systems. Knowledge gained by culture of organs and tissue explants', *Plant physiology* (ed. F. C. Steward), vol. V.B, chapter 6, Academic Press, New York and London.

Stuart, R. and Street, H. E. (1971) 'Studies on the growth in culture of plant cells. Further studies on the conditioning of culture media by suspensions of *Acer pseudoplatanus* L. cells', *J. exp. Bot.*, **22**, 96–106.

Sunderland, N. and Wells, B. (1968) 'Plastid structure and development in green callus tissues of *Oxalis dispar*', *Ann. Bot.*, **32**, 327–346.

Sutton-Jones, B. and Street, H. E. (1968) 'Studies on the growth in culture of plant cells III: Changes in fine structure during the growth of *Acer pseudoplatanus* L. cells in suspension culture', *J. exp. Bot.*, **19**, 114–118.

Torrey, J. G., Reinert, J., and Merkel, N. (1962) 'Mitosis in suspension culture of higher plant cells in synthetic medium', *Am. J. Bot.*, **49**, 420.

Tulett, A. J., Bagshaw, V., and Yeoman, M. M. (1969) 'Arrangement and structure of plastids in dormant and cultured tissue from Artichoke tubers', *Ann. Bot.*, **33**, 217–226.

Weibul, E. R. (1969) 'Stereological principles for morphometry in electron microscopic cytology', *Int. Rev. Cytol.*, **26**, 235–302.

Wheeler, H. and Hanchey, P. (1971) 'Pinocytosis and membrane dilation in uranyl-treated plant roots', *Science, N.Y.*, **171**, 68–71.

White, P. R. (1967) 'The promises and challenges of tissue culture for biology and for mankind', In: *Seminar on plant cell, tissue and organ cultures* (ed. B. M. Johri), pp. 12–19. U.G.C. Centre of Advanced Study in Botany, Plant Morphology and Embryology. Delhi.

Ultrastructure of cultured plant cells

Willison, J. H. M. and Cocking, E. C. (1969) 'Freeze-etching observations on tobacco leaves infected with tobacco mocaic virus', *J. gen. Virol.,* **4,** 229–233.

Willison, J. H. M., Grout, B., and Cocking, E. C. (1971) 'A mechanism for the pinocytosis of latex spheres by tomato fruit protoplasts', *Bioenergetics* **2,** 371–382.

Wilson, S. B., King, P. J., and Street, H. E. (1971) 'Studies on the growth in culture of plant cell suspensions', *J. exp. Bot.,* **22,** 177–207.

Withers, L. A. and Cocking, E. C. (1972) 'Fine structural studies on spontaneous and induced fusion of higher plant protoplasts', *J. Cell Sci.,* **11,** 59–75.

Yeoman, M. M. (1970) 'Early development in callus cultures', *Int. Rev. Cytol.,* **29,** 383–49.

9

Gland cells
E. Schnepf

Introduction

Glands are distinct groups of highly specialized cells, the function of which is the discharge of substances to the exterior (*exotropic*) or into special intercellular cavities (*endotropic*). They are often composed of *secretory cells* and different kinds of *auxiliary cells.*

'Secretion' is a common feature of living cells. It denotes transport to the exterior of the protoplast. The cell wall is a prominent secretion product. The term *excretion,* used in animal physiology to characterize the discharge of waste products, is controversial in botany (Schnepf, 1969a; Frey-Wyssling, 1970). An exact discrimination between secretion and excretion is either impossible or complicates the description of the processes and their intelligibility. At present we therefore try to avoid the term 'excretion'.

Secretion is preceded by the *ingestion* of raw material (which may take either the symplastic pathway in plasmodesmata, or apoplastic pathway in cell walls), and the synthesis of secretion product. Occasionally the discharge is not connected with a synthesis (e.g., water and ions). If the secreted material passes through the plasmalemma directly, the secretion process will be termed *eccrine.* Intracellular transmembrane transport, resulting in secretion vesicles followed by the extrusion of their contents by exocytosis, is called *granulocrine* secretion. Either form of secretion is said to be of the *merocrine* type.

Various cells (e.g., laticifers) synthesize inert substances which are not actually discharged, but deposited within the plasmalemma-bound part of the cell; these are storage cells or *excretion cells,* but not gland cells, since the actual extrusion process is absent. Further, if a group of such cells discharges its contents by lysis, it represents a gland, too (*holocrine* or lysigenous type). Thus, there are secretory cells which are not gland cells, and gland cells without an actually secretory character.

In contrast to many other specialized cells, there are several types of gland cell which exhibit dynamic ultrastructural modifications directly related to the secretory process. I shall, therefore, describe the development of the ultrastructure

of the gland cells, as well as the − possible − variations during the secretion
process, and not least, try to relate both to the secreted material and to the mode of
secretion. It is not intended to present an exhaustive survey but to demonstrate
typical gland cells by selection of favourable examples.

Different types of gland cell

Gland cells secreting mainly hydrophilic substances

Glands secreting hydrophilic substances are usually situated superficially, often
in the form of hairs or emergences; or they at least communicate with the exterior.
This group of glands is divided into mucilage glands, digestive glands (secreting
enzymes), nectaries (secreting sugars), salt glands, and hydathodes (secreting water).
It is impossible to make a strict distinction between these types since the secretion
is frequently a mixture of various substances.

Mucilage glands. The mucilages secreted by plant glands are usually polysaccharides, in contrast to animal mucilage glands which produce glycoproteins. The
first glands to be studied in detail with respect to the dynamic aspect of their ultrastructure will be the mucilage glands of carnivorous plants, especially of *Drosophyllu*
(see Schnepf, 1969a for review). *Drosophyllum lusitanicum* captures small insects by
means of mucilage. The leaves bear two kinds of gland: stalked glands which
secrete the mucilage; and sessile digestive glands. Apart from the stalk, the anatomy
of both glands is nearly identical.

In the head of the stalked glands (Fig. 9.7E) three cell layers cap a group of
short irregular tracheids which are connected by the vascular tissue of the stalk
with the xylem of the leaf. The cells of the two outermost layers have strong buttress
thickenings at their walls and a dense cytoplasm. The third layer consists of flat
cells with a large vacuole, their contiguous radial walls are cutinized. Thus, these
barrier cells block the apoplasm between the glandular cap and the leaf.

The highly viscous fluid secreted forms a globular droplet covering the cap of
the gland. Its volume can be measured microscopically. In experiments with
excised leaf parts in a moist chamber the dry weight of the secretion was
0.23%; 0.19% is represented by polysaccharide, the remainder, 0.04%, mainly by
ascorbic acid; proteins could not be found. The polysaccharide consists of
gluconic acid, galactose, arabinose, a small quantity of xylose, and traces of
rhamnose; uronic acids were not detected (Schnepf, 1963a). The application
of low concentrations of glucose (Schnepf, 1961b), or of the components of the
polysaccharide (Schnepf, 1963a), stimulates the secretion.

Ultrastructural research reveals the dynamics of the secretion process. The
secretory cells constituting the outer layers of the stalked, as well as the sessile,
glands were found to have numerous small wall protuberances, in addition to
the buttress thickenings, concentrated at all inner parts of the cell walls. They
characterize the gland cells as *transfer cells* (Gunning and Pate, 1969; see
chapter 13).

All secretory cells are similar in their fine structure, though one important

difference distinguishes the outer layer of the stalked glands. The secretory cells
have in common many well-developed mitochondria (the number of which is
about one and a half times as high in both inner layers as in the outer ones),
starch-free leucoplasts, an inconspicuous endoplasmic reticulum (ER), mainly in
the form of smooth surfaced vesicles, and a rather low content of ribosomes.

The difference pertains to the Golgi apparatus. The number of dictysomes was
found to be five times as high in the outer layer of the stalked glands as in the inner
one, and even twice as high as in the outer layer of the sessile glands. Only in the
cells of the outer layer of the stalked gland do rather voluminous Golgi vesicles
with a more or less opaque content occur (Schnepf, 1961a). This observation served
as a basis for the concept that the mucilage is secreted by means of the Golgi
apparatus, and that the secretion process is essentially restricted to the outer layer
of gland cells.

The dictyosomes (for a detailed description of the Golgi complex see chapter 3)
of the stalked glands of *Drosophyllum* are usually seen to be somewhat irregular
stacks of 4—5 Golgi cisternae. The polarity across the stack is represented by the
location of the Golgi vesicles and the electron opacity of their contents (Schnepf,
1965b), as well as by a gradient in the width of the cisternal lumen.

Vesicles with an opaque content are found adhering to Golgi cisternae and also
free in the cytoplasm. Without any poststaining the material bound within the
vesicle membrane exhibits various levels of contrast. When comparing the free
vesicles with the adherent ones a condensation is recognizable. The content stains
with lead; it is also possible to enhance its electron opacity by ruthenium red, which
is used in light microscopy to demonstrate pectins and similar acidic carbohydrates.

The rate of secretion was determined by estimating the amount of fluid
secreted per hour through a glandular surface area of 1 μm^2. There is a strict
correlation between this value and the ultrastructure of the Golgi apparatus;
both are influenced by different internal and external factors. All observations
confirm the idea that the Golgi vesicles are, in fact, secretion vesicles.

In young, immature glands the cells of the outer layer have a denser cytoplasm
and more ribosomes than in the active stage; the dictyosomes are smaller, Golgi
vesicles are not conspicuous. In old glands the secretory activity decreases, or
even ceases. The cytoplasm of the gland cells is reduced, while the vacuoles
enlarge. There are only few and small Golgi vesicles. The vesicles which are
considered to be secretion vesicles occur in the active stage only (Schnepf, 1961a).

The rate of secretion depends on the temperature; it reaches its maximum
at 32 °C, and the temperature coefficient is about 2. The number of secretion
vesicles counted in electron micrographs and related to a cytoplasmic area of
100 μm^2 increased from 15 to 73 when the temperature was raised from 14 °C to
29 °C for 2½ hours (Figs. 9.1 and 9.2). At the same time, the mean number of
Golgi cisternae per dictyosome decreased from 4.4 to 3.1. When taking the values
at 29 °C as 100%, the mucilage production at 14 °C is 33% and the number of
secretion vesicles is 21%. Considering that evaluation of the electron micrographs
was restricted to counting the vesicles, while their size and the density of their
contents were neglected, the relationship is impressive. Temperatures above

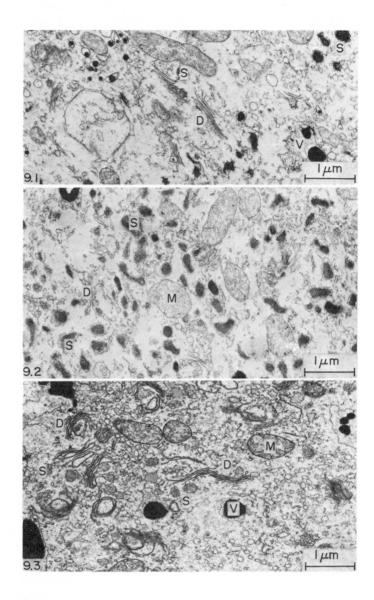

Figs. 9.1—9.3   Mucilage glands of *Drosophyllum lusitanicum* at different stages of activity.

Fig. 9.1   Low activity at 14°C.   Fig. 9.2 high activity after exposure to 29°C for 2½ hr.
Fig. 9.3 low activity after exposure to 29°C for 22 hr. D = dictyosomes, M= mitochondria,
S = Golgi apparatus derived secretion vesicles, V = vacuole. Osmium fixation.

32 °C reduce the number of secretion vesicles as well as the rate of secretion (Schnepf, 1961b).

As these experiments were performed with excised leaf parts in dark thermostats, photosynthetic formation of sugars (the raw material for the synthesis of mucilage) was precluded. As a consequence, in long duration experiments (Schnepf, 1961b) at temperatures near the optimum, the rate of secretion declined. At 28 °C it remained nearly constant for about 18 hours and then decreased markedly; at 20 °C the regression begins later (Schnepf, 1963b).

Corresponding modifications were observed in the Golgi elements. In glands fixed 1½ hours after the beginning of the experiment (during which the temperature was raised to 28 °C), 96 secretion vesicles per 100 μm^2 cytoplasmic area and 2.5 Golgi cisternae per dictyosome were found, whereas after 22 hours (Fig. 9.3) the figures were 24 and 4.5, respectively. Hence the secretory activity decreases from 100% to about 34%, and the number of secretion vesicles diminishes from 100% to 25%, while the number of Golgi cisternae per dictyosome, reduced during the previous phase of high secretory activity, is regenerated to a normal value (Schnepf, 1961b).

Withdrawal of water stops secretion immediately and completely. Glands from which water had been withheld for 1½–2 hours contained many secretion vesicles though their number was reduced as compared with the control gland (Schnepf, 1961b). However, it is noteworthy that all stages in the development of secretion vesicles were present. Thus it is believed that water shortage primarily affects the discharge process – and perhaps the maturation of the vesicles – and has only an indirect effect on their formation, if any.

Experiments with different metabolic inhibitors, or N_2-atmosphere, were undertaken to find which part of the secretion process is to be regarded as an active one. All inhibitors, when applied in usual concentrations, and also anaerobic conditions, prevented the appearance of a secretion droplet on the gland. The ultrastructure of the gland cell appeared modified accordingly (Schnepf, 1963b). After exposure to an N_2-atmosphere at 20 °C for 25 hours no secretion vesicles were detectable. Similar effects were observed with KCN (10^{-4} M and 10^{-3} M, 28 °C, 4 hr) and arsenite (10^{-2} M, 28 °C, 24 hr). The effects of KCN were reversible. The interference with the secretion process, marked by the lack of secretory vesicles, demonstrates that energy deficiency inhibits the formation of the secretory product rather than its discharge.

The strict interrelationship between the amount of secreted fluid and the number of secretion vesicles allows a rough calculation of the kinetics of the secretion process (Schnepf, 1961b). Assuming (a) that the size of the vesicles measured in electron micrographs is nearly the same as *in vivo*; (b) that the number of vesicles is not reduced during the fixation process; and (c) that the contents of the mature vesicles have nearly the same concentration as the discharged fluid on the gland surface, some kinetics can be calculated. It must be conceded, however, that the last assumption seems to be somewhat unrealistic, for a dilution in the water-saturated atmosphere must be taken into account.

The lifetime of the vesicles can be determined if the amount of fluid which passes through 1 μm^2 of gland surface area per minute is compared with the total volume of the secretion vesicles located under the corresponding surface area. The last value is determined by morphometric methods. At 28 °C the secretion intensity was 1.3 $\mu m^3/\mu m^2$/min. Below 1 μm^2 of surface area there was a total volume of secretion vesicles of 3.4 μm^3. To discharge 1.3 μm^3 of fluid per minute by a vesicle mass of 3.4 μm^3 a turnover rate of 3.4:1.3 = 2.6 minutes is necessary. Another calculation gave similar results.

As these data are based on uncertain postulates, they indicate at best the order of magnitude. Even this degree of accuracy is doubtful in the light of other results such as those provided by Northcote and Pickett-Heaps (1966). Using pulse-label incorporation of ^{3}H-glucose followed by autoradiography, these authors studied the secretion of the root cap slime. The distribution of the radioactivity at different times indicates that the lifetime of the secretion vesicles is about ten times as high as in the *Drosophyllum* glands. Further data in this respect are compiled in the contribution of Morré and Mollenhauer (chapter 3). On the other hand, the values obtained from *Drosophyllum* glands are supported by similar results obtained by studying secretion processes in some algae (Schnepf and Koch, 1966 – discharge of water; Brown, 1969 – scale production).

At the ultrastructural level we may distinguish several phases of mucilage secretion. It is obvious that the secretion vesicles derive from the dictyosomes. The secretion product can be demonstrated in the vesicles adhering to Golgi cisternae. There is sufficient evidence (for further data see chapter 3) to assume that the Golgi apparatus and the vesicles derived therefrom are the site of the mucilage synthesis.

However, an open question is that of the mode of sugar transport necessary for mucilage synthesis. Are the sugars provided by the ER via vesicles which become incorporated in the dictyosomes, or do they directly permeate the membranes of the Golgi cisternae and vesicles? As the contrast of the vesicle content is further modified after detachment, the synthetic processes appear to continue even in the free vesicles.

The detachment of secretion vesicles diminishes the amount of membranous material within the dictyosome. One role of the dictyosomes, therefore, seems to be the regeneration of membranes. In the case of mucilage secretion of *Drosophyllum,* it remains uncertain whether the vesicles or other elements of the ER are incorporated into the dictyosome or if the incorporation is limited to single membrane molecules or micelles.

Experiments affecting the rate of secretion revealed a reciprocal correlation between the number of secretion vesicles and the number of Golgi cisternae per dictyosome (Schnepf, 1963b). A very high rate of secretion results in a reduction in the number of Golgi cisternae within the stack. Consumption of the membrane material is greater than its regeneration. If the rate of secretion is lowered after a period of high activity, the dictyosomes enlarge, and the number of cisternae increases. This regeneration is remarkably insensitive to metabolic inhibitors.

Maturation of the contents of the secretion vesicles is associated with modifications of the vesicle membrane. This became evident in one fixation where the plasmalemma had split artificially. As a result, in cross sections it appeared as two lines separated by a gap of up to 20 nm. In these cells the membranes of the Golgi cisternae and of the adherent vesicles were of the usual thickness. The membranes of the detached vesicles, however, were modified in the same way as the plasmalemma; the gap became more evident as the vesicle content became more condensed (Schnepf, 1965b).

This transformation of the vesicle membrane, by which its structure becomes similar to the plasmalemma, seems to be a precondition for exocytosis. The membrane of the secretion vesicle fuses with the plasmalemma and becomes incorporated, while the secreted material is deposited as droplets in the extra-cytoplasmic space. After interruption of the water supply, such droplets accumulate within this space. A sufficient turgor pressure is apparently necessary to transport the secretion further.

As a consequence of extrusion, membrane material is added to the plasmalemma; a real increase, however, cannot be observed. Consequently, a backflow of membrane material must be assumed. Detecting no backward transport of intact membranes, I believe that the backflow is achieved by membrane micelles or single molecules which, possibly, are reincorporated into the membranes of the Golgi apparatus (Schnepf, 1969a), directly or via the endoplasmic reticulum.

Most of the vesicles empty into one of the inner cell walls. The secretion then moves within the apoplasm and, therefore, appears preferentially above the anticlinal walls (Schnepf, 1961a — observations in *Drosera*). The cuticle is traversed by pores which allow the mucilage to become free (Schnepf, 1965b).

In all other mucilage glands of plants studied adequately, the Golgi apparatus has been found to play a similar role in the secretion of polysaccharides to that shown for *Drosophyllum*; this is true of acidic polysaccharides as well as of neutral ones.

Many of these glands consist of special hairs. In spite of a great variety in shape and anatomy, their correspondence at the ultrastructural level is striking. This has been demonstrated by studies on glands of higher plants, mosses, and algae: in trapping slime glands of *Drosera* and *Pinguicula* (Fig. 9.7C) (Schnepf, 1961a); squamulae of young *Elodea* and *Potamogeton* shoots (Rougier, 1969); gland hairs of *Psychotria* (Horner and Lersten, 1968), *Pharbitis* (Healy, 1967), *Rumex* (Fig. 9.5 and Fig. 9.7D) and *Rheum* (Schnepf, 1968); and the inner glands of *Hibiscus* (Mollenhauer, 1967) discharging their products into intercellular cavities; in the subapical gland hairs of mosses (Bonnot and Hébant, 1970), and in glands of the red alga *Lomentaria* (Bouck, 1962) and the brown alga *Laminaria* (Schnepf, 1963d), the secretion follows the same course.

It seems noteworthy that all these glands except those of *Drosophyllum* and *Drosera* lack wall protuberances. They do not, therefore, represent transfer cells. Presumably, the enlargement of the plasmalemma in *Drosophyllum* gland cells is related to the uptake of digested material, or to the ingestion of raw material, rather than to the secretion of mucilage. In *Pinguicula,* wall protuberances occur in the sessile digestive glands but not in the stalked gland hairs which secrete mainly the trapping slime (Schnepf, 1961a, in contrast to Heslop-Harrison and Knox, 1971).

Barrier cells with cutinized walls occur neither in submersed plants nor in 'internal glands (including the trichomes of *Psychotria*), but only in those glands which are exposed to the open air.

The secretion product passes through the cuticle via pores in *Drosophyllum* (Schnepf, 1965b) and *Drosera* (Williams and Pickard, 1969), but directly in *Pinguicula* (Schnepf, 1961a). In the glands of *Rumex* and *Rheum* (Schnepf, 1968) the polysaccharide which covers the young leaves within the bud is first accumulated in a subcuticular space, and is then released through cuticular perforations which are presumably formed secondarily. The glandular trichomes in *Psychotria* lack a cuticle (Horner and Lersten, 1968).

Studies on the polysaccharide secretion in outer root cap cells of corn and wheat (Morré *et al.*, 1967) have confirmed the concept of a Golgi apparatus-mediated slime secretion. These workers have further extended our knowledge of the function of the Golgi apparatus in several details, such as the independence of the actual secretion process of protein synthesis. Moreover, several observations seem to point to a direct or indirect participation (via the Golgi apparatus, Mollenhauer, 1965) of the ER in the secretion process, as was also assumed for some mucilage glands (Horner and Lersten, 1968). As the secretory cells of the root cap do not form a distinct gland they will not be discussed in detail here.

Digestive glands. Glands secreting digestive enzymes are found in carnivorous plants (Lloyd, 1942). Closed *Nepenthes* pitchers already contain enzymes (Lüttge, 1964), whereas the digestive glands of *Drosera, Drosophyllum* and *Pinguicula* release them only after prey has been captured.

Digestive glands also absorb disintegrated material. A porous or incompletely agglutinated cuticle (Schnepf, 1965b – *Nepenthes*) facilitates its permeation. It is difficult to attribute the modifications observed in digestive glands either to the secretion or to the uptake, especially since it cannot be excluded that, even in the *Nepenthes* type, the secretion is also stimulated by the captured prey. Nevertheless, the so-called 'aggregation' of vacuoles, consisting of division and shrinkage processes (Lloyd, 1942), seems more probably to be related to absorption, as it is not restricted to the secretory cells.

A common feature observed in digestive glands is changes in the nuclear structure. These occur in the gland cells only and, therefore, seem to be related to enzyme secretion, though they are similarly found in the stalked glands of *Drosophyllum,* the digestive activity of which is believed to be low. In both gland cell layers of *Drosophyllum* glands (Schnepf, 1963c), the chromatin condenses. The nucleolus is gradually reduced, osmiophilic granules are observed in the karyolymph, and electron opaque material is found in the nuclear pores. Thereafter the number of polysomes attached to the ER increases, and the ER itself forms conspicuous complexes of rough surfaced cisternae and tubules. The mitochondria often swell, while the leucoplasts remain ultrastructurally unchanged. The Golgi apparatus shows such a diversity of structure that it has, as yet, been impossible to draw reasonable conclusions about its role.

In digestive glands of *Dionea* (Scala *et al.*, 1968; Schwab *et al.*, 1969) the production of large vesicles and coated vesicles by the dictyosomes is stimulated in a late phase of the digestive cycle, presumably in order to replenish the enzyme pool for the next cycle. The vacuoles are regarded as the site of enzyme storage. In contrast to the *Drosophyllum* glands, complex changes in plastidal fine structure and in the form and number of wall protuberances have been described; the formation of polysomes is also increased. As in the case of *Drosophyllum*, precise ideas on the ultrastructure of the secretion processes in *Dionea* have not yet been put forward.

In *Pinguicula* (Heslop-Harrison and Knox, 1971), different digestive enzymes were detected in the sessile as well as in the stalked glands; the proteolytic activity is mainly associated with the sessile glands. In unstimulated glands the enzymatic activity seems to be associated with the anticlinal walls between the secretory cells and, in part, with the vacuoles. Adequate stimulation leads to secretion of fluid within one hour and to discharge of the enzymes.

It seems reasonable to assume that the function of the organelles in plant digestive glands is similar to that of enzyme-secreting animal cells. An autoradiographic study by Figier (1969) on petiolar glands of *Mercurialis*, though not unambiguous, points in the same direction. Proteins, synthesized by ribosomes of the rough surfaced ER, migrate to the Golgi apparatus (and some other cell components) and are finally discharged into a subcuticular space.

Nectaries. Nectaries release a fluid with a high sugar content: the nectar. Frequently this process is performed by distinct gland cells, but in several plants a sap originating from sieve tubes moves to the exterior through intercellular spaces and modified stomata. In some of those nectaries, however, there is a 'nectary parenchyma' assumed to have a gland-like character (Frey-Wyssling and Häusermann, 1960).

Our discussion will be restricted to true secretory cells in morphologically defined glands. These, too, obtain the material to be secreted mainly from sieve tubes. The more the secretory organ is anatomically differentiated the more its secretion differs from the sieve tube sap (Lüttge, 1961).

The main sugars encountered in nectar are saccharose, glucose, fructose and various oligosaccharides. Invertases may modify the composition of the discharged fluid. De Fekete *et al.* (1967) presented a diagram summarizing the various transformations of sugars in nectaries.

The diversity of nectary types (Sperlich, 1939) corresponds to a diversity of ultrastructural features, and to a diversity of ideas on secretion mechanisms. They range from involvement of active sugar secretion, and secretion of enzymes transforming the sugars which migrate exclusively in the apoplasm (Vasiliev, 1969a), to reabsorption (Lüttge, 1961, 1969) of non-sugar components from the discharged sieve tube sap.

The glandular cells of the nectary trichomes of *Vicia faba* (Wrischer, 1962) and, especially, the epithelial cells of many of the so-called septal nectaries within the gynoecium of several Lilliiflorae (Schnepf, 1964a) have protuberances along their outer cell walls. Therefore, they are transfer cells (see chapter 13).

In the septal glands of *Gasteria* (Schnepf, 1964a), the wall protuberances

Gland cells

(Fig. 9.4) are of an extremely irregular shape and extend far into the cell. Thus, they enlarge the surface of the protoplast and the amount of the plasmalemma considerably. They are formed immediately before the secretory phase and are removed at its end.

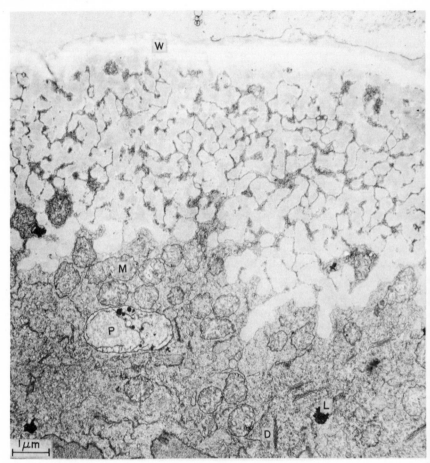

Fig. 9.4   Septal nectary of *Gasteria trigona,* apical part of a secretory cell in the active phase with many cell wall protuberances. D = dictyosome, L = lipoid droplet, M = mitochondrion, P = leucoplast, W = outer cell wall. Osmium fixation.

The cytoplasm contains many well-developed mitochondria and an extended ER in the form of rough surfaced cisternae, and smooth surfaced tubular and vesicular elements. The dictyosomes, though often surrounded by vesicles, show no pronounced activity during the secretory phase. Schnepf (1964a), in his ultrastructural research, could not detect distinct signs of a granulocrine secretion via vesicles derived from the ER or the Golgi apparatus. Therefore, he interpreted the wall protuberances as indicating an eccrine secretion, in which the plasmalemma is the site of active sugar transport; a participation of acid phosphatases, often found in high activity in nectaries, was assumed by several authors (e.g.,,

Figier, 1968). The extension of the plasmalemma might be related to reabsorptional processes as well.

In a recent autoradiographic study on septal nectaries of *Aloe*, Heinrich (personal communication), applying glucose-6-^{3}H and avoiding aqueous media during fixation and embedding procedures, obtained evidence in favour of a granulocrine secretion. He observed an accumulation of radioactive material in vacuolar structures, in vesicles and cisternae of the ER, and also in the cavity of the gynoecium into which the nectar is secreted. Moreover, these vacuoles and vesicles proved to be the site of glucose-6-phosphatase activity. Presumably these compartments split the sugar phosphates, releasing the phosphate into the cytoplasm, and discharge their contents by exocytosis.

These ideas are supported by the work of Mercer and Rathgeber (1962) on *Abutilon* nectary trichomes (Fig. 9.7F), revealing many vesicular elements; if secretion is inhibited by anaerobic conditions cisternae predominate. In cyathial nectaries of *Euphorbia pulcherrima* (Schnepf, 1964b), vesicular elements are common, too, but they are also found in inactive glands. In nectaries of *Diplotaxis* and *Ficaria,* Eymé (1967), on the other hand, observed Golgi vesicles during the secretory phase which are believed to participate in the secretion process. In other nectaries the Golgi apparatus is inactive and poorly developed (Vasiliev, 1969a; Zandonella, 1970).

Large crystalline aggregates of wound, smooth surfaced tubules which are connected with the ER, were observed by Eymé (1967) in several nectaries. They can hardly be related to the secretion process, as they occur in other cell types, too. A possible role of lomasomes (invaginations of the plasmalemma filled with membrane material) in transport is discussed by Eymé (1967), but their actual nature and function remain obscure. Corresponding to the high respiration rate of the glands (Ziegler, 1956), the mitochondria, in general, are numerous and well developed.

The contradictory results obtained from examining the ultrastructure of different nectaries allow no generalizing concept on its relationship to the specific secretion product. This parallels the diversity of nectary types mentioned above.

In the internal septal nectaries, the cuticle is often rudimentary (Schnepf, 1964a – *Gasteria*). In other glands the secretion passes through the cuticle by pores (Mercer and Rathgeber, 1962 – *Abutilon*) Cuticular pores were not found in other nectaries (e.g., Schnepf, 1964b – *Euphorbia*). Sperlich (1939) records examples of discharge through bursting cuticles. Nectary trichomes (Wrischer, 1962 – *Vicia*) generally have a basal barrier cell with a cutinized outer wall blocking the flow within the apoplasm.

Hydathodes. The diversity of hydathodes ranges from a simple system of intercellular spaces to complex true glands, and thus resembles that of nectaries; similarly, the secretion mechanisms vary greatly.

Some hydathodes (Sperlich, 1939) eliminate water directly from terminal tracheids and possess no glandular tissue. The release is caused by root pressure. In other plants, the terminal tracheids are in contact with a special parenchyma, the epithem, which may be provided with intercellular spaces (through which the water moves) or not (Häusermann and Frey-Wyssling, 1963). Frequently the epithem is enclosed by a sheath of suberinized cells, or of cells equipped with Casparian strips. Generally, the secretion passes through the epidermis via modified stomata. With respect to the site of the active transport (in the root or in the glands themselves), the epithem hydathodes are usually divided into passive and active ones. It is questionable, however, whether these types coincide with the anatomically defined types (with or without intercellular spaces). The high acid phosphatase activity found in epithems is believed to be related to a reabsorbing function of the cells (Häuserman and Frey-Wyssling, 1963).

The ultrastructure of epithem cells has been studied by Tucker and Hoefert (1968 – *Vitis*), and Perrin (1969 – *Saxifraga*; 1970a – *Saxifraga* and *Taraxacum*). In *Vitis* a normal organelle complement was found; chloroplasts (with few grana) and mitochondria are infrequent. The plastids of the *Saxifraga* and *Taraxacum* epithem cells contain remarkable aggregates of phytoferritin.

Other types of hydathode do not communicate directly with the tracheidal system. They consist of a single epidermal cell, of small cell groups, or of glandular hairs.

The trichome hydathodes – so far as they have been studied – seem to discharge various other substances together with water. Studying the glandular hairs of *Phaseolus multiflorus*, Perrin (1970b) even rejected the term hydathode for them. He observed material of unknown composition accumulated in subcuticular spaces of the apical gland cells, and, further, plasmalemma invaginations which might represent the final stage of an exocytotic secretion as well as small wall protuberances. Cuticular pores could not be detected.

In the course of the development of the glandular cells small vacuoles arise, enlarge, and fuse. During the secretory phase, cisternae of the ER form a layer beneath the plasmalemma, and their margins swell. The dictyosomes, the number of which increases during development, produce vesicles. As yet, the observations do not allow any idea of the morphology of the actual secretion process. Especially the function of straight tubular elements occurring in two types is obscure. They are arranged in conspicuous bundles with a core of opaque material. They were first observed by O'Brien (1967) in glandular hairs of *Phaseolus vulgaris.* During early stages of the gland development, the small type, 29 nm in diameter, arises from the ER and remains continuous with it (Steer and Newcomb, 1969). The larger tubules, 56–66 nm in diameter, have walls with a patterned substructure and contain an electron opaque material. They seem to replace gradually many of the smaller ones. Various experiments revealed that neither type is related to conventional microtubules. There are some indications that material originating from these tubules migrates into vacuoles. Similar tubular systems were also observed in *Coleus* leaf glands.

The trichome hydathodes of *Cicer* (Schnepf, 1965a), the secretion product of which contains some organic acids, are composed of a multicellular head of secretory cells and a stalk (Fig. 9.7A). Its most distal cell, the barrier cell, has a completely cutinized outer wall. At the apex of active glands the cuticle and the cutinized layer have separated from the cellulosic wall. Both are perforated by small pores which are not placed opposite each other, thus operating as a valve which does not open until the internal hydrostatic pressure is sufficient.

The secretory cells contain many mitochondria and small vacuoles. In active glands, the plasmalemma is removed from the cell walls, and the small extracytoplasmic space is filled with granular material. Vesicles are accumulated near the cell walls; their membrane structure exhibits a striking similarity to the plasmalemma. They are, therefore, believed to be in at least intermittent communication with the extracytoplasmic space, and their membranes to derive from the plasmalemma. Its enlargement is considered as indicating a kind of eccrine secretion.

Little is known about the role of the different glands in the hollow cavities of the subterranean leaves of *Lathraea*. They — or at least one type — discharge water. Most probably it is the peltate glands that are responsible for the water secretion. These each consist of a large central cell which is covered by four secretory cells. The cuticle is perforated centrally by a large pore. In the secretory cells the innermost parts of the anticlinal walls and the adjoining parts of the periclinal walls bear wall protuberances (Renaudin and Garrigues, 1967).

At times, the glandular trichomes discharge materials of unknown composition into small subcuticular spaces. Tightly packed wall protuberances (Schnepf, 1964c) along their outer walls, and many mitochondria, prove them to be transfer cells which might be active in secretion as well as in absorption. A barrier cell is not developed. Similar glands in the related *Odontites* function effectively in the discharge of water and certain solutes (Govier *et al.*, 1968).

Salt glands. A definite distinction between hydathodes and salt glands cannot be made, as the fluid secreted by hydathodes often contains rather high concentrations of salts, as well. The classical examples of salt glands are found in halophytes.

The salt glands of *Limonium* (Ziegler and Lüttge, 1966) are composed of 16 cells: 4 'secretory cells'; 4 'accessory cells'; 4 'inner' and 4 'outer cup cells'; in addition there are 4 collecting cells (Fig. 9.7G). The outer and inner parts of the cell walls between the collecting cells and the outer cup cells are completely cutinized, leaving a ring-like transfusion area of non-cutinized wall. Thus, a barrier blocks the apoplasm and forces all substances moving from the leaf via the collecting cells into the gland to pass through the plasmalemma of the outer cup cells in the region of the transfusion area, or to enter these cells through plasmodesmata which are found here in high number; that is, these substances take the symplastic pathway. The thick cuticle of the gland is traversed by four pores, each with a diameter of about 0.5 μm, one above each of the secretory cells.

The surface area of the cell walls of all glandular cells is enlarged by few small wall protuberances. A peculiar feature of the cytoplasm of the gland inner cells is, beside numerous mitochondria, extended ribbons of an electron opaque material.

Arisz *et al.* (1955) provided evidence that the salt secretion by glands is an active process. They assumed that the active phase is the accumulation of ions in the gland against a concentration gradient. Ziegler and Lüttge (1967) tried to follow the pathway of chloride ions in the leaves by autoradiography (at the light microscopical level) and by electron microscopic localization of silver precipitates. Regarding the limitations of the methods used, they concluded that the evidence for active concentration of chloride ions within the gland is unsatisfactory. The precipitations observed in different cell structures did not allow exact statements on the pathway or on the participation of the cell organelles in the secretion process.

In principle, the structure of the salt glands of *Tamarix* is similar, but less complex. They consist of only six secretory cells, arranged in three layers, and two highly vacuolated collecting cells. The barrier formed by cutinized outer and inner wall areas has the same consequence for the import of substances as discussed for *Limonium*; the high number of plasmodesmata in the transfusion area is also similar.

The secretory cells are characterized by many mitochondria and more or less elaborated wall protuberances (Thomson and Liu, 1967; Shimony and Fahn, 1968). The secreted material is composed of various salts, depending on the nutrition of the plant, and passes through the cuticle by pores. Occasionally, presumably in relation to their secretory activity, the secretory cells contain numerous small vacuoles distributed in the periphery of the cells. In position and structure they somewhat resemble the vesicles beneath the walls of *Cicer* glands described above. Usually they contain some electron opaque material but, in plants grown in a rubidium ion containing medium, large electron opaque aggregates were found in them. Therefore it is reasonable to assume that they accumulate the cations to be secreted. Presumably they release their contents into the apoplasm by fusion of their membranes with the plasmalemma. Close contact of both membranes resulting in a compound membrane were frequently observed (Thomson *et al.*, 1969). The origin of the vacuoles, as well as the significance of the compound membranes, remains an open question.

In the less complex salt glands of *Spartina foliosa*, which are composed of a cap cell and a basal cell, the secretion process seems to be eccrine. The basal cell has an extensive system of plasmalemma infoldings which extend from wall protuberances associated with mitochondria (Levering and Thomson, 1971).

Gland cells secreting mainly lipophilic substances

The lipophilic materials secreted by plant glands comprise a variety of different substances. The most important group is the terpenes, represented primarily by essential oils and resins. Waxes, fats, and flavonoid aglycons are frequently found as well. In general, the secretion is a mixture of such substances and even contains hydrophilic compounds, often in the form of gums.

In contrast to glands with hydrophilic secretions, which are predominantly merocrine, many of those secreting lipophilic substances are of the holocrine type. In some cases, a gland may start its secretion phase in the merocrine manner and end it in a holocrine one. Statements on the secretion mechanism have been based, up to the present, predominantly on comparative ultrastructural studies.

Exotropic merocrine glands. In a series of papers, Amelunxen (1964, 1965, 1967; Amelunxen *et al.,* 1969) studied the glands of *Mentha piperita.* The leaves bear two types of gland, both made up of a basal cell, a stalk cell (which acts as a barrier, due to its cutinized outer wall), and one (in the 'glandular hairs') or eight (in the 'glandular trichomes') secretory cells. Both gland types secrete various essential oils, such as menthone and menthol, though in glandular hairs oils could not be detected with the light microscope (Amelunxen *et al.,* 1969).

In the glandular trichomes the cytoplasm of young secretory cells in the pre-secretory phase resembles that of epidermal cells except that the ER is increased; the vacuoles frequently contain electron opaque material. The beginning of the secretion is marked by the formation of a subcuticular space. Its enlargement by the accumulation of the oil demonstrates the activity of the gland. During the secretion phase, the groundplasm is comparatively dense, the ER — which is increased further — has rather wide lumina. The dictyosomes remain scarce. The osmiophilia of the contents of the vacuoles gradually diminishes. At the end of the active stage the subcuticular space may reach such dimensions that the secretory cells loose their contact with the stalk and degenerate (Amelunxen, 1965).

The glandular hairs form only a very small subcuticular space. At this time they resemble the glandular trichomes in the structure of the ER and the density of the groundplasm; afterwards they develop numerous voluminous dictyosomes which are obviously active in the production of vesicles (Amelunxen, 1964, 1965). This activity suggests another secretion process during which hydrophilic material is released. In fact, the plasma volume is gradually reduced, and the developing extracytoplasmic space is filled by a flocculent material which may well have been discharged by Golgi vesicles. At least in *Salvia glutinosa,* where both types of gland also occur, the glandular hairs secrete mainly ruthenium red positive materials whereas the glandular trichomes — with an extended, smooth surfaced, tubular endoplasmic reticulum — discharge lipophilic substances (Schnepf, 1972).

In another paper, Amelunxen (1967) directed his attention to wavy bundles of fibres found in the gland hairs of *Mentha* as well as in other leaf cells. He believes these structures to be precursors of special oil vacuoles, but their nature and their involvement in real secretion processes has not yet been proved conclusively.

The glandular hairs of *Arctium lappa* (Schnepf, 1969b) consist of about sixteen cells arranged in two rows. They discharge a viscous oil into a subcuticular space capping the apical cells. The secretory nature of the gland cells is most prominent in the apical cells. It is especially evident from the remarkably developed ER: consisting of irregular elements, more cisterna-like and slightly studded with polysomes in young glands; and in the form of wound, smooth surfaced tubules in older ones (Fig. 9.6). The vacuoles are small and contain a more or less opaque material; mitochondria with an opaque matrix are numerous but relatively

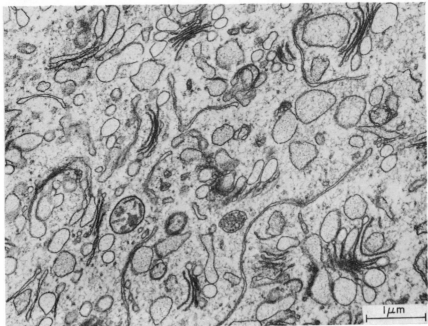

Fig. 9.5 Mucilage gland of *Rumex maximus* with abundant active dictyosomes. Potassium permanganate fixation.

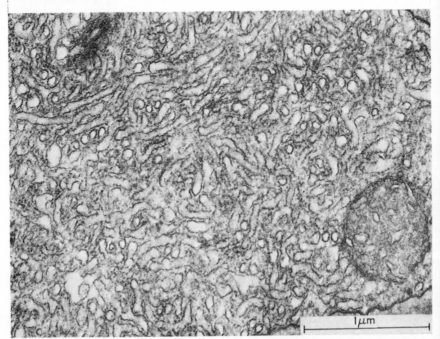

Fig. 9.6 Oil gland of *Arctium lappa* with extensive smooth surfaced tubular endoplasmic reticulum.

poorly provided with cristae. The leucoplasts are partially sheathed by elements of the ER.

The content of the voluminous subcuticular space is rather electron translucent. At the walls there are often stacks of weakly stained lamellae (about 3 nm in thickness) embedded in an opaque material. They look like wax films and, occasionally, are also found between the plasmalemma and the cell wall of secretory cells. A barrier of cutinized walls is not formed in the stalk.

Calceolaria flowers are pollinated by bees which collect a fatty oil (Vogel, 1971). This oil is secreted by glandular hairs. These consist of a short stalk, the distal cell of which has a completely cutinized outer wall, and a multicellular head. The oil is collected in a small apical subcuticular space before penetrating the cuticle. The most prominent cell components of active secretory cells (Schnepf, 1969c) are extended ER, made up of irregular, smooth surfaced cisternae and tubules; and leucoplasts, with poorly developed internal structures, which are more or less completely sheathed by elements of the ER. The rare dictyosomes do not produce vesicles. Immature glands, by contrast, are rich in rough surfaced ER and in active dictyosomes (which presumably secrete wall constituents).

Another type of secretion is produced by glandular hairs (Fig. 9.7B) of different species of *Primula*. Some glands discharge a whitish crystalline 'farina' mainly composed of various flavonoids; others a droplet of an oily nature which is also rich in flavonoids (Wollenweber and Schnepf, 1970). Some plants exclusively bear 'oil glands', others have both types distributed on different organs.

All glands are similar in consisting of a single secretory cell and a barrier cell. Again, in contrast to some former statements, in oil glands the secretion is not enclosed by the cuticle but only covered with a more rigid surface film.

With respect to the ultrastructure of the cytoplasm, both gland types are nearly identical. In mature glands, the most striking feature is again a crowded network of irregular, smooth surfaced tubules of the ER; cisternae sheathing the leucoplasts are not observed.

Buds are frequently covered with sticky or powdery masses which usually consist of lipophilic components, such as resins and flavonoid aglycones, as well as gums and mucilages. The complex composition of the secretion in *Alnus* buds is reflected in the fine structure of the secretory cells (Wollenweber *et al.*, 1971).

A dense groundplasm, with many ribosomes, marks young, still inactive, secretory cells of the multicellular glands. During maturation the ER proliferates to form a system of irregular, slightly dilated tubules and vesicles, predominantly of the smooth surfaced type. At the same time the Golgi apparatus is increased. In active glands the dictyosomes produce numerous secretion vesicles. Most probably they discharge the polysaccharidic components of the secretion. The leucoplasts are not sheathed by cisternae of the ER.

The heterogeneous material extruded accumulates in subcuticular spaces which at first arise above cleaving anticlinal walls. While they become more voluminous, the secretory cells gradually shrink.

Portions of the secretion product permeate the cuticle directly, while in other glands the cuticle bursts. At that time the secretory cells separate more and more. At last they degenerate while, in the stalk, cell walls cutinize and suberinize, and thus provide a new closure. In several details the bud glands of *Populus* (Vasiliev, 1970) resemble those of *Alnus.*

In all these examples it was impossible to detect the lipophilic secretion or its precursors within certain compartments. As a consequence, it is difficult to ascertain the site of its synthesis or the manner of its discharge. An exceptional case, perhaps, is the production of the stigma exudate of *Petunia* (Konar and Linskens, 1966; Kroh, 1967). The apical cell layers of the stigma may be regarded as a gland; its secretion consists mainly of an oily fat. The subepidermal cells release the exudate into intercellular spaces, the epidermal ones into subcuticular spaces. The cuticle ruptures, and the secretion droplets fuse to form a film.

The secretion, identified from its staining properties, is said to originate in cisternae of the ER which dilate at their margins. It is assumed that these dilatation develop into lipid droplets, and that active droplets (bounded by true membrane or not) penetrate the plasmalemma.

In the spadix appendices of the Aracea, *Typhonium,* which function as 'scent glands' or 'osmophores', the odorous oil seems to pass through the plasmalemma in the form of single molecules or small micelles, and only then to aggregate into droplets within the extracytoplasmic space (Schnepf, 1965b).

Endotropic merocrine glands. Endotropic merocrine glands, such as the resin glands of conifers, are similar to exotropic ones in many respects, but they discharge their products into schizogenous ducts or cavities. They are never separated from the apoplasm of the neighbouring tissue by cutinized or suberinize walls of barrier cells.

The resin ducts of *Pinus* are surrounded by a secretory epithelium. The epitheli cells contain a large number of leucoplasts, with poorly developed thylakoidal systems and sheathed by one or more cisternae of the ER (Wooding and Northcot 1965b). In resting periods, during winter, these sheathing cisternae fragment into vesicles (Vasiliev, 1970). Another remarkable structure in active secretory cells again is the densely packed, smooth surfaced tubular ER. Both, the ER and the leucoplasts, seem to be involved in resin production. Osmiophilic droplets, about 50 nm in diameter, are found in: the plastidal matrix; between the membranes of the plastidal envelope; between this envelope and the sheathing ER; in the sheathing and the cytoplasmic cisternae; in cisternae situated near the plasmalemma; and between the plasmalemma and the cell wall. Werker and Fahn (1968) succeeded in characterizing various lipophilic droplets by vital staining procedures.

Wooding and Northcote (1965b) consider this sequence to reflect the pathway of the secretion product, while Vasiliev is inclined to emphasize the role of the ER in terpenoid production, but acknowledges the support by the leucoplasts and, perhaps, by the mitochondria, too. The Golgi apparatus apparently has no function in this kind of secretion.

A transmembrane passage of entire terpenoid droplets is difficult to understand Nevertheless, Wooding and Northcote (1965b) present electron micrographs which seem to suggest such processes. During the permeation of the cell wall the particles apparently break down; they reaggregate in the resin duct.

Vasiliev (1969b, 1970) observed tightly sheathed leucoplasts, and a well-developed network of smooth surfaced tubular ER, in several other terpenoid glands, such as the excretion ducts of Umbelliferae; yet only in *Pinus* and *Hedera* glands did he recognize intracellular secretion droplets. Schnepf (1969e) also could not detect such particles in *Heracleum* and *Dorema* gland cells, but confirmed the results concerning the plastids: i.e., their sheath, and their crowded tubular internal structures and, especially, the ER. The glandular epithelial cells of oil ducts of *Solidago* also contain remarkable tubular ER and sheathed chloroplasts but, in contrast to the secretory cells, they form numerous osmiophilic lipoid droplets with a honeycomb-like surface within their groundplasm (Schnepf, 1969d).

It is obvious that all these differences are related to the plant species and to the peculiarities of the secretion product. Further studies are necessary to elucidate the method of secretion in all examples, and to find out if the process is uniform or not.

If the excretion is a mixture of lipophilic and polysaccharide components, such as in Umbelliferae, the latter seem to be secreted by the Golgi apparatus (Schnepf, 1969d).

Heinrich (1969) presented an unusual view of a merocrine essential oil gland with a schizogenous excretion cavity in *Ruta*. The gland produces methylketones. The secretory cells have a dense cytoplasm, with many ribosomes and conspicuous leucoplasts with tubular internal structures; but the ER is poorly developed, as compared to the glands described above, and does not sheath the leucoplasts. The methylketones accumulate in intercellular spaces. It is, as yet, impossible to correlate the fine structure with the function of the cell.

Holocrine glands. In holocrine glands, the production of the secretion is not followed by a true discharge process; instead, the secretory cells degenerate and lyse, and thus allow the accumulated material to coalesce to form voluminous droplets which may remain within the interior of the plant or finally reach the exterior. Frequently, the secretion phase is initiated by merocrine processes.

In *Citrus* and *Poncirus* glands (Heinrich, 1966), it is obvious that osmiophilic secretion products are synthesized by, or at least accumulated in, the plastids. Using the freeze-drying technique and applying different test substances, Heinrich (1970) was able to prove the volatile nature of the material within the plastids. These monoterpenes seem to be synthesized at or in the irregular tubular elements which form a crowded network within the plastids. The essential oils fuse to form droplets which increase in size until they more or less completely fill the plastids. Deposits of lipophilic material also occur in the cytoplasm, but are less voluminous.

In the course of development the secretory cells degenerate, the oil remains enclosed within the plastids for a long time. Finally, even the cell walls are lysed, and the lipid droplets fuse. The application of monoterpenes to young *Poncirus* glands results in their accumulation within schizogenous intercellular spaces and thus leads to a merocrine character in a lysigenous gland (Heinrich, 1969).

The 'spouting glands' of *Dictamnus* (Amelunxen and Arbeiter, 1967) have several features in common with the *Poncirus* glands. These glands secrete various essential oils; they occur on the inflorescences and different parts of the flowers and consist of a long multicellular stalk and a spherical glandular body which contains several layers of secretory cells around a central excretion cavity. The glandular body terminates in a trichome-like apex which easily breaks off, so allowing the oil to spout out. The cytoplasm of the secretory cells contains the usual organelle complement; ER is rather scarce, plastids are numerous. The essential oil accumulates in the plastids, gradually fills the spaces within the plastidal envelope and the thylakoids, and finally fuses to form large droplets while the plastids disintegrate. Thereafter the oil enters the cytoplasm where it is to be seen as foamy masses. It is possible that the cytoplasm itself also produces oil droplets during the last developmental phase. The thylakoidal system of the plastids is finally transformed into myelin-like figures which sometimes seem to be extruded into the vacuole. Finally, lysis of the cells takes place proceeding from the centre to the periphery. The peripheral cells also degenerate, though they do not produce considerable amounts of oil.

Another type of lysigenous essential oil gland, the multicellular glandular hairs of *Cleome,* was described by Amelunxen and Arbeiter (1969). Here, the plastids do not seem to play a role in secretion, although the ER in the form of a tubular network again appears to be involved. It is already well developed at the onset of oil production. The oil accumulates in the cytoplasm, at first appearing as membrane-bounded homogeneous droplets; later, in the lytic phase, these droplets become more granular, and finally they separate into the osmiophilic oil and less electron opaque vesicle-like elements.

Conclusions

The relationships between structure and function in plant gland cells have, as yet, been studied chiefly by comparative electron microscopy of different gland types solely. Only in a few favourable subjects, have ultrastructural analysis of different, experimentally influenced stages of secretory activity, application of radioactive tracers, or cytochemical methods extended our knowledge. A biochemical analysis of isolated glandular cell organelles is difficult because of the small size of the glands. Nevertheless, in some gland types it has been possible to elucidate the participation of certain cell organelles in the secretion process, while in many others the majority of the questions remain unanswered. Corresponding analyses of similar animal gland cells may provide more information.

The slime glands are relatively well understood, at least with respect to the later phases of the secretion process. The dictyosomes produce Golgi vesicles (Fig. 9.5) in which the polysaccharides are synthesized and transported. They are discharged by a typical exocytosis. These glands, thus best represent the granulocrine type. The subsequent pathway of the secretion — through the cell walls — depends on a sufficient turgor pressure in the secretory cells. The ingestion of the sugars used for the synthesis, and the possible role of the ER are not yet clear, whereas the function of the Golgi apparatus becomes increasingly well understood (see Morré and Mollenhauer: chapter 3).

In enzyme glands one would expect basically similar conditions to those

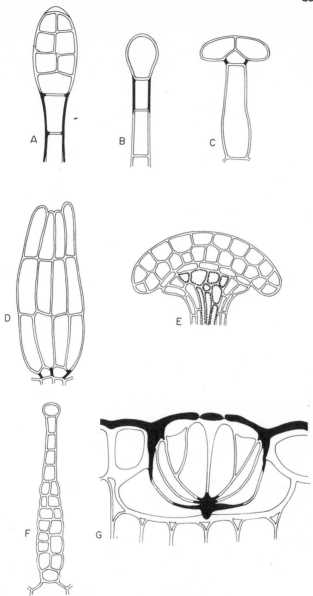

Fig. 9.7   Barrier cells in different plant glands.
(A) Trichome hydathode of *Cicer*.
(B) Flavonoid-oil gland of *Primula*.
(C) Stalked trapping slime gland of *Pinguicula*.
(D) Mucilage gland of *Rumex*.
(E) Stalked trapping slime gland of *Drosophyllum*.
(F) Trichome nectary of *Abutilon*.
(G) Salt gland of *Limonium*. Cutinized or suberinized wall parts are marked by thickened lines. A thin cuticle covering a normal cellulosic cell wall is not marked especially. (Adapted from Schnepf 1969a; courtesy of Springer Verlag, Vienna and New York.)

in the corresponding animal glands. In the secretion of sugars and salts, if perform
by the activity of superficial secretory cells and not by root or sieve tube pressure
active transport has been attributed to the plasmalemma (eccrine secretion) becau
of its enlargement by wall protuberances (Fig. 9.4) which are frequently found
(transfer cells, see Gunning and Pate: chapter 13). But reabsorption processes
might just as well be inferred from this cytological peculiarity. Moreover, recent
observations in favour of a granulocrine secretion accumulate in that, in several
glands, 'secretion vesicles' occur which are frequently attributed to formation fro*
the ER.

In most terpenoid, flavonoid, and fat glands, especially in those of the merocri*
type, an extended, smooth surfaced ER develops a tubular network and thus
represents the predominant cell organelle (Fig. 9.6); hence these cells strongly
resemble animal steroid producing cells (Christensen and Fawcett, 1961). An
involvement of the ER in the synthesis of sterols was shown biochemically by
Knapp *et al.* (1969). It is, therefore, quite reasonable to assume that it is the site
of (or, at least participates in) the synthesis of lipophilic secretions.

In many glands, the plastids (in general leucoplasts, sometimes also containing
packed tubules) obviously produce, or at least accumulate, lipophilic secretions.
Such glands often belong to the holocrine type; they lyse when the plastids
contain considerable amounts of lipoids. These materials are accumulated within
the thylakoidal lumen; thus, it may well be that the internal plastidal membranes
are involved in their synthesis. Common features of many secretory cells are
plastids sheathed by cisternae of the ER. These are found mainly in glands
secreting lipophilic substances, with the exception of flavonoid glands, but they
occasionally occur also in other types (Wooding and Northcote, 1965a). The
significance of this spatial relationship is, as yet, by no means understood.

Generally, lipophilic secretion products cannot be identified in electron
micrographs before they have been discharged. As a consequence, all ideas on the
transport and extrusion process are highly speculative.

The presence of rigid cell walls represents an important difference between pla*
and animal glands. In most exotropic glands, the secretion has to pass, not only
through cellulosic-pectinaceous walls (which are pathways rather than barriers),
but also through the cuticle. Either cuticular pores are preformed, or the cuticle
ruptures in the course of the secretion process, or the intact cuticle is sufficiently
permeable to allow the secretion to penetrate.

Peculiar structures of many glands are barrier cells with cutinized or suberinize*
cell walls, located in the glandular stalk or in a corresponding region (Fig. 9.7).
These separate the secretory cells from the apoplasm of the plant. They were
assumed to block a backflow of discharged material moving in the cell walls. But
the fact that they are found exclusively in glands exposed to the open air (and not
in submersed plants or internal glands), unless the latter are covered by a massive,
impermeable cuticle suggests another, perhaps more important, function: most
probably they do not block *secretion backflow* but *water outflow,* and thus
reduce the transpiration which is no doubt particularly high in glands having a
permeable cuticle.

Bibliography

Further reading

Mollenhauer, H. H. and Morré, D. J. (1966) 'Golgi apparatus and plant secretion', *A. Rev. Pl. Physiol.*, **17**, 27–46.

Schnepf, E. (1969) 'Sekretion und Exkretion bei Pflanzen', *Protoplasmatologia*, **VIII**, 8. Springer-Verlag, Vienna and New York.

References

Amelunxen, F. (1964) 'Elektronenmikroskopische Untersuchungen an den Drüsenhaaren von *Mentha piperita* L.', *Planta medica*, **12**, 121–139.

Amelunxen, F. (1965) 'Elektronenmikroskopische Untersuchungen an den Drüsenschuppen von *Mentha piperita* L.', *Planta medica*, **13**, 457–473.

Amelunxen, F. (1967) 'Einige Beobachtungen an den Blattzellen von *Mentha piperita* L.', *Planta medica*, **15**, 32–34.

Amelunxen, F. and Arbeiter, H. (1967) 'Untersuchungen an den Spritzdrüsen von *Dictamnus albus* L.', *Z. PflPhysiol.*, **58**, 49–69.

Amelunxen, F. and Aibeiter, H. (1969) 'Untersuchungen an den Drüsenhaaren von *Cleome spinosa* L.', *Z. PflPhysiol.*, **61**, 73–80.

Amelunxen, F., Wahlig, T., and Arbeiter, H. (1969) 'Uber den Nachweis des ätherischen Ols in isolierten Drüsenhaaren und Drüsenschuppen von *Mentha piperita* L.', *Z. PflPhysiol.*, **61**, 68–72.

Arisz, W. H., Camphuis, I. J., Heikens, H., and Tooren, A. J. van (1955) 'The secretion of the salt glands of *Limonium latifolium* Ktze.', *Acta bot. neerl.*, **4**, 322–338.

Bonnot, E.-J. and Hébant, C. (1970) 'Précisions sur la structure et le fonctionnement des cellules mucigènes de *Polytrichum juniperinum* Willd.', *C. r. hebd. Séanc. Acad. Sci., Paris, D* **271**, 53–55.

Bouck, G. B. (1962) 'Chromatophore development, pits, and other fine structure in the red alga, *Lomentaria baileyana* (Harv.) Farlow', *J. Cell Biol.*, **12**, 553–569.

Brown, R. M. Jr (1969) 'Observations on the relationship of the Golgi apparatus to wall formation in the marine chrysophycean alga, *Pleurochrysis scherffelii* Pringsheim', *J. Cell Biol.*, **41**, 109–123.

Campbell, R. (1972) 'Electron microscopy of the development of needles of *Pinus nigra* var. *maritima*', *Ann. Bot.*, **36**, 711–720.

Christensen, A. K. and Fawcett, D. W. (1961) 'The normal fine structure of opossum testicular interstitial cells', *J. biophys. biochem. Cytol.*, **9**, 653–670.

Dexheimer, J. (1972) 'Quelques aspects ultrastructuraux de la sécrétion de mucilage par les glandes digestives de *Drosera rotundifolia* L.', *C. r. hebd. Séanc. Acad. Sci., Paris, D* **275**, 1983–1986.

Eymé, J. (1967) 'Nouvelles observations sur l'infrastructure de tissus nectarigènes floraux', *Botaniste*, **50**, 169–183.

Fekete, M. A. R. de, Ziegler, H., and Wolf, R. (1967) 'Enzyme des Kohlenhydratstoffwechsels in Nektarien', *Planta*, **75**, 125–138.

Figier, J. (1968) 'Localisation infrastructurale de la phosphomonoestérase acide dans la stipule de *Vicia faba* L. au niveau du nectaire. Rôles possibles de cet enzyme dans les mécanismes de la sécrétion', *Planta*, **83**, 60–79.

Figier, J. (1969) 'Incorporation de glycine-^{3}H chez les glandes pétiolaires de *Mercurialis annua* L. Étude radioautographique en microscopie électronique', *Planta*, **87**, 275–289.

Figier, J. (1972) 'Localisation infrastructurale de la phosphatase acide dans les glandes pétiolaires d'*Impatiens holstii.* Rôles possibles de cette enzyme au cours des processus sécrétoires', *Planta*, **108**, 215–226.

Findlay, N. and Mercer, F. V. (1971) 'Nectar production in *Abutilon*. I. Movement of nectar through the cuticle', *Austr. J. biol. Sci.*, **24**, 647–656.

Findlay, N. and Mercer, F. V. (1971) 'Nectar production in *Abutilon*. II. Submicroscopic structure of the nectary', *Austr. J. biol. Sci.*, **24**, 657–664.

Gland cells

Findlay, N., Reed, M. L., and Mercer, F. V. (1971) 'Nectar production in *Abutilon*. III. Sugar secretion', *Austr. J. biol. Sci.*, **24**, 665–675.

Frey-Wyssling, A. (1970) 'Betrachtungen über die pflanzliche Stoffelimination', *Ber. schweiz. bot. Ges.*, **80**, 454–466.

Frey-Wyssling, A. and Häusermann, E. (1960) 'Deutung der gestaltlosen Nektarien', *Ber. schw bot. Ges.*, **70**, 150–162.

Govier, R. N., Brown, J. G. S., and Pate, J. S. (1968) 'Hemiparasitic nutrition in Angiosperms II. Root haustoria and leaf glands of *Odontites verna* (Bell.) Dum. and their relevance to the abstraction of solutes from the host', *New Phytol.*, **67**, 963–972.

Häusermann, E. and Frey-Wyssling, A. (1963) 'Phosphatase-Aktivität in Hydathoden', *Protoplasma*, **57**, 371–380.

Healey, P. L. (1967) 'The secretory trichome of *Pharbitis* – its ultrastructural and histochemical development', *Am. J. Bot.*, **54**, 638.

Heinrich, G. (1966) 'Licht- und elektronenmikroskopische Untersuchungen zur Genese der Exkrete in den lysigenen Exkreträumen von *Citrus medica*', *Flora*, **156A**, 451–456.

Heinrich, G. (1969) 'Elektronenmikroskopische Beobachtungen zur Entstehungsweise der Exkretbehälter von *Ruta graveolens, Citrus limon* und *Poncirus trifoliata*', *Öst. bot. Z.*, **117**, 397–403.

Heinrich, G. (1970) 'Elektronenmikroskopische Beobachtungen an den Drüsenzellen von *Poncirus trifoliata*; zugleich ein Beitrag zur Wirkung ätherischer Öle auf Pflanzenzellen und eine Methode zur Unterscheidung flüchtiger von nichtflüchtigen lipophilen Komponenten', *Protoplasma*, **69**, 15–36.

Heslop-Harrison, Y. and Knox, R. B. (1971) 'A cytochemical study of the leaf-gland enzymes of insectivorous plants of the genus *Pinguicula*', *Planta*, **96**, 183–211.

Horner, H. T. Jr and Lersten, N. R. (1968) 'Development, structure and function of secretory trichomes in *Psychotria bacteriophila* (Rubiaceae)', *Am. J. Bot.*, **55**, 1089–1099.

Knapp, F. F., Aexel, R. T., and Nicholas, H. J. (1969) 'Sterol biosynthesis in sub-cellular particles of higher plants', *Pl. Physiol, Lancaster*, **44**, 442–446.

Konar, R. N. and Linskens, H. F. (1966) 'The morphology and anatomy of the stigma of *Petunia hybrida*', *Planta*, **71**, 356–371.

Kroh, M. (1967) 'Bildung und Transport des Narbensekretes von *Petunia hybrida*', *Planta*, **77**, 250–260.

Levering, C. A. and Thomson, W. W. (1971) 'The ultrastructure of the salt gland of *Spartina foliosa*', *Planta*, **97**, 183–196.

Lloyd, F. E. (1942) *The carnivorous plants,* Ronald Press Comp., New York.

Lüttge, U. (1961) 'Über die Zusammensetzung des Nektars und den Mechanismus seiner Sekretion. I', *Planta*, **56**, 189–212.

Lüttge, U. (1964) 'Untersuchungen zur Physiologie der Carnivoren-Drüsen. I. Mitteilung. Die an den Verdauungsvorgängen beteiligten Enzyme', *Planta*, **63**, 103–117.

Lüttge, U. (1969) 'Aktiver Transport (Kurzstreckentransport bei Pflanzen)', *Protoplasmatolo* VIII, 7b, Springer-Verlag. Vienna and New York.

Mercer, F. V. and Rathgeber, N. (1962) 'Nectar secretion and cell membranes. Electron Microscopy, vol. 2: *5th. Int. Congr. Electron Microscopy Philadelphia*, WW11. Academic Press, New York and London.

Mollenhauer, H. H. (1965) 'Transition froms of Golgi apparatus secretion vesicles', *J. Ultrastruct. Res.*, **12**, 439–446.

Mollenhauer, H. H. (1967) 'The fine structure of mucilage secreting cells of *Hibiscus esculentus* pods', *Protoplasma*, **63**, 353–362.

Morré, D. J., Jones, D. D., and Mollenhauer, H. H. (1967) 'Golgi apparatus mediated polysaccharide secretion by outer root cap cells of *Zea mays*. I. Kinetics and secretory pathwa *Planta*, **74**, 286–301.

Northcote, D. H. and Pickett-Heaps, J. D. (1966) 'A function of the Golgi apparatus in polysaccharide synthesis and transport in the root-cap cells of wheat', *Biochem. J.*, **98**, 159–167.

O'Brien, T. P. (1967) 'Cytoplasmic microtubules in the leaf glands of *Phaseolus vulgaris*', *J. Cell Sci.*, **2**, 557–562.

Panessa, B. J. and Gennaro, J. F. Jr (1972) 'Ultrastructural changes during digestion in *Sarracenia purpurea*, as viewed by SEM, TEM and light microscopy', *J. Cell Biol.*, **55**, 199a

Bibliography

Perrin, A. (1969) 'Sur la présence et l'organisation de cristaux, renfermant des granules de constitution analogue à celle de la phytoferritine, dans le strome plastidial des cellules de l'épithème des hydathodes de *Saxifraga granulata* L. et *Saxifraga aizoon* Jacq.', *C. r. hebd. Séanc. Acad. Sci., Paris, D* **269**, 1521–1524.

Perrin, A. (1970a) Diversité des formes d'accumulation de la phytoferritine dans les cellules constituant l'épithème des hydathodes de *Taraxacum officinale* Weber et *Saxifraga aizoon* Jacq.', *Planta*, **93**, 71–81.

Perrin, A. (1970b) 'Organisation infrastructurale, en rapport avec les processus de sécrétion, des poils glandulaires (trichome-hydathodes) des *Phaseolus multiflorus*', *C. r. hebd. Séanc. Acad. Sci., Paris, D* **270**, 1984–1987.

Perrin, A. (1972) '*Contribution à l'étude de l'organisation et du fonctionnement des hydathodes: recherches anatomiques, ultrastructurales et physiologiques*', Thése, Lyon.

Ragetli, H. W. J., Weintraub, M., and Lo, E. (1972) 'Characteristics of *Drosera* tentacles. I. Anatomical and cytological detail', *Can. J. Bot.*, **50**, 159–168.

Renaudin, S. and Garrigues, R. (1967) 'Sur l'ultrastructure des glandes en bouclier de *Lathraea clandestina* L. et leur rôle physiologique', *C. r. hebd. Séanc. Acad. Sci., Paris, D* **264**, 1984–1987.

Renaudin, S. (1972) 'Sur l'aspect des glandes en bouclier de *Lathraea clandestina* L. en microscopie à balayage', *C. r. hebd. Séanc. Acad. Sci., Paris, D* **274**, 842–845.

Rougier, M. (1969) 'Mise en évidence de la fonction sécrétrice des squamules d'*Elodea canadensis* à l'aide de quelques techniques permettant la détection des polysaccharides en microscopie électronique', *J. Microscopie*, **8**, 82a.

Rougier, M. (1972) 'Étude cytochimique des squamules d'*Elodea canadensis*. Mise en évidence de leur sécrétion polysaccharidique et de leur activité phosphatasique acide', *Protoplasma*, **74**, 113–131.

Scala, J. Schwab, D., and Simmons, E. (1968) 'The fine structure of the digestive gland of Venus's flytrap', *Am. J. Bot.*, **55**, 649–657.

Schnepf, E. (1961a) 'Licht- und elektronenmikroskopische Beobachtungen an Insektivoren-Drüsen über die Sekretion des Fangschleimes', *Flora*, **151**, 73–87.

Schnepf, E. (1961b) 'Quantitative Zusammenhänge zwischen der Sekretion des Fangschleimes und den Golgi-Strukturen bei *Drosophyllum lusitanicum*', *Z. Naturf.*, **16b**, 605–610.

Schnepf, E. (1963a) 'Zur Cytologie und Physiologie pflanzlicher Drüsen. 1. Teil. Über den Fangschleim der Insektivoren', *Flora*, **153**, 1–22.

Schnepf, E. (1963b) 'Zur Cytologie und Physiologie pflanzlicher Drüsen. 2. Teil. Über die Wirkung von Sauerstoffentzug und von Atmungsinhibitoren auf die Sekretion des Fangschleimes von *Drosophyllum* und auf die Feinstruktur der Drüsenzellen', *Flora*, **153**, 23–48.

Schnepf, E. (1963c) 'Zur Cytologie und Physiologie pflanzlicher Drüsen. 3. Teil. Cytologische Veränderungen in den Drüsen von *Drosophyllum* während der Verdauung', *Planta*, **59**, 351–379.

Schnepf, E. (1963d) 'Golgi-Apparat und Sekretbildung in den Drüsenzellen der Schleimgänge von *Laminaria hyperborea*', *Naturwiss.*, **50**, 674.

Schnepf, E. (1964a) 'Zur Cytologie und Physiologie pflanzlicher Drüsen. 4. Teil. Licht- und elektronenmikroskopische Untersuchungen an Septalnektarien', *Protoplasma*, **58**, 137–171.

Schnepf, E. (1964b) 'Zur Cytologie und Physiologie pflanzlicher Drüsen. 5. Teil. Elektronen-mikroskopische Untersuchungen an Cyathialnektarien von *Euphorbia pulcherrima* in verschiedenen Funktionszuständen', *Protoplasma*, **58**, 193–219.

Schnepf, E. (1964c) 'Über Zellwandstrukturen bei den Köpfchendrüsen der Schuppenblätter von *Lathraea clandestina* L.', *Planta*, **60**, 473–482.

Schnepf, E. (1965a) 'Licht- und elektronenmikroskopische Beobachtungen an den Trichom-Hydathodem von *Cicer arietinum*', *Z. PflPhysiol.*, **53**, 245–254.

Schnepf, E. (1965b) 'Die Morphologie der Sekretion in pflanzlichen Drüsen', *Ber. dt. bot. Ges.*, **78**, 478–483.

Schnepf, E. (1968) 'Zur Feinstruktur der schleimsezernierenden Drüsenhaare auf der Ochrea von *Rumex* und *Rheum*', *Planta*, **79**, 22–34.

Schnepf, E. (1969a) 'Sekretion und Exkretion bei Pflanzen', *Protoplasmatologia*, VIII, 8. Springer-Verlag, Vienna and New York.

Schnepf, E. (1969b) 'Über den Feinbau von Öldrüsen. I. Die Drüsenhaare von *Arctium lappa*', *Protoplasma*, **67**, 185–194.

Gland cells

Schnepf, E. (1969c) 'Über den Feinbau von Öldrüsen. II. Die Drüsenhaare in *Calceolaria*-Blüten', *Protoplasma*, 67, 195–203.

Schnepf, E. (1969d) 'Über den Feinbau von Öldrüsen. III. Die Ölgänge von *Solidago canaden* und die Exkretschläuche von *Arctium lappa*', *Protoplasma*, 67, 205–212.

Schnepf, E. (1969e) 'Über den Feinbau von Öldrüsen. IV. Die Ölgänge von Umbelliferen: *Heracleum sphondylium* und *Dorema ammoniacum*', *Protoplasma*, 67, 375–390.

Schnepf, E. (1972) 'Tubuläres endoplasmatisches Reticulum in Drüsen mit lipophilen Ausscheidungen von *Ficus, Ledum* und *Salvia*', *Biochem. Physiol. Pfl.*, 163, 113–125.

Schnepf, E. and Koch, W. (1966) 'Über die Entstehung der pulsierenden Vacuolen von *Vacuolaria virescens* (Chloromonadophyceae) aus dem Golgi-Apparat', *Arch. Mikrobiol.*, 54, 229–236.

Schwab, D. W., Simmons, E., and Scala, J. (1969) 'Fine structure changes during function of the digestive gland of Venus's flytrap', *Am. J. Bot.*, 56, 88–100.

Shimony, C. and Fahn, A. (1968) 'Light- and electron-microscopical studies on the structure of salt glands of *Tamarix aphylla* L.', *J. Linn. Soc. (Bot.)*, 60, 283–288.

Sperlich, A. (1939) 'Exkretionsgewebe', In: *Handbuch der Pflanzenanatomie*. Bd. IV. Das trophische Parenchym. B (ed. K. Linsbauer), Gebr. Borntraeger Verlag, Berlin.

Steer, M. W. and Newcomb, E. H. (1969) 'Observations on tubules derived from the endoplasmic reticulum in leaf glands of *Phaseolus vulgaris*', *Protoplasma*, 67, 33–50.

Thomson, W. W., Berry W. L., and Liu, L. L. (1969) 'Localization and secretion of salt by th salt glands of *Tamarix aphylla*', *Proc. natn. Acad. Sci. U.S.A.*, 63, 310–317.

Thomson, W. W. and Liu, L. L. (1967) 'Ultrastructural features of the salt gland of *Tamarix aphylla* L.', *Planta*, 73, 201–220.

Tourte, Y. (1972) 'Les sécrétions mucilagineuses dans l'organe reproducteur femelle d'une fougère: Etude infrastructurale', *J. Microscopie*, 15, 377–394.

Tucker, S. C. and Hoefert, L. L. (1968) 'Ontogeny of the tendril in *Vitis vinifera*', *Am. J. Bot.*, 55, 1110–1119.

Unzelman, J. M. and Healey, P. L. (1972) 'Development and histochemistry of nuclear crystals in the secretory trichome of *Pharbitis nil*', *J. Ultrastruct. Res.*, 39, 301–309.

Vasiliev, A. E. (1969a) 'Submicroscopic morphology of nectary cells and problems of nectar secretion', (Russ.), *Akad. Nauk USSR, bot. J.*, 54, 1015–1031.

Vasiliev, A. E. (1969b) 'Some peculiarities of the endoplasmic reticulum in secretory cells of *Heracleum* sp.', (Russ.), *Akad. Nauk USSR, Citologia*, XI, 298–307.

Vasiliev, A. E. (1970) 'Uber die Lokalisation der Synthese von Terpenoiden in pflanzlichen Zellen', (Russ.), *Akad. Nauk USSR*, 5, 29–45.

Vasiliev, A. E. (1971) 'New data on the ultrastructure of the cells of flower nectary', (Russ.), *Akad. Nauk SSSR, Bot. Ž.*, 56, 1292–1306.

Vasiliev, A. E. (1972) 'The ultrastructure of the nectary cells of *Cucumber*', (Russ.), *Akad. Nauk SSSR, Citologia*, 14, 405–415.

Vogel, S. (1971) 'Olproduzierende Blumen, die durch ölsammelnde Bienen bestäubt werden', *Naturwiss.*, 58, 58.

Werker, E. and Fahn, A. (1968) 'Site of resin synthesis in cells of *Pinus halepensis* Mill', *Nature*, 218, 388–389.

Williams, S. E. and Pickard, B. G. (1969) 'Secretion, absorption and cuticle structure in *Drosera* tentacles', *Pl. Physiol.*, *Lancaster*, 44, Suppl. 5–6.

Wollenweber, E., Egger, K., and Schnepf, E. (1971) 'Flavenoid-Aglykone in *Alnus*-Knospen und die Feinstruktur der Drüsenzellen', *Biochem. Physiol. Pfl.*, 162, 193–201.

Wollenweber, E. and Schnepf, E. (1970) 'Vergleichende Untersuchungen über die flavonoiden Exkrete von "Mehl"- und "Öl"-Drüsen bei Primeln und die Feinstruktur der Drüsenzellen' *Z. PflPhysiol.*, 62, 216–227.

Wooding, F. B. P. and Northcote, D. H. (1965a) 'Association of the endoplasmic reticulum and the plastids in *Acer* and *Pinus*', *Am. J. Bot.*, 52, 526–531.

Wooding, F. B. P. and Northcote, D. H. (1965b) 'The fine structure of the mature resin canal cells of *Pinus pinea*', *J. Ultrastruct. Res.*, 13, 233–244.

Wrischer, M. (1962) Elektronenmikroskopische Beobachtungen an extrafloralen Nektarien von *Vicia faba* L.', *Acta bot. Croat.*, 20/21, 75–94.

Zandonella, P. (1970) 'Infrastructure des cellules du tissu nectarigène floral de quelques Caryophyllaceae', *C. r. hebd. Séanc. Acad. Sci.*, *Paris*, D 270, 1310–1313.

Ziegler, H. (1956) 'Untersuchungen über die Leitung und Sekretion der Assimilate', *Planta*, 47, 447–500.

Ziegler, H. and Lüttge, U. (1966) 'Die Salzdrüsen von *Limonium vulgare*. I. Mitteilung. Die Feinstruktur', *Planta*, 70, 193–206.

Ziegler, H. and Lüttge, U. (1967) 'Die Salzdrüsen von *Limonium vulgare*. II. Mitteilung. Die Lokalisierung des Chlorids', *Planta*, 74, 1–17.

10

Cambial cells
A. M. Catesson

Introduction

Cambia are the lateral meristems which produce the increase in thickness of both stem and root. Formerly, authors divided cambia into the vascular cambium, which gives rise to vascular tissues, and the cork cambium which produces cork. The cork cambium is now usually called *phellogen,* and the name *cambium* is preferentially used for vascular cambium. Phellogen is formed by short, identical cells while the vascular cambium is normally a complex tissue, composed of two cell types, each one with a definite role: the fusiform initials are elongated with tapering ends like a spindle, and give rise to the cells of the vertical system of phloem and xylem tracheids, vessels, sieve cells, fibres, and vertical parenchyma cells; the other kind of cells, the ray initials, are nearly isodiametric and produce only ray parenchyma cells, that is the elements of the horizontal system of phloem and xylem.

Although cambial anatomy and cell division have been studied for a long time, the cytology of cambial cells has been somewhat neglected. Bailey's pioneering work (1919–1930) had few competitors till the advent of the electron microscope and even the papers concerned with the ultrastructural study of the cambium cover only slightly more than half a dozen species of tree growing in the temperate regions. These species are pine, *Pinus strobus* (Srivastava and O'Brien, 1966; Murmanis, 1970, 1971), ash, *Fraxinus americana* (Srivastava, 1966), beech, *Fagus sylvatica* (Kidwai and Robards, 1969a and b; Robards and Kidwai, 1969a and b), lime tree, *Tilia americana* (Evert and Deshpande, 1970; Mia, 1970), and elm, *Ulmus americana* (Evert and Deshpande, 1970). We can add to these studies the light microscope observations made by Buvat on the locust tree, *Robinia pseudaccacia* (1956a) and by myself on the sycamore, *Acer pseudoplatanus* (Catesson, 1964). This work on *Robinia* and *Acer* was recently resumed at the ultrastructural level in our laboratory but the results are still unpublished. I may add to the list of publications on vascular cambium the two beautiful micrographs of *Robinia* cambium enclosed by Ledbetter and Porter in their book *Introduction to the fine structure of plants* (1970). The other papers published on vascular tissues are concerned mainly with xylem and phloem differentiation and deal

358

with the cambium only in passing (Hohl, 1960); Cronshaw and Wardrop, 1964; Kollmann and Schumacher, 1964; Cronshaw, 1965; Czaninski, 1968b, 1970; Zee and Chambers, 1969). More recently, a paper on the subject of *Pinus radiata* cambium has been published by Barnett (1971), while Philipson, Ward, and Butterfield (1971) have produced a book which, while providing an interesting and useful review of cambial anatomy and activity, deals only briefly with cambial ultrastructure.

Therefore, the plants studied are few and, furthermore, only stem cambium has been examined, except for the work of Zee and Chambers (1969). However, these few examples come from plants which differ greatly from a systematic point of view; gymnosperms and more or less advanced dicotyledons. But there has been no work on the cytology of anomalous cambia such as one can find in Chenopodiaceae and related families, or in various lianas, or in some monocotyledons. Further, no detailed cytological study has been made on the phellogen, and ultrastructural observations on cambium-like formations in tissue cultures are poor. Therefore, I shall have to limit the subject of this chapter essentially to the cell ultrastructure of stem cambia in arborescent plants.

The few ultrastructural studies on the cambium may be explained partly because it is difficult to isolate and fix this tissue properly. During winter, the cambium is poorly vacuolated and fixations are quite good, but ultrathin sections are easily torn along the xylem at the level of cambial cells. When the cambium is active, it is highly vacuolated and the cell organelles are poorly preserved by fixatives; a fixing medium suitable for differentiating xylem of phloem may not be adequate for cambium. As Kidwai and Robards (1969a) suggested, one must use different fixatives for different tissues, and also different fixatives for the same tissue at different times of the seasonal cycle.

The scarcity of ultrastructural studies on the cambium contrasts with the large number of papers concerned with cambial derivatives. This neglect of the cambium is deplorable, for anatomical and light microscope studies are not adequate to solve all problems. The first obstacle is the exact meaning of the word 'cambium': some authors insist that the term 'cambium' must be used for only one layer of cells, the true cambial initials which, alone, can give rise to cells in both phloem and xylem directions; others consider the cambium to be constituted by all the dividing cells, i.e., the cambial initials together with their derivatives. Ultrastructural studies should bring us more precise data and allow us to give a unanimous definition of the word 'cambium'. A second problem for which ultrastructural observations are needed is the origin and evolution of the cambium from primary meristems. In addition, a precise ultrastructural study will be useful to complement physiological work and biochemical analyses if we want to understand completely the mechanisms of seasonal activity in perennial plants.

These three subjects will be dealt with in this chapter but, first, we shall look at a general picture of cambial cells.

Portrait of active cambial cells and their derivatives

An example: the cambium of *Acer pseudoplatanus*

During the growing season, the cambial zone is 4–10 cells wide. This cambium appears non-storied in tangential sections, i.e., the fusiform cells are not arranged in regular rows and their ends overlap; in storied cambia, on the other hand, the fusiform initials occur in horizontal tiers so that their ends are all at approximatel the same level.

Fusiform initials have an average length of 320 μm in 2–3-year-old branches. These cells are much compressed radially; they are 2–4 μm thick and 15–20 μm wide. Cell walls are thin and punctuated by plasmodesmatal fields. The plasmalemma is more or less sinuous and is sometimes in contact with multivesicular bodies or isolated vesicles. These vesicles are either Golgi vesicles or pycnotic vesicles. Microtubules are numerous and often parallel to the cell axis. The resting nucleus occupies a more or less median position within the cell; it is rather voluminous and can reach 22 μm in length; it measures approximately 4 μm in tangential diameter and 2 μm in radial diameter. There are one to five nucleoli arranged in a vertical line, three being the average number. The fine structure of t nucleus is similar to that of the majority of plant cells.

Fusiform cells are highly vacuolated. *In vivo*, one can see one or two large vacuoles in each cell; their *p*H is slightly acid, as shown by their contents producing an orange-red colour with neutral red. A great number of anastomosing cytoplasmic strands extend across the vacuoles. These strands are often poorly preserved during fixation; they are however quite clear in Fig. 10.1A. In this micrograph, some vacuoles are present in the peripheral cytoplasm, but they are too small to be recognized *in vivo*. It is difficult to see their connections with the central vacuole on the micrographs. The peripheral cytoplasm is reduced to a thin layer 0.2–1 μm thick, while another thin layer surrounds the nucleus. Cytoplasm streaming can easily be seen in the cytoplasm, and in cytoplasmic strands. When fixation is good, the cytoplasm is rather dense: free ribosomes are abundant, and some occur singly, but they often aggregate into circular or spiral patterns. Endoplasmic reticulum generally appears in the form of typical rough cisternae. Dictyosomes of the Golgi apparatus are numerous and produce different kinds of vesicles. The larger ones are electron translucent, the smaller ones electron opaque (Fig. 10.2B).

Plastids and mitochondria are scattered in the ground cytoplasm. Mitochondria are abundant, ovoid or slightly elongated, with a mean length usually between 1 and 2 μm but sometimes reaching 3 μm; their width is around 0.5 μm. Mitochondria have a clear matrix and a few short cristae (Fig. 10.2B); this is a character common to all meristematic cells (see Clowes and Juniper, 1968, p. 163 Plastids, or rather proplastids, have a dense stroma, numerous vesicles and a few short tubules which, in many instances, can be seen to be continuous with the plastid inner membrane (Fig. 10.2B). There are generally no starch grains. Plastid are spherical, elongated, and sometimes curved. Constrictions can be seen frequen

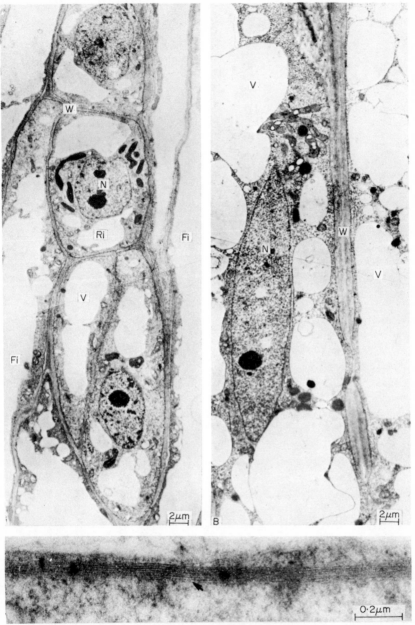

Fig. 10.1   Electron micrograph of *Acer pseudoplatanus* cambium. Fixation: glutaraldehyde-osmium tetroxide. Stained with potassium permanganate. N = nucleus, V = vacuole, W = cell wall.

(A) Tangential view of cambial cells in April. Ray initials (Ri) are in the middle of the micrograph, fusiform initial (Fi) on each side. Note the thin walls, the large vacuoles in the centre of the cells, and the small ones in the peripheral cytoplasm.

(B) Tangential view of fusiform initials in December, with thick walls and small vacuoles.

(C) Bundle of cytoplasmic fibrils (arrow) in a fusiform cell in January.

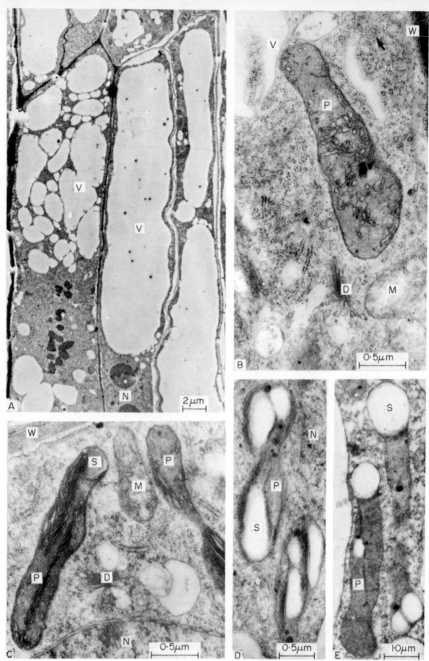

Fig. 10.2 Electron micrographs of *Acer pseudoplatanus* cells. Fixation: glutaraldehyde-osmium tetroxide; stained with lead citrate (Figs. A, B and C) or potassium permanganate (Figs. D and E). D = dictyosome, M = mitochondrion, N = nucleus, P = plastid, S = starch grain, V = vacuole, W = cell wall.

(A) Radial section of young procambial cells in April. Note the periclinal division in the cell on the left. (B and E). Details of fusiform cells either in April (B) or in December (E) showing the seasonal differences in plastid structure. Note the numerous cytoplasmic vesicles and the spiral arrangement of ribosomes (arrow) in April.

(C and D). Details of ray cells either in April (C) or in November (D). Plastid structure does not change much, but starch is more abundant in winter.

and can be followed through successive serial sections; they probably represent division images. Here and there, ferritin aggregates may be recognized (Catesson, 1966).

Osmiophilic globules are scarce during April and May. Some peroxysomes are present in the cells and their crystal-like core give the characteristic reaction of catalase (as we shall see later – p. 384).

Ray initials are clearly distinguished from fusiform initials by their smaller size. They are generally nearly isodiametric, their three dimensions ranging in size from 8 to 12 μm. However, some ray cells, just produced from a fusiform initial by transverse division, may be longer than this. Their ultrastructural features are rather similar to those of the fusiform cells, nuclei and plastids excepted. The nuclei are small, ovoid or spherical; they are 5–11 μm long and 3–5 μm wide. Ray cell plastids appear green *in vivo* and they contain chlorophyll. They are larger than those of fusiform cells, with the same high electron opacity. They contain small vesicles and stacks of thylakoids, but no mature grana; some plastoglobuli and starch grains may be present (Fig. 10.2C). Ray cell plastids are more highly differentiated than fusiform cell plastids. In *Robinia* xylem, there is also a higher degree of differentiation in horizontal parenchyma than in vertical parenchyma (Czaninski, 1968a, 1970).

Comparison with other species

The fine structure of cambial cells in other plants so far investigated is rather similar to that of sycamore cambial cells. The main differences concern the nucleus and plastid structure.

Nuclear shape, and the number of nucleoli in fusiform initials, differs from one plant to another. At first sight it seems that the more elongated the cells, the longer are the nuclei. The fusiform cells of dicotyledons contain smaller nuclei and fewer nucleoli than the homologous but longer cells of gymnosperms. In *Pinus strobus,* where the fusiform initial length varies between 870 and 4000 μm according to their age, nuclear dimensions oscillate in parallel between 63 and 82 μm; there are five or six nucleoli. These large nuclei do not seem to be polyploid (Bailey, 1920b).

Plastid fine structure varies greatly according to the species under investigation. In most plants, fusiform and ray cell plastids do not show any difference (contrary to *Acer* cells). In *Robinia,* however, one can find in the same ray cell proplastids with very few thylakoids, similar to those present in fusiform initials, and more elaborated plastids containing stacks of four or five lamellae. In *Salix* (Robards and Kidwai, 1969a), both kinds of plastid coexist in both types of initial, and plastoglobuli and starch grains are present. Plastids of *Fagus* (Kidwai and Robards, 1969b) are very similar. In *Ulmus* and in *Tilia* (Evert and Deshpande, 1970), plastids are abundant and highly variable in shape; the dense stroma contains one or two thylakoids, numerous vesicles, and starch grains. In *Pinus strobus* (Srivastava and O'Brien, 1966; Murmanis, 1971), plastids often appear curved, and sometimes extremely long; their size varies from 1.5 to 5 μm or more. The finely fibrillar stroma contains one to several thylakoids and elongated or circular vesicles. Electron opaque, intra-lamellar inclusions are sometimes present. Osmiophilic granules and starch grains occur in the stroma. In *Metasequoia* (Kollmann and Schumacher, 1964), plastids possess some elongated thylakoids.

Ferritin inclusions may be seen in cambial plastids in *Salix* (Robards and Humpherson, 1967) and *Robinia* (Czaninski, 1968a, and personal observations). In this latter species, ferritin molecules either form pseudocrystalline aggregates as in *Salix* and *Acer,* or are scattered in the stroma, so appearing as a cloud of dense specks.

In *Fraxinus americana* (Srivastava, 1966), cambial plastids differ greatly from those of the other species: they vary in shape and size; and the dense stroma contains some vesicles and a few lamellae, osmiophilic globules and starch grains. Their main feature is the presence of a dark, amorphous intra-lamellar inclusion. In the micrographs, these inclusions appear circular in face view, and in transverse sections are stacked like grana, but the partitions between saccules cannot be seen. To our knowledge, Srivastava is one of the first authors to describe this kind of plastid, and he suggested that the dense, osmiophilic material could play a role in the formation of new membranes as it disappears in phloem cells whose plastids have typical grana. Since Srivastava's paper, plastids with dense bodies have been described in various organs and plants (for the references, see Catesson, 1970), and recently Blackwell, Laetsch, and Hyde (1969) were able to show in aspen cultured cells that, upon exposure to light, the dense bodies, or thylakoid precursors, undergo a series of subdivisions and give rise to saccules and grana. This subdivision during proplastid differentiation explains the various shapes of inclusions found in cambial plastids of *Fraxinus*. As is frequent in other plastids with dense bodies (see Catesson, 1970), plastids of *Fraxinus* seem to contain two distinct lamellar systems: one constituted by the lamellae containing the dense material; the other by electron translucent vesicles and thylakoids. It should be interesting to compare the fine structure of cambial plastids with that of plastids from other meristematic tissues in *Fraxinus.*

Therefore, in cambial cells, proplastids or slightly differentiated plastids may belong either to the 'classical' type or to the 'dense inclusion' type according to the species. In arborescent plants, cambial plastids, even when they are true chloroplasts with stacks of lamellae, do not form grana. In herbaceous species, they may be typical chloroplasts with mature grana, as we have observed quite recently in carnation stem (unpublished results). The agranal structure of cambial plastids in trees is perhaps linked with the very small amount of light reaching the cambium across the bark. It is curious to note that, in spite of reduced illumination prolamellar bodies are quite infrequent in cambial plastids. Such prolamellar bodies have been seen only once: in a micrograph published by Cronshaw (1965) from *Acer rubrum* cambium.

Beyond the differences in nuclear and plastid fine structure, there are only slight dissimilarities between the cambial cells of the different plants which have been studied. These dissimilarities mainly relate to the presence or absence of coated vesicles, the quantities of osmiophilic granules, or the existence of peroxisomes. The latter have not always been identified, but they may be recognized in micrographs of ash and willow where they have a crystalline core. In the locust tree, peroxisomes without crystals are sometimes present, but on the whole they are rather infrequent.

In the great majority of investigated plants, the cytoplasm of cambial cells contains more or less abundant vesicles: some are undoubtedly pycnotic vesicles (Evert and Deshpande, 1970); others, produced by dictyosomes or endoplasmic reticulum, are probably linked with cell wall formation (Robards and Kidwai, 1969b); the origin and function of others remains unknown.

To our knowledge, only Zee and Chambers (1969) have published images of root cambial cells. According to these authors, in pea root the fine structure of the cambium is quite similar to that of ash or pine cambium as described by Srivastava (1966) and by Srivastava and O'Brien (1966).

The first cambial derivatives

It is not my purpose to discuss the differentiation of cambial cells into xylem or phloem, as these problems are the subject of the next chapters of this book. However, I want to give here a brief picture of the first steps of this evolution. The first indication of any differentiation is a marked enlargement of cambial derivatives which is accomplished primarily by radial expansion (Wilson, 1963; Cronshaw, 1965; Srivastava and O'Brien, 1966; Czaninski, 1968b, 1970). This extension is due to an increase in vacuolar volume without any conspicuous change in cell fine structure. Soon afterwards, deposition of new wall material occurs all around the cell concomitantly with an increase in the activity of endoplasmic reticulum, dictyosomes, and plasmalemma (Kollmann and Schumacher, 1964; Cronshaw, 1965; Srivastava and O'Brien, 1966; Czaninski, 1968b, 1970). From this point, the evolution of cambial derivatives differs greatly according to the kind of phloem or xylem cells that they will produce.

Ontogeny and evolution of the cambium

Origin of the cambium

The origin of the cambium is well known from an anatomical point of view and the reader is referred to anatomy textbooks (Esau, 1953; Fahn, 1969; Philipson, Ward, and Butterfield, 1971) as well as to the well-documented review of Philipson and Ward (1965). To my knowledge, there is no cytological study pertaining to the ontogeny of the root cambium. Only a few micrographs referring to root procambial cells have been published (Esau, 1965; Arsanto, 1970). These cells are rather long, with an elongated or lobate nucleus; their other ultrastructural characteristics are very similar to those of other primary meristematic cells.

In stems, cambia have a double origin: the greater part is initiated within the provascular strands and is called intrafascicular cambium; the other part originates in the ground tissue between the strands and forms the interfascicular cambium.

Intrafascicular cambium. Procambial strands arise from apical meristem activity.

Cambial cells

They are isolated and make a broken ring in the young internode. Primary vascular tissues differentiate within these strands which then become young vascular bundles. Some images of procambial cells of *Beta* have been published by Esau, Cheadle, and Gill (1966), who point out the abundance of dictyosomes and vesicles in those cells. Beyond this work, few details are known of procambial ultrastructure, and we shall return to our example, the sycamore.

In *Acer pseudoplatanus,* procambial cells are initiated by periclinal divisions of the deeper layers of the tunica (Catesson, 1964). They can be distinguished from the other meristematic cells only by their radially compressed shape (5 μm × 12 μm). The cytoplasm is rich in free ribosomes, and contains mitochondria, proplastids, and numerous small vacuoles. The nucleus generally has one voluminous nucleolus. Procambial cells elongate slightly as the internode is initiated. When it reaches a length of 80–100 μm, that is to say when the first sieve tubes differentiate, procambial cells are 15–20 μm in length. Free ribosomes are always abundant in the ground cytoplasm, dictyosomes are numerous, and the stroma of proplastids is dense and contains vesicles similar to those in the proplastids of mature fusiform initials. Small starch grains are sometimes present. Mitochondria have a clear matrix and are of variable size. There is a complex vacuolar net, as can be seen *in vivo,* and the shape of vacuoles in micrographs is highly variable. Nuclei are elongated, often with two or three nucleoli in line.

TABLE 10.1

Increase in Length of Procambial Cells During Growth of the Internode

internode length	0	100 μm	1 cm	5 cm	15 cm
procambial cell length	10 – 12 μm	15 – 20 μm	50 – 70 μm	200 μm	250 μm

Later there is an intensive cellular multiplication in the young bundles. Their increase in diameter is important, and their elongation may be very great, in keeping with internode elongation. Some of these internodes, the third for instance, may grow in less than four weeks from 100 μm to 15 cm. This increase in length is produced in part by cell elongation (Table 10.1), but chiefly by repeated transverse divisions (Table 10.2). Therefore there is here (in *Acer*) an *active* elongation of the procambium, while in *Pseudotsuga* there is only a *passive* growth due to cell expansion (Sterling, 1947).

In a very young bundle of sycamore (Fig. 10.3), when the first sieve tubes and tracheids are differentiated, one may recognize several regions (Catesson, 1964):

(a)   on the outside, the future pericycle fibres which are still unlignified (zone 1, Fig. 10.3);

(b)   along these fibres, the young phloem with the first sieve elements (zone 2,

TABLE 10.2

Percentage of the Three Types of Cell Division to Total Number of Mitoses in the Different Regions of a Young Vascular Bundle in an Internode 1 cm Long

	Transverse mitosis	Periclinal mitosis	Anticlinal mitosis
differentiating phloem	74	16	10
phloem mother cells	73	23	4
future cambial cells	50	36	14
differentiating xylem	61	30	9

(c) in the middle region, a complex zone which contains: *adjacent to the phloem,* a large band of small meristematic cells (8 μm × 15 μm) which mature to produce only phloem elements — I refer to them as phloem mother cells (zone 3, Fig. 10.3); *on the inside*, a zone of narrow, elongated, more vacuolated cells (8 μm × 60 μm) which will give rise to the cambium proper (zone 4);

(d) a large band of cells with tangential partitions, arranged in radial tiers; they are differentiating into parenchyma cells and vessels and constitute the primary metaxylem (zone 5);

(e) the inner region of the bundle which forms the protoxylem with annular or helical thickened tracheids (zone 6).

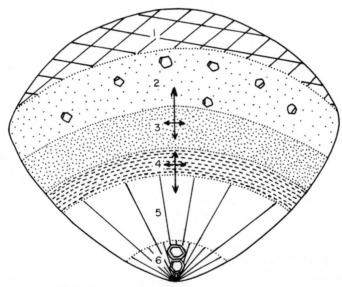

Fig. 10.3 Schematic drawing of a young vascular bundle seen in transverse section. (From Catesson, 1964.)

(1) future pericycle fibres.
(2) young phloem with functional sieve tubes.
(3) phloem mother cells.

(4) future cambial zone.
(5) differentiating xylem.
(6) protoxylem.

Cambial cells

The future cambial cells, which alone are of interest here, divide in all planes, but transverse mitoses predominate. However, periclinal divisions are more frequer in the future cambial zone than in other regions of the vascular bundle (Table 10.2 These cells (Fig. 10.2A) have elongated nuclei, 15–16 μm long, with two or three nucleoli in a line; their cytoplasm is dense and rich in ribosomes; dictyosomes are abundant and active. Vacuoles have many different appearances in the micrographs: sometimes there are one or two large vacuoles; alternatively a great number of small ones may be seen more or less fusing one with another. Mitochondria are generally globular, and some may attain a length of 3 μm. Proplastids are elongated or bell-shaped, and have a dense stroma with some isolated thylakoi and numerous vesicles.

When the internode reaches half its final length, the future cambial cells are about 200 μm long and their ends begin to taper off. Around this time transverse mitoses take place in some of these cells to give rise to the first ray initials. Thus, the middle region of the bundle acquires the characteristic appearance of a cambium with its two kinds of cell, fusiform and ray initials. The fine structure of these cells is then similar to that of mature cambium. However, the increase in length of fusiform initials goes on until the end of the internode growth, and even later on by intrusive growth. During the three or four weeks after the cessation of growth of the internode, the average length of fusiform initials increases from 250 to 320 μm. Nevertheless, in certain plants, cambial cells may b recognized long before the end of internode elongation (Sterling, 1946; Philipson and Ward, 1965).

Consequently, the cambium arises in a vascular bundle by means of *differentiation* of procambial cells into two kinds of peculiar cell. The cambium is not a residue of meristematic cells left by the differentiation of primary vascular tissues it is a distinct tissue with its own ontogeny.

Interfascicular cambium. The cambium which arises within the procambial strands forms a broken ring. The cells between the strands give rise to parenchyma cells. In the sycamore (Catesson, 1964), there are three or four tiers of parenchyma cells between the vascular bundles; these cells differentiate more slowly than adjoining cortical or medullary cells and they remain smaller in size. They are highly vacuolated, their mitochondria are short, and their plastids have a clear stroma, some stacks of lamellae and no vesicles. At the end of internode growth, these cells divide periclinally; the plastid stroma becomes denser, vesicles arise from the inner membrane, and these cells begin to function as ray initials.

During this period the fusiform initials adjoining the intrafascicular rays are, as we have seen, elongating: they intrude among radial cells and dissect the interfascicular rays into smaller rays. Some of the radial cells may also elongate and differentiate into fusiform initials, but this process is rather infrequent in sycamore. On the contrary, in other species, such as *Vicia faba, Teucrium scorodonia, Lycopus europeus* (personal observations) and in *Sequoia* (Sterling, 1946), a great number of fusiform cells arise from parenchyma cells in the interfascicular region. On the other hand, in *Rubus* (Resch, 1959) and in *Clematis vitalba* (personal data), interfascicular zones are not dissected by the intrusive growth of fusiform cells, and they remain conspicuous along the whole internode.

In conclusion, we can say that the cambium arises from the procambium, like the vascular tissues, by processes of cell growth, multiplication and differentiation; to these cells are added cells originating by means of a slight dedifferentiation from interfascicular parenchyma cells.

This cambial tissue differentiates progressively and can be recognized only during the last stages of internode growth when fusiform initials with characteristic tapering ends, and radial initials can be readily identified. Until this point, it is not possible to speak of a true cambium, and the tissues produced within the procambial strands, even if their cells form radial lines, are primary tissues. It would be better to keep the name 'secondary tissues' for those tissues produced by a well-defined cambium.

Aging of the cambium

We have seen that, in some species, the increase in length of fusiform cells goes on for some weeks after the termination of growth of the internode. In non-storied cambia, the average length of fusiform initials increases slowly during the whole life of the plant. Bailey (1923) showed that, in pine, fusiform cells of 60-year-old trunks reach a size of 4000 μm, while in young branches they do not exceed 870 μm in length. This intrusive growth partly contributes to the increase in girth of the cambium. In storied cambia, however, the fusiform initials remain of constant average length whatever the age of the plant may be.

The changes in length of fusiform initials are just one aspect of the profound modifications which take place in the cambial zone during the plant's life. Detailed studies of these transformations have been made during the last 30 years, but only from an anatomical viewpoint (see the reviews by Esau, 1953; Bannan, 1962, 1970; Srivastava, 1964; Philipson and Ward, 1965), and I shall refer here only to the results of these studies. Increase in girth is secured by three different means:

(a) increase in number of fusiform initials by anticlinal divisions;
(b) increase in number of ray initials and, in plants with multiseriate rays, a consecutive increase in width of vascular rays;
(c) intrusive elongation of fusiform initial in non-storied cambia.

In the majority of arborescent plants, the anticlinal divisions take place chiefly towards the end of radial growth and produce many more cells than are necessary for the increase in girth. This overproduction, to which is added the formation of fusiform cells from ray cells in Angiosperms, is counterbalanced by an important loss of fusiform cells: they divide transversely to give rise to ray cells, or they mature into phloem and xylem. On the contrary, in herbaceous species, there is no outright loss of fusiform initials, and the rate of production of new fusiform cells is just about adequate to keep pace with the diametral increase of the cambium (Cumbie, 1963, 1969; Walker and Cumbie, 1968).

Consequently, the life of the cambium may be indefinite although, in arborescent species at least, the initials are not permanent entities. Each initial functions as such only during a certain period; it differentiates into vascular tissues, or is converted into an initial of the other kind, and is then relieved by another initial of the same type which 'inherits' the initial function. Therefore the positions of the initials are continually shifting. This shifting induces important modifications in intercellular relationships, and the cambium is continually undergoing changes which point to a high degree of plasticity in this tissue. It is to be regretted that no data are available on ultrastructural evolution during these transformations. At the most, we may suppose that, in sycamore, the division of a fusiform cell into ray cells induces some differentiation of proplastids into chloroplasts, while the reverse process must occur when a ray initial elongates into a fusiform initial.

Cambial activity

Seasonal periodicity

In perennial plants from temperate regions cambial activity shows an annual rhythm linked with the succession of seasons. In tropical regions, cambial activity is rather dependent upon the amount of water available. When there is a dry season the cambium is inactive, except for plants whose roots can reach the levels that contain a certain amount of moisture even during the dry months (Fahn, 1969, p. 280). In the tropical rain forests, a great number of trees do not form annual rings; one-third of the species, however, exhibit a seasonal periodicity of cambial activity (Wilcox, 1962; Alvim, 1964).

Studies made on tropical plants are few compared to the abundant literature on seasonal activity of the cambium in temperate regions. There are many reviews on this subject (Bannan, 1962; Larson, 1962; Wilcox, 1962; Wort, 1962; Studhalter, Glock, and Agerter, 1963; Wareing, Hanney, and Digby, 1964; and Philipson, Ward, and Butterfield, 1971), and I shall point out only the more salient facts. In spring, a rather conspicuous swelling of the cells precedes the resumption of mitoses. The first divisions begin immediately below the terminal buds and spread to the bases of the branches and then downwards. The wave of cell divisions progresses down the stem very quickly in ring porous species, much more slowly in diffuse porous plants. In gymnosperms, the first mitoses take place in the cells adjacent to the xylem (Bannan, 1962); in ring porous angiosperms, there are no great differences between the onset of xylem and phloem production while, in diffuse porous angiosperms, the reactivation of phloic derivatives precedes cambial reactivation (see Davis and Evert, 1970). The initiation of cambial activity varies with species and regions. This activity slows down and eventually stops during the summer: in diffuse porous species, it ceases with the cessation of shoot growth; in ring porous species, summer wood production may continue for a long time after the end of apical growth (Wareing, Hanney, and Digby, 1964).

The annual cycle of cambial activity is related, like the bud cycle, to the antagonistic action of growth promotors (auxins, gibberellins . . .) and of growth inhibitors (abscissic acid . . .). The initiation of activity in spring is an auxin-related phenomenon originating in the developing buds (Larson, 1962; Wilcox, 1962; Wort, 1962; Wareing, Hanney, and Digby, 1964; Wareing and Phillips, 1970). However, in ring porous species, an auxin precursor, probably tryptophan, may be detected in cambial and cortical tissues before the swelling of the buds; this fact would explain the simultaneous resumption of cambial activity at all levels of the trunk and branches. It has also been shown that gibberellins play an important role in the onset of cambial activity, but that the control of this activity is auxin-dependent (Digby and Wareing, 1966b). Differentiation of cambial derivatives is also controlled by growth substances. It seems that a low level of auxin and a high level of gibberellin allows phloem production, while xylem production is stimulated by reverse proportions of these chemicals (Digby and Wareing, 1966a; see also Morey and Cronshaw, 1966, 1968; Robards, Davidson, and Kidwai, 1969; Lang, 1970). Auxin levels also affect the production of earlywood and late-wood: the latter is produced when the supply of auxin falls below a certain threshold. In ring porous species, the wood produced after the cessation of shoot growth is always latewood because the auxins synthesized by the leaves are sufficient to allow wood production, but not sufficient to permit earlywood differentiation (Digby and Wareing, 1966a). The study of tissue cultures has shown that wood formation is controlled by several substances (see Fosket and Torrey, 1969). At the end of the growth period, the level of auxin decreases while abscissic acid is synthesized in the leaves and then transferred to the buds (Addicott and Lyon, 1969; Wareing and Phillips, 1970). It is possible that, during this transfer which has been demonstrated particularly in sycamore (Phillips and Wareing, 1958), abscissic acid could first diffuse in the cambium of young internodes and then upwards to the buds and downwards to the older branches. This hypothesis would explain why, in this tree, divisions stop in young internodes before ceasing in the buds (Catesson, 1964). It would also explain why the wave of cessation of mitotic activity spreads progressively downwards (Cockerham, 1930).

An evident link also exists between growth substances and water availability. It is well known that cambial activity depends on water supply (Zahner, 1968). The connections between growth substances, water supply, and cambial activity are at present under examination (Little and Eidt, 1968, 1970).

Characteristics of cambial divisions

Cambial divisions were observed and described for the first time by Bailey (1919, 1920a). Vertical divisions may be either tangential (periclinal) or radial (anti-clinal). In the latter case, they are either truly vertical in storied cambia, or oblique (pseudo-transverse) in non-storied cambia. There are also true transverse mitoses. Periclinal and anticlinal divisions of the fusiform initials are rather peculiar on account of their great length. The phragmoplast extends over a long distance, sometimes several hundreds of micrometres, from the centre to the tips of the cell. In the light microscope, the phragmoplast is characterized by peripheral pyroninophilic fibres which appear like puffs at each end, and which Bailey called 'kinoplasmosomes' (1919). In these very large cells telophase is very long. According to Wilson (1964) the total time of division in pine would be around 24 hours: 5 hours for mitosis *sensu stricto*, and 19 hours for phragmoplast movement. The length of time between two successive divisions would be about 10 days, which agrees with Bannan's conclusions (1962). In these conditions, the rate of cell division in the cambium is much lower than in apical meristems where only 8 to 18 hours elapse between two successive mitoses (Baldovinos, 1953).

In a recent ultrastructural study, Evert and Deshpande (1970) showed that mitosis and cytokinesis in cambial cells are very similar to those of dividing cells in primary meristems (if one excepts the long period taken up by cell plate formation). However, an equatorial band of preprophase microtubules (chapter 6, p. 223) was seen only in ray cells. That this band was not encountered in fusiform cells could be due to its possible ephemeral existence. Cell plate formation begins at mid-anaphase when vesicles aggregate around groups of microtubules. These vesicles, in the first stages of cell plate formation, are smooth surfaced and derived from dictyosomes; some are electron opaque, others have clear contents. In more advanced stages, coated vesicles are also produced by dictyosomes and fuse with the young cell plate as the smooth vesicles also do. The fusion of coated vesicles with the cell plate continues long after the primary wall can be recognized. Very few microtubules traverse the cell plate during the last stages of development, while numerous tubules of endoplasmic reticulum can be seen crossing the plate at that time and primary wall plasmodesmata often enclose a tubular element of endoplasmic reticulum.

The problem of cambial initials: ultrastructural data and definition of the cambium

In 1873, Sanio expressed the idea that the cambium was a single layer of initials giving rise alternately to phloem cells on the outside and to xylem cells on the inside. But it was soon observed that division takes place also in the mother cells of vascular tissues. Therefore, several authors use the term 'cambium' to indicate the entire population of dividing cells between xylem and phloem (Priestley, 1930; Bannan, 1955; Wilson, 1964). This definition is clearly too broad, for mitosis may occur in already differentiated cells: for instance, the transverse divisions in parenchyma cells giving rise to future oxalate cells. Even periclinal divisions may take place during the last stages of differentiation of xylem or phloem. Figure 10.4A shows such a mitosis in a phloem parenchyma cell with thickened walls, and Figure 10.4B illustrates a periclinal division in a cell wedged between two differentiating vessels of the same radial tier (see also Fig. 11E, p. 277, and Figs. 12F and 12G, p. 278 in Catesson, 1964). As was emphasized by Bailey (in Wilson, 1964), cambial derivatives do not immediately lose their capacity for division, and the occurrence of mitoses in a broad zone does not mean that the whole zone is true cambium. Srivastava (also in Wilson, 1964) considers

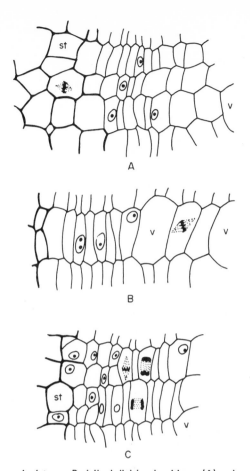

Fig. 10.4 *Acer pseudoplatanus.* Periclinal division in phloem (A) or in xylem (B) derivatives, and simultaneous anticlinal mitoses in the cambial zone (C). Phloem is on the left, xylem on the right. st = sieve tube, v = differentiating vessel.

that the cambial zone is at the centre of radial tiers where the cells are dividing at a faster rate, and I concur with this opinion. Mitotic counts, made in the cambial zone of sycamore, clearly show that the maximum number of divisions occurs in the central zone where the cells are narrow, RNA-rich, and where the cell walls are very thin (Fig. 10.5B), i.e., the restricted zone for which I proposed the term 'cambial zone' (Catesson, 1964). However, in some gymnosperms, the peak of mitotic activity seems to occur among the xylem mother cells (Bannan, 1962); therefore definition of the cambial zone is still difficult (see Philipson, Ward, and Butterfield, 1971).

Is it possible to point out clearly one initial layer among the dividing and redividing cells? Or, at least, to distinguish an initial cell in each radial tier? It is a controversial problem which was not solved by light microscope studies. Some authors who examined only gymnosperms believe that they can identify an initial cell in each radial tier (Bannan, 1955; Newman, 1956; Murmanis, 1969,

Cambial cells

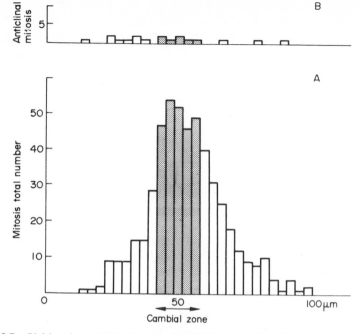

Fig. 10.5 Divisions in cambial cells and in their derivatives in May (*Acer pseudoplatanus*). 'Cambial Zone' is used here for the region where the cells are narrower, have the thinner walls and where the cytoplasm is very rich in ribosomes. The histograms were obtained from the examination of about 6000 horizontal tiers of cells by superposition of the 'cambial zones'. The total number of mitoses observed is shown in (A) and in (B) the number of anticlinal divisions present among them.

1970). On the other hand, Esau (1948) could not find any initial layer in *Vitis.* Esau later stated (in Wilson, 1964) that, as it is extremely difficult to determine which cell is the true initial, one can only speak of a cambial initial from a theoretical viewpoint to designate only the cell which is in a proper position to produce cells in both directions. The arguments given by Bannan (1962) to characterize cambial initials are open to discussion. This author considers that the existence of an uniseriate initiating row of cells is supported by the following arguments:

(a) the simultaneous appearance (or disappearance) of new radial files in both xylem and phloem;
(b) the extremely low proportion of pseudotransverse division outside the initiating layer;
(c) the fact that the initials and phloem mother cells are slightly shorter than xylem mother cells.

These conclusions are not valid for all species. For instance, in *Acer pseudoplatanus,* I found:

(a) simultaneous anticlinal divisions in the whole width of the cambial zone

374

(Fig. 10.4C, see also Catesson, 1964, Fig. 11E, p. 277 and 14C, p. 280); therefore the simultaneous origin of new tiers in phloem and xylem is not perforce produced by an anticlinal division of only one initial. Moreover, new radial files may be first produced by divisions in differentiating cambial derivatives, anticlinal divisions in the cambial zone proper taking place later (Fig. 10.4C);

(b)   an essentially similar rate of anticlinal mitoses in all layers of the cambial zone (Fig. 10.5A);

(c)   cell tips appearing in radial sections at the same level; differences were only visible in differentiating derivatives (Catesson, 1964, Fig. 19, p. 288).

It must be added, after these criticisms, that my observations, like those of Esau (1948), were made only on angiosperms. The authors who claim to be able to recognize the initial cell of each radial tier have worked with gymnosperms. More detailed studies, and precise comparisons between angiosperms and gymnosperms, must be made. Further, one may object that superposed images of mitoses mask the true appearance of each radial file, for Newman (1956) pointed out that the initial cell may not be at the same level in each tier. However, in spite of very detailed observations of a great number of sections, it was not possible in *Acer pseudoplatanus* to recognize the initial cell of each radial file (Catesson, 1964). In this species, all the cells of the zone where mitotic activity is at its maximum show the same characteristics:

(a)   a similar rate of division. Simultaneous anticlinal or periclinal mitosis in the same tier are frequent; sometimes six adjoining cells of the same radial file are dividing periclinally at the same time;

(b)   a similar concentration of RNA, higher than in differentiating derivatives;

(c)   identical structural characteristics.

For these reasons, I suggested that the cambium is really a multiseriate zone, all cells of which appear similar not only in their cytological characteristics but also in their properties of division.

The advent of electron microscope studies seemed to give us a useful tool to re-examine the problem of cambial initials. It was hoped that detailed ultrastructural studies would clearly elucidate the fine structural characteristics of initial cells, but this hope was not fulfilled. At the electron microscope level, the first signs of cell differentiation are revealed in cambial derivatives, but in each tier three or four cells remain absolutely similar between which it is impossible to distinguish the initial proper. Robards and Kidwai, after their study of *Salix fragilis* (1969a), explained that 'although the cambium proper is often regarded as a single layer of cells, it is impossible to distinguish between the cambium and the young differentiating elements on each side of it'. In *Pinus strobus*, Murmanis (1969, 1970, 1971) recognizes the initial cells by differences in thickness of tangential walls, but not by any ultrastructural characteristic, and Srivastava and O'Brien (1966) point out that 'although it is recognized that just one of these cells may be the fusiform initial *sensu stricto,* in practice it is impossible to distinguish between them with the techniques in use, even at the ultrastructural level'. Quite recently, Evert and Deshpande (1970) concluded from their studies that electron microscopes have not 'revealed any ultrastructural characteristics peculiar to any one layer of cells in the cambium'. Is it true that the cambium is formed by several layers of cells, each cell being endowed with the same meristematic properties, or are there really initial cells endowed with special properties? The problem remains unsolved, and we must hope that new detailed studies might provide us with a solution. For the moment, the notion of a cambial initial remains a theoretical idea.

Let us add that Sanio's image of a cambium (1873), producing alternately one cell on the xylem side and one cell on the phloem side, is not really true. The cambium may produce several daughter cells on the xylem side and then several cells on the phloem side. These derivatives redivide and differentiate more or less quickly into vascular tissues. However, xylem and phloem production are not equivalent. Xylem production is generally intensive but lasts only a short time; phloem production is less vigorous but takes place over a longer period. In sycamore, for example, there are two periods of phloem formation, April and June (Cokerham, 1930; Elliot, 1935), while the xylem is produced chiefly at the end of April and in May (Catesson, 1964).

The seasonal cycle of cambial cells

Physiological modifications

The seasonal transformations observed in the cambium of perennial plants are linked in part to the annual cycle of meristematic activity and additionally to the process of frost hardening at the beginning of winter. The annual variations of cambial activity are regulated, as we have seen, by the antagonist action of grow promotors and inhibitors. Cold hardening is marked by important physicochemica modifications of cambial cells.

One of the most striking of these modifications is the increase of osmotic pressure of the tissues. In sycamore the osmotic pressure, which averages 12 atm in spring, rises progressively from July until it reaches a maximal value of about 35 atm in January; then it decreases quickly during February and March. These variations are linked with temperature, for cambial osmotic pressure may reach 40 atm if December and January are very cold. Similar osmotic variations are observed in adjoining tissues, but they are less pronounced (Catesson, 1964). It is well known that the increase in osmotic pressure is one of the frost hardening related processes in plants. The cambium, a reserve of meristematic cells, has greater frost hardiness than adjoining tissues.

Variations in osmotic pressure are not directly accounted for by the modifications in water content of tissues. In sycamore, the water content of the cambium and young phloem is greatest in the spring. A second maximal water content, a little lower, is observed in December, while minima occur in September and January (Catesson, 1964). Similar results were obtained by Gibbs (1957) on Canadian trees.

The winter increase in osmotic pressure seems due to the accumulation of sugars in the vacuoles, for both phenomena vary in parallel. On the other hand, the concentration of soluble amino acids varies inversely with osmotic pressure (Le Saint and Catesson, 1966). In pine cambium, David (1967) also showed an antagonism between sugars and amino acids: from June to January, sugar concentration increases, amino acid concentration decreases; from February to May, the phenomenon is inverted. In locust tree bark, however, a winter increase in soluble proteins has been shown (Siminovitch and Chater, 1957).

Seasonal changes in fine structure

The physiological cycle of cambial cells produces seasonal variations in the organelles. The most striking transformations were observed by Bailey with the light microscope (1930). During the autumn, the central vacuole tends to divide into a number of small, independent vacuoles; cyclosis ceases; radial walls enlarge and take a characteristic beaded aspect in tangential view; and, at the same time, storage material is accumulated in the cells. Ultrastructural studies show that morphological modifications between active and resting periods are much more complex.

Cell walls and plasma membranes. In cambial cells, the tangential walls remain relatively thin throughout the year while radial walls strongly thicken between plasmodesmatal areas at the end of summer (Fig. 10.1B). Plasmodesmata appear to be less frequent in these radial walls, at least in some species (e.g., beech — Kidwai and Robards, 1969b), but this is opposite to the situation in pine (Murmanis, 1971). The thickening of walls is slow and progressive and does not take place during a definite period. The significance of this thickening is not clear: is it the sign of incipient cell differentiation, consecutive to the cessation of mitosis, or is it produced by the accumulation of storage polysaccharides?

Throughout the year the plasmalemma is sinuous. In pine, it is even more irregular in winter than in summer (Srivastava and O'Brien, 1966). During the active season, plasmalemma invaginations have often been interpreted, as we have seen, as pinocytotic vesicles or fusing Golgi vesicles. In winter, multivesicular bodies may be seen associated with the plasmalemma (Srivastava, 1966; Srivastava and O'Brien, 1966; personal observations on *Acer* and *Robinia*); in *Fagus* (Kidwai and Robards, 1969b), an electron opaque material is sometimes present outside the plasmalemma. All these images suggest that the plasma membrane remains active during winter, and this idea is strengthened, as we shall see later, by the existence of enzymic activities on the plasmalemma all the year round.

Vacuolar modifications. The seasonal cycle of cambial vacuoles has been well known since Bailey's work (1930). Vital observations show that the central vacuole, characteristic of actively dividing cambial initials, is broken down during September into elongated, irregular vacuoles which sometimes assume a 'myelin form', as Bailey called it. These elongated vacuoles coalesce into a complex, everchanging network. Numerous, small, globular vacuoles appear during November, while protoplasmic streaming ceases. This resting stage (Fig. 10.1B) persists until

February, when the small vacuoles fuse into a 'myelinic' network. This transform͏a
tion coincides exactly with the resumption of protoplasmic streaming. The sprin͏g
hydration of cortical tissues then begins, and the vacuolar network swells and
changes into a central vacuole traversed by numerous cytoplasmic strands. We h͏a
shown (Catesson, 1964) that these transformations are under the influence of
both temperature and osmotic pressure. For instance, the transition from
independent globular vacuoles to a 'myelinic' network may be obtained experimentally in winter either by a decrease in osmotic pressure or by an increase of
temperature (Table 10.3). But the reverse transformation, which normally occur͏s
in the autumn, cannot be obtained experimentally. It is possible that break-up
of the central vacuole is influenced by abscissic acid which is present in stem
tissues in autumn. A more detailed experimental study is necessary for our
better understanding of the vacuolar cycle.

TABLE 10.3

**Onset of Cytoplasmic Streaming and Vacuolar Changes as a Function of the
Temperature and Osmotic Pressure of the Medium (January)**

At the time of cutting, cambial cells possessed globular vacuoles and were not
streaming (adapted from Catesson, 1964)

temperature	mounting medium	time		
		30 min	1½ hr	4 hr
23 °C	levulose 24%	0	streaming + narrow, 'myelin-form' vacuoles	streaming ++ narrow, 'myelin-form' vacuoles
23 °C	levulose 6%	streaming + swollen vacuoles	streaming ++ large vacuoles	streaming +++ very large vacuoles
0 °C	levulose 24%	0	0	0
0 °C	levulose 6%	0	0 streaming ± globular vacuoles	streaming + 'myelin-form' vacuoles

This cycle involves great plasticity in the relationship between cytoplasm and
vacuoles, and the tonoplasts continually change their appearance. Ultrastructural
observations have strengthened the prior *in vivo* studies. During dormancy, ultra-
thin sections provide better images of various vacuolar shapes, and of relations
between vacuoles themselves. These are often irregular and arranged like the piec͏es
of a jig-saw puzzle, tonoplasts of two neighbouring vacuoles joining over a small
area (Fig. 10.1B). I have tried to study the influence of temperature and osmotic
changes at the ultrastructural level, but the results are deceiving as the thinness
of the sections cannot give an exact image of 'myelin form' vacuoles. In sections

they appear ovoid or elongated and differ only slightly from globular vacuoles. Moreover, the tonoplast is a very delicate membrane, easily damaged during fixation, and some images are difficult to interpret. However, a number of micrographs suggest the process of vacuolar fusion during the transitional stage (Fig. 10.6). Some vacuoles have small beaks at their periphery; these elongate, and two beaks, from different vacuoles, eventually join; they curve and hook one on the other like two cogs. The two adjacent tonoplasts protrude into one of the vacuoles and break a little later. This process may partly explain the great number of membrane remnants which are found in vacuoles (see also chapter 5, p. 193). Such images of vacuolar coalescence are also frequent in elongating cells of primary meristems (see Clowes and Juniper, 1968, p. 85; Nougarède, 1969, p. 339).

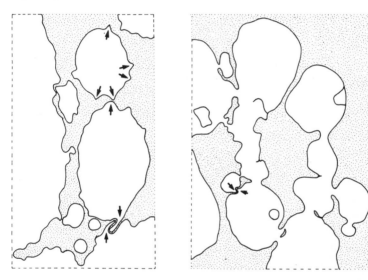

Fig. 10.6   Vacuole fusing during experimental transition from spherical to 'myelin form' vacuoles in winter (*Acer pseudoplatanus*). Note the formation of vacuolar beaks (arrows) before vacuolar confluence. The cytoplasm is stippled.

The presence of vacuolar beaks, of joining tonoplasts, and the existence of intravacuolar membranous material in normal cells during dormancy, suggest that these cells are not so completely dormant as they seem to be when observed *in vivo*. Very slow cytoplasmic movements may occur, producing slow vacuolar changes, even during the apparent rest period.

We have seen that sugars are accumulated in vacuoles during the winter. The vacuoles may also contain more or less abundant precipitates. The precipitates almost completely fill the vacuoles in *Robinia* in winter, and they disappear in spring. In *Acer* cambium, they are scarce in fusiform initials and much more abundant in vertical phloem parenchyma; in ray cells, where winter vacuoles are rather large, vacuolar precipitates are frequent but not very abundant; they also disappear in spring. These precipitates are probably tannins.

Mitochondrial cycle. Ultrastructural studies so far published do not mention seasonal mitochondrial changes, perhaps because the cambium was studied in active or resting conditions and not during transitional stages. The transitional steps, i.e., the resumption of activity or the onset of dormancy, are badly documented. However, light microscope studies by Buvat (1956a) showed mitochondrial changes during the annual cycle, and the results of our ultrastructural observations confirm Buvat's work.

In *Acer* for instance (Fig. 10.7) during the active growing season, mitochondria are globular or slightly elongated, exceptionally reaching 3 μm. Their length increases progressively, while the rate of division decreases during the summer: from September to November cross-sections of mitochondria range from 2 to 4 μm. They vary greatly in shape, frequently taking the appearance of an X, Y or Q. At the onset of dormancy, the reverse phenomenon takes place and, in January, mitochondria are again short and spherical; mitochondria as large as 3 μm are quite infrequent. A new period, characterized by elongated mitochondria, occurs at the time of spring hydration and cambial swelling, before the resumption of mitotic activity. During this brief period, mitochondria are variously shaped, sometimes with a beaded appearance, and they attain a length of 3–4 μm. This elongation seems to begin in the derivatives adjacent to the phloem of the preceding year, and then to spread to the cambial cells. During this cycle, mitochondrial fine structure remains the same: the matrix is always relatively translucent, and the cristae not very numerous. Similar mitochondrial changes, but much more pronounced, are visible in cambial derivatives and storage parenchyma cells of either phloem or xylem (Buvat, 1956b; Catesson, 1964; Czaninski, 1964, 1968a). In pine cambium, chain-like mitochondria have also been seen in autumn fixations (Murmanis, 1971).

Our first observations with the light microscope led us to think that several factors might relate to mitochondrial changes:

(a) *Temperature.* Cold induces a shortening of mitochondria (Genevès, 1955).

(b) *Osmotic pressure.* Buvat (1948) showed that the higher the tissue water content, the more elongated are the mitochondria, and he suggested that the mitochondrial cycle of cambial cells was related to the degree of cell hydration. However, in sycamore, the mitochondrial cycle is not parallel to the cycle of water content, but is rather closely related to the osmotic changes: when the osmotic pressure is low, under 25 atm, the average length of mitochondria varies between 2 and 4 μm; when the pressure rises above 25 atm, short mitochondria, less than 2 μm long, prevail. But the parallelism between osmotic pressure and mitochondrial length is not true during the height of the active period.

(c) *Mitotic activity.* During the peak of the active growing season, in April and May, mitochondria are always small, and their size appears to be linked to the relationship between the rate of mitochondrial division and the rate of mitosis (Catesson, 1964).

(d) *Protoplasmic streaming.* Buvat (1948) observed that slow protoplasmic streaming allows the formation of elongated mitochondria, while short mitochondria occur in cells with fast protoplasmic streaming. Measurements made by

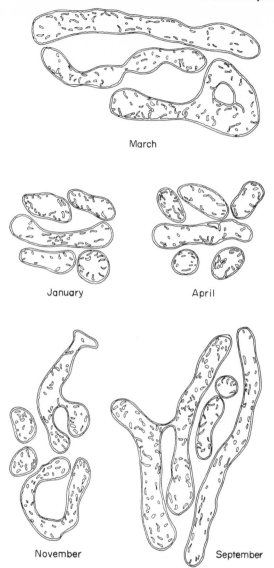

March

January   April

November   September

Fig. 10.7   Seasonal mitochondrial cycle in *Acer pseudoplatanus*. In January, mitochondria are short: they elongate in March, but in April and May they become short again; they elongate a second time in summer, only to shorten again in autumn.

Thimann and Kaufman (1957), in pine cambium, showed that cyclosis is fastest during the peak of mitotic activity, i.e., when we see the shortest mitochondria.

(e) *Growth hormones.* It is possible that the antagonist action between growth promotors and inhibitors could have an effect on mitochondrial morphology, but there are no useful data available on this subject.

It is clear that the mitochondrial cycle must be studied in species other than

Robinia and *Acer,* and that the hypotheses formed to explain this cycle must be checked by experimental studies.

Plastid evolution. Here again, as for the mitochondrial cycle, the lack of detailed data does not permit us to extend to other species the plastid modifications that I have observed in *Acer.* In papers on cambial ultrastructure, the only changes reported concern the presence or absence of starch, but these changes may be in opposite directions according to the plant under study. In pine, for instance, starch is frequent in spring but quite scarce in winter (Srivastava and O'Brien, 1966; Murmanis, 1971), while the opposite is observed in the fusiform cells of sycamore.

In this last species, beyond seasonal changes in starch content, seasonal modifications of the lamellar system occur and these modifications are more conspicuous in fusiform than in radial initials. In April and May the plastids cont numerous vesicles (Figs. 10.2B and C). While mitotic activity slows down, the vesicles decrease in number and are replaced by a few short thylakoids. At the beginning of winter (Fig. 10.2E), plastids have various sizes and shapes: they are round, elongated, curved, and sometimes star-shaped. They are narrow, not much more than 1 μm wide, but their length may reach 7 or 8 μm. The stroma is very dense with a few short thylakoids, isolated or in pairs. Some plastids contain plastoglobuli and small starch grains. The latter are so small that they cannot be recognized with light microscope techniques. In contrast to the cambia initials, undifferentiated derivatives may be recognized by the greater number of thylakoids in their plastids. This is a great help to delimiting the cambium proper in winter when there are only two layers of cells with quite undifferentiated proplastids. In March, thylakoids appear to vesiculate, while in April we again find the plastids described in the first part of this chapter. A similar cycle, but less conspicuous, can be recognized in ray cells: vesicles are present in plastids during April and May but disappear in summer. Starch is present throughout the year in these cells (Figs. 10.2C and D).

For the time being, there is no explanation of this seasonal cycle in plastids. It may be linked to different rates of plastid multiplication.

Golgi apparatus and endoplasmic reticulum. In some species, like ash (Srivastava, 1966), the Golgi apparatus does not show any variation. In willow (Robards and Kidwai, 1969a), in pine (Murmanis, 1971) and in sycamore, dictyosomes are clearly less numerous in winter and are characterized by a total absence of separated dictyosome vesicles. In March, in sycamore, the tips of Golgi cisternae swell and vesicle production begins. These vesicles are abundant during the active growth season and decrease in number at the onset of dormancy During this cycle, the number of cisternae remains constant, around 4 or 5. In spring, the resumption of Golgi activity is accompanied by a great increase in the number of dictyosomes both in willow and in sycamore. In these two species we can really speak of a winter rest for the Golgi apparatus.

Endoplasmic reticulum is generally abundant in cambial cells. During the active season, it forms ribosome-studded cisternae parallel to the cell walls. In winter, rough endoplasmic reticulum is often less abundant than in summer although it persists in some species (e.g., sycamore). In other species smooth reticulum is much more frequent (e.g., willow − Robards and Kidwai, 1969a; pine − Srivastava and O'Brien, 1966; or locust tree − personal data). Winter smooth reticulum occurs as elongated cisternae or numerous separate vesicles.

If reticulum ribosomes are less frequent during winter, free ribosomes are abundant throughout the year. They form polysomes in spring but are isolated in winter (Srivastava, 1966; Robards and Kidwai, 1969a; personal data).

It seems likely that seasonal variations of the Golgi apparatus and endoplasmic reticulum activity are closely related to the influence of growth substances. Indole acetic acid increases protein and RNA synthesis and allows the formation of new dictyosomes and ribosomes (Nitsch, 1968). Likewise, auxins and cytokinins prevent senescence (Sacher, 1968), while abscissic acid inhibits DNA and RNA synthesis (Wareing, Good, and Manuel, 1968; Shih and Rappaport, 1970), and accelerates RNase activity (Pilet, 1970). It is possible that dormancy implies an extensive repression of RNA (Wareing and Phillips, 1970). These observations can be compared to aging phenomena. After 24 hours of aging on water, Jerusalem artichoke slices show an increase in the number of ribosomes and of endoplasmic reticulum profiles; ribosomes are no longer scattered but gathered in polysomes or attached to the reticulum cisternae; dictyosomes remain infrequent and their number increases only if 2,4-D is added (Fowke and Setterfield, 1968). In potato slices aged on water, we observed the same phenomena, but there was a conspicuous increase in dictyosome number even without 2,4-D. These morphological changes were related to an increase in protein and lipid synthesis in the microsomal fraction (Ben Abdelkader, 1969; Ben Abdelkader and Catesson, 1970). Aging phenomena, responses to a trauma, are somewhat similar to cambial changes at the onset of meristematic activity. It would again be interesting to possess a better understanding of the action of growth substances on cellular ultrastructure.

Different inclusions and storage material. In autumn, cambial cells elaborate storage material, and various inclusions appear in the cytoplasm. Lipid droplets may be present throughout the year, but in some species they increase in size and number in autumn as in *Salix* (Robards and Kidwai, 1969a), in *Pinus* (Murmanis, 1971) and in *Robinia*. In this latter plant, osmiophilic inclusions are specially abundant in winter. In sycamore, we followed this accumulation of lipids by histochemical techniques (Catesson, 1964). Some authors have suggested that the winter increase in lipids may be related to frost hardiness (e.g., Parker, 1957).

Some species, like beech (Kidwai and Robards, 1969b) or willow (Robards and Kidwai, 1969a), accumulate storage proteins in the autumn which form dense, membrane-bound bodies 0.75−2.0 μm in diameter. Histochemical tests for protein gave a strong positive reaction in such bodies. In other species, bundles of fine filaments may appear in the cytoplasm (Fig. 10.1C). These filaments are very straight and paired; the space between two filaments of a pair varies between 12 and 15 nm according to the species. In the absence of transverse views of these bundles we cannot say if these pairs of filaments represent longitudinal sections

383

of fine tubules. If this were the case, the tubules would have half the diameter of ordinary microtubules and would be narrower than most P-protein tubules of sieve tubes (Cronshaw and Esau, 1968). Such cambial bundles of filaments have been described in pine (Srivastava and O'Brien, 1966), and in ash (Srivastava, 1966), and we have found them in fusiform cells of sycamore. In this species, 2 to 10 fibril pairs are stacked in bundles parallel to the cell axis. These bundles have a very rigid appearance and can reach a length of 13 μm in sections; their true length is probably greater, as the plane of the sections is generally not parallel to the bundle axis. In the same cell, several parallel bundles may coexist. In sycamore, these bundles are present from October to March. They disappear at the end of March which suggests that they may represent a storage protein, but this hypothesis needs to be investigated by cytochemical methods. Parthasarathy and Mühlethaler (1972) recently discussed the possible function of such microfilaments in relation to cytoplasmic streaming, and questioned the fact that the cambial cells in which Srivastava and O'Brien reported the presence of bundles of filaments were truly active. In sycamore cambium, these filaments are commonly found in dormant, non-streaming cells, as I have established by joint light and electron microscopical observations.

Cytochemical studies

Biochemical work on the cambium is difficult for it is impossible to separate the narrow cambial zone from the adjacent tissues. The more precise data available concern at least the cambial zone and the young phloem. For this reason, the recent techniques of ultrastructural cytochemistry open an interesting field of investigation as they give us a tool to visualize *in situ* the products of cell metabolism. They should enable us to characterize the cambium in contrast to the adjoining tissues, and to follow chemical processes during differentiation of cambial derivatives. They will perhaps also give us, at last, an answer to the problem of the definition of cambial initials. But ultrastructural cytochemistry of cambial cells is only just beginning.

Peroxidatic activities. We have previously seen that the presence of catalase may be recognized in cambial peroxisomes. Catalase is characterized by a peroxidatic activity at pH 9 which can be inhibited by aminotriazole. Cambial peroxisomes, as also phloem peroxisomes, of sycamore show this activity (Czanins and Catesson, 1970). In these cells other peroxidatic activities may be demonstrat throughout the year in the cell wall and along the plasmalemma; they exist at pH's varying between 5 and 9, and are inhibited by potassium cyanide and by heating at 100 °C. Other than the reactions on the cell wall and plasmalemma, onl slight, irregular peroxidatic activities may be present in some endoplasmic reticulum cisternae, at least under our experimental conditions. Cambial cells react similarly to storage parenchyma cells, and differ from companion cells and vessel associated cells where a vacuolar peroxidatic activity can be demonstrated (Czaninski and Catesson, 1969, 1970). Cambial cells differ also from primary meristematic cells where peroxidatic activity has been shown in young vacuoles and, less frequently, in ergastoplasmic profiles and dictyosomes (Marty, 1969;

Poux, 1969). Even when cambial vacuoles are small, during winter, vacuolar peroxidatic activity is very infrequent, and then always slight. But, here again, we need data from other species before we can draw general conclusions from these first observations.

Phosphatasic activities. Reactions at pH 7.2 with ATP as substrate were obtained by Robards and Kidwai (1969c) in beech cambium. Lead deposits were found in nuclei and mitochondria, and on either side of the plasmalemma. Fragmentary personal observations gave us similar results with thiaminepyrophosphate as substrate at pH 7. These results, as those obtained for peroxidatic activities, point to the important role of the plasmalemma in cambial cells all the year round.

At pH 5, with β-glycerophosphate as substrate, results vary according to the season. During the active growth period, lead precipitates are essentially localized in the endoplasmic reticulum cisternae in cambial cells and in the adjacent tissues, as we observed in sycamore and locust tree (Catesson and Czaninski, 1967). In winter, there is no reaction inside the cells; a heavy lead phosphate deposit occurs between the plasmalemma and the cell wall, so emphasizing the outlines of cambial and phloem cells (Catesson and Czaninski, 1968). This does not seem to be an artifact related to difficulties in substrate penetration as, in winter, diaminobenzine (used for research on peroxidatic activities) enters easily into the cells though its molecular weight (360) is slightly higher than that of β-glycerophosphate (306). The significance of the results obtained with β-glycerophosphate during the resting period are still difficult to explain without additional biochemical data. Is there a real decrease in acid phosphatase content, or is there an inactivation of the enzyme which would thus become undetectable cytochemically?

Conclusion

Study of the cambium holds great interest, not only for the cytologist, but also for the physiologist or the forestry worker, as cambium is the tissue which produces the increase in girth of plants by giving rise to new vascular tissues, in particular to wood.

Cytological and cytochemical approaches must be associated with physiological methods so as to enable us to obtain a clear and comprehensive picture of cambial role and activity. But ultrastructural studies are few and cytochemical work is just beginning. Numerous questions are either still not clarified or remain completely unanswered. The differences in cambial activity between gymnosperms and angiosperms, and between trees and herbaceous plants remains to be elucidated. The relations between the hormonal cycle and seasonal changes in ultrastructure must be examined thoroughly. We have almost no ultrastructural data on cambial growth, differentiation and aging. Anomalous cambia have been neglected, which is to be regretted, for it is possible that their study could have given us valuable pointers to understanding cambial cell physiology. Cytochemical techniques appear to be useful tools for studying cambial physiology and biochemistry at the cell level, as the cambium cannot be isolated as a single tissue without some vascular tissue attached to it. The first results obtained by cytochemical methods underline the particular characteristics of the cambium as opposed to primary meristems.

Therefore, we can conclude by saying that our knowledge on cambial ultrastructure, and its relationships to plant development, are still fragmentary, and that many problems remain to be investigated.

Bibliography

Further reading

(a) General

Boureau, E. (1956) *Anatomie Végétale*, vol. 2, pp. 354–378, Presses Universitaires de Fran Paris.

Esau, K. (1953) *Plant anatomy*, pp. 125–135; 380–391; 489–498, John Wiley and Sons, N York.

Esau, K. (1960) *Anatomy of seed plants*, pp. 114–121; 190–202; 238–257, John Wiley, N York.

Fahn, A. (1969), *Plant anatomy*, Oxford, pp. 272–285, Pergamon Press.

Philipson, W. R., Ward, J. M., and Butterfield, B. G. (1971) *The vascular cambium – its development and activity*, Chapman and Hall, London.

(b) Cell structure

Catesson, A. M. (1964) 'Origine, fonctionnement et variations cytologiques saisonnières du cambium de l'*Acer pseudoplatanus* L. (Acéracées)', *Annls. Sci. nat. (Bot.)*, 12ème série, 5, 229–498.

Evert, R. F. and Deshpande, B. P. (1970) 'An ultrastructural study of cell division in the cambium', *Am. J. Bot.*, 57, 942–951.

Kidwai, P. and Robards, A. W. (1969b) 'On the ultrastructure of resting cambium of *Fagus sylvatica* L.', *Planta*, 89, 361–368.

Robards, A. W. and Kidwai, P. (1969a) 'A comparative study of the ultrastructure of resting and active cambium of *Salix fragilis* L.', *Planta*, 84, 239–249.

Srivastava, L. M. (1966) 'On the fine structure of the cambium of *Fraxinus americana* L.', *J. Cell Biol.*, 31, 79–93.

Srivastava, L. M. and O'Brien, T. P. (1966) 'On the ultrastructure of cambium and its vascul derivatives. I – Cambium of *Pinus strobus*', *Protoplasma*, 61, 257–276.

References

Addicott, F. T. and Lyon, J. L. (1969) 'Physiology of abscissic acid and related substances', *A. Rev. Pl. Physiol.*, 20, 139–164.

Alvim, P. de T. (1964) 'Tree growth periodicity in tropical climates', In: *The formation of wood in forest trees* (ed. M. H. Zimmerman), pp. 479–495, Academic Press Inc., New York.

Arsanto, J. P. (1970) 'Infrastructure et différenciation du protophloème dans les jeunes racines du Sarrasin (*Polygonum fagopyrum*, Polygonacée)', *C. r. hebd. Séanc. Acad. Sci., Paris*, 270, 3071–3074.

Bailey, I. W. (1919) 'Phenomena of cell division in the cambium of arborescent Gymnosperms and their cytological significance', *Proc. natn. Acad. Sci.*, 5, 283–285.

Bailey, I. W. (1920a) 'The formation of the cell plate in the cambium of the higher plants', *Proc. natn. Acad. Sci.*, **6**, 197–200.

Bailey, I. W. (1920b) 'The cambium and its derivative tissues: III – A reconnaissance of cytological phenomena in the cambium', *Am. J. Bot.*, **7**, 417–434.

Bailey, I. W. (1923) 'The cambium and its derivative tissues: IV – The increase in girth of the cambium', *Am. J. Bot.*, **10**, 499–509.

Bailey, I. W. (1930) 'The cambium and its derivative tissues: V – A reconnaissance of the vacuome in living cells', *Z. Zellforsch. Mikrosk. Anat.*, **10**, 651–682.

Baldovinos, G. (1953) 'Growth of the root tip', In: *Growth and differentiation in plants* (ed. W. E.'Lommis), Iowa State College Press, Ames, Iowa.

Bannan, M. W. (1955) 'The vascular cambium and radial growth of *Thuya occidentalis* L.', *Can. J. Bot.*, **33**, 113–138.

Bannan, M. W. (1962) 'The vascular cambium and tree-ring development', In: *Tree growth* (ed. T. T. Koslowski), pp. 3–21, The Ronald Press, New York.

Bannan, M. W. (1970) 'A survey of cell length and frequency of multiplicative divisions in the cambium of Conifers', *Can. J. Bot.*, **48**, 1585–1590.

Barnett, J. R. (1971) 'Winter activity in the cambium of *Pinus radiata*', *N.Z. J. Forest. Sci.*, **1**, 208–222.

Ben Abdelkader, A. (1969) 'Influence de la "survie" (ageing) sur la biosynthèse des phospholipides dans les "microsomes" de tubercule de Pomme de terre', *C. r. hebd. Séanc. Acad. Sci., Paris*, **268**, 2406–2409.

Ben Abdelkader, A. and Catesson, A. M. (1970) 'Ageing of potato slices: morphological modifications and changes in lipid metabolism of organelles', *Abstracts of the European Biological Session, International Association of Gerontology*, Leeds, April 1970.

Blackwell, S. J., Laetsch, W. M., and Hyde, B. B. (1969) 'Development of chloroplast fine structure in aspen tissue culture', *Am. J. Bot.*, **56**, 457–463.

Buvat, R. (1948) 'Recherches sur les effets cytologiques de l'eau', *Rev. Cytol. et Cytophys. végét.*, **10**, 1–36.

Buvat, R. (1956a) 'Variations saisonnières du chondriome dans le cambium de *Robinia pseudoacacia*', *C. r. hebd. Séanc. Acad. Sci., Paris*, **243**, 1908–1911.

Buvat, R. (1956b) 'Variations saisonnières du chondriome dans les cellules parenchymateuses du phloème de *Robinia pseudoacacia*', *C. r. hebd. Séanc. Acad. Sci., Paris*, **243**, 2127–2130.

Catesson, A. M. (1964) 'Origine, fonctionnement et variations cytologiques saisonnières du cambium de l'*Acer pseudoplatanus* L. (Acéracées).', *Annls. Sci. nat. (Bot.)*, 12ème série, **5**, 229–498.

Catesson, A. M. (1966) 'Présence de phytoferritine dans le cambium et les tissus conducteurs de la tige de Sycomore, *Acer pseudoplatanus*', *C. r. hebd. Séanc. Acad. Sci., Paris*, **262**, 1070–1073.

Catesson, A. M. (1970) 'Evolution des plastes de Pomme au cours de la maturation du fruit. Modifications ultrastructurales et accumulation de ferritine', *J. Microscopie*, **9**, 949–974.

Catesson, A. M. and Czaninski, Y. (1967) 'Mise en évidence d'une activité phosphatasique acide dans le réticulum endoplasmique des tissus conducteurs de Robinier et de Sycomore', *J. Microscopie*, **6**, 509–514.

Catesson, A. M. and Czaninski, Y. (1968) 'Localisation ultrastructurale de la phosphatase acide et cycle saisonnier dans les tissus conducteurs de quelques arbres', *Bull Soc. fr. Physiol. Végét.*, **14**, 165–173.

Clowes, F. A. L. and Juniper, B. E. (1968) *Plant Cells*, Blackwell, Oxford.

Cockerham, G. (1930) 'Some observations on cambial activity and seasonal starch content in Sycamore (*Acer pseudoplatanus*)', *Proc. Leeds phil. lit. Soc.*, **2**, 64–80.

Cronshaw, J. (1965) 'The organization of cytoplasmic components during the phase of cell wall thickening in differentiating cambial derivatives of *Acer rubrum*', *Can. J. Bot.*, **43**, 1401–1407.

Cronshaw, J. and Esau, K. (1968) 'P-Protein in the phloem of *Curcurbita* I. The development of P-Protein bodies', *J. Cell Biol.*, **1**, 25–39.

Cronshaw, J. and Wardrop, A. B. (1964) 'Organization of cytoplasm in differentiating xylem of *Pinus radiata*', *Aust. J. Bot.*, **12**, 15–23.

Cumbie, B. G. (1963) 'The vascular cambium and xylem development in *Hibiscus lasiocarpus*', *Am. J. Bot.*, **50**, 944–951.

Cambial cells

Cumbie, B. G. (1969) 'Developmental changes in the vascular cambium of *Polygonum lapathifolium*', *Am. J. Bot.*, **56**, 139–146.

Czaninski, Y. (1964) 'Variations saisonnières du chondriome dans les cellules du parenchyme ligneux vertical du *Robinia pseudo-acacia*', *C. r. hebd. Séanc. Acad. Sci., Paris*, **258**, 679–682.

Czaninski, Y (1968a) 'Etude du parenchyme ligneux du Robinier (parenchyme à réserves et cellules associées aux vaisseaux) au cours du cycle annuel', *J. Microscopie*, **7**, 145–164.

Czaninski, Y. (1968b) 'Etude cytologique de la différenciation du bois de Robinier. I. Différenciation des vaisseaux', *J. Microscopie*, **7**, 1051–1068.

Czaninski, Y. (1970) 'Etude cytologique de la différenciation du bois de Robinier. II. Différenciation des cellules du parenchyme (cellules à réserves et cellules associées aux vaisseaux)', *J. Microscopie*, **9**, 389–406.

Czaninski, Y. and Catesson, A. M. (1969) 'Localisation ultrastructurale d'activités peroxydasiques dans les tissus conducteurs végétaux au cours du cycle annuel', *J. Microscopie*, **8**, 875–888.

Czaninski, Y. and Catesson, A. M. (1970) 'Activités peroxydasiques d'origines diverses dans les cellules d'*Acer pseudoplatanus* (tissus conducteurs et cellules en culture)', *J. Microscopie*, **9**, 1089–1102.

David, A. (1967) 'Relations entre l'activité histogène, in situ et in vitro, du cambium du tronc du Pin maritime et la teneur en glucides solubles et acides aminés libres du liber-cambium', *C. r. hebd. Séanc. Acad. Sci., Paris*, **265**, 1602–1605.

Davis, J. D. and Evert, R. F. (1970) 'Seasonal cycle of phloem development in woody vines', *Bot. Gaz.*, **131**, 128–138.

Digby, J. and Wareing, P. F. (1966a) 'The effect of applied growth hormones on cambial division and the differentiation of cambial derivatives', *Ann. Bot.*, **30**, 539–548.

Digby, J. and Wareing, P. F. (1966b) 'The relationship between endogenous hormone level in the plant and seasonal aspects of cambial activity', *Ann. Bot.*, **30**, 607–622.

Elliot, J. H. (1935) 'Seasonal changes in the development of the phloem of the Sycamore *Acer pseudoplatanus* L.', *Proc. Leeds Phil. lit. Soc.*, **3**, 55–67.

Esau, K. (1948) 'Phloem structure in the grapevine, and its seasonal changes', *Hilgardia*, **18**, 217–296.

Esau, K. (1953) *Plant anatomy*, John Wiley, New York.

Esau, K. (1965) 'Anatomy and cytology of *Vitis* phloem', *Hilgardia*, **37**, 17–72.

Esau, K., Cheadle, V. I., and Gill, R. H. (1966) 'Cytology of differentiating tracheary elements. I. Organelles and membranes systems', *Am. J. Bot.*, **53**, 756–764.

Evert, R. F. and Deshpande, B. P. (1970) 'An ultrastructural study of cell division in the cambium', *Am. J. Bot.*, **57**, 942–951.

Fahn, A. (1969) *Plant anatomy*, Pergamon Press, Oxford.

Fosket, D. E. and Torrey, J. G. (1969) 'Hormonal control of cell proliferation and xylem differentiation in cultured tissues of *Glycine max* var. Biloxi', *Pl. Physiol. (Lancaster)*, **44**, 871–880.

Fowke, L. C. and Setterfield, G. (1968) 'Cytological responses in Jerusalem artichoke tuber slices during aging and subsequent auxin treatment', In: *Biochemistry and physiology of plant growth substances* (ed. F. Wightman and G. Setterfield), pp. 581–602. The Runge Press, Ottawa.

Genevès, L. (1955) 'Recherches sur les effets cytologiques du froid', *Rev. Cytol. et Biol. Végét.*, **16**, 1–207.

Gibbs, R. D. (1957) 'Patterns in the seasonal water content of trees', In: *The physiology of forest trees* (ed. K. V. Thimann), pp. 43–69, The Ronald Press, New York.

Hohl, H. R. (1960) 'Über die submikroskopische Struktur normaler und hyperplastscher Gewebe von *Datura stramonium* L.', *Bull. Soc. bot. Suisse*, **70**, 395–439.

Kidwai, P. and Robards, A. W. (1969a) 'The appearance of differentiating vascular cells after fixation in different solutions', *J. exp. Bot.*, **20**, 664–670.

Kidwai, P. and Robards, A. W. (1969b) 'On the ultrastructure of resting cambium of *Fagus sylvatica* L.', *Planta*, **89**, 361–368.

Kollmann, R. and Schumacher, W. (1964) 'Über die Feinstruktur des phloems von *Metasequoia glyptostroboides* und seine jahreszietlichen veränderungen. V-Mitteilung: Die Differenzierung der Siebzellen im Verlaufe einer Vegetations-periode', *Planta*, **63**, 155–190.

Bibliography

Lang, A. (1970) 'Gibberellins: structure and metabolism', *A. Rev. Pl. Physiol.*, **21**, 537–570.

Larson, P. R. (1962) 'Auxin gradients and the regulation of cambial activity', In: *Tree growth* (ed. T. T. Koslowski), pp. 97–117, The Ronald Press, New York.

Ledbetter, M. C. and Porter, K. R. (1970) *Introduction to the fine structure of plant cells*, pp. 86–92, Springer-Verlag, Berlin.

Le Saint, A. M. and Catesson, A. M. (1966) 'Variations simultanées des teneurs en eau, en sucres solubles, en acides aminés et de la pression osmotique dans le phloème et le cambium de Sycomore pendant les périodes de repos apparent et de reprise de la croissance', *C. r. hebd. Séanc. Acad. Sci., Paris*, **263**, 1463–1466.

Little, C. H. A. and Eidt, D. C. (1968) 'Effect of abscissic acid on budbreak and transpiration in woody species', *Nature*, **220**, 498–499.

Little, C. H. A. and Eidt, D. C. (1970) 'Relationship between transpiration and cambial activity in *Abies balsamea*', *Can. J. Bot.*, **48**, 1027–1028.

Marty, F. (1969) 'Caractérisation cytochimique infrastructurale de peroxysomes (= 'microbodies' *sensu stricto*) chez *Euphorbia characias*', *C. r. hebd. Séanc. Acad. Sci., Paris*, **268**, 1388–1391.

Mia, A. J. (1970) 'Fine structure of active, dormant and aging cambial cells in *Tilia americana*', *Wood Sci.*, **3**, 34–42.

Morey, P. R. and Cronshaw, J. (1966) 'Induced structural changes in cambial derivatives of *Ulmus americana*', *Protoplasma*, **62**, 76–85.

Morey, P. R. and Cronshaw, J. (1968) 'Developmental changes in the secondary xylem of *Acer rubrum* induced by various auxins and 2, 3, 5 tri-iodobenzoic acid', *Protoplasma*, **65**, 287–313.

Murmanis, L. (1969) 'Vascular cambium with emphasis on the location of the initial cell in *Pinus strobus* L.', *Abstr. XIth International Botanical Congress, Seattle*, p. 155.

Murmanis, L. (1970) 'Locating the initial in the vascular cambium of *Pinus strobus* by electron microscopy', *Wood Sci. Techn.*, **4**, 1–17.

Murmanis, L. (1971) 'Structural changes in the vascular cambium of *Pinus strobus* L. during an annual cycle', *Ann. Bot.*, **35**, 133–141.

Newman, I. V. (1956) 'Pattern in meristems of vascular plants. I. Cell partition in living apices and in the cambial zone in relation to the concepts of initial cells and apical cells', *Phytomorphology (India)*, **6**, 1–19.

Nitsch, J. P. (1968) 'Studies on the mode of action of auxins, cytokinins and gibberellins at the subcellular level', In: *Biochemistry and physiology of plant growth substances* (ed. F. Wightman and G. Setterfield), pp. 563–580, The Runge Press, Ottawa.

Nougarède, A. (1969) *Biologie vegétale. I. Cytologie*, Masson et Cie, Paris.

Parker, J. (1957) 'Seasonal changes in some chemical and physical properties of living cells of *Pinus ponderosa* and their relation to freezing resistance', *Protoplasma*, **48**, 147–163.

Parthasarathy, M. V. and Mühlethaler, K. (1972) 'Cytoplasmic microfilaments in plant cells', *J. Ultrastruct. Res.*, **38**, 46–62.

Phillips, I. D. J. and Wareing, P. F. (1958) 'Studies in dormancy of Sycamore. I. Seasonal changes in the growth substances contents of the shoot', *J. exp. Bot.*, **9**, 350–364.

Philipson, W. R. and Ward, J. M. (1965) 'The ontogeny of the vascular cambium in the stem of seed plants', *Biol. Rev.*, **40**, 534–579.

Philipson, W. R., Ward, J. M., and Butterfield, B. G. (1971) *The vascular cambium – its development and activity*, Chapman and Hall, London.

Pilet, P. E. (1970) 'The effect of auxin and abscissic acid on the catabolism of RNA', *J. exp. Bot.*, **21**, 446–451.

Poux, N. (1969) 'Localisation d'activités enzymatiques dans les cellules du méristème radiculaire de *Cucumis sativus* L. II. Activité peroxydasique', *J. Microscopie*, **8**, 855–866.

Priestley, J. H. (1930) 'Studies in the physiology of cambial activity. II. The concept of sliding growth', *New Phytol.*, **29**, 96–140.

Resch, A. (1959) 'Über Leptombündel und isolierte Siebrohren sowie deren Korrelationen zu den Übrigen Leitungsbahnen in der Sprossachse', *Planta*, **1–2**, 467–515.

Robards, A. W., Davidson, E., and Kidwai, P. (1969) 'Short-term effects of some chemicals on cambial activity', *J. exp. Bot.*, **20**, 912–921.

Robards, A. W. and Humpherson, P. G. (1967) 'Phytoferritin in plastids of the cambial zone of Willow', *Planta*, **76**, 169–178.

Cambial cells

Robards, A. W. and Kidwai, P. (1969a) 'A comparative study of the ultrastructure of resting and active cambium of *Salix fragilis* L.', *Planta*, **84**, 239–249.

Robards, A. W. and Kidwai, P. (1969b) 'Vesicular involvement in differentiating plant vascular cells', *New Phytol.*, **68**, 343–350.

Robards, A. W. and Kidwai, P. (1969c) 'Cytochemical localization of phosphatase in differentiating secondary vascular cells', *Planta*, **87**, 227–238.

Sacher, J. A. (1968) 'Senescence: effects of auxin and kinetin on RNA and protein synthesis in subcellular fractions of fruit and leaf tissue sections', In: *Biochemistry and physiology of plant growth substances* (ed. F. Wightman and G. S. Setterfield), pp. 1457–1477, The Runge Press, Ottawa.

Sanio, K. (1873) 'Anatomie der gemeinen Kiefer (*Pinus silvestris*)', *Jb. wiss. Bot.*, **9**, 50–126.

Shih, C. Y. and Rappaport, L. (1970) 'Regulation of bud rest in tubers of potato, *Solanum tuberosum* L. – VII. Effect of abscissic and gibberellic acids on nucleic acid synthesis in excised buds', *Pl. Physiol. (Lancaster)*, **45**, 33–36.

Siminovitch, D. and Chater, A. P. J. (1957) 'Biochemical processes in the living bark of the black locust tree in relation to frost hardiness and the seasonal cycle', In: *The physiology of forest trees* (ed. K. V. Thimann), pp. 219–250, The Ronald Press, New York.

Srivastava, L. M. (1964) 'Anatomy, chemistry and physiology of bark', In: *International Review of Forestry Research* (ed. J. A. Romberger and P. Mikola), vol. 1, pp. 203–277, Academic Press Inc., New York.

Srivastava, L. M. (1966) 'On the fine structure of the cambium of *Fraxinus americana* L.', *J. Cell Biol.*, **31**, 79–93.

Srivastava, L. M. and O'Brien, T. P. (1966) 'On the ultrastructure of cambium and its vascular derivatives. I – Cambium of *Pinus strobus*', *Protoplasma*, **61**, 257–276.

Sterling, C. (1946) 'Growth and vascular development in the shoot apex of *Sequoia sempervirens* (Lamb.) Endl. III – Cytological aspects of vascularisation', *Am. J. Bot.*, **33**, 35–45.

Sterling, C. (1947) 'Organization of the shoot apex of *Pseudotsuga taxifolia* (Lamb.) Britt. II – Vascularisation', *Am. J. Bot.*, **34**, 272–280.

Studhalter, R. A., Glock, W. S., and Agerter, S. R. (1963) 'Tree growth', *Bot. Rev.*, **29**, 245–365.

Thimann, K. V. and Kaufman, D. (1957) 'Cytoplasmic streaming in the cambium of white pine', In: *The physiology of forest trees* (ed. K. V. Thimann), pp. 479–492, The Ronald Press, New York.

Walker, N. E. and Cumbie, B. G. (1968) 'Anticlinal division in the vascular cambium of *Hibiscus*', *Trans. Missouri Acad. Sci.*, **2**, 66–72.

Wareing, P. F., Good, J., and Manuel, J. (1968) 'Some possible physiological roles of abscissic acid', In: *Biochemistry and physiology of plant growth substances* (ed. F. Wightman and G. Setterfield), pp. 1561–1579, The Runge Press, Ottawa.

Wareing, P. F., Hanney, C. E. A., and Digby, J. (1964) 'The role of endogenous hormones in cambial activity and xylem differentiation', In: *The formation of wood in forest trees* (ed. M. H. Zimmerman), pp. 323–344, Academic Press Inc., New York.

Wareing, P. F. and Phillips, I. D. J. (1970) *The control of growth and differentiation in plants*, Pergamon Press, Oxford.

Wilcox, H. W. (1962) 'Cambial growth characteristics', In: *Tree growth* (ed. T. T. Koslowski), pp. 57–88, The Ronald Press, New York.

Wilson, B. F. (1963) 'Increase in cell wall surface area during enlargement of cambial derivatives in *Abies concolor*', *Am. J. Bot.*, **50**, 95–102.

Wilson, B. F. (1964) 'A model for cell production by the cambium of conifers', In: *The formation of wood in forest trees* (ed. M. H. Zimmerman), pp. 19–36, Academic Press Inc., New York.

Wort, D. J. (1962) 'Physiology of cambial activity', In: *Tree growth* (ed. T. T. Koslowski), pp. 89–95, The Ronald Press, New York.

Zahner, R. (1968) 'Water deficits and growth of trees', In: *Water deficits and plant growth* (ed. T. T. Koslowski), vol. 2, pp. 191–254, Academic Press Inc., New York.

Zee, S. Y. and Chambers, T. C. (1969) 'Development of the secondary phloem of the primary root of *Pisum*', *Aust. J. Bot.*, **17**, 199–214.

11

Phloem differentiation and development
J. Cronshaw

Introduction

Early experiments on phloem transport

Research on the transport systems of plants has a long history dating back to the middle of the seventeenth century when William Harvey's exciting discovery of the circulation of blood in animals stimulated other scientists to look for similar circulatory systems in plants. A seventeenth-century Italian physician, Marcello Malphigi, performed experiments by removing a ring-like portion of bark from trees and observed that swelling subsequently occurred above the ring. Malphigi concluded that the removal of the ring of bark had caused a blocking of the flow of nutrients down the trunk of the tree and that the accumulation of nutrients above the ring had induced the stimulation of growth. Further experiments by early forest botanists confirmed Malphigi's results and gradually it became accepted that long distance transport of nutrients takes place in phloem tissue which in the trunks of trees makes up the inner part of the bark.

Discovery of the sieve element

The concept of the phloem as the primary long distance food-conducting system of the plant became firmly established in the middle of the nineteenth century with the discovery of the sieve element by Theodore Hartig. It is the presence of sieve elements that characterizes phloem tissue. Hartig also made the first analysis of sieve tube sap, the solution that exudes when the bark of a tree is cut deeply enough so that the actively conducting phloem is reached. Hartig's analysis of phloem exudate revealed that it contained a high concentration of sugar (up to 33 per cent) and also small amounts of minerals and nitrogenous substances.

The mass flow mechanism

At about the time that Hartig discovered the sieve element another German forest botanist, Ernest Münch, was making observations on phloem exudation. Münch observed that exudate was produced by red oak trees following an incision in the phloem. Because new incisions made above or below the primary wound did not deliver any exudate, Münch concluded that the transporting phloem cells were delicate structures and were readily blocked. From his observations Münch proposed a mechanism to explain the transport of sugars through the phloem based on a physical model. The model is extremely simple and consists of two osmometers connected to each other by a tube and immersed in the same, or in different, solutions. The first osmometer contains a solution considerably more concentrated than the surrounding solution. The second osmometer contains a solution which is considerably less concentrated than the surrounding solution. Within this system there is an osmotic movement of water into the first osmometer and out of the second osmometer so that solution is consequently carried through the connecting tube in a bulk or mass flow. The flow continues until the concentration of both solutions is equal. Münch suggested that the living plant contains a system analogous to the physical model, and that the continued production of assimilates at specific sites of the plant while these assimilates are utilized in other parts of the plant, maintains a difference in concentrations. These conditions result in a flow through the connecting phloem cells whenever and wherever there is a gradient of concentrations. This proposed mechanism for the movement of nutrients in the phloem is called the 'mass flow mechanism'. The mechanism requires that solution can move through the phloem sieve elements without much physical resistance to the flow. This is the critical consideration in the application of the Münch model to the plant. The mechanism will not work in the plant unless it contains structures that will permit the flow of solution through them.

Other mechanisms of transport

Structural studies of the phloem have indicated the possibility that there may not be a pathway of low resistance through the sieve elements. For this and other reasons alternate mechanisms of transport have been postulated that do not require the mass movement of solution. These mechanisms are based on a type of diffusional phenomenon and have in common the facts that the phloem cells play an active role in the movement, and that the nutrients move independently from the water.

Although there is no general agreement among plant physiologists as to the mechanism of long distance phloem transport, a vast body of physiological data about the substances that move, and about the rate of transport, has been accumulated. The substances that move are sugars, principally sucrose, together with small quantities of minerals, amino acids, and hormones. In the case of virus infected plants, intact virus particles are also transported at the same rate as the nutrients. These substances are transported through the phloem at 50–200 cm/hr (Weatherley and Johnson, 1968), and although higher concentrations can occur, the sieve tube sap usually contains approximately a 10% solution of sugar. This rate of transport is far in excess of that which could take place by simple diffusion or by cytoplasmic streaming, and to explain it we must invoke some kind of active mechanism.

Possible phloem transport mechanisms have been discussed recently in an excellent review by Weatherley and Johnson (1968) who concluded that 'With progress in studies of the fine structure of the phloem the question of how the sieve tube works has become more and not less puzzling'.

Structure in relation to function

Whatever the mechanism of transport, it must be compatible with structural observations of phloem cells. For this reason the phloem has been intensively studied by structural techniques. The *in vivo* structure of phloem cells must be precisely determined in order to speculate on mechanisms explaining the translocation process. Unfortunately, even though a great deal of research effort has been expended and the electron microscope has revealed a sieve element of unsuspected complexity, we do not have a complete picture of phloem structure. The most controversial feature is the one which would allow us to accept or reject the mass flow hypothesis: namely, whether or not the sieve element connections have a structure preventing a bulk flow of materials through the phloem conduits.

In considering the structure and differentiation of phloem, first let us turn our attention to the structure of the phloem as a whole, then to the structure and differentiation of sieve elements, returning later to the controversy of the sieve element connections and the mechanism of translocation.

Phloem tissue

Phloem is a complex tissue consisting of numerous cells which vary in structure in different plant species. We have seen that it is characterized by the presence of a highly specialized cell type, the sieve element. These cells receive their name from the presence of numerous relatively small perforations in their walls which are aggregated into regions known as sieve areas. The contents of contiguous sieve elements are in communication through the perforations. Sieve elements are elongated cells arranged longitudinally one above the other. In flowering plants these longitudinal series are particularly well developed and are called sieve tubes.

In addition to sieve elements, phloem tissue contains various kinds of parenchyma cells and may also contain thick walled sclereids and fibres. Sieve elements function in conjunction with these surrounding cells and are in dynamic equilibrium with them. Thus, water and food materials are moved in and out of the sieve elements by way of the parenchyma cells. Any parenchyma cell can function by loading the sieve element with sugar or by removing sugar from the sieve element, the direction of movement at any one time depending on the sugar content of the parenchyma cells relative to that of the sieve element.

Certain parenchyma cells are closely associated with the sieve elements and are known as companion cells in angiosperms, and albuminous cells in gymnosperms. Companion cells are derived from the same meristematic cell as their associated sieve element. They are characterized by prominent plasmodesmatal connections with the sieve element, dense cytoplasm, and usually an absence of starch. Albuminous cells stain strongly with certain dyes. In recent years it has been recognized that there may be a gradation of cells between 'typical' companion cells, and albuminous cells and other phloem parenchyma cells. For a detailed discussion see Esau (1969).

Detailed anatomical studies of the phloem of minor veins of peas (Gunning *et al.*, 1968) led to the discovery of specialized parenchyma cells of the phloem known as transfer cells (chapter 13). Transfer cells have wall protuberances which may extend into the cell for a considerable distance and may branch to form a labyrinth. They have a variety of forms and have now been described in numerous situations where one expects high rates of solute flow over short distances. Pate and Gunning (1969) have speculated that transfer cells in the phloem play a specialized role in loading and unloading sieve elements with nutrients.

The sieve element

The sieve element has already been introduced as the cell type specialized for long distance transport of nutrients and as the characteristic cell of phloem tissue. Specialized perforations of the wall of the sieve element are arranged in areas, the sieve areas, which vary in size in different kinds of plants. In the lower vascular plants and gymnosperms the pores are generally small and rather uniform in size. In the angiosperms larger pores are found, and these are usually more irregular in size. It is thought that there has been an evolutionary specialization of sieve areas, the sieve areas of angiosperms with large pores representing an evolutionarily advanced condition. In the sieve elements of angiosperms, the end walls have highly specialized sieve areas with larger pores. These end walls are referred to as sieve plates and serve to connect sieve elements with one another. Thus, in the angiosperms, numerous sieve elements are connected end to end by walls bearing sieve plates to form long tubes — the sieve tubes. Sieve plates may, however, also occur on lateral walls connecting contiguous sieve tubes.

In addition to the specialized wall structure, the sieve element has a highly specialized protoplast with many unusual features. Unlike the protoplast of xylem tracheary elements, the protoplast of the sieve element is retained in the mature cell. The contribution of this protoplast to the functioning of the cell in long distance transport has remained a subject for speculation since the nineteen thirties. Two contrasting views have been put forward, one emphasizing the uniqueness of the protoplast and ascribing a *passive* role of the cell in translocation; the other emphasizing normal cell features and ascribing to the cell an *active* role in nutrient transport.

Several features of the sieve element protoplast remain controversial. Sieve elements contain specific protein components, and a difficult unsolved problem is a determination of the distribution of this protein component in the mature cell. There are also many unsolved questions about the formation, molecular structure and properties of these protein components, about their aggregation into specific bodies, and about their subsequent disaggregation during the differentiation of the cell. Another unsolved problem is a determination of the structure to function relationship of the unique endoplasmic reticulum system of the sieve elements. Other problems solved to the satisfaction of most but not all botanists are those concerned with the fate of the nucleus and vacuoles during the development of the sieve elements.

A major reason why many questions about the structure of sieve elements remain unanswered is that cutting of the sieve elements or chemical alteration of the membrane with fixatives, when preparing material for electron microscopy, is certain to release the hydrostatic pressure within. This pressure has been estimated to be as high as 30 atm (Weatherley, 1967), and it is obvious that a release of this pressure would cause large disturbances in the structural arrangement of the sieve element subcellular components. Even with the best techniques available, and taking all possible precautions, it is impossible to be certain that such changes have not occurred. This problem of the release of hydrostatic pressure is, of course, a problem of mature functioning sieve elements. Young sieve elements at the early stages of differentiation have not developed a hydrostatic pressure and are much more amenable to study by cytological techniques. For this reason one of the best ways to study the structural organization of the sieve element is to build up a picture of the undifferentiated cell and to follow the changes that occur during differentiation.

Sieve element differentiation

Early stages of development

Since the introduction of reliable methods of fixation of plant cells for electron microscopy using aldehydes (Ledbetter and Porter, 1963), there have been several excellent descriptions of the differentiation of sieve elements. Detailed studies have been made of *Pisum* (pea) (Bouck and Cronshaw, 1965), *Acer* (Northcote and Wooding, 1966), *Nicotiana* (tobacco) (Cronshaw and Esau, 1967), *Cucurbita* (Cronshaw and Esau, 1968a, b; Evert, Murmanis, and Sachs, 1966), *Coleus* (Steer and Newcomb, 1969) and *Glycine* (soybean) (Wergin and Newcomb, 1970). The structure of the sieve element at the earliest stages of development is essentially similar in all cases. They are first recognizable by their wall structure. Before visible changes from the meristematic state occur in the protoplast a region of wall material is deposited which gives the wall a thicker appearance than that of adjacent cells. The thickness of the wall may be judged from the position of the middle lamella, the wall to the sieve element side being thicker than that on the parenchyma cell side. The thicker region of wall, known as the nacré wall, is adjacent to the plasma membrane and is frequently more electron opaque than other wall regions.

The plasma membrane of the young sieve element has the usual triple layered structure consisting of two electron-opaque lines separated by a less electron opaque region. Early in the development of the sieve element, however, changes occur in the structure of the plasma membrane so that the two electron opaque lines appear thicker and stain more heavily than their counterparts in adjacent cells. The plasma membrane also develops asymmetry, the electron opaque line to the wall side appearing thicker than the electron opaque line on the inside. This asymmetry may be an indication of high cellulose synthetase active at the outer surface of the plasma membrane. Sieve elements at this early stage of differentiation are rapidly depositing large quantities of wall material, and several workers have suggested that cellulose synthetase activity is associated with the outer region of the plasma membrane. Other structural features of cells at this stage indicative of a high rate of wall deposition are a wavy appearance of the plasma membrane, presumably due to a high rate of fusion of dictyosome derived vesicles bringing wall substances to the surface, and numerous peripheral microtubules. The peripheral microtubules are especially abundant in sieve elements at early stages of differentiation, and in favourable sections can be seen to be attached to the plasma membrane with an electron opaque substance. At later stages of differentiation, when cell wall deposition has ceased, microtubules are seen only rarely.

The young sieve element protoplast contains the usual components of a meristematic cell (Fig. 11.1). The cytoplasm is dense but has a vacuolar system which may consist of a single large vacuole or, more commonly, several smaller ones. The nucleus is often elongated and lobed. Other components, the mitochondria, plastids, dictyosomes, ribosomes, and cisternae of endoplasmic reticulum show a relatively uniform distribution throughout the cell. The high density of these organelles probably indicates intense synthetic activity at these early stages of differentiation.

During later stages, the protoplast is drastically remodeled, with many radical changes taking place. A new component, P-protein, is fabricated, the endoplasmic reticulum is reorganized, changes occur in plastids and mitochondria, and the nucleus, vacuole system, dictyosomes, and ribosomes are disassembled and disappear. At the same time changes occur in the cell wall and specialized cell-to-cell connections are formed. The changes in these subcellular systems will now be examined in detail.

P-proteins (slime)

For many years it has been known that proteinaceous accumulations occur in phloem cells. Formerly these proteinaceous components were known as slime, and aggregations of them as slime bodies and slime plugs. Slime bodies were observed in young phloem cells and slime plugs were acknowledged as artifacts formed by the accumulation of slime at sieve plates following the rupture of sieve tubes. After extensive examination of these components in various differentiating and mature phloem cells at the fine structural level, the terms P-protein and P-protein bodies were introduced to describe them (Esau and Cronshaw, 1967). The main reason for the introduction of these new terms was that

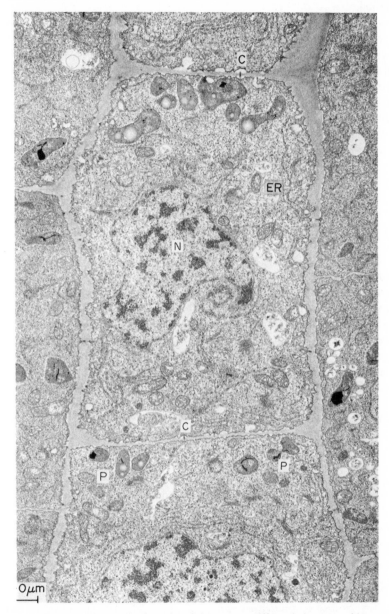

Fig. 11.1 Electron micrograph of a series of three young differentiating protophloem sieve elements from a tobacco root together with adjacent parenchyma cells. Note the normal nucleus (N), endoplasmic reticulum (ER) and the random distribution of organelles in the dense cytoplasm. In the oldest cell are small aggregates of P-protein (P). Electron-lucent callose (C) is present on the end walls around the plasmodesmata. ×7500.

the former term slime was misleading in that plant slimes are usually carbohydrate in nature.

In numerous fine structural studies P-proteins have been resolved into amorphous, fibrillar, tubular, and crystalline components. They are particularly characteristic of sieve elements and have been studied in great detail because of their possible involvement in the translocation process and the functioning of the sieve elements. It is well established by observation at the light microscope level of certain dicotyledonous species that dense bodies known as P-protein bodies or slime bodies originate in the cytoplasm of sieve elements at early stages of their differentiation and that they disaggregate as the cells mature. It is the component parts of these bodies, together with other cell components, that form the plugs known as slime plugs produced at the sieve plates when a sieve element is ruptured. In some cases, slime bodies are retained in the mature sieve elements, and this phenomenon has been described for several species of the subfamily Papilionacea of the Leguminosae and for some sieve elements of *Cucurbita.*

Studies with the electron microscope have shown that P-proteins have many morphological forms. They may be in the form of small granules, fibrils, tubules or crystalline components and there are usually developmental relationships between these forms. The most commonly encountered form of P-protein is a tubular one. Tubular P-proteins have been found in *Acer* (Northcote and Wooding, 1966), *Primula* (Tamulevich and Evert, 1966), *Nicotiana* (Cronshaw and Esau, 1967; Parthasarathy and Mühlethaler, 1969; Wooding, 1969; Cronshaw and Anderson, 1969), *Cucurbita* (Cronshaw and Esau, 1968a; Parthasarathy and Mühlethaler, 1969), *Coleus* (Steer and Newcomb, 1969), *Glycine* (Wergin and Newcomb, 1970), *Pisum* (Zee, 1968), *Beta* (Esau, Cronshaw, and Hoeffert, 1967), *Ulmus* (Evert and Deshpande, 1969) and *Mimosa* (Esau, 1971). In some species the formation of tubules is preceded by the formation of a fibrous P-protein, and although this has not been recognized in all the plants studied it may be a general phenomenon. In mature sieve elements the P-protein is often in the form of beaded fibrils, and apparently these are derived from the tubular form (Cronshaw and Esau, 1967; Parthasarathy and Mühlethaler, 1969; Steer and Newcomb, 1969). In the papilionaceous legumes a crystalline form of P-protein has been observed which is an intermediate aggregation between a tubular form and dispersed beaded fibrils (Wergin and Newcomb, 1970).

Formation of P-protein

When first evident in differentiating sieve elements, P-protein may be in the form of tubules as in tobacco (Cronshaw and Esau, 1967), fibrils as in *Cucurbita* (Cronshaw and Esau, 1968a), or as finely granular material as in soybean (Wergin and Newcomb, 1970) and pea (Bouck and Cronshaw, 1965). The first suggestion of a mechanism of P-protein formation was by Bouck and Cronshaw (1965). They observed in pea that small regions of granular P-protein were closely associated with cisternae of endoplasmic reticulum and suggested that the endoplasmic reticulum may be directly involved in the synthesis of P-protein.

Newcomb (1967) observed vesicles bearing spinelike projections in cells located adjacent to protophloem sieve elements in *Phaseolus* root tips. Newcomb termed these structures spiny vesicles and, because they were frequently associated with P-protein, suggested that they may play a role in its formation. Subsequent work showed associations between spiny vesicles and P-protein in *Cucurbita* (Cronshaw and Esau, 1968a), *Coleus* (Steer and Newcomb, 1969), *Nicotiana* (Esau and Gill, 1970b) and sugar beet (Esau and Gill, 1970a), and appears to confirm a functional relationship between the two cell components.

Other observations have shown a close association of apparently active dictyosomes adjacent to developing regions of P-proteins, suggesting that these organelles are involved in P-protein formation. In view of the fact that developing regions of P-proteins are not membrane bound, and dictyosome derived products are usually transferred through a membrane, this latter association may be fortuitous. It may be that the apparently active dictyosomes are involved in the packaging of materials for transport to the developing cell wall. It has also been suggested that P-proteins are formed by the numerous free ribosomes and polysomes that closely surround the developing regions of P-protein. Thus, although numerous observations and suggestions have been made, a determination of the precise mechanisms of P-protein formation must await further research.

P-protein bodies

As development of the sieve element proceeds, the small accumulations of P-protein expand to form large bodies known as P-protein bodies. The morphology of the P-protein components and their arrangement into P-protein bodies varies in different kinds of plants.

The P-protein bodies of *Nicotiana* and *Coleus* are similar and have been studied in detail (Cronshaw and Esau, 1967; Steer and Newcomb, 1969). In the young sieve elements of these species the components of the P-protein bodies are well-formed tubules of undetermined length and about 23 nm in width (Fig. 11.3). The bodies appear to increase in size by the addition of more tubules to these units. There is usually one major P-protein body per cell and this increases in size until it is approximately the same size or perhaps even greater than that of the nucleus. It is important to note that the P-protein bodies are not membrane bound. The bodies are ellipsoidal in shape, usually elongated in the same direction as the differentiating sieve element. The P-protein tubules are arranged more or less parallel to one another so that a cross-section of the body cuts the tubules in transection. This means that the length of the tubules is in the same direction as the long axis of the ellipsoidal bodies. In some cases the bodies contain crystalline regions of tubules; and in these regions the tubules have a hexagonal close-packed arrangement. Within the P-protein bodies, the individual tubules have stained, spoke-like projections which appear to connect adjacent tubules. These connections between adjacent tubules may be responsible for the structural integrity of the bodies at this stage.

In *Cucurbita* the situation is apparently more complex. In the young developing cells, the P-protein has more than one morphological form and is organized
into P-protein bodies of two types, one of which is larger than the other
(Cronshaw and Esau, 1968a). There are a large number of these bodies per cell,
and these develop in the peripheral cytoplasm. The larger type of P-protein body
contains the bulk of the P-protein in the form of fine fibrils. This type of
P-protein body is first recognized in the cytoplasm as a small aggregate of fine
fibrils intermixed with ribosomes, dictyosomes and cisternae of endoplasmic
reticulum. Presumably these accumulations are the regions of synthesis of
P-protein. As the deposition of fibrous P-protein continues, the aggregates
become larger and the other cell organelles are excluded from the centre of
the accumulation. The second type of P-protein body found in *Cucurbita* is
smaller and consists of an assembly of distinct tubules which are similar
morphologically to the P-protein tubules found in *Nicotiana* and *Coleus*. These
smaller tubular P-protein bodies occur adjacent to the larger type of P-protein
body. Within them the tubules usually have a characteristic arrangement, which
in certain sections is clearly quadrangular. In the larger bodies the fibrous
P-protein component may become modified to show a range of structures. Often
these structures have an electron opaque outer region and an electron translucent
central region, giving them the appearance of small tubules. The variability of
structures found in the large P-protein bodies indicates that there is a developmental sequence from the fine fibrils, which are formed initially, to the small
tubules. These small tubules are morphologically distinct from the common
type of P-protein tubules found in *Nicotiana, Coleus,* and the small bodies of
Cucurbita.

In the papilionaceous legumes large unique bodies have been described. In
1891, Strasburger described large refractive structures in the sieve elements of
Robinia. Since that time similar structures have been described from several
species and have been variously called irregular bodies (Bouck and Cronshaw,
1965); slime bodies (Wark and Chambers, 1965); crystalline inclusions (Zee,
1968); flagellar inclusions (LaFlèche, 1966) and mucoid bodies (Gerola,
Lombardo, and Catara, 1969). These bodies are now known to be P-protein
bodies. Recently, Wergin and Newcomb (1970), working with soybeans, have
described in detail P-protein formation and possible relationships of crystalline
P-proteins with other P-protein components. In the soybean, P-protein first appears
in a cell as a finely granular material which accumulates among ribosomes, vesiculate
endoplasmic reticulum, and numerous dictyosomes. Among these granular
masses, bundles of tubules appear which then aggregate and align themselves
longitudinally in the sieve element. By further transformations these bundles
of tubules are converted into an electron opaque crystalline P-protein body.
This body continues to grow by the aggregation of more material and at
maturity may be as long as $15-30\ \mu$m. The main part of the body is square in
cross-section and at either end it is terminated by sinuous 'tails'. These characteristic tails were the structures observed by earlier workers who described
the bodies as flagellar inclusions (LaFlèche, 1966).

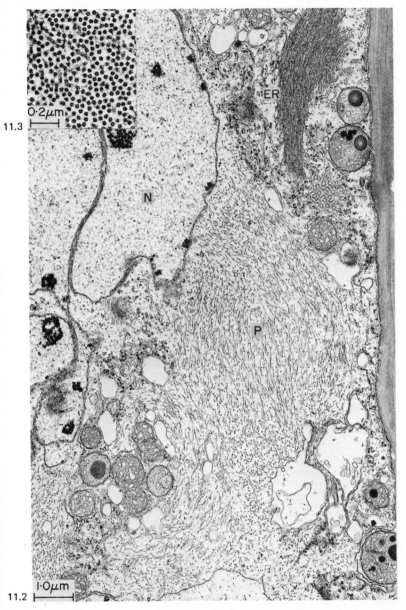

Fig. 11.2   Electron micrograph of a portion of a sieve element differentiating in a tobacco pith culture. The P-protein (P) is dispersing, the endoplasmic reticulum (ER) is reorganizing and the nucleus (N) is degenerating. x13 500.

Fig. 11.3   *Nicotiana tabaccum.* Electron micrograph of a portion of a P-protein body showing P-protein tubules in transection. x50 000.

P-protein bodies are thus membraneless assemblages of proteinaceous components, and it is pertinent to ask what maintains the integrity of these bodies. There is no surrounding membrane and, except for the stained projections between the components, they are apparently free to move out in the cytoplasm. However, the integrity of the P-protein bodies is maintained until a certain stage in the differentiation of the sieve elements, when disaggregation of the P-protein bodies takes place.

Disaggregation of P-protein bodies

The stage of differentiation of the sieve element when the P-protein bodies are fully formed is followed by one in which a series of rapid and unique events take place. These include the disassembly of the nucleus, vacuolar systems, dictyosomes, and ribosomes, changes in the form of the endoplasmic reticulum, and the disaggregation of the P-protein bodies. The first indication of these changes is usually a loosening of the tubular or fibrillar components of the P-protein bodies. P-protein components then start to move apart from one another and gradually become dispersed throughout the cytoplasm of the sieve element (Fig. 11.2). At this stage the lumen of the cell is drastically altered by the series of events mentioned above, which includes a hydration of the cytoplasm due mainly to the breakdown of the vacuolar system of the cell.

At the time of dispersal of the P-protein body in tobacco and *Coleus,* beaded fibrils appear in the cytoplasm associated with the disaggregating tubules. It has been postulated that these beaded fibrils are derived from the P-protein tubules (Cronshaw and Esau, 1967). In the soybean the large crystalline P-protein bodies disperse into a mass of fine striated fibrils which resemble the striated fibrils of tobacco and *Coleus.* In tobacco the degree of change from one form of P-protein to another appears to be varied, and tubular forms and beaded fibrillar forms may be present at the same time in mature sieve elements (Cronshaw and Anderson, 1969).

From observations such as these, it is postulated that P-proteins can undergo conformational changes, and these will now be discussed in relation to the structure of the P-protein components.

Structure and conformational changes of P-proteins

We have seen that P-proteins may be granular, fibrillar, tubular or crystalline. The most commonly occurring form, however, appears to be the tubular one. In several species of plants, observations of P-proteins have suggested that tubular P-proteins may be converted to a fibrillar type. What is the molecular architecture of the P-protein components and how could conformation changes take place? While trying to answer these questions researchers have repeatedly compared P-protein tubules with the conventional microtubules found in plant and animal cells. Microtubules have a diameter of about 24 nm which is similar to that of P-protein tubules. The apparent lumen of the microtubules is, however, larger than that of P-protein tubules. Microtubules have been shown to be built up of protein subunits approximately 4 nm x 8 nm in size. These subunits are arranged in a shallow helix (8–10 degree pitch) in such a way that there appear to be thirteen linear strands (Porter, 1966). Parthasarathy and Mühlethaler (1969) have suggested that P-protein tubules are also formed from a helical arrangement of protein subunits. These authors presented alternative models, one based on a single helix, the other based on two helices. They suggested that the number of subunits arranged around the tubule was either six or twelve. It was further suggested that, by a stretching and partial unwinding of the helices in the tubule, it could be converted into structures that would appear as striated fibrils in the electron microscope. The striations would actually be the place where two helices cross. The evidence on which these proposed models was based consisted of observations of cross striation of varying pitch in longitudinal views of tubules and on images obtained of transectional views of P-protein tubules using a photographic image reinforcement technique.

Cronshaw (1970a) presented evidence that P-protein components may have a tubular structure for part of their length and a beaded fibrillar structure for part of their length, unequivocally demonstrating the interconvertibility of the two morphological forms. Evidence was also presented of P-protein components with an appearance suggesting a two-helix structure. It was suggested that these results supported the two-helix model originally proposed by Parsatharathy and Mühlethaler (1969). Recent work in our laboratory has provided further evidence in support of this model. Two helices have been clearly demonstrated in P-protein components extracted from tobacco plants and negatively stained. We have also applied the image reinforcement technique to transectional views of P-protein tubules and determined that there are six subunits arranged around the tubule.

The proteinaceous nature of P-protein components was determined originally by cytochemical studies at the light microscope level (Cronshaw and Esau, 1967). Aggregations of P-protein stain strongly with mercuric bromophenol blue, a characteristic reaction of proteins. Kollmann, Dorr, and Kleinig (1970) have confirmed the proteinaceous nature by direct chemical analysis of filaments extracted from *Cucurbita* and *Nicotiana*. These filaments had measurements similar to the P-protein components in sectioned material. Cronshaw (1970b) also isolated P-protein components from *Cucurbita, Cucumis,* and *Nicotiana*. Measurements of the diameter of negatively stained extracted fibrils were in agreement with measurements made on sectioned material. The isolated fibrils could be disrupted by agents known to disrupt the tertiary structure of proteins.

The proteinaceous nature of the P-protein components thus appears to have been substantiated, and evidence is accumulating suggesting that P-protein tubules have a structure based on two tightly wound helices of subunits with six subunits around the tubule. Most of the various electron microscope images that have been obtained of P-proteins can be explained in terms of this structure and of forms that would be derived by a loosening and unwinding of the helices.

P-protein function

It is pertinent to ask, 'What do we know of the function of P-proteins?' At the present time this is a difficult question to answer. Large amounts of this material are synthesized in the phloem cells, chiefly in the sieve elements, but also in companion cells and parenchyma cells. It has been postulated that, since the sieve element is a cell specialized for the long distance transport of nutrients, this proteinaceous component is most probably functional in this transport. To date, however, no one has put forward a mechanism of long distance transport of nutrients involving the P-protein component which seems feasible. Perhaps the large surface area of P-proteins serves to distribute metabolically active molecules. Or perhaps the P-protein components are contractile filaments which could generate the motive force for phloem transport.

The one known function of the P-protein is that, if the hydrostatic pressure of functional sieve elements is suddenly released, the P-protein component is capable of rapidly plugging pores in the sieve plate. This rapid clogging stops the flow of nutrients from the phloem and is essentially a wound-healing mechanism. If the pores of the sieve plate were not plugged in this way the plant would suffer a drastic loss of nutrients which would continue to flow down through the phloem tissue in response to a hydrostatic pressure generated at the sites of phloem loading of nutrients.

Membrane systems of the differentiating sieve element

We have seen that the young sieve elements have a vacuolar system, normal profiles of endoplasmic reticulum, and numerous dictyosomes. At the time of the disaggregation of the P-protein bodies the vacuolar membrane apparently changes its permeability properties and starts to break down. This causes a hydration of the cytoplasm which consequently appears less dense in the electron microscope. The breakdown of the vacuolar membrane has been questioned by Evert and co-workers (Tamulevich and Evert, 1966; Evert, Murmanis, and Sachs, 1966; Evert and Deshpande, 1969) who believe that a vacuole continues to exist in the cytoplasm throughout the differentiation of the sieve element. To account for the disparity between their view and that of other authors, Evert and his co-workers have suggested that the vacuole membrane of the sieve element is extremely fragile and is easily destroyed when phloem tissue is prepared for electron microscopy. However, electron microscopists have described vacuoles in young differentiating sieve elements of all species thus far examined. Evert and co-workers' view would mean that the vacuolar membrane at least changes during differentiation to become more sensitive to the fixation procedures of electron microscopy. At the present time the bulk of evidence appears to support the more widely held view that the vacuolar membrane does indeed change its permeability properties and lose its integrity during differentiation.

Early in the development of the sieve element, cisternae of endoplasmic reticulum become associated with the sites of future pores of the sieve plate. This association will be discussed in relation to the formation of the sieve plate pores.

At the stage of differentiation of the sieve element when the P-protein bodies start to disperse and the vacuolar system changes, the endoplasmic reticulum starts to reorganize. This reorganization was first described for sieve elements of *Pisum* by Bouck and Cronshaw (1965). These authors considered the differentiated or changed form of endoplasmic reticulum to be sufficiently distinct from the endoplasmic reticulum of other cells that it merited a new term, thus they called it the *sieve tube reticulum*. Later, Srivastava and O'Brien (1966) suggested the term *sieve element reticulum* so that it would be applicable to sieve elements in all plant taxa and not only to those having sieve tubes, i.e., the angiosperms.

In the sieve elements of *Pisum,* changes in the organization of the endoplasmic reticulum are first apparent by the formation of stacks of cisternae. At first, pairs or triplets of endoplasmic reticulum cisternae are seen. These stacks of cisternae do not have associated ribosomes. Between the cisternae is a fibrous dark-staining component which resembles fibrils of P-protein. Stacks of various sizes become associated with the nuclear envelope. During this process the contents of the nucleus disintegrate and ultimately the chromatin and nucleoli can no longer be recognized within the nucleus. Shortly after the association of stacks of cisternae or vesicles along the nuclear envelope, the nuclear envelope itself breaks down. The stacks continue to enlarge by the addition of further cisternae until sometimes massive aggregates are formed (Fig. 11.2). Prior to, and after the breakdown of the intact nucleus the aggregates of endoplasmic reticulum organize into loose patches along the inner surface of the plasma membrane.

In addition to the stacked system of cisternae, individual cisternae come to form a layer just inside the plasma membrane and closely associated with it. The final arrangement of sieve element reticulum thus consists of a peripheral network of single cisternae close to the plasma membrane in addition to regions of stacked cisternae. A similar sequence of events has been described in tobacco (Cronshaw and Anderson, 1971), in *Cucurbita* (Esau and Cronshaw, 1968) and in bean (Esau and Gill, 1971). Other forms of aggregated ER have been described. Behnke (1968) has described stacked convoluted endoplasmic reticulum in sieve elements of *Dioscorea* which assumes the form of a highly ordered lattice, the component elements of which are tubules.

What is the function of the reorganized system of endoplasmic reticulum? The answer to this question remains pure speculation. Perhaps the system is one of storage of the fibrous material which occurs within and between the cisternae. Perhaps it is a storage site for various enzymes or other molecules necessary for the functioning of the cell.

At the time of reorganization of the membranous components of the endoplasmic reticulum, the dictyosomes and ribosomes of the cell are disassembled and disappear. Thus, the mature sieve element lacks its synthetic machinery. Not only does it lose its information centre, the nucleus; it loses also the sites of translation, the ribosomes, and its secretory system, the dictyosomes.

Plastids and the mitochondria

Mitochondria are present in young sieve elements and remain apparently unmodified throughout the differentiation of the cell. Some authors have described degenerative changes in mitochondria (Esau and Cheadle, 1965), but these are probably due to fixation artifacts.

The plastids of young sieve elements contain a dense electron opaque matrix and have little differentiation of their membrane systems. They may contain small osmiophilic globules, occasionally a crystal, and sometimes starch. During differentiation the plastids enlarge to some extent and the matrix becomes less electron opaque. The internal membrane systems, however, remain poorly developed.

The sieve elements of most dicotyledons commonly contain plastids which elaborate a form of starch with a specific reddish-brown staining reaction with iodine. A detailed study of the sieve element plastids of bean and their starch deposits has been made recently by Palevitz and Newcomb (1970). Cytochemical studies showed that the sieve element starch is composed of highly branched molecules with numerous α-1,6 linkages. Some dicotyledenous species have sieve element plastids that do not typically accumulate starch. The plastids of these species have a proteinaceous component in the form of a fibrous ring. This type of plastid was first described in *Beta* (Esau, 1965). A recent survey by Behnke (1969b) has described the *Beta* type plastid in seventeen species.

In the plastids of monocotyledons, crystalline proteinaceous bodies are often found. In a recent survey of monocotyledonous plastids Behnke (1969a) has described crystalline proteinaceous inclusions within plastids of numerous species. Starch may or may not be present.

Differentiation of the sieve area pores

Apart from the fact that the nature of the contents of the sieve area pores still confounds us, the structural features of sieve area pores, and the changes that occur during their formation, appear to be well established. Initially, the pore sites are plasmodesmata, and each plasmodesma develops into a pore. Early in the development of the sieve element, cisternae of endoplasmic reticulum become closely applied to the plasma membrane at the pore site region. This is followed by the appearance of the carbohydrate callose (a β 1,3-linked glucan) as an electron translucent substance outside the plasma membrane at the pore site region. Callose is known to be deposited as a reaction to wounding in a variety of cells and to be rapidly synthesized in sieve elements in response to injury or other disturbance (see review by Eschrich, 1971). The deposition of callose which takes place during the development of sieve plate pores is, however, not thought to be a wound effect or artifact but an integral part of normal cell differentiation.

Callose continues to be deposited at the developing pore sites. Where the callose is deposited it appears to replace previously deposited wall material, so forming opposing callose masses. These masses are cone-shaped pads which, as development proceeds, penetrate deeper and deeper into the wall, replacing disappearing wall material. Callose deposition continues until the opposing callose masses fuse and form cylinders in the wall surrounding the plasmodesmata. At about the time of fusion of the opposing callose masses the plasmodesmata enlarge in the region of the middle lamella to produce a structure known as a median nodule. The plasmodesmata further enlarge from the median nodule with the dissolution of callose until, finally, a large pore is formed with the plasma membrane continuous through it from cell to cell. Residual callose is nearly always observed as a thin walled cylinder around the pore.

Pore formation thus involves synthetic events and the formation of callose and degradation phenomena involving the enzymatic removal of wall material. Presumably, the enzymes involved in the degradation of callose and other wall materials are located in the cell wall. How are these enzymes moved out of the cells? How are their activities localized? Where are the enzymes located for the synthesis of callose? At the present time these questions remain unanswered. It has been suggested that the cisternae of endoplasmic reticulum which become associated with the pore sites early in development may play a part in the control of pore development, but the evidence for this is only that of morphological association. Clearly the answers to the questions must await further research.

Sieve element plasmodesmata

Sieve elements are connected to adjacent parenchyma cells by plasmodesmata which have a usual structure consisting of a canal lined with plasma membrane and a central tubule. During the differentiation of sieve elements the plasmodesmatal canal often widens on the sieve element side and becomes branched on the parenchyma cell side. A cylinder of callose may be associated with the widened canal on the sieve element side. Branched plasmodesmata are known to be characteristic of connections between sieve elements and companion cells (Esau and Cheadle, 1965) and have been described in minor veins between sieve elements and any contiguous parenchyma cell (Esau, 1967).

Contents of the sieve area pores

A proper consideration of the nature of the contents of the sieve area pores must take into account the question of possible artifacts produced by the release of pressure from the sieve elements. Plugging of the sieve plate pores by P-protein and occasional strands of endoplasmic reticulum has been described by many workers, and the key question is: Does this plugging represent an *in vivo* condition? Early observations by several researchers (Hepton *et al.*, 1955; Schumacher and Kollman, 1959; Kollman, 1960), indicated that the pores were plugged with electron opaque material. In contrast to these results Esau and Cheadle (1965), using potassium permanganate as a fixative, demonstrated 'open' pores in *Cucurbita.* Later it was realized that potassium permanganate does not preserve some cell components and, even if the pores were plugged, potassium permanganate fixation may have dissolved out the plugging material.

When glutaraldehyde and acrolein were demonstrated to yield excellent fixation of phloem tissue (Bouck and Cronshaw, 1965), plugged sieve area pores were described for many species (see Cronshaw and Anderson, 1969). The validity of these observations was questioned when Esau, Cronshaw, and Hoefert (1967) showed that in sugar beet plants infected with the beet yellows virus, some sieve plate pores were occupied by virus particles instead of P-protein. These observations were interpreted as indicating a mass movement of the virus particles, together with nutrients, through the sieve tubes. It was suggested that the sieve plate pores could be plugged by either P-protein or virus particles on the release of hydrostatic pressure. These observations led to a series of experiments to try to determine the *in vivo* condition of the sieve plate pores.

In tobacco, with usual methods of fixation (that is, cutting specimens from the plant for fixation or by fixing whole plants using glutaraldehyde or formaldehyde-glutaraldehyde mixture), the pores were universally plugged with P-protein (Cronshaw and Anderson, 1969). With the rapidly penetrating fixative acrolein, however, some pores were observed that did not have dense plugs of P-protein. Similarly, if whole plants were rapidly frozen in liquid nitrogen and then fixed in acrolein, formaldehyde-glutaraldehyde, or glutaraldehyde, some unplugged pores were observed (Cronshaw and Anderson, 1969). Also, if very thin slices of tissue were used so that fixatives had only a short distance to penetrate, pores were observed that were either unplugged or loosely filled with P-protein (Anderson and Cronshaw, 1970a).

Several investigators have attempted to reduce the hydrostatic pressure of the sieve elements before fixation. Rouschal (1941) pretreated phloem with glycerine in order to reduce the pressure in sieve tubes. Using this technique, he was able to prevent both slime plug formation and plugging of sieve plate pores. Anderson and Cronshaw (1970a) tried wilting tobacco and bean plants before fixation with formaldehyde-glutaraldehyde. Good fixation was apparently obtained as judged by the lack of plasmolysis of the cells and the nature of the organelles. In several wilted plants the P-protein was rather evenly distributed throughout the lumen of the sieve element, and gave the appearance of being in a three-dimensional network that was continuous through the sieve area pores. P-protein was not compacted in the pores. There was no evidence that rapid flow of the sieve element contents had occurred in these sieve elements, and it was concluded that the electron micrographs obtained most probably depicted the *in vivo* condition of the cells (Fig. 11.4). It should be mentioned, however, that other plants treated in a similar way had evidence of rapid flow in the sieve elements and pores plugged with P-protein. Anderson and Cronshaw (1970a) suggested that perhaps plants that were wilted too little or too much would still be sensitive to fixation damage. It was interesting that there was a correlation between the amount of callose in the cylinders around the pores and the degree of plugging. The wilted plants with unplugged pores had less callose

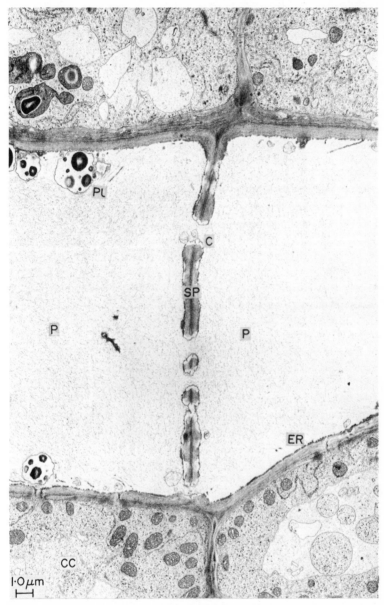

Fig. 11.4   Electron micrograph of portions of sieve elements and adjacent cells in a wilted plant of tobacco. The sieve-plate pores are large and there is little callose. P-protein is not compacted in the pores. Details: C = callose, CC = companion cell, Pl = plastid, P = P-protein, SP = sieve plate, ER = endoplasmic reticulum. x6000.

than the wilted plants with plugged pores. As it is generally agreed that callose deposition on sieve plates takes place in response to wounding, this could be further evidence that the unplugged situation represented the *in vivo* condition. Plugs of P-protein in sieve area pores may be further compacted by the formation of wound callose (Currier, 1957).

Precisely what are the effects of excess flow due to the release of hydrostatic pressure from the sieve elements? In attempting to answer this question some interesting observations were made by Anderson and Cronshaw (1969). Specimens had been cut from tobacco plants, releasing the hydrostatic pressure in the sieve element, and were processed for electron microscopy. When these specimens were examined near the original cut surfaces, dense plugs were found on the sieve plates, consisting of P-protein, sieve element reticulum, and starch grains from ruptured plastids. Aggregates of P-protein, and in some instances cisternae of endoplasmic reticulum, formed extended strands on the 'downstream' side of the sieve plate. Starch grains smaller in diameter than the sieve plate pores were able to pass through them; however, of considerable interest was the observation that starch grains larger than the pore diameter could block the pores, and these pores contained little or no P-protein. It can be argued that, in the rapid flow of content upon the cutting of a sieve tube, if a large starch grain lodges over a pore it can block it, preventing the flow of P-protein into the pore which so remains without a P-protein plug. On the other hand, Spanner and co-workers (Siddiqui and Spanner, 1970) have countered this argument by suggesting that the starch grains coming from the peripherally located plastids would arrive at the sieve plates after the P-protein from the central region of the lumen. Spanner goes on to suggest that perhaps the large starch grains are responsible for the removal of P-protein plugs originally present in the pores. The fact that large starch grains may block empty pores and have large amounts of P-protein on the 'upstream' side of them would seem to negate Spanner's argument. Also, in some cases, medium-sized starch grains occupy a pore together with densely compacted P-protein. These observations would thus again seem to suggest that sieve plate pores *in vivo* are unplugged.

Further evidence in favour of unplugged pores comes from observations of sieve elements which were differentiated in tissue culture. Sieve elements can be induced to differentiate in isolated nodules in tobacco callus (Wooding, 1969) or in tobacco pith cultures (Cronshaw and Anderson, 1971). One would expect that these sieve elements, which are not part of a continuous transport system of a plant, could be fixed without the artifacts associated with the release of excess hydrostatic pressure. It was found that differentiation of the cell wall into sieve areas and the differentiation of the protoplasmic contents of the sieve elements in tissue culture closely resemble those in intact plants. With regard to the sieve plate pores both the plugged and unplugged condition was found. Comparing these results with those obtained from tobacco plants fixed by similar procedures where the sieve plate pores are universally plugged, we can argue that the unplugged pores most probably represent the condition in the living cells. Most probably the induced sieve elements had generated some hydrostatic pressure as they matured, and the plugged pores in those cells where they occurred were a fixation artifact.

The results from any of the individual situations described, do not in themselves provide a definitive answer regarding the nature of the contents of the sieve plate pores. Taken together, however, they are compatible with the suggestion that the P-protein is arranged in a three-dimensional network in the lumen of the sieve element, the network extending from cell to cell through the sieve plate pores. Other cell components would occupy a peripheral position. Plugged pores would result from disturbance of the sieve tube either by cutting or chemical alteration of the membranes by fixatives. This type of arrangement of the P-protein and other components would be compatible with a mass flow type of mechanism for translocation through the sieve element conduits.

Recent results from several laboratories support the arrangement of P-protein in sieve elements and sieve plate pores just described. Johnson (1968) prepared freeze-etch replicas of *Nymphoides* phloem and observed that the sieve plate pores contained loosely arranged P-protein. Similar results were obtained using chemically fixed and sectioned preparations. Shih and Currier (1969) fixed intact cotton plants with acrolein-glutaraldehyde and observed unplugged pores in the sieve plates. Behnke (1971) fixed *Aristolochia* phloem in cold glutaraldehyde-acrolein and observed that P-protein filaments, tubular endoplasmic reticulum, and sometimes a filamentous component derived from plastid crystalloids extended through the sieve plate pores without plugging them. Thus, although the early observations of electron microscopists revealed plugged pores and seemed to be incompatible with the Münch mass flow hypothesis, recent observations and experiments have provided evidence that differentiated phloem sieve elements have a structure that would permit the rapid translocation of substances through them by a pressure flow mechanism.

Bibliography

Selected further reading

Crafts, A. S. and Crisp, C. E. (1971) *Phloem transport in plants,* Freeman, San Francisco.
Esau, K. (1969) 'The phloem', In: *Handbuch der Pflanzenanatomie,* Band V, Teil 2, Borntraeger, Berlin.
Northcote, D. H. and Wooding, F. B. P. (1968) 'The structure and function of phloem tissue', *Sci. Prog., Oxf.,* **56**, 35–58.
Weatherley, P. R. and Johnson, R. P. C. (1968) 'The form and function of the sieve tube: A problem in reconciliation', *Int. Rev. Cytol.,* **24**, 149–192.

References

Anderson, R. and Cronshaw, J. (1969) 'The effects of pressure release on the sieve plate pores of *Nicotiana'*, *J. Ultrastruct. Res.,* **29**, 50–59.
Anderson, R. and Cronshaw, J. (1970a) 'Sieve plate pores in tobacco and bean', *Planta,* **91**, 173–180.
Anderson, R. and Cronshaw, J. (1970b) 'Sieve element pores in *Nicotiana* pith culture', *J. Ultrastruct. Res.,* **32**, 458–471.
Behnke, H. D. (1968) 'Zum Aufban gitterartiger Membranstrukturen im Siebelementplasma von *Dioscorea'*, *Protoplasma,* **66**, 287–310.

Phloem differentiation and development

Behnke, H. D. (1969a) 'Die Siebrohren-Plastiden der Monokotyledonen', *Planta*, 84, 174–184.

Behnke, H. D. (1969b) 'Uber Siebrohren-Plastiden der Caryophyllales', *Planta*, 89, 275–283.

Behnke, H. D. (1971) 'The contents of the sieve-plate pores in *Aristolochia*', *J. Ultrastruct. Res.*, 36, 493–498.

Bouck, G. B. and Cronshaw, J. (1965) 'The fine structure of differentiating sieve tube elements', *J. Cell Biol.*, 25, 79–96.

Cronshaw, J. (1970a) 'The P-protein components of sieve elements', *Septième Congrès International de Microscopie Électronique, Grenoble, 1970*, 429–430.

Cronshaw, J. (1970b) 'Electron microscopy of phloem', Royal Microscopical Society Micro 70 Conference Programme – *Proc. R. microsc. Soc.*

Cronshaw, J. and Anderson, R. (1969) 'Sieve plate pores of *Nicotiana*', *J. Ultrastruct. Res.*, 27, 134–138.

Cronshaw, J. and Anderson, R. (1971) 'Phloem differentiation in tobacco pith culture , *J. Ultrastruct. Res.*, 34, 244–259.

Cronshaw, J. and Esau, K. (1967) 'Tubular and fibrillar components of mature and differentiating sieve elements', *J. Cell Biol.*, 34, 801–816.

Cronshaw, J. and Esau, K. (1968a) 'P-protein in the phloem of *Cucurbita*. I. Development of P-protein bodies', *J. Cell Biol.*, 38, 25–39.

Cronshaw, J. and Esau, K. (1968b) 'P-protein in the phloem of *Cucurbita*. II. The P-protein of mature sieve elements', *J. Cell Biol.*, 38, 292–303.

Currier, H. B. (1957) 'Callose substance in plant cells', *Am. J. Bot.*, 44, 478–488.

Esau, K. (1965) 'Fixation images of sieve element plastids in *Beta*', *Proc. natn. Acad. Sci., U.S.A.*, 54, 429–437.

Esau, K. (1967a) 'Minor veins in *Beta* leaves: structure related to function', *Proc. Am. phil. Soc*', 111, 219–233.

Esau, K. (1969) 'The phloem', In: *Handbuch der Pflanzenanatomie*, Band V, Teil 2, Gebrüder Borntraeger, Stuttgart.

Esau, K. (1971) 'Development of P-protein in sieve elements of *Mimosa pudica*', *Protoplasma*, 73, 225–238.

Esau, K. and Cheadle, V. I. (1965) 'Cytologic studies on phloem', *Univ. Calif. Publs. Bot.*, 36, 253–344.

Esau, K. and Cronshaw, J. (1967) 'Tubular components in cells of healthy and tobacco mosaic virus-infected *Nicotiana*', *Virology*, 33, 26–35.

Esau, K. and Cronshaw, J. (1968) 'Endoplasmic reticulum in the sieve element of *Cucurbita*', *J. Ultrastruct. Res.*, 23, 1–14.

Esau, K., Cronshaw, J., and Hoefert, L. L. (1967) 'Relation of beet yellows virus to the phloem and to movement in the sieve tube', *J. Cell Biol.*, 32, 71–87.

Esau, K. and Gill, R. H. (1970a) 'A spiny cell component in the sugar beet', *J. Ultrastruct. Res.*, 31, 444–455.

Esau, K. and Gill, R. H. (1970b) 'Observations on spiny vesicles and P-protein in *Nicotiana tabacum*', *Protoplasma*, 69, 373–388.

Esau, K. and Gill, R. H. (1971) 'Aggregation of endoplasmic reticulum and its relation to the nucleus in a differentiating sieve element', *J. Ultrastruct. Res.*, 34, 144–158.

Eschrich, W. (1971) 'Biochemistry and fine structure of phloem in relation to transport', *A. Rev. Pl. Physiol.*, 21, 193–214.

Evert, R. F., Murmanis, L., and Sachs, I. B. (1966) 'Another view of ultrastructure of *Cucurbita* phloem', *Ann. Bot.*, 30, 563–585.

Evert, R. F. and Deshpande, B. P. (1969) 'Electron microscope investigation of sieve-element ontogeny and structure in *Ulmus americana*', *Protoplasma*, 68, 403–432.

Gerola, F. M., Lombardo, G., and Cataro, A. (1969) 'Histological localization of citrus infectious variegation virus (CVV) in *Phaseolus vulgaris*', *Protoplasma*, 67, 319–326.

Gunning, B. E. S., Pate, J. S., and Briarty, L. (1968) 'Specialized "transfer cells" in minor veins of leaves and their possible significance in phloem translocation', *J. Cell Biol.*, 37, C7–C12.

Hepton, C. E. L., Preston, R. D., and Ripley, G. W. (1955) 'Electron microscope observations on the structure of the sieve plates in *Cucurbita*', *Nature, Lond.*, 176, 868–870.

Johnson, R. P. C. (1968) 'Microfilaments in pores between frozen-etched sieve elements', *Planta*, 81, 314–332.

Bibliography

Kollman, R. (1960) 'Untersuchungen über das Protoplasma der Siebröhren von *Passiflora coerulea* II Elektronenoptische Untersuchunger', *Planta*, 55, 67–107.

Kollman, R., Dorr, I., and Kleinig, H. (1970) 'Protein filaments-structural components of the phloem exudate', *Planta*, 95, 86–94.

LaFlèche, D. (1966) 'Ultrastructure et cytochimie des inclusions flagellées des cellules criblées de *Phaseolus vulgaris'*, *J. Microscopie*, 5, 493.

Ledbetter, M. C. and Porter, K. R. (1963) 'A "microtubule" in plant cell fine structure', *J. Cell Biol.*, 19, 239.

Newcomb, E. H. (1967) 'A spiny vesicle in slime-producing cells of the bean root', *J. Cell Biol.*, 35, C17–C22.

Northcote, D. H. and Wooding, F. B. P. (1966) 'Development of sieve tubes in *Acer pseudoplatanus'*, *Proc. R. Soc.*, B 163, 524.

Palevitz, B. A. and Newcomb, E. H. (1970) 'A study of sieve element starch using sequential enzymatic digestion and electron microscopy', *J. Cell Biol.*, 45, 383–398.

Parthasarathy, M. V. and Mühlethaler, K. (1969) 'Ultrastructure of protein tubules in differentiating sieve elements', *Cytobiologie*, 1, 17–36.

Pate, J. S. and Gunning, B. E. S. (1969) 'Vascular transfer cells in angiosperm leaves a taxonomic and morphological survey', *Protoplasma*, 68, 135–156.

Porter, K. R. (1966) 'Cytoplasmic microtubules and their functions', In: *Ciba Symp.*, pp. 308–345, Churchill, London.

Rouschal, E. (1941) 'Untersuchungen über die Protoplasmatik und Funktion der Siebröhren', *Flora, Jena*, 35, 135–200.

Schumacher, W. and Kollman, R. (1959) 'Zur Anatomie des Siebröbrenplasmas bei *Passiflora coerulea'*, *Ber. dt. bot. Ges.*, 72, 176–179.

Shih, C. Y. and Currier, H. B. (1969) 'Fine structure of phloem cells in relation to translocation in the cotton seedling', *Am. J. Bot.*, 56, 464–472.

Siddiqui, A. W. and Spanner, D. C. (1970) 'The state of the pores in functioning sieve plates', *Planta*, 91, 181–189.

Srivastava, L. M. and O'Brien, T. P. (1966) 'On the ultrastructure of cambium and its vascular derivatives. II. Secondary phloem of *Pinus strobus* L.', *Protoplasma*, 61, 277–293.

Steer, M. W. and Newcomb, E. H. (1969) 'Development and dispersal of P-protein in the phloem of *Coleus blumei* Benth', *J. Cell Sci.*, 4, 155–169.

Strasburger, E. (1891) *Ueber den Bau und die Verrichtungen der Leitungsbahnen in den Pflancen: Histologische Beitrage*, Heft III, Gustav Fischer, Jena.

Tamulevich, S. R. and Evert, R. F. (1966) 'Aspects of sieve element ultrastructure in *Primula obconica'*, *Planta*, 69, 319–337.

Wark, M. C. and Chambers, T. C. (1965) 'I. The sieve element ontogeny', *Aust. J. Bot.*, 13, 171–183.

Weatherley, P. E. (1967) 'The mechanism of sieve tube translocation: observation, experiment and theory', *Adv. Sci.*, 18, 571–577.

Weatherley, P. E. and Johnson, R. P. C. (1968) 'The form and function of the sieve tube: A problem in reconciliation', *Int. Rev. Cytol.*, 24, 149–192.

Wergin, W. P. and Newcomb, E. H. (1970) 'Formation and dispersal of crystalline P-protein in sieve elements of soybean (Glycine Max L.)', *Protoplasma*, 71, 365–388.

Wooding, F. B. P. (1969) 'P-protein and microtubular systems in *Nicotiana* callus phloem', *Planta*, 85, 284–298.

Zee, S. Y. (1968) 'Ontogeny of cambium and phloem in the epicotyl of *Pisum sativum'*, *Aust. J. Bot.*, 16, 419–426.

12

Primary vascular tissues
T. P. O'Brien

Introduction

With few exceptions, the formation of a new organ in any vascular plant is accompanied by the differentiation of a vascular system within it. In general, the first organs of a new plant are those of the embryo and, as more organs are added by the growth and differentiation of the products of the embryonic meristems, new vascular tissues differentiate within the new organs. Linkages are established between the new vascular tissue and that of the more mature organs, creating a continuous network of vascular tissue which ramifies throughout the whole of the plant body.

In the vascular cryptogams, and in many flowering plants, there comes a time when no new vascular tissue is formed within mature parts of the plant body. Such plants are said to exhibit *primary growth* and their vascular system consists solely of primary vascular tissues. In woody plants, the phase of primary growth, especially in the stem and root, is followed by the development of a lateral meristem, the vascular cambium. The growth of these cells is predominantly but not exclusively radial (Wardrop, 1964), and the derivatives of this meristem form the secondary plant body within which secondary vascular tissues differentiate. Though the formation and activity of the vascular cambium pose many interesting questions (see Philipson, Ward, and Butterfield, 1971; also chapter 10), I shall be concerned in this chapter almost exclusively with primary vascular tissues.

Primary vascular tissues are classified into two types — *primary xylem* and *primary phloem.* Each of these tissues in characterized by the presence of one highly differentiated cell type, the *tracheary elements* (*tracheids* and *vessel members*) of the xylem, and *sieve elements* (*sieve cells* and *sieve tube members*) of the phloem. These specialized elements are always accompanied by some parenchyma cells and, in many instances, supportive tissue elements (e.g., fibres and collenchyma cells) are also present. It has been proposed for more than a century that the major function of vascular tissue is the conduction of nutrients and water — the transpiration stream moving in the xylem, the translocation stream in the phloem. In many cases this primary function of conduction is clearly combined with a supportive role for the vascular tissues, demonstrated most elegantly in the familiar girder construction of the vascular bundles of grass leaves (see Haberlandt, 1914). These functions of conduction and support have been attributed generally to the more highly differentiated cell-types within the vascular tissue, the role of the accompanying parenchyma cells being something of a mystery. In the secondary vascular system, abundant deposits of starch, lipids, and tannins have led easily to an acceptance of these cells as depots for nutrients and/or waste products. However, such deposits are usually less abundant in the parenchyma cells of the primary body, and it is only in recent years that good evidence has been obtained that these cells may regulate the composition of the conducting streams (see, e.g., Carr, 1966; Pate and Wallace, 1964; Pate et al., 1965).

Patterns

Many of us first become aware of vascular tissues in plants by studying venation patterns in photosynthetic organs, and perhaps you can still remember the first occasion you picked up the fragile skeleton of a partially decomposed autumn leaf. Though a glance at a venation pattern will often suffice to identify a plant as a fern, monocotyledon or dicotyledon, and though the details of the pattern may be important aids in the identification of genera or species (as, e.g., in many ferns, *Acacia,* or *Eucalyptus*), it is sobering to realize that so little is known about the functional significance of these very different venation patterns. What *are* the essential functional differences between dichotomous, reticulate, and striate venation patterns?

Similar problems are posed by the extraordinary variations in vascular patterns encountered in stems and roots of different vascular plants (see Foster and Gifford, 1959; Sporne, 1962; Bierhorst, 1971). The *stelar theory,* developed especially by van Tieghem and Douliot (1866), was an attempt to produce a rational basis for interpreting these different vascular patterns. The subject is well discussed by Foster and Gifford (p. 50, *et seq.*) and Esau (1965a, p. 370, *et seq.*), and will not be treated in detail here. However, Esau's conclusions are important. On p. 371 she states:

Contributors to the physiological anatomy have made little or no use of the stelar concept . . . and in much of the later literature the stele is used as a convenient abbreviation for the vascular system.

On p. 377, she sets out, with characteristic insight, why this is so.

The segregation of the different tissues of the plant body varies in distinctness . . . the presence or absence of an anatomic delimitation of cortex, endodermis, pericycle, medullary rays, leaf gaps and pith constitutes a variation in the relative distribution of the vascular and ground tissues. On the one hand are the plant axes with an almost diagrammatic division into cortex, vascular cylinder, and pith (if present), and with distinctly differentiated endodermis and pericycle;

on the other, are the axes having no sharp boundary between the vascular and fundamental tissues and lacking a pericycle. In the extreme condition, the vascular system is dispersed to such an extent that no cortex or pith can be delimited (stems of many monocotyledons).

In short, the comparative anatomy of plant axes reveals numerous patterns of vascularization which, although they can be classified, are not well understood in functional terms. We do not yet understand the significance of different arrangements of vascular tissue. What is the significance in angiosperms of alternating strands of phloem and xylem in the root compared with their collateral or bicollateral arrangement in the stem? What is the significance of the polyarch condition in some herbaceous monocotyledonous roots compared with the diarch state found in most gymnosperms? Does a solenostele perform vascular functions that are different from those performed by a protostele or a dictyostele? What are the functions of an endodermis — how do we explain its universality in roots, its presence in the stems of most vascular cryptogams and its absence (with a few exceptions; O'Brien and Carr, 1970 and references therein) in the stems of most angiosperms? It is a sad reflection on the state of our knowledge that the answers we must give to these questions are almost entirely speculative.

The vascular pattern of any mature organ is the result of both a temporal and spatial sequence of cellular differentiation. The mature pattern does not emerge all at once but unfolds slowly as the organ develops. A median longitudinal section of an organ such as a root is a display in space of the sequence of cytodifferentiation through which the various cells produced in the meristem will pass in time. From a study of serial transverse sections of such an organ one can construct the sequence of vascular differentiation. Studies of this kind have been made on a variety of organs from different vascular plants (see Esau, 1965b) and permit one to draw some important conclusions.

Firstly, in most organs, future vascular tissue (termed *procambium* or *provascular tissue*) can usually be identified and distinguished from neighbouring cells quite early in the life of the organ (Golub and Wetmore, 1948; Sharman and Hitch, 1967; Hitch and Sharman, 1968, 1971). In the light microscope, this identification is based on cell shape (procambial cells are often longer and narrower than their neighbours), position, and 'more strongly stained cytoplasm'. It is easy to deduce that the long, narrow shape of procambial cells is the result of more frequent longitudinal and less frequent transverse divisions in the provascular strand, but the origin of the 'more strongly stained cytoplasm' is not clear. It is due in part to differences in the degree of vacuolation (Fig. 12.1) but it may also be due to real differences in stainable protein and RNA in the cytoplasm. I am unaware of any careful, comparative explorations of the histochemistry and fine structure of procambial and non-procambial cells.

Secondly, in the normal course of events, vascular cells differentiate only within the procambium (vascular cells that form in tissue culture or in wounded tissue may sometimes provide exceptions to this rule). Thus the pattern formed by the provascular tissue normally determines the pattern of the vascular system of the mature plant body. This point should be borne firmly in mind. Faced with

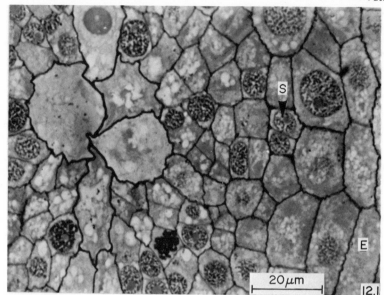

Fig. 12.1 Transverse section of the procambium in the root of the fern *Polystichum proliferum* (R. Br.) C. Presl. The root is diarch and its centre is occupied by the large cells of the future metaxylem. The endodermis (E) and first future sieve elements (S) can be identified. Toluidine blue staining.

the complexities of the vascular networks found in some organs (e.g., a maize cob, a wheat node, a palm stem, or a bracken rhizome) it is easy to forget that the pattern of the mature vascular network is simply the result of further cellular differentiation (producing the cell types of xylem and phloem) within the network already 'blocked out' partly or completely as procambium. Where a fusion exists between two or more mature strands, the fusion was made possible by the prior 'fusion' of differentiating strands of procambium.

Thirdly, given that the gross outlines of the mature vascular system of the primary plant body are determined by the pattern in which provascular tissue differentiates, the detailed structure of the mature vascular system depends upon the positions in which the elements of the xylem and phloem differentiate within the provascular tissue. It is at this level that we encounter many of the most fascinating differences in vascular patterns between the same organ in different species (e.g., diarch versus polyarch steles in roots) or among different organs of the same species (e.g., strands of xylem and phloem on alternate radii in roots versus collateral or bicollateral strands in stems).

Unfortunately, the relationship between the meristem (the procambium) and its derivatives is not as simple as it might appear from a consideration of these three points alone. In Esau's words (1965a, p. 377):

The differentiation of the primary vascular tissues from the procambium follows various developmental patterns. The maturation of the first vascular elements in the procambial strand or cylinder may occur while the procambium is still actively dividing, or it may happen after most of the divisions have been completed and the procambium clearly shows the outline and the internal pattern of the future vascular system (Wetmore, 1943). The former relation is commonly found in the aerial parts of the seed plants . . . (the latter) is characteristic of stems of many lower vascular plants and of most roots.

Again, on p. 381:

The differentiation of the vascular elements occurs simultaneously in more than one direction, both transversely and longitudinally and the various steps in tissue formation overlap at the same levels of the axis. Each developmental stage — the formation of procambium, the differentiation of phloem, and the differentiation of xylem — presents special aspects.

The root is the simplest case to understand. Procambium is evident quite close to the apex of the root (Fig. 12.2). About 200 μm further away from the apex, the first mature sieve elements of the protophloem make their appearance at the edge of the stele. At about this level, the rate of extension increases rapidly for about 200–300 μm and then falls to zero and is accompanied by the early stages of differentiation of the xylem and phloem elements. The first fully mature xylem elements (the tracheary elements of the protoxylem) also appear at the margins of the stele, but on radii alternate to those occupied by the protophloem sieve elements. Thus, in roots, the bulk of vascular differentiation (the maturation phase of metaxylem and metaphloem) is accompanied by very little extension.

One may summarize these facts for roots by stating that the differentiation of procambium, phloem and xylem are continuous and acropetal (towards the root apex), with maturation of the protophloem preceding that of the protoxylem. In the radial direction, differentiation is centripetal, metaxylem and metaphloem forming internally to the protoxylem and protophloem.

In the stem, matters are more variable and less clear cut. Esau (1965a, p. 383) states that:

Several conifers and dicotyledons with vascular tissues organized into leaf-trace systems had procambium differentiating acropetally and continuously from existing vascular tissue in the stem toward the apex and in most the procambium was identifiable beneath the youngest leaf primordia (Esau 1942; Lawalreé, 1948; McGahan, 1955).

In grasses, procambium also differentiates acropetally into developing leaf primordia but, as Hitch and Sharman point out (1968, p. 153):

In a vegetative shoot, all the procambial strands originate in the leaf primordia independently of the vascular system of the older parts of the plant and extend downwards from their first points of origin.

Thus, in contrast to the situation in some conifers and dicotyledons, the differentiation of procambium in the vegetative apex of grasses is discontinuous

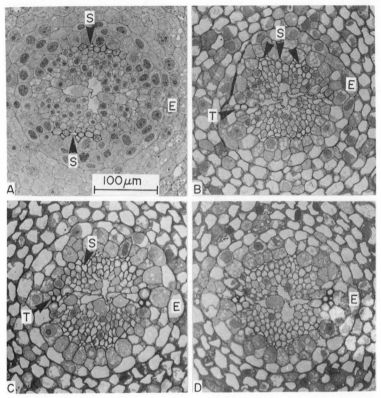

Fig. 12.2   Maturation of vascular elements in the stele of the root of the fern *Polystichum proliferum* (R. Bv.) C. Presl. (A) is at the level of the first, fully differentiated sieve elements (S) whereas at (B), about 12 sieve elements are differentiated, and lignified walls can be discerned on two differentiating tracheary elements (T). Note the extensive vacuolation of the endodermis (E) and cortical tissues that occurs in the levels between (A) and (B). In (C) the protoxylem tracheary elements have differentiated, while in (D) three mature tracheary elements are formed and the earliest sieve tubes have been partly obliterated. (A–D) are all at the same magnification, and stained by toluidine blue.

and bidirectional, procambial strands differentiating both acropetally and basipetally from their initial sites of origin in the disk of insertion of the leaf primordia. I suggest that the presence of these basipetally differentiating strands of procambium in the grass shoot is responsible in part for the extraordinary complexity of vascular connections in grass nodes (see Chrysler, 1906; Hitch and Sharman, 1968, 1971; O'Brien and Zee, 1971). As each new primordium produces its basipetally differentiating strands of various rank (mid-rib, first, second, third order laterals), these strands must either fuse with or pass between the sets of previously differentiated procambial strands from the three or four primordia that have preceded it (see Fig. 12.3), building as they go quite complex networks. It seems likely that the complexity of vascular connections described in some other monocotyledonous axes (e.g., palms; Tomlinson, 1970) can also be correlated with the presence of basipetally differentiating complexes of leaf-trace procambium.

In all cases that have been studied with sufficient precision and by adequate methods, maturation of protophloem sieve elements in stems and leaves appears to precede that of protoxylem tracheary elements by many hours, and in some cases by several days (Esau, 1965b; Jacobs, 1970; Swift and O'Brien, 1971).

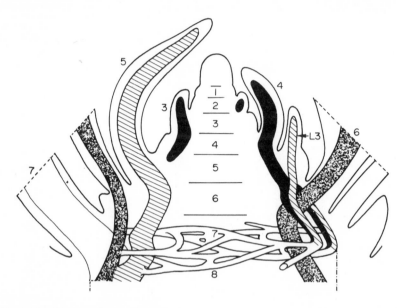

Fig. 12.3   Diagram of an ideal median longitudinal section, cut in the plane of the leaves, of the apex of a *Dactylis glomerata* L. shoot. It is reconstructed to scale from a series of transverse sections and shows the course and extent of the median procambial strands and the branches of the youngest nodal plexus linking these. The numbers 1–8 are the primordia and their corresponding discs of insertion; L3 is the third lateral strand of primordium 5. Note the basipetal course of differentiation of the median strands (cf. 3 and 4) and the 'obstruction' of the strand in 4 by the strand in 6. (Redrawn from Hitch and Sharman, 1968.)

The direction of maturation of these elements within the procambium is still open to some discussion. A considerable body of evidence supported the view (Esau, 1965b) that the maturation of protophloem sieve elements was continuous and acropetal, in contrast to the maturation of protoxylem tracheary elements which occurred in isolated loci related to leaf initiation, and spread bidirectionally from these loci upwards into the growing leaf and downwards into the stem. However, detailed studies of vascular differentiation in *Coleus*, employing around-the-clock sampling of tissues, revealed a second locus of tracheary element maturation in the stem (Jacobs and Morrow, 1957) and a second locus of sieve element maturation in the leaf (Jacobs and Morrow, 1967). Thus, the pattern of differentiation of the first sieve tubes in the *Coleus* leaf is remarkably similar to that of the tracheary elements. The first sieve element differentiates on the outer side of the procambial strand at a locus near the base of the leaf, a locus opposite the region where the first tracheary element will form in the same strand, albeit some days later (Jacobs, 1970).

Figure 12.4 summarizes the results of a recent detailed study of vascular differentiation in the wheat embryo (Swift and O'Brien, 1971). In this embryo, as in many others, only procambial strands are present in the mature seed, further differentiation of vascular elements accompanying germination. Such a system provides one with an ideal opportunity to study the course of xylem and phloem maturation in a normal plant that is entirely free of any influence that can be ascribed to pre-existing vascular tissue. The root system was not studied, but four separate loci of sieve element maturation were found (Fig. 12.4, 1–4). Locus 1 (in the scutellum) and locus 3 (in the mid-rib of the primary leaf) appear to be similar in their behaviour (bidirectional differentiation) to the isolated loci reported for *Coleus* leaf. Loci 2 and 4 are similar to the loci that differentiate within the *Coleus* stem with an acropetal sequence of differentiation.

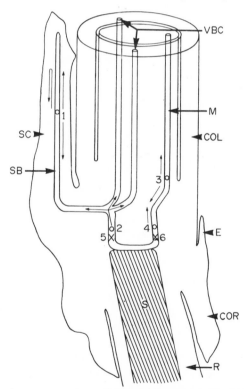

Fig. 12.4 Diagram of part of a wheat seedling, showing the six postulated loci of vascular differentiation. Each locus of sieve element differentiation is indicated by a circle, each locus of tracheary element differentiation by a cross. The arrows associated with loci 1–4 show the directions of sieve element maturation. The directions of tracheary element maturation from loci 5 and 6 are the same as those shown for sieve element maturation from loci 2 and 4 respectively. COL = coleoptile, COR = coleorhiza, E = epiblast, M = midrib bundle of first foliage leaf, R = radicle, S = stele of radicle, SB = central scutellar bundle, SC = scutellum, VBC = vascular bundles of the coleoptile. (from Swift and O'Brien, 1971.)

Curiously, no separate locus could be found in the coleoptile, which appears to differentiate both protophloem and protoxylem in an acropetal direction. However, we could find no evidence for four loci of tracheary element maturation, the simplest interpretation of our observations being that there were just two loci (Fig. 12.4, 5 and 6) from which differentiation is initially acropetal.

Many students react to this kind of discussion with an impatient 'so what? Does it really matter in what sequence the plant constructs its "plumbing" system?' While I am unaware of any evidence which demonstrates clearly that the precise sequence of vascular differentiation in a new organ is important, I consider it likely that this will prove to be the case when we come to know more about the functions of sieve elements (and especially about the very much neglected protophloem sieve elements; see Shah and Daniel, 1971) and what moves in them and in protoxylem tracheary elements. Most organs (some embryos may provide exceptions) must be very dependent for their nutrition during the early stages of vascularization upon other parts of the plant. This fact, so obvious for roots, is often forgotten in considering vascular differentiation within shoot apices and their products (leaves, stems, buds, floral organs). It is most unlikely that, at the time protoxylem and protophloem differentiate within any organ produced by a shoot apex, that organ is making any significant contribution to its dry matter increment. If sieve elements speed up the transport of sugars, amino acids, vitamins, hormones, and other metabolites needed for growth of young primordia, then it is certainly conceivable that it could make a great deal of difference to the growth rate of a new primordium if it is vascularized with sieve elements continuously and acropetally, rather than having to wait until basipetal differentiation links it with pre-existing sieve elements. Similar considerations apply to the maturation of tracheary elements if one takes into account the time that stomata differentiate within the epidermis of any foliar organ and the organ's sensitivity to moisture stress. I believe that some thought along these lines might convince the sceptical that the precise manner in which the 'plumbing' is assembled may have considerable physiological significance.

It should be clear from the preceding discussion that a complete understanding of the vascular pattern of any mature organ requires a detailed answer to three questions. What controls the differentiation of procambial cells? What controls the direction in which procambial strands differentiate and determines whether two discrete loci of procambial tissue will fuse or remain separate from one another? Finally, what controls the spatial and temporal sequences of maturation of vascular elements within a procambial strand? In my view we are a long way from having detailed answers to any of these questions.

The answer to the first question lies partly in the control of the plane of cytokinesis. Knowledge of the fine structure of division in a variety of cells is beginning to demonstrate the role of the microtubule and microtubule organizing-centres in the division process, but these studies have given no clue at all about the control of the plane of cytokinesis. It is possible to state categorically (see chapter 6) that the plane of cytokinesis is *not* determined by the presence or orientation of a spindle or by the presence or orientation of micro-tubules in the cytokinetic structure.

The orderly development of procambial strands in embryos, the maintenance of procambium development in root and shoot apices severed from contact with mature vascular elements, and the formation of cells with procambial character-istics in tissue cultures (Esau, 1965b; Cutter, 1969), all indicate that the induc-tion of procambial cells is *not* due to influences exerted by pre-existing mature vascular tissues. Furthermore, the factors that control the direction of differentia-tion of procambium and the fusion of procambial strands appear to be entirely unknown.

The answer to the third question remains as mysterious as ever. A considerable body of experimental evidence (see Wetmore, de Maggio, and Rier, 1964; Esau, 1965b; Cutter, 1969; Jacobs, 1970) has shown that auxin and sugar are critical variables in the induction and differentiation of vascular tissues in plants. However, while this work shows clearly that the amount of vascular tissue that is formed and its composition (xylem or phloem only, or both) are controlled by the concen-trations of auxin and sugar supplied, there is no explanation for the different positions occupied by tracheary elements and sieve elements in different organs. In particular there is as yet no indication of what controls the positions occupied by protophloem and protoxylem within procambial strands.

Differentiation of tracheary elements

In the primary xylem, maturation of tracheary elements produces pipes within an otherwise cellular tissue, pipes whose walls are fashioned from cell walls deposited in part by the differentiating tracheary elements. Where parenchyma cells abut the pipes, the cell walls of the parenchyma cells also contribute to the pipe wall. Wherever two tracheary elements abut one another, either at lateral walls or at a cross wall, the lumina of adjacent elements are interconnected at maturity. In tracheids, this luminal connection is always across a remnant of a pair of primary walls which are modified during maturation of the cells to produce a cellulose-rich residue, the 'pit membrane' (see Fig. 12.6). Where adjacent vessel members are in contact, some part of the luminal connection is absolute, all trace of the pre-existing pair of primary walls being destroyed. Thus, adjacent vessel members are interconnected, at least somewhere along their zone of contact, by a true pore across which no pit membrane extends. The areas that contain these pores are called *perforation plates* and the set of inter-connected vessel members is termed a vessel.

If one examines preparations of the primary xylem in internodes that have been stained to reveal the lignified walls of the tracheary elements, one can usually identify a number of different kinds of wall pattern (see Fig. 12.5, and Bierhorst, 1960; Bierhorst and Zamora, 1965), ranging from annular to helical, reticulate and pitted arrangements of the lignified wall. In general, annular or helical thickenings are found in the protoxylem, especially if the organ concerned undergoes considerable extension growth, whereas reticulate and pitted elements are more characteristic of situations where extension is minimal, e.g., xylem formed at nodes or in internodes after extension is completed. Since lignification of a cellulose-rich wall renders it brittle and inextensible, it has been recognized for many years that pipes constructed of annular and helical elements are ideal for maintaining their structure within a rapidly extending organ. Such elements are extensible because they can stretch between the bands of lignified thickening and will not collapse when stretched passively by the growth of the adjacent living cells unless the bands tilt. The

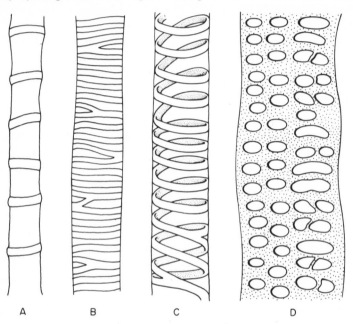

A   B   C   D

Fig. 12.5   Selected patterns of wall thickening in primary xylem tracheary elements. (A) Stretched annular; (B) reticulate; (C) helical; (D) pitted.

walls of pitted elements, once lignified, are inextensible, and there has been a considerable amount of discussion about the role of these elements in controlling organ growth-rate. It is possible that the differentiation of this kind of element within the xylem is a cause of the declining growth-rate of extending organs, but most workers believe that these cells are a response to the declining growth-rate (see Stafford, 1948). Treatments that initiate a reduced growth-rate in an extending organ certainly increase the frequency of pitted elements (Goodwin, 1942). The fact that reticulate elements form in tissue culture (Earle and Torrey, 1965) and in wound parenchyma (O'Brien and McCully, 1969; Hepler and Fosket, 1971) where maturation is not likely to be accompanied by a reduction in growth rate, leads me to believe that the pitted element is more likely to be a response to declining growth-rate than a cause of it. The suggestion by Bierhorst and Zamora (1965) that pitted elements are actually derived ontogenetically from helical elements by the deposition of additional wall material between the

424

gyres of the helical thickenings should be tested by a thorough ontogenetic study using thin sections for, if their ideas are substantiated, they would provide strong additional support for this view of the pitted element.

How are these bands of wall thickening deposited, and what is the fine structure of cells engaged in depositing such elaborate wall patterns? Early observations (Crüger, 1855; Sinnott and Block, 1945; see discussion in Cronshaw and Bouck, 1965) suggested that the cytoplasm developed a pattern which mirrored that of the wall thickenings prior to their deposition. Studies based on $KMnO_4$ and OsO_4 fixation failed to clarify the cytological basis of these early observations but, in all the cases where glutaraldehyde/OsO_4 fixation has been used (Hepler and Newcomb, 1964; Cronshaw, 1965; Cronshaw and Bouck, 1965; Esau, Cheadle, and Gill, 1966; Pickett-Heaps, 1966, 1967; Cronshaw, 1967; see also chapter 6), microtubules, lying above the developing wall thickenings and parallel to them, have been demonstrated in the cortical cytoplasm. It is well known (Roelofsen, 1959, p. 118) that cellulose microfibrils are oriented parallel to the axis of these bands of thickening and these cells provide the best evidence available for a correlation between microtubule orientation and microfibril orientation. Unfortunately, it is not easy to establish the significance of this correlation of orientations. In only one case is there a reasonably complete picture of the behaviour of the microtubules during the whole sequence of wall formation (Pickett-Heaps, 1966). Initially, microtubules are spaced out evenly along the lateral walls and oriented circumferentially around the cell. In later stages they group into bands of the same orientation (36 regularly spaced, consecutive groups were found in one cell) that alternate with cisternae of ER which lie very close to the cell membrane. The bands of thickening appear to develop beneath these groups of microtubules which remain associated with them throughout their formation.

Autoradiography of tissue fed glucose-6-^{3}H showed only weak incorporation into the wall up until the time the bands of thickening began to appear. From that time onwards, incorporation into the bands was marked. The experiment did not demonstrate conclusively whether label was incorporated into the wall between the bands or only into the bands, though some of Pickett-Heaps' illustrations suggest that the latter may be the case. A small percentage of dictyosomes was also labelled, but most of the cytoplasmic label appeared to be localized to the cisternae of ER that penetrated between the bands. It must be emphasized that those experiments did not demonstrate that any of the cytoplasmic label was in polysaccharides.

In a later paper, Pickett-Heaps (1968) stressed that there is an association between cytoplasmic 'vesicles' and the band of thickenings, an association evident from the time the microtubules first group into bands. Despite a careful study of his and other workers' micrographs, I do not find it easy to interpret the nature of these vesicles, nor am I entirely convinced that this association is always present (cf. Pickett-Heaps, 1966 with 1968). There is no doubt that dictyosomes are abundant in these cells and that sections of dictyosomes sometimes show an appearance similar to that of the vesicles near the wall (e.g., his Fig. 12, 1968), but I find it impossible to decide from published micrographs whether this appearance is due to sections of perforated dictyosome cisternae, dictyosome vesicles, or elements of tubular ER that lie close to the dictyosomes. In particular, I find that the 'vesicles' (his Fig. 11) which Pickett-Heaps (1968) suggests are fusing with one another and perhaps with the developing thickenings are very similar to sections of tubular ER cisternae reported in the cell cortex in a wide variety of plant cells (O'Brien, 1972).

This difficulty is just one example of a problem faced by all who wish to relate cytoplasmic fine structure to a particular aspect of cytodifferentiation. It is relatively easy to establish by electron microscopy that differentiating tracheary elements are rich in ER and dictyosomes (Porter, 1961; Hepler and Newcomb, 1963; Esau *et al.;* 1966), and that they contain a highly ordered distribution of microtubules (Pickett-Heaps, 1966). However, one cannot establish the structure of the fixed cell-surface without an analysis of serial sections or an examination of stereoscopic pairs, and these techniques have not been used to study differentiating tracheary elements. Furthermore, even if this were to be done, it has to be admitted that we know very little about the temporal stability of the associations between organelles and membrane systems visualized in thin sections. The relationship between the static image and the dynamic reality is at best uncertain: vesicles do not come armed with little arrows that tell us which way they were going at the time of fixation! Unfortunately, pulse-chase experiments with labelled metabolites, monitored by EM autoradiography, rarely lead to an unequivocal answer about the structural basis of a proposed or known biochemical pathway since the interpretation is usually complicated by uncertainties about the size of precursor pools, multiple sites of tracer incorporation *in vivo*, and the chemical nature of the labelled macromolecule demonstrated by autoradiography (O'Brien, 1972).

Perhaps the greatest difficulty that confronts the electron microscopist is establishing cause and effect between the structure observed and the functions being carried out by the cell. The very well-documented correlation (Newcomb, 1969) between orientation of wall microfibrils and cortical microtubules in differentiating primary xylem tracheary elements provides an excellent example of this dilemma. Pickett-Heaps (1967) has shown that, if colchicine is applied to these cells before the bands start to form, the treated cells lack microtubules and no bands form. However, if colchicine is added after the microtubules have grouped and the bands have started to develop, most of the microtubules disappear and the regular pattern of band development is destroyed. Thickenings form, but they are quite irregular in size and may be combined into giant bands. However, there is no evidence that the deposition of cellulose microfibrils within the bands becomes randomized, since microfibrils still appear to be deposited parallel to the axis of the irregular bands. On the other hand, Hepler and Fosket (1971), in a study of wound-vessel formation in colchicine-treated *Coleus* stems, report that colchicine treatment does destroy microfibril orientation within the irregular wall-thickenings. Thus, these experiments with this type of wall formation suggest that, although the microtubules play a part in determining the gross pattern in which deposition of wall microfibrils takes place, the orientation of the wall microfibrils may not always be determined by the microtubules.

One of the most striking features of the many published electron micrographs of differentiating tracheary elements in the primary xylem is the paucity of plasmodesmatal connections between adjacent tracheary elements or between tracheary elements and vascular parenchyma cells. There is no evidence that the gyres of thickening in these cells are deposited between areas that contain primary pit fields rich in plasmodesmata. One does encounter occasionally what might be termed a 'half-plasmodesma' extending from a parenchyma cell to the middle lamella region of an adjacent tracheary element. However, these are infrequent and encountered just as often beneath a gyre of thickening as between adjacent gyres (see Cronshaw and Bouck, 1965; O'Brien and Thimann, 1967). It seems likely that the procambial cells destined to form tracheary elements either have few plasmodesmata or else lose them during the early stages of cytodifferentiation. This point is certainly worth studying, since the comparative fine structure of procambial cells is so poorly understood. In view of the speed with which differentiating tracheary elements become labelled by glucose, leucine, phenylalanine, tyrosine, and cinnamic acid (Pickett-Heaps, 1966, 1968), and the entirely heterotrophic nature of these cells (I am unaware of any report of chloroplasts in differentiating tracheary elements), the paucity of plasmodesmata raises important questions about the significance of symplastic versus apoplastic pathways of nutrition for these cells. Viewed with this in mind, the gyres of thickening wall may actually serve an additional function, namely to increase the surface area of the cell membrane/apoplast boundary.

A careful ontogenetic study of pitted tracheary elements is an important task for the future. If Bierhorst and Zamora (1965) are correct in their view that these elements are derived ontogenetically from annular, helical, or reticulate elements, one would expect to find bands of microtubules in these cells during the early stages of their differentiation. It will be fascinating to discover what happens when the additional wall material is added between the gyres to produce the pitted element. Is this later deposition accompanied by an oriented set of microtubules? Are both sets present simultaneously? Does lignification of the gyres begin before deposition of the extra wall? Are the later wall-additions affected by colchicine? Since the differentiation of pitted elements can be regulated by controlling the growth-rate of the tissue, it may even be possible to reveal an ultrastructural reflection of this altered growth-rate in the cytoplasm of these cells.

To the best of my knowledge, primary xylem tracheary elements, unlike those of the secondary xylem which are lignified throughout the compound middle lamella and secondary wall, are lignified only in the gyres of thickening. This observation is based on the detection of strongly and metachromatically stained layers of primary wall beneath lignified gyres of thickening wherever two tracheary elements abut one another (see O'Brien and Thimann, 1967). In favourable circumstances, a similar layer of wall, whose staining reactions with cationic dyes are probably due to the presence of carboxyl groups in the wall, can also be detected beneath lignified gyres where they abut parenchyma cells, though this demonstration requires thin sections (0.5–2 μm thick; O'Brien and McCully, 1969, Fig. 108). Casual study of the xylem in members representative of all of the major groups of vascular plants suggests that this is a general phenomenon in the primary xylem, though some pitted elements may provide exceptions. This observation indicates that the cell controls very precisely just what part of the wall is to be lignified, and raises the question of how this is achieved.

Staining reactions of thin sections (O'Brien and McCully, 1969, Fig. 104), patterns of autofluorescence (Hepler et al., 1970), the distribution of electron opaque regions after $KMnO_4$ fixation, and autoradiographic experiments with tritiated cinnamic acid, a very specific precursor of lignin in the system under study (Pickett-Heaps, 1968), all suggest that lignification begins in the gyre just where it abuts the original primary wall of the procambial cell (the situation in pitted elements has not been studied). Lignification progresses steadily inwards as the gyre develops in thickness, though it lags somewhat behind the increase in gyre thickness. As the growth-rate of the gyres declines, the lignification is eventually completed.

The simplest explanations of this observation are that the enzyme complex responsible for lignification is added to the gyre in an active form shortly after a layer of gyre is synthesized or else it is added in an inactive form at the time the gyre layer is deposited and is activated shortly afterwards. In either case, there is a strong possibility that the enzyme complex is extracellular. What fraction of the total process of lignification can be carried out by this postulated extracellular complex is not clear but, as Pickett-Heaps points out, it is likely that all steps necessary to incorporate cinnamic acid into lignin can be carried out extracellularly. In the wheat coleoptile, he found that tritiated cinnamic acid was incorporated into lignin of tracheary elements that lacked cytoplasm, even when the tracer was supplied for as little as 15 minutes prior to fixation. Unfortunately, the attractive simplicity of this interpretation is marred somewhat by the fact that Pickett-Heaps, using EM autoradiography, detected significant amounts of cytoplasmic label apparently overlying the ER in cells that retained their protoplast during the period of cinnamic acid labelling. Extraction experiments showed that essentially all the label was in phenolic materials, and one must therefore conclude that a metabolite of cinnamic acid was present in the cytoplasm. However, this metabolite might not be associated with lignification of the wall. Alternatively, one might propose that the ER is the site of synthesis of the enzyme complex that carried out lignification (once the complex is incorporated into the wall), and that the cytoplasmic label is the result of the synthesis of a small amount of lignin by the nascent enzyme complex just prior to its discharge to the wall. Perhaps there is a way of testing these different possibilities?

The colchicine experiments add one additional interesting fact. Unfortunately, Pickett-Heaps (1967) does not discuss the matter but, as far as I can tell from an inspection of his micrographs, the deformed bands of thickening deposited in colchicine-treated cells *are lignified.* Certainly, those deposited in colchicine-treated wound xylem are lignified (Hepler and Fosket, 1971). This indicates that control of the distribution of the enzyme complex that carried out lignification is coupled to the processes that synthesize the polysaccharide framework of the gyres. In particular, it makes it very unlikely that microtubules play a role in 'directing' or 'shuttling' membrane-bound precursors of the active lignification complex to their appropriate sites in the wall. This point, and the orientation of the cellulose microfibrils in gyres deposited by colchicine-treated cells, deserve very careful study.

There is pressing need for a technique that will differentiate reliably between a lignified wall and an unlignified wall in an electron micrograph. The fact that the primary wall of the procambial cell remains unlignified while that of a cambial cell becomes lignified everywhere except at pit fields is a fascinating difference between the tracheary elements of the primary and secondary xylem. Hepler *et al.* (1970) have studied lignification in tracheary elements that differentiate from parenchyma cells in wounded *Coleus* internodes. They equate the electron opaque regions of the wall after $KMnO_4$ fixation with the lignified regions of the wall and demonstrate that, on this basis, the primary wall of the parenchyma cell which forms the tracheary element is lignified beneath the gyres of reticulate thickening. Hepler *et al.* (1970) recognize that their results differ from those obtained by studying other systems but their results serve to point out an important problem. Either the primary walls of these 'wound tracheary elements' are unlignified, making them similar to those of other primary tracheary elements that have been described, and the electron opacity in the primary wall after $KMnO_4$ fixation is due to something other than lignin, or they are lignified and behaving like secondary-xylem tracheary elements, a perfectly reasonable alternative since these cells differentiate their gyres of wall thickening without undergoing surface growth. Given that in other cells, starch, mucilage, and vacuolar deposits may all be electron opaque after $KMnO_4$ fixation (O'Brien, 1972), it is clear that high electron opacity after $KMnO_4$ fixation cannot be equated reliably with the presence of lignin.

I have stressed this question of the control of the pattern of lignification for, as we shall see presently, this pattern determines what areas of the pipe wall are destroyed by hydrolases during the death of the tracheary element and may also determine the passive permeability of the pipe wall to dissolved solutes.

The complete and rapid destruction of the protoplasts of tracheary elements at maturity is one of the most remarkable and constant aspects of their differentiation, an aspect which has received very little attention. We know almost nothing about the biochemistry of this process and even less about its control, and yet it is an absolutely essential step in the construction of a fluid-filled pipe within an otherwise cellular mass of tissue. Some would deny that there is any problem by regarding the destruction as the inevitable outcome of 'turnover of cytoplasmic constituents' should synthesis of new constituents stop. However, this patently evasive answer has never been an adequate explanation for the disappearance of

primary wall constituents at vessel perforation plates, for there is very little
evidence that cell wall polysaccharides 'turn over' at appreciable rates.

Although good, direct evidence to support the view is lacking, I believe that
protoplast destruction is brought about by the action of hydrolases whose activity
is stimulated in the protoplast during the closing stages of maturation of the
tracheary element. Destruction of the protoplast and its cell membrane lead
eventually to an exposure of the cell walls of the tracheary element to a 'soup'
which degrades all areas of cell wall that it contacts unless these areas are protected against the action of the enzymes it contains. O'Brien and Thimann (1967)
studying annular and reticulate elements in coleoptile xylem, identified two
classes of substance that rendered a wall resistant to degradation: lignin; and
an unidentified component of the middle lamella, probably pectic acid. They
found that the primary wall of the procambial cell was degraded to a meshwork
of fine fibrils (when viewed with the electron microscope) which they called a
'hydrolysed wall'. Such walls were birefringent and negative to the periodic
acid-Schiff's (PAS) reaction. Where two tracheary elements abutted one
another, the primary walls of both procambial cells were degraded to hydrolysed
walls except where they lay beneath the gyres of lignified thickening (Fig. 12.6).

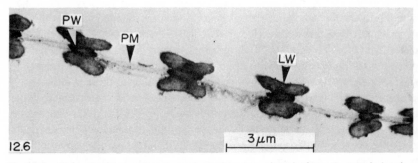

Fig. 12.6   The wall of two adjacent vessel members in a wheat leaf bundle, seen in longitudinal section. LW = gyres of lignified wall, PM = 'pit membrane', a hydrolysed remnant of
the primary walls (PW) that separated these cells before protoplast destruction. The primary
walls are still detectable beneath the lignified gyres.

Such hydrolysed walls between gyres of thickening correspond to 'pit membranes'
of tracheids. On the other hand, where the tracheary elements abutted parenchyma
cells, the procambial cell wall of the tracheary element was degraded to a
hydrolysed wall only between the gyres of thickening, hydrolysis stopping at
the middle lamella region of the adjacent parenchyma cell wall.

These observations were confirmed in the primary xylem of a variety of other
organs and were extended to the secondary xylem of willow by O'Brien (1970).
Adjacent vessel members of willow are connected by a large perforation and also
at numerous bordered pits. The 'pit membrane' of the bordered pit consists of a
hydrolysed wall identical in all respects to those encountered in the primary
xylem, whereas the large apertures of the perforation plates are devoid of any
trace of wall material. In secondary xylem tracheary elements, the regions of the

primary wall of the cambial cells that become the pit membranes and perforation plates are, of course, the only regions of the primary wall that remain unlignified.

It has been known for many years that the margo of the pit membrane in the bordered pits of conifer tracheids consists solely of cellulose microfibrils (see Bailey, 1957). O'Brien and Thimann (1967) suggested that this region was part of a hydrolysed wall and that the enrichment of the torus with pectic acid protected it from degradation. We had not, at that time, seen the excellent micrographs of Liese (1965) that demonstrate most clearly that this is indeed the case. In these pits, the margo and the compound middle lamella that lies beneath the torus thickening consist solely of networks of fine fibrils, identical in appearance to other hydrolysed walls. The non-cellulosic polysaccharides have been lost and the hydrolysed part of the wall contrasts sharply with the non-hydrolysed torus thickenings. This observation, coupled with the birefringence of the hydrolysed wall, suggests that it is rich in cellulose. At first sight, this appears to conflict with the PAS-negativity of these walls, but careful study of the PAS reaction in other walls that consist largely of cellulose showed that these are also PAS-negative (O'Brien and Thimann, 1967), despite the commonly held view that cellulose is PAS-positive (see Jensen, 1962, p. 190).

Sheldrake and Northcote (1968) have demonstrated that the guttation fluid of *Avena* coleoptiles is rich in a variety of sugars, including uronic acids, galactose, xylose, and arabinose. Although the presence of these sugars does not prove that they are products of wall hydrolysis, it is certainly consistent with such a view. The guttation fluid was also rich in amino acids and contained a trace of ribose. Acid phosphatase, ribonuclease, deoxyribonuclease and protease activity were also detected. Unfortunately, these results do not prove that the enzymes and nutrients are products of protoplast hydrolysis in the tracheary elements, since guttation fluid is likely to be enriched with materials secreted from the storage reserves in the endosperm, scutellum, and coleoptile. The authors looked for, but failed to detect, cellulase, xylanase, or mannanase activity. However, using a more sensitive assay, Sheldrake (1970) did detect cellulase activity in guttation fluid of barley, wheat, and oat coleoptiles.

It must be admitted that, at the present time, there are serious gaps in our understanding of the processes of protoplast and wall degradation in the tracheary element. The morphological evidence suggests that the protoplast is destroyed before the wall is degraded and that the non-cellulosic polysaccharides are removed from any area of wall that is not protected either by lignification, or by some as yet unidentified component of the middle lamella, probably pectic acid (O'Brien and Thimann, 1967; O'Brien, 1970; Zee and O'Brien, 1970). The hydrolases present in guttation fluid are consistent with the idea of an enzymatic destruction of the protoplast, but the only wall-degrading hydrolase that has been identified in the guttation fluid is cellulase. If the hydrolysed walls and pit membranes consist of cellulose, it is difficult to understand why they are not degraded completely by the cellulase. It is equally difficult to explain how cellulose-rich hydrolysed walls can persist in the pit membranes of bordered pits or along lateral walls in primary xylem vessels when such membranes have been removed completely from the perforation plates of the same vessels.

I believe that the cellulase activity in the tracheary element may never be sufficient to degrade the unprotected primary walls completely and that the total loss of the wall in the perforation plate may be partly mechanical. Characteristically, perforations occur when a large area of wall is unlignified. If the non-cellulosic polysaccharides are removed and the cellulose microfibrils partly degraded by enzyme attack, perhaps the flow of the transpiration stream is sufficient to rupture the relatively unsupported hydrolysed wall.

In all of the cases that have been studied in thin sections, lignification of a wall appears to render it completely resistant to degradation during maturation of the tracheary element. However, fungi are known that can destroy cellulose in lignified walls and Bierhorst and Zamora (1965) suggest that lignified thickenings can be degraded during the formation of perforation plates in some species. A careful study of the fine structure of such elements during their ontogeny should be most rewarding.

It is interesting to reflect upon the remarkable adaptiveness of wall hydrolysis in the case of the annularly thickened tracheary element. These cells are characteristic of tissues undergoing rapid elongation. To function as a pipe, essential in an organ undergoing rapid growth, these cells must lose their contents but at that moment they would also lose the capacity to plasticize the extension of their unlignified primary walls in the regions between the bands of thickening. In 1970, I summarized my view of this problem as follows:

What would happen in a rapidly elongating organ to the primary wall of a protoxylem tracheary element once its protoplast was destroyed? It is unlikely that the adjacent living cells could plasticise the wall of the dead tracheary element At best, the dead protoxylem would act to check the growth rate of the extending organ, and at worst the wall might rupture, perhaps tilting the rings of thickening and collapsing the all-important water conduit. But this problem is solved because the non-cellulosic polysaccharides are removed from the primary wall *between* the bands of thickening, releasing the cellulose microfibrils in those regions from interaction with the non-cellulosic substances. Thus, precisely at the time when the vital activity of the tracheary element is unable to plasticise the wall, the 'concrete is removed from the reinforcing steel', and the resistance of the wall to passive stretching must largely disappear. The lignified bands of thickening remain, anchored to similar bands in neighbouring cells or to walls of living vascular cells. These bands prevent collapse of the passively stretching element, giving time for the differentiation of a new file of tracheary elements.

Bierhorst and Zamora (1965) have suggested that all primary xylem tracheary elements are constructed initially as annular or reticulate elements and that pitted elements are derived ontogenetically from such elements by the addition of more wall material between the primary gyres. If one also accepts that the perforation plates of vessels have evolved from the pit membranes of tracheids by the development of wider spacings between the lignified walls where tracheary elements contact one another, it becomes clear that, to understand the construction of tracheary elements, we must understand what controls the pattern of the primary gyres, the secondary infilling, and the process of lignification. The remarkable adaptiveness of the annular element suggests that the procambial cell 'recognizes' in some manner that it is part of a rapidly extending organ and programmes the synthesis of an annular element accordingly. If Bierhorst and Zamora (1965) are correct, this programme can be 'corrected' if the extension rate of the organ drops. Additional lignified wall is added to create the structurally rigid pitted element which is much better adapted to resist the bending stresses found in mature aerial organs. The secondary xylem tracheary element shows structural adaptations similar to those of the pitted primary element.

What can be said in summary about the differentiation of primary xylem tracheary elements? As the procambial cell is stimulated to differentiate into a tracheary element, it initiates the synthesis of cell-wall material, rich in cellulose, in a pattern that is related to the conditions of growth of the organ of which the cell is a part. This pattern of cell wall deposition is reflected in the distribution of cytoplasmic microtubules that lie just beneath the cell membrane in the areas of cell-wall formation. If the population of microtubules is destroyed with colchicine before the bands of thickening are initiated, no bands form, but if the colchicine is not added until the bands are initiated, their deposition continues in an irregular manner. It seems likely, therefore, that the microtubules determine two things: the initial sites at which the specialized wall is deposited, and the orientation of the cellulose microfibrils in the specialized wall. Shortly after the bands are initiated, lignification begins and is carried out apparently by an extracellular enzyme complex whose distribution parallels the deposit of cellulose-rich wall. When the lignified bands have been constructed, the protoplast is destroyed, probably by autolysis, and about the same time, unlignified areas of wall lose their non-cellulosic polysaccharides. In tracheids, these degraded areas of unlignified wall persist as 'pit-membranes', a cellulose-rich remnant of wall material, but in vessel-members, some part of this wall residue is completely removed, creating the perforations of the perforation plate.

The reactions of neighbouring parenchyma cells

Since 1964, we have examined the xylem in a wide variety of organs embedded in glycol methacrylate, sectioned at $1-2$ μm (Feder and O'Brien 1968) and stained with the cationic dye toluidine blue O. In such sections, walls rich in acidic polyuronides (pectic acids) stain strongly and metachromatically (a deep reddish-purple), contrasting sharply with the green or aqua-blue shades of adjacent lignified walls (see O'Brien et al., 1964; Feder and O'Brien, 1968). To date we have found no exception to the rule that, wherever a parenchyma cell abuts a hydrolysed wall of a tracheary element, in either the primary or the secondary xylem, the wall of the parenchyma cell is strongly and metachromatically stained by toluidine blue O. Unfortunately, because the dye is metachromatic, the intensity and colour of staining cannot be used to measure the amount of acidic polysaccharides in the wall (see Kelly, 1956; Baker, 1958). Nonetheless, it is likely that the increased staining of such wall layers is due to their enrichment with acidic polyuronides or at least to a change in the structure of the wall that allows the acidic polyuronides to be stained more readily. These layers of wall that stain strongly with toluidine blue O are also strongly PAS-positive, but one cannot decide at present whether the PAS-positivity is due to an increased content of polyuronides or to an enrichment with other non-cellulosic polysaccharides. This could be checked by careful staining of sections from which the bulk of the acidic polyuronides had been extracted, leaving the other non-cellulosic polysaccharides relatively intact.

In 1967, O'Brien and Thimann encountered images in electron micrographs of parenchyma cells (their Figs. 33 and 40) that were consistent with the view that 'middle lamella material' was being added to the parenchyma cell wall, but only on the side where it abutted the hydrolysed wall of a tracheary element. This response is seen most dramatically where ray parenchyma cells abut secondary xylem elements (Schmid, 1965; O'Brien, 1970). In these cases, a complete layer of wall, rich in polyuronides and strongly stained both by toluidine blue O and the PAS reaction, is added to the wall of the ray cell. The deposition of this new layer of wall is greatest at the pits (which lack plasmodesmata) where the ray cell is in contact with the lumen of the tracheary element. The new wall may bulge into the lumen of the tracheary element and is seen to be continuous with similar wall material that has been deposited inside the lignified secondary wall of the ray cell. In willow, no trace of a hydrolysed wall can be detected where the wall of the parenchyma cell bulges into the lumen of the tracheary element and the electron micrographs suggest that the additional, protective wall-material is added fast enough to prevent hydrolysis of the primary wall of the tracheary element in that region (see Figs. 8 and 9, O'Brien, 1970).

Tyloses, outgrowths from ray parenchyma cells which may partly or completely occlude the lumina of old tracheary elements, have been recognized as a common feature of the secondary xylem for many years (see Esau, 1965). That they are very common in primary xylem, especially in the protoxylem, seems to be less widely appreciated (but see Calvin, 1967). Fig. 12.7 is a diagram of a small section of the protoxylem (a stretched annular element) from the mid-rib of a wheat leaf. Tyloses of this kind occurred 20 times in a distance of 2 cm of leaf length. Very little flow of the transpiration stream takes place through such an element! It is a remarkable fact that wherever tyloses occur in either primary or secondary xylem, the tyloses appear to develop by the growth of a primary wall that has been enriched with acidic polyuronides (see Foster, 1967; O'Brien, 1970). Is it possible that enrichment of a wall with acidic polyuronides somehow makes it easy for the cell to undergo intrusive growth into a cavity? Does the fact that the tips of pollen tubes growing *in vitro* (Dashek and Rosen, 1966) and the tips of root hairs growing *in vivo* are rich in acidic polyuronides offer support for this suggestion?

In 1967, O'Brien and Thimann pointed out that the fine structure of xylem parenchyma cells (abundance of rough ER and mitochondria) indicated that these cells might be important in accumulating solutes, both organic and inorganic, from the transpiration stream, and perhaps secreting other metabolites, such as vitamins and hormones, to it. It was demonstrated, subsequently, that such cells, in mature, non-growing internodes and leaves of pea, can accumulate a variety of amino acids when fed bleeding sap via the transpiration stream (Pate and O'Brien, 1968). Careful investigation of the fine structure of the vascular tissues of the pea leaf and internode led to the discovery of vascular transfer cells (Gunning *et al.*, 1968), cells with wall ingrowths of a variety of forms now described from many situations, non-vascular as well as vascular, where one expects high rates of solute flux over short distances. This subject is treated in detail in chapter 13, but I wish to emphasize one or two aspects here. In all of the transfer cells so far described, the wall thickenings which characterize them are unlignified, stain strongly and metachromatically with toluidine blue O and are strongly PAS-positive. In the electron microscope, such walls have an open, porous appearance, and it is obvious that this kind of wall is very similar to that which one finds

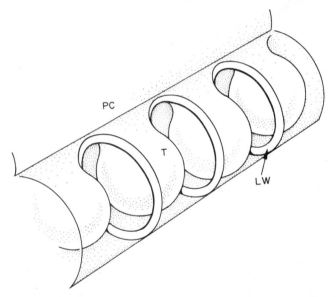

Fig. 12.7   Diagram of tyloses (T) that have protruded from a xylem parenchyma cell (PC) into the lumen of a stretched, annular vessel member in the protoxylem of a wheat leaf. LW = gyres of lignified wall of the vessel member.

wherever parenchyma cells abut the transpiration stream of tracheary elements. Indeed, for the particular case of xylem parenchyma cells, one can find examples which span a complete spectrum: smooth walls; smooth walls that bulge into the lumen of the tracheary element; walls covered with tiny asperities; walls with complex flanges; walls with labyrinthine ingrowths. In each case, the walls have identical staining reactions and very similar fine structure. Furthermore, the development of wall ingrowths in xylem parenchyma cells is confined always to those regions of the parenchyma cell that abut the *unlignified* wall of the tracheary element. This fact, coupled with the fact that wall ingrowths of transfer

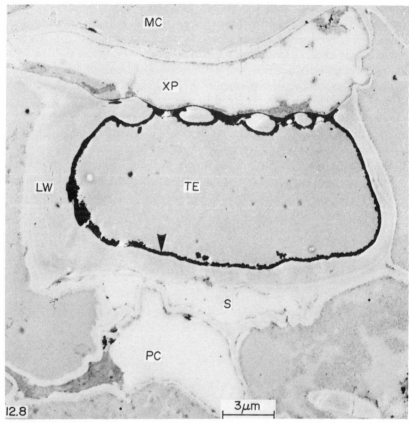

Fig. 12.8 Electron micrograph of a transverse vein from a wheat leaf fed 1% ferric chloride, and treated with 1% silver nitrate before dehydration in ethanol and embedding in epoxy resin. The vein consists in TS of a tracheary element (TE) and sieve element (S), with adjacent xylem (XP) and phloem (PC) parenchyma cells. A mesophyll cell (MC) abuts the xylem parenchyma cell. Note that the electron-opaque deposit (presumably derived from silver chloride) lines the lumen of the tracheary element and penetrates for a short distance into the wall of the parenchyma between the lignified bands. The lignified wall itself (LW) is unstained. Identical results were obtained when leaves fed ferric chloride were immersed in ammonium sulphide, suggesting that neither the ferric salt nor the chloride ions readily permeate through the lignified wall of the vessel. Note: the specimen has not been treated with OsO_4, or salts of uranium or lead.

cells are never lignified, suggests to me that solute uptake by xylem parenchyma cells is negligible except where non-lignified wall abuts the transpiration stream. I suggest that passage of solutes through lignified free space is very much slower than its passage through unlignified regions of the free space. If uptake (or secretion) of solutes by the cell membranes of xylem parenchyma cells is limited to those areas which do not underlie lignified thickenings, one begins to see the 'need' for the extra membrane/unlignified free-space contact created in xylem parenchyma transfer cells at sites of high solute flux.

Unfortunately, I am unaware of any direct evidence to support the view that lignified walls are relatively impermeable to solutes, although I suspect that a careful search of the literature on wood preservation might turn up such evidence. However, there are two pieces of evidence that suggest that the permeability of lignified walls to ions is different to that of the protective walls of xylem parenchyma cells and transfer cell wall-ingrowths. Colloidal lanthanum solutions do not penetrate into lignified walls (Gunning and Pate, 1969; Zee and O'Brien, 1970), but they do penetrate readily into transfer cell wall-thickenings. Similarly, we have been able to show that ferric chloride solutions will not penetrate into lignified walls but do penetrate into, and are retained by, xylem parenchyma cell walls where they abut tracheary elements (Fig. 12.8). Both of these observations are open to objections; the ionic species used are colloidal, the particle-size distribution is unknown, and one cannot be certain that absence of iron or lanthanum from the lignified wall is due to poor penetration rather than to selective loss during processing of the tissue for microscopy. Despite such objections, these two observations, taken at their face value, do support my contention that the lignified free-space is relatively impermeable, at least to positively charged ions. One can note in passing that, if lignified free-space is always relatively impermeable to solutes, the role of symplastic connections in the nutrition of heterotrophic sclerenchyma and lignified epidermal cell-types assumes a new significance.

Conclusions

Although I have not covered all the topics that a complete discussion of primary vascular tissues would require (e.g., maturation of the phloem has been omitted entirely since it is treated in chapter 11), I believe that the analysis has gone far enough for all to grasp what is known and what is urgently in need of further study. It is patently clear to me that we know almost nothing about the differentiation or the fine structure of the procambium, and yet it is at this stage of vascular development that the vascular network is created. We appear to be equally ignorant about what controls the sites within the procambial strands at which phloem and xylem elements mature. Although it is clear that auxin and sugar are involved in determining the composition of a vascular strand, we know nothing about what controls the position of mature vascular elements within the strands.

We know much about the structure of mature tracheary elements, but very little about the fine structure of different types of tracheary element during differentiation. It is clear that microtubules play some role in regulating the pattern of the wall but no one knows what they do. The sequence of annular to reticulate to pitted element is beautifully adaptive but we have not the slightest idea how the cell determines which kind of wall to deposit nor, in detail, how the cell deposits gyres of thickening, lignifies them, and then hydrolyses its protoplast and unprotected primary wall. And when it comes to stating the functions of xylem parenchyma cells with their strangely modified walls — why, it is only a few years since we have begun to recognize that these cells are something more than what is 'left over' in vascular bundles after all the 'interesting cells' have been classified.

Acknowledgements

I am grateful to Mr P. F. Lumley and Dr Katherine Esau for helpful comments on the manuscript, and to Dr J. Kuo, Mrs R. Clark, and Mrs C. Rosser for their assistance with the illustrations.

Bibliography

References

Bailey, I. W. (1957) 'Die Struktur der Tüpfelmembranen bei den Tracheiden der Koniferen', *Holz Roh-u Werkst.*, **15**, 210–213.

Baker, J. R. (1958) *Principles of biological microtechnique*, Methuen and Co. Ltd., London.

Bierhorst, D. W. (1960) 'Observations on tracheary elements', *Phytomorphology*, **10**, 249–30

Bierhorst, D..W. (1971) *Morphology of vascular plants*, The Macmillan Co., New York.

Bierhorst, D. W. and Zamora, P. M. (1965) 'Primary xylem elements and element associations of angiosperms', *Am. J. Bot.*, **52**, 657–710.

Calvin, C. L. (1967) 'The vascular tissues and development of sclerenchyma in the stem of the mistletoe *Phoradendron flavescens*', *Bot. Gaz.*, **128**, 35–59.

Carr, D. J. (1966) 'Metabolic and hormonal regulation of growth and development', In: *Trends in plant morphogenesis* (ed. E. G. Cutter), pp. 253–283, Longmans, Green and Co., London.

Chrysler, M. A. (1906) 'The nodes of grasses', *Bot. Gaz.*, **4**, 1–16.

Cronshaw, J. (1965) 'Cytoplasmic fine structure and cell development in differentiating xylem elements', In: *Cellular ultrastructure of woody plants* (ed. W. A. Côté, Jr), pp. 99–124, Syracuse Univ. Press, Syracuse, N.Y.

Cronshaw, J. (1967) 'Tracheid differentiation in tobacco pith cultures', *Planta*, **72**, 78–90.

Cronshaw, J. and Bouck, G. B. (1965) 'The fine structure of differentiating xylem elements', *J. Cell Biol.*, **24**, 415–431.

Crüger, H. (1855) 'Zur Entwicklungsgeschicte der Zellenwand', *Bot. Ztg.*, **13**, 601–613, 617–629.

Cutter, E. G. (1969) *Plant anatomy: experiment and interpretation*, Edward Arnold Ltd., London.

Dashek, W. V. and Rosen, W. G. (1966) 'Electron microscopical localization of chemical components in the growth zone of lily pollen tubes', *Protoplasma*, **61**, 192–204.

Earle, E. D. and Torrey, J. G. (1965) 'Morphogenesis in cell colonies grown from *Convolvulus* cell suspensions plated on synthetic media', *Am. J. Bot.*, **52**, 891–899.

Esau, K. (1942) 'Vascular differentiation in the vegetative shoot of *Linum*. I. The procambium', *Am. J. Bot.*, **29**, 738–747.

Esau, K. (1965a) *Plant anatomy* (2nd ed.), John Wiley and Sons, Inc., New York.

Esau, K. (1965b) *Vascular differentiation in plants*, Holt, Rinehart and Winston, New York.

Esau, K., Cheadle, V. I., and Gill, R. H. (1966) 'Cytology of differentiating tracheary element I. Organelles and membrane systems', *Am. J. Bot.*, **53**, 756–764.

Feder, N. and O'Brien, T. P. (1968) 'Plant microtechnique: some principles and new methods', *Am. J. Bot.*, **55**, 123–142.

Foster, A. S. and Gifford, E. M., Jr (1959) *Comparative morphology of vascular plants*, W. H. Freeman and Co., San Francisco.

Foster, R. C. (1967) 'Fine structure of tyloses in three species of the *Myrtaceae*', *Aust. J. Bot.*, **15**, 25–34.

Golub. S. J. and Wetmore, R. H. (1948) 'Studies of development in the vegetative shoot of *Equisetum arvense* L. II. The mature shoot', *Am. J. Bot.*, **35**, 767–781.

Goodwin, R. H. (1942) 'On the development of xylary elements in the first internode of *Avena* in dark and light', *Am. J. Bot.*, **29**, 818–828.

Gunning, B. E. S. and Pate, J. S. (1969) ' "Transfer cells" – plant cells with wall ingrowths, specialized in relation to short distance transport of solutes – their occurrence, structure, and development', *Protoplasma*, **68**, 107–133.

Gunning, B. E. S., Pate, J. S., and Briarty, L. G. (1968) 'Specialized "transfer cells" in minor veins of leaves and their possible significance in phloem translocation', *J. Cell Biol.*, **37**, C7–C12.

Haberlandt, G. (1914) *Physiological plant anatomy,* MacMillan and Co., London.

Hepler, P. K. and Fosket, D. E. (1971), The role of microtubules in vessel member differentiation in *Coleus. Protoplasma* **72**, 213–236.

Hepler, P. K. and Newcomb, E. H. (1963) 'The fine structure of young tracheary xylem elements arising by redifferentiation of parenchyma in wounded *Coleus* stem', *J. exp. Bot.*, **14**, 496–503.

Hepler, P. K. and Newcomb, E. H. (1964) 'Microtubules and fibrils in the cytoplasm of *Coleus* cells undergoing secondary wall deposition', *J. Cell Biol.*, **20**, 529–533.

Hepler, P. K., Fosket, D. E., and Newcomb, E. H. (1970) 'Lignification during secondary wall formation in *Coleus*: an electron microscopic study', *Am. J. Bot.*, **57**, 85–96.

Hitch, P. A. and Sharman, B. C. (1968) 'Initiation of procambial strands in leaf primordia of *Dactylis glomerata* L. as an example of a temperate herbage grass', *Ann. Bot.*, **32**, 153–164.

Hitch, P. A. and Sharman, B. C. (1971) 'The vascular pattern of festucoid grass axes, with particular reference to nodal plexi', *Bot. Gaz.*, **132**, 38–56.

Jacobs, W. P. (1970) 'Regeneration and differentiation of sieve tube elements', *Int. Rev. Cytol.*, **28**, 239–273.

Jacobs, W. P. and Morrow, I. B. (1957) 'A quantitative study of xylem development in the vegetative study of xylem development in the vegetative shoot apex of *Coleus*', *Am. J. Bot.*, **44**, 823–842.

Jacobs, W. P. and Morrow, I. B. (1967) 'A quantitative study of sieve-tube differentiation in vegetative shoot apices of *Coleus*', *Am. J. Bot.*, **54**, 425–431.

Jensen, W. A. (1962) *Botanical histochemistry,* W. H. Freeman and Co., San Francisco.

Kelly, J. W. (1956) 'The metachromatic reaction', *Protoplasmatologia*, II, D2. 98 pp.

Lawalreé, A. (1948) 'Histogénèse florale et végétative chez quelques Composées', *Cellule*, **52**, 215–294.

Liese, W. (1965) 'The fine structure of bordered pits in softwoods', In: *The ultrastructure of woody plants* (ed. W. A. Côté, Jr), pp. 271–290, Syracuse Univ. Press, Syracuse, N.Y.

McGahan, M. W. (1955) 'Vascular differentiation in the vegetative shoot of *Xanthium chinense*', *Am. J. Bot.*, **42**, 132–140.

Newcomb, E. H. (1969) 'Plant microtubules', *A. Rev. Pl. Physiol.*, **14**, 43–64.

O'Brien, T. P. (1970) 'Further observations on hydrolysis of the cell wall in the xylem', *Protoplasma*, **69**, 1–14.

O'Brien, T. P. (1972) 'The cytology of cell wall formation in some eukaryotic cells', *Bot. Rev.*, **38**, 87–118.

O'Brien, T. P. and Carr, D. J. (1970) 'A suberized layer in the cell walls of the bundle sheath of grasses', *Aust. J. biol. Sci.*, **23**, 275–287.

O'Brien, T. P., Feder, N., and McCully, M. E. (1964) 'Polychromatic staining of plant cell walls by toluidine blue O', *Protoplasma*, **59**, 367–373.

O'Brien, T. P. and McCully, M. E. (1969) *Plant structure and development: a pictorial and physiological approach,* The MacMillan Co., New York.

O'Brien, T. P. and Thimann, K. V. (1967) 'Observations on the fine structure of the oat coleoptile. III. Correlated light and electron microscopy of the vascular tissues', *Protoplasma*, **63**, 443–478.

O'Brien, T. P. and Zee, S.-Y. (1971) 'Vascular transfer cells in the vegetative nodes of wheat', *Aust. J. biol. Sci.*, **24**, 207–217.

Pate, J. S. and O'Brien, T. P. (1968) 'Microautoradiographic study of the incorporation of labelled amino acids into insoluble compounds of the shoot of a higher plant', *Planta*, **78**, 60–71.

Pate, J. S. and Wallace, W. (1964) 'Movement of assimilated nitrogen from the root system of the field pea (*Pisum arvense* L.)', *Ann. Bot.*, **28**, 83–99.

Primary vascular tissues

Pate, J. S., Walker, J., and Wallace, W. (1965) 'Nitrogen-containing compounds in the shoot system of *Pisum arvense* L. II. The significance of amino-acids and amides released from nodulated roots', *Ann. Bot.* **29**, 475–493.

Philipson, W. R., Ward, J. M., and Butterfield, B. G. (1971) *The vascular cambium,* Chapman and Hall Ltd., London.

Pickett-Heaps, J. D. (1966) 'Incorporation of radio-activity into wheat xylem walls', *Planta,* **71**, 1–14.

Pickett-Heaps, J. D. (1967) 'The effects of colchicine on the ultrastructure of dividing plant cells, xylem wall deposition and distribution of cytoplasmic microtubules', *Devl. Biol.,* **15**, 206–236.

Pickett-Heaps, J. D. (1968) 'Xylem wall deposition. Radioautographic investigations using lignin precursors', *Protoplasma,* **65**, 181–205.

Porter, K. R. (1961) 'The endoplasmic reticulum: some current interpretations of its form and functions', In: *Biological structure and function,* vol. I (eds. T. W. Goodwin and O. Lindberg), pp. 127–155, Academic Press Inc., New York.

Roelofsen, P. A. (1959) 'The plant cell wall', In: *Encyclopaedia of plant anatomy,* vol. III, pt. 4. Gebruder-Borntraeger, Berlin–Nikolassee.

Schmid, R. (1965) 'The fine structure of pits in hardwood', In: *Cellular ultrastructure of woody plants* (ed. W. A. Côté, Jr), Syracuse Univ. Press, Syracuse, N.Y.

Shah, J. J. and Daniel, P. (1971) 'Protophloem sieve elements in the adventitious root of *Pennisetum typhoides* S. and H.', *La Cellule,* **68**, 259–266.

Sharman, B. C. and Hitch, P. A. (1967) 'Initiation of procambial strands in leaf primordia of bread wheat, *Triticum aestivum* L.', *Ann. Bot.,* **31**, 229–243.

Sheldrake, A. R. (1970) 'Cellulase and cell differentiation in *Acer pseudoplatanus*', *Planta.* **95**, 167–178.

Sheldrake, A. R. and Northcote, D. H. (1968) 'Some constituents of xylem sap and their possible relationship to xylem differentiation', *J. exp. Bot.,* **19**, 681–689.

Sinnott, E. W. and Bloch, R. (1945) 'The cytoplasmic basis of intercellular patterns in vascular differentiation', *Am. J. Bot.,* **32**, 151–156.

Sporne, K. R. (1962) *The morphology of pteridophytes,* Hutchinson and Co. Ltd., London.

Stafford, H. A. (1948) 'Studies on the growth and xylary development of *Phleum pratense* seedlings in darkness and in light', *Am. J. Bot.,* **35**, 706–715.

Swift, J. G. and O'Brien, T. P. (1971) 'Vascular differentiation in the wheat embryo', *Aust. J. Bot.,* **19**, 63–91.

Tomlinson, P. B. (1970) 'Monocotyledons – towards an understanding of their morphology and anatomy', *Adv. Bot. Res.,* **3**, 208–290.

Van Tieghem, P. and Douliot, H. (1886) 'Sur la polystélie', *Annls Sci. Nat. (Bot),* Ser. 7. **3**, 275–286.

Wardrop, A. B. (1964) 'The structure and formation of the cell wall in xylem', In: *The formation of wood in forest trees* (ed. M. H. Zimmerman), pp. 87–134, Academic Press Inc., New York.

Wetmore, R. H. (1943) 'Leaf-stem relationships in the vascular plants', *Torreya,* **43**, 16–28.

Wetmore, R. H., de Maggio, A. E., and Rier, J. P. (1964) 'Contemporary outlook on the differentiation of vascular tissues', *Phytomorphology,* **14**, 203–217.

Zee, S.-Y. and O'Brien, T. P. (1970) 'A special type of tracheary element associated with "xylem discontinuity" in the floral axis of wheat', *Aust. J. biol. Sci.,* **23**, 783–791.

13

Transfer cells
Brian E. S. Gunning and John S. Pate

The transfer cell concept

Transport of solutes over short distances is one of many vital processes which
transcend taxonomic differences between organisms. The cells of both plants and
animals must pass dissolved substances between their protoplasts and the extracellular environment, whether external to the organism, or that internal 'free-
space', the *apoplast*. They also pass material from cell to cell, the interconnected
protoplasts together constituting the *symplast*. Apoplast and symplast are
botanical terms, but the concepts apply equally well to animals: in fact some of
the most striking examples of homology between the two Kingdoms are to be
found amongst the structural adaptations that relate to transport of solutes
within the symplast, and between symplast and apoplast.

The supreme arbiter of these transport processes is, of course, the plasma
membrane. In keeping with the ubiquity of its functions, it looks much the
same in all cells and organisms, and can develop equivalent local specializations
in plants and animals. Examples of short distance transport processes found in
plants, mediated by plasma membrane specializations, and familiar as processes
to zoologists, include: endocytosis of small particles (Mayo and Cocking, 1969a,
b), large particles (e.g., bacteria from infection threads in legume nodules, Dixon,
1969), and perhaps solutes (MacRobbie, 1970; Costerton and MacRobbie, 1970);
passage of solutes within the symplast *via* low resistance cell-to-cell junctions
(plasmodesmata, see, e.g., Tyree, 1970); and restriction of solute transport
within the apoplast by tight junctions (the Casparian strip, suberized lamellae,
etc., Ledbetter, 1968; Bonnett, 1968; Pate, Gunning, and Briarty, 1969;
Codaccioni, 1970; O'Brien and Carr, 1970). These homologies serve to emphasize
that a final example described below, and much more germane to our subject,
is far from being an isolated case.

The most spectacular of the permanent adaptations of the plasma membrane
of animal cells are the various plications which amplify its surface area. Most
zoologists would agree that the more intensively a cell engages in the absorption
or secretion of solutes, the more highly developed will be the invaginations or

outgrowths of its plasma membrane. The view that plant cells can develop an adaptation that is structurally and functionally equivalent to these is crucial to the transfer cell concept.

It is only rarely that the plasma membrane of the plant cell can form outward extensions (Rowley and Flynn, 1971). The cell wall usually precludes such configurations. However, cells in which invaginations or infoldings increase the surface area of the plasma membrane are now known to be common in plants. The membrane is usually moulded around inwardly directed projections of cell wall material, and these 'wall ingrowths' constitute an easily detected diagnostic feature for the cell type. We first applied the term transfer cell to certain phloem parenchyma cells in minor veins of leaves (Gunning, Pate, and Briarty, 1968) and later, having become impressed in the course of extensive surveys by the widespread occurrence of cells with wall ingrowths, we suggested that all cells so endowed should be called transfer cells (Gunning and Pate, 1969a).

In applying the term in this wide context, the implication we wished to convey was not that a totally 'new' type of cell had been discovered, but that many of the old, familiar cell types can have in common a specialization related to short-distance transport processes. Gland cells, epidermal cells, phloem parenchyma cells, xylem parenchyma cells, companion cells, pericycle cells, and others, can all produce a distinctive 'wall-membrane apparatus' consisting of wall ingrowths and amplified plasma membrane. Our argument was, and is, that this phenomenon is sufficiently common and important to make it convenient to have a term that cuts across the conventional classification of plant cell types. 'Transfer cell' seemed at the time to have appropriate connotations, and the choice has been to some extent justified by subsequent results from our own and other laboratories. There are situations, to be discussed later, where the word 'cell' is inappropriate. Clearly, no matter what anatomical situation is being considered, the feature deserving the greatest emphasis is the possession of a wall-membrane apparatus.

In the sections that follow we shall describe the wall-membrane apparatus of the transfer cell and explore possible relationships between its structure and its function, developing the theme that it represents a functional 'module' that can be brought into use in numerous situations in the plant, its versatility enabling it to perform a variety of tasks falling within the category of symplast – apoplast exchanges. Some of the many possible roles for the apparatus will then be analysed in terms of the physiology of the plant. Having established that the potential for the development of the apparatus is deep-rooted in the plant genome, we shall conclude by considering the morphogenetic factors that bring about the realization of this potential.

The wall-membrane apparatus

Cell wall ingrowths are not difficult to detect. Superb drawings may be found in the early literature, and anyone who wishes to see a wall labyrinth has only to prepare some hand-cut sections with a razor blade of, say, a *Tradescantia* node;

when stained with toluidine blue the massed wall ingrowths become visible even at low magnification. We have found optical microscopy of $1-2$ μm sections of material fixed in aldehydes and embedded in glycol methacrylate (Feder and O'Brien, 1968) to be an especially useful technique. The improvement in resolution as compared with thicker sections allows individual wall ingrowths to be seen, and it has consequently been feasible to undertake large scale surveys of different tissues and plants. Some examples of these preparations are shown in Figs. 13.1—13.4. Conventional transmission electron microscopy of ultra-thin sections adds further details of fine structure (e.g., Fig. 13.13), and high voltage electron microscopy provides especially dramatic images of sections that are in the same range of thickness as those used in optical microscopy, but are, of course, viewed with much greater resolution (e.g., Fig. 13.5). The third dimension can be brought into the picture either by making reconstructions from serial ultra-thin sections (Fig. 13.6, same material as Fig. 13.13) or by looking at cytoplasm-free preparations in the scanning electron microscope (Fig. 13.7, see also Briarty (1971) and Idle (1971)).

Structure and composition

The wall ingrowths are continuous with the secondary wall and may be regarded as a specialized form of it: the middle lamella does not penetrate into them. They consist of microfibrils that usually appear randomly arranged, and are probably embedded in a matrix. They can be stained using the periodic acid-Schiff reaction, with alcian blue, or with the polychromatic basic dye toluidine blue (Figs. 13.1— 13.4). Their metachromasy with the latter is intense at pH 4 or above, but diminishes in the region of pH 3 and only very weak staining is observed at pH values lower than this. Methylation with thionyl chloride in methanol (Stoward, 1967) or methanolic hydrochloric acid (Barka and Anderson, 1963) reduces their metachromasy, and subsequent saponification at least partially restores it, without abolishing the periodic acid-Schiffs reactivity. The metachromatic reaction is frequently found to be poor, but can usually be intensified if the sections are first rinsed for a few seconds in very dilute (0.01 N) hydrochloric acid and then in water.

The likely interpretation of these histochemical observations is that the wall ingrowths contain carbohydrates with carboxyl groups open to methylation and capable of becoming blocked under natural conditions by counter-ions that can be displaced with acid. It seems reasonable to ascribe these properties to the matrix in which the microfibrils are presumed to lie. An additional property of the matrix can be suggested from the electron-lucent space consistently visible between the ingrowths and the plasma membrane (e.g., Fig. 13.13). Fluorescence microscopy of specimens stained with aniline blue at high pH to reveal callose (obvious in sieve plates) indicates that this substance is not present in the space, which is more likely to be an artifact produced by shrinkage of the matrix during specimen processing. Tufts of fibrillar material are often seen bridging the space, again suggesting that the ingrowth and the plasma membrane were originally in intimate contact, but have become separated.

443

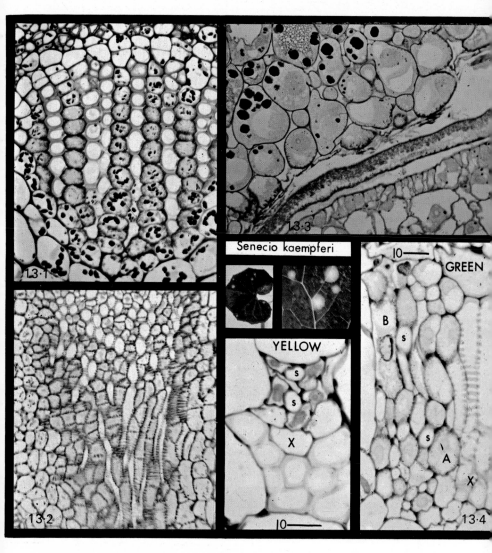

Fig. 13.1   Leaf trace from a node of *Trifolium repens,* showing transfer cells between the rows of xylem elements (lignified thickenings stained blue-green). The magnification (x400) is too low to resolve individual wall ingrowths, but every parenchyma cell that backs onto the xylem is seen to have a thick, fuzzy, purple-stained wall, which is in fact a dense carpet of wall ingrowths.

Fig. 13.2   Leaf trace from a node of *Trollius europaeus*. This is one example where without colour contrast it is easy to mistake wall ingrowths (purple) in transfer cells for lignified thickenings (blue-green) in xylem elements. The wall ingrowths are thick flanges—a form not commonly encountered outside the grasses. Magnification x170.

Fig. 13.3   Wall ingrowths in a developing *Vicia sepium* seed. A thin layer of endosperm lies between tissue of the young cotyledon (lower right) and the endothelium at the inner face of the inner integument. Wall ingrowths are seen in the endothelium and on both faces of the endosperm; this embryo was too young to have developed epidermal transfer cells on its cotyledons. Some crushed nucellar cells are included. Magnification x270.

Fig. 13.4   Minor vein transfer cells in variegated leaves of *Senecio kaempferi*. Minor veins in yellow parts of the leaf have few wall ingrowths in their phloem cells (lower left, s = sieve element, X = xylem). Veins in green parts of the leaf (right-hand side) have both A- and B-type transfer cells. Scale markers: 10 μm.

    Staining (Figs. 13.1—13.4): Periodic acid-Schiff reaction, counterstained with toluidine blue.

Certain cytoplasmic specializations accompany the wall-membrane apparatus. The most notable is the abundance of mitochondria, always provided with numerous cristae. Cisternae of endoplasmic reticulum can be so abundant as to be the most conspicuous element of the cytoplasm (Wooding and Northcote, 1965; Eymé and Suire, 1967; Dörr, 1968; Fahn and Rachmilevitz, 1970). Even when present in a less well-developed form they are of interest in that they frequently lie close to the involuted plasma membrane, a fact not readily appreciated from looking at single ultra-thin sections, but more apparent when sequences of adjacent sections are studied (Gunning, 1970). Whilst it is easy to envisage a role for mitochondria in a cell thought to engage in active transport of solutes across its plasma membrane, the function (if any) of the endoplasmic reticulum is more enigmatic. Mediation of intracellular transport is one possibility, especially the delivery of solutes to or from the plasma membrane.

Function

To go into precise details of transfer cell function, and particularly the nature of the transported solutes, would be to anticipate later considerations of the versatility of the wall-membrane apparatus. We deal here only with general aspects of function, applicable to all types of transfer cell.

The simplest hypothesis concerning the function of the transfer cell wall-membrane apparatus is that, *ceteris paribus,* the greater the surface area of a barrier — such as the plasma membrane — that is limiting transport, the greater will be the total flux across it. This would be so whether transport occurs by passive diffusion or by active processes. There are, as yet, no direct tests of this as far as transfer cells are concerned, and we look enviously at, for example, data concerning toad oocytes, where increases in the total permeability to water that take place during development are significantly and positively correlated with up to eleven-fold increases in surface area brought about by the formation of microvilli, the trans-membrane flux per unit area remaining approximately constant (Dick, Dick, and Bradbury, 1970).

The actual amplification factor by which wall ingrowths increase the surface area of the transfer cell plasma membrane (as compared with the same cell imagined to be devoid of ingrowths) depends on several factors. The frequency and geometry of the ingrowths are of obvious importance, and their distribution must also be taken into account. Quite commonly the ingrowths are restricted to a specific part of the cell periphery, that is, the cell is polarized. Amplification factors up to a ceiling of about 20 have been recorded when measurements are restricted to that part of the wall bearing the ingrowths (Gunning, Pate, and Green, 1970). Comparable data for cells with microvilli in the animal kingdom are 26 for mouse intestinal epithelium, and 129 for *Ascaris* intestine (Béguin, 1966). Twenty-fold amplification in a transfer cell is equivalent to a pattern of hexagonally close-packed microvilli, each 0.75 μm long, 0.1 μm in diameter, and with centre-to-centre spacing 0.12 μm (amplification factor 22.3): such a pattern is in no way unrealistic.

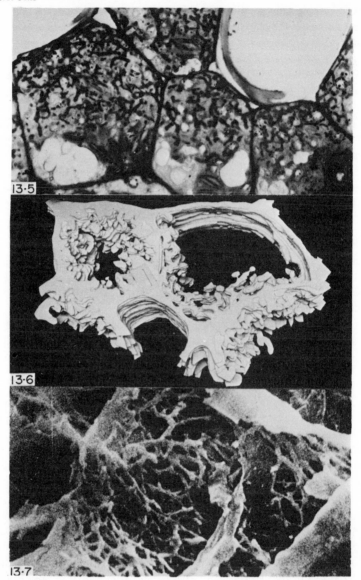

Fig. 13.5   High voltage electron micrograph of a 0.5 μm section of xylem parenchyma transfer cells in the cotyledonary node of *Lamium*. Labyrinths of wall ingrowths extend about two-thirds of the way across the cytoplasm of the transfer cells from the walls nearest the lumina of the tracheids (empty spaces, top left and right). Chloroplasts and vacuoles can be seen in the generally dense cytoplasm. Magnification x5000. (We thank Dr V. E. Cosslett for the use of his high voltage electron microscope.)

Fig. 13.6   Photograph of a model made at magnification x150 000 using cut out sheets of polystyrene to represent the disposition of the cell walls and wall ingrowths in each successive section of a sequence showing minor vein phloem transfer cells in a *Senecio vulgaris* leaf. On the left: A-type cell; on the right: B-type, lower centre: sieve element. Scale marker

446

It would be wrong to assume that surface area is the only factor to be considered. There is as yet no information on whether or not the plasma membrane has special transport properties in the vicinity of the ingrowths, and a similar paucity of facts exists concerning movement of substances within the wall ingrowths themselves. In fact, one powerful criticism of the transfer cell concept is that although wall ingrowths can be conceived of as being beneficial via their surface area effects, it can be argued just as forcibly that since they lengthen the diffusion path to or from a large fraction of the plasma membrane, they are likely to be more of a hindrance than a help. Added to this, all that is known about the dimensions of unstirred layers at the surface of plant cells (Dainty, 1969) supports the view that the ingrowths will be stagnant backwaters in which diffusion could well be rate limiting. A closer look at wall ingrowths as a milieu for transport phenomena is clearly indicated.

It is known that lanthanum hydroxide colloid can penetrate the matrix between the microfibrils of wall ingrowths in transfer cells (Gunning and Pate, 1969a; Zee and O'Brien, 1971a). The presence of open channels, larger in dimensions than any that may exist in lignified thickenings (into which the lanthanum probe does not penetrate) is indicated. Taken together with the histochemical evidence we can therefore picture the wall ingrowths as being penetrated by a network of channels, probably larger than 2 nm in diameter, and lined with fixed anionic charges. These charges will not only have an appreciable ion-exchange capacity, but will also create favourable conditions for electro-osmosis. Thus a potential difference existing between the cytoplasm and the surrounding apoplast could drive the coupled transport of solutes and water within the ingrowth itself. The data and calculations presented by Tyree (1969) indicate that electro-osmotic phenomena may be of considerable importance in transport in multicellular plants, and the idea is attractive in the present context in that it provides a mechanism for mass-flow within the wall ingrowths, thereby overcoming the hazards of the stagnant unstirred layers. Electro-osmosis is not, however, the only possible means of generating such a mass flow.

Mass-flow could be a consequence of the geometry of the wall-membrane apparatus, according to the 'standing gradient osmotic flow' hypothesis developed by animal physiologists to account not only for absorptive transport phenomena in epithelia of intestine, gall bladder and kidney (Diamond and Bossert, 1967), and secretory processes in salivary glands and Malpighian tubes, but also for

---

on top of model 1 μm. The wall labyrinths of the A-cell are relatively non-polarized; the B-cell is polarized, with ingrowths absent from the wall that faces outwards from the vein to the mesophyll. An ultra-thin section of the same leaf material is shown in Fig. 13.13. (Courtesy of I. Marks.)

Fig. 13.7   Scanning electron micrograph of part of a transversely sectioned vascular strand in a *Trifolium* root nodule, showing pericycle transfer cells with their wall labyrinths, sandwiched between smooth-walled conducting cells (upper left) and part of an endodermal cell (lower right). Magnification x5400. (Courtesy of Dr L. G. Briarty, reproduced with permission from *J. Microscopy*, **94**, 181, 1971).

coupled absorption and secretion at the basal and apical surfaces respectively of certain animal cells (Oschman and Berridge, 1971). In all of these cases the cell surfaces develop elaborate plications – microvilli, basal infoldings, and canaliculi. All have in common the possession of long, narrow, fluid-filled spaces bounded by the plasma membrane, open at one end and closed at the other. They all transport fluids that range from isotonic to a few times isotonic, and, as with transfer cells, there is considerable diversity in the nature of the solutes concerned. Having already drawn an analogy between the wall-membrane apparatus of transfer cells and surface specializations of animal cells, we now take it one stage further and apply the standing gradient hypothesis.

Absorptive situations (Fig. 13.8A). Solute pumps located in or on the plasma membrane withdraw osmotically active solutes from the lumen of the wall

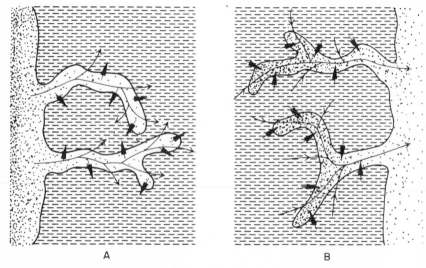

A   B

Fig. 13.8 Diagrams showing how the standing gradient osmotic flow hypothesis can be applied to the wall-membrane apparatus of transfer cells. (A) Absorptive situation, (B) Secretory situation.

 The solid line represents the plasma membrane, dashes the cytoplasm, and the density of the stippling is high or low where the concentration of solutes within the cell wall and the wall ingrowths is high or low respectively. Heavy arrows represent solute pumps, fine arrows the direction of water movement.

ingrowth (heavy arrows). This raises the water potential of the luminal fluid and lowers that of the cytoplasm, thereby establishing a gradient down which water will flow into the cell (fine arrows). As a consequence there are two mechanisms by which solutes enter the wall ingrowths from the surrounding apoplast or external environment (as the case may be): one is diffusion down their own concentration gradients, and the other is that they are swept in with the flow of water. In the steady state, active solute absorption maintains a standing osmotic gradient, with the highest water potential at the inner, closed end of the ingrowth. The continual intake of water and solutes must be balanced by their concomitant removal, and for this there are a number of possibilities. Water could be exported at other parts of the cell boundary where the potential gradient is suitable, solutes could enter intracellular storage compartments, and both solutes and water could leave the cell by 'trans-symplastic streaming' (Tyree, 1970) via plasmodesmata. It is a characteristic of transfer cells that they are connected to underlying cells by plasmodesmata, so that in absorptive situations they can be viewed as the gateway to whole tracts of the symplast, rather than as cells supplying their own individual needs.

Secretory situations (Fig. 13.8B). Solute pumps located in or on the plasma membrane pass osmotically active solutes from the cytoplasm into the lumen of the wall ingrowths (heavy arrows). This lowers the water potential of the luminal fluid and raises that of the cytoplasm, thereby establishing a gradient down which water will flow out of the cell (fine arrows). Solutes enter the apoplast or the external environment surrounding the cell by diffusion down their own concentration gradients and in the mass flow of water. In the steady state, active solute secretion maintains a standing osmotic gradient, with the lowest water potential at the inner, closed end of the ingrowth. The continual exit of water and solute must be balanced by their concomitant intake, either across the plasma membrane, from intracellular storage compartments, or via plasmo-desmata. The presence of plasmodesmata on what a zoologist would call the basal surface of secretory transfer cells allows them to be regarded as exit ports for their symplastic hinterland.

A precedent for applying the standing osmotic gradient hypothesis to plant material has been set by Anderson, Aikman, and Meiri (1970), who use it to account for the characteristics of exudation of fluid from cut roots, taking either the stele or the xylem as a long, narrow channel, closed at one end by the root meristem, and in which the standing gradient is established. Compared with this, and with many of the channels found in animal tissues, transfer cell wall ingrowths are short and hence less favourable for gradient formation. It is not often that they exceed a few μm in length (e.g., in *Anthoceros* vaginula up to 10 μm (Fig. 8 in Gunning and Pate, 1969a), in *Lamium* node xylem parenchyma up to 5–7 μm (Fig. 13.5)). Segel (1970), however, advises that the cross-sectional area–perimeter ratio of the channel is of great importance, and that if it is low enough, even quite short channels can be operative. He also shows that there is only a weak requirement for the solute pumps to be concentrated towards the closed end of the channels. Once again we must confess to having insufficient evidence on which to base sound judgements or even opinions on these matters. One approach that should, however, be feasible is to look for solute-dependent ATPase enzymes in the vicinity of the wall ingrowths, for judging by zoological parallels such as avian salt glands (Abel, 1969; Ernst and Philpott, 1970) and Malpighian tubules (Oschman and Berridge, 1971), their presence is to be predicted. The only fundamental difference between absorptive and secretory situations lies in the vectorial properties of the, as yet hypothetical, solute pumps.

Transfer cells in the physiology of the plant

We have so far explored the functional potentialities of the wall-membrane apparatus in general terms only. The fact that it occurs in many different anatomical situations has, however, been mentioned, and it is logical to proceed now to examine its diverse roles in the life of the plant. We wish to avoid needless repetition of a recent review of the subject (Gunning and Pate, 1969a), and propose to focus attention on just two families of transfer cells – those in reproductive structures, and those in the vascular system of the plant. These are selected partly because a large amount of new information has accrued, and partly because of the contrasts they offer in terms of function. The first type is likely to be involved in local transport processes, whereas the second, being associated with the long-distance conducting channels of the plant, may influence or be influenced by, processes occurring at a considerable distance, and thus require that aspects of the physiology of the whole plant be taken into account in considerations of their function.

Since the amount of direct physiological information that can be called upon in assessing transfer cell function is limited, it is necessary to fall back on the classical procedures of physiological plant anatomy and attribute functional significance to structural features such as the location of transfer cells in the organ or tissue, and the position of the wall-membrane apparatus with respect to adjacent cells. Knowledge of the time at which the wall ingrowths appear can also be useful. The case histories which follow are therefore compiled largely by weaving information on spatial organization and temporal relationships into a pattern expressed in physiological terms.

Transfer cells and reproduction

From the most primitive archegoniate plants to the modern flowering plants it is basic to the organization of the reproductive cycle for one nuclear generation to be borne on, and to be more or less dependent for its nutrition on, a parent generation of differing nuclear phase. Since cytoplasmic contact is almost invariably lacking between such generations (see Israel and Sagawa, 1964; Schulz and Jensen, 1971), all materials exchanged between the partners must pass across the intervening apoplast, and it seems logical to ascribe complementary physiological attributes to the donor and receptor cells adjoining this boundary. At the same time constraints of one sort or another may operate against the efficient transfer of materials from the host to the dependent generation. The more obvious cases of this are displayed in the mosses and ferns where contact is limited to a relatively small and confined placental region and, as the sporophyte grows, and its requirements for water and solutes increase, an ever decreasing proportion of its bulk and surface consists of absorptive tissues. In the higher plants, also, physical restriction to exchange may be imposed through the development of some sort of impervious envelope, be it cuticle, layer of suberized cells, or air gap, shrouding all but a part of the boundary between the generations. In these situations the new generation is totally dependent on the

450

tissues of its parent and, being inactive in transpiration, there is no possibility, as there is in mosses and ferns, of a sympathetic flow of water bringing in solutes.

Having this general background in mind, it is not surprising to find that many of the various types of 'placental' junctions found in the reproductive organs of plants exhibit highly specialized features and frequently possess structural equipment that enhances the area of membrane surfaces between one generation and another. The elaborate cellular interdigitations of the placenta of certain liverworts and ferns (Gunning and Pate, 1969a, b), the elongate suspensors of gymnosperms and the haustorial endospermic and antipodal extensions from the embryo sacs of flowering plants (see Wardlaw, 1955; Davis, 1966) bear witness to such compensatory influences at the macroscopic level, while, in these selfsame situations and elsewhere, the occurrence of the wall-membrane apparatus characteristic of transfer cells, is an equivalent, if not more effective, means of amplification of contact area operating at the ultrastructural level.

It is the purpose of the sections that follow to chronicle the discovery, description, and function of the wall-membrane apparatus known now to occur in the reproductive apparatus of the major groups of land plants.

The placenta of bryophytes and ferns. This region, where the haustorial foot of the sporophyte is embedded in nutritive tissues of its parent gametophyte, has double significance in the transfer cell story, for it was here that some of the first descriptions were made of what we now recognize as transfer cells, and here, also, that we might seek, in evolutionary terms, the birthplace of the cell type.

The earlier anatomists, among them Bower (1908), Goebel (1900) and Haberlandt (1914), were quick to appreciate the value, in terms of exposure of absorptive surface, of the globose, anchor shaped, or digitate sporophytic feet of many mosses and liverworts, but the discovery of the more subtle microscopic features of this region was not made until some years later when Lorch (1925, 1931) recorded in detailed drawings the deposits of wall material in the tangential outer walls of the outermost cells of the foot of many moss sporophytes and the opposed gametophyte cells of the vaginula. He concluded that the deposits were of a cellulosic nature and were penetrated by fine protoplasmic strands, both findings later fully confirmed by Blaikley (1933). The last author deduced that the cytoplasmic filaments within the wall thickening might have an important function in the efficient absorption of water by the sporophyte from the vaginula of the gametophyte, and in a most elegant series of illustrations she depicted the dense cytoplasm of the placental cells and their wall ingrowths.

The first electron micrographs of the wall labyrinths in the foot of mosses were published by Maier (1967), working with *Polytrichum,* and Eymé and Suire (1967), studying *Mnium.* Since then positive identifications of placental transfer cells by optical and electron microscopy have been extended to hornworts and liverworts (e.g., *Anthoceros* (Gunning and Pate, 1969a), *Phaeoceros* (F. Burr, personal communication), *Sphaerocarpos* (Kelley, 1969) and to many other mosses (K. Maier, personal communication). In the ferns placental transfer cells were described in *Adiantum* and *Polypodium* by Gunning and Pate (1969b) and, more recently, we have found them in a horsetail (*Equisetum*) and a club moss (*Lycopodium*). Negative findings are reported for a species of *Sphagnum* (K. Maier, personal communication) and for *Pellia* and *Selaginella* (present authors unpublished), although in the latter two instances only one age of sporophyte was examined, possibly too young to carry transfer cells. It already looks, then, as if transfer cells are very common on both sides of the junction between the two generations (see Fig. 13.9A) and their combined effect may be interpreted as a coupled 'push-pull' system for the solute transfer. Indeed, Bopp and Weniger (1971) recently suggest that a system of this kind operates in respect of water transport across the placenta of *Funaria*. Observing that a fluid collects in the vaginula of the gametophyte after removal of the sporophyte, they conclude that the wall labyrinth of the gametophyte secretes a solution and that the correspondi wall labyrinth lining the sporophyte foot is specially adapted for the efficient absorption of this solution from the vaginula. Evidence from other sources would indicate that the gametophyte may not only provide water, but also certain organic solutes to its parasitic partner. For instance, Paolillo and Bazzaz (1968), studying carbon assimilation of isolated sporophytes of *Polytrichum* and *Funaria,* showed that neither appeared to be self-sufficient through its own photosynthesis, and, in fact, in *Polytrichum* they failed to observe assimilation beyond the compensation point. If this proves to be general, the sporophyte must be markedly dependent on its host for photosynthate: presumably some form of carbohydrate would be received from the gametophyte, although the studies on soluble nitrogen and amino acid levels in aging sporophytes and gametophytes of certain mosses by Guyomarc'h (1969) point to organic compounds of nitrogen as also being involved in transport to the sporophyte. Critical studies of translocation using labelled tracers are obviously called for. It might be thought that the nutritional relationship between the two partners would be specific, but Arnaudow (1925) showed that young sporophytes removed from their parent vaginula and inserted into the vaginula of an unrelated species survived and grew.

Sporophyte-gametophyte junctions of gymnosperms. Knowledge is almost

---

Fig. 13.9   Diagrams showing where the wall-membrane apparatus can occur in reproductive cells and tissues.

(A) Bryophyte sporophyte (S) = gametophyte, (G) = junction.

(B) Angiosperm embryo sac with synergids (Sy) and antipodals (An).

(C) Young embryo (Em) with endosperm (Es, stippled), suspensor (Su), and nucellus (N).

(D) Old embryo (Em) of, e.g., *Vicia,* showing cotyledon epidermis (C.Ep), endosperm (Es, stippled), endothelium (End) at the inner face of the integuments (Is), and small suspensor (Su). For the sake of clarity the wall ingrowths on the inner and outer faces of the endosperm (see C) have been omitted. See also Fig. 13.3.

(E) Old embryo (Em) of, e.g., *Phaseolus,* with wall ingrowths restricted to the micropylar part of the endosperm, and to the large suspensor (Su). Other lettering as in D.

(F) Grass seed, showing aleurone transfer cells (ATC) at that part of the aleurone boundary layer (Al) of the endosperm (Es, stippled) overlaying the seed stalk at the hilar pad. Apart from this region the embryo (Em) and endosperm are surrounded by a cuticle (Cu).

(G) Seed with perisperm (Per), e.g., *Mesembryanthemum.* Other lettering as in D.

(Fig. 13.9 reproduced with permission from Pate and Gunning, 'Transfer cells', *A. Rev. Pl. Physiol.,* **23,** 1972).

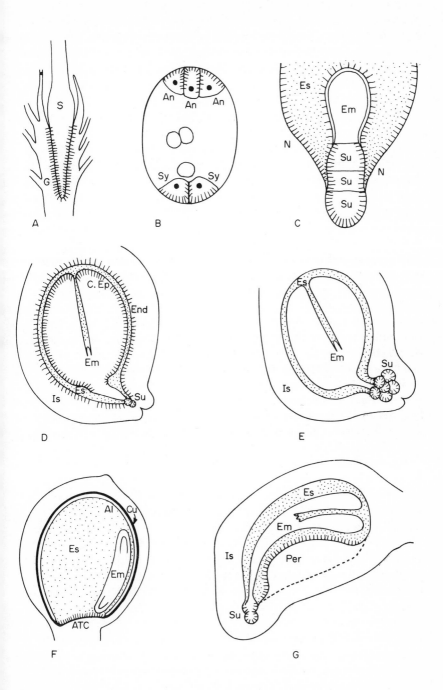

A

B

C

D

E

F

G

totally lacking in this direction. Our own observations of the sporophyte-megagametophyte interface in *Larix, Pinus, Sequoia,* and *Picea* have failed to find transfer cells, and the only structure we know that resembles a wall-membrane apparatus occurs on the inner wall of the megagametophyte (female prothallus) of *Cryptomeria,* where diminutive wall projections are just visible by optical microscopy. Electron microscopy will be required to investigate this situation further.

Junctions between generations in the angiosperm ovule. The mature embryo sac of the flowering plant, a small few-celled megagametophyte, provides several instances where wall labyrinths have been reported. Most common of these is the 'filiform apparatus' of the synergids, consisting of a complex of ingrowths extending inwards from the outer walls of these cells (see Fig. 13.9B). The strategic placing of this apparatus against the micropylar end of the embryo sac, and its very widespread occurrence (see Van der Pluijm, 1964; Van Went and Liskens, 1967), has led authors such as Jensen (1965) to propose that it has a generalized function in facilitating transfer from sporophytic tissues to the embryo sac, and that it might also direct and possibly nourish the entering pollen tube. The apparatus and its parent synergids rarely persist for long after fertilization, consistent within a transient and specific function in the events of reproduction.

At the other end of the embryo sac, equivalent wall specializations have been encountered in the antipodal cells (see Fig. 13.9B) in at least two genera (*Zea* (Diboll and Larsen, 1966; Diboll, 1968) and *Linum* (Vazart, 1968, 1969)), and these antipodal transfer cells appear to continue functioning, and may even enlarge and divide after fertilization has taken place. Consequently an extended nutritional role in seed development is to be suspected. Other species are known (e.g., *Piper* and *Gramineae*) where a massive proliferation of antipodal tissue occurs after fertilization, and these may provide further examples of transfer cells.

The act of fertilization, and presumably also the later growth of the embryo, prompts a series of developmental changes in the various protective, nutritive, and storage tissues surrounding the young sporophyte. Earliest and most spectacular among these is the enormous enlargement of the embryo sac wall, usually at the expense of part or all of the nucellus, and creating the space in which the embryo and endosperm are accommodated. Even at an early stage when the contained endosperm is still in a free nuclear condition, inwardly projecting wall papillae may form on the boundary wall of the endosperm. Certain regions of this boundary wall seem to be particularly well endowed with ingrowths, and elaborate labyrinths may develop as enlargement and differentiation of the endosperm takes place. Commonly in legumes (e.g., *Pisum* (Marinos, 1970), *Vicia, Phaseolus, Spartium,* and *Lathyrus* (Pate and Gunning, unpublished)), a continuous investment of ingrowths clothes the micropylar region of the outer endosperm wall, and this investment continues on the inner boundary of the endosperm where it adjoins the suspensor cells and the radicle region of the embryo (see Figs. 13.3, 13.9C). Marinos (1970) has commented in detail on these opposing zones of wall ingrowths on the endosperm faces of *Pisum,* and notes that, in the region of the radicle, longer strands of wall material may connect across from the outer to the inner of these boundary walls of the endosperm, anchoring the lower end of the embryo and its suspensor in the micropylar end of the embryo sac. Outside the legumes other examples known to us of the wall membrane apparatus being associated with endosperm are those provided by Vazart and Vazart (1966) for *Linum,* by Diboll (1968) for *Zea* and by Schulz and Jensen (1971) for *Capsella.*

Virtually all embryos, of course, develop a suspensor, this varying greatly from species to species in size, complexity and longevity. Since it can extend into sporophytic tissues beyond the micropylar end of the embryo sac, the suspensor is regarded by many authors as a haustorial structure, and in some species it is presumed to function in this fashion until relatively late in the life of the seed. Zones of wall ingrowths, largely on the outer faces of the suspensor, have been reported, either for the basal cell alone (e.g., in *Capsella,* Schulz and Jensen, 1969), or for all of the suspensor cells (Figs. 13.9C, E) (e.g., in *Phaseolus* (Clutter and Sussex, 1968; Schnepf and Nagl, 1970), *Pisum* (Marinos, 1970), *Mesembryanthemum, Epilobium, Antirrhinum, Vicia, Lathyrus, Schrophularia* (Pate and Gunning, unpublished data)). Transfer cells *par excellence* are to be found in the suspensor of *Phaseolus vulgaris,* which grows to as much as 0.3 mm in diameter as the embryo develops.

The later stages of seed development usually witness the deposition of ergastic substances of various sorts within one or more clearly defined regions of the seed and, measured quantitatively in terms of assimilate transport, the establishment of these reserves probably represents the most important single process in the whole of seed development. It is not surprising to find, therefore, that where structural layers or cytoplasmic discontinuities impede such transport, zones of transfer cells have been evolved. In those species which do not develop perisperm, and in which endospermic tissues are only of ephemeral importance, the cells of the embryo cotyledons eventually act as the major repository for reserves. There are two obvious routes by which assimilates could reach these cells: one across the cotyledon epidermis from the surrounding compartment, and the other a symplastic pathway within the embryo, originating in the suspensor and radicle. There is no evidence favouring either of these routes, but in several genera (e.g., *Vicia, Lathyrus, Pisum*) epidermal transfer cells develop on the abaxial faces of the cotyledons and their petioles, the wall ingrowths appearing when the embryo first completely fills the embryo sac, as the last reserves of the endosperm are exhausted (Fig. 13.9D), and as the first protein bodies appear in the sub-epidermal cells of the cotyledon (Graham and Gunning, 1970).

In contrast with the situation already described for suspensor transfer cells, the epidermal transfer cells of the embryo are physically separated from their presumed source of assimilates in the integument by layers of crushed and partly digested nucellus and endosperm (Fig. 13.3). In view of their condition these layers probably offer little obstacle to the inward flow of assimilates, and rate limitation in transfer is more likely at sites of supply and release from the donor tissues of the sporophytic integuments. In some of these legume genera the integuments carry an extensive vascular network (Corner, 1951; Eames, 1961), and, in addition, a specialized layer of supposedly nutritive tissue — the endothelium — forms at the inner face of the inner integument. The cells in this layer may in turn be transformed into transfer cells, each cell exhibiting zones of wall ingrowths facing the embryo cavity (e.g., in species of *Vicia* and *Lathyrus,* Pate and Gunning, unpublished, and see Fig. 13.3), and since epidermal transfer cells are also present, a 'push-pull' export—import system may be envisaged, strikingly similar, in fact, to that already described for the bryophyte and fern placenta.

The two genera *Vicia* and *Lathyrus* may well exemplify the ultimate in specialization for direct uptake by the embryo cotyledons, for, as we have seen, outer and inner endosperm layers bear the wall-membrane apparatus early in embryo development while, when the embryo is fully enlarged, a similar structural specialization is added to the cotyledon epidermis and endothelium. It is of interest to note that both of the above genera possess a relatively small suspensor while, on the other hand, in *Phaseolus,* which possesses a very well-developed suspensor apparatus, epidermal and endothelial transfer cells appear to be lacking (cf. Figs. 13.9D, E). It would clearly be of great interest to make comparative studies of nutrition amongst these genera.

There are other groups of flowering plants in which the cotyledons store little if any excess reserves. Instead, it is the endosperm or the perisperm (a derivative of the nucellus) which lays down the main energy source of the seed. We are still woefully ignorant of how the flow of assimilates to these unrelated and sometimes competing sinks in the seed is governed.

A case where transfer cells are almost certainly implicated in nutrition of both embryo and endosperm is to be found in certain members of the *Gramineae,* where several workers have recorded a zone of specialized aleurone cells confined to the region of the hilar pad, or placenta, these cells developing labyrinths of wall material on their outer tangential and radial walls (Kiesselbach and Walker (1952) for maize (*Zea*); Rost (1970), Rost and Lersten (1971) for yellow foxtail (*Setaria*); and Zee and O'Brien (1971b) for millet (*Echinochloa*)). These cells, termed 'aleurone transfer cells' by Rost (1970), represent, essentially, a highly specialized layer of outer endosperm. Facing the nucellar projection at the chalazal end of the seed, lying close to the vascular tissue of the pedicel, and occupying the sole, confined gap in the cuticular layer of the seed coat, this zone of transfer cells must surely represent the major, if not the only, port of entry for assimilates destined for the embryo or inner tissues of the endosperm, and, by analogy with the situations described earlier, one would suggest an absorptive function for the outward facing wall-membrane apparatus (Fig. 13.9F). Contrasting millet and wheat, Zee and O'Brien (1971b) make the interesting point that the gap in the seed coat of the latter is relatively extensive, and the underlying aleurone layer does not develop wall ingrowths; their interpretation is that surface area amplification occurs only where there is a 'structural bottleneck'.

The only case discovered so far of the wall-membrane apparatus being associated with perisperm is found in the ice plant (*Mesembryanthemum calycinum*) where the outer endospermal layer facing the perisperm develops wall ingrowths (Fig. 13.9G). The significance of the placement of this apparatus vis-à-vis the nutritional interrelationships of perisperm and endosperm has yet to be appreciated.

Sporophyte-microgametophyte junctions in angiosperms. The male side of the reproductive cycle of seed-bearing plants is unusual in having two distinct heterotrophic stages in its nutrition, the first of these relating to the initiation and growth of microspores (pollen grains) in the microsporangium (pollen sacs) of the parent sporophyte, the second concerning the germination of the pollen and the growth of the microgametophyte (pollen tube) in the ovary of the same or another sporophyte. In both of these stages of nutrition we find examples relevant to the transfer cell concept.

During the formation of the pollen grains, two instances of structures enhancing the area of contact between microsporangium and microspores have been recorded — the rather sparse invaginations described for the inner wall of the tapetum of *Paeonia* by Marquardt, Barth, and von Rahden (1968) and, on the recipient microspores, the cytoplasmic evaginations noted recently for *Nuphar* and *Epilobium* by Rowley and Flynn (1971). We hesitate, at least for the present, to classify either of these as 'transfer cells', but the enlarged membrane apparatus in each case provides pertinent homologies with the cell type.

The pollen tube enters into a highly specific relationship with the pistil into which it grows. Rosen (1968) has reviewed the main features of this relationship, and of special significance here are the studies from his laboratory of the morphological changes that take place in the pollen tube as it becomes established in the sporophyte (see Welk, Millington, and Rosen, 1965; Rosen and Gawlik, 1966). Pollen tubes growing in incompatible pistils, or *in vitro*, have a single compartmented cap at the top of the tube but, when developing in a compatible ovary, the tip of the pollen tube develops numerous deep cytoplasmic embayments containing pinocytic vesicles, a feature which Rosen regards as being characteristic of the heterotrophic condition. The message that such a structure carries in terms of surface area relationships is obvious.

In many ovaries the pollen tube follows a predestined intercellular route down the style and into the ovary. Frequently this entire pathway is lined with cytologically distinct stigmatoid cells or transmitting tissue. Such tissue is generally held to be responsible for the release of chemotropically-active substances and nutrients.

Quite typical transfer cells have been observed in these tissues. In the style itself (e.g., in *Muscari* (present authors, unpublished)) the wall-membrane apparatus develops on faces of the stigmatoid cells in contact with the pollen tube, while lower down in the ovary transmitting tissue takes the form of a secretory epithelium of transfer cells (e.g., in *Lilium* (Rosen and Thomas, 1970) and in *Fritillaria* (present authors, unpublished)). Wall ingrowths in transmitting tissue of the second of these genera develop prior to the arrival of the pollen tubes; here, at least, physical contact between the generations is not essential for development of transfer cells.

Our experience of surveying for transfer cells in other anatomical situations leads us to suspect that the few examples listed above represent but the tip of the iceberg. If this is the case a most interesting area will have emerged for morphogenetic investigations on the cell type.

Transfer cells in the vascular system

As the search for transfer cells progressed, it became evident that the development of these specialized cells in the vascular system is widespread and common, and we have come to regard the association of transfer cells and the long-distance transport systems as being of fundamental significance in a conceptual appraisal of the cell type. In the sections that follow, we deal with the morphological and functional characteristics of vascular transfer cells in each of their three main locations.

Stem vascular transfer cells

If measured in terms of overall abundance and sheer breadth of distribution within the major taxa, transfer cells in the xylem of stems emerge so far as the most important of the genre. We shall refer to them simply as 'xylem transfer cells'. A survey of 190 species, including clubmosses, horsetails, ferns, gymnosperms, and representatives of 58 angiosperm families (Gunning, Pate, and Green, 1970), showed them to be present in all of these groups except the clubmosses. It is predicted that they exist in the majority of monocotyledons and dicotyledons, and it would be fascinating to discover whether they also exist in the fossil remains of extinct forms held to be important in the evolution of present-day vascular plants.

It soon became apparent during the survey that the distribution of xylem transfer cells in the shoot, and their spatial relationships with other categories of cell, conform to a standard pattern. Firstly, the cells are obviously modified xylem parenchyma. Secondly, they develop ingrowths only on those walls which abut on, or are closest to, conducting elements of the xylem. Thirdly, they are most prominent in the nodes (Fig. 13.10), regardless of whether cotyledons, scale leaves, foliage leaves, bracts, or floral structures are subtended. Nevertheless, the best examples are usually found in nodes bearing the larger leaves, and here, the size and intricacy of the wall-membrane apparatus often exceeds that in other types of transfer cell in the same plant (Figs. 13.1, 13.2, 13.5). Fourthly, they are restricted to foliar traces, provided that the departure of these traces creates gaps in the vascular cylinder — witness their absence from clubmosses, which are microphyllus and have leaf traces but no leaf gaps. A marked waning in frequency and intensity of development of wall ingrowths is observable along the files of xylem transfer cells in the adaxial face of a foliar trace as it is followed out past the leaf gap and into the petiole, or, at its proximal end, back into the parent stem. This applies whether the nodes are uni-, tri-, or multi-lacunar. We have found instances, for example the cotyledonary traces of many dicotyledonous seedlings, where the zone of transfer cells does not extend for more than a few millimetres beyond the ends of the leaf gap. This is not always the case, however, for vestiges of the cell type may sometimes be traced much further beyond the node. In the stems of many legumes, sections through the middle of an internode will reveal xylem transfer cells in some of the bundles, these being the lower ends of traces serving foliar organs higher up the stem. In other species, especially ferns, xylem transfer cells may spill over from the distal end of a leaf trace into the veins of the leaf: we will comment further on this in the next section.

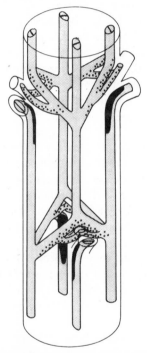

Fig. 13.10 Diagrammatic representation of the vascular system of two nodes and the inter-vening internode in a simple shoot bearing opposite-decussate leaves. The zone on the leaf traces where xylem transfer cells occur is shown in solid black; regions where phloem transfer cells may be found in the branch traces and at the upper margins of the leaf gaps are stippled.

It is quite common for nodes which bear xylem transfer cells also to possess transfer cells in their phloem. These 'phloem transfer cells' are modified vascular parenchyma and, like the xylem transfer cells, are distributed in the stem in a highly characteristic fashion. Surprising as it may seem at first sight, the phloem transfer cell appears to be best developed in parts of the vascular network where xylem transfer cells are absent or inconspicuous: they rarely occur in leaf traces, for example, but are seen along the margins of the leaf gap and in the traces to an axillary bud, these two regions rarely, if ever, carrying xylem transfer cells (see Fig. 13.10). A special instance has been recorded for young epigeal seedlings of dicotyledons in which the departing traces to the cotyledons carry superlative examples of xylem transfer cells but no phloem transfer cells, whilst the reverse holds true for the vasculature of the plumule where strands are well endowed with phloem transfer cells even before they differentiate xylem elements (Pate, Gunning, and Milliken, 1970). In monocotyledons as well, it also seems rare for the two types of transfer cell to coexist within the same region of a vascular strand. Thus, O'Brien and Zee (1971) state that two of the three types of vascular strand which they recognize as the vegetative node of wheat possess only phloem transfer cells, whilst in the third, xylem transfer cells predominate.

Bearing in mind the clearly defined pattern of distribution of xylem transfer cells within nodes, and the strategic positioning of their wall-membrane apparatus in relation to the lumina of the accompanying vessels and tracheids, there are strong grounds for postulating that one of their main functions is to foster local and intense exchange of solutes between the xylem sap and the symplast or, more specifically, between the transpiration stream entering the base of a leaf and the tissues and organs adjacent to the departing leaf trace. There is already considerable evidence to suggest that solutes can pass quite freely between the xylem conduits and the surrounding living tissues of a stem (reviewed by Pate, 1971), but it remains to be seen whether any especially large fluxes of solutes can occur in nodes where transfer cells occur. Pate, Gunning, and Milliken (1970) provide some experimental evidence in this connection, suggesting that in young seedlings the collars of xylem transfer cells in the cotyledonary strands traversing the hypocotyl, and the ancillary investments of phloem transfer cells in the same region, together play a vital role in the nutrition of the plumule, particularly in early stages before it has developed functional connections with the vascular network interconnecting cotyledons and root. By feeding tritiated leucine through the cut base of the hypocotyl of young, transpiring seedlings, they show that transfer cells at the cotyledonary node become greatly enriched with the radiosubstrate, and that the young plumule, a structure only weakly active in transpiration, acquire a high and uniform density of labelling. By contrast, the cotyledons become only weakly labelled during the time course of such an experiment, suggesting that the seedling transfer cells are very effective in collecting the introduced amino acid, and are largely responsible for channelling so much of it toward the young plumule A further type of experiment in which xylem sap exuding from seedlings decapitated either above or below the zone of hypocotyl transfer cells is collected and analysed for its main nitrogenous constituent, nitrate, shows that the concentration of this solute may decrease by as much as 50–60 per cent as the xylem sap runs the gauntlet of the transfer cells. Conversely, the concentration of glutamine, another nitrogenous solute, present in lesser quantities, increases in the xylem from base to top of the hypocotyl, showing that the solute fluxes are not always to the benefit of the symplast. If extended to analyses of other solutes an experimental inventory of this kind might tell us much about the versatility of this type of transfer cell.

We know that spectacular sets of nodal transfer cells exist in the stem of older plants of many species, and it might be envisaged that these have as their prime function the supply of solutes (minerals, amino acids, amides, etc.) that are normally carried in the xylem to any growing structure that constitutes a sink but is not itself capable of transpiration. Axillary meristems, which often lie only a few cells above a carpet of transfer cells on the adaxial face of the departing leaf trace, are obvious candidates to benefit from the activities of nodal transfer cells, although the potential value of such a system might not be realized until relatively late in the life of a node, and then only if a weakening or abolition of apical dominance allowed a bud to break its dormancy. However, an equally important role for the nodal transfer cells might be in the nutrition of the apical meristem, for it is conceivable that materials carried in the traces serving the crown of young leaves below the apex might be absorbed and diverted through the symplast towards the apex. In a similar manner flowers and fruits might also receive significant elements of their diet from xylem and phloem transfer cells. Anatomical grounds in support of this may be found in a study by Zee and O'Brien (1971a) of transfer cells underlying the placenta in the nodal region of the wheat spikelet.

The architecture of the node suggests another, not entirely unrelated, function of transfer cells. In the region of the leaf gap, xylem and phloem from adjacent strands run facing each other and in close proximity (Fig. 13.10) and, whether or not the relevant transfer cells develop, an opportunity must surely exist for a direct exchange of solutes between the xylem of one strand and phloem of another. Here, indeed, may lie the *raison d'être* for the complementary patterns of distribution of phloem and xylem transfer cells, these together facilitating a mixing of sets of assimilates originating in roots and in leaves, while at the same time providing the means whereby cross traffic of solutes can be encouraged between the different orthostichies and types of appendage, including fruits.

It is somewhat discouraging to find how little factual evidence exists concerning solute transport in and around nodes. Anatomists have long been aware of the possible physiological significance of their complex vascular systems (Dormer, 1955), and the new observations on transfer cells indicate more clearly than ever that nodes must be traffic control centres. Herein lies a most interesting challenge for the plant physiologist.

Leaf vascular transfer cells. Specialized phloem parenchyma bearing wall ingrowths were first depicted, for the vein of a broad bean leaf, by Ziegler (1965), and the subsequent discovery of virtually identical cells in the minor veins of leaves of several other species of flowering plant (Gunning, Pate, and Briarty, 1968) spurred us on to a further search for such cells, now extended to over 1000 species, including clubmosses, horsetails, ferns, and gymnosperms and representatives of some 250 families of dicotyledons and monocotyledons. It has emerged that the phloem transfer cell of minor veins is a relatively rare structure in comparison with its counterpart in the xylem of the node, being absent from all taxa except the angiosperms, within which it is common only in certain predominantly herbaceous families of dicotyledons. With one exception (*Zosteraceae*) minor veins of leaves of monocotyledons appear to lack transfer cells. This class of transfer cell seems to have been exploited relatively late in the evolution of vascular plants, in fact, in those advanced orders whose phloem bears sieve tubes as opposed to sieve cells. The only instance at hand which conflicts with this supposition is the horsetail, *Equisetum,* species of which exhibit transfer cells in the phloem regions of the traces leading to their scale leaves, but even in this exceptional genus transfer cells are not visible within the scale leaves themselves.

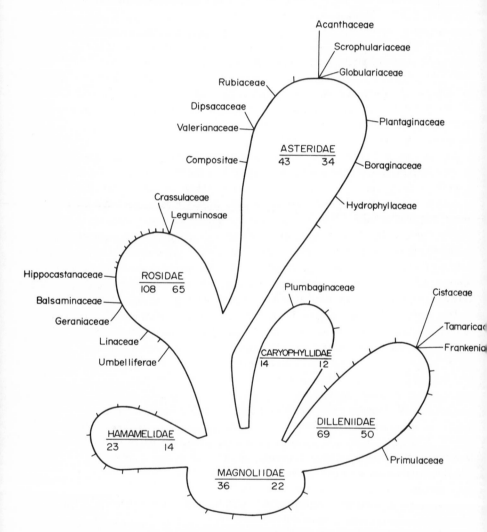

Fig. 13.11   Position of families with representatives bearing minor vein phloem transfer cells in the classification scheme proposed for the Dicotyledons by Cronquist (1968). Each sub-class, represented by a lobe in the diagram, is named, and the numbers beneath the names show, on the left, the total number of families in the sub-class, and, on the right, the number of these families surveyed for presence or absence of this type of transfer cell.

We now have examined sufficient numbers of the dicotyledons to advance a somewhat tentative picture of the overall distribution of minor vein transfer cells among the families and orders of this group of plants (see Pate and Gunning, 1969). This is summarized in Fig. 13.11, in which the 'positive' families (in which at least one member bears leaf transfer cells) are listed against the sub-classes recognized in a classification scheme proposed by Cronquist (1968). It is seen that these 'positive' families are restricted to the four more advanced sub-classes, and that most of these families contain species which are of a predominantly herbaceous habit. Furthermore, judging from the somewhat scattered distribution of the positive families, it seems possible that leaf transfer cells may have been exploited on several isolated instances during the evolution of the dicotyledons, although the occurrence of these cells in closely related groupings (e.g., within the three families of legumes, and within the *Rubiaceae–Valerianaceae–Dipsacaceae–Compositae* complex) would suggest that once the habit of bearing these cells had evolved it was perpetuated during later evolution of a family or group of related families. In the *Compositae,* for example, every one of the sixty-odd genera we have examined has been shown to possess most prolific sets of vein transfer cells, and one is tempted to conclude from this that the habit of bearing these cells may have been inherited piecemeal from an ancestral stock.

Morphological study of the transfer cells of leaf minor veins has shown that four different types can be distinguished, two, designated A- and B-type transfer cells, are located within the phloem, the other two, C- and D-type transfer cells, occupy the xylem region of the vein.

The A-type cell turns out to be by far the most common and might well be regarded as the 'type species'. It is associated specifically with the sieve elements, and we have strong reasons for regarding it as being anatomically equivalent to a modified companion cell (Pate and Gunning, 1969). When present in company with other types of transfer cell it is immediately recognizable: by its strict partnership with the sieve elements; by its especially dense cytoplasm; and by the fact that, in contrast to the other types of transfer cells, its prolific wall ingrowths extend right round its cell periphery, avoiding only the numerous pit areas where plasmodesmata connect it to its associated sieve element (see Figs. 13.4, 13.6, 13.13).

The other phloem-based transfer cell — the B-type cell — appears to represent a modification of phloem parenchyma cells other than companion cells, and accordingly it is more widely scattered throughout a vein. Nevertheless, a relationship with the conducting channels of the phloem is to be suspected since the zones of wall ingrowths of the B-type cell are restricted to those walls oriented towards the sieve elements or their associated companion or A-type cells. In consequence, B-type cells may exhibit several discrete zones of wall ingrowths, each occupying a region of the cell adjacent to a conducting channel of the phloem (Figs. 13.6 and 13.13 contrast the A- and B-type cells).

The two other types of vein transfer cell are much less common and conspicuous than the A- and B-type cells. One, the C-type cell, is clearly a modified xylem parenchyma cell. Its wall-membrane apparatus lines walls shared with tracheids or vessels and we regard it as equivalent to the xylem transfer cells already described for the stem — in certain ferns and dicotyledons (e.g., *Galium*), xylem transfer cells extend out along the leaf trace into the finest veins of the leaf. However, in such instances wall ingrowth development declines markedly as one approaches the finer veins, so that it is likely that the stem, not the leaf vein, is the primary seat of this class of transfer cell. Moreover, when present in a minor vein, the small number of these C-type cells and the sparsity of their wall ingrowths indicate that they might have little real impact on the overall functioning of the minor vein. The final class of transfer cell, the D-type cell, clearly does have its true home within the leaf vein, since it is a modified bundle sheath cell and is always restricted to veins of small diameter supplying the islands of mesophyll in the leaf. Its zone or zones of wall ingrowths are usually very sparse, each being oriented towards the nearest xylem elements, and judging from the rarity of the cell type and the fact that its cytoplasm is not noticeably different from that of unspecialized sister cells in the bundle sheath, it is unlikely that it has any important role of its own in the functioning of the vein.

We now have at hand a somewhat complex picture regarding the distribution of these four categories of vein transfer cells within the species, genera, and families so far examined. The scheme below embraces every combination of transfer cells so far encountered:

(a)　*in ferns:* C alone.

(b)*　*in dicotyledons:*

A alone; A + C; A + C + D (together accounting for 62% of the positive genera examined).

A + B; A + B + C; A + B + D; A + B + C + D (together accounting for 35% of the positive genera examined).

B alone (3% of positive genera examined).

(c)*　in monocotyledons (*Zosteraceae* only): a phloem transfer cell which cannot be classified as either A or B.

* See Pate and Gunning (1969).

In every instance it would appear that the combination of transfer cells is species-specific, although closely related species of a genus may be sometimes found to bear contrasting combinations of transfer cells (e.g., species of *Plantago* with B, A + B, or no transfer cells, Pate and Gunning, 1969). Needless to say, we know virtually nothing of the morphological significance of these various groupings of types of transfer cell in a vein, let alone of the subtle physiological flavours that each might well impart. Yet one is swayed by the fact that more than one-half of the genera examined have A-type cells as their only leaf transfer cells, encouraging the belief that this cell might carry out a more important role than the others in the physiology of the minor vein.

Before turning to the possible function of transfer cells in minor veins some facts must be gleaned concerning the timing of their development in the leaf. In contrast to the situation in the phloem of nodes where transfer cells become recognizable almost at the procambial stage of a vascular strand, the presumptive transfer cells of the minor vein do not develop their specialized walls until relatively late in the life of the leaf, in fact, not until after stomata have been perforated, air spaces have appeared in the mesophyll, and the conducting elements of xylem and phloem have differentiated. Measured in terms of leaf expansion, the appearance of wall ingrowths in the various types of transfer cells is found to coincide with the leaf having reached about two-thirds to three-quarters of its final area (I. Marks, unpublished data). Of greater significance is the relationship that exists between transfer cell development and the export activity of a leaf. We have studied the state of transfer cell development simultaneously with the capacity, if any, to export photosynthates, by feeding different ages of leaf number six of *Pisum arvense* with $^{14}CO_2$, and taking samples for cytological examination after radioactive solutes had been extracted. The results (see Fig. 13.12) are that wall ingrowths first become visible by optical

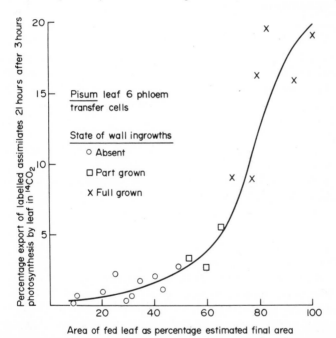

Fig. 13.12   Development of minor vein transfer cell wall ingrowths and of the capacity to export products of photosynthesis from a leaf. Each point represents an individual leaf: the upright axis shows its capacity to export, the horizontal axis its area relative to the fully expanded condition, and the state of its transfer cell wall ingrowths is symbolized to indicate whether they are absent, part, or full grown.

microscopy just as the leaf commences to export, and that the further growth of the ingrowths roughly parallels the build-up in export activity as the leaf matures. We have also found that wall ingrowths fail to reach a normal size and number in leaves grown in darkness (*Pisum*), in the leaves of albino mutants (*Pisum*), or in the non-green regions of variegated leaves (species of *Sedum, Impatiens,* and *Senecio* (Fig. 13.4)). Admittedly all of this evidence is circumstantial, but it does seem to point rather strongly towards a positive association between the specialized attributes of the transfer cell and the export of solutes from the leaf.

From what we know of the general physiology of the mature leaf, its value in term of translocation to the rest of the plant stems from its ability to generate through its photosynthesis an exportable surplus of assimilates, and from its capacity in its transpiration to attract, in excess of its current requirements, inorganic and organic solutes moving in the xylem from the root system. Solutes from either of these sources must be loaded on to the sieve tubes if they are to be exported from the leaf and, judging from the strategic positioning of the vein parenchyma, it can hardly be disputed that its cells must be directly involved in some manner in this loading process. What is at issue is how this transfer to the sieve elements is accomplished, and in what manner, if any, minor veins equipped with transfer cells function differently from those bearing normal parenchyma cells.

Whatever interpretation we place on the function of the different types of leaf transfer cell, there is no escaping the fact that the presence of each and every one augments the area of contact between the apoplastic and symplastic compartments of the vein over and above that possible in veins with a similar number of conventional parenchyma cells of comparable dimensions. We adhere to our earlier contention (Pate and Gunning, 1969) that all leaf transfer cells, regardless of type or placement, are specialized in at least one particular direction, that of absorbing solutes from the extracytoplasmic region of the minor vein. Fig. 13.13 epitomizes this concept of intraveinal retrieval, the arrows overlying the electron micrograph depicting the proposed directions of solute flux, from the xylem to the extracellular compartment of the vein, to the population of transfer cells and thence, via plasmodesmatal connections, to the sieve tubes. For the sake of simplicity, transport activities in relation to the passage of photosynthate to the veins are not included in the figure, and it is depicted, possibly quite wrongly, that the main extracellular source of solutes to the system is the transpiration stream.

Minor veins bearing transfer cells are thus pictured as acting in a kidney-like fashion, filtering off and concentrating certain of the inorganic and organic solutes delivered in the xylem and exporting these out of the leaf without them ever leaving the confines of the vein. Considering the enormous combined length of minor veins in a leaf and the correspondingly great area of contact that exists within them between xylem and phloem, intraveinal recycling of solutes may prove to be both widespread and quantitatively important in the functioning of leaves, and it is more likely that such an activity is carried out efficiently when transfer cells are present. In fact, retrieval might be particularly effective in the many species where transfer cells can be seen to bridge the gap between files of xylem and sieve elements (see Fig. 13.13). Arguing that the greater the surface for absorption within the vein, the greater its capacity for solute retrieval, those species

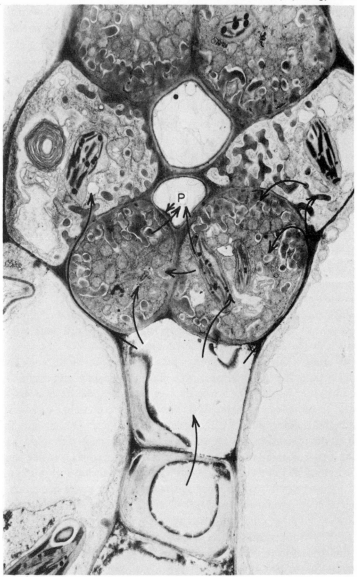

Fig. 13.13   Transverse section of a minor vein in a leaf of *Senecio vulgaris* (see also Fig. 13.6). The phloem region, in the top half of the picture, shows six parenchyma cells centred around two sieve elements. Pairs of companion cells modified as A-type transfer cells lie both above and below the sieve elements. There is a B-type transfer cell to each side. The A- and the B-types differ strikingly in the relative density of their cytoplasm, in the polarity of their zones of wall ingrowths, and in their connections with the sieve elements: plasmodesmata (P) between sieve elements and A-types are common, but are very rare between sieve elements and B-types. In the lower half of the picture two xylem elements occupy the centre, while to each side there are parts of two large bundle sheath cells. Arrows indicate some of the possible routes for intra-veinal retrieval of solutes delivered to the apoplast in the xylem (see text). Magnification x8000.

carrying all four types of transfer cell in their veins might well represent the ultimate in efficiency for, in them, the absorptive apparatus of the transfer cells extends to almost every intercellular boundary within a vein (e.g., see Fig. 2, Pate and Gunning, 1969).

As in the case of their specialized counterparts, the vein transfer cells are integrated with the sieve elements, bundle sheath, and mesophyll by plasmodesmatal connections so that it is justifiable to implicate them in the other important activity of vein parenchyma, namely the ferrying of photosynthetically fixed assimilates via the symplast into the translocatory channel of the vein. In this respect, the transfer cell is certainly no better adapted to load the sieve elements than is a conventional parenchyma cell of the vein, and, indeed, one might argue that the penalty that is paid for developing zones of wall ingrowths specializing in exchange between apoplast and symplast is to decrease the boundary contacts available for symplastic transfer through the plasmodesmata. Unfortunately, there is no information as yet on the relative frequency of plasmodesmata on the walls of normal parenchyma and transfer cells in leaf veins, so that it is not possible to check this line of reasoning, but it is logical to suggest that a compromise has been reached during the evolution of the transport system of leaves, and it would not be unexpected if one were to find many subtle differences and compensatory mechanisms of a structural or physiological nature regulating the balance between apoplastic exchange and symplastic transfer within leaf minor veins in different species.

In any event, a vein carrying transfer cells would appear to be much better equipped structurally than a vein not so endowed in dealing with assimilates leaking from the mesophyll cells into the free space of the leaf. In several species this extracytoplasmic compartment has been shown to contain a variety of solutes, including inorganic ions (Bernstein, 1971), and also sugars and other organic solutes (Jacobson, 1971) suspected to have originated from the mesophyll. At least one group of workers (Kursanov and his colleagues (see Kursanov, 1968; Kursanov and Brovchenko, 1969)) have proposed that absorption by veins of photosynthetically produced substances leaking to the free space of the leaf may be an important mechanism for loading the sieve elements. It would be fascinating to know whether those herbaceous species, in whose leaf veins transfer cells abound, possess a free space system which is regularly enriched with assimilates, thus providing a milieu ripe for exploitation by transfer cells. Pursuing this line of reasoning further, one might propose that the very conditions prompting the appearance of transfer cells in phloem of minor veins may be realized only where the possession of sieve tubes allows the retrieval and export of extracellular solutes from mesophyll and the vein to be accomplished with a high level of efficiency. In herbaceous dicotyledons, where it is well known that extremely fast rates of production, turnover, and cycling of assimilates take place, the potential carrying capacities of sieve tubes may be exploited in fullest measure, so that it comes as no surprise to find that, in this category of plants, phloem transfer cells are particularly common. Nevertheless, the virtual absence of phloem transfer cells in veins of leaves of the predominantly

herbaceous monocotyledons is certainly a puzzling anomaly, since there do
not appear to be any grounds for suspecting that their translocatory activities
are in any way less efficient than in comparable dicotyledons. Nor, indeed, is the
propensity for transfer cell development lacking in these same monocotyledonous
species for, as we have seen, many of them are especially well endowed with both
xylem and phloem transfer cells in the vasculature of their stems.

Root vascular transfer cells. Having demonstrated that vascular transfer cells
occur quite commonly in the above-ground parts of plants, it is of some interest
to record that we have searched widely but virtually to no avail for comparable
cells within the xylem and phloem of roots. In fact, the only instances where
root transfer cells have been discovered is in a few genera of ferns (e.g., *Azolla*),
where the pericycle cells adjacent to the xylem develop wall ingrowths, albeit
sparsely, and only on walls adjoining tracheids.

If this handful of somewhat specialized species proves to be the only group
of plants bearing root transfer cells, it is necessary to ask why the habit was
not perpetuated or developed widely during the later evolution of plants and
why, in particular, it did not become established in roots of those families of
herbaceous angiosperms which carry a veritable riot of these cells in their shoots.
Two possible reasons come to mind: the first that the vascular parenchyma of
roots might differ fundamentally from that of the shoot vascular system in lacking
the ability to develop the wall-membrane apparatus; the second that the propensity
for transformation into transfer cells does exist in root vascular parenchyma but
that physiological conditions, or morphogenetic impulses for such transformation,
are lacking. As shown below, we have grounds to reject the first and favour
the second alternative.

One situation in which the wall-membrane apparatus appears in roots is in
tissue infected with nematode worms belonging to the family Heteroderidae. Wall
ingrowths are one of many cytological oddities of the 'giant cells' that form in
the stele in the vicinity of females of the root knot nematode, *Meloidogyne* spp.
(Bird, 1961; Paulson and Webster, 1970). The giant cells represent either modified
stelar parenchyma or modified xylem elements, most strikingly diverted out of
their normal developmental pathway. They are considered to be a source of
nutrient for the parasite as it grows in an intercellular pocket to form a large
egg-containing cyst. In mapping the distribution of wall ingrowths we have found
(Dennis and Gunning, unpublished) that the giant cells of tomato roots can be
polarized with respect to both adjacent xylem and adjacent phloem. On the one
hand they are analogous to stem xylem parenchyma transfer cells, with ingrowths
on that part of their wall backing on xylem elements, and on the other they
resemble both B-type transfer cells of minor vein phloem and the transfer cells
of *Cuscuta* haustoria (Kollmann and Dörr, 1969) in possessing ingrowths on the
wall nearest the sieve elements. We find that the same two types of polarity
occur in potato roots infected with the cyst nematode, *Heterodera rostockiensis*.
In this case, there is a most remarkable variation: whereas with *Meloidogyne* the
giant cells near the head of the nematode are merely multinucleate, with *Heterodera* they are syncytial, and the giant cells, freely interconnected by cell wall

dissolution, in one region have a wall-membrane apparatus against xylem elements in another against phloem tissue, while intervening regions have smooth walls. This is remarkably versatile behaviour for what is to all intents and purposes a single protoplast.

As usual, the interpretation of these structural features has to be speculative. The analogies with xylem transfer cells, B-type transfer cells, and *Cuscuta* haustorial transfer cells, might suggest that the dual wall-membrane apparatus of the giant cells aids retrieval of solutes from both xylem and phloem, providing a balanced diet of nitrogenous materials, minerals and sugars for growth of the cells and ultimately for that of the worm which feeds on them. In some respects the giant cells resemble the transfer cells mentioned earlier in this article for the suspensors of certain higher plant embryos (e.g., *Phaseolus*): both are endo-polyploid; both are exceptionally large cells; both develop wall ingrowths against the surrounding tissues; and both relate to the nutrition of 'foreign' parasites, either nematode or young sporophyte.

Returning to the question of whether roots are or are not capable of producing transfer cells, the giant cells, though admittedly part of a pathological condition and quite clearly exceptional, do indicate that root cells do not lack the capacity for wall ingrowth formation. What is needed is an appropriate physiological stimulus and this, we would suggest, is absent in the normal root.

The second situation where transfer cells are encountered in roots may also be regarded as a pathological condition. This is the root nodules of leguminous plants infected by the nitrogen-fixing organism *Rhizobium,* a situation dealt with at length by Pate, Gunning, and Briarty (1969) and, for sake of brevity, only the salient features will be mentioned here. The transfer cells of nodules are modified stelar parenchyma, occurring throughout the peripheral vascular network of the nodule as concentric rings of pericycle cells surrounding the central xylem and phloem of each vascular strand. They differ from the vascular transfer cells described so far in that their zones of wall ingrowths are not necessarily oriented towards the conducting elements, nor indeed can the cells be associated specifically with either the xylem or phloem elements (Fig. 13.7). Admittedly, the most prolific sets of wall ingrowths often lie alongside the xylem conduits, but in many cases equally dense tufts of ingrowths adjoin the common boundary between adjacent transfer cells, sometimes considerably distant, in fact, from the nearest conducting tissues. In short, the surface area amplification within the strands is towards the stelar apoplast in general, whether it consists of xylem elements or intercellular free space.

The vascular bundles running through a nodule are each enclosed by an endodermis with Casparian thickenings, lying immediately outside the pericycle transfer cells and appearing to form a 'tight junction' between the apoplast of the vascular bundle and the surrounding apoplast of the nodule. We have proposed a model of functioning for the vascular strand in which the principal activity of the transfer cells is thought to be the selective secretion to the bundle apoplast of amides and amino acids. These solutes diffuse down their concentration gradient in the symplast from the donor bacterial tissue, through the endodermis, and into

470

the transfer cells. The secreted solutes lower the water potential of the bundle apoplast, leading to influx of water across the endodermis, and thereby flushing the secreted amino compounds out of the nodule to the rest of the plant via the xylem. Evidence supporting these suggestions comes from the fact that legume nodules bleed copiously from their cut xylem if detached from a root and placed with their distal, apical end in water. Analysis of the exuded xylem sap shows it to contain amino acids and amides typical of xylem transport, some of them at concentrations up to ten times that in the nitrogen-fixing bacterial tissue of the nodule. The legume nodule is, in effect, a gland, secreting into the apoplast in the manner of an endocrine gland in animals. Apoplastic conduits — xylem and serum — conduct away the secreted material in both. By concentrating its products prior to export in this remarkable manner, the root nodule differs basically from the root, and the osmotically operated system motivated by the transfer cells gives the nodule an economy of solvent in xylem transport approaching that in phloem.

Morphogenesis of transfer cells

Prior to the development of their distinctive wall-membrane apparatus, transfer cells are latent, and look like their non-specialized counterparts. This unique morphogenetic event is considered now, not so much by presenting new information as by taking observations that have already been described, and seeking their significance with respect to the mechanisms that regulate the formation of wall ingrowths.

Morphogenesis at the level of species and tissues

The marked variation between plant species, and the constancy within species, implies that the form and anatomical distribution of the wall-membrane apparatus are genetically determined traits. The possibility that all plants possess the necessary genetic information is not so unlikely as it may seem. Firstly, there are many anatomical sites to be searched before it can be stated that any species of flowering plant is devoid of transfer cells; as far as we are aware this has never been done. Secondly, the presence of transfer cells in sporophyte—gametophyte junctions of bryophytes and ferns speaks of an ancient evolutionary origin for the traits.

Data on taxonomic distribution of the types of transfer cell are inadequate for full discussion of these matters, but enough is known about vascular transfer cells to illustrate some generalizations. For instance, if a species realizes its potential for formation of transfer cells in its minor veins, there is a very high probability (no exceptions have been found thus far) that it also does so in its nodal xylem parenchyma. The converse does not apply: within the dicotyledons nodal transfer cells are found in more sub-classes; within a sub-class such as the Rosidae they occur in more orders; within orders like the Polemoniales, Lamiales, Dipsacales, Rosales, or Umbellales they are found in more families; and in families such as the Umbelliferae and Leguminosae they are present in more genera. It is true, however, that the taxa in which nodal transfer cells occur in the absence of their fellows in minor veins are often quite closely related to the taxa possessing both types.

Flowering plants play many variations on this theme of expressing or repressing their potential morphogenetic capacity to make wall ingrowths in a variety of anatomical sites. Table 13.1 makes the point for a group of legume species.

TABLE 13.1

Development of Selected Types of Transfer Cell in Some Legumes

	Location							
	Reproductive				Vascular			
							Minor vein phloem	
	Endosperm	Integumentary endothelium	Cotyledon epidermis	Floral nectary	Foliar trace xylem parench.	Root nodule pericycle	A type	B type
Vicia sepium	●	●	●	●	●	●	●	○
Lathyrus odoratus	●	●	●	○	●	●	●	○
Lupinus albus	●	○	○	☆	●	●	●	●
Pisum arvense	●	○	○	○	●	●	●	○
Spartium junceum	●	○	○	☆	●	○	●	○
Phaseolus multiflorus	●	○	○	○	●	○	○	○

● Developed. ☆ Not applicable. ○ Not developed.

At one extreme there are plants like *Vicia* spp., with wall ingrowths in root nodules, in both xylem and phloem of nodal traces leading to cotyledons, stipules and leaves, in the phloem of leaf minor veins, stipules and pods, in floral and extra-floral nectaries, in synergids, on inner and outer faces of the endosperm, in the epidermis of the developing cotyledons, and in the integumentary endothelium. In view of this diversity it is extraordinary that B-type transfer cells are missing, though the unspecialized phloem parenchyma cells which could develop into them (and do so in, e.g., *Lupinus*) are present. There is no accounting for this and the numerous similar quirks of morphogenesis that emerge from comparisons of the distribution of transfer cells in different species.

Two isolated and remarkable examples help us to come to a very tentative conclusion. One has been introduced on page 469, where the stele of the flowering plant root is described as being devoid of transfer cells under normal conditions, yet, given an appropriate stimulus, namely infection with a nematode, the cells can produce the wall-membrane apparatus. The second example is the behaviour of sycamore cells grown in tissue culture to form cell aggregates. The outermost cells of each aggregate develop wall ingrowths (Davey and Street, 1971). They are in direct contact with the culture medium and must in physiological terms be not unlike two other sites for transfer cells, the epidermis of submerged leaves of water plants (Gunning and Pate, 1969a), and the epidermis of embryos immersed in endospermic or endothelial fluid (p. 455).

Plant cells are said to be totipotent. The above two examples, taken together with all the other available data, strikingly indicate that, among the genetic information underlying totipotency, there is a portion concerned with the formation of the transfer cell wall-membrane apparatus. Whether it is under positive or negative control is debatable; a stimulus may be required either to de-repress the morphogenetic potential, or to elicit its expression. Judging by the widely scattered taxonomic distribution of transfer cells — surely not to be regarded as multiple, parallel evolution — the potential could well be present in all plants, at any rate in all flowering plants, though perhaps latent. Superimposed on this there is the established fact that realization of the potential in one anatomical site in a particular species does not necessarily mean that it will be realized in others. We have on several occasions stated the desirability of carrying out the ultimate, but exceedingly difficult, test of the transfer cell hypothesis, that is, to see whether variations in taxonomic and anatomical occurrence of the cells reflect corresponding variations in evolutionary selection pressures seated in physiological priorities. Particularizing from the data in Table 13.1, is solute transport in *Phaseolus* nodules or minor veins non-limiting? Was it limiting in *Vicia* and was this behind the realization of the potential for formation of its wall-membrane apparatus?

Morphogenesis at the level of the transfer cell

Under normal conditions the development of the wall-membrane apparatus is one event in an orderly sequence taking place both within the transfer cell and in the surrounding tissues. This is apparent from developmental studies of minor veins (p. 461ff.), nodes (Pate, Gunning, and Milliken, 1970) and glands (e.g., Fahn and Rachmilevitz, 1970). But this orderly sequence can be interrupted. A leaf kept under conditions of low light intensity fails to complete transfer cell formation in its veins (Gunning, Pate, and Briarty, 1968). A seedling grown in the dark, or with cotyledons removed or darkened, or in the absence of carbon dioxide, fails to develop nodal xylem parenchyma transfer cells (Pate, Gunning, and Milliken, 1970). The leaf and the node can be reprieved from this state of incomplete development, e.g., by restoring carbon dioxide supply to the seedling.

We have likened the development of the wall-membrane apparatus to the adaptive formation of an enzyme in response to provision of its substrate (Pate, Gunning, and Milliken, 1970). To take the above three examples: minor vein transfer cells become recognizable when the leaf first balances its budget and starts to export, that is to say, the amplified membrane surface appears when solutes destined to be exported become available (p. 465ff.). The experiments on seedlings can be interpreted in the same manner — the manipulations are likely to reduce the amount of solute in the xylem sap, and the xylem-associated transfer cells fail to develop; they do complete their development if the solute supply is restored (Pate, Gunning, and Milliken, 1970). In the gland, commencement of nectar secretion is concomitant with the development of the wall labyrinth (e.g., Fahn and Rachmilevitz, 1970). The parallel between the formation of the wall-membrane apparatus and of adaptive enzymes is given added strength by some additional facts. Firstly, membrane transport proteins of microorganisms can be formed adaptively and the amount formed per cell varies according to the quantity of substrate and nutrient that is available (Pardee, 1968). Secondly, avian salt glands not only develop more (Na^+-K^+)—ATPase activity when the birds are salt-stressed, but the secretory cells also produce much more extensive plasma membrane infoldings (Ernst and Ellis, 1969).

The real problems arise when the morphogenetic process is examined in detail. One's first impression on looking at micrographs of transfer cells is that they have synthesized fragments and strands of wall material in a totally disorganized fashion. Chaotic though it may appear, closer inspection shows that it is organized chaos. Signs of organization that betray the existence of control systems manifest themselves in a variety of ways.

Firstly, there is a measure of specificity in the shape of the ingrowths. For example, in Fig. 13.5 the ingrowths of *Lamium* xylem parenchyma transfer cells are shown to be more or less constant in thickness, whether measured at their point of origin on the cell wall, or at a remote extremity deep in the cytoplasm. Then again, *Lamium* transfer cell wall ingrowths are quite different in shape from some of the others that we illustrate (compare Fig. 13.5 with Figs. 13.2 and 13.3). Though it is necessary to be aware of the phenomenon of modulation in the intensity of development of the wall ingrowths, in which some cells within a given plant are more, or less, well endowed than others, it is clear that some plant species characteristically produce short, papillate ingrowths, others long and filiform types, perhaps branched and interconnected, and others robust flanges: there is, in short, species-specificity. Presumably the morphology of the ingrowths is genetically determined, but if this is so, one must add the proviso that variation from tissue to tissue within a given plant is possible. One striking example of tissue-specific differences in the form of transfer cell wall ingrowths is seen in the shoot of rock-rose (*Helianthemum* spp.) where xylem and phloem parenchyma transfer cells are totally unlike one another, with branched filiform ingrowths and flanges respectively (compare Fig. 23 of Gunning, Pate, and Green, 1970 with Figs. 17 and 18 of Pate and Gunning, 1969). The grass *Lolium* has nodal transfer cells with flanges or trabeculae (as in other grasses), while the giant cells induced in its root by infection with the nematode *Heterodera* develop papillate ingrowths (Dennis and Gunning, unpublished observations). We have no idea how the cell moulds wall ingrowths of defined shape; nor how a single genome can give rise to tissue-specific differences.

The other manifestations of control concern the position and abundance of wall ingrowths. Some transfer cells, for instance, are quite as polarized as those animal cells described as having 'apical' microvilli or 'basal' infoldings. In the case of xylem parenchyma transfer cells the clustering of the ingrowths on the wall between the xylem element and the transfer cell (Figs. 13.1, 13.5) suggests strongly that a localized stimulus for their formation comes from the xylem. This does not, however, help to explain how the cell carries out the deposition of wall material on one particular face. Even more puzzling are vein phloem parenchyma cells, where a polarized B-type cell can lie alongside, and be connected by plasmodesmata to, a relatively non-polarized A-type (see Figs. 13.6, 13.13). Once again, the operation of tissue-specific and cell-specific genetic programmes is evident.

As to abundance, modulation of the intensity of development of the wall-membrane apparatus has already been described. The best example, in xylem parenchyma transfer cells, can be seen by following a leaf trace from an internode through a node and out into the petiole (see p. 459). It could be envisaged that the total surface area amplification could be regulated by metering either the amount of wall precursor or the amount of membrane constituents available to the cell. The latter would be sufficient if the membrane embodies within it mechanisms for the production of specified quantities of the fabric of the wall ingrowths.

A somewhat different type of control has been uncovered by quantitative estimation of surface area amplification factors for A-type transfer cells in minor vein phloem. It is common for these cells to have amplification factors in the range 2—5, with pronounced inter-specific variation. Within a species or an individual there is also variation in the size of the cells, seen as differences in diameter when they are viewed in transverse section. Despite this variation the individual cells show remarkable constancy in their amplification factors, the simplest interpretation being that the larger the diameter of the cell, the larger the wall surface on which ingrowths can develop. Maintenance of a constant density per unit area of ingrowths of standard size accounts for the observation (Marks and Gunning, unpublished).

The morphogenesis of transfer cells clearly presents as many problems for the student of cell differentiation as the cells themselves do for plant physiologists. We do not understand the basis of their taxonomic and anatomical distribution. Where they do occur, we can do no more than recognize that regulation of their development is temporal, morphological, spatial, and quantitative. Hopefully, recognition will be the forerunner of greater understanding.

More than three-quarters of a century ago, Haberlandt, the great proponent of physiological plant anatomy, wrote:

Since the correlation of structure with function is always evident as regards both the gross anatomy and the histology of the plant body, a similar connection between the morphological features of protoplasm and the functions of the individual protoplast would doubtless be demonstrable, were it not that most of the structures concerned are of ultramicroscopic dimensions.

In basing suggestions for the role of transfer cells in the life of the plant on what is known about their gross anatomy, their histology, and their now-demonstrable ultramicroscopic features, we have followed Haberlandt's precept that structure is correlated with function. Ramifications of the story have touched upon many branches of plant science, from taxonomy and evolution to cell and plant physiology and morphogenesis, and at every turn the prevailing dearth of relevant knowledge has been all too obvious. There is here such an opportunity for collaboration between cytologists and physiologists as can have arisen only seldom since the days of Haberlandt: by their very existence transfer cells present a challenge; that challenge must be met, and in so doing we may all learn something more about plants.

Postscript

Many publications dealing with transfer cells have appeared since this chapter was written. A general review has appeared (Pate, J. S. and Gunning, B. E. S. (1972), 'Transfer Cells', *Ann. Rev. Pl. Physiol.*, **23**, 173–196), and the following additional references are selected and cited here because of their relevance to points raised in the chapter. Adenosine triphosphatase has been detected in the wall-membrane apparatus (see p. 449) (Maier, K. and Maier, U. (1972) 'Localizatic of Beta-Glycerophosphatase and Mg^{++} - activated Adenosine triphosphatase in a Moss Haustorium, and the Relation of These Enzymes to the Cell Wall Labyrinth', *Protoplasma*, **75**, 91–112.). The absorptive hyphae of the phloem-feeding parasite *Cuscuta*, with their remarkable endoplasmic reticulum, have been described in detail (see pp. 445 and 469) (Dörr, I. (1972) 'Der Anschluss der *Cuscuta*-Hyphen an die Siebrohren ihrer Wirtspflanzen', *Protoplasma*, **75**, 167–184.). Our picture of the evolutionary origin of xylem transfer cells (see pp. 458 and 471) has been modified by the discovery that cells adjacent to the 'hydroids' (equivalent to xylem) in the conducting tissue of a moss can develop wall ingrowths (Hébant, C. (1970) 'A new look at the Conducting Tissues of Mosses (Bryopsida): their Structure, Distribution, and Significance', *Phytomorphology*, **20**, 390–410). The nematode-induced transfer cells discussed in p. 469 have been thoroughly investigated (Jones, M. G. K. and Northcote, D. H. (1972) 'Nematode – Induced Syncytium – a Multinucleate Transfer Cell,' *J. Cell Sci.*, **10**, 789–809; Jones, M. G. K. and Northcote, D. H. (1972) 'Transfer Cells induced in *Coleus* roots by the Root-Knot Nematode *Meloidogyne arenaria*,' *Protoplasma*, **75**, 381–420).

Bibliography

References

Abel, J. H. (1969) 'Electron microscopic demonstrations of adenosine triphosphate phospho-hydrolase activity in herring gull salt glands', *J. Histochem. Cytochem.*, **17**, 570–584.
Anderson, W. P., Aikman, D. P., and Meiri, A. (1970) 'Excised root exudation – a standing-gradient osmotic flow', *Proc. R. Soc.*, B **174**, 445–458.
Arnaudow, N. (1925) 'Über Transplantieren von Moosembryonen', *Flora*, **118/9**, 17–26.

Barka, T. and Anderson, P. J. (1963) *Histochemistry, theory, practice and bibliography*, Harper and Row, New York.

Béguin, F. (1966) 'Étude au microscope électronique de la cuticle et de ses structure associées chez quelques cestodes. Essai d'histologie comparée', *Z. Zellforsch. mikrosk. Anat.*, **72**, 30–46.

Bernstein, L. (1971) 'Method for determining solutes in the cell wall of leaves', *Pl. Physiol. Lancaster*, **47**, 361–365.

Bird, A. F. (1961) 'The ultrastructure and histochemistry of a nematode-induced giant cell', *J. Cell Biol.*, **11**, 701–715.

Blaikley, N. M. (1933) 'The structure of the foot in certain mosses and in *Anthoceros laevis*', *Trans. Roy. Soc. Edinb.*, **57**, 699–709.

Bonnett, H. T. (1968) 'The root endodermis: Fine structure and function', *J. Cell Biol.*, **37**, 199–205.

Bopp, M. and Weniger, H. P. (1971) 'Wassertransport vom Gametophyten zum Sporophyten bei Laubmoosen', *Z. PflPhysiol.*, **64**, 190–198.

Bower, F. O. (1908) *The origin of a land flora. A theory based upon the facts of alternation*, Macmillan, London.

Briarty, L. G. (1971) 'A method for preparing living plant cell walls for scanning electron microscopy', *J. Microscopy*, **94**, 181–184.

Clutter, M. E. and Sussex, I. M. (1968) 'Ultrastructural development of bean embryo cells containing polytene chromosomes', *J. Cell Biol.*, **39**, 26a.

Codaccioni, M. (1970 'Présence de cellules de type endodermique dans la tige de quelques Dicotylédones', *C. r. hebd. Séanc. Acad. Sci., Paris*, **271**, 1515–1517.

Corner, E. J. H. (1951) 'The leguminous seed', *Phytomorphology*, **1**, 117–150.

Costerton, J. W. F. and MacRobbie, E. A. C. (1970) 'Ultrastructure of *Nitella translucens* in relation to ion transport', *J. exp. Bot.*, **21**, 535–542.

Cronquist, A. (1968) *The evolution and classification of flowering plants*, Nelson, Ltd., London and Edinburgh.

Dainty, J. (1969) 'The water relations of plants', In: *Physiology of plant growth and development* (ed. M. B. Wilkins), McGraw-Hill, London.

Davey, M. R. and Street, H. E. (1971) 'Studies on the growth in culture of plant cells. IX. Additional features of the fine structure of *Acer pseudoplatanus* L. cells cultured in suspension', *J. exp. Bot.*, **22**, 90–95.

Davis, G. L. (1966) *Systematic embryology of the angiosperms*, John Wiley, New York.

Diamond, J. M. and Bossert, W. H. (1967) 'Standing-gradient osmotic flow', *J. Gen. Physiol.*, **50**, 2061–2083.

Diboll, A. G. (1968) 'Fine structural development of the megametophyte of *Zea mays* following fertilization', *Am. J. Bot.*, **55**, 787–806.

Diboll, A. G. and Larson, D. A. (1966) 'An electron microscopic study of the mature megagametophyte in *Zea mays*', *Am. J. Bot.*, **53**, 391–402.

Dick, E. G., Dick, D. A. T., and Bradbury, S. (1970) 'The effect of surface microvilli on the water permeability of single toad oocytes', *J. Cell Sci.*, **6**, 451–476.

Dixon, R. O. D. (1969) 'Rhizobia (with particular reference to relationships with host plants)', *Ann. Rev. Microbiol.*, **23**, 137–158.

Dormer, K. J. (1955) '*Asarum europaeum* – a critical case in vascular morphology', *New Phytol.*, **54**, 338–342.

Dörr, I. (1968) 'Feinbau der Kontakte zwischen *Cuscuta* – Hyphen und den siebröhren ihrer Wirtspflanzen', *Vortr. Ges. Geb., Bot. N.F.*, **2**, 24–26.

Eames, E. J. (1961) *Morphology of the angiosperms*, McGraw-Hill, New York.

Ernst, S. A. and Ellis, R. A. (1969) 'The development of surface specialization in the secretory epithelium of the avian salt gland in response to osmotic stress', *J. Cell Biol.*, **40**, 305–321.

Ernst, S. A. and Philpott, C. W. (1970) 'Preservation of Na^+-K^+-activated and Mg^{++}-activated adenosine triphosphatase activities of avian salt gland and teleost gill with formaldehyde as fixative', *J. Histochem. Cytochem.*, **18**, 251–262.

Eymé, J. and Suire, C. (1967) 'Au sujet de l'infrastructure des cellules de la région placentaire de *Mnium cuspidatum* Hedw. (Mousse Bryale acrocarpe)', *C. r. hebd. Séanc. Acad. Sci., Paris*, **265**, 1788–1791.

Fahn, A. and Rachmilevitz, T. (1970) 'Ultrastructure and nectar secretion in *Lonicera japonica'*, In: *New research in plant anatomy* (supplement to vol. 63, *Bot. J. Linn. Soc.*), Academic Press, London.

Feder, N. and O'Brien, T. P. (1968) 'Plant microtechnique: some principles and new methods', *Am. J. Bot.*, **55**, 123–144.

Goebel, K. (1900) *Organography of plants, especially of the Archegoniatae and Spermaphyta* (Engl. Translation), Oxford University Press, Oxford.

Graham, T. A. and Gunning, B. E. S. (1970) 'Localization of legumin and vicilin in bean cotyledon cells using fluorescent antibodies', *Nature, Lond.*, **228**, 81–82.

Gunning, B. E. S. (1970) 'Electron microscopy of transfer cells associated with the xylem of leaf traces', *Proc. R. microsc. Soc.*, **5**, 202–203.

Gunning, B. E. S. and Pate, J. S. (1969a) ' "Transfer Cells" – Plant cells with wall ingrowths, specialized in relation to short distance transport of solutes – their occurrence, structure and development', *Protoplasma*, **68**, 107–113.

Gunning, B. E. S. and Pate, J. S. (1969b) 'Cells with wall ingrowths (Transfer Cells) in the placenta of ferns', *Planta*, **87**, 271–274.

Gunning, B. E. S., Pate, J. S., and Briarty, L. G. (1968) 'Specialized "Transfer Cells" in minor veins of leaves and their possible significance in phloem translocation', *J. Cell Biol.*, **37**, C7–12.

Gunning, B. E. S., Pate, J. S., and Green, L. W. (1970) 'Transfer cells in the vascular system of stems: taxonomy, association with nodes and structure', *Protoplasma*, **71**, 147–171.

Guyomarc'h, C. (1969) 'Sur les variations de l'azote soluble au cours du développement du sporophyte chez deux Mousses Bryales', *C. r. hebd. Séanc. Acad. Sci., Paris*, **268**, 2339–2342.

Haberlandt, G. (1914) *Physiological plant anatomy*, 4th ed. (trans.), Macmillan & Co., London.

Idle, D. B. (1971) 'Preparation of plant material for scanning electron microscopy', *J. Microscopy*, **93**, 77–79.

Israel, H. W. and Sagawa, Y. (1964) 'Post-pollination ovule development in *Dendrobium* orchids. II. Fine structure of the nucellar and archesporial phases', *Caryologia*, **17**, 301–316.

Jacobson, S. L. (1971) 'A method for extraction of extracellular fluid: use in development of a physiological saline for Venus's-flytrap', *Can. J. Bot.*, **49**, 121–127.

Jensen, W. A. (1965) 'The ultrastructure and histochemistry of the synergids of cotton', *Am. J. Bot.*, **52**, 238–256.

Kelley, C. (1969) 'Wall projections in the sporophyte and gametophyte of *Sphaerocarpos'*, *J. Cell Biol.*, **41**, 910–914.

Kiesselbach, T. A. and Walker, E. R. (1952) 'Structure of certain specialized tissues in the kernel of corn', *Am. J. Bot.*, **39**, 561–569.

Kollmann, R. and Dörr, I. (1969) 'Strukturelle Grundlagen des zwischenzelligen Stoffaustausches', *Ber. dt. bot., Ges.*, **82**, 415–425.

Kursanov, A. L. (1968) 'Free space and transport of metabolites in parenchymal tissues', In: *Transport and distribution of matter in cells of higher plants* (eds. K. Mothes, E. Müller, A. Nelles, and D. Newmann), Akademie-Verlag, Berlin.

Kursanov, A. L. and Brovchenko, B. (1969) 'Free space as an intermediate zone between photosynthesizing and conducting cells of leaves', *Fiziol. Rast.*, **16**, 965–972.

Ledbetter, M. C. (1968) 'The casparian strip: A site of a functional tight junction in plant roots', *J. Cell Biol.*, **37**, 39A.

Lorch, W. (1925) 'Über die Saugzellen im Fusse und in der Vaginula bei den Laubmoosen', *Ber. dt. bot. Ges.*, **43**, 120–126.

Lorch, W. (1931) 'Anatomie der Laubmoose', In: Linsbauer, *Handbuch der Pflanzenanatomie* vol. VII/I Borntraeger, Berlin.

MacRobbie, E. A. C. (1970) 'Quantized fluxes of chloride to the vacuole of *Nitella translucens*' *J. exp. Bot.*, **21**, 355–364.

Maier, K. (1967) 'Wandlabyrinthe im Sporophyten von *Polytrichum'*, *Planta*, **77**, 108–126.

Marinos, N. G. (1970) 'Embryogenesis of the Pea (*Pisum sativum*) I. The cytological environment of the developing embryo', *Protoplasma*, **70**, 261–279.

Marquardt, H., Barth, O. M., and von Rahden, U. (1968) 'Zytophotometrische und elektronenmikroskopische Beobachtungen über die Tapetumzellen in den Antheren von *Paeonia tenuifolia'*, *Protoplasma*, **65**, 407–421.

Mayo, M. A. and Cocking, E. C. (1969a) 'Pinocytic uptake of polystyrene latex particles by isolated tomato fruit protoplasts', *Protoplasma*, **68**, 223–230.

Mayo, M. A. and Cocking, E. C. (1969b) 'Detection of pinocytic activity using selective staining with phosphotungstic acid', *Protoplasma*, **68**, 231–236.

O'Brien, T. P. and Carr, D. J. (1970) 'A suberized layer in the cell walls of the bundle sheath of grasses', *Aust. J. biol. Sci.*, **23**, 275–287.

O'Brien, T. P. and Zee, S.-Y. (1971) 'Vascular transfer cells in the vegetative nodes of wheat', *Aust. J. biol. Sci.*, **24**, 207–217.

Oschman, J. L. and Berridge, M. J. (1971) 'The structural basis of fluid secretion', *Fed. Proc.*, **30**, 49–56.

Paolillo, D. J. and Bazzaz, F. A. (1968) 'Photosynthesis in sporophytes of *Polytrichum* and *Funaria*', *The Bryologist*, **71**, 335–343.

Pardee, A. B. (1968) 'Membrane transport proteins', *Science, N.Y.*, **162**, 632–637.

Pate, J. S. (1971) 'Movement of nitrogenous solutes in plants', In: *Nitrogen–15 in soil plant studies*, International Atomic Energy Agency, Vienna.

Pate, J. S. and Gunning, B. E. S. (1969) 'Vascular transfer cells in Angiosperm leaves: a taxonomic and morphological survey', *Protoplasma*, **68**, 135–156.

Pate, J. S., Gunning, B. E. S., and Briarty, L. G. (1969) 'Ultrastructure and functioning of the transport system of the leguminous root nodule', *Planta*, **85**, 11–34.

Pate, J. S., Gunning, B. E. S., and Milliken, F. F. (1970) 'Function of transfer cells in the nodal region of stems, particularly in relation to the nutrition of young seedlings', *Protoplasma*, **71**, 313–334.

Paulson, R. E. and Webster, J. M. (1969) 'Giant cell formation in tomato roots caused by *Meloidogyne incognita* and *Meloidogyne hapla* (Nematoda) infection. A light and electron microscope study', *Can. J. Bot.*, **48**, 271–276.

Rosen, W. G. (1968) 'Ultrastructure and physiology of pollen', *A. Rev. Pl. Physiol.*, **19**, 435–462.

Rosen, W. G. and Gawlik, S. R. (1966) 'Relation of lily pollen tube fine structure to pistil incompatibility and mode of nutrition', In: *Electron microscopy* (Proc. VI Intern. Micros., Kyoto), Maruzen, Tokyo.

Rosen, W. G. and Thomas, H. R. (1970) 'Secretory cells of lily pistils. I. Fine structure and function', *Am. J. Bot.*, **57**, 1108–1114.

Rost, T. L. (1970) 'Fine structure of the aleurone layer in yellow foxtail grass (*Setaria lutescens*)', *Am. J. Bot.*, **57**, 738.

Rost, T. L. and Lersten, N. R. (1971) 'Transfer aleurone cells in *Setaria lutescens* (Gramineae)', *Protoplasma*, **71**, 403–408.

Rowley, J. R. and Flynn, J. J. (1971) 'Migration of lanthanum through the pollen wall', *Cytobiol.*, **3**, 1–12.

Schnepf, E. and Nägl, W. (1970) 'Uber einige Strukturbesonderheiten der Suspensorzellen von *Phaseolus vulgaris*', *Protoplasma*, **69**, 133–143.

Schulz, S. R. and Jensen, W. A. (1969) '*Capsella* embryogenesis: The suspensor and the basal cell', *Protoplasma*, **67**, 139–163.

Schulz, S. R. and Jensen, W. A. (1971) '*Capsella* embryogenesis: the chalazal proliferating tissue', *J. Cell Sci.*, **8**, 201–227.

Segel, L. A. (1970) 'Standing-gradient flows driven by active solute transport', *J. theor. Biol.*, **29**, 233–250.

Stoward, P. J. (1967) 'The histochemical properties of some periodate-reactive mucosubstances of the pregnant Syrian hamster before and after methylation with methanolic thionyl chloride', *Jl. R. microsc. Soc.*, **87**, 77–103.

Tyree, M. T. (1969) 'The thermodynamics of short-distance translocation in plants', *J. exp. Bot.*, **20**, 341–349.

Tyree, M. T. (1970) 'The symplast concept. A general theory of symplastic transport according to the thermodynamics of irreversible processes', *J. theor. Biol.*, **26**, 181–214.

Van der Pluijm, J. E. (1964) 'An electron microscopic investigation of the filiform apparatus in the embryo sac of *Torenia fournieri*', In: *Pollen physiology and fertilization*, 6–16, North-Holland Publishing Co., Amsterdam.

Van Went, J. and Linskens, H. F. (1967) 'Die Entwicklung des sogenannten "Fadenapparates" im Embryosack von *Petunia hybrida*', *Genet. Breeding Res.*, **37**, 51–56.

Vazart, B. and Vazart, J. (1966) 'Infrastructure du sac embryonnaire du lin (*Linum usitatissimum* L.)', *Rev. Cytol. Biol. Végétales*, **24**, 251–266.

Transfer cells

Vazart, J. (1968) 'Infrastructure de l'ovule du lin, *Linum usitatissimum* L. Le complex antipodal', *C. r. hebd. Séanc. Acad. Sci., Paris,* **266,** 211–213.

Vazart, J. (1969) 'Organization et ultrastructure du sac embryonnaire du lin (*Linum usitatissimum*)', *Rev. Cytol. Biol. Végétales,* **32,** 227–240.

Wardlaw, C. W. (1955) *Embryogenesis in Plants,* Methuen & Co. Ltd., London.

Welk, M., Millington, W. F., and Rosen, W. G. (1965) 'Chemotropic activity and the pathway of the pollen tube in lily', *Am. J. Bot.,* **52,** 774–781.

Wooding, F. B. P. and Northcote, D. H. (1965) 'An anomalous wall thickening and its possible role in the uptake of stem-fed tritiated glucose by *Pinus pinea', J. Ultrastruct. Res.,* **12,** 463–472.

Zee, S.-Y. and O'Brien, T. P. (1971a) 'Vascular transfer cells in the wheat spikelet', *Aust. J. biol. Sci.,* **24,** 35–49.

Zee, S.-Y. and O'Brien, T. P. (1971b) 'Aleurone transfer cells and other structural features of the spikelet of millet', *Aust. J. biol. Sci.,* **24,** 391–395.

Ziegler, H. (1965) 'Die Physiologie pflanzlicher Drüsen', *Ber. dt. bot. Ges.,* **78,** 466–477.

14

Reproduction in flowering plants
William A. Jensen

Introduction

In 1824, Giovanni Amici, Italian mathematician, astronomer, and microscope maker, saw the pollen of *Portulaca oleracea* germinate on the stigma and thus became the first man to see a pollen tube. Three years later, a young French botanist, Brongniart, made a more complete study of pollen tubes and misunderstood their true structure and function, thus establishing a pattern to be followed by many plant embryologists over the next 140 years. A great many brilliant observations and correct conclusions were made during this period, but as many controversies flared and problems remained unresolved.

By 1900, the general pattern of reproduction in flowering plants had been established. Pollen was known to consist of two cells: the vegetative, that would produce the tube; and the generative, that would form two male gametes or sperms. Pollen was known to germinate on the stigma, grow down the style, and carry the male gametes to the embryo sac contained within the ovule. The 'normal' embryo sac was established as a seven celled, eight nucleate structure: the egg and two additional cells, the synergids, at one end; three cells called the antipodals, at the other; and a large binucleate central cell in between. Somehow, the pollen tube discharged into the embryo sac, one male gamete nucleus fusing with the egg nucleus to form the zygote and the second fusing with the two nuclei of the central cell (the polar nuclei) to give rise to the primary endosperm nucleus. This is the famous double fertilization of angiosperms independently discovered by Sergius Nawaschin and Leon Guignard in 1898–99.

The first half of this century saw many of the details of the structure and development of pollen, pollen tubes, embryo sacs, and embryogenesis described. The microscope and histological techniques were pushed to their limits and a great body of literature accumulated, which has been masterfully summarized by P. Maheshwari (1950). Yet, many fundamental questions remained unanswered. What was the nature of the generative cell? Indeed, was it a true cell at all? Were the male gametes cells or only naked nuclei? What was the condition of the cells of the embryo sac? Were they true cells? If so, were they surrounded by walls or not? How did the pollen tube locate the embryo sac? How did it enter the embryo sac and what was the method of discharge? What was the role of the synergids, if any, in pollen tube discharge? How did one male nucleus always enter the egg and the second the central cell? Were the male gametes different in size or shape? Did they move? How did the nuclei fuse? What changes occurred in the zygote between the time it was formed and the first division? What directed the early development of the embryo?

These are but some of the questions that remained unanswered. Many still remain outstanding, but ultrastructural studies have answered many. It is only fair to point out that many observations yielding the correct answers were made with the light microscope, but in almost every case contrary data was also collected that clouded the issue to the point where the question remained open. In addition, as one may easily predict, the ultrastructural studies have uncovered more questions, many of which remain unanswered. To illustrate the present stage of knowledge in this area of botany, I would like to examine some of these questions and the answers we now have.

Structural problems in angiosperm reproduction

What is the ultrastructure of the pollen grain?

The general morphology of the pollen grain was well known before the advent of the electron microscope (Maheshwari, 1950). Indeed, a great and extensive literature exists on pollen, particularly the wall of the pollen grain that is so useful in identifying the plant producing the pollen (Erdtman, 1952). I will not attempt to review this literature, to which the electron microscope has contributed much, but will confine my discussion to the inside of the grain.

The cytoplasm of the vegetative cell is extremely dense in the mature grain, there are no large vacuoles, and what vacuoles are present are small (Bopp-Hassenkamp, 1960; Diers, 1963b; Sassen, 1964; Dexheimer, 1965b; Larson, 1965; Maruyama, 1966; Jensen, Fisher, and Ashton, 1968; Chardard, 1969; Vazart, 1969 Cocucci and Jensen, 1970). Plastids of varying degrees of complexity have been seen in pollen; these are, of course, storage plastids, and elaborate and highly characteristic amyloplasts are found in many pollen grains. Numerous dictyosom and mitochondria are also present in all pollen.

The endoplasmic reticulum (ER) is one of the most interesting of the various cell parts found in pollen (Maruyama, 1966; Chardard, 1969). Present in large amounts it frequently has large cisternal regions filled with dense material known to be protein and suspected of containing carbohydrate. The ER is found in many forms in pollen: in cotton, for example, the ER forms spherical pockets (Fisher, Jensen, and Ashton, 1968) which are lined with lipid droplets and packed with dictyosome vesicles containing carbohydrate. The surface of the ER is completely covered with ribosomes spaced so closely that it is difficult to resolve them individually. All in all, the ER is a remarkable storage structure admirably evolved with regard to the function of pollen.

The vegetative nucleus is usually large, and frequently highly lobed or indented

(Diers, 1963b; Sassen, 1964; Jensen, Fisher, and Ashton, 1968), and chromatin is usually not seen condensed. No nucleoli can be seen in many species, and, if they are present, they are small; an exception appears to be the orchids where large, persistent nucleoli are found (Cocucci and Jensen, 1970).

What is the nature of the generative cell?

The ultrastructure of the generative cell is of considerable interest because it will divide to form the two male gametes or sperms. On the basis of early light microscope observations questions were raised as to whether or not the generative nucleus seen in the pollen grain was surrounded by a true cell. While this point had been settled, at least for some species, with additional light microscope observations, it was one of the first attacked with the electron microscope. The work of Bopp-Hassenkamp (1960), Diers (1963a), and Sassen (1964) showed the generative cell to be a true cell, although one greatly reduced in structure. Further studies have extended our knowledge of generative cells in numerous species and have indicated some of the variations in structure that can be found.

All generative cells are surrounded by a plasma membrane and a cell wall, although the thickness of the wall may vary greatly. In some species it is thin or incomplete (Maruyama, Gay, and Kaufman, 1965; Fisher, Jensen, and Ashton, 1968; Burgess, 1970), while in others it is thick and may be clearly visible with the light microscope (Sassen, 1964; Cocucci and Jensen, 1968; Vazart, 1969). Histochemical work (Jensen, Fisher, and Ashton, 1968) at both the light and electron microscope level has shown that this wall is polysaccharide in nature.

The cytoplasm of the generative cell is poor in organelles (Jensen, Fisher, and Ashton, 1968; Chardard, 1969; Burgess, 1970): most such cells contain small amounts of ER, some ribosomes, a few dictyosomes, assorted single membrane bound vesicles, and a handfull of mitochondria. Plastids have not been found in the generative cells of most species so far examined, however one exception is that of *Pelargonium* in which amyloplasts are clearly present (Lombardo and Gerolay, 1968). Available evidence suggests that plastids are excluded during the formation of the generative cell (Angold, 1968; Mepham and Lane, 1970).

The nucleus of the generative cell is usually a compact structure containing condensed chromatin and a nucleolus. Collections of microtubules have been found at the end of the nucleus in at least one species (Jensen, Fisher, and Ashton, 1968); and they are commonly seen in the cytoplasm, sometimes in clusters near the periphery of the cell (Hoefert, 1969; Burgess, 1970).

What happens upon germination of the pollen?

The pollen germinates shortly after it lands on the surface of a compatible stigma. There are two types of stigmatic surfaces: those that are wet and those that are dry (Vasil and Johri, 1964). Linskens and his co-workers have made careful studies of the petunia stigma (Konar and Linskens, 1966a, b; Linskens, 1969), which is a wet type, while I and my co-workers (Jensen and Fisher, 1969) have examined cotton, which has a dry surface.

In petunia, the exudate is secreted via the dictyosome vesicles and first accumulates between the wall and the cuticle (Kroh, 1964; Konar and Linskens, 1966a, b). The exudate was analysed and shown to contain both carbohydrates and oils. The cells are rich in cytoplasm containing large amounts of ER. In contrast, the stigmatic cells of cotton are highly vacuolate, contain relatively little cytoplasm, and produce no exudate (Jensen and Fisher, 1969); instead, they appear to degenerate on the day that the surface is receptive to pollen.

At germination, a tube grows from one of the previously differentiated pores in the wall of the pollen grain. The wall that is produced is continuous with the innermost layer, or intine, of the wall surrounding the pollen grain (Larson, 1965). There have been few detailed studies of the ultrastructural events associated with the early development of the pollen tube wall (Larson, 1965; Crang and Miles, 1969); however, there are some excellent studies of the growth of the pollen tube particularly at the tip (Rosen et al., 1964; Dashek and Rosen, 1966; Rosen, 1969).

Physiological and biochemical studies have shown that the early growth of the pollen tube is controlled by messenger RNA present in the pollen prior to germination (Mascarenhas, 1965), and only after growth has started is more messenger RNA produced by the vegetative nucleus. The same studies showed that ribosomes are not synthesized by the vegetative nucleus in snapdragons. This is compatible with the observations that the vegetative nucleus lacks a nucleolus and that the ER in the pollen grain is literally packed with ribosomes (Jensen, Fisher, and Ashton, 1968). Such observations also fit well with the fact that most pollen tubes live for only a matter of hours and grow at high rates of speed (Linskens, 1969); one exception being the pollen tubes of many orchids that may take weeks or even months to reach the embryo sac. As noted earlier, vegetative nuclei of such pollen contain nucleoli and it would be interesting to determine if such pollen tubes synthesize new ribosomes (Cocucci and Jensen, 1969c).

The nature of the wall of the pollen tube, especially at the tip where growth is occurring, has been studied intensively, particularly by Rosen and his co-workers (Rosen et al., 1964; Rosen, 1969). Their observations make the point that the condition of the wall is different in tubes growing *in vitro,* as opposed to those growing *in vivo* (Rosen and Gawlik, 1966). The latter appear to have wall thickening similar to transfer walls near the tip, while the *in vitro* tubes have only thin walls. The walls of all pollen tubes are not necessarily thin; those of cotton, for example, are thick and multilayered (Jensen and Fisher, 1970), while those of lily appear relatively thin.

The cytoplasm of the pollen tube near the growing tip is filled with cell organelles (Dexheimer, 1965a; Jensen and Fisher, 1970) which, at the very tip, are found in characteristic arrangements (Rosen et al., 1964; Dashek and Rosen, 1966). Prominent among these are membrane-bound vesicles which are filled with polysaccharide material and are involved in cell wall synthesis while the tube is growing, thus marking the extent of pollen tube discharge in the embryo sac (Jensen and Fisher, 1968b). The careful work of Woude, Morré, and Bracker (1971) has not only characterized them on the ultrastructural level but isolated them and analysed them chemically.

In the older parts of the pollen tube the cytoplasm is pressed to the wall by a large vacuole and plugs of wall material form, sealing off sections of the tube (Jensen and Fisher, 1970). In cotton some portions of the ER become extended with protein (Jensen and Fisher, 1970). The cytoplasm trapped behind the plug degenerates (Jensen and Fisher, 1970).

What is the relationship of the pollen tube to the cells of the stigma and style?

We saw earlier that stigmas come in two conditions, either wet or dry. Styles also come in two forms either open (as in *Lilium*) or solid (as in cotton or petunia).

In the case of the flowers with open styles the pollen tubes germinate on the stigmatic hairs and grow down the surface on these hairs to the mucilage-filled canal (Yamada, 1965) in the centre of the style. This canal is lined with special cells that develop highly thickened walls in a characteristic 'transfer type' cell wall thickening (Rosen and Thomas, 1970). The work of Rosen and his associates has clearly demonstrated the transfer of material from the canal cells to the pollen tubes (Kroh, *et al.*, 1970).

In the case of the solid styles the picture is slightly more complicated because the tube must grow into the tissue of the flower. In the case of *Petunia*, it has been shown that the pollen tube first penetrates the cuticle of the stigmatic hair cells (Konar and Linskens, 1966a, b). The exudate from the stigmatic hairs accumulates between the wall and the cuticle and it is in this layer that the pollen tube tip grows. The tube continues to grow in the wall layers, and when it reaches the style it encounters the special cells of the transmitting tissue which are elongate, and thick-walled, and between which the pollen tube grows. In cotton, also a solid style flower, the picture is somewhat different (Jensen and Fisher, 1969). In this case the pollen tube first grows along the surface of the stigmatic cells, penetrating the cuticle only at the base of the cells; it then grows in the large intercellular spaces that are present between the stigmatic cells and then through the wall layers of the transmitting cells which have thick, loosely organized walls. There is evidence (Jensen and Fisher, 1970) that the pollen tube grows through the third layer of these transmitting cells of the style rather than the middle lamella. As yet there is no clear evidence that the pollen tubes in a solid style absorb substances from the transmitting cells, but it would be difficult to believe that they are not receiving some material from these cells. At least in the case of cotton there seems to be no clear relationship between the growth of the pollen tube and degeneration of the transmitting cells.

What is the condition of the sperm?

As the pollen tube grows down the style, the generative cell divides, producing two sperms or male gametes. (In some species, particularly in the grasses, the generative cell divides during the formation of the pollen grains so that the sperm are present in the pollen.)

Despite their obvious importance to the whole process of reproduction in flowering plants, few sperm have been seen with the electron microscope. There are published reports for the sperms of cotton (Jensen and Fisher, 1968a), *Petunia*

(Sassen, 1964), barley (Cass, 1973) and sugar beet (Hoefert, 1969), the latter two specimens having sperms that are present in the pollen. The general conclusion, from what observations we have, is that sperms are true cells possessing a plasma membrane but apparently lacking a cell wall, at least in the pollen tube. The angiosperm sperms are simple cells, each consisting of a nucleus, usually with condensed chromatin, and a nucleolus. The cytoplasm that surrounds this nucleus is small in amount and contains only a few cell organelles. The mitochondria that are present seem greatly reduced in internal structure and to date no plastid or plastid-like structures have been seen. A few dictyosomes may be present, as well as small amounts of ER and some ribosomes. Numerous single membrane-bound vesicles are seen sometimes containing dense material within them.

Characteristic of the sperm cytoplasm is the presence of microtubules. In the sperms of sugar beet and barley the tubules are in clusters of two or three in a pattern around the periphery of the cell running longitudinally to the long axis of the cell, while, in cotton, there are random microtubules in the periphery of the cell but these are not grouped into clusters. There are several obvious interpretations for such clusters of microtubules: the first, suggested by Hoefert (1969), is that they are involved in a possible movement of the sperm and may account for the amoeboid motion ascribed to some plant sperm on the basis of light microscope observation (Vazart, 1958) (although this observation has been recently challenged by Navashin, 1969; Cass, 1973); the second suggestion (Cass, 1973) is that they are skeletal elements and accounts for the elongate appearance of the sperm in the pollen and pollen tube; finally, as suggested by Burgess (1970), they may have had a role in the division of the generative cell. At present there is not enough evidence available to suggest which possibility is the correct one.

With regard to the absence of plastids or plastid-like structures in the sperm, it is important to point out that the sperms of such species as *Pelargonium*, in which amyloplasts have been seen in a generative cell and in which cytoplasmic plastid inheritance is known to occur, have not yet been seen with the electron microscope. This is clearly an important area of research that has been overlooked and requires additional investigation.

The electron microscope observations also bear on an old contention with regard to sperm. This is, that the two sperm cells are different in some morphological feature such as size or density of cytoplasm. This difference in size is critical to the argument of how double fertilization can occur, and, to date, no differences in ultrastructure have been seen between the two sperms. They appear to be equal in size, organization and content.

What is the pollen tube growing toward?

The ultimate destination of the pollen tube, of course, is the egg of the embryo sac, which is located in the ovule consisting of the outer and inner integuments, the nucellus, and the embryo sac or megagametophyte. The condition of these

various tissues varies greatly from species to species (Maheshwari, 1950). In
Capsella, for example (Schulz and Jensen, 1968a), the nucellus has disappeared
in the mature ovule, while in cotton the nucellus is a massive structure of
thirty or more cell layers surrounding the embryo sac (Jensen, 1963). The
ultrastructure of the nucellar cells in cotton has been studied in detail (Jensen,
1965a), and a column of cells was found through which the pollen tube grows
on its way to the embryo sac; these cells differentiate and degenerate before
the pollen tube reaches them.

The embryo sac itself can contain varying numbers of cells and varying numbers
of nuclei (Mahashwari, 1950). The most general case of the so-called 'normal'
embryo sac is an eight nucleate, seven celled structure: at the micropylar end are
three cells of the egg apparatus — the egg itself and the two synergids; at the
chalazal end are the three antipodal cells. Separating them is the large central
cell with its two polar nuclei. The bulk of ultrastructural studies thus far conducted have been with embryo sacs of the normal type that include *Torinia,*
cotton, *Capsella,* corn (maize) and orchid. On the basis of these ultrastructural
studies we can answer a number of questions concerning the cells of the embryo
sac. Let us start by considering the ultrastructure of the various cells in the
embryo sac at maturity but before fertilization.

What is the ultrastructure of the synergids?

The great majority of embryo sacs have two synergids. These are large cells that
are occupied at their micropylar end by a structure known as the filiform
apparatus and at their chalazal end by one or more conspicuous vacuoles. The
nucleus of the cell is located either in the centre or toward the micropylar end.

The nature of the filiform apparatus and the wall surrounding the synergid
have long been a source of discussion among plant embryologists. Research with
the electron microscope has made it clear that the filiform apparatus is an extension of the wall of the synergid (Jensen, 1963, 1965b; Pluijm, 1964; Godineau,
1966; Van Went and Linskens, 1967; Diboll, 1968; Schulz and Jensen, 1968a;
Vazart, 1969; Cocucci and Jensen, 1969a; Van Went, 1970a). The filiform apparatus
is shaped in the form of a number of finger-like projections that extend into the
cell. The structure of this wall is unusual in that it contains a number of elements
including a twisted cylindrical unit surrounded by a puffier wall material. Preliminary histochemical investigations indicate that the wall is extremely rich in
pectins and hemicellulose and contains little cellulose (Jensen, 1963). In *Petunia*
the filiform apparatus appears less elaborate and more fleeting in its appearance,
developing only late in the maturation of the cell and decreasing after fertilization (Van Went and Linskens, 1967). One of the effects of the filiform apparatus
is to greatly increase the area of the plasma membrane.

The synergid is surrounded by walls that become progressively thinner from
the micropylar to the chalazal end. In cotton, corn, and other plants (Jensen,
1963, 1965; Godineau, 1966; Diboll, 1968; Vazart, 1969; Van Went, 1970a), the wall
disappears about half way along this distance and the chalazal end of the cell is
surrounded only by a plasma membrane and not an additional wall. In some

plants, such as *Capsella* (Schulz and Jensen, 1968a) and orchid (Cocucci and Jensen, 1969a), the wall is present around the entire cell but becomes honey-combed at the chalazal end.

The nuclei of the synergids are large and contain a prominent nucleolus. The cytoplasm is dense and contains numerous mitochondria, storage plastids, dense amounts of ER, many dictyosomes, and numerous ribosomes and vesicles. The synergid in cotton is about 100 μm long and there are three distinct populations of dictyosomes found in various parts of the cell (Jensen, 1965b). The cytoplasm of the synergid gives all the appearances of being that of a highly active cell. Histochemical observations have indicated large amounts of protein and nucleic acids in them and there is evidence of high protein synthesis (Jensen, 1965b; Schulz and Jensen, 1968a). One of the curious aspects of the synergids resides in the vacuoles, which have been shown by microincineration studies in cotton, to contain large amounts of ash (Jensen, 1965b). It has not yet been possible to analyse the nature of the ash but, because of the incineration conditions used, it is highly likely that it is either phosphorus or calcium. We will return to the synergids when we consider the events associated with fertilization and pollen tube discharge. It is clear, however, that the synergids are highly active cells and are not simply the evolutionary remains of more elaborate structures. While all the things they are doing are not completely clear they must be carrying out many functions and they obviously have an important and vital role in the development and functioning of the embryo sac.

What have ultrastructural studies revealed concerning the egg?

The egg of the angiosperm actually appears more variable than the synergids in terms of its ultrastructure. Some, as is the case for cotton (Jensen, 1965b), are highly vacuolate cells with a small number of mitochondria, amyloplasts, and dictyosomes. The ribosomes that are present are usually few in number and not associated into elaborate polysomes. There is a small amount of ER present and much of this closely appressed to the plasma membrane. This general pattern is also true for *Petunia* (Van Went, 1970b). In other eggs, such as corn (Diboll, 1968), the egg has relatively few, small vacuoles and a rich, dense cytoplasm filled with mitochondria, plastids and dictyosomes. Still other eggs are more or less inter-mediate, as is the case of *Capsella* (Schulz and Jensen, 1968b) and *Linum* (Vazart, 1969) where there are prominent vacuoles but the cell is not as highly vacuolate as it is in cotton. The ultrastructure of some half a dozen eggs are known and they give the general impression of cells which are not metabolically active. It is not clear whether they should be considered dormant, as they contain quite a large number of organelles and seem to be high in protein and nucleic acids, yet they surely are less active in appearance when compared with neighbouring synergids.

Again the egg has a wall surrounding it at the micropylar end, and either no wall at the chalazal end or a wall which is honeycomb-like in appearance. Only in *Epidendrum* (Cocucci and Jensen, 1969a) has a thick wall been observed surrounding the egg, but even here there are notable gaps in the wall at the chalazal end of the cell.

What of the ultrastructure of the central cell?

The central cell is one of the neglected cells of the embryo sac, although it is the progenitor of the endosperm that has excited much interest over the years. In fact, it is generally ignored as a cell in the classical literature. In the normal embryo sac it is a binucleate cell, and these two nuclei are usually termed the polar nuclei. They may fuse, remain separate or be partially fused before fertilization occurs. In the early literature, the cell in which these nuclei occur is usually unnamed and is assumed to be the remains of the cytoplasm of the embryo sac left over from the formation of the other cells that are present (Mahashwari, 1950). It is, however, a perfectly good cell and an interesting one.

The central cell is surrounded and clearly demarked by its own plasma membrane. The wall surrounding the central cell is quite complex and highly variable: in some cases, such as *Capsella* (Schulz and Jensen, 1969), it has extensions from the wall that are like small filiform apparatus projections; in other species, such as cotton, the wall is generally smooth (Jensen, 1965b). There is little wall that can be identified as being associated with the central cell around the chalazal end of the egg or synergids that project into it. The central cell is usually occupied by a large vacuole which presses the cytoplasm toward the periphery of the cell. This cytoplasm is rich in ER and ribosomes and contains numerous mitochondria, amyloplasts or, in some cases, chloroplasts and dictyosomes (Jensen, 1963; Schulz and Jensen, 1969; Van Went, 1970b).

The polar nuclei are usually the largest in the embryo sac, and frequently contain enormous nucleoli (Jensen, 1963; Vazart, J., 1969; Van Went, 1970b). There are numerous connections between the ER and the nuclear envelope of the polar nuclei. The polar nuclei in cotton contain many unusual ultrastructural features: they contain small ribosome-like granules that are fixed in permanganate and are therefore not true ribosomes (Jensen, 1963); they also apparently contain vesicles within the nuclear envelope and small satellite nucleoli associated with the large ones. Finally, in contrast to the egg, the cytoplasm of the central cell is active in appearance.

What is the ultrastructure of the antipodals?

Antipodals have been studied in such things as corn (Diboll and Larsen, 1966), *Capsella* (Schulz and Jensen, 1971), and *Epidendrum* (Cocucci and Jensen, 1969a). They are often not present in the mature embryo sac as in the case of cotton. In some species they are small cells with relatively thick walls and a dense cytoplasm filled with ribosomes (Cocucci and Jensen, 1969a) In the case of corn (Diboll, 1968) they continue to divide for some time so that their numbers may be quite large. The antipodals of corn have a developmental sequence through which they progress which has been detailed by Diboll (1968). We will not discuss at length the antipodals as they are not an immediate participant in the act of fertilization in angiosperms. Instead, let us turn our attention back to the pollen tube and consider what happens as it appears near the embryo sac.

How does the pollen tube enter the embryo sac?

It is in this area of the entrance and discharge of the pollen tube that the electron microscope observations have contributed much to the clarification of observations made with the light microscope. The electron microscope observations themselves, however, have raised numerous questions, and many of the events associated with pollen tube entrance and discharge into the embryo sac remain only partially explained. I would, therefore, like to describe briefly what we know happens at this stage and present some of the problems that remain.

The pollen tube grows across the funiculus, down the micropyle and through the nucellus — if that tissue is still present at this stage. The pollen tube continues to grow into the filiform apparatus of one of the two synergids and then into the cytoplasm of that same synergid. Once in the synergid, the pollen tube grows for a short distance and then discharges. The pollen tube discharge includes, besides the cytoplasm, the tube nucleus and the two sperms. This same pattern of pollen tube growth and discharge has been found in every species thus far examined with the electron microscope (Jensen, 1963; Pluijm, 1964; Diboll, 1968; Schulz and Jensen, 1968a; Jensen and Fisher, 1968b; Cocucci and Jensen, 1969b; Van Went, 1970a). It seems reasonable to assume at this point that it is, in fact, the general pattern, and that any deviations from it must now be considered minor. Before these observations with the electron microscope, four patterns were ascribed to pollen tube discharge in embryo sacs: (a) the pollen tube destroyed the synergid on contact and discharged into the embryo sac; (b) a pollen tube grew between the egg and synergid and discharged into the embryo sac; (c) the pollen tube grew between the egg or synergid and the embryo sac wall, and discharged into the embryo sac; and (d) the pollen tube grew into one of the synergids and discharged into that synergid. The correct observations had been made by Wylie (1941) with the light microscope but this was considered a minor variation on the general pattern. The first description given was considered the usual one. Finally, it must be pointed out that in several cases, as in *Petunia*, the light microscope observations described the pollen tube as growing between the wall of the embryo sac and the synergid, and discharging directly into the embryo sac, while the electron microscope observations place petunia with all the others in having the pollen tube discharge into the synergid. It should also be pointed out that the earlier observations tended to ignore the cellular nature of the central cell and use 'central cell' as synonymous with the embryo sac. This clearly leads to problems in understanding how the sperm nuclei can enter the egg and the central cell.

There is no question that the general pattern I have stated above is true on the basis of electron microscope observations so far made, yet it is equally true that many questions and problems remain unanswered. Let us look at some of these before proceeding to other aspects of embryogenesis in angiosperms.

What directs the pollen tube into one of the two synergids?

The two synergids and their filiform apparatus completely fill the micropylar end of the embryo sac. The egg is arranged distal to the end of the synergid so that the tip of the egg does not reach the micropylar end of the embryo sac itself. The tip of the pollen tube must therefore meet one of the synergids, but why does it grow into only one? In *Torenia* (Pluijm, 1964) the pollen tube apparently grows between the two synergids for a while before discharging into one of them, but in all other species examined it grows directly through the filiform apparatus of one synergid. Is this purely by chance or is there some directing influence? My co-workers and I (Jensen and Fisher, 1968b) believe that one of the two synergids begins to degenerate before the pollen tube enters it. We have made observations on cotton which suggest that there are changes initiated in one of the two synergids before the pollen tube reaches it. As the second synergid is still in an unaltered condition after the pollen tube has discharged, it seems only reasonable to conclude that the synergid which had begun to degenerate earlier was the one entered by the pollen tube. Observations by Cass and Jensen (1970) point to the same phenomena in barley, except that here the synergid will degenerate without pollination. Again in barley the pollen tube grows into the degenerated synergid. Mogensen (1972) reports a similar phenomenon in *Quercus*. There are other observations which suggest that the growth of the pollen tube during this final stage is not at random. In *Epidendrum* (Cocucci and Jensen, 1969b), for example, the pollen tube grows directly through the micropyle to the micropylar end of the embryo sac, the nucellus having degenerated during the final stages of maturation of the embryo sac. In this species the pollen tube forms a cap over the end of the embryo sac and then a smaller tube grows through the filiform apparatus of one of the two synergids and discharges into it. This hardly sounds like a random phenomenon. Observations on cotton also suggest that the compound which is present in the vacuoles of the synergids is released some time during the growth of the pollen tube to the embryo sac (Jensen and Fisher, 1968b). Whether this substance, or some other, is acting as a chemotropic guide to the pollen tube during its final stages of development remains unanswered, yet it is a reasonable hypothesis to pursue. Chemotropic substances have been demonstrated at various stages of the pollen tube development and it seems logical that such substances may be involved with this final growth of the pollen tube (Mascarenhas and Machlis, 1962; Rosen, 1962). *Torenia* (Pluijm, 1964) and *Petunia* (Van Went, 1970a) do not show such changes. It might be that such changes do occur in these species but at a chemical rather than a morphological level. It would be interesting to see if there are any differences that can be detected in the two synergids before pollen tube entrance in these species.

The work of Chao (1971) suggests that, in *Paspalum,* the pollen tubes are growing down a path of periodic acid—Schiff's reacting material that leads them to the embryo sac and the synergids.

What happens to the synergid after the entrance of the pollen tube?

The cytoplasm of the synergid rapidly undergoes degeneration, one of the significant features of which is the breakdown of the synergid plasmalemma (Jensen and Fisher, 1968b). The plasmalemma of cells adjacent to the degenerating synergid is not affected. Apparently the cytoplasm of the degenerating synergid rapidly gels and forms a semi-solid mass, and becomes marked by organelles such as mitochondria and plastids which have enormously thickened, blackened outer membranes even while the inner membranes may still appear to be normal. The ER becomes fragmented and rounds up into small pockets. In general, the cytoplasm becomes extremely dense in appearance and stains heavily with all dyes. One of the reasons for this is the large number of aldehyde groups that are apparently produced during this stage (Jensen and Fisher, 1968b).

The synergid nucleus is normally pressed to one side of the cell while the vegetative nucleus is pushed to a similar position or chalazally located. These two nuclei form the so-called X bodies which have long plagued the embryological literature (Pluijm, 1964; Fisher and Jensen, 1969; Van Went, 1970c).

There are some additional unanswered questions. One of these is, why does the pollen tube stop growing after it enters the synergid? The pollen tube has been growing vigorously for many millimetres and, in some cases, centimetres of tissue and now, when it reaches the embryo sac, it penetrates the synergid, and it suddenly stops within a matter of micrometres. Not only does it stop but it discharges. While ultrastructural studies in the cells cannot provide the answers to these questions they can provide some working hypotheses. One of the points that comes out of the ultrastructural work is that, until the tip of the pollen tube enters the cytoplasm of the synergid, it has been growing through cell walls or through air. Even in the case of the solid style flowers and those which have large nucelli, the pollen tube does not actually disrupt or enter any cell along the way. Only when it reaches the synergid does it come in contact with the cytoplasm of the cell. Moreover, the cell may be already in a stage of degeneration or be chemically altered in preparation for the ultimate breakdown which occurs. In either case, the tip of the pollen tube is entering a completely different milieu than it has seen before. In addition, the work of Linskens and his co-workers (Linskens and Schrauwen, 1966; Stanley and Linskens, 1967) suggests that the oxygen tension in the embryo sac is quite low, thus accounting for the discharge of the pollen tube because, when the pollen tubes are placed in similar oxygen tension, they stop growing and disrupt. An alternate explanation is that the pollen tube, as it comes into the synergid, is inhibited from growing, possibly by a reaction of the enzymes in the cell wall in the tip of the pollen tube and further elongation is prevented. The pressure that has been continually driving the pollen tube on toward greater growth in the style now increases as cell wall expansion is inhibited and the weakest point of the pollen tube ruptures, resulting in the discharge of the contents. At least in some pollen tubes the process is complicated by the fact that the discharge is not from the tip of the tube but from the pore located along the side of the pollen tube. This is true for cotton (Jensen, 1963; Jensen and Fisher, 1968b), and the same general mechanism would still be employed except that in this case the weakest point is determined by another growth condition in the development of the pollen tube or by a localized site of enzyme activity as the pollen tube growth ceases. Finally, in cotton and some other species, after the pollen tube has discharged, a plug of wall material forms in the pore itself so that further discharge is not possible (Jensen and Fisher, 1968b). A similar plugging of the pollen tube occurs in some species that have thin wall pollen tubes by the collapsing, and thus sealing off, of the end of the tube. This seems to be the mechanism present in *Capsella* (Schulz and Jensen, 1968a).

What do the sperm look like in the synergid?

This question cannot be answered on the ultrastructural level because no one has yet seen the sperms in the synergid, or any other part of the embryo sac, with the electron microscope; however, sperm have been seen in the synergid, or what is commonly termed the pollen tube discharge, with the light microscope. Such studies have revealed little of the structure of these cells except that they indicate that they are true cells at this time (Cass and Jensen, 1970). What has been observed with the electron microscope are sperm nuclei in both the egg and the central cell (Jensen and Fisher, 1967). In all cases only the nuclei have been seen. No cytoplasm or plasma membrane is seen associated with the sperm nuclei. Moreover, there are no reports of observations with the electron microscope of degenerate sperm cytoplasm in either the egg or the central cell.

How do the sperm and nuclei enter the egg and the central cell?

This remains one of the major unanswered questions of angiosperm embryology. However, based on the observations reported earlier, a hypothesis can be formulated (Fig. 14.1). According to this hypothesis, the sperm nuclei are discharged into the degenerating synergid, the plasma membrane of which has already disappeared. The sperm are transported by the force of the discharge into contact with the plasma membrane of the egg and the central cell. The plasma membranes of the sperm and the egg or central cells fuse and the sperm nucleus moves into the cytoplasm of either the egg or central cell (Jensen and Fisher, 1968b; Linskens, 1968).

This hypothesis is attractive because it explains the observations so far made and provides explanations for several other phenomenon, such as male cytoplasmic inheritance. There are well-documented cases of male cytoplasmic inheritance in some plants, while it seems equally certain that there are other plants with little male cytoplasmic inheritance. According to this hypothesis, in those plants in which male cytoplasmic heredity was common we would expect the cytoplasm of the sperm, including organelles such as mitochondria and plastids (if there are plastids present), to enter the egg along with the sperm nucleus. In those species in which there is little male cytoplasmic inheritance, only the nucleus would pass through the bridge formed by the fusion of the egg and sperm plasma membrane.

Reproduction in flowering plants

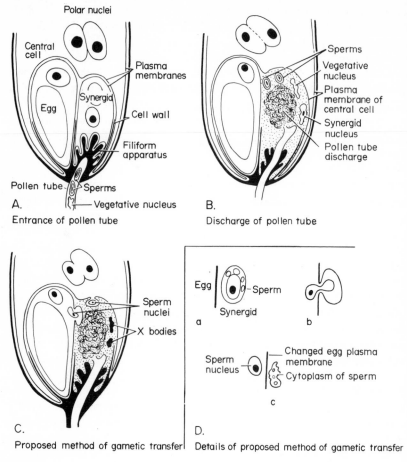

Fig. 14.1 Diagrammatic summary of the events associated with double fertilization in angiosperms.

In addition, this hypothesis allows us to think of a mechanism to explain double fertilization. The problem here is how to explain the fact that one sperm nucleus enters the egg while the second enters the central cell. Differences in the morphology or physiology of the two sperm have been invoked in thoughts on double fertilization but these seem unrealistic in terms of the current observation of the physiology and morphology of the sperms. Instead, this hypothesis allows one to think about fertilization membranes. All that is required for double fertilization to occur is for the membrane that fuses with the sperm membrane to be altered as soon as that fusion begins. If one sperm were to make contact first with the egg, a reaction would occur in the egg plasma membrane that made it unreceptive to fusing with the second sperm. Consequently, the second sperm would only fuse with the plasma membrane of the central cell.

It must be made clear that this is strictly a hypothesis and that until the sperm have been seen in the synergid and in the act of fusing with either the egg or central cell we cannot be certain that it is a valid explanation of the facts. It does, however, provide a basis for continued observation and experimentation with the hope of further elucidation of this major unanswered question.

What happens after the sperm nuclei enter the egg and central cell?

The sperm nucleus is apparently carried to the egg nucleus or the polar nuclei by cytoplasmic streaming. There is no evidence that within the egg or central cell the sperm move by self-locomotion, although there is no evidence that they do not. From their structure, however, it would appear unlikely that this is the case and that the nuclei, which are relatively small in relation to the nuclei of the egg or the central cell, are simply carried by cytoplasmic streaming to the respective egg and polar nuclei. In cotton, and many other plants, the sperm nucleus in the egg apparently reaches the egg nucleus before the sperm nucleus in the central cell reaches the polar nuclei (Jensen, 1963; Jensen and Fisher, 1967). However, the fusion of the egg and the sperm nucleus is much slower than that of the sperm nucleus and the polar nuclei.

How does nuclear fusion take place?

In the case of cotton (Jensen, 1964) and *Petunia* (Van Went, 1970c) the following sequence of events has been shown to occur. The nuclei come into immediate proximity with each other. If there are strands of ER projecting from the nuclear membrane they will first fuse and then apparently shorten so that the outer nuclear membranes come into contact. Where there is little ER the outer nuclear membranes simply come into contact directly. In either case the nuclear membranes fuse, forming continuous bridges of membranes between the two nuclei. These bridges shorten, and eventually the inner nuclear membranes come into contact and they fuse. Thus a series of bridges is formed between the two nuclei, which begin to enlarge and fuse, starting from the centre of the region between the two nuclei, and coalesce in an outward pattern. Thus, the cytoplasm and organelles trapped between the two nuclei are squeezed out towards the periphery. Eventually, the fusion is complete, and little morphological evidence of the fusion remains except that the last of the bridges may be present for some time.

The same pattern is followed in the case of the fusion of the sperm nucleus with the two polar nuclei. In cotton, the two polar nuclei begin to fuse before fertilization but remain only partially fused until the male nucleus is present and then all three fuse quite rapidly (Jensen and Fisher, 1967).

In the case of cotton (Jensen and Fisher, 1967), it would seem that while the sperm nucleus in the egg reaches the egg nucleus before the sperm nucleus in the central cell reaches the polar nuclei, the fusion of the egg and sperm nucleus takes a matter of hours, while the fusion of the egg and polar nuclei take only a matter of minutes.

This type of nuclear fusion has been shown to be true not only in higher plants but also in algae and in animals. Whether this is true for all cases of sperm nucleus and egg nucleus fusion in angiosperms remains to be seen. Gerassimova-Navashina (1960, 1961) has summarized the light microscopic observations on gametic nuclear fusion in plants to indicate that there are three types, depending upon the state of the nucleus. All of the species so far examined with the electron microscope fall into the same type and it will be of interest to see, when the observations are extended, if there are different modes of nuclear fusion found in different species.

What happens following the formation of the zygote nucleus?

There may be a number of changes in the zygote nucleus itself. Again, in the case of cotton (Jensen and Fisher, 1967), a number of small nucleoli may be produced in the sperm end, and these fuse into one so that the final zygote nucleus has only two nucleoli — one from the egg and one from the sperm. In *Capsella* the nucleolus of the sperm fuses with that of the egg so that the zygote has only a single new nucleolus (Schulz and Jensen, 1968b).

More spectacular changes occur in the cytoplasm of the zygote. What actually happens depends a great deal on the morphology of the egg before fertilization. The egg of cotton is a large vacuolate cell and one of the main changes involves the decrease in cell size of the zygote in relation to that of the egg (Jensen, 1968). The zygote has only half the volume of the egg and, after a reduction in size has taken place, is a cell with a relatively dense cytoplasm with a vacuole at the micropylar end. The nucleus is located at the chalazal end of the cell surrounded by a shell of mitochondria and amyloplasts. The mechanism of this reduction in size of the zygote is not known, but I have suggested that it is a result of coming to equilibrium between the osmotic pressure in the egg in relation to the rapidly developing endosperm that surrounds it.

Such shrinkage in cell size appears to be related to the fact that the cotton egg is a highly vacuolate cell. In *Capsella* (Schulz and Jensen, 1968b) where the vacuole is initially much smaller in size, the zygote actually increases in length before it divides. In this case the vacuole again is in the micropylar end of the cell and the egg nucleus is surrounded by cytoplasm at the chalazal end. In *Epidendrum* (Cocucci and Jensen, 1969b), on the other hand, there appears to be no change in cell size whatever, either increasing or decreasing. This would also seem to be the case for corn (Diboll. 1968).

Associated with the formation of the zygote nucleus is the appearance of a large polysome complex within the cytoplasm of the zygote. In cotton (Jensen, 1968c), these polysomes may contain as many as 30—40 individual ribosomes, and appear associated with the membranes of plastids and occasionally mito-chondria. Large polysome complexes have been found in every zygote studied, although usually not of the heroic proportions found in cotton. In cotton, there is a clear second generation of ribosomes that appears in the zygote before its division. These appear as individual ribosomes or as small polysomes and exist side by side with large polysomes. One reasonable interpretation of this data is that a long-lived messenger RNA is released by the new zygote nucleus which ties up most of the ribosomes that are originally present in the egg. Then a new set of ribosomes are produced by the zygote nucleus itself. Preliminary studies indicate that these polysomes can be traced through the early stages of embryo development and are present in the young globular embryo but disappear at the differentiation of the heart stage. Obviously, these observations have intriguing possibilities in relation to information and differentiation and should be followed with biochemical and cytochemical studies.

There are other changes which occur in the zygote as well. One of these is the formation of starch in the amyloplasts (Jensen, 1963). While starch is present in the egg cell of cotton, it is found only in small amounts. After the zygote is formed and the plastids are found clustered around the outside of the nuclear membrane, starch synthesis begins and conspicuous starch grains are found surrounding the nucleus. An increase in starch in the zygote also occurs in *Epidendrum* and *Capsella*.

An unusual occurrence is found in cotton. During the period in which the zygote is shrinking, but before it divides, a special ER is formed (Jensen, 1968b) which contains numerous tubules. These are quite different in appearance from the microtubules found free in the cytoplasm of the zygote near the periphery of the cell. The ER tubules are slightly larger in diameter, are wavy in appearance, and branch and fuse. They are continuous with a membrane of the ER. The chalazal end of the zygote is filled with this tube containing ER and the tubes can even be seen in the nuclear membrane. No function has yet been ascribed to these microtubules in the ER, and they have not been found in other zygotes examined with the electron microscope. Interestingly, they have been found in other organisms, primarily animals, including man. No explanation for their presence in any system has yet been given.

The dictyosomes become quite active and a wall around the entire zygote is formed. There is no sign of organelle degeneration in the zygote. Neither are there any indications of cell organelles being produced by the nuclear membrane as have been reported for ferns.

What happens at the division of the zygote?

The first division of the zygote is always unequal, resulting in the formation of a small terminal cell and a large basal cell (Maheshwari, 1950). In most cases, the small terminal cell goes on to produce through repeated cell divisions the embryo proper, while the large basal cell divides to form the suspensor.

The two cells look quite different with both the light and electron microscope. In general, the small terminal cell stains darkly for the nucleic acids and protein, while the larger basal cell stains relatively lightly (Jensen, 1968a). One exception to this is in the case of *Capsella* (Schulz and Jensen, 1968b) where, for a short period, the basal cell actually stains more intensely than the terminal cell, but the situation is soon reversed. At the ultrastructural level there are also marked differences in the electron density of the two cells. The terminal cell appears to contain a higher number of ribosomes per unit volume than the basal cells. The nucleus of the terminal cell is surrounded by a shell of mitochondria and plastids while the organization in the basal cell appears more casual. The terminal cell also appears to contain less ER than the basal cell and, in the case of cotton, most of the tube containing ER is found in the basal cell rather than the terminal.

Once cell division begins in the zygote, and continues in the resulting embryon cells, the starch which accumulated in the zygote is rapidly utilized and cannot be seen either on the light or electron microscope level (Jensen, 1968a).

What happens after these first divisions?

The subsequent course in the development of the terminal and basal cell depends somewhat upon the species. Unfortunately, we have little ultrastructural work on early embryo development.

One series of studies has been carried out on *Capsella,* which is the traditional material for plant embryologists (Schulz and Jensen, 1968b, c; 1969). In this case the basal cell continues to undergo a limited number of divisions, resulting in a large basal cell connected to a series of smaller suspensor cells that in turn join the developing embryo. The basal cell is much larger than the other cell and contains a conspicuous central vacuole area. The wall at the micropylar end of the basal cell develops what appears to be a smaller version of a filiform apparatus and is clearly the transfer wall. The nucleus becomes highly lobed and suspended in a strand of cytoplasm transversing the large vacuole. (In similar basal cells of other species (Clutter and Sussex, 1969; Nagl, 1969) it has been shown that they are highly polyploid and the chromosomes actually seem to be in a polytene situation.) From the general size and staining of the nucleus of the basal cell of *Capsella* one would reasonably conclude that it too is highly polyploid. The cytoplasmic matrix of the cell darkens in the late globular stage of embryo development and histochemical staining for protein becomes intense. The basal cell remains active after the suspensor cytoplasm has degenerated.

The suspensor cells in *Capsella* are more vacuolate and contain more ER and dictyosomes, but fewer ribosomes, and stain less intensely for protein and nucleic acid than the cells of the embryo. The end walls of the suspensor cells contain numerous plasmodesmata. There are no plasmodesmata in the walls separating the suspensor from the embryo sac. The lower suspensor cells fuse with the embryo sac wall, and a lateral wall of the lower and middle suspensor cells produce finger-like extensions into the endosperm. At the heart stage of embryo development the suspensor cells begin to degenerate and gradually lose their ability to stain for protein and nucleic acids.

These observations suggest that the suspensor and basal cell function as an embryonic root in the absorption and translocation of nutrients from the integuments to the developing embryo. This would appear a logical result of their position in development. It also agrees with such embryos as cotton, where the suspensor is much reduced and represents only a pyramid-shaped cluster of some four to six cells at the base of the developing embryo. The traditional view has been that the suspensor pushes the embryo into the developing endosperm and thus aids the absorption of material from the endosperm by the embryo. The ultrastructure of the basal cells and the suspensor cells of *Capsella* do not support this type of function but do support the concept of transport.

The suspensor cells of *Phaseolus vulgaris* have also been studied with the electron microscope (Schnepf and Nagl, 1970). These cells also show wall protuberances, highly endopolyploid nuclei, and plastids of varying, often bizarre, shapes. They also contain large amounts of smooth tubular ER. What happens in the development of the terminal cells into the embryo?

Again, the most detailed published work on embryo development is of *Capsella*, where the embryo was studied from the terminal cell of the three celled embryo through a globular embryo consisting of some 32–64 cells (Schulz and Jensen, 1968c). As I noted earlier, during the early development of the terminal cell, starch and lipid deposits that had accumulated disappear. Cell size and the number of organelles per cell decreases with repeated divisions through the formation of the globular embryo. This is also true for the cotton embryo, where the cell size is drastically reduced during early embryogenesis. The cells contain few dictyosomes and little ER but are rich in ribosomes grouped into small polysomes. In the case of cotton the same general description is true, but the large polysomes that were present in the zygote persist through the first several generations of cell divisions and then gradually disappear.

The embryo stains intensely for protein and nucleic acids. The walls are extremely thin and no starch is present during these early stages. One of the surprising observations with regard to the *Capsella* embryo was that there were no apparent ultrastructural or histochemical differences in the cells of the protoderm, procambium, and ground meristem until the formation of the heart-shaped embryo.

The observations, which have so far been made on the ultrastructural level with regard to embryos, are indeed scanty, but several lines of speculation are suggested. The reduction in cell size and number of organelles can be interpreted as representing a shift in importance from the single cell to masses of interacting cells with the increasing age of the embryo. The minimum cell size is reached immediately before the stage of cell differentiation that proceeds rapidly from the late globular embryo onwards (Pollack and Jensen, 1964). It is almost as if the minimum size is a critical determining factor in the expression of the differentiation of the cells composing the cell mass of the embryo.

The lack of clear differences in type, number, or distribution of cell organelles between the different developing tissues suggests that position effect, that is where a given cell is in the embryo with relation to the surrounding cells, is more important than internal structural modifications in the future development of that cell. Clearly the programming that is occurring is being expressed on a molecular level and is only slowly translated into ultrastructural differences that appear much later in the development of the embryo.

Conclusions

Finally, what can we conclude from all the questions we have been asking concerning embryogenesis in flowering plants?

The first, obvious conclusion is that we need a great deal more data. We need basic data from observations made by people looking at more stages of development of more species of angiosperms. Next, we need more data that combines dynamic and analytical studies with the ultrastructural work. Histochemical and cytochemical studies at both the light and electron microscope level show promise in extending the morphological observations and tying them in with the physiological data which are available from culture experiments. We need to collect such basic data as a general composition of the various cells involved in embryogenesis and, in particular, the distribution of various important classes of enzymes. Wherever possible such a study should involve quantitative analyses of the systems through the use of techniques that involve both the isolation of cell parts by various microprocedures and other types of procedure based on absorption phenomena under the microscope.

The modern study of plant embryogenesis has just begun. The results that are available suggest that ultrastructural procedures offer a way of obtaining data that will help us to understand fundamental problems in cell and organism development. The next decades should be exciting ones indeed in the field of plant embryogenesis.

Bibliography

Further reading

Erdtman, C. (1952) *Pollen morphology and plant taxonomy,* The Chronica Botanica Co., Waltham. A summary of light microscope pollen analysis by one of the founders of the field.

Linskens, H. F. (1969) 'Fertilization Mechanisms in Higher Plants', In: *Fertilization* (eds. C. B. Metz and A. Monroy), vol. II, pp. 189–253. Academic Press Inc., New York. Summary of the literature to about 1965 (despite the publication date of 1969).

Maheshwari, P. (1950) *An Introduction to the embryology of angiosperms,* McGraw-Hill, New York. The classic work on traditional light microscope embryology.

Rosen, W. B. (1969) 'Ultrastructure and physiology of pollen', *A. Rev. Pl. Physiol.,* **19,** 435–462. An excellent review of various aspects of pollen biology.

Steffen, K. (1963) 'Fertilization', In: *Recent advances in embryology of angiosperms* (ed. P. Maheshwari), pp. 105–143. Int. Soc. Plant Morphologists, Delhi. A summary of the light microscope data on fertilization in angiosperms.

Vasil, I. K. and Johri, B. M. (1964) 'The style, stigma, and pollen tube', *Phytomorphology,* **14,** 352–369. An excellent paper on the light microscope morphology of these parts.

References

Angold, R. E. (1968) 'The formation of the generative cell in the pollen grain of *Endymion non-scriptus* (L.)', *J. Cell Sci.,* **3,** 573–578.

Bopp-Hassenkamp, G. (1960) 'Elektronenmikroskopische Untersuchungen an Pollenschläuchen ziveier *Liliaceen', Z. Naturf.,* **18,** 562–566.

Burgess, J. (1970) 'Cell shape and mitotic spindle formation in the generative cell of *Endymion non-scriptus', Planta,* **95,** 92–85.

Cass, D. D. (1973) 'An ultrastructural and Nomarski-interference study of the sperm of barley,' *Can. J. Bot.,* **51,** 601–605.

Cass, D. D. and Jensen, W. A. (1970) 'Fertilization in barley', *Am. J. Bot.,* **57,** 62–70.

Chao, C.-Y. (1971) 'A periodic acid-Schiff's substance related to the directional growth of pollen tube into embryo sac in *Paspalum* ovules', *Am. J. Bot.,* **58,** 649–654.

Chardard, R. (1969) 'Aspects infrastructure de la maturation des grains de pollen de quelques Orchidacées', *Rev. Cytol. Biol. vég.,* **32,** 67–100.

Clutter, M. E. and Sussex, I. M. (1969) 'Ultrastructural development of bean embryo cells containing polytene chromosomes', *J. Cell Biol.,* **39,** 26a.

Cocucci, A. and Jensen, W. A. (1969a) 'Orchid embryology, The mature megagametophyte of *Epidendrum scutella',* *Kurtziana,* **5,** 23–38.

Cocucci, A. and Jensen, W. A. (1969b) 'Orchid embryology. Megagametophyte of *Epidendrum scutella* following fertilization', *Am. J. Bot.,* **56,** 629–640.

Cocucci, A. and Jensen, W. A. (1969c) 'Orchid embryology. Pollen tetrads of *Epidendrum scutella* in the anther and on the stigma', *Planta,* **84,** 215–229.

Crang, R. E. and Miles, G. B. (1969) 'An electron microscope study of germinating *Lychnis alba* pollen', *Am. J. Bot.,* **56,** 398–405.

Dashek, W. V. and Rosen, W. G. (1966) 'Electron microscopical localization of chemical components in the growth zone of Lily pollen tube', *Protoplasma,* **61,** 192–204.

Dexheimer, J. (1965a) 'Sur les structures cytoplasmiques dans les tubes polliniques de *Lobelia erinus',* *C. r. hebd. Séanc. Acad. Sci., Paris,* **261,** 507–508.

Dexheimer, J. (1965b) 'Sur les structures cytoplasmiques dans les grains de pollen de *Lobelia erinus',* *C. r. hebd. Séanc. Acad. Sci., Paris,* **260,** 6963–6965.

Diboll, A. G. (1968) 'Fine structure development of the megagametophyte of *Zea mays* following fertilization', *Am. J. Bot.,* **55,** 787–806.

Diboll, A. G. and Larsen, D. A. (1966) 'An electron microscope study of the mature megagametophyte in *Zea mays', Am. J. Bot.,* **53,** 391–402.

Diers, L. (1963a) 'Elektronenmikroskopische Beobactungen an der vegetatinen Zelle *Oenothera hookeri', Z. Naturf.,* **18,** 562–566.

Diers, L. (1963b) 'Elektronenmikroskopische Beobactungen an der vegetatinen Zeile im auskeimenden Pollenkorn von *Oenothera hookeri', Z. Naturf.* **18,** 1092–1097.

Fisher, D. B. and Jensen, W. A. (1969) 'Cotton embryogenesis: The identification, as nuclei, of the X-bodies in the degenerate synergid', *Planta,* **84,** 122–133.

Fisher, D. B., Jensen, W. A., and Ashton, M. E. (1968) 'Histochemical studies of pollen: Storage pockets in the endoplasmic reticulum (ER)', *Histochemie,* **13,** 169–182.

Gerassimova-Navashina, H. (1960) 'A contribution to the cytology of fertilization in flowering plants', *Nucleus,* **3,** 111–120.

Gerassimova-Navashina, H. (1961) 'Fertilization and events leading up to fertilization, and their bearing on the origin of angiosperms', *Phytomorphology,* **11,** 139–146.

Godineau, J. C. (1966) 'Ultrastructure du sac embryonnaire du *Crepis tectorum,* less cellules du pole micropylaire', *C. r. hebd. Séanc. Acad. Sci., Paris,* **263,** 852–855.

Hoefert, L. (1969) 'Ultrastructure of *Beta* pollen', I. Cytoplasmic constituents. *Am. J. Bot.,* **56,** 363–368.

Jensen, W. A. (1963) 'Cellular differentiation in embryos', *Brookhaven symposium in biology,* **16,** 179–202.

Jensen, W. A. (1964) 'Observations on the fusion of nuclei in plants', *J. Cell Biol.,* **23,** 669–672.

Jensen, W. A. (1965a) 'The composition and ultrastructure of the nucellus in cotton', *J. Ultrastruct., Res.,* **13,** 112–128.

Jensen, W. A. (1965b) 'The ultrastructure and histochemistry of the synergids of cotton', *Am. J. Bot.,* **52,** 238–256.

Jensen, W. A. (1965c) 'The ultrastructure and composition of the egg and central cell of cotton', *Am. J. Bot.,* **52,** 781–797.

Jensen, W. A. (1968a) 'Cotton embryogenesis: The zygote', *Planta,* **79,** 346–366.

Jensen, W. A. (1968b) 'Cotton embryogenesis: The tube-containing endoplasmic reticulum', *J. Ultrastruct., Res.,* **22,** 296–302.

Jensen, W. A. (1968c) 'Cotton embryogenesis: Polysome formation in the zygote', *J. Cell Biol.,* **36,** 403–406.

Jensen, W. A. (1969) 'Cotton embryogenesis: Pollen tube development in the nucellus', *Can. J. Bot.,* **47,** 383–385.

Jensen, W. A. (1972) 'The embryo sac and fertilization in angiosperms,' Harold L. Lyon Arboretum Lecture No. 3. Honolulu, Hawaii, 32 pp.

Jensen, W. A. and Fisher, D. B. (1967) 'Cotton embryogenesis: Double fertilization', *Phytomorphology,* **17,** 261–269.

Jensen, W. A. and Fisher, D. B. (1968a) 'Cotton embryogenesis: The Sperm', *Protoplasma,* **65,** 277–286.

Jensen, W. A. and Fisher, D. B. (1968b) 'Cotton embryogenesis: The entrance and discharge of the pollen tube in the embryo sac', *Planta, 78,* 158–183.

Jensen, W. A. and Fisher, D. B. (1969) 'Cotton embryogenesis: The tissues of the stigma and style and their relation to the pollen tube', *Planta, 84,* 97–121.

Jensen, W. A. and Fisher, D. B. (1970) 'Cotton embryogenesis: The pollen tube in the stigma and style', *Protoplasma, 69,* 215–235.

Jensen, W. A., Fisher, D. B., and Ashton, M. E. (1968) 'Cotton embryogenesis: The pollen cytoplasm', *Planta, 81,* 206–228.

Konar, R. N. and Linskens, H. F. (1965) 'Some observations on the stigmatic exudate in *Petunia hybrida', Naturwissenschaften, 52,* 625.

Konar, R. N. and Linskens, H. F. (1966a) 'The morphology and anatomy of the stigma of *Petunia hybrida', Planta, 71,* 356–371.

Konar, R. N. and Linskens, H. F. (1966b) 'Physiology and biochemistry of the stigmatic fluid of *Petunia hybrida', Planta, 71,* 372–387.

Kroh, M. (1964) 'An electron microscope study of the behavior of Cruciferae pollen after pollination', In: *Pollen physiology and fertilization* (ed. H. F. Linskens), pp. 221–224, North-Holland Publ. Co., Amsterdam.

Kroh, M. (1967) 'Fine structure of *Petunia* pollen germinated *in vivo', Rev. Paleo. Palyn.,* **3,** 197–203.

Kroh, M., Miki-Hirosige, H., Rosen, W., and F. Loewus (1970) 'Incorporation of label into pollen tube walls from myo-inositol-labeled *Lilium longiflorum* pistils', *Pl. Physiol., Lacnaster,* **45,** 457–471.

Larson, D. (1965) 'Fine structural changes in the cytoplasm of germinating pollen', *Am. J. Bot., 52,* 139–154.

Linskens, H. F. (1968) 'Egg-sperm interactions in higher plants', *Accademia Nazionale dei Lincei,* **365,** 48–60.

Linskens, H. F. and Schrauwen, T. (1966) 'Measurements of oxygen tension changes in the style during pollen tube growth', *Planta,* **71,** 98–106.

Lombardo, G. and Gerola, F. M. (1968) 'Cytoplasmic inheritance and ultrastructure of the malegenerative cell of higher plants', *Planta, 82,* 105–110.

Marinos, N. G. (1970) 'Embryogenesis of the pea (*Pisum sativum*). I. The cytological environment of the developing embryo', *Protoplasma,* **70,** 261–279.

Maruyama, K. (1966) 'Behavior of membrane system in the cell during cell divisions of microsporogenesis in *Tradescantia paludosa* III. Postmeiotic mitosis', *Cytologia,* **31,** 257–269.

Maruyama, K., Gay, H., and Kaufmann, B. P. (1965) 'The nature of the wall between generative cell and vegetative nuclei in the pollen grain of *Tradescantia paludosa', Am. J. Bot., 52,* 605–610.

Mascarenhas, J. P. (1965) 'Pollen tube growth and ribonucleic acid synthesis by vegetative and generative nuclei of *Tradescantia', Am. J. Bot., 52,* 605–610.

Mascarenhas, J. P. and Machlis, L. (1962) 'The hormonal control of the directional growth of pollen tubes', *Vitamins and hormones,* **20,** 347–371.

Mepham, R. H. and Lane, G. R. (1970) 'Observations on the fine structure of developing microspores of *Tradescantia bractiata', Protoplasma,* **70,** 1–20.

Mogensen, H. L. (1972) 'Fine structure and composition of the egg apparatus before and after fertilization in *Quercus gambelii:* The functional ovule', *Am. J. Bot..* 59. 931–941.

Nagl, W. (1969) 'Banded polytene chromosomes in the legume *Phaseolus vulgaris', Nature, Lond.,* **221,** 70–71.

Navashin, M. S. (1969) 'On the nature of the movement of the generative cell in the pollen tube and the problem of the localization of cell elements', *Rev. Cytol. Biol. vég.,* **32,** 141–148.

Pluijm, J. E. van der (1964) 'An electron microscopic investigation of the filiform apparatus in the embryo sac of *Torenia fournieri',* In: *Pollen physiology and fertilization* (ed. H. F. Linskens), pp. 8–16, North-Holland Publ. Co., Amsterdam.

Pollock, E. G. and Jensen, W. A. (1964) 'Cell development during early embryogenesis in *Capsella* and *Gossypium', Am. J. Bot., 51,* 915–921.

Rosen, W. G. (1962) 'Cellular chemotropism and chemotaxis', *Q. Rev. Biol.,* **37,** 242–259.

Rosen, W. G. and Gawlik, S. R. (1966) 'Fine structure of lily pollen tubes following various fixation and staining procedures', *Protoplasma,* **61,** 181–191.

Rosen, W. G., Gawlik, S. R., Dashek, W. V., and Siegesmund, K. A. (1964) 'Fine structure and cytochemistry of *Lilium* pollen tubes', *Am. J. Bot.*, **51**, 61–71.

Rosen, W. G. and Thomas, H. R. (1970) 'Secretory cells of lily pistils. I. Fine structure and function', *Am. J. Bot.*, **57**, 1108–1114.

Sassen, M. M. A. (1964) 'Fine structure of *Petunia* pollen grain and pollen tube', *Acta bot. neerl.*, **13**, 175–181.

Schnepf, E. and Nagl, N. (1970) 'Uber einige Strukturbesonderheiten der Suspensorzellen von *Phaseolus vulgaris*', *Protoplasma*, **69**, 133–143.

Schulz, R. and Jensen, W. A. (1968a) '*Capsella* embryogenesis: The synergids before and after fertilization', *Am. J. Bot.*, **55**, 541–552.

Schulz, R. and Jensen, W. A. (1968b) '*Capsella* embryogenesis: The egg, zygote, and young embryo', *Am. J. Bot.*, **55**, 807–819.

Schulz, R. and Jensen, W. A. (1968c) '*Capsella* embryogenesis: The early embryo', *J. Ultrastruct., Res.*, **22**, 376–392.

Schulz, P. and Jensen, W. A. (1969) '*Capsella* embryogenesis. The suspensor and basal cell', *Protoplasma*, **67**, 139–163.

Schulz, P. and Jensen, W. A. (1971) '*Capsella* embryogenesis: The chalazal proliferating tissue', *J. Cell Biol.*, **8**, 201–227.

Stanley, R. G. and Linskens, H. F. (1967) 'Oxygen tension as a control mechanism in pollen tube rupture', *Science, N.Y.*, **157**, 833–834.

Van Went, J. F. (1970a) 'The ultrastructure of the synergids of *Petunia*', *Acta Bot. neerl.*, **19**, 121–132.

Van Went, J. F. (1970b) 'The ultrastructure of the egg and central cell of *Petunia*', *Acta bot. neerl*, **19**, 313–322.

Van Went, J. F. (1970c) 'The ultrastructure of the fertilized embryo sac of *Petunia*', *Acta bot. neerl.*, **19**, 468–480.

Van Went, J. F. and Linskens, H. F. (1967) 'Die Entwicklung des sogenannten "Fadenapparates" in Embryosack von *Petunia hybrida*', *Genet. Breeding Res.*, **37**, 51–56.

Vazart, B. (1958) 'Differenciation des cellules sexuelles et fecondation chez les Phanérogames', *Protoplasmatologia*, **7**, 1–158.

Vazart, B. (1969) 'Structure et évolution de la cellule génératrice du Lin, *Linum usitatissimum* L., au cours des premiers stades de la maturation du pollen', *Rev. Cytol. Biol. Vég.*, **32**, 101–114.

Vazart, J. (1969) 'Organisation et ultrastructure du sac embryonnaire du Lin (*Linum usitalissimum* L.)', *Rev. Cytol. Biol. Vég.*, **32**, 227–240.

Vazart, B. and Vazart, J. (1966) 'Infrastructure du sac embryonnaire du Lin (*Linum usitatissimum* L.)', *Rev. Cytol. Biol. Vég.*, **24**, 251–266.

Welk, Sr. M., Millington, W. F., and Rosen, W. G. (1965) 'Chemotropic activity and the pathway of the pollen tube in lily', *Am. J. Bot.*, **52**, 774–781.

Woude, W. J. van der, Morré, D. J., and Bracker, C. E. (1971) 'Isolation and characterization of secretory vesicles in germinated pollen of *Lilium longiflorum*', *J. Cell Sci.*, **8**, 331–351.

Wylie, R. B. (1941) 'Some aspects of fertilization in *Vallisneria*', *Am. J. Bot.*, **28**, 169–174.

Yamada, Y. (1965) 'Studies on the histological and cytological changes in the tissues of pistil after pollination', *Jap. J. Bot.*, **19**, 69–82.

15

Motile male gametes of plants
Dominick J. Paolillo

Introduction

The study of spermatogenesis in plants with swimming sperms has had a long and colourful history. The early light microscopists recognized the interesting features of spermatogenesis and worked diligently to elucidate them. They realized that, coincident with the onset of sperm formation, the flagellar apparat makes its appearance *de novo* and a series of dramatic morphological changes occurs as the sperms are prepared for their eventual release. Here, it was thought one had the opportunity to detect homologies between plant and animal cells, and this has proved to be the case. Cytologists were also interested in knowing what the sperms were made of so that they could speculate intelligently on the contribution the sperm brings to the egg. For example, does the sperm contain a plastid that becomes part of the cytoplasmic inheritance of the zygote?

As one who has worked for many hours with light and electron microscopes, I find myself somewhat awed at the accuracy of certain early observations on spermatogenesis, and on these points we find ourselves only confirming the reports of our predecessors and relating their observations to the realm of ultrastructure that they could never hope to see. There is also a generation of microscopists who helped construct the plant sciences with their light microscop and who have lived on to take their places at the consoles of electron microscop in various countries of the world. Inasmuch as we know these people we can understand how the light microscope was put to such good use, for they have taught us also that the ultimate in resolution lies not in the formula of the lens b in the minds of men who are willing to work with whatever tools of observation are at their disposal.

They have taught us the continuity of light and electron microscopy and, having tried myself to repeat some of their light microscopic observations, I am in a position to affirm the diligence of their efforts. The challenge to electron microscopists is to bring our minds to as sharp a focus as our pictures, that we may exceed by one order of magnitude the resolution of our instruments, just as our predecessors have done with the light microscope. I make these remarks now because it is not practical to mention in each case which of the currently acceptable ideas were first formulated by light microscopists. Nor is it practical to mention each of the light microscopic observations that are relevant to the interpretation of our present knowledge. I will simply state at the outset that without the outstanding light microscopic literature on the subject, our task of interpreting the structure of the sperms of plants would be much harder. However, electron microscopic investigations do allow us to settle some of the disputes that could not be settled by light microscopy. And they allow us to ask new questions that further excite our interest.

There remains for this introduction only the delimitation of the scope of this review. My own investigations of spermatogenesis are concentrated on the hairy-cap moss, *Polytrichum,* where we have found it absolutely necessary to apply developmental information to obtain a positive identification of the organelles found in the maturing sperms. Of course, many of the parts identified in *Poly-trichum* sperms are also components of other sperms, and a common groundplan of construction can be discovered. However, differences in arrangement, number, and condition of parts also come to light. Our approach then, will be both comparative and developmental.

Most of the emphasis will be placed on the components common to swimming sperms of green plants and on the highlights of the developmental changes that radically alter the cellular condition of the maturing sperm cells. While we shall be concerned with plants ranging from algae through gymnosperms, we will at the same time restrict our principal considerations to plants in which the sperms are the only swimming cells in the life history. Among the algae, *Nitella* and *Chara* will be considered throughout the text. Other algae will be treated separately to introduce the comparison between sperms and zoospores.

Centriole origin and replication

The components of the mature sperms are markedly altered from their original condition in the spermatogenous cells. But of all the components, only the flagellar apparatus seems to have no direct continuity with organelles in the cells of the young antheridium. The question of the origin of flagella and associated structures was a continuous source of controversy in the literature based on light microscopy, for neither the time of origin nor the nature of the entities perceived could be agreed upon by all workers (see, especially, Lepper, 1956, for review of this topic). One of the objectives of this chapter, then, is to list the achievements of electron microscopists in their attempts to unravel the mysteries of the flagellar apparatus that could not be solved with the light microscope. The first part of these achievements relates to the origin of centrioles that later become the basal bodies of the flagella.

A *de novo* origin of centrioles

It has been suggested by numerous authors that, when the sperms are the only motile cells a plant has, the centrioles that become the basal bodies originate *de novo* during spermatogenesis. My own thoughts on the matter are conditioned by our still unpublished work on *Polytrichum*. A *de novo* origin of centrioles is difficult to prove. I can only report that a diligent search for centrioles in young antheridia produces no positive result. In *Polytrichum,* the first indication of centriole formation occurs in cells that are entering the prophase of the ultimate division in the spermatogenous cells. Suire (1970) reports that the centrioles appear after the final division in the liverwort, *Pellia.* Centrioles participate in the final division of *Marchantia* (Kreitner, 1970) and *Phaeoceros* (Moser, 1970).

It is the habit of botanists to call the last two generations of spermatogenous cells by special names when describing bryophytes. The cells of the penultimate generation are called 'androcyte mother cells'. The cells formed by the ultimate cytokinesis are the 'androcytes'. The term androcyte has been used in preference to spermatid to dissociate the spermatogenous cells from any context of meiosis. The terms spermatid and androcyte are otherwise equivalent because the androcytes metamorphose into motile male gametes, or sperms. Thus one can report that, in bryophytes, the centrioles first appear probably in the androcyte mother cells or androcytes, depending on the species, and these cells are many generations removed from meiosis.

The entities first detected during development in *Polytrichum* are procentrioles in the sense of Gall (1961) in that they possess a hub and cartwheel but lack the nine triplet fibres. However, the triplets become evident before the mitosis is complete. The procentrioles graduate, then, to centrioles in a structural sense, and furthermore they become attached to spindle fibres (see later). Procentrioles and centrioles are easily detected in the androcyte mother cells of *Polytrichum,* first in a perinuclear position and later in the polar positions with respect to the plane of mitotic division. They are not found in either of these locations during earlier generations of the spermatogenous cells. Hence, procentrioles are either absent, or extraordinarily well hidden and non-polar in early spermatogenous cells. It does not appear that we are dealing with preformed entities tucked alongside the nucleus as in the vegetative hyphal tips of the fungus *Allomyces,* at the onset of zoosporogenesis (Renaud and Swift, 1964). Therefore it is appropriate to propose a *de novo* origin of centrioles in *Polytrichum,* by which I imply specifically that procentrioles as structural entities appear in the androcyte mother cells when they are absent in earlier cell generations.

Admittedly, all contemporary studies, including our own, are subject to shortcomings in sampling. We have, however, the fortunate circumstance that, during the last division in *Polytrichum,* the spermatogenous cells are not uniform. Those in the basal region are found to be further along in their development than those at the tip. Thus, one antheridium at the crucial stage offers information from a time *segment* in development rather than from a single point in the process.

There is no extensive electron microscopic literature on the early stages of centriole formation in plants. In the fern *Marsilea* and the cycad *Zamia* (Mizukami and Gall, 1966) a *de novo* origin of centrioles is postulated using the same arguments that we have applied to *Polytrichum.* As the final division

is approached (*Zamia*), or before the penultimate division (*Marsilea*), large cytoplasmic bodies that are called 'blepharoplasts' become visible. Electron micrographs show that a blepharoplast consists of procentrioles arranged in a sphere.
In the late stages of their existence the spheres become hollow to the extent that
the procentrioles become restricted to the periphery. While the procentrioles
remain in aggregates, each constitutes only the hub and spokes, or cartwheel
(Gibbons and Grimstone, 1960) of a centriole that is yet to mature. When the
procentrioles disperse into the cytoplasm, the triplet fibres appear and, thus,
centrioles become recognizable. There is no doubt that these centrioles become
the basal bodies of the flagella as spermatogenesis proceeds. Continuity at this
end of the ontogenetic sequence cannot be refuted. It is continuity of centrioles
with any pre-existing organelle in early spermatogenous cells that seems to be
lacking. Of course, one can still favour the less attractive argument that the
conspicuousness of procentrioles and eentrioles as the final mitosis nears is
simply a function of their entering a replicative phase.

Centriole replication in plants

Moser and Kreitner (1970) established an 'end-to-end' or linear replication of
centrioles in the androcyte mother cells of the bryophytes *Anthoceros* (actually
Phaeoceros, according to Moser, 1970) and *Marchantia*. In each of the androcytes
there is a complex structure consisting of two centrioles with their cartwheels
on a common hub. The double nature of this entity is revealed by a discontinuity
of the triplets in the median region. Perkins (1970) has named this linear configuration a 'bicentriole' because it differs from the typical diplosome in lacking the
right-angled orientation of the two components. The attachment of the two
parts is by way of the proximal ends (cartwheel ends) of the centrioles. The tilt
of the triplet fibres is, accordingly, reversed in the two halves of the bicentriole.
The free ends of the bicentriole are destined to grow out as the flagella when
the centrioles are converted to basal bodies.

Once the bicentriole 'shears' into two parts, the bryophytic androcyte has its
full complement of two centrioles. This shearing, or binary fission, can be called
a centriole multiplication, but it remains obvious that the multiplication should
not be confused with the replication of centrioles, that must precede the
multiplication. It is to be inferred, however, that the replication is itself linear,
and the bicentriole is the terminal stage in the replication.

Bicentrioles have been reported for the fungus *Albugo* in vegetative hyphae
and during zoospore formation by Berlin and Bowen (1964) who also concluded
that the linear arrangement of two centrioles must be a stage in centriole
replication (Berlin and Bowen, 1965). Tourte and Hurel-Py (1967) have suggested
linear replication of centrioles in the fern *Pteridium*. Here, as in *Marsilea,* the
formation of centrioles (or at least, procentrioles) precedes the penultimate
mitosis. Here, also, there are numerous centrioles to account for the numerous
basal bodies seen at later stages. The replication of procentrioles of *Marsilea* and
Zamia should be reinvestigated for the occurrence of linear replication. This
question is of more than passing interest because one needs to determine how

widespread linear replication is when comparing plants and animals. While Turner (1966) reported bicentrioles in the liverwort *Reboulia,* he preferred to speculate that the replication is by 'budding' at right angles to existing centrioles in the alga *Nitella* during spermatogenesis (Turner, 1968). He did not view the actual replication in *Nitella* however, and the photograph of a bicentriole in his later paper on *Nitella* (Turner, 1970) remains unexplained. Pickett-Heaps' (1968) study of *Chara*, and other charophycean algae, also indicates a perpendicular relationship between the two centrioles in each spermatogenous cell. It is clear from the work of Moser and Kreitner (1970) and Perkins (1970) that the linear arrangement in a bicentriole does not preclude subsequent rearrangements. More must be learned about centriole replication in the charophytes before they are confirmed to be different from other green plants in this regard.

It has already come to light that the linear replication of centrioles is not restricted to bryophytes, ferns, and fungi. Perkins (1970) reported *de novo* origin of procentrioles and their linear replication during Meiosis I in the protist *Labyrinthula,* during zoosporulation. After the bicentriole shears into two parts, a typical diplosome is formed by a repositioning of components. Additional replication is presumed to be by orthagonal budding because only typical diplosomes, not bicentrioles, were seen subsequently. The original procentrioles form in granular aggregations of dense material in the cytoplasm, by extrapolation of a synthetic process that falls short of centriole formation during vegetative mitoses. In *Labyrinthula,* as in *Polytrichum,* the centrioles are formed during a particular mitotic cycle, have a polar behaviour including attachment to the spindle fibres, and become basal bodies in the normal course of development.

Homologies among centrioles

Before any further comparisons are drawn, it is necessary to reiterate that the androcyte mother cells in the bryophyte antheridium are many cell generations removed from meiosis. While this situation is not universal among plants it is certainly the typical case in cryptogams. I raise this point because it is necessary to sort out extraneous variables when one pursues fundamental comparisons. The ultimate and penultimate mitoses in the spermatogenous cells of cryptogamic plants are thought to be typical mitoses, like those of preceding divisions. Therefore, neither the formation of centrioles nor their configuration during replication is a function of meiosis or any other atypical mitosis. In plants with swimming sperms, the formation and replication of centrioles should be considered, instead, a function of the impending assembly of the flagellar apparatus.

For this reason, various authors have been reluctant to apply the name centriole (or, in the light microscopic literature, centrosome) to the entities that produce flagella in plants (e.g., Suire, 1970). Of course, the greater part of the controversy is based on light microscopic observations (Lepper, 1956). One might be inclined to suggest that the entire argument should simply be set aside because new bases for defining and describing the relevant entities are available because of electron microscopy. Before taking that suggestion seriously, one should consider that Webber (1901), for example, insisted on the name 'blepharoplast' for the cytoplasmic entities that give rise to the ciliated band during spermatogenesis in *Zamia*. He used this distinctive terminology in spite of his understanding that a homology between blepharoplasts and the centrosomes of animal cells might eventually be proven beyond reasonable doubt. Now it has been shown that the blepharoplasts, as designated by Webber in *Zamia,* consist of aggregates of procentrioles and that the latter become eventually the basal bodies of flagella (Mizukami and Gall, 1966). But rather than showing that 'blepharoplasts' and 'centrosomes' are homologous, what electron microscopy actually demonstrates is that the *units* comprising these diverse aggregations are homologous. There is no pressing need to go beyond this point in comparing an animal diplosome (centrosome) and a blepharoplast of *Zamia*. What Webber was striving for in his terminology was a way of characterizing *distinctive behaviour* without posing the question of homology. This remains a legitimate goal and one that we accept tacitly and routinely in electron microscopy. For example, Perkins (1970) and others have pointed out that any structural distinction between centrioles and basal bodies is a function of ontogeny, because the former become the latter. Once this simple fact is accepted, there is no need to eliminate either term, because each describes a fundamental entity in one of its various aspects or stages of development.

Similarly, there is no need to root out the other distinctive terms that have been applied in recognition of distinctive developmental or behavioural patterns. That electron microscopes show us the smaller entities is poor excuse to ignore the larger. In the interim, however, electron microscopists must continue to examine light microscopic entities to determine whether they are composed of more fundamental organelles that one cannot recognize in the light microscope. In passing, we can note that the arguments over homologies between the flagellar apparatus of plants and those of animals have their parallel in the question of whether or not plants possess a structure homologous to the Golgi apparatus of animals. This comparison is rendered apropos by the fact that the Golgi apparatus functions in spermatogenesis of animals to form the acrosome. While the component Golgi bodies, or dictyosomes, are definitely present in plant cells, no aggregation of dictyosomes ever functions in the androcyte to form an acrosome. We shall return to that point later; for now we must continue with the developmental rather than comparative aspects of our subject.

The evidence from *Polytrichum*

Because a single antheridium of *Polytrichum* contributes information on a segment of the developmental process, any antheridium at the crucial stage of development should tell us a great deal about centriole formation and replication. In one such antheridium, the spermatogenous cells of the distal region showed granular aggregations of dense material in the cytoplasm that resembled what is reported to precede procentriole formation in *Labyrinthula* (Perkins, 1970) and in ciliated epithelium of the rhesus monkey oviduct (Anderson and Brenner, 1971). The connection between these aggregations and procentriole formation in *Polytrichum* is suggested by microtubules converging in the area, as is the case later, when procentrioles are recognizable. Further down the antheridium, procentrioles were present in perinuclear locations. Present evidence leaves undecided the question of whether there is a sequence of first one and then a second procentriole, or whether both emerge together.

The subsequent stages outlined below are sequenced according to the observation that the overall length increases from the relatively short procentrioles that are first seen, to the bicentrioles that are ultimately obtained, and that, as the replicating procentrioles age, they come to resemble the bicentriole more and more. These simplifying assumptions are consistent with the overall maturation gradient in the antheridium and with the evidence that the bicentriole is the initial state of centrioles in the androcytes formed by the final cytokinesis in the antheridium.

The two procentrioles become arranged side by side, with an opaque material between them (Fig. 15.1). Longitudinal sections show that this opaque material is more or less restricted to the future mid-region of the developing bicentriole (Fig. 15.2). The bicentrioles become longer, dissociate from one another and lie at various angles with reference to each other (Fig. 15.3). Subsequently, one bicentriole migrates to each pole. Throughout the replication and at the poles, the bicentrioles are associated with microtubules, presumably the earliest components of the mitotic spindle. During the replication triplets are absent, but a single fibril, presumed tentatively to be the A fibril, is found in each of the positions that are occupied by triplets after the bicentrioles attain their polar positions in the cell. During the poleward migration, a developing bicentriole continues to show opaque material associated with its mid-region (Fig. 15.4). This material eventually disappears. The fully developed bicentriole is a symmetrical double structure with a continuous hub and cartwheel that fall short of each end.

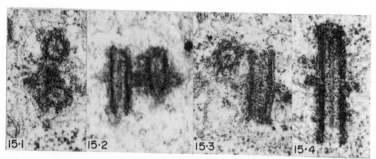

Figs. 15.1—15.4 Procentrioles and centrioles during replication in *Polytrichum*. 15.1 Transversely sectioned procentrioles. 15.2 Longitudinally sectioned, replicating procentrioles. 15.3 Developing bicentrioles after side-by-side positions have been abandoned. 15.4 Asymmetric bicentriole during poleward migration. Magnification: x40 000.

The present evidence leaves unclear exactly how the replication proceeds. However, because a large number of our micrographs show the developing bicentriole to be asymmetric (Fig. 15.4), I conclude, tentatively, that the replication is by linear growth from the end of a pre-existing procentriole.

Perkins (1970) visualized a more or less symmetrical bilateral growth of the bicentriole in *Labyrinthula*. If the differences are real, perhaps they result from the fact that procentriole origin and replication are combined into one process in *Labyrinthula*, whereas they are temporarily separated in *Polytrichum*.

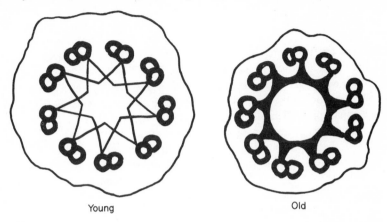

Young   Old

Fig. 15.5 Comparison of transverse sections in young vs. old transition regions in *Polytrichum* flagella.

Two points need emphasis before the next stages of development are described. Firstly, the origin of procentrioles from granular aggregates suggested here for *Polytrichum* and established for *Labyrinthula* by Perkins (1970) resembles the 'acentriolar' formation of procentrioles in the rhesus monkey ciliated epithelium (Anderson and Brenner, 1971). The authors who studied that process postulate that organization of a procentriole from a granular aggregate requires the influence of an existing centriole or of a deuterosome, a dense sphere found at the base of the procentriole. It seems impossible to identify any analogous organizing influence in *Polytrichum* and *Labyrinthula*. Secondly, these and other recent electron microscopic studies make *de novo* origin of centrioles an acceptable hypothesis, and another example of Nature's diversity.

If we here attempt to refine our concept of *de novo* origin, it seems reasonable to require that the formation of a procentriole occurs within an aggregation of substances accumulated specifically for that purpose, and in the absence of existing centrioles or, at least, remotely located from them. Thus, the orthogonal budding typical of diplosomes is excluded by this concept. Whether 'acentriolar' formation of procentrioles in the rhesus monkey oviduct should also be excluded is a debatable point, but one that is not crucial to our present subject.

The blepharoplast of bryophytes

When Webber (1901) applied the term blepharoplast in the description of spermatogenesis in *Zamia*, he designated only the spherical clusters of procentrioles.

The dispersal of the procentrioles is visualized as an abrupt change in the light microscope. The continuity of events is recognized more clearly at the electron microscopic level, where the emphasis can be placed on the development of individual basal bodies. But, as a matter of fact, the flagellar apparatus in *Zamia* is so large as to pose a number of serious problems for analysis by thin sections (Norstog, 1967, 1968). In bryophytes only two flagella are formed and, accordingly, a single bicentriole is sufficient to supply the basal bodies that are needed. The centrioles are always immersed in the cytoplasm and appear together as no more than a granule in the light microscope. Thus, there is no abrupt change in their appearance when cilliogenesis begins, and light microscopists naturally described subsequent stages as the developmental stages in the history of the blepharoplast (Wilson, 1911; Allen, 1917; Weier, 1931).

A further complication in relating light and electron microscopic descriptions comes from the response of the materials to different fixatives. Electron microscopists are so used to the excellent fixation produced routinely by glutaraldehyde that this quality of fixation is simply taken for granted. Potassium permanganate, however, does not even fix flagella or basal bodies, and has only a restricted use in the study of spermatogenesis, for example, in distinguishing one kind of microtubule from another (Diers, 1967; Kreitner, 1970). An analogous situation must exist with light microscope fixatives, because a careful reading of the relevant papers indicates that some of the components of the androcytes and sperms must have been missing from the preparations that were examined, but not necessarily the same components in every case.

Beginning with the studies of Heitz (1959, 1960) it was evident that the developing basal bodies of bryophytes are associated with two other cytoplasmic structures. One of these Heitz readily identified as a mitochondrion, the other remained more mysterious and elicited no direct comparison with other cytoplasmic organelles. He named this structure '*die Dreiergruppe*' because it appears to consist of three layers. Carothers and Kreitner (1967) later detected four layers in this structure in *Marchantia,* and accordingly coined the term '*Vierergruppe*'. When it was subsequently determined that the number of layers might vary among species, the generic term 'multilayered structure' (MLS) was proposed to avoid further proliferation of terms while facilitating comparisons (Paolillo *et al.,* 1968a, b).

Once attention was given to the three-dimensional aspects of the MLS it became evident that parts of the MLS had been detected in earlier studies (Satô, 1954, 1956; Manton, 1957), because the first layer of the MLS was found to consist of microtubules (Paolillo, 1965). The comparisons had failed to suggest themselves to Heitz largely because he did not assemble adequate materials for a three-dimensional reconstruction. Furthermore, both Satô and Manton had described mature sperms rather than androcytes and the gap between the two remained to be bridged. Finally, Satô had used shadowed mounts rather than sections. However, Manton (1957) had already sensed that what she saw as the fibrous band in *Sphagnum* was the same as the filamentous appendage revealed in a variety of bryophytes by Satô (1954, 1956). Thus the initial steps in the 'comparative microanatomy' of plant sperms had been taken and it only remained for developmental studies to follow suit.

Initial attempts at reconstructing the MLS (Paolillo, 1965) were only partly successful. More elegant reconstructions quickly followed as the application of superior embedding techniques markedly improved the results (Carothers and Kreitner, 1967, 1968; Suire, 1970). In their reconstruction of the entire blepharoplast of *Marchantia,* Carothers and Kreitner (1968) clearly illustrated the components that comprise the blepharoplast in the bryophytes. What the light microscopists had described as a filament growing over the nucleus (e.g., Wilson, 1911) was the layer of microtubules, extending itself rearward as the androcyte ages. The granule from which the filament grows consists of the two developing basal bodies, the anterior ends of the microtubules, all the underlying elements of the MLS, and (sooner or later) mitochondrial structures as well. There is little wonder that disagreements emerged among light microscopic descriptions. In fact, the electron microscopic evidence indicates that the history of the blepharoplast varies somewhat with the species, and its three-dimensional shape, size, etc. is related to the ultimate size and shape of the sperm.

In spite of specific variations in structure, in all cases the complete MLS exists only at the anterior of the nucleus in the developing sperms. While the number of microtubules in the first layer varies, it appears that the separation of microtubules into groups is a common feature, and the microtubules are more numerous where the MLS is complete than they are over the nucleus. Hence, there is specialization of the microtubules according to their relative location in the first layer of the MLS. Only some of the tubules attain great length. In some instances not all the microtubules persist in the mature sperm (Suire, 1970).

Virtually nothing is known of the origin of the MLS. We know that the MLS is present very early in the androcytes of *Polytrichum* but stages in its formation have not been observed. Suire (1970) suggests that the first of the layers to appear in *Pellia* is the layer of microtubules. Clearly, the microtubules are at first short, but if they grow from the anterior region (assuming the MLS is a synthetic region for the microtubules) then the microtubules must be pushed over the nucleus so that they change their relative position with respect to the nuclear envelope along their entire length.

In bryophytes, only the microtubular layer of MLS persists in an easily recognizable form in the mature sperm. The other layers are either altered or disappear. One tends to ascribe morphogenetic rather than permanent structural roles to the MLS because it is a transient entity. The genesis of microtubules is one possible role for the MLS, but it seems more important to me to emphasize that the MLS effectively polarizes the androcyte, and thereby contributes an ordering, or organizing, influence to the metamorphosis of the androcyte into a sperm. The complete complement of layers in the MLS is found only at the anterior of the nucleus, and the presence of microtubules over the nuclear envelope is realized only after the relationship of MLS and nucleus is established.

One can suggest then, that the juxtaposition of MLS and nucleus determines the anterior of the nucleus, which is approximately spherical and entirely uncondensed prior to its association with the MLS. The MLS is polarized front to rear by virtue of the fact that the two developing basal bodies lie on top of the microtubules with their cartwheel ends forward, with respect to the ultimate configuration of the sperm. Finally, the MLS seems to attract the mitochondria to the anterior of the developing sperm, thus playing a role in what the light microscopists called the formation of the apical body (e.g., Weier, 1931).

Some sort of microtubular sheath or appendage is found in all the plant sperms investigated so far with electron microscopy. Because the microtubules are part of the MLS, the obvious question to ask next is whether or not the MLS is found in all cases. Of course one realizes that developmental studies provide the most definitive answer to this question, especially if the MLS proves to be entirely absent or transient.

The general occurrence of the MLS in bryophytes cannot be disputed (Heitz, 1959, 1960; Paolillo, 1965; Carothers and Kreitner, 1967, 1968; Suire, 1970). Also, as far as we know, throughout these plants the MLS is a transient structure, except for the layer of microtubules. In the fern *Pteridium* and in the horsetail *Equisetum* the MLS persists into the mature sperm (Duckett and Bell, 1969; Bell *et al.*, 1971). According to Bell *et al.* (1971) the MLS occupies the first 1½ gyres in a *Pteridium* sperm and the first 2½ gyres in *Equisetum.* Norstog (1967, 1968) has identified an MLS in sperms of *Zamia.* It appears then, that the MLS is of widespread occurrence, but its shape and exact structure, and its ability to persist as a structural element in the mature sperm varies with the species.

An MLS does not occur in the charophytes (Pickett-Heaps, 1968; Turner, 1968). Instead, a dense region of cytoplasm attends the region of the developing basal bodies and the anterior of the microtubules, which develop in spite of the absence of an MLS, as such. This divergence from bryophytes is interesting because of the recurring question of whether the evolution of charophytes has anything to do with the evolution of bryophytes (Manton, 1970). It is evident that some algae are capable of forming an MLS in motile cells. *Coleochaete* is reported to form such a structure in its zoospores (McBride, 1971). Again, this matter is of some interest in a comparative sense, because this alga has also been mentioned in the past with regard to the origin of a land flora. While contemporary students do not seek a direct origin from the algae for the bryophytes, it remains a legitimate goal to develop our ultrastructural evidence for whatever it can contribute to the discussion of the broader topics of comparative morphology.

Another aspect of the possible comparisons relates to a more detailed consideration of the MLS. In mosses, the MLS is reported to be three-layered (Paolillo, 1965; Paolillo *et al.*, 1968a; Bonnott, 1967) whereas it appears to be four-layered in hepatics (Carothers and Kreitner, 1967; Suire, 1970). It is premature to place much emphasis on this difference, for one cannot readily resolve the ultrastructural events into 'hepatic' behaviour vs. 'moss' behaviour. However, I return to the question of fine structure of the MLS to make one final comment on the basis of observations reported so far. In my initial attempt to reconstruct the MLS (Paolillo, 1965) I made a point which has been somewhat neglected in more recent, more elegant reconstructions. In an electron micrograph it is not always intuitively obvious whether the real structure is represented by electron translucent or electron opaque areas in the micrographs. Accordingly, I suggested that light areas were structural and the dark areas matrix in the lower layers of the MLS. Although I thought that the second layer, for example, consisted of electron translucent columns, it is more likely that there are closely packed subspherical particles. Although all published accounts give indication of substructuring in the electron translucent components of the layers of the MLS, little attention is given to this fact, and the electron opaque 'fins' of the second layer, for example, are taken as the real structure. Of course, the electron translucent and opaque areas must somehow be integrated to give the organelle that actually exists. But what is at stake here is where the structural entity or subunit of structure resides. I point to the parallel with the microtubules of the first layer. Whereas a microtubule appears as a dark circle in transverse sections at low magnification, at higher magnification it is a series of electron translucent areas in a circle of dark matrix (Ledbetter and Porter, 1964). The clear areas are the real subunits of microtubular structure (Pease, 1963). More attention should be given to parallel considerations in the other layers of the MLS.

The highlights of sperm maturation

Spermatogenous cells have the characteristics of ordinary meristematic cells when they are first delineated and through their multiplication phase. With the origin of centrioles and the cessation of multiplication, a far-reaching metamorphosis begins. We can call these changes the maturation of the sperm. In bryophytes, maturation occurs after a short or lengthy multiplication phase that builds the contents of the multicellular antheridium. In charophytes, long filaments are produced during the multiplication, and each cell of a filament becomes a sperm, eventually. In the cycads and *Gingko* divisions are few, and only two sperms are formed in each pollen tube. Although there is no way to generalize on the history of the spermatogenous cells because of species variation, comparative microanatomy of the sperms is a promising area of study, and this offers hope, also, that comparative developmental studies will be rewarding.

Because of the small number of developmental studies that fully elucidate maturation of the sperms of plants, any generalizations on this process are best restricted to the broadest comparisons. The descriptions that follow are most directly applicable to the biflagellated sperms of charophytes and bryophytes. Publications on the details of development in multiflagellated sperms are not yet available. We shall approach the subject by enumerating the principal components of the maturing sperms and the principal changes that occur in these components as the sperms mature.

The nucleus

Each sperm has a nucleus. At the beginning of maturation the nucleus is nearly spherical, uncondensed, and has the usual porous envelope. By the time the sperm is released, its nucleus is elongated, condensed, and essentially devoid of nucleoplasm. The envelope is non-porous, and there is no opportunity for nuclear-cytoplasmic interaction in the ordinary sense, with the cytoplasm being largely eliminated.

Condensation of the nucleus accompanies a change in shape of the organelle, and both of these changes are correlated with the locations of extranuclear structures. The anterior end of the nucleus is the first to change, and this is the region of the nucleus adjacent to the blepharoplast, or its analogue. Plant sperms are built on a helical plan. The direction of the coil is a property of the sperm that is consistent with the construction of the basal bodies (see later). Hence the anterior of the nucleus and the directional aspects of nuclear elongation are co-ordinated with the location and polarity of the developing basal bodies. The dorsal side, or outer curvature of the helix, seems to be determined in the nucleus by the growth of the microtubules along the nuclear envelope. Thus, whatever symmetry the nucleus develops is, as one might expect, a result of co-ordinated development that 'fits' the nucleus to the other components of the sperm. As Turner (1970) has shown in *Nitella,* disrupting microtubular systems with colchicine prevents the elongation and coiling of the nucleus.

Condensation of the nuclear chromatin follows any number of patterns. In *Polytrichum,* dense chromatic regions of large dimensions appear within the nucleoplasm, and other regions of density occur adjacent to the nuclear envelope (Paolillo *et al.,* 1968a, b; Genevès, 1970). Within the nucleus the dense masses are rich in DNA and are of chromosomal dimensions (Genevès, 1970). In *Conocephalum* (Satô, 1962), *Pellia* (Suire, 1970) and *Marchantia* (Kreitner, 1970) the condensation proceeds first through the formation of numerous threads of variable size. Satô (1962) reports the smallest size for the fibrils (17–20 nm), Kreitner (1970) gives a larger size (20–25 nm), and Suire (1970) reports a diameter of 30 nm for these chromatin fibrils. Accounts on the substructure of the fibrils differ markedly. In *Phaeoceros* (Moser, 1970) condensation of the chromatin occurs first along the part of the nuclear membrane that contacts the microtubules that elongate rearward from the MLS. In *Nitella* (Turner, 1968) and *Chara* (Pickett-Heaps, 1968) early peripheral disposition of chromatin is followed by dispersal of chromatin, formation of fibrils, and overall condensation.

Microtubules

Where an MLS is present, the microtubules are the first layer of the MLS and their extension rearward is closely associated with the nuclear envelope, which appears to adhere to the microtubules. Hence, the tubules are a stabilizing influence on one region of the nuclear envelope, and it is also reasonable to assume that the microtubules contribute structurally to the coiled shape of the mature sperm. There is no concrete evidence for a locomotory function for microtubules in plant sperms.

The microtubules that persist in the mature sperms have been called the filamentous appendage (Satô, 1954, 1956) the fibrous band (Manton, 1957, 1959a) the manchette (Pickett-Heaps, 1968) and the spline (Carothers and

Kreitner, 1968). One cannot justify replacing all these terms with a single specific term. Circumstances vary so profoundly from study to study that the terms are not synonymous, or at least they carry strongly different connotations in their original uses. Filamentous appendage and spline were both coined for descriptions of liverworts, but the former referred to microtubules attached to the mature nucleus while the latter was used to describe microtubules in relation to their development as part of the MLS. Fibrous band, as first applied by Manton (1957) for *Sphagnum* denotes a small number of microtubules appressed to the nucleus. A larger number of microtubules is found under the plasmalemma in the cytoplasm. It is not yet known whether all or any of these tubules are related developmentally to an MLS. In *Pellia,* Suire (1970) recognizes a 'languette transitoire' and a 'languette definitive' because more tubules are produced in association with the MLS than are conserved in the mature sperm. In contrast to *Sphagnum* sperms, in the sperms of *Pellia* microtubules that are not directly associated with the nucleus do not persist; but neither is the final association of microtubules and nucleus as intimate as it is, for example, in *Polytrichum* (Paolillo *et al.,* 1968b), *Marchantia* (Kreitner, 1970), or *Phaeoceros* (Moser, 1970). In fact, in *Pellia* the band of microtubules may stand 'on edge' on the dorsal surface of the nucleus (Manton, 1970). The term manchette was used for the assemblage of microtubules in *Chara.* In both *Chara* (Pickett-Heaps, 1968) and *Nitella* (Turner, 1968) the microtubules develop without the other layers characteristic of an MLS, so one cannot claim that an MLS exists in the charophytes. But even comparing the two charophytes one notes that in *Chara* the microtubules completely surround the anterior of the sperm while in *Nitella* they do not (Moestrup, 1970).

The charophyte sperms further show that the microtubules align other organelles with the nucleus. The mitochondria take up a position in linear alignment under the microtubules anterior to the nucleus. Behind the nucleus, again in linear array beneath the microtubules, one finds the plastids. Thus, the microtubules seem to have a capacity for 'organizing' or otherwise influencing their immediate environment. Another aspect of this phenomenon may be manifested in the maturing sperms of *Pellia.*

We have mentioned that in *Pellia* there are more microtubules in the MLS than there are in the mature sperm (Suire, 1970). The supernumerary microtubules do not grow rearward over the nucleus. Instead, they are diverted to a tubular system originating from the endoplasmic reticulum and lying parallel to the elongating nucleus. One can only speculate on cause and effect in this case, but if the extension of microtubules goes beyond the limits of the permanent organelles, perhaps the microtubules organize the cytoplasm immediately beneath them in a way parallel to their effect on the nucleus. In *Pellia* this influence is manifested in the association of microtubules and cytoplasmic membranes. In the charophytes the extension of microtubules beyond the rear of the nucleus differentiates the cytoplasm of that region, and the plastids come to occupy that location. In all the plant sperms studied to date the region anterior to the nucleus is overlain with microtubules, and here the mitochondria are to be found.

Whereas the studies already available indicate diverse developmental histories for the microtubular assemblages of plant sperms, they also point to the concept that the microtubules have morphogenetic as well as structural roles. Their presence seems to determine the course of nuclear elongation and condensation, and they contribute structurally to the helical form of the sperm. They seem effectively to align and link the components of a sperm, and they may contribute to the differentiation that orders those components into their final relative positions. It has also been suggested that microtubules play a role in the conduction of metabolites (Manton, 1964b).

Finally, we must emphasize that the bands of microtubules are internally differentiated. Disposition of microtubules into groups at early stages of maturation can be correlated with their final dispositions in the mature sperm. For example, in *Polytrichum* (Paolillo *et al.,* 1968a, b) two groups of six and four tubules, respectively, are formed. Of the group of six, a single microtubule lies on top of the mitochondrion where the latter is adjacent to the anterior of the nucleus, while the other five are on top of the nucleus itself. Only three of these five are long enough to go beyond the length of the mitochondrion, where they are rejoined by their wayward member and continue rearward on the nucleus. The group of four tubules is lateral to the nucleus, and quite short, seeming only to hold the tip of the nucleus against the mitochondrion. The ten microtubules thus accounted for assume an approximation of their final postures early in the maturation of the sperm as the organelles adjust to each other. Moreover, the underlying layers of the MLS extend further rearward under the six tubules as opposed to the four in early stages of sperm maturation. I relate these details to offer some indication of the precision of integration among co-ordinated developments during maturation. As published accounts show, the details of the process are a function of the species under study. From all cases, however, we derive the understanding that something is at work to determine the location, length, and, ultimately, the function of microtubules in the sperm. The microtubules are not an undifferentiated strand or rope, but rather an internally differentiated band that is constructed under precise controls that we do not understand. It is evident, however, that, depending on their location in the band, the microtubules are associated with different organelles or with different aspects of the same organelle.

Plastids and mitochondria

The plastids and mitochondria differ in their behaviour in different species, but some overall patterns are discernable. In mosses, the mitochondria coalesce around the plastid to form a 'Nebenkern' (Paolillo, 1965; Bonnott, 1967). The Nebenkern divides into two parts, with a substantial portion of the mitochondrial mass migrating to the interior side of the developing blepharoplast. The remainder accompanies the plastids into the more or less lateral-posterior cytoplasmic remnant. The liverwort *Pellia* follows this sequence of developments (Suire, 1970) except that the plastid and mitochondria seem to be more firmly attached to the concave surface of the nucleus than in the mosses.

In the liverwort *Marchantia* (Kreitner, 1970), fusion of the mitochondria does not involve the plastid, and only one small mitochondrion accompanies the plastid to the rear of the sperm. While the anterior mitochondrion becomes part of the developing apical complex, the small mitochondrion becomes lodged between the nucleus and the plastid. This complex tailpiece is integrated into the body of the sperm, being aligned with the other components under the microtubules, as in the charophytes. The liverwort *Sphaerocarpus* (Diers, 1967) has sperms that seem to be organized like those of *Marchantia*, and a similar situation applies in the hornwort *Phaeoceros* (Moser, 1970), although here there is no mitochondrion posterior to the nucleus. In the charophytes, migration of the mitochondria to the anterior and of plastids to the posterior seems to occur without massive fusions.

While the maturing sperms of *Chara* (Pickett-Heaps, 1968) and *Nitella* (Turner, 1968) each have a number of plastids, the rule among bryophytes is one plastid per cell. In spite of the report of two plastids per androcyte in *Polytrichum* by Genevès (1967) I remain convinced that there is a single, polymorphic plastid in the early androcyte. The tendency for mitochondria to cluster around the plastid is manifested even before the final cytokinesis. The two types of organelles come into close association. Sun (1964) misidentified the two components of this complex Nebenkern, largely because his fixations did not preserve the boundary between the mitochondria and the plastid, but also because he did not follow the development of the Nebenkern. Of course, the plastid of the androcyte is not a chloroplast, but it does resemble the plastids of a meristematic cell. In *Polytrichum*, the plastids are dense and internally simple throughout the development of the spermatogenous tissue (Paolillo *et al.*, 1968a, b). In *Pellia* (Suire, 1970), the density of the plastid increases abruptly when the androcytes are formed. As the sperm matures in *Polytrichum*, the plastid comes to resemble the plastids of albino tissues, with little internal differentiation and a degenerate, vacuolate appearance. The plastids in mature sperms of *Chara, Nitella, Marchantia,* and *Phaeoceros* might be more accurately characterized as amyloplasts because of their large starch content.

Thus the history of the plastids varies with the species. The plastids may be numerous (*Chara, Nitella, Pteridium, Marsilea, Equisetum, Zamia*) or single (*Polytrichum, Bryum, Pellia, Marchantia, Phaeoceros*). The plastid may act as a focus for mitochondrial fusions (mosses and *Pellia*) or not. The plastid(s) may be integrated into the sperm body (*Chara, Nitella, Marchantia, Phaeoceros*) or not. In the multiflagellated sperms of *Equisetum*, the ferns, and *Zamia*, the plastids evidently lie free in the cytoplasmic vesicle, which is more substantial than in the biflagellated sperms. Again, they are amyloplasts in the mature sperms.

The large sperms of *Zamia* pose a special problem for two reasons. Firstly, their remarkable size makes them hard to reconstruct from thin sections and electron micrographs (Norstog, 1967, 1968). Secondly, of all the sperms studied so far, those of *Zamia* resemble the swimming cells of algae the most closely. It is ironic that the sperms of a gymnosperm should show this resemblance while those of lower archegoniates do not (see later).

In all plant sperms the anterior is occupied by mitochondria. Microtubules overlie the mitochondria, with or without the other layers of an MLS, depending on the species. Likewise, the degree of fusion of mitochondria varies with the species. On one extreme (*Chara, Nitella*) a number of mitochondria persist, linearly arranged (Pickett-Heaps, 1968; Turner, 1968). On the other (mosses, liverworts) there is a single mitochondrion that becomes streamlined into a definite shape that is characteristic of the species. In *Polytrichum* (Paolillo *et al.*, 1968a, b) the apical mitochondrion elongates and lies beside the nucleus for much of its length. The apical mitochondrion of *Marchantia* is blunt and spatulate (Kreitner, 1970), overlapping the nucleus for only a short distance, while the apical mitochondrion in *Phaeoceros* does not overlap the nucleus at all (Moser, 1970).

The cristae of the mitochondria are unusually saccate during the early stages in maturation, and this condition is reputed to imply active respiration in these organelles (Manton, 1970). This condition does not always persist to maturity, however. If the cristae become platelike their orientation may be adjusted with respect to the position of other components in the sperm, particularly the basal bodies and the nucleus (Paolillo *et al.*, 1968a, b).

The basal bodies and flagella

When the sperms begin to mature they contain centrioles with the typical cartwheel plus nine triplet fibres. As the centrioles begin the transformation to basal bodies they are reoriented with relation to each other. In *Nitella* (Turner, 1968) and *Chara* (Pickett-Heaps, 1968) a spindle-shaped striated body, perhaps of contractile nature, seems to assist in this reorientation. The spindle is attached to both centrioles, and Pickett-Heaps (1968) has compared it to striated components of ciliary roots of other algal cells (e.g., in *Oedogonium*, Hoffman and Manton, 1962, 1963; Ringo, 1967). A similar component develops between the centrioles of the *Phaeoceros* androcyte (Moser, 1970) as they rotate into position before the growth of the flagella.

The following account gives the principal changes occurring in the basal bodies of bryophyte sperms, based on the available literature (Paolillo *et al.*, 1968a, b; Kreitner, 1970; Moser, 1970; Suire, 1970). The basal bodies change their length and continue to adjust their positions relative to each other and to the other organelles, as the sperm matures. If the basal bodies develop to different lengths, the shorter lies anterior to the longer in its insertion into the sperm. Each basal body is asymmetrical in several ways: (a) the proximal end has the cartwheel, and the opposite end, which was the free end in a bicentriole, grows out as the flagellum; (b) all the triplets tilt in the same direction, and this direction is the same for both basal bodies; and (c) certain of the triplets are longer than the others at the proximal end of the basal body. The third of these features generates additional differences among species, for while the phenomenon of triplet extension seems to be widespread (no exception discovered so far) it is not the same triplet that is the longest in each case for the anterior basal body. On the other hand, in the posterior basal body the three triplets closest to the underlying microtubules are uniformly the longest. It seems reasonable that the differential lengths of triplets alter the shape of the basal bodies for a better fit into the body of the mature sperm.

The basal body is connected to the flagellum (which has nine doublets plus two central singlets) through a stellate transition region of the kind encountered in the algae (Lang, 1963; Manton, 1964a). Other details vary so much from species to species that it is best to deal with them only in the context of the original research papers. However, the stellate transition region can be found in both biflagellated and multiflagellated sperms. Likewise, in both biflagellated and multiflagellated sperms the basal bodies are inserted into the sperms on top of the microtubules.

The flagella of sperms must be approached as objects for developmental study. For example, the basal bodies continue to change in some species with the loss of the cartwheel structure (Moestrup, 1970). In *Polytrichum* other developmental changes occur, including the alteration of the transition region and the formation of a helical wrapper around the central elements of the axoneme (Paolillo, 1967). The latter is probably present in other bryophytes as well (see figures in Diers, 1967; Kreitner, 1970; Suire, 1970) and is remarkably like the helical wrapper reported for the central complex of the flagellum of the flatworm by Silveria and Porter (1964).

Thus, the flagellum is a source of comparative as well as developmental information, and one can begin to construct how developmental changes can be translated into phylogenetic changes. For example, consider the stellate transition region: the immature flagellum of *Polytrichum* compares favourably with the mature flagella of certain algae in this regard. But subsequent alterations make the transition region different if *two mature* flagella are compared. The transition regions in young and mature flagella of *Polytrichum* are illustrated in Fig. 15.5. This contrast is not much different from that given by Mignot (1967) for euglenoid vs. chlorophycean transition regions, the latter of which has, of course, the stellate configuration in transverse section. Hence, divergent types of transition regions can be obtained; at least theoretically, by extrapolation, or serial development from the more usual type. Subsequently, the usual pattern might disappear altogether as developmental steps are dropped out of the process and the end point is reached more directly. The same model can be applied with regard to the helical wrapper around the central fibres of the axoneme. In *Polytrichum* this wrapper comes late in development and surrounds the two central fibrils. In the flatworm flagellum the two central fibrils are absent (Silveria and Porter, 1964). This situation could have come about by the simplification of a flagellum of the type seen in *Polytrichum,* with the formation of the central fibrils being dropped out of the developmental process and the development of the wrapper being emphasized all the more. My goal here is simply to point out how comparisons among flagella must be made, with a developmental point of view in mind. If we actually do believe that there are certain fundamental features of flagellar construction (e.g., 9 + 2 pattern) then we should be willing to make the appropriate developmental studies that will place the deviant types (e.g., 9 + '0') in proper perspective.

In this regard, we come again to the conclusion that the sperms of the gymnosperm *Zamia* (Norstog, 1967, 1968) are remarkably like the motile algal cells in organization. For, although the basal bodies in *Zamia* are long, their organization at maturity retains the obvious similarities to the algal basal bodies, and the stellate transition region persists.

Before leaving the subject of basal bodies and flagella, we must return to the fact that the asymmetry of the basal bodies is consistent with the coiling of the nucleus. This is easiest to understand in the biflagellated sperms. As Fig. 15.6

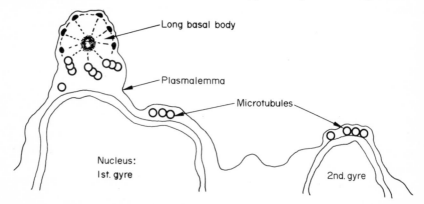

Fig. 15.6 Relative positioning of basal bodies and successive gyres of the nucleus, with the 'A' fibrils of the triplets in the clockwise configuration.

shows, if the A fibril of the triplet is viewed in the clockwise position the successive gyres of the nucleus will be to the right in the section. When the orientation of the section in three-dimensional space is known, the 'handedness' of the coil of the sperm can be determined. If the description given above turns out to be a view from the basal body toward the flagellum, the basal body is said to have the clockwise configuration (Gibbons and Grimstone, 1960) and the sperm is a left-handed helix.

We have reassembled 73 sperms of *Polytrichum* from serial sections and five of these (unpublished data) cannot be made to fit the left-handed helix that is supposedly universal among plant sperms (Bell *et al.*, 1971). If these results can be further substantiated, the occurrence of the counter-clockwise enantiomorphic basal bodies will be established, because a right-handed sperm coil could contain only counter-clockwise basal bodies. The only previously confirmed configuration for basal bodies and flagella is the clockwise configuration (Gibbons, 1961, 1963; Satir, 1965). The source of the enantiomorphy might be sought in variability in the synthesis of B and C fibrils during centriole replication. Alternatively, there might be occasional variability in which end of the centrioles grow the flagella. This seems less attractive because all the bicentrioles we have observed to date are strongly polarized.

Other organelles

Androcytes contain the full complement of cellular components except for
large vacuoles. Among those not yet discussed we must list the endoplasmic
reticulum and ribosomes. Both of these components are at their peak in the
androcyte and they wane as the sperm matures. While the Golgi bodies, or
dictyosomes are, likewise, active in the androcyte, they do not cluster together
to form a Golgi complex as is the case in animal sperms. There is no acrosome
formed as such, and the Golgi bodies generally disappear as the sperms mature.
A possible analogue for the acrosome is suggested by our evidence from *Polytrichum* (Paolillo *et al.*, 1968b) and by the study of *Marchantia* (Kreitner, 1970),
where it appears that the lower layers of the MLS may persist in an altered form
at the very tip of the sperm. Assigning any role to this remnant is speculative,
however, and its only real resemblance to an acrosome is its position in the
sperm, not its origin or structure.

General comparisons

Any broad comparisons must take several forms because our subject matter is
so diverse. However, I shall limit our discussion to three questions: (a) what are
the similarities and contrasting features among sperms of the charophytes,
pteridophytes and lower gymnosperms?, (b) how do they compare with the
sperms of oogamous algae, some of which also have zoospores (non-sexual,
motile cells) in the life history? and (c) how do they compare with animal
sperms?

Comparisons within the charophytes and land plants

The first of our questions merely requires a synopsis based on the descriptive
aspects of this review so far. The more or less common features of the developmental sequences that have been outlined include the growth of microtubules and
flagella, the elongation and condensation of the nucleus, the migration of all
or most of the mitochondrial mass to the anterior end of a maturing sperm, and
the migration of the plastid(s) rearward. Thus, the basic parts of the sperms, when
released, are easy to enumerate: nucleus, apical mitochondrion (or mitochondria),
basal bodies and flagella, microtubules, plastid(s), and residual cytoplasm essentially
depleted of ribosomes and membrane systems, in the more specialized cases.

The studies now available for bryophytes indicate that the sperms may be
as diverse in the details of their structure as the gametophytes that bear them.
The shapes and relative sizes of parts, the numbers and dispositions of microtubules, the positioning of basal bodies, etc., all vary with the species. The
plastid may be associated with mitochondria or not, or both kinds of organelle
can be integrated into the tail of the sperm or left suspended in the residual
cytoplasm. In contrast to the sperms of bryophytes, those of the ferns, *Equisetum,*
and the cycad *Zamia* have an MLS at maturity, have more numerous flagella,
and more abundant cytoplasm. Of course, the overall size is greater for these

multiflagellated sperms than for the sperms of bryophytes. The MLS is a transient structure in bryophytes and is absent in the charophytes.

Comparisons with algae

The second general comparison I wish to make is between the sperms we have discussed so far and those of algae other than *Chara* and *Nitella*. Moestrup (1970) and Manton (1970) have made the salient comparisons that one can draw among these algae, by reference to the resemblance of sperms to zoospores. *Prasiola* (Manton and Friedmann, 1959), *Oedogonium* (Hoffman and Manton, 1963), Cystoseiraceae (Manton, 1964b), *Dictyota* (Manton, 1959b) and *Lithodesmium* (Manton and Von Stosch, 1966) have sperms that resemble zoospores. The nuclei are rounded rather than elongate, and the cells are more or less generalized, with a nearly complete roster of organelles. However, the sperms are not necessarily the same size as zoospores. In *Oedogonium* the sperms are smaller than zoospores of the same genus and, in addition, the flagellar bases are not anchored in the nucleus as they are in the zoospore (Hoffman and Manton, 1963). In *Dictyota* sperms the plastid is at best vestigal (Manton, 1959b) or absent (Manton, 1970). The sperm of *Lithodesmium* has no Golgi body (Manton and Von Stosch, 1966) but otherwise resembles a zoospore.

Moestrup (1970) regards this assemblage of forms as contrasting in sperm morphology with *Vaucheria,* Fucaceae (Manton and Clarke, 1951, 1956; Manton, Clarke, and Greenwood, 1953), charophytes, ferns and fern allies, and bryophytes. With the exception of Fucaceae and the addition of *Zamia* this assemblage represents the plants we have dealt with so far.

Special attention has been given to the Fucaceae by Manton and her co-workers and they have attempted to determine whether the sperms of the *Fucus* type are primitive or derived. They speculate that the latter is the case (Manton and Clarke, 1956). In the *Fucus* sperm the nucleus is elongated and the components of the sperm are reduced to the minimum and are closely appressed to each other. The anterior of the cell is ornamented with a 'proboscis' that is a flattened structure containing numerous, recurved microtubules. Comparative studies lead to the conclusion that these microtubules are homologous to the microtubular flagellar 'root' seen in other sperms and in zoospores (Manton, 1964b). The flagellar roots are attached to the flagellar bases and their location and extent in the cell varies with the species. It can be demonstrated that they often contact mitochondria in ways that suggest more than casual relationships between the root and these potential sources of chemical energy. Hence, it is suggested that the roots are used for metabolic rather than structural purposes in these cells (Manton, 1964b).

Of course, our interest in these roots is stimulated by their resemblance to the assemblages of microtubules found in sperms of land plants and the charophytes. Manton has already called attention to this comparison, and one might suggest at the same time that the microtubules in all sperms, function in conduction of metabolites as well as for structural purposes. It remains for us here to lay out some simple speculations on broad comparative grounds.

It can be argued that in algae with zoospores and sperms, the two types of swimming cells represent the same genetic information channelled down two separate lines of specialization within the same species. In forms like *Fucus,* the sperms have been maintained in the life history but the zoospores have been dropped out. In the land flora, the same situation holds true. One can reason that all swimming cells are fair game for comparison, and this would reach all the way to the vegetative cells in the life histories of some unicells. These speculations are a reiteration of the commonly accepted notion that swimming cells have been progressively restricted and specialized in plant evolution. The ultimate loss of all tendency to produce swimming cells occurs in higher gymnosperms and angiosperms (see also, Manton, 1970 for similar arguments).

It remains conceivable that the microtubules seen in sperms of land plants are an alternative expression of the inherited genetic information that is currently used to produce microtubular roots attached to the flagellar bases of algal sperms and zoospores. It should be realized, however, that among the algae the forms of these roots are as diverse as the arrangements of microtubules in the sperms of land plants.

Where the MLS comes into the picture is hard to say. Sperms of *Chara* and *Nitella* seem to fit with other algal sperms in the absence of an MLS, but it is reported that an MLS is present in zoospores of *Coleochaete* (McBride, 1971). Perhaps the MLS is a property of one or a few major lines of plant evolution. However, layered striated components are well known for flagellar roots in the algae. These are found in addition to the microtubules as are the other layers of the MLS in land plants. Examples include the layered component in the dinoflagellate *Amphidium* (Dodge and Crawford, 1968), a paracrystalline component in euglenoids (Mignot, 1966) and striated components as in *Chlamydamonas* (Ringo, 1967) and *Oedogonium* (Hoffman and Manton, 1962, 1963). The details are so diverse that speculation on the homologies between the MLS on the one hand and microtubular roots plus striated or layered components on the other is not in order.

Developmental comparisons between algae in general and land plants in general are not made possible by the few studies that contain developmental information on the algae (Manton and Von Stosch, 1966; Moestrup, 1970). In *Vaucheria* (Moestrup, 1970) a large mass of cytoplasm is cleaved to form numerous gametes. A nucleus, two attached flagella, a few large mitochondria and a Golgi body are included in each sperm but plastids, for example, are left behind in the communal cytoplasm. This loss of components has its counterpart in the sperms of land plants in the shedding of cytoplasmic vesicles by individual sperms. In *Lithodesmium* the flagella appear before meiosis is complete (meiosis is coincident with spermatogenesis in this case) but there is only one flagellum per cell and it has a 9 + '0' structure, with no central complex whatsoever (Manton and Von Stosch, 1966).

The sperms of *Zamia* offer a final topic for evolutionary speculations. The evidence available (Norstog, 1967, 1968) indicates that, of all the sperms of land plants studied so far by electron microscopy, it is the *Zamia* sperm that most nearly resembles a multiflagellated algal zoospore. The apical region of the cell is differentiated by the flagellar apparatus, not unlike the situation in *Oedogonium* zoospores (Hoffman and Manton, 1962). The nucleus is rounded and located in the basal region of the cell, which contains numerous other organelles as well. The sperms are less 'streamlined' than those of the bryophytes or ferns. Perhaps the environment in which the sperms of *Zamia* swim (pollen tubes and nucellus) is either a permissive one or actually favours this type of sperm. The nucleus never has to be hardened off for a trip 'out of doors', since it is placed right on the doorstep of the egg cell inside the developing ovule.

Comparisons with animal sperms

Our final line of comparison is the one that attracted the attention of the light microscopists who were in search of fundamental homologies between the cells of plants and animals (for example, Bowen, 1926, 1927; Weier, 1931; Eyme, 1954; Lepper, 1956). Here, we will confine ourselves to the sperms of bryophytes, although the arguments can be extended to other land plants by virtue of the comparisons already made.

Considerable debate was devoted to the question of whether or not the apical region of the bryophyte sperm was occupied by the homologue of the acrosome found in animal sperms. Because the light microscope was incapable of resolving the closely juxtaposed parts as discrete entities, the Nebenkern of mosses gave the impression that the plastid was involved in 'acrosome' formation, and was, therefore, the homologue of the animal Golgi apparatus (Weier, 1931). We know now that there is no acrosome formed in bryophytes and that the Golgi bodies simply remain dispersed. Furthermore, there is no recognizable homologue of the plastid in animal cells.

It has further been revealed that the common feature of construction in the sperms of plants and animals is that the flagella are inserted near mitochondria. In the metazoan sperm this is typically the mid-region, but in plant sperms it is anterior of the cell. Of special interest is the microgamete of the sporozoan *Eimeria*, which has apical flagella and an apical mitochondrion derived by fusion of smaller mitochondria (Cheissin, 1965) as do the bryophytes. The sperms of flatworms are built along the same lines, and the mitochondrion at the anterior of a sperm is similarly derived (Silveria and Porter, 1964). In the simplest possible comparison, the sperms of bryophytes, *Eimeria*, and the flatworms are 'folded' versions of an idealized metazoan sperm (Fig. 15.7). The comparison is strengthened by the parallels between mitochondrial fusions in both plants and animals, in the so-called Nebenkern formation (De Robertis *et al.*, 1965).

Thus, parts of the cell that appear to be homologous from plants to animals on structural and/or physiological grounds are homologous on behavioural grounds as well. The 'mission oriented' activity of the sperms of both plants and animals has probably been important in preserving this similar behaviour over the aeons in plant and animal cells. By examining the parts of the mature sperm, and by tracing the origin of these parts, the homologies of plant and animal centrioles (thus, basal bodies and flagella), nuclei, and mitochondria appear evident. That the plant Golgi bodies (dictyosomes) fail to perform in the formation of an acrosome does not bespeak a fundamental disparity in the Golgi bodies of plants vs. animals. Rather, it indicates that plants have gone about solving the problems of penetration of the female gametes in a different way from animals.

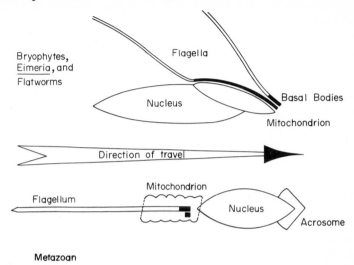

Fig. 15.7 Comparison of idealized metazoan sperm and the 'folded' sperms of bryophytes, *Eimeria,* and flatworms.

Sperms and fertilization

There are no studies on fertilization in plants that match the elegance of contemporary studies on fertilization in animals. Manton and Friedmann (1959) have shown the importance of the flagella in attaching the sperms of *Prasiola* to the eggs. Diers (1967) reports that the *Sphaerocarpus* sperms lose all components except the headpiece and the nucleus as they descend the neck canal of the archegonium. In contrast to this, Yuasa (1952) reported on the basis of light microscopy that the plastid of the *Marchantia* sperm enters the egg with the nucleus.

The studies by Duckett and Bell (1971, 1972) on the fern *Pteridium* indicate that the plastids of the sperms are lost in the neck of the archegonium. They suggest that flexing actions of the microtubules in the sperms are responsible for the movements of the sperms into the venter of the archegonium, and that the microtubules cause the sperms to coil again before they enter the egg. One to six sperms enter an egg, but all of these are sequestered in a chamber at the top of the egg until one sperm enters the cytoplasm proper. The sperms are thought to more or less carve their way into the egg and its cytoplasm by use of the so-called border rim, a modification of the anterior edge of the first 1½ gyres of the sperm. The multilayered structure is part of the border rim. Breakdown of flagellar membranes begins in the neck of the archegonium, but the axonemes and the paraphernalia to which they are attached are not cast off from a sperm nucleus until the sperm has entered the egg cytoplasm. Polyspermy is supposed to be prevented by rapid physiological changes in the egg rather than by the secretion of new membranes. Duckett (1972) reports that a rapid buildup of membrane-bound ribosomes in pentagonal arrays follows fertilization. This change may be symptomatic of rapid metabolic alterations in the egg, due to fertilization.

Final comments

While the subject of spermatogenesis is hardly exhausted by this review, or for that matter by current research, I hope I have shown that sufficient data are available to plan further research. The questions of the origin of centrioles and enantiomorphy certainly deserve further study, and the sample of organisms that are studied developmentally must be diversified. The lycopods, which contain both biflagellated and multiflagellated forms, are likely candidates here, and *Gingko* deserves special examination among the gymnosperms. Sperms need to be further investigated physiologically, after the fashion of Rice and Laetsch (1967), with a watch being kept for ultrastructural correlations. More imaginative functions for the components of the sperms must be investigated also; for example, the suggestion of an 'acrosomal' function for the proboscis microtubules in *Fucus* (Manton, 1970). Obviously, fertilization and zygote development need further study.

These investigations are more easily suggested than they are accomplished. But for investigators who would like to work on some of Nature's most intricate creations, the sperms of plants offer ample opportunities for additional investigations.

Bibliography

References

Allen, C. E. (1917) 'The spermatogenesis of *Polytrichum juniperinum*', *Ann. Bot.*, **31**, 269–292.
Anderson, R. G. W. and Brenner, R. M. (1971) 'The formation of basal bodies (centrioles) in the rhesus monkey oviduct', *J. Cell Biol.*, **50**, 10–34.
Bell, P. R., Duckett, J. G., and Myles, D. (1971) 'The occurrence of a multilayered structure in the motile spermatozoids of *Pteridium aquilinum*', *J. Ultrastruct. Res.*, **34**, 181–189.
Berlin, J. D. and Bowen, C. C. (1964) 'Centrioles in the fungus *Albugo candida*', *Am. J. Bot.*, **51**, 650–652.
Berlin, J. D. and Bowen, C. C. (1965) 'Mitosis and zoospore formation in *Albugo*', *Am. J. Bot.*, **52**, 613 (Abstract).
Bonnot, E. J. (1967) 'Le plan d'organisation fondamental de la spermatide de *Bryum capillare* (L.) Hedw.', *C. r. hebd. Séanc. Acad. Sci., Paris*, D, **265**, 958–961.
Bowen, R. H. (1926) 'Chondriosomes and Golgi apparatus in plant cells', *Science, N.Y.*, **87**, 188–190.
Bowen, R. H. (1927) 'A preliminary report on the structural elements of the cytoplasm in plant cells', *Biol. Bull.*, **53**, 179–196.

Carothers, Z. B. and Kreitner, G. L. (1967) 'Studies of spermatogenesis in the Hepaticae I. Ultrastructure of the Vierergruppe in *Marchantia*', *J. Cell Biol.*, **33**, 43–51.

Carothers, Z. B. and Kreitner, G. L. (1968) 'Studies of spermatogenesis in the Hepaticae II. Blepharoplast structure in the spermatid of *Marchantia*', *J. Cell Biol.*, **36**, 603–616.

Cheissin, E. M. (1965) 'Electron microscopic study of microgametogenesis in two species of coccidia from rabbit (*Eimeria magna* and *E. intestinalis*)', *Acta Protozool.*, **3**, 215–224.

De Robertis, E. D. P., Nowinski, W. W., and Saez, F. A. (1965) *Cell biology*, p. 172, W. B. Saunders Co., Philadelphia.

Diers, L. (1967) 'Der Feinbau des Spermatozoids von *Sphaerocarpus donnellii* Aust. (Hepaticae)', *Planta*, **72**, 119–145.

Dodge, J. D. and Crawford, R. M. (1968) 'Fine structure of the dinoflagellate *Amphidium carteri* Hulbert', *Protistologica*, **4**, 231–242.

Duckett, J. G. (1972) 'Pentagonal arrays of ribosomes in fertilized eggs of *Pteridium aquilinum* (L) Kuhn', *J. Ultrastruct. Res.* **38**, 390–397.

Duckett, J. G. and Bell, P. R. (1969) 'The occurrence of a multilayered structure in the sperm of a Pteridophyte', *Planta*, **89**, 203–211.

Duckett, J. G. and Bell, P. R. (1971) 'Studies on fertilization in archegoniate plants. I. Changes in the structure of the spermatozoids of *Pteridium aquilinum* (L) Kuhn during entry into the archegonium', *Cytobiologie*, **4**, 421–436.

Duckett, J. G. and Bell, P. R. (1972) 'Studies on fertilisation in archegoniate plants. II. Egg penetration in *Pteridium aquilinum* (L) Kuhn', *Cytobiologie*, **6**, 35–50.

Eyme, J. (1954) 'Recherches cytologiques sur la mousses', *Le Botaniste*, **38**, 1–166.

Gall, J. G. (1961) 'Centriole replication. A study of spermatogenesis in the snail *Viviparus*', *J. biophys. biochem. Cytol.*, **10**, 163–193.

Genevès, L. (1967) 'Sur le groupement des plastes et des mitochondries pendant la differentiation du spermatozoide de *Polytrichum formosum* (Bryacées)', *C. r. hebd. Séanc. Acad. Sci., Paris, D*, **265**, 1679–1682.

Genevès, L. (1970) 'Différenciation de l'appareil nucléaire a l'echelle infrastructurale, pendant la spermatogenèse, chez une bryophyte (*Polytrichum formosum*)', In: *Comparative spermatology* (ed. B. Baccetti), pp. 159–168. *Proc. Int. Symp. Rome and Siena*, 1–5 July 1969.

Gibbons, I. R. (1961) 'Structural asymmetry in cilia and flagella', *Nature, Lond.*, **190**, 1128–1129.

Gibbons, I. R. (1963) 'A method for obtaining serial sections of known orientation from single spermatozoa', *J. Cell Biol.*, **16**, 626–629.

Gibbons, I. R. and Grimstone, A. V. (1960) 'On flagellar structure in certain flagellates', *J. biophys. biochem. Cytol.*, **7**, 697–716.

Heitz, E. (1959) 'Electronenmikroskopische Untersuchungen über zwei auffallende Strukturen an der Geisselbasis der Spermatiden von *Marchantia polymorpha*, *Preissia quadrata*, *Sphaerocarpus donnellii*, *Pellia fabroniana* (Hepaticae)', *Z. Naturf.*, **14b**, 399–401.

Heitz, E. (1960) 'Uber die Geisselstruktur sowie die Dreiergruppe in den Spermatiden der Leber- und Laubmosse', *Proc. Eur. Reg. Conf. Electron Microscopy*, Delft, **2**, 934–937.

Hoffman, L. R. and Manton, I. (1962) 'Observations on the fine structure of the zoospore of *Oedogonium cardiacum* with special reference to the flagellar apparatus', *J. exp. Bot.*, **13**, 443–449.

Hoffman, L. R. and Manton, I. (1963) 'Observations on the fine structure of *Oedogonium*. II. The spermatozoid of *O. cardiacum*', *Am. J. Bot.*, **50**, 455–463.

Kreitner, G. L. (1970) *The ultrastructure of spermatogenesis in the liverwort, 'Marchantia polymorpha'*, Ph.D. thesis, University of Illinois.

Lang, N. J. (1963) 'An additional ultrastructural component of flagella', *J. Cell Biol.*, **19**, 631–634.

Ledbetter, M. C. and Porter, K. R. (1964) 'Morphology of microtubules of plant cells', *Science, N.Y.*, **144**, 872–874.

Lepper, R., Jr (1956) 'The plant centrosome and the centrosome-blepharoplast homology', *Bot. Rev.*, **22**, 375–417.

Manton, I. (1957) 'Observations with the electron microscope on the cell structure of the antheridium and spermatozoid of *Sphagnum*', *J. exp. Bot.*, **8**, 382–400.

Manton, I. (1959a) 'Observations on the microanatomy of the spermatozoid of the bracken fern (*Pteridium aquilinum*)', *J. biophys. biochem. Cytol.*, **6**, 413–417.

Motile male gametes of plants

Manton, I. (1959b) 'Observations on the internal structure of the spermatozoid of *Dictyota*', *J. exp. Bot.*, **10**, 448–461.

Manton, I. (1964a) 'The possible significance of some details of flagellar bases in plants', *Jl. R. microsc. Soc.*, **82**, 279–285.

Manton, I. (1964b) 'A contribution towards understanding of "the primitive fucoid" ', *New Phytol.*, **63**, 244–254.

Manton, I. (1970) 'Plant spermatozoids', In: *Comparative spermatology* (ed. B. Baccetti), pp. 143–158, *Proc. Int. Symp. Rome and Siena*, 1–5 July 1969.

Manton, I. and Clarke, B. (1951) 'An electron microscope study of the spermatozoid of *Fucus serratus*', *Ann. Bot.*, **15**, 461–471.

Manton, I. and Clarke, B. (1956) 'Observations with the electron microscope on the internal structure of the spermatozoid of *Fucus*', *J. exp. Bot.*, **7**, 416–432.

Manton, I., Clarke, B., and Greenwood, A. D. (1953) 'Further observations with the electron microscope on spermatozoids in the brown algae', *J. exp. Bot.*, **4**, 319–329.

Manton, I. and Friedmann, I. (1959) 'Gametes, fertilization, and zygote development in *Prasiola stipitata*', *Nova Hedwigia*, **1**, 443–462.

Manton, I. and Von Stosch, H. A. (1966) 'Observations on the fine structure of the male gamete of the centric diatom *Lithodesmium undulatum*', *Jl. R. microsc. Soc.*, **85**, 119–134.

McBride, G. E. (1971) 'The flagella base in *Coleochaete* and its evolutionary significance', *J. Phycol.*, **7**, S.13. (Abstract.)

Mignot, J. P. (1966) 'Structure et ultrastructure de quelques Euglénomonadines', *Protistologica*, **2**(3), 51–117.

Mignot, J. P. (1967) 'Affinités de Euglénomonadines et de Chloromonadines. Remarques sur la systematique de Euglénida', *Protistologica*, **3**, 25–60.

Mizukami, I. and Gall, J. (1966) 'Centriole replication II. Sperm formation in the fern, *Marsilea*, and the cycad, *Zamia*', *J. Cell Biol.*, **29**, 97–111.

Moestrup, Ø. (1970) 'On the fine structure of the spermatozoids of *Vaucheria sesculpicaria* and on the later stages in spermatogenesis', *J. mar. biol. Ass. U.K.*, **50**, 513–523.

Moser, J. W. (1970) *An ultrastructural study of spermatogenesis in 'Phaeoceros laevis' subsp. 'carolinianus'*, Ph.D. thesis, University of Illinois.

Moser, J. W. and Kreitner, G. L. (1970) 'Centrosome structure in *Anthoceros laevis* and *Marchantia polymorpha*', *J. Cell Biol.*, **44**, 454–458.

Norstog, K. (1967) 'Fine structure of the spermatozoid of *Zamia* with special reference to the flagellar apparatus', *Am. J. Bot.*, **54**, 831–840.

Norstog, K. (1968) 'Fine structure of the spermatozoid of *Zamia:* Observations on the microtubule system and related structures', *Phytomorphology*, **18**, 350–356.

Paolillo, D. J., Jr (1965) 'On the androcyte of *Polytrichum*, with special reference to the Dreiergruppe and the limosphere (Nebenkern)', *Can. J. Bot.*, **43**, 669–676.

Paolillo, D. J., Jr (1967) 'On the structure of the axoneme in flagella of *Polytrichum juniperinum*', *Trans. Am. microsc. Soc.*, **86**, 428–433.

Paolillo, D. J., Jr, Kreitner, G. L., and Reighard, J. A. (1968a) 'Spermatogenesis in *Polytrichum juniperinum* I. The origin of the apical body and elongation of the nucleus', *Planta*, **78**, 226–247.

Paolillo, D. J., Jr, Kreitner, G. L., and Reighard, J. A. (1968b) 'Spermatogenesis in *Polytrichum juniperinum* II. The mature sperm', *Planta*, **78**, 248–261.

Pease, D. C. (1963) 'The ultrastructure of flagellar fibrils', *J. Cell Biol.*, **18**, 313–326.

Perkins, F. O. (1970) 'Formation of centriole and centriole-like structures during meiosis and mitosis in *Labyrinthula* sp. (Rhizopodea, Labyrinthulida). An electron microscope study', *J. Cell Sci.*, **6**, 629–653.

Pickett-Heaps, J. D. (1968) 'Ultrastructure and differentiation in *Chara* (*fibrosa*) IV. Spermatogenesis', *Aust. J. biol. Sci.*, **21**, 655–690.

Renaud, F. L. and Swift, H. (1964) 'The development of basal bodies and flagella in *Allomyces*', *J. Cell Biol.*, **23**, 339–354.

Rice, H. V. and Laetsch, W. M. (1967) 'Observations on the morphology and physiology of *Marsilea* sperm', *Am. J. Bot.*, **54**, 856–866.

Ringo, D. L. (1967) 'Flagellar motion and fine structure of the flagellar apparatus in *Chlamydomonas*', *J. Cell Biol.*, **33**, 543–571.

Satir, P. (1965) 'Structure and function in cilia and flagella. Facts and problems', In: *Protoplasmatologia*, No. 3E, Springer-Verlag, Vienna.

Bibliography

Satô, S. (1954) 'On the filamentous appendage, a new structure of the spermatozoid of *Conocephalum conicum* discovered by means of the electron microscope', *J. Hattori Bot. Lab.*, 12, 113–115.

Satô, S. (1956) 'The filamentous appendage in the spermatozoids of Hepaticae as revealed by the electron microscope', *Bot. Mag. Tokyo*, 69, 435–438.

Satô, S. (1962) 'Electron microscope studies on the cell structure during spermatogenesis in *Conocephalum conicum*', *J. Hattori Bot. Lab.*, 25, 154–162.

Silveria, M. and Porter, K. R. (1964) 'The spermatozoids of flatworms and their microtubular systems', *Protoplasma*, 59, 240–266.

Suire, C. (1970) 'Recherches cytologiques sur deux hépatiques: *Pellia epiphylla* (L.) Corda (Metzgeriale) et *Radula complanata* (L.) Durn. (Jungermanniale). Ergastome, sporogénèse et spermatogénèse', *Le Botaniste*, 53, 125–392.

Sun, C. N. (1964) 'Fine structure of the spermatozoid of two mosses with special reference to the so called "Nebenkern" ', *Protoplasma*, 58, 663–666.

Tourte, Y. and Hurel-Py, G. (1967) 'Ontogénie et ultrastructure de l'appareil cinetique des spermatozoids du *Pteridium aquilinum* L.', *C. r. hebd. Séanc. Acad. Sci., Paris, D*, 265, 1289–1292.

Turner, F. R. (1966) *Changes in cellular organization during spermatogenesis*, Ph.D. thesis, University of Texas.

Turner, F. R. (1968) 'An ultrastructural study of plant spermatogenesis. Spermatogenesis in *Nitella*', *J. Cell Biol.*, 37, 370–393.

Turner, F. R. (1970) 'The effects of colchicine on spermatogenesis in *Nitella*', *J. Cell Biol.*, 46, 220–234.

Webber, H. J. (1901) 'Spermatogenesis and fecundation of *Zamia*', *U.S.D.A. Dept. of Agric. Bureau of Plant Industry Bulletin No. 2*, Govt. Printing Office, Washington.

Weier, T. E. (1931) 'A study of the moss plastid after fixation by mitochondrial, osmium and silver techniques II. The plastid during spermatogenesis in *Polytrichum commune* and *Catherinaea undulata*', *La Cellule*, 41, 51–84.

Wilson, M. (1911) 'Spermatogenesis in the Bryophyta', *Ann. Bot.*, 25, 415–459.

Yuasa, A. (1952) 'Studies in the cytology of Pteridophyta, XXIX. Does the spermatozoid accompany the plastid in fertilization?', *Cytologia*, 16, 347–351.

Index

Acetic acid, effect on interphase nucleus, 2
Acid phosphatase:
  cytochemical localization of, 183
  in provacuoles, 187
Acidic polyuronides, 434
ADP/O ratio, 60, 61
Albuminous cells, 393
Aldehyde fixation for electron microscopy,
  219, 395
Aleurone grains, 183, 191, 203
  enzymes of, 199
  location of reserve proteins in, 192
  mobilization in, 204
  origin and development, 191–192,
    193
  protein accumulation in, 199
Aleurone transfer cells, 456
Aleurone vacuoles, fusion into single central
    vacuole, 203
Algae:
  advanced mitotic systems in, 231–236
  cell walls in, 266
  sperms of, 524–526
  unusual mitotic systems in, 236–238
  zoospores of, 525
Allium cepa, 13
Allium porrum, 12, 13, 14, 18, 22
  interphase nucleus in, 6
  nucleoli, 4
Amyloplasts in zygote, 497
Anaphase, 20
  microtubules in, 228
Androcyte mother cells, 506, 507, 508
Androcytes, 506, 523
Annulate lamellae, 91, 95
Antheridium, 509
Anticlinal divisions, 369, 371
Antipodal cells, 454
  ultrastructure, 489
Apoplast, 441

ATP:
  synthesis, 59
  synthesis sites, 60
Autolysis, 209
  in mesophyll of wilting *Ipomöa purpurea*
    corolla, 207
Autophagy, 196, 197
  in meristem cells, 194
Axillary meristems, 461

Bernhard's regressive staining technique, 16
Bicentrioles, 507, 510, 512
  binary fission of, 507
  symmetrical bilateral growth in
    Labyrinthula, 511
Biochemical differentiation, 196–199
  apparent correlation with lysosomal
    activity, 197, 201
Biological membranes, structural models of,
    64
Bivalent cations, effect on nuclear structures, 3
Blepharoplasts, 507, 508, 509, 516
  of bryophytes, 511–515
  components, 513
  microtubules in, 513, 516
  multilayered structure in, *see* Multilayered
    structure
  response of material to different
    fixatives, 512
Buds, lipophilic coating, 347

Calceolaria, fatty oil secreted by, 347
Callose, 406, 408, 409
  deposition, 407
Callus, 310–315
  induction of, 310
  origination of, 310
  plasmodesmata in, 319
  polysaccharide components, in sycamore,
    318